Problem Books in Mathematics

T0189895

Edited by K.A. Bencsáth
P.R. Halmos

Problem Books in Mathematics

Series Editors: K.A. Bencsáth and P.R. Halmos

Pell's Equation
by *Edward J. Barbeau*

Polynomials
by *Edward J. Barbeau*

Problems in Geometry
by *Marcel Berger, Pierre Pansu, Jean-Pic Berry, and Xavier Saint-Raymond*

Problem Book for First Year Calculus
by *George W. Bluman*

Exercises in Probability
by *T. Cacoullos*

Probability Through Problems
by *Marek Capiński and Tomasz Zastawniak*

An Introduction to Hilbert Space and Quantum Logic
by *David W. Cohen*

Unsolved Problems in Geometry
by *Hallard T. Croft, Kenneth J. Falconer, and Richard K. Guy*

Berkeley Problems in Mathematics (3rd ed.)
by *Paulo Ney de Souza and Jorge-Nuno Silva*

Problem-Solving Strategies
by *Arthur Engel*

Problems in Analysis
by *Bernard R. Gelbaum*

Problems in Real and Complex Analysis
by *Bernard R. Gelbaum*

Theorems and Counterexamples in Mathematics
by *Bernard R. Gelbaum and John M.H. Olmsted*

Exercises in Integration
by *Claude George*

(continued after index)

Paulo Ney de Souza   Jorge-Nuno Silva

Berkeley Problems
in Mathematics

Third Edition

 Springer

Paulo Ney de Souza
Department of Mathematics
University of California at Berkeley
Berkeley, CA 94720-3840
USA
desouza@math.berkeley.edu

Jorge-Nuno Silva
Departamento de Matemática
Faculdade de Ciências de Lisboa
1700 Lisboa
Portugal
jnsilva@lmc.fc.ul.pt

Series Editors:
Katalin A. Bencsáth
Mathematics
School of Science
Manhattan College
Riverdale, NY 10471
USA
katalin.bencsath@manhattan.edu

Paul R. Halmos
Department of Mathematics
Santa Clara University
Santa Clara, CA 95053
USA
phalmos@scuacc.scu.edu

Cover illustration: Tracy Hall, University of California at Berkeley

Mathematics Subject Classification (2000): 26-06, 26Bxx, 34Axx, 30Axx

Library of Congress Cataloging-in-Publication Data
de Souza, Paulo Ney.
    Berkeley problems in mathematics / Paulo Ney de Souza, Jorge Nuno-Silva
        p.   cm.— (Problem books in mathematics)
    Includes bibliographical references and index.
    ISBN 0-387-20429-6 (alk. paper) — ISBN 0-387-00892-6 (pbk : alk. paper)
    1. Mathematics—Problems, exercises, etc.   2. University of California, Berkeley. Dept. of
    Mathematics—Examinations.   I. Silva, Jorge-Nuno.   II. Title.   III. Series.
    QA43.D38—2004
    510'.76—dc22   2003063347

ISBN 0-387-20429-6 (hardcover)   Printed on acid-free paper.
ISBN 0-387-00892-6 (softcover)

Printed in the United States of America.   (EB)

9 8 7 6 5 4 3 2

springeronline.com

para Aline, Laura, Manuel & Stephanie

Preface

In 1977 the Mathematics Department at the University of California, Berkeley, instituted a written examination as one of the first major requirements toward the Ph.D. degree in Mathematics. This examination replaced a system of standardized Qualifying Exams. Its purpose was to determine whether first-year students in the Ph.D. program had mastered basic mathematics well enough to continue in the program with a reasonable chance of success.

Historically, any one examination is passed by approximately half of the students taking it and students are allowed three attempts. Since its inception, the exam has become a major hurdle to overcome in the pursuit of the degree and, therefore, a measure of the minimum requirements to successful completion of the program at Berkeley. Even though students are allowed three attempts, most would agree that the ideal time to complete the requirement is during the first month of the program rather than in the middle or end of the first year. This book was conceived on this premise, and its intent is to publicize the material and aid in the preparation for the examination during the undergraduate years, when one is deeply involved with the material that it covers.

The examination is now offered twice a year in the second week of each semester, and consists of 6 hours of written work given over a 2-day period with 9 problems each (10 before 1988). Students select 6 of the 9 problems (7 of 10 before 1988). Most of the examination covers material, mainly in analysis and algebra, that should be a part of a well-prepared mathematics student's undergraduate training. This book is a compilation of the more than 1000 problems which have appeared on the Prelims during the last few decades and currently make up a collection which is a delightful field to plow through, and solutions to most of them.

When Berkeley was on the Quarter system, exams were given three times a year: Spring, Summer, and Fall. Since 1986, the exams have been given twice a year, in January and September.

From the first examination through Fall 1981, the policy was: two attempts allowed; each examination 6 hours; total 14/20 problems. From Winter 1982 through Spring 1988, the policy was: two attempts allowed; each examination 8 hours; total 14/20 problems. Starting Fall 1988, the policy was: three attempts allowed; each examination 6 hours; total 12/18 problems. In all cases, the examination must be passed within 13 months of entering the Ph.D. program.

The problems are organized by subject and ordered in increasing level of difficulty, within clusters. Each one is tagged with the academic term of the exam in which it appeared using abbreviations of the type **Fa87** to designate the exam given in the **Fall** semester of **1987**. Problems that have appeared more than once have been merged and show multiple tags for each exam. Sometimes the merge required slight modifications in the text (a few to make the problem correct!), but the original text has been preserved in an electronic version of the exams (see Appendix A). Other items in the Appendices include the syllabus, passing scores for the exams and a Bibliography used throughout the solutions.

Classifying a collection of problems as vast as this one by subjects is not an easy task. Some of the problems are interdisciplinary and some have solutions as varied as Analysis and Number Theory (1.1.18 comes to mind!), and the choices are invariably hard. In most of these cases, we provide the reader with an alternative classification or pointers to similar problems elsewhere.

We would like to hear about other solutions to the problems here and comments on the existing ones. They can be sent by e-mail to the authors.

This project started many years ago, when one of us (PNdS) came to Berkeley and had to go through the lack of information and uncertainties of the exam and got involved with a problem solving group. First thanks go to the group's members: Dino Lorenzini, Hung The Dinh, Kin-Yin Li, and Jorge Zubelli, and then to the Prelim Workshop leaders, many of whose names escape us now but the list includes, besides ourselves, Matthew Wiener, Dmitry Gokhman, Keith Kearnes, Geon Ho Choe, Mike May, Eliza Sachs, Ben Lotto, Ted Jones, David Cruz-Uribe, Jonathan Walden, Saul Schleimer and Howard Thompson; and also to the many people we have discussed these problems with like Don Sarason, George Bergman, Reginald Koo, D. Popa, C. Costara, József Sándor, Elton Hsu, Enlin Pan, Bjorn Poonen, Assaf Wool and Jin-Gen Yang. Many thanks to Debbie Craig for swift typesetting of many of the problems and to Janet Yonan for her help with the archeological work of finding many of the old and lost problem sets, and finally to Nefeli's for the best coffee west of Rome, we would not have survived without it!

We thank also the Department of Mathematics and the Portuguese Studies Program of UC Berkeley, MSRI, University of Lisbon, CMAF, JNICT, PRAXIS XXI, FEDER, project PRAXIS/2/2.1/MAT/125/94 and FLAD, which supported one of the authors on the Summers of 96, 97, 2000 and 2002, and CNPq grant 20.1553/82-MA that supported the other during the initial phase of this project.

This is a project that could not have been accomplished in any typesetting system other than TeX. The problems and solutions are part of a two-pronged database that is called by sourcing programs that generate several versions (working, final paper version, per-exams list, and the on-line HTML and PDF versions) from a single source. Silvio Levy's TeX support and counseling was a major resource backing our efforts and many thanks also to Noam Shomron for help with non-standard typesetting.

Berkeley, California   Paulo Ney de Souza
August 2003   desouza@math.berkeley.edu

                                                              Jorge Nuno Silva
                                                     jnsilva@lmc.fc.ul.pt

Contents

Part I

Problems

Part 1

Problems

1
Real Analysis

1.1 Elementary Calculus

Problem 1.1.1 (Fa87) *Prove that $(\cos\theta)^p \leqslant \cos(p\theta)$ for $0 \leqslant \theta \leqslant \pi/2$ and $0 < p < 1$.*

Problem 1.1.2 (Fa77) *Let $f : [0, 1] \to \mathbb{R}$ be continuously differentiable, with $f(0) = 0$. Prove that*

$$\sup_{0\leqslant x\leqslant 1} |f(x)| \leqslant \sqrt{\int_0^1 (f'(x))^2 \, dx}.$$

Problem 1.1.3 (Sp81) *Let $f(x)$ be a real valued function defined for all $x \geqslant 1$, satisfying $f(1) = 1$ and*

$$f'(x) = \frac{1}{x^2 + f(x)^2}.$$

Prove that

$$\lim_{x\to\infty} f(x)$$

exists and is less than $1 + \dfrac{\pi}{4}$.

Problem 1.1.4 (Sp95) *Let $f, g: [0, 1] \to [0, \infty)$ be continuous functions satisfying*

$$\sup_{0\leqslant x\leqslant 1} f(x) = \sup_{0\leqslant x\leqslant 1} g(x).$$

Prove that there exists $t \in [0, 1]$ with $f(t)^2 + 3f(t) = g(t)^2 + 3g(t)$.

Problem 1.1.5 (Fa86) *For f a real valued function on the real line, define the function Δf by $\Delta f(x) = f(x+1) - f(x)$. For $n \geqslant 2$, define $\Delta^n f$ recursively by $\Delta^n f = \Delta(\Delta^{n-1} f)$. Prove that $\Delta^n f = 0$ if and only if f has the form $f(x) = a_0(x) + a_1(x)x + \cdots + a_{n-1}(x)x^{n-1}$ where $a_0, a_1, \ldots, a_{n-1}$ are periodic functions of period 1.*

Problem 1.1.6 (Fa00) *Suppose that $f : \mathbb{R} \to \mathbb{R}$ is a nonconstant function such that $f(x) \leqslant f(y)$ whenever $x \leqslant y$. Prove that there exist $a \in \mathbb{R}$ and $c > 0$ such that $f(a+x) - f(a-x) \geqslant cx$ for all $x \in [0, 1]$.*

Problem 1.1.7 (Fa81) *Either prove or disprove (by a counterexample) each of the following statements:*

1. *Let $f : \mathbb{R} \to \mathbb{R}$, $g : \mathbb{R} \to \mathbb{R}$ be such that*

$$\lim_{t \to a} g(t) = b \quad and \quad \lim_{t \to b} f(t) = c.$$

   Then

$$\lim_{t \to a} f(g(t)) = c.$$

2. *If $f : \mathbb{R} \to \mathbb{R}$ is continuous and U is an open set in $\mathbb{R}$, then $f(U)$ is an open set in $\mathbb{R}$.*

3. *Let f be of class C^∞ on the interval $(-1, 1)$. Suppose that $|f^{(n)}(x)| \leqslant 1$ for all $n \geqslant 1$ and all x in the interval. Then f is real analytic; that is, it has a convergent power series expansion in a neighborhood of each point of the interval.*

Problem 1.1.8 (Fa97) *Prove that for all $x > 0$, $\sin x > x - \dfrac{x^3}{6}$.*

Problem 1.1.9 (Su81) *Let*

$$y(h) = 1 - 2\sin^2(2\pi h), \qquad f(y) = \frac{2}{1 + \sqrt{1-y}}.$$

Justify the statement

$$f(y(h)) = 2 - 4\sqrt{2}\pi \, |h| + O(h^2)$$

where

$$\limsup_{h \to 0} \frac{O(h^2)}{h^2} < \infty.$$

Problem 1.1.10 (Fa82)   1. *Prove that there is no continuous map from the closed interval $[0, 1]$ onto the open interval $(0, 1)$.*

2. *Find a continuous surjective map from the open interval $(0, 1)$ onto the closed interval $[0, 1]$.*

3. *Prove that no map in Part 2 can be bijective.*

Problem 1.1.11 (Sp99) *Suppose that f is a twice differentiable real function such that $f''(x) > 0$ for all $x \in [a, b]$. Find all numbers $c \in [a, b]$ at which the area between the graph $y = f(x)$, the tangent to the graph at $(c, f(c))$, and the lines $x = a$, $x = b$, attains its minimum value.*

Problem 1.1.12 (Fa94, Sp98) *Find the maximum area of all triangles that can be inscribed in an ellipse with semiaxes a and b, and describe the triangles that have maximum area.*
Note: See also Problem 2.2.2.

Problem 1.1.13 (Fa93) *Let f be a continuous real valued function on $[0, \infty)$. Let A be the set of real numbers a that can be expressed as $a = \lim_{n\to\infty} f(x_n)$ for some sequence (x_n) in $[0, \infty)$ such that $\lim_{n\to\infty} x_n = \infty$. Prove that if A contains the two numbers a and b, then contains the entire interval with endpoints a and b.*

Problem 1.1.14 (Su81) *Show that the equation*

$$x \left(1 + \log \left(\frac{1}{\varepsilon \sqrt{x}}\right)\right) = 1, \quad x > 0, \quad \varepsilon > 0,$$

has, for each sufficiently small $\varepsilon > 0$, exactly two solutions. Let $x(\varepsilon)$ be the smaller one. Show that

1. *$x(\varepsilon) \to 0$ as $\varepsilon \to 0+$;*

yet for any $s > 0$,

2. *$\varepsilon^{-s} x(\varepsilon) \to \infty$ as $\varepsilon \to 0+$.*

Problem 1.1.15 (Sp82) *Suppose that $f(x)$ is a polynomial with real coefficients and a is a real number with $f(a) \neq 0$. Show that there exists a real polynomial $g(x)$ such that if we define p by $p(x) = f(x)g(x)$, we have $p(a) = 1$, $p'(a) = 0$, and $p''(a) = 0$.*

Problem 1.1.16 (Su84) *Let $p(z)$ be a nonconstant polynomial with real coefficients such that for some real number a, $p(a) \neq 0$ but $p'(a) = p''(a) = 0$. Prove that the equation $p(z) = 0$ has a nonreal root.*

Problem 1.1.17 (Fa84, Fa97) *Let f be a C^2 function on the real line. Assume f is bounded with bounded second derivative. Let*

$$A = \sup_{x \in \mathbb{R}} |f(x)|, \quad B = \sup_{x \in \mathbb{R}} |f''(x)|.$$

Prove that

$$\sup_{x \in \mathbb{R}} |f'(x)| \leqslant 2\sqrt{AB}.$$

Problem 1.1.18 (Fa90) *Find all pairs of integers a and b satisfying $0 < a < b$ and $a^b = b^a$.*

Problem 1.1.19 (Sp92) *For which positive numbers a and b, with $a > 1$, does the equation $\log_a x = x^b$ have a positive solution for x?*

Problem 1.1.20 (Sp84) *Which number is larger, π^3 or 3^π ?*

Problem 1.1.21 (Sp94) *For which numbers a in $(1, \infty)$ is it true that $x^a \leqslant a^x$ for all x in $(1, \infty)$?*

Problem 1.1.22 (Sp96) *Show that a positive constant t can satisfy*

$$e^x > x^t \quad for\ all \quad x > 0$$

if and only if $t < e$.

Problem 1.1.23 (Su77) *Suppose that $f(x)$ is defined on $[-1, 1]$, and that $f'''(x)$ is continuous. Show that the series*

$$\sum_{n=1}^{\infty} \left(n \left(f \left(\frac{1}{n} \right) - f \left(-\frac{1}{n} \right) \right) - 2f'(0) \right)$$

converges.

Problem 1.1.24 (Fa96) *If f is a C^2 function on an open interval, prove that*

$$\lim_{h \to 0} \frac{f(x + h) - 2f(x) + f(x-h)}{h^2} = f''(x) .$$

Problem 1.1.25 (Su85)   *1. For $0 \leqslant \theta \leqslant \frac{\pi}{2}$, show that*

$$\sin \theta \geqslant \frac{2}{\pi} \theta .$$

  2. By using Part 1, or by any other method, show that if $\lambda < 1$, then

$$\lim_{R \to \infty} R^\lambda \int_0^{\frac{\pi}{2}} e^{-R \sin \theta}\, d\theta = 0.$$

Problem 1.1.26 (Su78) *Let $f : \mathbb{R} \to \mathbb{R}$ be continuous. Suppose that $\mathbb{R}$ contains a countably infinite subset S such that*

$$\int_p^q f(x)\, dx = 0$$

if p and q are not in S. Prove that f is identically 0.

Problem 1.1.27 (Fa89) *Let the function f from $[0, 1]$ to $[0, 1]$ have the following properties:*

- f is of class C^1;

- $f(0) = f(1) = 0$;

- f' is nonincreasing (i.e., f is concave).

Prove that the arclength of the graph of f does not exceed 3.

Problem 1.1.28 (Sp93) Let f be a real valued C^1 function on $[0, \infty)$ such that the improper integral $\int_1^\infty |f'(x)|dx$ converges. Prove that the infinite series $\sum_{n=1}^\infty f(n)$ converges if and only if the integral $\int_1^\infty f(x)dx$ converges.

Problem 1.1.29 (Su82) Let E be the set of all continuous real valued functions $u : [0, 1] \to \mathbb{R}$ satisfying

$$|u(x) - u(y)| \leqslant |x - y|, \quad 0 \leqslant x, y \leqslant 1, \quad u(0) = 0.$$

Let $\varphi : E \to \mathbb{R}$ be defined by

$$\varphi(u) = \int_0^1 \left(u(x)^2 - u(x) \right) dx.$$

Show that φ achieves its maximum value at some element of E.

Problem 1.1.30 (Fa87) Let S be the set of all real C^1 functions f on $[0, 1]$ such that $f(0) = 0$ and

$$\int_0^1 f'(x)^2 dx \leqslant 1.$$

Define

$$J(f) = \int_0^1 f(x)\, dx.$$

Show that the function J is bounded on S, and compute its supremum. Is there a function $f_0 \in S$ at which J attains its maximum value? If so, what is f_0?

Problem 1.1.31 (Fa82, Fa96) Let f be a real valued continuous nonnegative function on $[0, 1]$ such that

$$f(t)^2 \leqslant 1 + 2 \int_0^t f(s)\, ds$$

for $t \in [0, 1]$. Show that $f(t) \leqslant 1 + t$ for $t \in [0, 1]$.

Problem 1.1.32 (Sp96) Suppose φ is a C^1 function on $\mathbb{R}$ such that

$$\varphi(x) \to a \quad and \quad \varphi'(x) \to b \quad as \quad x \to \infty.$$

Prove or give a counterexample: b must be zero.

Problem 1.1.33 (Su77) *Show that*

$$F(k) = \int_0^{\pi/2} \frac{dx}{\sqrt{1 - k\cos^2 x}}$$

$0 \leqslant k < 1$, *is an increasing function of k.*

Problem 1.1.34 (Fa79) *Given that*

$$\int_{-\infty}^{\infty} e^{-x^2} dx = \sqrt{\pi} \,,$$

find $f'(t)$ explicitly, where

$$f(t) = \int_{-\infty}^{\infty} e^{-tx^2} dx, \qquad t > 0.$$

Problem 1.1.35 (Fa80) *Define*

$$F(x) = \int_{\sin x}^{\cos x} e^{(t^2 + xt)} dt.$$

Compute $F'(0)$.

Problem 1.1.36 (Fa95) *Let $f : \mathbb{R} \to \mathbb{R}$ be a nonzero C^∞ function such that $f(x)f(y) = f\left(\sqrt{x^2 + y^2}\right)$ for all x and y and that $f(x) \to 0$ as $|x| \to \infty$.*

1. *Prove that f is an even function and that $f(0)$ is 1.*

2. *Prove that f satisfies the differential equation $f'(x) = f''(0)xf(x)$, and find the most general function satisfying the given conditions.*

Problem 1.1.37 (Fa01) *Let S be the set of continuous real-valued functions on $[0, 1]$ such that $f(x)$ is rational whenever x is rational. Prove that S is uncountable.*

1.2   Limits and Continuity

Problem 1.2.1 (Sp02) *Let f be a bounded, continuous, real-valued function on $\mathbb{R}^2$. Define the function g on $\mathbb{R}$ by*

$$g(x) = \int_{-\infty}^{\infty} \frac{f(x, t)}{1 + t^2} dt \,.$$

Prove that g is continuous.

Problem 1.2.2 (Fa90) *Suppose that f maps the compact interval I into itself and that*

$$|f(x) - f(y)| < |x - y|$$

for all x, y ∈ I, x ≠ y. Can one conclude that there is some constant M < 1 such that, for all x, y ∈ I,

$$|f(x) - f(y)| \leqslant M|x - y|?$$

Problem 1.2.3 (Sp90) *Let the real valued function f on [0, 1] have the following two properties:*

- *If [a, b] ⊂ [0, 1], then f ([a, b]) contains the interval with endpoints f (a) and f (b) (i.e., f has the Intermediate Value Property).*

- *For each c ∈ ℝ, the set f⁻¹(c) is closed.*

Prove that f is continuous.

Problem 1.2.4 (Fa00) *Let f : ℝ → ℝ be uniformly continuous with f (0) = 0. Prove: there exists a positive number B such that |f (x)| ⩽ 1 + B|x|, for all x.*

Problem 1.2.5 (Sp83, Sp01) *Suppose that f is a continuous function on ℝ which is periodic with period 1, i.e., f (x + 1) = f (x). Show:*

1. *The function f is bounded above and below and achieves its maximum and minimum.*

2. *The function f is uniformly continuous on ℝ.*

3. *There exists a real number x_0 such that*

$$f(x_0 + \pi) = f(x_0).$$

Problem 1.2.6 (Sp77) *Let h : [0, 1) → ℝ be a function defined on the half-open interval [0, 1). Prove that if h is uniformly continuous, there exists a unique continuous function g : [0, 1] → ℝ such that g(x) = h(x) for all x ∈ [0, 1).*

Problem 1.2.7 (Fa99) *Let f be a continuous real valued function on [0, ∞) such that $\lim_{x \to \infty} f(x)$ exists (finitely). Prove that f is uniformly continuous.*

Problem 1.2.8 (Sp84) *Prove or supply a counterexample: If the function f from ℝ to ℝ has both a left limit and a right limit at each point of ℝ, then the set of discontinuities of f is, at most, countable.*

Problem 1.2.9 (Fa78) *Let f : ℝ → ℝ satisfy f (x) ⩽ f (y) for x ⩽ y. Prove that the set where f is not continuous is finite or countably infinite.*

Problem 1.2.10 (Su85, Fa96) *A function* $f : [0, 1] \to \mathbb{R}$ *is said to be upper semicontinuous if given* $x \in [0, 1]$ *and* $\varepsilon > 0$, *there exists a* $\delta > 0$ *such that if* $|y - x| < \delta$, *then* $f(y) < f(x) + \varepsilon$. *Prove that an upper semicontinuous function f on* $[0, 1]$ *is bounded above and attains its maximum value at some point* $p \in [0, 1]$.

Problem 1.2.11 (Su83) *Prove that a continuous function from* $\mathbb{R}$ *to* $\mathbb{R}$ *which maps open sets to open sets must be monotonic.*

Problem 1.2.12 (Fa91) *Let* f *be a continuous function from* $\mathbb{R}$ *to* $\mathbb{R}$ *such that* $|f(x) - f(y)| \geqslant |x - y|$ *for all x and y. Prove that the range of f is all of* $\mathbb{R}$.
Note: See also Problem 2.1.8.

Problem 1.2.13 (Fa81) *Let* f *be a continuous function on* $[0, 1]$. *Evaluate the following limits.*

  1.

$$\lim_{n \to \infty} \int_0^1 x^n f(x)\, dx .$$

  2.

$$\lim_{n \to \infty} n \int_0^1 x^n f(x)\, dx .$$

Problem 1.2.14 (Fa88, Sp97) *Let* f *be a function from* $[0, 1]$ *into itself whose graph*

$$G_f = \{(x, f(x)) \mid x \in [0, 1]\}$$

is a closed subset of the unit square. Prove that f is continuous.
Note: See also Problem 2.1.2.

Problem 1.2.15 (Sp89) *Let* f *be a continuous real valued function defined on* $[0, 1] \times [0, 1]$. *Let the function g on* $[0, 1]$ *be defined by*

$$g(x) = \max \{f(x, y) \mid y \in [0, 1]\}.$$

Prove that g is continuous.

Problem 1.2.16 (Fa01) *Let the function* $f : \mathbb{R} \to \mathbb{R}$ *be bounded on bounded sets and have the property that* $f^{-1}(K)$ *is closed whenever K is compact. Prove f is continuous.*

1.3   Sequences, Series, and Products

Problem 1.3.1 (Su85) *Let* $A_1 \geqslant A_2 \geqslant \cdots \geqslant A_k \geqslant 0$. *Evaluate*

$$\lim_{n \to \infty} \left(A_1^n + A_2^n + \cdots + A_k^n\right)^{1/n} .$$

Note: See also Problem 5.1.11.

Problem 1.3.2 (Sp96) *Compute*

$$L = \lim_{n \to \infty} \left(\frac{n^n}{n!} \right)^{1/n}.$$

Problem 1.3.3 (Sp92) *Let* $x_0 = 1$ *and*

$$x_{n+1} = \frac{3 + 2x_n}{3 + x_n}, \qquad n \geqslant 0.$$

Prove that $x_\infty = \lim_{n \to \infty} x_n$ *exists, and find its value.*

Problem 1.3.4 (Fa97) *Define a sequence of real numbers* (x_n) *by*

$$x_0 = 1, \qquad x_{n+1} = \frac{1}{2 + x_n} \quad \text{for} \quad n \geqslant 0.$$

Show that (x_n) *converges, and evaluate its limit.*

Problem 1.3.5 (Fa89, Sp94) *Let* α *be a number in* $(0, 1)$. *Prove that any sequence* (x_n) *of real numbers satisfying the recurrence relation*

$$x_{n+1} = \alpha x_n + (1 - \alpha)x_{n-1}$$

has a limit, and find an expression for the limit in terms of α, x_0 *and* x_1.

Problem 1.3.6 (Fa92) *Let* k *be a positive integer. Determine those real numbers* c *for which every sequence* (x_n) *of real numbers satisfying the recurrence relation*

$$\frac{1}{2}(x_{n+1} + x_{n-1}) = cx_n$$

has period k *(i.e.,* $x_{n+k} = x_n$ *for all* n*).*

Problem 1.3.7 (Sp84) *Let* a *be a positive real number. Define a sequence* (x_n) *by*

$$x_0 = 0, \quad x_{n+1} = a + x_n^2, \quad n \geqslant 0.$$

Find a necessary and sufficient condition on a *in order that a finite limit* $\lim_{n \to \infty} x_n$ *should exist.*

Problem 1.3.8 (Sp03) *Let* x_n *be a sequence of real numbers so that*

$$\lim_{n \to \infty} (2x_{n+1} - x_n) = x.$$

Show that $\lim_{n \to \infty} x_n = x$.

Problem 1.3.9 (Sp00) *Let* a *and* x_0 *be positive numbers, and define the sequence* $(x_n)_{n=1}^{\infty}$ *recursively by*

$$x_n = \frac{1}{2}\left(x_{n-1} + \frac{a}{x_{n-1}} \right).$$

Prove that this sequence converges, and find its limit.

Problem 1.3.10 (Fa95) *Let x_1 be a real number, $0 < x_1 < 1$, and define a sequence by $x_{n+1} = x_n - x_n^{n+1}$. Show that $\liminf\limits_{n \to \infty} x_n > 0$.*

Problem 1.3.11 (Fa80) *Let $f(x) = \frac{1}{4} + x - x^2$. For any real number x, define a sequence (x_n) by $x_0 = x$ and $x_{n+1} = f(x_n)$. If the sequence converges, let x_∞ denote the limit.*

1. *For $x = 0$, show that the sequence is bounded and nondecreasing and find $x_\infty = \lambda$.*

2. *Find all $y \in \mathbb{R}$ such that $y_\infty = \lambda$.*

Problem 1.3.12 (Fa81) *The Fibonacci numbers $f_1, f_2, \ldots$ are defined recursively by $f_1 = 1$, $f_2 = 2$, and $f_{n+1} = f_n + f_{n-1}$ for $n \geqslant 2$. Show that*

$$\lim_{n \to \infty} \frac{f_{n+1}}{f_n}$$

exists, and evaluate the limit.
Note: See also Problem 7.5.20.

Problem 1.3.13 (Fa79) *Prove that*

$$\lim_{n \to \infty} \left(\frac{1}{n+1} + \frac{1}{n+2} + \cdots + \frac{1}{2n} \right) = \log 2.$$

Problem 1.3.14 (Sp02) *For n a positive integer, let H_n denote the nth partial sum of the harmonic series*

$$H_n = \sum_{j=1}^{n} \frac{1}{j}.$$

Let $k > 1$ be an integer. Prove that

$$\log k - \frac{C}{n} < H_{nk} - H_n < \log k \qquad (n = 1, 2, \ldots),$$

where $\log k$ is the natural logarithm of k, and C is a constant.

Problem 1.3.15 (Sp90) *Suppose $x_1, x_2, x_3, \ldots$ is a sequence of nonnegative real numbers satisfying*

$$x_{n+1} \leqslant x_n + \frac{1}{n^2}$$

for all $n \geqslant 1$. Prove that $\lim\limits_{n \to \infty} x_n$ exists.

Problem 1.3.16 (Sp93) *Let (a_n) and (ε_n) be sequences of positive numbers. Assume that $\lim_{n \to \infty} \varepsilon_n = 0$ and that there is a number k in $(0, 1)$ such that $a_{n+1} \leqslant ka_n + \varepsilon_n$ for every n. Prove that $\lim\limits_{n \to \infty} a_n = 0$.*

Problem 1.3.17 (Fa83) *Prove or disprove (by giving a counterexample), the following assertion: Every infinite sequence $x_1, x_2, \ldots$ of real numbers has either a nondecreasing subsequence or a nonincreasing subsequence.*

Problem 1.3.18 (Su83) *Let $b_1, b_2, \ldots$ be positive real numbers with*

$$\lim_{n \to \infty} b_n = \infty \quad and \quad \lim_{n \to \infty} (b_n / b_{n+1}) = 1.$$

Assume also that $b_1 < b_2 < b_3 < \cdots$. Show that the set of quotients $(b_m/b_n)_{1 \leqslant n < m}$ is dense in $(1, \infty)$.

Problem 1.3.19 (Sp81) *Which of the following series converges?*

1.

$$\sum_{n=1}^{\infty} \frac{(2n)!(3n)!}{n!(4n)!}.$$

2.

$$\sum_{n=1}^{\infty} \frac{1}{n^{1+1/n}}.$$

Problem 1.3.20 (Fa91) *Let $a_1, a_2, a_3, \ldots$ be positive numbers.*

1. *Prove that $\sum a_n < \infty$ implies $\sum \sqrt{a_n a_{n+1}} < \infty$.*

2. *Prove that the converse of the above statement is false.*

Problem 1.3.21 (Su80, Sp97) *For each $(a, b, c) \in \mathbb{R}^3$, consider the series*

$$\sum_{n=3}^{\infty} \frac{a^n}{n^b (\log n)^c}.$$

Determine the values of (a, b, c) for which the series

1. *converges absolutely;*

2. *converges but not absolutely;*

3. *diverges.*

Problem 1.3.22 (Sp91) *For which real numbers x does the infinite series*

$$\sum_{n=1}^{\infty} \frac{\sqrt{n+1} - \sqrt{n}}{n^x}$$

converge?

Problem 1.3.23 (Fa94) *For which values of the real number a does the series*

$$\sum_{n=1}^{\infty} \left(\frac{1}{n} - \sin \frac{1}{n} \right)^a$$

converge?

Problem 1.3.24 (Sp91) *Let A be the set of positive integers that do not contain the digit 9 in their decimal expansions. Prove that*

$$\sum_{a \in A} \frac{1}{a} < \infty;$$

that is, A defines a convergent subseries of the harmonic series.

Problem 1.3.25 (Sp89) *Let $a_1, a_2, \ldots$ be positive numbers such that*

$$\sum_{n=1}^{\infty} a_n < \infty.$$

Prove that there are positive numbers $c_1, c_2, \ldots$ such that

$$\lim_{n \to \infty} c_n = \infty \quad and \quad \sum_{n=1}^{\infty} c_n a_n < \infty.$$

Problem 1.3.26 (Fa90) *Evaluate the limit*

$$\lim_{n \to \infty} \cos \frac{\pi}{2^2} \cos \frac{\pi}{2^3} \cdots \cos \frac{\pi}{2^n}.$$

1.4   Differential Calculus

Problem 1.4.1 (Su83) *Outline a proof, starting from basic properties of the real numbers, of the following theorem: Let $f : [a, b] \to \mathbb{R}$ be a continuous function such that $f'(x) = 0$ for all $x \in (a, b)$. Then $f(b) = f(a)$.*

Problem 1.4.2 (Sp84) *Let $f(x) = x \log \left(1 + x^{-1} \right), 0 < x < \infty$.*

1. Show that f is strictly monotonically increasing.

2. Compute $\lim f(x)$ as $x \to 0$ and $x \to \infty$.

Problem 1.4.3 (Sp85) *Let $f(x), 0 \leqslant x < \infty$, be continuous, differentiable, with $f(0) = 0$, and that $f'(x)$ is an increasing function of x for $x \geqslant 0$. Prove that*

$$g(x) = \begin{cases} f(x)/x, & x > 0 \\ f'(0), & x = 0 \end{cases}$$

is an increasing function of x.

Problem 1.4.4 (Su79, Fa97) *1. Give an example of a differentiable function* $f : \mathbb{R} \to \mathbb{R}$ *whose derivative* f' *is not continuous.*

2. Let f *be as in Part 1. If* $f'(0) < 2 < f'(1)$, *prove that* $f'(x) = 2$ *for some* $x \in [0, 1]$.

Problem 1.4.5 (Sp90) *Let* $y : \mathbb{R} \to \mathbb{R}$ *be a* C^∞ *function that satisfies the differential equation*

$$y'' + y' - y = 0$$

for $x \in [0, L]$, *where* L *is a positive real number. Suppose that* $y(0) = y(L) = 0$. *Prove that* $y \equiv 0$ *on* $[0, L]$.

Problem 1.4.6 (Su85) *Let* $u(x)$, $0 \leqslant x \leqslant 1$, *be a real valued* C^2 *function which satisfies the differential equation*

$$u''(x) = e^x u(x).$$

1. Show that if $0 < x_0 < 1$, *then* u *cannot have a positive local maximum at* x_0. *Similarly, show that* u *cannot have a negative local minimum at* x_0.

2. Now suppose that $u(0) = u(1) = 0$. *Prove that* $u(x) \equiv 0$, $0 \leqslant x \leqslant 1$.

Problem 1.4.7 (Sp98) *Let* K *be a real constant. Suppose that* $y(t)$ *is a positive differentiable function satisfying* $y'(t) \leqslant Ky(t)$ *for* $t \geqslant 0$. *Prove that* $y(t) \leqslant e^{Kt}y(0)$ *for* $t \geqslant 0$.

Problem 1.4.8 (Sp77, Su82) *Suppose* f *is a differentiable function from the reals into the reals. Suppose* $f'(x) > f(x)$ *for all* $x \in \mathbb{R}$, *and* $f(x_0) = 0$. *Prove that* $f(x) > 0$ *for all* $x > x_0$.

Problem 1.4.9 (Sp99) *Suppose that* f *is a twice differentiable real-valued function on* $\mathbb{R}$ *such that* $f(0) = 0$, $f'(0) > 0$, *and* $f''(x) \geqslant f(x)$ *for all* $x \geqslant 0$. *Prove that* $f(x) > 0$ *for all* $x > 0$.

Problem 1.4.10 (Sp87) *Show that the equation* $ae^x = 1 + x + x^2/2$, *where* a *is a positive constant, has exactly one real root.*

Problem 1.4.11 (Su78, Fa89) *Suppose* $f : [0, 1] \to \mathbb{R}$ *is continuous with* $f(0) = 0$, *and for* $0 < x < 1$, f *is differentiable and* $0 \leqslant f'(x) \leqslant 2f(x)$. *Prove that* f *is identically* 0.

Problem 1.4.12 (Sp84) *Let* $f : [0, 1] \to \mathbb{R}$ *be continuous function, with* $f(0) = f(1) = 0$. *Assume that* f'' *exists on* $0 < x < 1$, *with* $f'' + 2f' + f \geqslant 0$. *Show that* $f(x) \leqslant 0$ *for all* $0 \leqslant x \leqslant 1$.

Problem 1.4.13 (Sp85) *Let* v_1 *and* v_2 *be two real valued continuous functions on* $\mathbb{R}$ *such that* $v_1(x) < v_2(x)$ *for all* $x \in \mathbb{R}$. *Let* $\varphi_1(t)$ *and* $\varphi_2(t)$ *be, respectively, solutions of the differential equations*

$$\frac{dx}{dt} = v_1(x) \quad and \quad \frac{dx}{dt} = v_2(x)$$

for $a < t < b$. If $\varphi_1(t_0) = \varphi_2(t_0)$ for some $t_0 \in (a, b)$, show that $\varphi_1(t) \leqslant \varphi_2(t)$ for all $t \in (t_0, b)$.

Problem 1.4.14 (Fa83, Fa84) *Prove or supply a counterexample: If f and g are C^1 real valued functions on $(0, 1)$, if*

$$\lim_{x \to 0} f(x) = \lim_{x \to 0} g(x) = 0,$$

if g and g' never vanish, and if

$$\lim_{x \to 0} \frac{f(x)}{g(x)} = c,$$

then

$$\lim_{x \to 0} \frac{f'(x)}{g'(x)} = c.$$

Problem 1.4.15 (Fa00) *Let f be a real-valued differentiable function on $(-1, 1)$ such that $f(x)/x^2$ has a finite limit as $x \to 0$. Does it follow that $f''(0)$ exists? Give a proof or a counterexample.*

Problem 1.4.16 (Sp90, Fa91) *Let f be an infinitely differentiable function from $\mathbb{R}$ to $\mathbb{R}$. Suppose that, for some positive integer n,*

$$f(1) = f(0) = f'(0) = f''(0) = \cdots = f^{(n)}(0) = 0.$$

Prove that $f^{(n+1)}(x) = 0$ for some x in $(0, 1)$.

Problem 1.4.17 (Sp00) *Let f be a positive function of class C^2 on $(0, \infty)$ such that $f' \leqslant 0$ and f'' is bounded. Prove that $\lim_{t \to \infty} f'(t) = 0$.*

Problem 1.4.18 (Sp86) *Let f be a positive differentiable function on $(0, \infty)$. Prove that*

$$\lim_{\delta \to 0} \left(\frac{f(x + \delta x)}{f(x)} \right)^{1/\delta}$$

exists (finitely) and is nonzero for each x.

Problem 1.4.19 (Sp88) *Suppose that $f(x)$, $-\infty < x < \infty$, is a continuous real valued function, that $f'(x)$ exists for $x \neq 0$, and that $\lim_{x \to 0} f'(x)$ exists. Prove that $f'(0)$ exists.*

Problem 1.4.20 (Sp88) *For each real value of the parameter t, determine the number of real roots, counting multiplicities, of the cubic polynomial $p_t(x) = (1 + t^2)x^3 - 3t^3 x + t^4$.*

Problem 1.4.21 (Sp91) *Let the real valued function f be defined in an open interval about the point a on the real line and be differentiable at a. Prove that if*

(x_n) is an increasing sequence and (y_n) is a decreasing sequence in the domain of f, and both sequences converge to a, then

$$\lim_{n \to \infty} \frac{f(y_n) - f(x_n)}{y_n - x_n} = f'(a).$$

Problem 1.4.22 (Fa86) *Let f be a continuous real valued function on $[0, 1]$ such that, for each $x_0 \in [0, 1)$,*

$$\limsup_{x \to x_0^+} \frac{f(x) - f(x_0)}{x - x_0} \geqslant 0.$$

Prove that f is nondecreasing.

Problem 1.4.23 (Sp84) *Let I be an open interval in $\mathbb{R}$ containing zero. Assume that f' exists on a neighborhood of zero and $f''(0)$ exists. Show that*

$$f(x) = f(0) + f'(0) \sin x + \frac{1}{2} f''(0) \sin^2 x + o(x^2)$$

($o(x^2)$ denotes a quantity such that $\dfrac{o(x^2)}{x^2} \to 0$ as $x \to 0$).

Problem 1.4.24 (Sp84) *Prove that the Taylor coefficients at the origin of the function*

$$f(z) = \frac{z}{e^z - 1}$$

are rational numbers.

Problem 1.4.25 (Sp79) *Give an example of a function $f : \mathbb{R} \to \mathbb{R}$ having all three of the following properties:*

- *$f(x) = 0$ for $x < 0$ and $x > 2$,*

- *$f'(1) = 1$,*

- *f has derivatives of all orders.*

Problem 1.4.26 (Sp99) *Prove that if n is a positive integer and α, ε are real numbers with $\varepsilon > 0$, then there is a real function f with derivatives of all orders such that*

1. *$|f^{(k)}(x)| \leqslant \varepsilon$ for $k = 0, 1, \ldots, n - 1$ and all $x \in \mathbb{R}$,*

2. *$f^{(k)}(0) = 0$ for $k = 0, 1, \ldots, n - 1$,*

3. *$f^{(n)}(0) = \alpha$.*

Problem 1.4.27 (Su83) *Let* $f : \mathbb{R} \to \mathbb{R}$ *be continuously differentiable, periodic of period* 1, *and nonnegative. Show that*

$$\frac{d}{dx}\left(\frac{f(x)}{1 + cf(x)}\right) \to 0 \quad (\text{as } c \to \infty)$$

uniformly in x.

Problem 1.4.28 (Su81) *Let* $I \subset \mathbb{R}$ *be the open interval from* 0 *to* 1. *Let* $f : I \to \mathbb{C}$ *be* C^1 *(i.e., the real and imaginary parts are continuously differentiable). Suppose that* $f(t) \to 0$, $f'(t) \to C \neq 0$ *as* $t \to 0+$. *Show that the function* $g(t) = |f(t)|$ *is* C^1 *for sufficiently small* $t > 0$ *and that* $\lim\limits_{t \to 0+} g'(t)$ *exists, and evaluate the limit.*

Problem 1.4.29 (Fa95) *Let* $f : \mathbb{R} \to \mathbb{R}$ *be a* C^∞ *function. Assume that* $f(x)$ *has a local minimum at* $x = 0$. *Prove there is a disc centered on the* y *axis which lies above the graph of* f *and touches the graph at* $(0, f(0))$.

1.5   Integral Calculus

Problem 1.5.1 (Fa98) *Let* f *be a real function on* $[a, b]$. *Assume that* f *is differentiable and that* f' *is Riemann integrable. Prove that*

$$\int_a^b f'(x)\, dx = f(b) - f(a).$$

Problem 1.5.2 (Sp98) *Using the properties of the Riemann integral, show that if* f *is a non-negative continuous function on* $[0, 1]$, *and* $\int_0^1 f(x)dx = 0$, *then* $f(x) = 0$ *for all* $x \in [0, 1]$.

Problem 1.5.3 (Fa90) *Suppose* f *is a continuous real valued function. Show that*

$$\int_0^1 f(x)x^2\, dx = \frac{1}{3}f(\xi)$$

for some $\xi \in [0, 1]$.

Problem 1.5.4 (Sp77) *Suppose that* f *is a real valued function of one real variable such that*

$$\lim_{x \to c} f(x)$$

exists for all $c \in [a, b]$. *Show that* f *is Riemann integrable on* $[a, b]$.

Problem 1.5.5 (Sp78) *Let* $f : [0, 1] \to \mathbb{R}$ *be Riemann integrable over* $[b, 1]$ *for all* b *such that* $0 < b \leqslant 1$.

1. *If* f *is bounded, prove that* f *is Riemann integrable over* $[0, 1]$.

2. *What if f is not bounded?*

Problem 1.5.6 (Su81) *Let* $f : \mathbb{R} \to \mathbb{R}$ *be continuous, with*

$$\int_{-\infty}^{\infty} |f(x)| \, dx < \infty.$$

Show that there is a sequence (x_n) *such that* $x_n \to \infty$, $x_n f(x_n) \to 0$, *and* $x_n f(-x_n) \to 0$ *as* $n \to \infty$.

Problem 1.5.7 (Su85) *Let*

$$f(x) = e^{x^2/2} \int_x^{\infty} e^{-t^2/2} \, dt$$

for $x > 0$.

1. *Show that* $0 < f(x) < \dfrac{1}{x}$.

2. *Show that* $f(x)$ *is strictly decreasing for* $x > 0$.

Problem 1.5.8 (Su84) *Let* $\varphi(s)$ *be a* C^2 *function on* $[1, 2]$ *with* φ *and* φ' *vanishing at* $s = 1, 2$. *Prove that there is a constant* $C > 0$ *such that for any* $\lambda > 1$,

$$\left| \int_1^2 e^{i\lambda x} \varphi(x) \, dx \right| \leqslant \frac{C}{\lambda^2}.$$

Problem 1.5.9 (Fa85) *Let* $0 \leqslant a \leqslant 1$ *be given. Determine all nonnegative continuous functions* f *on* $[0, 1]$ *which satisfy the following three conditions:*

$$\int_0^1 f(x) \, dx = 1,$$

$$\int_0^1 x f(x) \, dx = a,$$

$$\int_0^1 x^2 f(x) \, dx = a^2.$$

Problem 1.5.10 (Fa85, Sp90) *Let* f *be a differentiable function on* $[0, 1]$ *and let*

$$\sup_{0 < x < 1} |f'(x)| = M < \infty.$$

Let n *be a positive integer. Prove that*

$$\left| \sum_{j=0}^{n-1} \frac{f(j/n)}{n} - \int_0^1 f(x) \, dx \right| \leqslant \frac{M}{2n}.$$

Problem 1.5.11 (Fa83) *Let $f : [0, \infty) \to \mathbb{R}$ be a uniformly continuous function with the property that*

$$\lim_{b \to \infty} \int_0^b f(x)\, dx$$

exists (as a finite limit). Show that

$$\lim_{x \to \infty} f(x) = 0.$$

Problem 1.5.12 (Fa86) *Let f be a real valued continuous function on $[0, \infty)$ such that*

$$\lim_{x \to \infty} \left(f(x) + \int_0^x f(t)\, dt \right)$$

exists. Prove that

$$\lim_{x \to \infty} f(x) = 0.$$

Problem 1.5.13 (Sp83) *Let $f : \mathbb{R}_+ \to \mathbb{R}_+$ be a monotone decreasing function, defined on the positive real numbers with*

$$\int_0^\infty f(x)\, dx < \infty.$$

Show that

$$\lim_{x \to \infty} x f(x) = 0.$$

Problem 1.5.14 (Fa90, Sp97) *Let f be a continuous real valued function satisfying $f(x) \geqslant 0$, for all x, and*

$$\int_0^\infty f(x)\, dx < \infty.$$

Prove that

$$\frac{1}{n} \int_0^n x f(x)\, dx \to 0$$

as $n \to \infty$.

Problem 1.5.15 (Sp87) *Evaluate the integral*

$$I = \int_0^{1/2} \frac{\sin x}{x}\, dx$$

to an accuracy of two decimal places; that is, find a number I^* such that $|I - I^*| < 0.005$.

Problem 1.5.16 (Fa87) *Show that the following limit exists and is finite:*

$$\lim_{t \to 0^+} \left(\int_0^1 \frac{dx}{(x^4 + t^4)^{1/4}} + \log t \right).$$

Problem 1.5.17 (Fa95) *Let f and f' be continuous on $[0, \infty)$ and $f(x) = 0$ for $x \geqslant 10^{10}$. Show that*

$$\int_0^\infty f(x)^2 dx \leqslant 2\sqrt{\int_0^\infty x^2 f(x)^2 dx} \sqrt{\int_0^\infty f'(x)^2 dx} .$$

Problem 1.5.18 (Fa88) *Let f be a continuous, strictly increasing function from $[0, \infty)$ onto $[0, \infty)$ and let $g = f^{-1}$. Prove that*

$$\int_0^a f(x)\, dx + \int_0^b g(y)\, dy \geqslant ab$$

for all positive numbers a and b, and determine the condition for equality.

Problem 1.5.19 (Sp94) *Let f be a continuous real valued function on $\mathbb{R}$ such that the improper Riemann integral $\int_{-\infty}^\infty |f(x)|\, dx$ converges. Define the function g on $\mathbb{R}$ by*

$$g(y) = \int_{-\infty}^\infty f(x) \cos(xy)\, dx .$$

Prove that g is continuous.

Problem 1.5.20 (Fa99) *Let f and g be continuous real valued functions on $\mathbb{R}$ such that $\lim_{|x| \to \infty} f(x) = 0$ and $\int_{-\infty}^\infty |g(x)| dx < \infty$. Define the function h on $\mathbb{R}$ by*

$$h(x) = \int_{-\infty}^\infty f(x - y)g(y) dy .$$

Prove that $\lim_{|x| \to \infty} h(x) = 0$.

Problem 1.5.21 (Sp88) *Prove that the integrals*

$$\int_0^\infty \cos x^2\, dx \quad and \quad \int_0^\infty \sin x^2\, dx$$

converge.

Problem 1.5.22 (Fa85) *Let $f(x)$, $0 \leqslant x \leqslant 1$, be a real valued continuous function. Show that*

$$\lim_{n \to \infty} (n + 1) \int_0^1 x^n f(x)\, dx = f(1).$$

Problem 1.5.23 (Su83, Sp84, Fa89) *Compute*

$$\int_0^\infty \frac{\log x}{x^2 + a^2} dx$$

where $a > 0$ is a constant.

Problem 1.5.24 (Sp85) *Show that*

$$I = \int_0^\pi \log(\sin x)\, dx$$

converges as an improper Riemann integral. Evaluate I.

Problem 1.5.25 (Sp02) *Prove that $\int_0^\infty \dfrac{\sin x}{\sqrt{x}}\, dx$ converges as an improper Riemann integral, but that $\int_0^\infty \dfrac{|\sin x|}{\sqrt{x}}\, dx = \infty$.*

1.6   Sequences of Functions

Problem 1.6.1 (Fa84) *Prove or supply a counterexample: If f is a nondecreasing real valued function on $[0, 1]$, then there is a sequence of continuous functions on $[0, 1]$, $\{f_n\}$, such that for each $x \in [0, 1]$,*

$$\lim_{n \to \infty} f_n(x) = f(x).$$

Problem 1.6.2 (Fa77, Sp80) *Let $f_n : \mathbb{R} \to \mathbb{R}$ be differentiable for each $n = 1, 2, \ldots$ with $|f_n'(x)| \leqslant 1$ for all n and x. Assume*

$$\lim_{n \to \infty} f_n(x) = g(x)$$

for all x. Prove that $g : \mathbb{R} \to \mathbb{R}$ is continuous.

Problem 1.6.3 (Fa87) *Suppose that $\{f_n\}$ is a sequence of nondecreasing functions which map the unit interval into itself. Suppose that*

$$\lim_{n \to \infty} f_n(x) = f(x)$$

pointwise and that f is a continuous function. Prove that $f_n(x) \to f(x)$ uniformly as $n \to \infty$, $0 \leqslant x \leqslant 1$. Note that the functions f_n are not necessarily continuous.

Problem 1.6.4 (Fa85) *Let f and f_n, $n = 1, 2, \ldots$, be functions from $\mathbb{R}$ to $\mathbb{R}$. Assume that $f_n(x_n) \to f(x)$ as $n \to \infty$ whenever $x_n \to x$. Show that f is continuous. Note: The functions f_n are not assumed to be continuous.*

Problem 1.6.5 (Sp99) *Suppose that a sequence of functions $f_n : \mathbb{R} \to \mathbb{R}$ converges uniformly on $\mathbb{R}$ to a function $f : \mathbb{R} \to \mathbb{R}$, and that $c_n = \lim_{x \to \infty} f_n(x)$ exists for each positive integer n. Prove that $\lim_{n \to \infty} c_n$ and $\lim_{x \to \infty} f(x)$ both exist and are equal.*

Problem 1.6.6 (Sp81)   *1. Give an example of a sequence of C^1 functions*

$$f_k : [0, \infty) \to \mathbb{R}, \quad k = 0, 1, 2, \ldots$$

such that $f_k(0) = 0$ for all k, and $f'_k(x) \to f'_0(x)$ for all x as $k \to \infty$, but $f_k(x)$ does not converge to $f_0(x)$ for all x as $k \to \infty$.

2. State an extra condition which would imply that $f_k(x) \to f_0(x)$ for all x as $k \to \infty$.

Problem 1.6.7 (Fa84) *Show that if f is a homeomorphism of $[0, 1]$ onto itself, then there is a sequence $\{p_n\}$, $n = 1, 2, 3, \ldots$ of polynomials such that $p_n \to f$ uniformly on $[0, 1]$ and each p_n is a homeomorphism of $[0, 1]$ onto itself.*

Problem 1.6.8 (Sp95) *Let $f_n : [0, 1] \to [0, \infty)$ be a continuous function, for $n = 1, 2, \ldots$. Suppose that one has*

$$(*) \quad f_1(x) \geqslant f_2(x) \geqslant f_3(x) \geqslant \cdots \quad \text{for all } x \in [0, 1].$$

Let $f(x) = \lim_{n \to \infty} f_n(x)$ and $M = \sup_{0 \leqslant x \leqslant 1} f(x)$.

1. Prove that there exists $t \in [0, 1]$ with $f(t) = M$.

2. Show by example that the conclusion of Part 1 need not hold if instead of $(*)$ we merely know that for each $x \in [0, 1]$ there exists n_x such that for all $n \geqslant n_x$ one has $f_n(x) \geqslant f_{n+1}(x)$.

Problem 1.6.9 (Fa82) *Let $f_1, f_2, \ldots$ be continuous functions on $[0, 1]$ satisfying $f_1 \geqslant f_2 \geqslant \cdots$ and such that $\lim_{n \to \infty} f_n(x) = 0$ for each x. Must the sequence $\{f_n\}$ converge to 0 uniformly on $[0, 1]$?*

Problem 1.6.10 (Sp78) *Let $k \geqslant 0$ be an integer and define a sequence of maps*

$$f_n : \mathbb{R} \to \mathbb{R}, \quad f_n(x) = \frac{x^k}{x^2 + n}, \quad n = 1, 2, \ldots.$$

For which values of k does the sequence converge uniformly on $\mathbb{R}$? On every bounded subset of $\mathbb{R}$?

Problem 1.6.11 (Sp81) *Let $f : [0, 1] \to \mathbb{R}$ be continuous and $k \in \mathbb{N}$. Prove that there is a real polynomial $P(x)$ of degree $\leqslant k$ which minimizes (for all such polynomials)*

$$\sup_{0 \leqslant x \leqslant 1} |f(x) - P(x)|.$$

Problem 1.6.12 (Fa79, Fa80) *Let $\{P_n\}$ be a sequence of real polynomials of degree $\leqslant D$, a fixed integer. Suppose that $P_n(x) \to 0$ pointwise for $0 \leqslant x \leqslant 1$. Prove that $P_n \to 0$ uniformly on $[0, 1]$.*

Problem 1.6.13 (Su85) *Let f be a real valued continuous function on a compact interval $[a, b]$. Given $\varepsilon > 0$, show that there is a polynomial p such that $p(a) = f(a)$, $p'(a) = 0$, and $|p(x) - f(x)| < \varepsilon$ for $x \in [a, b]$.*

Problem 1.6.14 (Sp95) *For each positive integer n, define $f_n : \mathbb{R} \to \mathbb{R}$ by $f_n(x) = \cos nx$. Prove that the sequence of functions $\{f_n\}$ has no uniformly convergent subsequence.*

Problem 1.6.15 (Fa03) *Let $C_{[0,1]}$ denote the space of continuous functions on $[0, 1]$. Define*

$$d(f, g) = \int_0^1 \frac{|f(x) - g(x)|}{1 + |f(x) - g(x)|} dx.$$

1. Show that d is a metric on $C_{[0,1]}$.

2. Show that $(C_{[0,1]}, d)$ is not a complete metric space.

Problem 1.6.16 (Fa86) *The Arzelà–Ascoli Theorem asserts that the sequence $\{f_n\}$ of continuous real valued functions on a metric space Ω is precompact (i.e., has a uniformly convergent subsequence) if*

(i) Ω is compact,

(ii) $\sup \|f_n\| < \infty$ (where $\|f_n\| = \sup\{|f_n(x)| \mid x \in \Omega\}$),

(iii) the sequence is equicontinuous.

Give examples of sequences which are not precompact such that: (i) and (ii) hold but (iii) fails; (i) and (iii) hold but (ii) fails; (ii) and (iii) hold but (i) fails. Take Ω to be a subset of the real line. Sketch the graph of a typical member of the sequence in each case.

Problem 1.6.17 (Sp01) *Let the functions $f_n : [0, 1] \to [0, 1]$ ($n = 1, 2, \ldots$) satisfy $|f_n(x) - f_n(y)| \leqslant |x - y|$ whenever $|x - y| \geqslant 1/n$. Prove that the sequence $\{f_n\}_{n=1}^\infty$ has a uniformly convergent subsequence.*

Problem 1.6.18 (Fa92) *Let $\{f_n\}$ be a sequence of real valued C^1 functions on $[0, 1]$ such that, for all n,*

$$|f_n'(x)| \leqslant \frac{1}{\sqrt{x}} \quad (0 < x \leqslant 1),$$

$$\int_0^1 f_n(x) \, dx = 0.$$

Prove that the sequence has a subsequence that converges uniformly on $[0, 1]$.

Problem 1.6.19 (Fa96) *Let M be the set of real valued continuous functions f on $[0, 1]$ such that f' is continuous on $[0, 1]$, with the norm*

$$\|f\| = \sup_{0 \leqslant x \leqslant 1} |f(x)| + \sup_{0 \leqslant x \leqslant 1} |f'(x)|.$$

Which subsets of M are compact?

Problem 1.6.20 (Su80) *Let* (a_n) *be a sequence of nonzero real numbers. Prove that the sequence of functions* $f_n : \mathbb{R} \to \mathbb{R}$

$$f_n(x) = \frac{1}{a_n}\sin(a_n x) + \cos(x + a_n)$$

has a subsequence converging to a continuous function.

Problem 1.6.21 (Sp82) *Let* $\{f_n\}$ *be a sequence of continuous functions from* $[0, 1]$ *to* $\mathbb{R}$. *Suppose that* $f_n(x) \to 0$ *as* $n \to \infty$ *for each* $x \in [0, 1]$ *and also that, for some constant* K, *we have*

$$\left| \int_0^1 f_n(x)\,dx \right| \leqslant K < \infty$$

for all n. Does

$$\lim_{n \to \infty} \int_0^1 f_n(x)\,dx = 0 ?$$

Problem 1.6.22 (Sp82, Sp93) *Let* $\{g_n\}$ *be a sequence of twice differentiable functions on* $[0, 1]$ *such that* $g_n(0) = g_n'(0) = 0$ *for all n. Suppose also that* $|g_n''(x)| \leqslant 1$ *for all n and all* $x \in [0, 1]$. *Prove that there is a subsequence of* $\{g_n\}$ *which converges uniformly on* $[0, 1]$.

Problem 1.6.23 (Fa93) *Let* K *be a continuous real valued function defined on* $[0, 1] \times [0, 1]$. *Let* F *be the family of functions* f *on* $[0, 1]$ *of the form*

$$f(x) = \int_0^1 g(y)K(x, y)\,dy$$

with g a real valued continuous function on $[0, 1]$ *satisfying* $|g| \leqslant 1$ *everywhere. Prove that the family* F *is equicontinuous.*

Problem 1.6.24 (Fa78) *Let* $\{g_n\}$ *be a sequence of Riemann integrable functions from* $[0, 1]$ *into* $\mathbb{R}$ *such that* $|g_n(x)| \leqslant 1$ *for all n, x. Define*

$$G_n(x) = \int_0^x g_n(t)\,dt.$$

Prove that a subsequence of $\{G_n\}$ *converges uniformly.*

Problem 1.6.25 (Su79) *Let* $\{f_n\}$ *be a sequence of continuous real functions defined* $[0, 1]$ *such that*

$$\int_0^1 (f_n(y))^2\,dy \leqslant 5$$

for all n. Define $g_n : [0, 1] \to \mathbb{R}$ *by*

$$g_n(x) = \int_0^1 \sqrt{x + y}\,f_n(y)\,dy.$$

1. *Find a constant $K \geqslant 0$ such that $|g_n(x)| \leqslant K$ for all n.*

2. *Prove that a subsequence of the sequence $\{g_n\}$ converges uniformly.*

Problem 1.6.26 (Su81) *Let $\{f_n\}$ be a sequence of continuous functions defined from $[0, 1] \to \mathbb{R}$ such that*

$$\int_0^1 (f_n(x) - f_m(x))^2 \, dx \to 0 \quad as \quad n, m \to \infty.$$

Let $K : [0, 1] \times [0, 1] \to \mathbb{R}$ be continuous. Define $g_n : [0, 1] \to \mathbb{R}$ by

$$g_n(x) = \int_0^1 K(x, y) f_n(y) \, dy.$$

Prove that the sequence $\{g_n\}$ converges uniformly.

Problem 1.6.27 (Fa82) *Let $\varphi_1, \varphi_2, \ldots, \varphi_n, \ldots$ be nonnegative continuous functions on $[0, 1]$ such that the limit*

$$\lim_{n \to \infty} \int_0^1 x^k \varphi_n(x) \, dx$$

exists for every $k = 0, 1, \ldots$. Show that the limit

$$\lim_{n \to \infty} \int_0^1 f(x) \varphi_n(x) \, dx$$

exists for every continuous function f on $[0, 1]$.

Problem 1.6.28 (Sp83) *Let $\lambda_1, \lambda_2, \ldots, \lambda_n, \ldots$ be real numbers. Show that the infinite series*

$$\sum_{n=1}^{\infty} \frac{e^{i\lambda_n x}}{n^2}$$

converges uniformly over $\mathbb{R}$ to a continuous limit function $f : \mathbb{R} \to \mathbb{C}$. Show, further, that the limit

$$\lim_{T \to \infty} \frac{1}{2T} \int_{-T}^{T} f(x) \, dx$$

exists.

Problem 1.6.29 (Sp85) *Define the function ζ by*

$$\zeta(x) = \sum_{n=1}^{\infty} \frac{1}{n^x}.$$

Prove that $\zeta(x)$ is defined and has continuous derivatives of all orders in the interval $1 < x < \infty$.

Problem 1.6.30 (Sp85) *Let f be continuous on* $\mathbb{R}$, *and let*

$$f_n(x) = \frac{1}{n} \sum_{k=0}^{n-1} f\left(x + \frac{k}{n}\right).$$

Prove that $f_n(x)$ *converges uniformly to a limit on every finite interval* $[a, b]$.

Problem 1.6.31 (Sp87) *Let* f *be a continuous real valued function on* $\mathbb{R}$ *satisfying*

$$|f(x)| \leqslant \frac{C}{1 + x^2},$$

where C is a positive constant. Define the function F on $\mathbb{R}$ *by*

$$F(x) = \sum_{n=-\infty}^{\infty} f(x + n).$$

1. *Prove that F is continuous and periodic with period* 1.

2. *Prove that if G is continuous and periodic with period* 1, *then*

$$\int_0^1 F(x)G(x)\,dx = \int_{-\infty}^{\infty} f(x)G(x)\,dx.$$

Problem 1.6.32 (Sp79) *Show that for any continuous function* $f : [0, 1] \to \mathbb{R}$ *and* $\varepsilon > 0$, *there is a function of the form*

$$g(x) = \sum_{k=0}^{n} C_k x^{4k}$$

for some $n \in \mathbb{Z}$, *where* $C_0, \ldots, C_n \in \mathbb{Q}$ *and* $|g(x) - f(x)| < \varepsilon$ *for all x in* $[0, 1]$.

1.7   Fourier Series

Problem 1.7.1 (Sp80) *Let* $f : \mathbb{R} \to \mathbb{R}$ *be the unique function such that* $f(x) = x$ *if* $-\pi \leqslant x < \pi$ *and* $f(x + 2n\pi) = f(x)$ *for all* $n \in \mathbb{Z}$.

1. *Prove that the Fourier series of f is*

$$\sum_{n=1}^{\infty} \frac{(-1)^{n+1} 2 \sin nx}{n}.$$

2. *Prove that the series does not converge uniformly.*

3. *For each* $x \in \mathbb{R}$, *find the sum of the series.*

Problem 1.7.2 (Su81) *Let $f : \mathbb{R} \to \mathbb{R}$ be the function of period 2π such that $f(x) = x^3$ for $-\pi \leqslant x < \pi$.*

1. *Prove that the Fourier series for f has the form $\sum\limits_{1}^{\infty} b_n \sin nx$ and write an integral formula for b_n (do not evaluate it).*

2. *Prove that the Fourier series converges for all x.*

3. *Prove*

$$\sum_{n=1}^{\infty} b_n^2 = \frac{2\pi^6}{7}.$$

Problem 1.7.3 (Su82) *Let $f : [0, \pi] \to \mathbb{R}$ be continuous and such that*

$$\int_0^{\pi} f(x) \sin nx \, dx = 0$$

for all integers $n \geqslant 1$. Is f identically 0?

Problem 1.7.4 (Sp86) *Let f be a continuous real valued function on $\mathbb{R}$ such that*

$$f(x) = f(x+1) = f\left(x + \sqrt{2}\right)$$

for all x. Prove that f is constant.

Problem 1.7.5 (Sp88) *Does there exist a continuous real valued function $f(x)$, $0 \leqslant x \leqslant 1$, such that*

$$\int_0^1 xf(x) \, dx = 1 \quad and \quad \int_0^1 x^n f(x) \, dx = 0$$

for $n = 0, 2, 3, 4, \ldots$? Give an example or a proof that no such f exists.

Problem 1.7.6 (Fa80) *Let g be 2π-periodic, continuous on $[-\pi, \pi]$ and have Fourier series*

$$\frac{a_0}{2} + \sum_{n=1}^{\infty} (a_n \cos nx + b_n \sin nx).$$

Let f be 2π-periodic and satisfy the differential equation

$$f''(x) + kf(x) = g(x)$$

where $k \neq n^2, n = 1, 2, 3, \ldots$. Find the Fourier series of f and prove that it converges everywhere.

Problem 1.7.7 (Su83) *Let f be a twice differentiable real valued function on $[0, 2\pi]$, with $\int_0^{2\pi} f(x)dx = 0 = f(2\pi) - f(0)$. Show that*

$$\int_0^{2\pi} (f(x))^2 \, dx \leqslant \int_0^{2\pi} (f'(x))^2 \, dx.$$

Problem 1.7.8 (Fa81) *Let f and g be continuous functions on $\mathbb{R}$ such that $f(x+1) = f(x)$, $g(x+1) = g(x)$, for all $x \in \mathbb{R}$. Prove that*

$$\lim_{n \to \infty} \int_0^1 f(x)g(nx)\, dx = \int_0^1 f(x)\, dx \int_0^1 g(x)\, dx.$$

1.8 Convex Functions

Problem 1.8.1 (Sp81) *Let $f : [0, 1] \to \mathbb{R}$ be continuous with $f(0) = 0$. Show there is a continuous concave function $g : [0, 1] \to \mathbb{R}$ such that $g(0) = 0$ and $g(x) \geqslant f(x)$ for all $x \in [0, 1]$.*
Note: A function $g : I \to \mathbb{R}$ is concave *if*

$$g(tx + (1-t)y) \geqslant tg(x) + (1-t)g(y)$$

for all x and y in I and $0 \leqslant t \leqslant 1$.

Problem 1.8.2 (Sp82) *Let $f : I \to \mathbb{R}$ (where I is an interval of $\mathbb{R}$) be such that $f(x) > 0$, $x \in I$. Suppose that $e^{cx} f(x)$ is convex in I for every real number c. Show that $\log f(x)$ is convex in I.*
Note: A function $g : I \to \mathbb{R}$ is convex *if*

$$g(tx + (1-t)y) \leqslant tg(x) + (1-t)g(y)$$

for all x and y in I and $0 \leqslant t \leqslant 1$.

Problem 1.8.3 (Sp86) *Let f be a real valued continuous function on $\mathbb{R}$ satisfying the* mean value inequality *below:*

$$f(x) \leqslant \frac{1}{2h} \int_{x-h}^{x+h} f(y)\, dy, \quad x \in \mathbb{R}, \quad h > 0.$$

Prove:

1. *The maximum of f on any closed interval is assumed at one of the endpoints.*

2. *f is convex.*

2
Multivariable Calculus

2.1   Limits and Continuity

Problem 2.1.1 (Fa94) *Let the function* $f : \mathbb{R}^n \to \mathbb{R}^n$ *satisfy the following two conditions:*

(i) $f(K)$ *is compact whenever* K *is a compact subset of* $\mathbb{R}^n$.

(ii) *If* $\{K_n\}$ *is a decreasing sequence of compact subsets of* $\mathbb{R}^n$, *then*

$$f\left(\bigcap_{1}^{\infty} K_n\right) = \bigcap_{1}^{\infty} f(K_n).$$

Prove that f *is continuous.*

Problem 2.1.2 (Sp78) *Prove that a map* $g : \mathbb{R}^n \to \mathbb{R}^n$ *is continuous only if its graph is closed in* $\mathbb{R}^n \times \mathbb{R}^n$. *Is the converse true?*
Note: See also Problem 1.2.14.

Problem 2.1.3 (Su79) *Let* $U \subset \mathbb{R}^n$ *be an open set. Suppose that the map* $h : U \to \mathbb{R}^n$ *is a homeomorphism from* U *onto* $\mathbb{R}^n$, *which is uniformly continuous. Prove* $U = \mathbb{R}^n$.

Problem 2.1.4 (Sp89) *Let* f *be a real valued function on* $\mathbb{R}^2$ *with the following properties:*

1. *For each* y_0 *in* $\mathbb{R}$, *the function* $x \mapsto f(x, y_0)$ *is continuous.*

2. *For each x_0 in $\mathbb{R}$, the function $y \mapsto f(x_0, y)$ is continuous.*

3. *$f(K)$ is compact whenever K is a compact subset of $\mathbb{R}^2$.*

Prove that f is continuous.

Problem 2.1.5 (Sp91) *Let f be a continuous function from the ball $B_n = \{x \in \mathbb{R}^n \mid \|x\| < 1\}$ into itself. (Here, $\|\cdot\|$ denotes the Euclidean norm.) Assume $\|f(x)\| < \|x\|$ for all nonzero $x \in B_n$. Let x_0 be a nonzero point of B_n, and define the sequence (x_k) by setting $x_k = f(x_{k-1})$. Prove that $\lim x_k = 0$.*

Problem 2.1.6 (Su78) *Let N be a norm on the vector space $\mathbb{R}^n$; that is, $N : \mathbb{R}^n \to \mathbb{R}$ satisfies*

$$N(x) \geqslant 0 \ \text{and} \ N(x) = 0 \ \text{only if} \ x = 0,$$
$$N(x + y) \leqslant N(x) + N(y),$$
$$N(\lambda x) = |\lambda| N(x)$$

for all $x, y \in \mathbb{R}^n$ and $\lambda \in \mathbb{R}$.

1. *Prove that N is bounded on the unit sphere.*

2. *Prove that N is continuous.*

3. *Prove that there exist constants $A > 0$ and $B > 0$, such that for all $x \in \mathbb{R}^n$, $A\|x\| \leqslant N(x) \leqslant B\|x\|$.*

Problem 2.1.7 (Fa97) *A map $f : \mathbb{R}^m \to \mathbb{R}^n$ is proper if it is continuous and $f^{-1}(B)$ is compact for each compact subset B of $\mathbb{R}^n$; f is closed if it is continuous and $f(A)$ is closed for each closed subset A of $\mathbb{R}^m$.*

1. *Prove that every proper map $f : \mathbb{R}^m \to \mathbb{R}^n$ is closed.*

2. *Prove that every one-to-one closed map $f : \mathbb{R}^m \to \mathbb{R}^n$ is proper.*

Problem 2.1.8 (Sp83) *Suppose that $F : \mathbb{R}^n \to \mathbb{R}^n$ is continuous and satisfies*

$$\|F(x) - F(y)\| \geqslant \lambda \|x - y\|$$

for all $x, y \in \mathbb{R}^n$ and some $\lambda > 0$. Prove that F is one-to-one, onto, and has a continuous inverse.
Note: See also Problem 1.2.12.

2.2 Differential Calculus

Problem 2.2.1 (Sp93) *Prove that $\dfrac{x^2 + y^2}{4} \leqslant e^{x+y-2}$ for $x \geqslant 0$, $y \geqslant 0$.*

Problem 2.2.2 (Fa98) *Find the minimal value of the areas of hexagons circumscribing the unit circle in $\mathbb{R}^2$.*
Note: See also Problem 1.1.12.

Problem 2.2.3 (Sp03) *Define $f : \mathbb{R}^2 \to \mathbb{R}^2$ by $f(x, 0) = 0$ and*

$$f(x, y) = \left(1 - \cos \frac{x^2}{y}\right) \sqrt{x^2 + y^2}$$

for $y \neq 0$.

1. *Show that f is continuous at $(0, 0)$.*

2. *Calculate all the directional derivatives of f at $(0, 0)$.*

3. *Show that f is not differentiable at $(0, 0)$.*

Problem 2.2.4 (Fa86) *Let $f : \mathbb{R}^2 \to \mathbb{R}$ be defined by:*

$$f(x, y) = \begin{cases} x^{4/3} \sin(y/x) & \text{if } x \neq 0 \\ 0 & \text{if } x = 0. \end{cases}$$

Determine all points at which f is differentiable.

Problem 2.2.5 (Sp00) *Let F, with components $F_1, \ldots, F_n$, be a differentiable map of $\mathbb{R}^n$ into $\mathbb{R}^n$ such that $F(0) = 0$. Assume that*

$$\sum_{j,k=1}^{n} \left| \frac{\partial F_j(0)}{\partial x_k} \right|^2 = c < 1.$$

Prove that there is a ball B in $\mathbb{R}^n$ with center 0 such that $F(B) \subset B$.

Problem 2.2.6 (Fa02) *Let p be a polynomial over $\mathbb{R}$ of positive degree. Define the function $f : \mathbb{R}^2 \to \mathbb{R}^2$ by $f(x, y) = (p(x + y), p(x - y))$. Prove that the derivative $Df(x, y)$ is invertible for an open dense set of points (x, y) in $\mathbb{R}^2$.*

Problem 2.2.7 (Fa02) *Find the most general continuously differentiable function $g : \mathbb{R} \to (0, \infty)$ such that the function $h(x, y) = g(x)g(y)$ on $\mathbb{R}^2$ is constant on each circle with center $(0, 0)$.*

Problem 2.2.8 (Sp80, Fa92) *Let $f : \mathbb{R}^n \to \mathbb{R}^n$ be continuously differentiable. Assume the Jacobian matrix $(\partial f_i / \partial x_j)$ has rank n everywhere. Suppose f is proper; that is, $f^{-1}(K)$ is compact whenever K is compact. Prove $f(\mathbb{R}^n) = \mathbb{R}^n$.*

Problem 2.2.9 (Sp89) *Suppose f is a continuously differentiable map of $\mathbb{R}^2$ into $\mathbb{R}^2$. Assume that f has only finitely many singular points, and that for each positive number M, the set $\{z \in \mathbb{R}^2 \mid |f(z)| \leqslant M\}$ is bounded. Prove that f maps $\mathbb{R}^2$ onto $\mathbb{R}^2$.*

Problem 2.2.10 (Fa81) *Let f be a real valued function on $\mathbb{R}^n$ of class C^2. A point $x \in \mathbb{R}^n$ is a critical point of f if all the partial derivatives of f vanish at x; a critical point is nondegenerate if the $n \times n$ matrix*

$$\left(\frac{\partial^2 f}{\partial x_i \partial x_j}(x) \right)$$

is nonsingular.

Let x be a nondegenerate critical point of f. Prove that there is an open neighborhood of x which contains no other critical points (i.e., the nondegenerate critical points are isolated).

Problem 2.2.11 (Su80) *Let $f : \mathbb{R}^n \to \mathbb{R}$ be a function whose partial derivatives of order $\leqslant 2$ are everywhere defined and continuous.*

1. *Let $a \in \mathbb{R}^n$ be a critical point of f (i.e., $\frac{\partial f}{\partial x_j}(a) = 0$, $i = 1, \ldots, n$). Prove that a is a local minimum provided the Hessian matrix*

$$\left(\frac{\partial^2 f}{\partial x_i \partial x_j} \right)$$

   is positive definite at $x = a$.

2. *Assume the Hessian matrix is positive definite at all x. Prove that f has, at most, one critical point.*

Problem 2.2.12 (Fa88) *Prove that a real valued C^3 function f on $\mathbb{R}^2$ whose Laplacian,*

$$\frac{\partial^2 f}{\partial x^2} + \frac{\partial^2 f}{\partial y^2},$$

is everywhere positive cannot have a local maximum.

Problem 2.2.13 (Fa01) *Let the function u on $\mathbb{R}^2$ be harmonic, not identically 0, and homogeneous of degree d, where $d > 0$. (The homogeneity condition means that $u(tx, ty) = t^d u(x, y)$ for $t > 0$.) Prove that d is an integer.*

Problem 2.2.14 (Su82) *Let $f : \mathbb{R}^3 \to \mathbb{R}^2$ and assume that 0 is a regular value of f (i.e., the differential of f has rank 2 at each point of $f^{-1}(0)$). Prove that $\mathbb{R}^3 \setminus f^{-1}(0)$ is arcwise connected.*

Problem 2.2.15 (Sp87) *Let the transformation T from the subset $U = \{(u, v) \mid u > v\}$ of $\mathbb{R}^2$ into $\mathbb{R}^2$ be defined by $T(u, v) = (u + v, u^2 + v^2)$.*

1. *Prove that T is locally one-to-one.*

2. *Determine the range of T, and show that T is globally one-to-one.*

Problem 2.2.16 (Fa91) *Let f be a C^1 function from the interval $(-1, 1)$ into $\mathbb{R}^2$ such that $f(0) = 0$ and $f'(0) \neq 0$. Prove that there is a number ε in $(0, 1)$ such that $\| f(t) \|$ is an increasing function of t on $(0, \varepsilon)$.*

Problem 2.2.17 (Fa80) *For a real 2×2 matrix*

$$X = \begin{pmatrix} x & y \\ z & t \end{pmatrix},$$

let $\| X \| = x^2 + y^2 + z^2 + t^2$, and define a metric by $d(X, Y) = \| X - Y \|$. Let $\Sigma = \{ X \mid \det(X) = 0 \}$. Let

$$A = \begin{pmatrix} 1 & 0 \\ 0 & 2 \end{pmatrix}.$$

Find the minimum distance from A to Σ and exhibit an $S \in \Sigma$ that achieves this minimum.

Problem 2.2.18 (Su80) *Let $S \subset \mathbb{R}^3$ denote the ellipsoidal surface defined by*

$$2x^2 + (y - 1)^2 + (z - 10)^2 = 1.$$

Let $T \subset \mathbb{R}^3$ be the surface defined by

$$z = \frac{1}{x^2 + y^2 + 1}.$$

Prove that there exist points $p \in S$, $q \in T$, such that the line $\overline{pq}$ is perpendicular to S at p and to T at q.

Problem 2.2.19 (Sp80) *Let P_2 denote the set of real polynomials of degree $\leqslant 2$. Define the map $J : P_2 \to \mathbb{R}$ by*

$$J(f) = \int_0^1 f(x)^2 \, dx.$$

Let $Q = \{ f \in P_2 \mid f(1) = 1 \}$. Show that J attains a minimum value on Q and determine where the minimum occurs.

Problem 2.2.20 (Su79) *Let X be the space of orthogonal real $n \times n$ matrices. Let $v_0 \in \mathbb{R}^n$. Locate and describe the elements of X, where the map*

$$f : X \to \mathbb{R}, \qquad f(A) = \langle v_0, A v_0 \rangle$$

takes its maximum and minimum values.

Problem 2.2.21 (Fa78) *Let $W \subset \mathbb{R}^n$ be an open connected set and f a real valued function on W such that all partial derivatives of f are 0. Prove that f is constant.*

Problem 2.2.22 (Sp77) *In* $\mathbb{R}^2$, *consider the region* $\mathcal{A}$ *defined by* $x^2 + y^2 > 1$. *Find differentiable real valued functions* f *and* g *on* $\mathcal{A}$ *such that* $\partial f/\partial x = \partial g/\partial y$ *but there is no real valued function* h *on* $\mathcal{A}$ *such that* $f = \partial h/\partial y$ *and* $g = \partial h/\partial x$.

Problem 2.2.23 (Sp77) *Suppose that* $u(x, t)$ *is a continuous function of the real variables* x *and* t *with continuous second partial derivatives. Suppose that* u *and its first partial derivatives are periodic in* x *with period* 1, *and that*

$$\frac{\partial^2 u}{\partial x^2} = \frac{\partial^2 u}{\partial t^2}.$$

Prove that

$$E(t) = \frac{1}{2} \int_0^1 \left(\left(\frac{\partial u}{\partial t}\right)^2 + \left(\frac{\partial u}{\partial x}\right)^2 \right) dx$$

is a constant independent of t.

Problem 2.2.24 (Su77) *Let* $f(x, t)$ *be a* C^1 *function such that* $\partial f/\partial x = \partial f/\partial t$. *Suppose that* $f(x, 0) > 0$ *for all* x. *Prove that* $f(x, t) > 0$ *for all* x *and* t.

Problem 2.2.25 (Fa77) *Let* $f : \mathbb{R}^n \to \mathbb{R}$ *have continuous partial derivatives and satisfy*

$$\left| \frac{\partial f}{\partial x_j}(x) \right| \leqslant K$$

for all $x = (x_1, \dots, x_n)$, $j = 1, \dots, n$. *Prove that*

$$|f(x) - f(y)| \leqslant \sqrt{n} K \|x - y\|$$

(where $\|u\| = \sqrt{u_1^2 + \cdots + u_n^2}$ *).*

Problem 2.2.26 (Fa83, Sp87) *Let* $f : \mathbb{R}^n \setminus \{0\} \to \mathbb{R}$ *be a function which is continuously differentiable and whose partial derivatives are uniformly bounded:*

$$\left| \frac{\partial f}{\partial x_i}(x_1, \dots, x_n) \right| \leqslant M$$

for all $(x_1, \dots, x_n) \neq (0, \dots, 0)$. *Show that if* $n \geqslant 2$, *then* f *can be extended to a continuous function defined on all of* $\mathbb{R}^n$. *Show that this is false if* $n = 1$ *by giving a counterexample.*

Problem 2.2.27 (Sp79) *Let* $f : \mathbb{R}^n \setminus \{0\} \to \mathbb{R}$ *be differentiable. Suppose*

$$\lim_{x \to 0} \frac{\partial f}{\partial x_j}(x)$$

exists for each $j = 1, \dots, n$.

1. *Can* f *be extended to a continuous map from* $\mathbb{R}^n$ *to* $\mathbb{R}$?

2. *Assuming continuity at the origin, is f differentiable from $\mathbb{R}^n$ to $\mathbb{R}$?*

Problem 2.2.28 (Sp82) *Let $f : \mathbb{R}^2 \to \mathbb{R}$ have directional derivatives in all directions at the origin. Is f differentiable at the origin? Prove or give a counterexample.*

Problem 2.2.29 (Fa78) *Let $f : \mathbb{R}^n \to \mathbb{R}$ have the following properties: f is differentiable on $\mathbb{R}^n \setminus \{0\}$, f is continuous at 0, and*

$$\lim_{p \to 0} \frac{\partial f}{\partial x_i}(p) = 0$$

for $i = 1, \ldots, n$. Prove that f is differentiable at 0.

Problem 2.2.30 (Su78) *Let $U \subset \mathbb{R}^n$ be a convex open set and $f : U \to \mathbb{R}^n$ a differentiable function whose partial derivatives are uniformly bounded but not necessarily continuous. Prove that f has a unique continuous extension to the closure of U.*

Problem 2.2.31 (Fa78)   *1. Show that if $u, v : \mathbb{R}^2 \to \mathbb{R}$ are continuously differentiable and $\dfrac{\partial u}{\partial y} = \dfrac{\partial v}{\partial x}$, then $u = \dfrac{\partial f}{\partial x}$, $v = \dfrac{\partial f}{\partial y}$ for some $f : \mathbb{R}^2 \to \mathbb{R}$.*

2. *Prove there is no $f : \mathbb{R}^2 \setminus \{0\} \to \mathbb{R}$ such that*

$$\frac{\partial f}{\partial x} = \frac{-y}{x^2 + y^2} \quad and \quad \frac{\partial f}{\partial y} = \frac{x}{x^2 + y^2}.$$

Problem 2.2.32 (Su79) *Let $f : \mathbb{R}^3 \to \mathbb{R}$ be such that*

$$f^{-1}(0) = \{v \in \mathbb{R}^3 \mid \|v\| = 1\}.$$

Suppose f has continuous partial derivatives of orders $\leqslant 2$. Is there a $p \in \mathbb{R}^3$ with $\|p\| \leqslant 1$ such that

$$\frac{\partial^2 f}{\partial x^2}(p) + \frac{\partial^2 f}{\partial y^2}(p) + \frac{\partial^2 f}{\partial z^2}(p) \geqslant 0 ?$$

Problem 2.2.33 (Sp92) *Let f be a differentiable function from $\mathbb{R}^n$ to $\mathbb{R}^n$. Assume that there is a differentiable function g from $\mathbb{R}^n$ to $\mathbb{R}$ having no critical points such that $g \circ f$ vanishes identically. Prove that the Jacobian determinant of f vanishes identically.*

Problem 2.2.34 (Fa83) *Let $f, g : \mathbb{R} \to \mathbb{R}$ be smooth functions with $f(0) = 0$ and $f'(0) \neq 0$. Consider the equation $f(x) = t g(x)$, $t \in \mathbb{R}$.*

1. *Show that in a suitably small interval $|t| < \delta$, there is a unique continuous function $x(t)$ which solves the equation and satisfies $x(0) = 0$.*

2. *Derive the first order Taylor expansion of $x(t)$ about $t = 0$.*

Problem 2.2.35 (Sp78) *Consider the system of equations*

$$3x + y - z + u^4 = 0$$
$$x - y + 2z + u = 0$$
$$2x + 2y - 3z + 2u = 0$$

1. *Prove that for some $\varepsilon > 0$, the system can be solved for (x, y, u) as a function of $z \in [-\varepsilon, \varepsilon]$, with $x(0) = y(0) = u(0) = 0$. Are such functions $x(z)$, $y(z)$ and $u(z)$ continuous? Differentiable? Unique?*

2. *Show that the system cannot be solved for (x, y, z) as a function of $u \in [-\delta, \delta]$, for all $\delta > 0$.*

Problem 2.2.36 (Sp81) *Describe the two regions in (a, b)-space for which the function*

$$f_{a,b}(x, y) = ay^2 + bx$$

restricted to the circle $x^2 + y^2 = 1$, has exactly two, and exactly four critical points, respectively.

Problem 2.2.37 (Fa87) *Let u and v be two real valued C^1 functions on $\mathbb{R}^2$ such that the gradient ∇u is never 0, and such that, at each point, ∇v and ∇u are linearly dependent vectors. Given $p_0 = (x_0, y_0) \in \mathbb{R}^2$, show that there is a C^1 function F of one variable such that $v(x, y) = F(u(x, y))$ in some neighborhood of p_0.*

Problem 2.2.38 (Fa94) *Let f be a continuously differentiable function from $\mathbb{R}^2$ into $\mathbb{R}$. Prove that there is a continuous one-to-one function g from $[0, 1]$ into $\mathbb{R}^2$ such that the composite function $f \circ g$ is constant.*

Problem 2.2.39 (Su84) *Let $f : \mathbb{R} \to \mathbb{R}$ be C^1 and let*

$$u = f(x)$$
$$v = -y + xf(x).$$

If $f'(x_0) \neq 0$, show that this transformation is locally invertible near (x_0, y_0) and the inverse has the form

$$x = g(u)$$
$$y = -v + ug(u).$$

Problem 2.2.40 (Su78, Fa99) *Let $M_{n \times n}$ denote the vector space of real $n \times n$ matrices. Define a map $f : M_{n \times n} \to M_{n \times n}$ by $f(X) = X^2$. Find the derivative of f.*

Problem 2.2.41 (Su82) *Let $M_{2 \times 2}$ be the four-dimensional vector space of all 2×2 real matrices and define $f : M_{2 \times 2} \to M_{2 \times 2}$ by $f(X) = X^2$.*

1. *Show that f has a local inverse near the point*

$$X = \begin{pmatrix} 1 & 0 \\ 0 & 1 \end{pmatrix}.$$

2. *Show that f does not have a local inverse near the point*

$$X = \begin{pmatrix} 1 & 0 \\ 0 & -1 \end{pmatrix}.$$

Problem 2.2.42 (Fa80) *Show that there is an $\varepsilon > 0$ such that if A is any real 2×2 matrix satisfying $|a_{ij}| \leqslant \varepsilon$ for all entries a_{ij} of A, then there is a real 2×2 matrix X such that $X^2 + X^t = A$, where X^t is the transpose of X. Is X unique?*

Problem 2.2.43 (Sp96) *Let $M_{2\times2}$ be the space of 2×2 matrices over $\mathbb{R}$, identified in the usual way with $\mathbb{R}^4$. Let the function F from $M_{2\times2}$ into $M_{2\times2}$ be defined by*

$$F(X) = X + X^2.$$

Prove that the range of F contains a neighborhood of the origin.

Problem 2.2.44 (Fa78) *Let $M_{n\times n}$ denote the vector space of $n \times n$ real matrices. Prove that there are neighborhoods U and V in $M_{n\times n}$ of the identity matrix such that for every A in U, there is a unique X in V such that $X^4 = A$.*

Problem 2.2.45 (Sp79, Fa93) *Let $M_{n\times n}$ denote the vector space of $n \times n$ real matrices for $n \geqslant 2$. Let $\det : M_{n\times n} \to \mathbb{R}$ be the determinant map.*

1. *Show that $\det$ is C^∞.*

2. *Show that the derivative of $\det$ at $A \in M_{n\times n}$ is zero if and only if A has rank $\leqslant n - 2$.*

Problem 2.2.46 (Fa83) *Let $F(t) = \big(f_{ij}(t)\big)$ be an $n\times n$ matrix of continuously differentiable functions $f_{ij} : \mathbb{R} \to \mathbb{R}$, and let*

$$u(t) = \operatorname{tr}\left(F(t)^3\right).$$

Show that u is differentiable and

$$u'(t) = 3\operatorname{tr}\left(F(t)^2 F'(t)\right).$$

Problem 2.2.47 (Fa81) *Let $A = \big(a_{ij}\big)$ be an $n \times n$ matrix whose entries a_{ij} are real valued differentiable functions defined on $\mathbb{R}$. Assume that the determinant $\det(A)$ of A is everywhere positive. Let $B = \big(b_{ij}\big)$ be the inverse matrix of A. Prove the formula*

$$\frac{d}{dt}\log\left(\det(A)\right) = \sum_{i,j=1}^{n} \frac{da_{ij}}{dt} b_{ji}.$$

Problem 2.2.48 (Sp03)   1. *Prove that there is no continuously differentiable, measure-preserving bijective function $f : \mathbb{R} \to \mathbb{R}_+$.*

2. *Find an example of a continuously differentiable, measure-preserving bijective function $f : \mathbb{R} \times \mathbb{R} \to \mathbb{R} \times \mathbb{R}_+$.*

2.3   Integral Calculus

Problem 2.3.1 (Sp78) *What is the volume enclosed by the ellipsoid*

$$\frac{x^2}{a^2} + \frac{y^2}{b^2} + \frac{z^2}{c^2} = 1?$$

Problem 2.3.2 (Sp78) *Evaluate*

$$\iint_A e^{-x^2-y^2}\, dx dy,$$

where $A = \{(x, y) \in \mathbb{R}^2 \mid x^2 + y^2 \leqslant 1\}$.

Problem 2.3.3 (Sp98) *Given the fact that* $\int_{-\infty}^{\infty} e^{-x^2} dx = \sqrt{\pi}$, *evaluate the integral*

$$I = \int_{-\infty}^{\infty} \int_{-\infty}^{\infty} e^{-(x^2+(y-x)^2+y^2)} dx\, dy .$$

Problem 2.3.4 (Fa86) *Evaluate*

$$\iint_{\mathcal{R}} (x^3 - 3xy^2)\, dx dy,$$

where

$$\mathcal{R} = \{(x, y) \in \mathbb{R}^2 \mid (x + 1)^2 + y^2 \leqslant 9, \quad (x - 1)^2 + y^2 \geqslant 1\}.$$

Problem 2.3.5 (Fa98) *Let* $\varphi(x, y)$ *be a function with continuous second order partial derivatives such that*

1. $\varphi_{xx} + \varphi_{yy} + \varphi_x = 0$ *in the punctured plane* $\mathbb{R}^2 \setminus \{0\}$,

2. $r\varphi_x \to \dfrac{x}{2\pi r}$ *and* $r\varphi_y \to \dfrac{y}{2\pi r}$ *as* $r = \sqrt{x^2 + y^2} \to 0$.

Let C_R *be the circle* $x^2 + y^2 = R^2$. *Show that the line integral*

$$\int_{C_R} e^x (-\varphi_y\, dx + \varphi_x\, dy)$$

is independent of R, and evaluate it.

Problem 2.3.6 (Sp80) *Let* $S = \{(x, y, z) \in \mathbb{R}^3 \mid x^2 + y^2 + z^2 = 1\}$ *denote the unit sphere in* $\mathbb{R}^3$. *Evaluate the surface integral over* S:

$$\iint_S (x^2 + y + z)\, dA.$$

Problem 2.3.7 (Sp81) *Let* $\vec{i}$, $\vec{j}$, *and* $\vec{k}$ *be the usual unit vectors in* $\mathbb{R}^3$. *Let* $\vec{F}$ *denote the vector field*

$$(x^2 + y - 4)\vec{i} + 3xy\vec{j} + (2xz + z^2)\vec{k}.$$

1. *Compute* $\nabla \times \vec{F}$ *(the curl of* $\vec{F}$*).*

2. *Compute the integral of* $\nabla \times \vec{F}$ *over the surface* $x^2 + y^2 + z^2 = 16$, $z \geqslant 0$.

Problem 2.3.8 (Sp91) *Let the vector field* F *in* $\mathbb{R}^3$ *have the form*

$$F(r) = g(\|r\|)r \qquad (r \neq (0, 0, 0)),$$

where g *is a real valued smooth function on* $(0, \infty)$ *and* $\|\cdot\|$ *denotes the Euclidean norm. (* F *is undefined at* $(0, 0, 0)$*.) Prove that*

$$\int_C F \cdot ds = 0$$

for any smooth closed path C *in* $\mathbb{R}^3$ *that does not pass through the origin.*

Problem 2.3.9 (Fa91) *Let* B *denote the unit ball of* $\mathbb{R}^3$, $B = \{r \in \mathbb{R}^3 \mid \|r\| \leqslant 1\}$. *Let* $J = (J_1, J_2, J_3)$ *be a smooth vector field on* $\mathbb{R}^3$ *that vanishes outside of* B *and satisfies* $\nabla \cdot \vec{J} = 0$.

1. *For* f *a smooth, scalar-valued function defined on a neighborhood of* B, *prove that*

$$\int_B (\nabla f) \cdot \vec{J} \, dxdydz = 0.$$

2. *Prove that*

$$\int_B J_1 \, dxdydz = 0.$$

Problem 2.3.10 (Fa94) *Let* D *denote the open unit disc in* $\mathbb{R}^2$. *Let* u *be an eigenfunction for the Laplacian in* D; *that is, a real valued function of class* C^2 *defined in* $\overline{D}$, *zero on the boundary of* D *but not identically zero, and satisfying the differential equation*

$$\frac{\partial^2 u}{\partial x^2} + \frac{\partial^2 u}{\partial y^2} = \lambda u,$$

where λ *is a constant. Prove that*

$$(*) \qquad \iint_D |\text{grad } u|^2 \, dxdy + \lambda \iint_D u^2 dxdy = 0,$$

and hence that $\lambda < 0$.

Problem 2.3.11 (Fa03) *Let* $\lambda, a \in \mathbb{R}$, *with* $a < 0$. *Let* $u(x, y)$ *be an infinitely differentiable function defined on an open neighborhood of closed unit disc* $\mathcal{D}$ *such that*

$$\frac{\partial^2 u}{\partial x^2} + \frac{\partial^2 u}{\partial y^2} = \lambda u \qquad \text{in int}(\mathcal{D})$$

$$D_n u = au \qquad \text{in } \partial \mathcal{D}.$$

Here $D_n u$ *denotes the directional derivative of* u *in the direction of the outward unit normal. Prove that if* u *is not identically zero in the interior of* $\mathcal{D}$ *then* $\lambda < 0$.

Problem 2.3.12 (Sp92) *Let* f *be a one-to-one* C^1 *map of* $\mathbb{R}^3$ *into* $\mathbb{R}^3$, *and let* J *denote its Jacobian determinant. Prove that if* x_0 *is any point of* $\mathbb{R}^3$ *and* $Q_r(x_0)$ *denotes the cube with center* x_0, *side length* r, *and edges parallel to the coordinate axes, then*

$$|J(x_0)| = \lim_{r \to 0} r^{-3} \text{vol} \left(f(Q_r(x_0)) \right) \leqslant \limsup_{x \to x_0} \frac{\|f(x) - f(x_0)\|^3}{\|x - x_0\|^3}.$$

Here, $\| \cdot \|$ *is the Euclidean norm in* $\mathbb{R}^3$.

3
Differential Equations

3.1 First Order Equations

Problem 3.1.1 (Fa87) *Find a curve* C *in* $\mathbb{R}^2$, *passing through the point* $(3, 2)$, *with the following property: Let* $L(x_0, y_0)$ *be the segment of the tangent line to* C *at* (x_0, y_0) *which lies in the first quadrant. Then each point* (x_0, y_0) *of* C *is the midpoint of* $L(x_0, y_0)$.

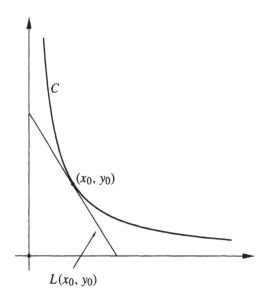

Problem 3.1.2 (Su78) *Solve the differential equation* $g' = 2g$, $g(0) = a$, *where a is a real constant.*

Problem 3.1.3 (Fa93, Fa77) *Let n be an integer larger than 1. Is there a differentiable function on* $[0, \infty)$ *whose derivative equals its* n^{th} *power and whose value at the origin is positive?*

Problem 3.1.4 (Sp78)   *1. For which real numbers* $\alpha > 0$ *does the differential equation*

$$\frac{dx}{dt} = x^\alpha, \qquad x(0) = 0,$$

have a solution on some interval $[0, b]$, $b > 0$?

2. For which values of α *are there intervals on which two solutions are defined?*

Problem 3.1.5 (Sp78) *Consider the differential equation*

$$\frac{dx}{dt} = x^2 + t^2, \quad x(0) = 1.$$

1. Prove that for some $b > 0$, *there is a solution defined for* $t \in [0, b]$.

2. Find an explicit value of b having the property in Part 1.

3. Find a $c > 0$ *such that there is no solution on* $[0, c]$.

Problem 3.1.6 (Fa78) *Solve the differential equation*

$$\frac{dy}{dx} = x^2 y - 3x^2, \qquad y(0) = 1.$$

Problem 3.1.7 (Sp93) *Prove that every solution* $x(t)$ $(t \geqslant 0)$ *of the differential equation*

$$\frac{dx}{dt} = x^2 - x^6$$

with $x(0) > 0$ *satisfies* $\lim\limits_{t \to \infty} x(t) = 1$.

Problem 3.1.8 (Sp80) *Consider the differential equation*

$$x' = \frac{x^3 - x}{1 + e^x}.$$

1. Find all its constant solutions.

2. Discuss $\lim\limits_{t \to \infty} x(t)$, *where* $x(t)$ *is the solution such that* $x(0) = \dfrac{1}{2}$.

Problem 3.1.9 (Su77, Su80, Sp82, Sp83) *Prove that the initial value problem*

$$\frac{dx}{dt} = 3x + 85 \cos x, \qquad x(0) = 77,$$

has a solution $x(t)$ defined for all $t \in \mathbb{R}$.

Problem 3.1.10 (Fa82) *Let $f : \mathbb{R} \to \mathbb{R}$ be a continuous nowhere vanishing function, and consider the differential equation*

$$\frac{dy}{dx} = f(y).$$

1. *For each real number c, show that this equation has a unique, continuously differentiable solution $y = y(x)$ on a neighborhood of 0 which satisfies the initial condition $y(0) = c$.*

2. *Deduce the conditions on f under which the solution y exists for all $x \in \mathbb{R}$, for every initial value c.*

Problem 3.1.11 (Sp79) *Find all solutions to the differential equation*

$$\frac{dy}{dx} = \sqrt{y}, \qquad y(0) = 0.$$

Problem 3.1.12 (Sp83) *Find all solutions $y : \mathbb{R} \to \mathbb{R}$ to*

$$\frac{dy}{dx} = \sqrt{y(y-2)}, \qquad y(0) = 0.$$

Problem 3.1.13 (Su83) *Find all real valued C^1 solutions $y(x)$ of the differential equation*

$$x\frac{dy}{dx} + y = x \qquad (-1 < x < 1).$$

Problem 3.1.14 (Sp84) *Consider the equation*

$$\frac{dy}{dx} = y - \sin y.$$

Show that there is an $\varepsilon > 0$ such that if $|y_0| < \varepsilon$, then the solution $y = f(x)$ with $f(0) = y_0$ satisfies

$$\lim_{x \to -\infty} f(x) = 0.$$

Problem 3.1.15 (Fa84) *Consider the differential equation*

$$\frac{dy}{dx} = 3xy + \frac{y}{1+y^2}.$$

Prove

1. *For each* $n = 1, 2, \ldots$, *there is a unique solution* $y = f_n(x)$ *defined for* $0 \leqslant x \leqslant 1$ *such that* $f_n(0) = 1/n$.

2. $\lim\limits_{n \to \infty} f_n(1) = 0$.

Problem 3.1.16 (Fa85) *Let* $y(t)$ *be a real valued solution, defined for* $0 < t < \infty$, *of the differential equation*

$$\frac{dy}{dt} = e^{-y} - e^{-3y} + e^{-5y}.$$

Show that $y(t) \to +\infty$ *as* $t \to +\infty$.

Problem 3.1.17 (Sp01) *Consider the differential-delay equation given by* $y'(t) = -y(t - t_0)$. *Here, the independent variable* t *is a real variable, the function* y *is allowed to be complex valued, and* t_0 *is a positive constant. Prove that if* $0 < t_0 < \pi/2$ *then every solution of the form* $y(t) = e^{\lambda t}$, *with* λ *complex, tends to* 0 *as* $t \to +\infty$.

Problem 3.1.18 (Fa86) *Prove the following theorem, or find a counterexample: If* p *and* q *are continuous real valued functions on* $\mathbb{R}$ *such that* $|q(x)| \leqslant |p(x)|$ *for all* x, *and if every solution* f *of the differential equation*

$$f' + qf = 0$$

satisfies $\lim\limits_{x \to +\infty} f(x) = 0$, *then every solution* f *of the differential equation*

$$f' + pf = 0$$

satisfies $\lim\limits_{x \to +\infty} f(x) = 0$.

Problem 3.1.19 (Fa86) *Discuss the solvability of the differential equation*

$$(e^x \sin y)(y')^3 + (e^x \cos y)y' + e^y \tan x = 0$$

with the initial condition $y(0) = 0$. *Does a solution exist in some interval about* 0? *If so, is it unique?*

Problem 3.1.20 (Fa92) *Let* f *and* g *be positive continuous functions on* $\mathbb{R}$, *with* $g \leqslant f$ *everywhere. Assume the initial value problem*

$$\frac{dx}{dt} = f(x), \quad x(0) = 0,$$

has a solution defined on all of $\mathbb{R}$. *Prove that the initial value problem*

$$\frac{dx}{dt} = g(x), \quad x(0) = 0,$$

also has a solution defined on all of $\mathbb{R}$.

Problem 3.1.21 (Sp95) *Let* $f : \mathbb{R} \to \mathbb{R}$ *be a bounded continuously differentiable function. Show that every solution of* $y'(x) = f(y(x))$ *is monotone.*

3.2   Second Order Equations

Problem 3.2.1 (Sp97) *Suppose that* $f''(x) = (x^2 - 1)f(x)$ *for all* $x \in \mathbb{R}$, *and that* $f(0) = 1$, $f'(0) = 0$. *Show that* $f(x) \to 0$ *as* $x \to \infty$.

Problem 3.2.2 (Sp77) *Find the solution of the differential equation*

$$y'' - 2y' + y = 0,$$

subject to the conditions

$$y(0) = 1, \quad y'(0) = 1.$$

Problem 3.2.3 (Fa77) *Find all solutions of the differential equation*

$$\frac{d^2 x}{dt^2} - 2\frac{dx}{dt} + x = \sin t$$

subject to the condition $x(0) = 1$ *and* $x'(0) = 0$.

Problem 3.2.4 (Su79) *Let* $x : \mathbb{R} \to \mathbb{R}$ *be a solution to the differential equation*

$$5x'' + 10x' + 6x = 0.$$

Prove that the function $f : \mathbb{R} \to \mathbb{R}$,

$$f(t) = \frac{x(t)^2}{1 + x(t)^4}$$

attains a maximum value.

Problem 3.2.5 (Su84) *Let* $x(t)$ *be the solution of the differential equation*

$$x''(t) + 8x'(t) + 25x(t) = 2\cos t$$

with initial conditions $x(0) = 0$ *and* $x'(0) = 0$. *Show that for suitable constants* α *and* δ,

$$\lim_{t \to \infty} (x(t) - \alpha \cos(t - \delta)) = 0.$$

Problem 3.2.6 (Fa79, Su81, Fa92) *Let* $y = y(x)$ *be a solution of the differential equation* $y'' = -|y|$ *with* $-\infty < x < \infty$, $y(0) = 1$ *and* $y'(0) = 0$.

1. *Show that* y *is an even function.*

2. *Show that* y *has exactly one zero on the positive real axis.*

Problem 3.2.7 (Fa95) *Determine all real numbers* $L > 1$ *so that the boundary value problem*

$$x^2 y''(x) + y(x) = 0, \quad 1 \leqslant x \leqslant L$$
$$y(1) = y(L) = 0$$

has a nonzero solution.

Problem 3.2.8 (Fa83) *For which real values of p does the differential equation*

$$y'' + 2py' + y = 3$$

admit solutions $y = f(x)$ with infinitely many critical points?

Problem 3.2.9 (Sp87) *Let p, q and r be continuous real valued functions on $\mathbb{R}$, with $p > 0$. Prove that the differential equation*

$$p(t)x''(t) + q(t)x'(t) + r(t)x(t) = 0$$

is equivalent to (i.e., has exactly the same solutions as) a differential equation of the form

$$\big(a(t)x'(t)\big)' + b(t)x(t) = 0,$$

where a is continuously differentiable and b is continuous.

Problem 3.2.10 (Fa93) *Let the function $x(t)$ $(-\infty < t < \infty)$ be a solution of the differential equation*

$$\frac{d^2x}{dt^2} - 2b\frac{dx}{dt} + cx = 0$$

such that $x(0) = x(1) = 0$. (Here, b and c are real constants.) Prove that $x(n) = 0$ for every integer n.

Problem 3.2.11 (Sp93) *Let k be a positive integer. For which values of the real number c does the differential equation*

$$\frac{d^2x}{dt^2} - 2c\frac{dx}{dt} + x = 0$$

have a solution satisfying $x(0) = x(2\pi k) = 0$?

Problem 3.2.12 (Fa01) *Consider the second-order linear differential equation*

$$\frac{d^2x}{dt^2} + p\frac{dx}{dt} + qx = 0. \tag{$*$}$$

Here, the independent variable t varies over $\mathbb{R}$, the unknown function x is as-sumed to be real valued, and p and q are continuous functions on $\mathbb{R}$. Assume that the solutions of $(*)$ are defined for all t (which is actually guaranteed by the theory), and that the solution set is translation invariant: if f is a solution and s is a real number, then the function $g(t) = f(t + s)$ is also a solution. Prove that p and q are constant.

Problem 3.2.13 (Sp85) *Let $h > 0$ be given. Consider the linear difference equa-tion*

$$\frac{y((n+2)h) - 2y((n+1)h) + y(nh)}{h^2} = -y(nh), \quad n = 0, 1, 2, \ldots.$$

(Note the analogy with the differential equation $y'' = -y$.)

1. *Find the general solution of the equation by trying suitable exponential substitutions.*

2. *Find the solution with $y(0) = 0$ and $y(h) = h$. Denote it by $S_h(nh)$, $n = 1, 2, \ldots$.*

3. *Let x be fixed and $h = \dfrac{x}{n}$. Show that*

$$\lim_{n \to \infty} S_{\frac{x}{n}}\left(\frac{nx}{n}\right) = \sin x \,.$$

3.3   Higher Order Equations

Problem 3.3.1 (Su78) *Let E be the set of functions $f : \mathbb{R} \to \mathbb{R}$ which are solutions to the differential equation $f''' + f'' - 2f = 0$.*

1. *Prove that E is a vector space and find its dimension.*

2. *Let $E_0 \subset E$ be the subspace of solutions g such that $\lim_{t \to \infty} g(t) = 0$. Find $g \in E_0$ such that $g(0) = 0$ and $g'(0) = 2$.*

Problem 3.3.2 (Fa98) *Find a function $y(x)$ such that $y^{(4)} + y = 0$ for $x \geqslant 0$, $y(0) = 0$, $y'(0) = 1$ and $\lim_{x \to \infty} y(x) = \lim_{x \to \infty} y'(x) = 0$.*

Problem 3.3.3 (Sp87) *Let V be a finite-dimensional linear subspace of $C^\infty(\mathbb{R})$ (the space of complex valued, infinitely differentiable functions). Assume that V is closed under D, the operator of differentiation (i.e., $f \in V \Rightarrow Df \in V$). Prove that there is a constant coefficient differential operator*

$$L = \sum_{k=0}^{n} a_k D^k$$

such that V consists of all solutions of the differential equation $Lf = 0$.

Problem 3.3.4 (Fa94)   1. *Find a basis for the space of real solutions of the differential equation*

$$(*) \qquad \sum_{n=0}^{7} \frac{d^n x}{dt^n} = 0.$$

2. *Find a basis for the subspace of real solutions of $(*)$ that satisfy*

$$\lim_{t \to +\infty} x(t) = 0.$$

Problem 3.3.5 (Sp94)   *1. Suppose the functions $\sin t$ and $\sin 2t$ are both solutions of the differential equation*

$$\sum_{k=0}^{n} c_k \frac{d^k x}{dt^k} = 0,$$

where $c_0, \ldots, c_n$ are real constants. What is the smallest possible order of the equation? Write down an equation of minimum order having the given functions as solutions.

2. Will the answers to Part 1 be different if the constants $c_0, \ldots, c_n$ are allowed to be complex?

Problem 3.3.6 (Sp95) *Let $y : \mathbb{R} \to \mathbb{R}$ be a three times differentiable function satisfying the differential equation $y''' - y = 0$. Suppose that $\lim_{x \to \infty} y(x) = 0$. Find real numbers a, b, c, and d, not all zero, such that $ay(0) + y'(0) + cy''(0) = d$.*

3.4   Systems of Differential Equations

Problem 3.4.1 (Sp79) *Consider the system of differential equations:*

$$\frac{dx}{dt} = y + tz$$

$$\frac{dy}{dt} = z + t^2 x$$

$$\frac{dz}{dt} = x + e^t y.$$

Prove there exists a solution defined for all $t \in [0, 1]$, such that

$$\begin{pmatrix} 1 & 2 & 3 \\ 4 & 5 & 6 \\ 7 & 8 & 9 \end{pmatrix} \begin{pmatrix} x(0) \\ y(0) \\ z(0) \end{pmatrix} = \begin{pmatrix} 0 \\ 0 \\ 0 \end{pmatrix}$$

and also

$$\int_0^1 \left(x(t)^2 + y(t)^2 + z(t)^2 \right) dt = 1.$$

Problem 3.4.2 (Su79, Fa79, Fa82, Su85) *Find all pairs of C^∞ functions $x(t)$ and $y(t)$ on $\mathbb{R}$ satisfying*

$$x'(t) = 2x(t) - y(t), \qquad y'(t) = x(t).$$

Problem 3.4.3 (Su80) *Consider the differential equations*

$$\frac{dx}{dt} = -x + y, \qquad \frac{dy}{dt} = \log(20 + x) - y.$$

Let $x(t)$ and $y(t)$ be a solution defined for all $t \geq 0$ with $x(0) > 0$ and $y(0) > 0$. Prove that $x(t)$ and $y(t)$ are bounded.

Problem 3.4.4 (Sp81) *Consider the system of differential equations*

$$\frac{dx}{dt} = y + x(1 - x^2 - y^2)$$
$$\frac{dy}{dt} = -x + y(1 - x^2 - y^2).$$

1. *Show that for any x_0 and y_0, there is a unique solution $(x(t), y(t))$ defined for all $t \in \mathbb{R}$ such that $x(0) = x_0$, $y(0) = y_0$.*

2. *Show that if $x_0 \neq 0$ and $y_0 \neq 0$, the solution referred to in Part 1 approaches the circle $x^2 + y^2 = 1$ as $t \to \infty$.*

Problem 3.4.5 (Sp84) *Show that the system of differential equations*

$$\frac{d}{dt} \begin{pmatrix} x \\ y \\ z \end{pmatrix} = \begin{pmatrix} 0 & 1 & 0 \\ 2 & 0 & 0 \\ 0 & 0 & 3 \end{pmatrix} \begin{pmatrix} x \\ y \\ z \end{pmatrix}$$

has a solution which tends to ∞ as $t \to -\infty$ and tends to the origin as $t \to +\infty$.

Problem 3.4.6 (Sp91) *Let $x(t)$ be a nontrivial solution to the system*

$$\frac{dx}{dt} = Ax,$$

where

$$A = \begin{pmatrix} 1 & 6 & 1 \\ -4 & 4 & 11 \\ -3 & -9 & 8 \end{pmatrix}.$$

Prove that $\|x(t)\|$ is an increasing function of t. (Here, $\| \cdot \|$ denotes the Euclidean norm.)

Problem 3.4.7 (Su84) *Consider the solution curve $(x(t), y(t))$ to the equations*

$$\frac{dx}{dt} = 1 + \frac{1}{2}x^2 \sin y$$
$$\frac{dy}{dt} = 3 - x^2$$

with initial conditions $x(0) = 0$ and $y(0) = 0$. Prove that the solution must cross the line $x = 1$ in the xy plane by the time $t = 2$.

Problem 3.4.8 (Fa80) *Consider the differential equation $x'' + x' + x^3 = 0$ and the function $f(x, x') = (x + x')^2 + (x')^2 + x^4$.*

1. *Show that f decreases along trajectories of the differential equation.*

2. *Show that if $x(t)$ is any solution, then $(x(t), x'(t))$ tends to $(0, 0)$ as $t \to \infty$.*

Problem 3.4.9 (Fa84) *Consider the differential equation*

$$\frac{dx}{dt} = y, \quad \frac{dy}{dt} = -ay - x^3 - x^5, \quad \text{where } a > 0.$$

1. Show that

$$F(x, y) = \frac{y^2}{2} + \frac{x^4}{4} + \frac{x^6}{6}$$

decreases along solutions.

2. Show that for any $\varepsilon > 0$, there is a $\delta > 0$ such that whenever $\|(x(0), y(0))\| < \delta$, there is a unique solution $(x(t), y(t))$ of the given equations with the initial condition $(x(0), y(0))$ which is defined for all $t \geqslant 0$ and satisfies $\|(x(t), y(t))\| < \varepsilon$.

Problem 3.4.10 (Fa83) *1. Let $u(t)$ be a real valued differentiable function of a real variable t which satisfies an inequality of the form*

$$u'(t) \leqslant au(t), \quad t \geqslant 0, \quad u(0) \leqslant b,$$

where a and b are positive constants. Starting from first principles, derive an upper bound for $u(t)$ for $t > 0$.

2. Let $x(t) = (x_1(t), x_2(t), \ldots, x_n(t))$ be a differentiable function from $\mathbb{R}$ to $\mathbb{R}^n$ which satisfies a differential equation of the form

$$x'(t) = f(x(t)),$$

where $f : \mathbb{R}^n \to \mathbb{R}^n$ is a continuous function. Assuming that f satisfies the condition

$$\langle f(y), y \rangle \leqslant \|y\|^2, \quad y \in \mathbb{R}^n$$

derive an inequality showing that the norm $\|x(t)\|$ grows, at most, exponentially.

Problem 3.4.11 (Fa81) *Consider an autonomous system of differential equations*

$$\frac{dx_i}{dt} = F_i(x_1, \ldots, x_n),$$

where $F = (F_1, \ldots, F_n) : \mathbb{R}^n \to \mathbb{R}^n$ is a C^1 vector field.

1. Let U and V be two solutions on $a < t < b$. Assuming that

$$\langle DF(x)z, z \rangle \leqslant 0$$

for all x, z in $\mathbb{R}^n$, show that $\|U(t) - V(t)\|^2$ is a decreasing function of t.

2. Let $W(t)$ be a solution defined for $t > 0$. Assuming that

$$\langle DF(x)z, z \rangle \leqslant -\|z\|^2,$$

show that there exists $C \in \mathbb{R}^n$ such that

$$\lim_{t \to \infty} W(t) = C.$$

Problem 3.4.12 (Fa81) *Let* $V : \mathbb{R}^n \to \mathbb{R}$ *be a* C^1 *function and consider the system of second order differential equations*

$$x_i''(t) = f_i(x(t)), \qquad 1 \leqslant i \leqslant n,$$

where

$$f_i = -\frac{\partial V}{\partial x_i}.$$

Let $x(t) = (x_1(t), \ldots, x_n(t))$ *be a solution of this system on a finite interval* $a < t < b$.

1. *Show that the function*

$$H(t) = \frac{1}{2}\langle x'(t), x'(t) \rangle + V(x(t))$$

   is constant for $a < t < b$.

2. *Assuming that* $V(x) \geqslant M > -\infty$ *for all* $x \in \mathbb{R}^n$, *show that* $x(t)$, $x'(t)$, *and* $x''(t)$ *are bounded on* $a < t < b$, *and then prove all three limits*

$$\lim_{t \to b} x(t), \quad \lim_{t \to b} x'(t), \quad \lim_{t \to b} x''(t)$$

   exist.

Problem 3.4.13 (Sp86) *For* λ *a real number, find all solutions of the integral equations*

$$\varphi(x) = e^x + \lambda \int_0^x e^{(x-y)} \varphi(y)\, dy, \qquad 0 \leqslant x \leqslant 1,$$

$$\psi(x) = e^x + \lambda \int_0^1 e^{(x-y)} \psi(y)\, dy, \qquad 0 \leqslant x \leqslant 1.$$

Problem 3.4.14 (Sp86) *Let* V *be a finite-dimensional vector space (over* $\mathbb{C}$ *) of* C^∞ *complex valued functions on* $\mathbb{R}$ *(the linear operations being defined pointwise). Prove that if* V *is closed under differentiation (i.e.,* $f'(x)$ *belongs to* V *whenever* $f(x)$ *does), then* V *is closed under translations (i.e.,* $f(x + a)$ *belongs to* V *whenever* $f(x)$ *does, for all real numbers* a *).*

Problem 3.4.15 (Fa99) *Describe all three dimensional vector spaces* V *of* C^∞ *complex valued functions on* $\mathbb{R}$ *that are invariant under the operator of differentiation.*

Problem 3.4.16 (Fa88) *Let the real valued functions* $f_1, \ldots, f_{n+1}$ *on* $\mathbb{R}$ *satisfy the system of differential equations*

$$f_{k+1}' + f_k' = (k+1)f_{k+1} - kf_k, \quad k = 1, \ldots, n$$
$$f_{n+1}' = -(n+1)f_{n+1}.$$

Prove that for each k,

$$\lim_{t \to \infty} f_k(t) = 0.$$

Problem 3.4.17 (Fa91) *Consider the vector differential equation*

$$\frac{dx(t)}{dt} = A(t)x(t)$$

where A *is a smooth* $n \times n$ *function on* $\mathbb{R}$. *Assume* A *has the property that* $\langle A(t)y, y \rangle \leqslant c\|y\|^2$ *for all* y *in* $\mathbb{R}^n$ *and all* t, *where* c *is a fixed real number. Prove that any solution* $x(t)$ *of the equation satisfies* $\|x(t)\| \leqslant e^{ct}\|x(0)\|$ *for all* $t > 0$.

Problem 3.4.18 (Sp94) *Let* W *be a real* 3×3 *antisymmetric matrix, i.e.,* $W^t = -W$. *Let the function*

$$X(t) = \begin{pmatrix} x_1(t) \\ x_2(t) \\ x_3(t) \end{pmatrix}$$

be a real solution of the vector differential equation $\dfrac{dX}{dt} = WX$.

1. *Prove that* $\|X(t)\|$, *the Euclidean norm of* $X(t)$, *is independent of* t.

2. *Prove that if* v *is a vector in the null space of* W, *then* $X(t) \cdot v$ *is independent of* t.

3. *Prove that the values* $X(t)$ *all lie on a fixed circle in* $\mathbb{R}^3$.

Problem 3.4.19 (Sp80) *For each* $t \in \mathbb{R}$, *let* $P(t)$ *be a symmetric real* $n \times n$ *matrix whose entries are continuous functions of* t. *Suppose for all* t *that the eigenvalues of* $P(t)$ *are all* $\leqslant -1$. *Let* $x(t) = (x_1(t), \ldots, x_n(t))$ *be a solution of the vector differential equation*

$$\frac{dx}{dt} = P(t)x.$$

Prove that

$$\lim_{t \to \infty} x(t) = 0.$$

Problem 3.4.20 (Sp89) *Let*

$$A = \begin{pmatrix} 0 & 0 & 0 & 0 \\ 1 & 0 & 0 & 0 \\ 0 & 1 & 0 & 0 \\ 0 & 0 & 1 & 0 \end{pmatrix}, \qquad B = \begin{pmatrix} 0 & 1 & 0 & 0 \\ 0 & 0 & 1 & 0 \\ 0 & 0 & 0 & 1 \\ 0 & 0 & 0 & 0 \end{pmatrix}$$

Find the general solution of the matrix differential equation $\dfrac{dX}{dt} = AXB$ *for the unknown* 4×4 *matrix function* $X(t)$.

$$\begin{pmatrix} 0 & 0 & 0 & 0 \\ 0 & 0 & 0 & 0 \\ 0 & 0 & 0 & 0 \\ 0 & 0 & 0 & 0 \end{pmatrix} \qquad \begin{pmatrix} 0 & 0 & 0 & 0 \\ 0 & 0 & 0 & 0 \\ 0 & 0 & 0 & 0 \\ 0 & 0 & 0 & 0 \end{pmatrix}$$

4
Metric Spaces

4.1 Topology of $\mathbb{R}^n$

Problem 4.1.1 (Fa02) *Let S be a subset of $\mathbb{R}$. Let C be the set of points x in $\mathbb{R}$ with the property that $S \cap (x - \delta, x + \delta)$ is uncountable for every $\delta > 0$. Prove that $S \setminus C$ is finite or countable.*

Problem 4.1.2 (Sp03) *Let $A \subseteq \mathbb{R}$ be uncountable.*

1. *Show that A has at least one accumulation point.*

2. *Show that A has uncountably many accumulation points.*

Problem 4.1.3 (Fa02) *Let $\{x(i, j) \mid i, j \in \mathbb{N}\}$ be a doubly indexed set in a complete metric space (X, ρ) such that*

$$\rho(x(i, j), x(k, \ell)) \leqslant \min \left\{ \max \left\{ \frac{1}{i}, \frac{1}{k} \right\}, \max \left\{ \frac{1}{j}, \frac{1}{\ell} \right\} \right\}$$

Prove that the iterated limits $\lim_{i \to \infty} \lim_{j \to \infty} x(i, j)$, $\lim_{j \to \infty} \lim_{i \to \infty} x(i, j)$ exist and are equal.

Problem 4.1.4 (Sp01) *Let T_0 be the interior of a triangle in $\mathbb{R}^2$ with vertices A, B, C. Let T_1 be the interior of the triangle whose vertices are the midpoints of the sides of T_0, T_2 the interior of the triangle whose vertices are the midpoints of the sides of T_1, and so on. Describe the set $\cap_{n=0}^{\infty} T_n$.*

Problem 4.1.5 (Sp86, Sp94, Sp96, Fa98) *Let K be a compact subset of $\mathbb{R}^n$ and $\{B_j\}$ a sequence of open balls that covers K. Prove that there is a positive number*

ε such that each ε-ball centered at a point of K is contained in one of the balls B_j.

Problem 4.1.6 (Su81) *Prove or disprove: The set $\mathbb{Q}$ of rational numbers is the intersection of a countable family of open subsets of $\mathbb{R}$.*

Problem 4.1.7 (Fa77) *Let $X \subset \mathbb{R}$ be a nonempty connected set of real numbers. If every element of X is rational, prove X has only one element.*

Problem 4.1.8 (Su80) *Give an example of a subset of $\mathbb{R}$ having uncountably many connected components. Can such a subset be open? Closed?*

Problem 4.1.9 (Fa00) *Let $f_n : \mathbb{R}^k \to \mathbb{R}^m$ be continuous $(n = 1, 2, \ldots)$. Let K be a compact subset of $\mathbb{R}^k$. Suppose $f_n \to f$ uniformly on K. Prove that*

$$S = f(K) \cup \bigcup_{n=1}^{\infty} f_n(K) \text{ is compact.}$$

Problem 4.1.10 (Sp83) *Show that the interval $[0, 1]$ cannot be written as a countably infinite disjoint union of closed subintervals of $[0, 1]$.*

Problem 4.1.11 (Su78, Sp99, Sp03) *Let X and Y be nonempty subsets of a metric space M. Define*

$$d(X, Y) = \inf\{d(x, y) \mid x \in X, y \in Y\}.$$

1. *Suppose X contains only one point x, and Y is closed. Prove*

$$d(X, Y) = d(x, y)$$

   for some $y \in Y$.

2. *Suppose X is compact and Y is closed. Prove*

$$d(X, Y) = d(x, y)$$

   for some $x \in X$, $y \in Y$.

3. *Show by example that the conclusion of Part 2 can be false if X and Y are closed but not compact.*

Problem 4.1.12 (Sp82) *Let $S \subset \mathbb{R}^n$ be a subset which is uncountable. Prove that there is a sequence of distinct points in S converging to a point of S.*

Problem 4.1.13 (Fa89) *Let $X \subset \mathbb{R}^n$ be a closed set and r a fixed positive real number. Let $Y = \{y \in \mathbb{R}^n \mid |x - y| = r \text{ for some } x \in X\}$. Show that Y is closed.*

Problem 4.1.14 (Sp92, Fa99) *Show that every infinite closed subset of $\mathbb{R}^n$ is the closure of a countable set.*

Problem 4.1.15 (Fa86) *Let* $\{U_1, U_2, \ldots\}$ *be a cover of* $\mathbb{R}^n$ *by open sets. Prove that there is a cover* $\{V_1, V_2, \ldots\}$ *such that*

1. $V_j \subset U_j$ *for each* j;

2. *each compact subset of* $\mathbb{R}^n$ *is disjoint from all but finitely many of the* V_j.

Problem 4.1.16 (Sp87) *A standard theorem states that a continuous real valued function on a compact set is bounded. Prove the converse: If* K *is a subset of* $\mathbb{R}^n$ *and if every continuous real valued function on* K *is bounded, then* K *is compact.*

Problem 4.1.17 (Su77) *Let* $A \subset \mathbb{R}^n$ *be compact,* $x \in A$; *let* (x_i) *be a sequence in* A *such that every convergent subsequence of* (x_i) *converges to* x.

1. *Prove that the entire sequence* (x_i) *converges.*

2. *Give an example to show that if* A *is not compact, the result in Part 1 is not necessarily true.*

Problem 4.1.18 (Fa89) *Let* $X \subset \mathbb{R}^n$ *be compact and let* $f : X \to \mathbb{R}$ *be continuous. Given* $\varepsilon > 0$, *show there is an* M *such that for all* $x, y \in X$,

$$|f(x) - f(y)| \leqslant M|x - y| + \varepsilon.$$

Problem 4.1.19 (Su78) *Let* $\{S_\alpha\}$ *be a family of connected subsets of* $\mathbb{R}^2$ *all containing the origin. Prove that* $\bigcup_\alpha S_\alpha$ *is connected.*

Problem 4.1.20 (Fa79) *Consider the following properties of a function* $f : \mathbb{R}^n \to \mathbb{R}$:

1. f *is continuous.*

2. *The graph of* f *is connected in* $\mathbb{R}^n \times \mathbb{R}$.

Prove or disprove the implications $1 \Rightarrow 2, 2 \Rightarrow 1$.

Problem 4.1.21 (Sp01) *Let* U *be a nonempty, proper, open subset of* $\mathbb{R}^n$. *Construct a function* $f : \mathbb{R}^n \to \mathbb{R}$ *that is discontinuous at each point of* U *and continuous at each point of* $\mathbb{R}^n \setminus U$.

Problem 4.1.22 (Sp82) *Prove or give a counterexample: Every connected, locally pathwise connected set in* $\mathbb{R}^n$ *is pathwise connected.*

Problem 4.1.23 (Sp81) *The set of real* 3×3 *symmetric matrices is a real, finite-dimensional vector space isomorphic to* $\mathbb{R}^6$. *Show that the subset of such matrices of signature* $(2, 1)$ *is an open connected subspace in the usual topology on* $\mathbb{R}^6$.

Problem 4.1.24 (Fa78) *Let* $M_{n \times n}$ *be the vector space of real* $n \times n$ *matrices, identified with* $\mathbb{R}^{n^2}$. *Let* $X \subset M_{n \times n}$ *be a compact set. Let* $S \subset \mathbb{C}$ *be the set of all numbers that are eigenvalues of at least one element of* X. *Prove that* S *is compact.*

Problem 4.1.25 (Su81) *Let $S\mathbb{O}(3)$ denote the group of orthogonal transformations of $\mathbb{R}^3$ of determinant 1. Let $Q \subset S\mathbb{O}(3)$ be the subset of symmetric transformations $\neq I$. Let P^2 denote the space of lines through the origin in $\mathbb{R}^3$.*

1. *Show that P^2 and $S\mathbb{O}(3)$ are compact metric spaces (in their usual topologies).*

2. *Show that P^2 and Q are homeomorphic.*

Problem 4.1.26 (Fa83) *Let m and n be positive integers, with $m < n$. Let $M_{m \times n}$ be the space of linear transformations of $\mathbb{R}^m$ into $\mathbb{R}^n$ (considered as $n \times m$ matrices) and let L be the set of transformations in $M_{m \times n}$ which have rank m.*

1. *Show that L is an open subset of $M_{m \times n}$.*

2. *Show that there is a continuous function $T : L \to M_{m \times n}$ such that $T(A)A = I_m$ for all A, where I_m is the identity on $\mathbb{R}^m$.*

Problem 4.1.27 (Fa91) *Let $M_{n \times n}$ be the space of real $n \times n$ matrices. Regard it as a metric space with the distance function*

$$d(A, B) = \sum_{i,j=1}^{n} |a_{ij} - b_{ij}| \qquad \left(A = (a_{ij}), B = (b_{ij}) \right).$$

Prove that the set of nilpotent matrices in $M_{n \times n}$ is a closed set.

Problem 4.1.28 (Sp00) *Let S be an uncountable subset of $\mathbb{R}$. Prove that there exists a real number t such that both sets $S \cap (-\infty, t)$ and $S \cap (t, \infty)$ are uncountable.*

4.2   General Theory

Problem 4.2.1 (Sp02) *Let (X, ρ) and (Y, σ) be metric spaces. Assume that:*

1. *$f, f_1, f_2, \ldots$ are bijective functions of X onto Y with inverses $g, g_1, g_2, \ldots$*

2. *g is uniformly continuous*

3. *$f_n \to f$ uniformly as $n \to \infty$.*

Prove that $g_n \to g$ uniformly as $n \to \infty$.

Problem 4.2.2 (Fa99) *Let $E_1, E_2, \ldots$ be nonempty closed subsets of a complete metric space (X, d) with $E_{n+1} \subset E_n$ for all positive integers n, and such that $\lim_{n \to \infty} \operatorname{diam}(E_n) = 0$, where $\operatorname{diam}(E)$ is defined to be*

$$\sup\{d(x, y) \mid x, y \in E\}.$$

Prove that $\bigcap_{n=1}^{\infty} E_n \neq \emptyset$.

Problem 4.2.3 (Fa00) *Let A be a subset of a compact metric space (X, d). Assume that, for every continuous function $f : X \to \mathbb{R}$, the restriction of f to A attains a maximum on A. Prove that A is compact.*

Problem 4.2.4 (Fa93) *Let X be a metric space and (x_n) a convergent sequence in X with limit x_0. Prove that the set $C = \{x_0, x_1, x_2, ...\}$ is compact.*

Problem 4.2.5 (Sp79) *Prove that every compact metric space has a countable dense subset.*

Problem 4.2.6 (Fa80) *Let X be a compact metric space and $f : X \to X$ an isometry. Show that $f(X) = X$.*

Problem 4.2.7 (Sp97) *Let M be a metric space with metric d. Let C be a nonempty closed subset of M. Define $f : M \to \mathbb{R}$ by*

$$f(x) = \inf\{d(x, y) \mid y \in C\}.$$

Show that f is continuous, and that $f(x) = 0$ if and only if $x \in C$.

Problem 4.2.8 (Su84) *Let $C^{1/3}$ be the set of real valued functions f on the closed interval $[0, 1]$ such that*

1. *$f(0) = 0$;*

2. *$\|f\|$ is finite, where by definition*

$$\|f\| = \sup\left\{\frac{|f(x) - f(y)|}{|x - y|^{1/3}} \mid x \neq y\right\}.$$

Verify that $\|\cdot\|$ is a norm for the space $C^{1/3}$, and prove that $C^{1/3}$ is complete with respect to this norm.

Problem 4.2.9 (Sp00) *Let $\{f_n\}_{n=1}^{\infty}$ be a uniformly bounded equicontinuous sequence of real-valued functions on the compact metric space (X, d). Define the functions $g_n : X \to \mathbb{R}$, for $n \in \mathbb{N}$ by*

$$g_n(x) = \max\{f_1(x), \ldots, f_n(x)\}.$$

Prove that the sequence $\{g_n\}_{n=1}^{\infty}$ converges uniformly.

Problem 4.2.10 (Sp87) *Let $\mathcal{F}$ be a uniformly bounded, equicontinuous family of real valued functions on the metric space (X, d). Prove that the function*

$$g(x) = \sup\{f(x) \mid f \in \mathcal{F}\}$$

is continuous.

Problem 4.2.11 (Fa91) *Let X and Y be metric spaces and f a continuous map of X into Y. Let $K_1, K_2, \ldots$ be nonempty compact subsets of X such that $K_{n+1} \subset K_n$ for all n, and let $K = \bigcap K_n$. Prove that $f(K) = \bigcap f(K_n)$.*

Problem 4.2.12 (Fa92) *Let* (X_1, d_1) *and* (X_2, d_2) *be metric spaces and* $f : X_1 \to X_2$ *a continuous surjective map such that* $d_1(p, q) \leqslant d_2(f(p), f(q))$ *for every pair of points* p, q *in* X_1.

1. *If* X_1 *is complete, must* X_2 *be complete? Give a proof or a counterexample.*

2. *If* X_2 *is complete, must* X_1 *be complete? Give a proof or a counterexample.*

4.3   Fixed Point Theorem

Problem 4.3.1 (Fa79) *An accurate map of California is spread out flat on a table in Evans Hall, in Berkeley. Prove that there is exactly one point on the map lying directly over the point it represents.*

Problem 4.3.2 (Fa87) *Define a sequence of positive numbers as follows. Let* $x_0 > 0$ *be any positive number, and let* $x_{n+1} = (1 + x_n)^{-1}$. *Prove that this sequence converges, and find its limit.*

Problem 4.3.3 (Su80) *Let* $f : \mathbb{R} \to \mathbb{R}$ *be monotonically increasing (perhaps discontinuous). Suppose* $0 < f(0)$ *and* $f(100) < 100$. *Prove* $f(x) = x$ *for some* x.

Problem 4.3.4 (Su82, Sp95) *Let* K *be a nonempty compact set in a metric space with distance function* d. *Suppose that* $\varphi \colon K \to K$ *satisfies*

$$d(\varphi(x), \varphi(y)) < d(x, y)$$

for all $x \neq y$ *in* K. *Show there exists precisely one point* $x \in K$ *such that* $x = \varphi(x)$.

Problem 4.3.5 (Fa82) *Let* K *be a continuous function on the unit square* $0 \leqslant x, y \leqslant 1$ *satisfying* $|K(x, y)| < 1$ *for all* x *and* y. *Show that there is a continuous function* $f(x)$ *on* $[0, 1]$ *such that we have*

$$f(x) + \int_0^1 K(x, y) f(y) \, dy = e^{x^2}.$$

Can there be more than one such function f?

Problem 4.3.6 (Fa88) *Let* g *be a continuous real valued function on* $[0, 1]$. *Prove that there exists a continuous real valued function* f *on* $[0, 1]$ *satisfying the equation*

$$f(x) - \int_0^x f(x - t) e^{-t^2} \, dt = g(x).$$

Problem 4.3.7 (Su84) *Show there is a unique continuous real valued function* $f : [0, 1] \to \mathbb{R}$ *such that*

$$f(x) = \sin x + \int_0^1 \frac{f(y)}{e^{x+y+1}} \, dy.$$

Problem 4.3.8 (Fa85, Sp98) *Let (M, d) be a nonempty complete metric space. Let S map M into M, and write S^2 for $S \circ S$; that is, $S^2(x) = S(S(x))$. Suppose that S^2 is a* strict contraction; *that is, there is a constant $\lambda < 1$ such that for all points $x, y \in M, d\left(S^2(x), S^2(y)\right) \leqslant \lambda d(x, y)$. Show that S has a unique fixed point in M.*

5

Complex Analysis

5.1 Complex Numbers

Problem 5.1.1 (Fa77) *If a and b are complex numbers and $a \neq 0$, the set a^b consists of those complex numbers c having a logarithm of the form $b\alpha$, for some logarithm α of a. (That is, $e^{b\alpha} = c$ and $e^{\alpha} = a$ for some complex number α.) Describe set a^b when $a = 1$ and $b = 1/3 + i$.*

Problem 5.1.2 (Su77) *Write all values of i^i in the form $a + bi$.*

Problem 5.1.3 (Sp85) *Show that a necessary and sufficient condition for three points a, b, and c in the complex plane to form an equilateral triangle is that*

$$a^2 + b^2 + c^2 = bc + ca + ab.$$

Problem 5.1.4 (Fa86) *Let the points a, b, and c lie on the unit circle of the complex plane and satisfy $a + b + c = 0$. Prove that a, b, and c form the vertices of an equilateral triangle.*

Problem 5.1.5 (Sp77) *1. Evaluate $P_{n-1}(1)$, where $P_{n-1}(x)$ is the polynomial*

$$P_{n-1}(x) = \frac{x^n - 1}{x - 1}.$$

2. *Consider a circle of radius 1, and let $Q_1, Q_2, \ldots, Q_n$ be the vertices of a regular n-gon inscribed in the circle. Join Q_1 to $Q_2, Q_3, \ldots, Q_n$ by segments of a straight line. You obtain $(n-1)$ segments of lengths $\lambda_2, \lambda_3, \ldots, \lambda_n$.*

Show that

$$\prod_{i=2}^{n} \lambda_i = n.$$

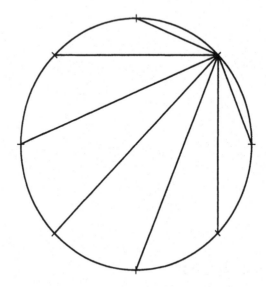

Problem 5.1.6 (Sp03) *Prove that for each integer $n \geq 0$ there is a polynomial $T_n(x)$ with integer coefficients such that the identity*

$$2 \cos nz = T_n(2 \cos z)$$

holds for all z.

Problem 5.1.7 (Sp90) *Let $z_1, z_2, \ldots, z_n$ be complex numbers. Prove that there exists a subset $J \subset \{1, 2, \ldots, n\}$ such that*

$$\left| \sum_{j \in J} z_j \right| \geq \frac{1}{4\sqrt{2}} \sum_{j=1}^{n} |z_j|.$$

Problem 5.1.8 (Sp94) *Let $a_1, a_2, \ldots, a_n$ be complex numbers. Prove that there is a point x in $[0, 1]$ such that*

$$\left| 1 - \sum_{k=1}^{n} a_k e^{2\pi i k x} \right| \geq 1.$$

Problem 5.1.9 (Fa82) *Let a and b be complex numbers whose real parts are negative or 0. Prove the inequality $|e^a - e^b| \leq |a - b|$.*

Problem 5.1.10 (Fa95) *Let A be a finite subset of the unit disc in the plane, and let $N(A, r)$ be the set of points at distance $\leq r$ from A, where $0 < r < 1$. Show* *that the length of the boundary $N(A, r)$ is, at most, C/r for some constant C independent of A.*

Problem 5.1.11 (Su82) *For complex numbers $\alpha_1, \alpha_2, \ldots, \alpha_k$, prove*

$$\limsup_n \left| \sum_{j=1}^k \alpha_j^n \right|^{1/n} = \sup_j |\alpha_j|.$$

Note: See also Problem 1.3.1.

5.2 Series and Sequences of Functions

Problem 5.2.1 (Fa95) *Show that*

$$(1+z+z^2+\cdots+z^9)(1+z^{10}+z^{20}+\cdots+z^{90})(1+z^{100}+z^{200}+\cdots+z^{900})\cdots = \frac{1}{1-z}$$

for $|z| < 1$.

Problem 5.2.2 (Fa94) *Suppose the coefficients of the power series*

$$\sum_{n=0}^{\infty} a_n z^n$$

are given by the recurrence relation

$$a_0 = 1, \quad a_1 = -1, \quad 3a_n + 4a_{n-1} - a_{n-2} = 0, \quad n = 2, 3, \ldots.$$

Find the radius of convergence of the series and the function to which it converges in its disc of convergence.

Problem 5.2.3 (Fa03) *Show that the differential equation*

$$f''(z) = zf(z), \qquad f(0) = 1, \qquad f'(0) = 1$$

has an unique entire solution in the complex plane.

Problem 5.2.4 (Fa93) *Describe the region in the complex plane where the infinite series*

$$\sum_{n=1}^{\infty} \frac{1}{n^2} \exp\left(\frac{nz}{z-2}\right)$$

converges. Draw a sketch of the region.

Problem 5.2.5 (Su77) *Let f be an analytic function such that*

$$f(z) = 1 + 2z + 3z^2 + \cdots \quad \text{for } |z| < 1.$$

Define a sequence of real numbers $a_0, a_1, a_2, \ldots$ by

$$f(z) = \sum_{n=0}^{\infty} a_n (z+2)^n.$$

What is the radius of convergence of the series

$$\sum_{n=0}^{\infty} a_n z^n?$$

Problem 5.2.6 (Sp77) *Let the sequence $a_0, a_1, \ldots$ be defined by the equation*

$$1 - x^2 + x^4 - x^6 + \cdots = \sum_{n=0}^{\infty} a_n (x-3)^n \quad (0 < x < 1).$$

Find

$$\limsup_{n \to \infty} \left(|a_n|^{\frac{1}{n}} \right).$$

Problem 5.2.7 (Su78) *Suppose the power series*

$$\sum_{n=0}^{\infty} a_n z^n$$

converges for $|z| < R$ where z and the a_n are complex numbers. If $b_n \in \mathbb{C}$ are such that $|b_n| < n^2 |a_n|$ for all n, prove that

$$\sum_{n=0}^{\infty} b_n z^n$$

converges for $|z| < R$.

Problem 5.2.8 (Sp99) *Let $b_1, b_2, \ldots$ be a sequence of real numbers such that $b_k \geqslant b_{k+1}$ for all k and $\lim_{k \to \infty} b_k = 0$. Prove that the power series $\sum_{k=1}^{\infty} b_k z^k$ converges for all complex numbers z such that $|z| \leqslant 1$ and $z \neq 1$.*

Problem 5.2.9 (Sp79) *For which $z \in \mathbb{C}$ does*

$$\sum_{n=0}^{\infty} \left(\frac{z^n}{n!} + \frac{n^2}{z^n} \right)$$

converge?

Problem 5.2.10 (Su79) *Show that*

$$\sum_{n=0}^{\infty} \frac{z}{\left(1+z^2\right)^n}$$

converges for all complex numbers z exterior to the lemniscate

$$\left|1+z^2\right| = 1.$$

Problem 5.2.11 (Su82) *Determine the complex numbers z for which the power series*

$$\sum_{n=1}^{\infty} \frac{z^n}{n^{\log n}}$$

and its term by term derivatives of all orders converge absolutely.

Problem 5.2.12 (Su84) *Suppose*

$$f(z) = \sum_{n=0}^{\infty} a_n z^n$$

has radius of convergence $R > 0$. Show that

$$h(z) = \sum_{n=0}^{\infty} \frac{a_n z^n}{n!}$$

is entire and that for $0 < r < R$, there is a constant M such that

$$|h(z)| \leqslant M e^{|z|/r}.$$

Problem 5.2.13 (Sp01) *Let the power series $\sum_{n=0}^{\infty} c_n z^n$, with positive radius of convergence R, represent the function f in the disk $|z| < R$. For $k = 0, 1, \ldots$ let s_k be the k-th partial sum of the series $s_k(z) = \sum_{n=0}^{k} c_n z^n$. Prove that*

$$\sum_{k=0}^{\infty} |f(z) - s_k(z)| < \infty$$

for each z in the disk $|z| < R$.

Problem 5.2.14 (Sp85) *Let $R > 1$ and let f be analytic on $|z| < R$ except at $z = 1$, where f has a simple pole. If*

$$f(z) = \sum_{n=0}^{\infty} a_n z^n \qquad (|z| < 1)$$

is the Maclaurin series for f, show that $\lim_{n \to \infty} a_n$ exists.

Problem 5.2.15 (Fa95) *Find the radius of convergence R of the Taylor series about $z = 1$ of the function $f(z) = 1/(1 + z^2 + z^4 + z^6 + z^8 + z^{10})$. Express the answer in terms of real numbers and square roots only.*

Problem 5.2.16 (Sp78) *Prove that the uniform limit of a sequence of complex analytic functions is complex analytic. Is the analogous theorem true for real analytic functions?*

Problem 5.2.17 (Su79) *Let $g_n(z)$ be an entire function having only real zeros, $n = 1, 2, \ldots$ Suppose*

$$\lim_{n \to \infty} g_n(z) = g(z)$$

uniformly on compact sets in $\mathbb{C}$, with g not identically zero. Prove that $g(z)$ has only real zeros.

Problem 5.2.18 (Sp86) *Let $f, g_1, g_2, \ldots$ be entire functions. Assume that*

1. *$|g_n^{(k)}(0)| \leqslant |f^{(k)}(0)|$ for all n and k;*

2. *$\lim_{n \to \infty} g_n^{(k)}(0)$ exists for all k.*

Prove that the sequence $\{g_n\}$ converges uniformly on compact sets and that its limit is an entire function.

Problem 5.2.19 (Sp92) *Find a Laurent series that converges in the annulus $1 < |z| < 2$ to a branch of the function $\log\left(\dfrac{z(2 - z)}{1 - z}\right)$.*

5.3   Conformal Mappings

Problem 5.3.1 (Fa77) *Consider the following four types of transformations:*

$$z \mapsto z + b, \quad z \mapsto 1/z, \quad z \mapsto kz \quad (\text{where } k \neq 0),$$

$$z \mapsto \frac{az + b}{cz + d} \quad (\text{where } ad - bc \neq 0).$$

Here, z is a variable complex number and the other letters denote constant complex numbers. Show that each transformation takes circles to either circles or straight lines.

Problem 5.3.2 (Fa78) *Give examples of conformal maps as follows:*

1. *from $\{z \mid |z| < 1\}$ onto $\{z \mid \Re z < 0\}$,*

2. *from $\{z \mid |z| < 1\}$ onto itself, with $f(0) = 0$ and $f(1/2) = i/2$,*

3. *from $\{z \mid z \neq 0, 0 < \arg z < \frac{3\pi}{2}\}$ onto $\{z \mid z \neq 0, 0 < \arg z < \frac{\pi}{2}\}$.*

Problem 5.3.3 (Sp83) *A fractional linear transformation maps the annulus $r < |z| < 1$ (where $r > 0$) onto the domain bounded by the two circles $|z - \frac{1}{4}| = \frac{1}{4}$ and $|z| = 1$. Find r.*

Problem 5.3.4 (Sp80) *Does there exist an analytic function mapping the annulus*

$$A = \{z \mid 1 \leqslant |z| \leqslant 4\}$$

onto the annulus

$$B = \{z \mid 1 \leqslant |z| \leqslant 2\}$$

and taking $C_1 \to C_1$, $C_4 \to C_2$, where C_r is the circle of radius r?

Problem 5.3.5 (Su80) *Exhibit a conformal map from $\{z \in \mathbb{C} \mid |z| < 1, \Re z > 0\}$ onto $\mathbb{D} = \{z \in \mathbb{C} \mid |z| < 1\}$.*

Problem 5.3.6 (Sp90) *Find a one-to-one conformal map of the semidisc*

$$\left\{ z \in \mathbb{C} \mid \Im z > 0, \ \left| z - \frac{1}{2} \right| < \frac{1}{2} \right\}$$

onto the upper half-plane.

Problem 5.3.7 (Fa97) *Conformally map the region inside the disc given by $\{z \in \mathbb{C} \mid |z - 1| \leqslant 1\}$ and outside the disc $\{z \in \mathbb{C} \mid |z - \frac{1}{2}| \leqslant \frac{1}{2}\}$ onto the upper half-plane.*

Problem 5.3.8 (Sp95) *Prove that there is no one-to-one conformal map of the punctured disc $G = \{z \in \mathbb{C} \mid 0 < |z| < 1\}$ onto the annulus $A = \{z \in \mathbb{C} \mid 1 < |z| < 2\}$.*

Problem 5.3.9 (Fa02) *Let n be a positive integer. Find a group of linear fractional transformations of $\mathbb{C}$ that fixes the two points $z = 1$ and $z = -1$ and has order n.*

5.4   Functions on the Unit Disc

Problem 5.4.1 (Fa02) *Let α be a number in $(0, \pi/2)$. Prove that the function $f(z) = e^{-1/z}$ is uniformly continuous in $S_\alpha = \{z \mid 0 < |z| \leqslant 1, |\text{Arg } z| \leqslant \alpha\}$, a sector of the complex plane.*

Problem 5.4.2 (Fa82) *Let a and b be nonzero complex numbers and $f(z) = az + bz^{-1}$. Determine the image under f of the unit circle $\{z \mid |z| = 1\}$.*

Problem 5.4.3 (Su83, Fa96) *Let f be analytic on and inside the unit circle $C = \{z \mid |z| = 1\}$. Let L be the length of the image of C under f. Show that $L \geqslant 2\pi |f'(0)|$.*

Problem 5.4.4 (Sp80) *Let*

$$f(z) = \sum_{n=0}^{\infty} c_n z^n$$

be analytic in the disc $\mathbb{D} = \{z \in \mathbb{C} \mid |z| < 1\}$. *Assume* f *maps* $\mathbb{D}$ *one-to-one onto a domain* G *having area* A. *Prove*

$$A = \pi \sum_{n=1}^{\infty} n|c_n|^2.$$

Problem 5.4.5 (Su83) *Compute the area of the image of the unit disc* $\{z \mid |z| < 1\}$ *under the map* $f(z) = z + z^2/2$.

Problem 5.4.6 (Sp80) *Let*

$$f(z) = \sum_{n=0}^{\infty} a_n z^n$$

be an analytic function in the open unit disc $\mathbb{D}$. *Assume that*

$$\sum_{n=2}^{\infty} n|a_n| \leqslant |a_1| \quad with \quad a_1 \neq 0.$$

Prove that f *is injective.*

Problem 5.4.7 (Sp03) *Let* $f(z)$ *be a function that is analytic in the unit disc* $\mathbb{D} = \{|z| < 1\}$. *Suppose that* $|f(z)| \leqslant 1$ *in* $\mathbb{D}$. *Prove that if* $f(z)$ *has at least two fixed points* z_1 *and* z_2, *then* $f(z) = z$ *for all* $z \in \mathbb{D}$.

Problem 5.4.8 (Su85) *For each* $k > 0$, *let* X_k *be the set of analytic functions* $f(z)$ *on the open unit disc* $\mathbb{D}$ *such that*

$$\sup_{z \in \mathbb{D}} \left\{ (1 - |z|)^k |f(z)| \right\}$$

is finite. Show that $f \in X_k$ *if and only if* $f' \in X_{k+1}$.

Problem 5.4.9 (Sp88) *Let the function* f *be analytic in the open unit disc of the complex plane and real valued on the radii* $[0, 1)$ *and* $[0, e^{i\pi\sqrt{2}})$. *Prove that* f *is constant.*

Problem 5.4.10 (Fa91) *Let the function* f *be analytic in the disc* $|z| < 1$ *of the complex plane. Assume that there is a positive constant* M *such that*

$$\int_0^{2\pi} |f'(re^{i\theta})|\, d\theta \leqslant M, \qquad (0 \leqslant r < 1).$$

Prove that

$$\int_{[0,1)} |f(x)|\, dx < \infty.$$

Problem 5.4.11 (Fa78) *Suppose $h(z)$ is analytic in the whole plane, $h(0) = 3 + 4i$, and $|h(z)| \leqslant 5$ if $|z| < 1$. What is $h'(0)$?*

Problem 5.4.12 (Fa98) *Let f be analytic in the closed unit disc, with $f(-\log 2) = 0$ and $|f(z)| \leqslant |e^z|$ for all z with $|z| = 1$. How large can $|f(\log 2)|$ be?*

Problem 5.4.13 (Fa79, Fa90) *Suppose that f is analytic on the open upper half-plane and satisfies $|f(z)| \leqslant 1$ for all z, $f(i) = 0$. How large can $|f(2i)|$ be under these conditions?*

Problem 5.4.14 (Fa85) *Let $f(z)$ be analytic on the right half-plane $H = \{z \mid \Re z > 0\}$ and suppose $|f(z)| \leqslant 1$ for $z \in H$. Suppose also that $f(1) = 0$. What is the largest possible value of $|f'(1)|$?*

Problem 5.4.15 (Su82) *Let $f(z)$ be analytic on the open unit disc $\mathbb{D} = \{z \mid |z| < 1\}$. Prove that there is a sequence (z_n) in $\mathbb{D}$ such that $|z_n| \to 1$ and $(f(z_n))$ is bounded.*

Problem 5.4.16 (Sp93) *Let f be an analytic function in the unit disc, $|z| < 1$.*

1. *Prove that there is a sequence (z_n) in the unit disc with $\lim_{n\to\infty} |z_n| = 1$ and $\lim_{n\to\infty} f(z_n)$ exists (finitely).*

2. *Assume f nonconstant. Prove that there are two sequences (z_n) and (w_n) in the disc such that $\lim_{n\to\infty} |z_n| = \lim_{n\to\infty} |w_n| = 1$, and such that both limits $\lim_{n\to\infty} f(z_n)$ and $\lim_{n\to\infty} f(w_n)$ exist (finitely) and are not equal.*

Problem 5.4.17 (Fa81, Sp89, Fa97) *Let f be a holomorphic map of the unit disc $\mathbb{D} = \{z \mid |z| < 1\}$ into itself, which is not the identity map $f(z) = z$. Show that f can have, at most, one fixed point.*

Problem 5.4.18 (Fa87) *If $f(z)$ is analytic in the open disc $|z| < 1$, and $|f(z)| \leqslant 1/(1 - |z|)$, show that*

$$|a_n| = \left| \frac{f^{(n)}(0)}{n!} \right| \leqslant (n+1)\left(1 + \frac{1}{n}\right)^n < e(n+1).$$

Problem 5.4.19 (Sp88)   1. *Let f be an analytic function that maps the open unit disc, $\mathbb{D}$, into itself and vanishes at the origin. Prove that $|f(z) + f(-z)| \leqslant 2|z|^2$ in $\mathbb{D}$.*

2. *Prove that the inequality in Part 1 is strict, except at the origin, unless f has the form $f(z) = \lambda z^2$ with λ a constant of absolute value one.*

Problem 5.4.20 (Sp91) *Let the function f be analytic in the unit disc, with $|f(z)| \leqslant 1$ and $f(0) = 0$. Assume that there is a number r in $(0, 1)$ such that $f(r) = f(-r) = 0$. Prove that*

$$|f(z)| \leqslant |z| \left| \frac{z^2 - r^2}{1 - r^2 z^2} \right|.$$

Problem 5.4.21 (Sp85) *Let $f(z)$ be an analytic function that maps the open disc $|z| < 1$ into itself. Show that $|f'(z)| \leqslant 1/(1 - |z|^2)$.*

Problem 5.4.22 (Sp87, Fa89) *Let f be an analytic function in the open unit disc of the complex plane such that $|f(z)| \leqslant C/(1 - |z|)$ for all z in the disc, where C is a positive constant. Prove that $|f'(z)| \leqslant 4C/(1 - |z|)^2$.*

5.5   Growth Conditions

Problem 5.5.1 (Sp03) *Let f be an entire function such that $\Re f(z) \geqslant -2$ for all $z \in \mathbb{C}$. Show that f is constant.*

Problem 5.5.2 (Fa90) *Let the function f be analytic in the entire complex plane, and suppose that $f(z)/z \to 0$ as $|z| \to \infty$. Prove that f is constant.*

Problem 5.5.3 (Fa99) *Let the rational function f in the complex plane have no poles for $\Im z \geqslant 0$. Prove that*

$$\sup\{|f(z)| \mid \Im z \geqslant 0\} = \sup\{|f(z)| \mid \Im z = 0\}.$$

Problem 5.5.4 (Fa97) *Let f be an entire function such that, for all z, $|f(z)| = |\sin z|$. Prove that there is a constant C of modulus 1 such that $f(z) = C \sin z$.*

Problem 5.5.5 (Fa79, Su81) *Suppose f and g are entire functions with $|f(z)| \leqslant |g(z)|$ for all z. Prove that $f(z) = cg(z)$ for some constant c.*

Problem 5.5.6 (Sp02) *Let p and q be nonconstant complex polynomials of the same degree whose zeros lie in the open disk $|z| < 1$. Prove that if $|p(z)| = |q(z)|$ for $|z| = 1$ then $q = \lambda p$ for a unimodular constant λ.*

Problem 5.5.7 (Fa98) *Let f be an entire function. Define $\Omega = \mathbb{C} \setminus (-\infty, 0]$, the complex plane with the ray $(-\infty, 0]$ removed. Suppose that for all $z \in \Omega$, $|f(z)| \leqslant |\log z|$, where $\log z$ is the principal branch of the logarithm. What can one conclude about the function f?*

Problem 5.5.8 (Sp97) *Let f and g be two entire functions such that, for all $z \in \mathbb{C}$, $\Re f(z) \leqslant k\Re g(z)$ for some real constant k (independent of z). Show that there are constants a, b such that*

$$f(z) = ag(z) + b.$$

Problem 5.5.9 (Sp02) *Let the continuous real–valued function φ on the complex plane satisfy the Sub–Mean–Value Property: for any point z_0,*

$$\varphi(z_0) \leq \frac{1}{2\pi} \int_0^{2\pi} \varphi(z_0 + re^{i\theta})d\theta \qquad (0 < r < 1).$$

Prove that φ obeys the Maximum Modulus Principle: the maximum of φ over any compact set K is attained on the boundary of K.

Problem 5.5.10 (Su78) *Let $f : \mathbb{C} \to \mathbb{C}$ be an entire function and let $a > 0$ and $b > 0$ be constants.*

1. *If $|f(z)| \leqslant a\sqrt{|z|} + b$ for all z, prove that f is a constant.*

2. *What can one prove about f if*

$$|f(z)| \leqslant a|z|^{5/2} + b$$

for all z?

Problem 5.5.11 (Fa90) *Let the function f be analytic in the entire complex plane and satisfy*

$$\int_0^{2\pi} |f(re^{i\theta})|\, d\theta \leqslant r^{17/3}$$

for all $r > 0$. Prove that f is the zero function.

Problem 5.5.12 (Fa96) *Does there exist a function f, analytic in the punctured plane $\mathbb{C} \setminus \{0\}$, such that*

$$|f(z)| \geqslant \frac{1}{\sqrt{|z|}}$$

for all nonzero z?

Problem 5.5.13 (Fa91) *Let the function f be analytic in the entire complex plane and satisfy the inequality $|f(z)| \leqslant |\Re z|^{-1/2}$ off the imaginary axis. Prove that f is constant.*

Problem 5.5.14 (Fa01) *Let the nonconstant entire function f satisfy the conditions*

1. *$f(0) = 0$;*

2. *for each $M > 0$, the set $\{z \mid |f(z)| < M\}$ is connected.*

Prove that $f(z) = cz^n$ for some constant c and positive integer n.

5.6 Analytic and Meromorphic Functions

Problem 5.6.1 (Sp96) *Let $f = u + iv$ be analytic in a connected open set D, where u and v are real valued. Suppose there are real constants a, b and c such that $a^2 + b^2 \neq 0$ and*

$$au + bv = c$$

in D. Show that f is constant in D.

Problem 5.6.2 (Sp88) *True or false: A function $f(z)$ analytic on $|z - a| < r$ and continuous on $|z - a| \leqslant r$ extends, for some $\delta > 0$, to a function analytic on $|z - a| < r + \delta$? Give a proof or a counterexample.*

Problem 5.6.3 (Fa80) *Do there exist functions $f(z)$ and $g(z)$ that are analytic at $z = 0$ and that satisfy*

1. $f(1/n) = f(-1/n) = 1/n^2, n = 1, 2, \ldots,$

2. $g(1/n) = g(-1/n) = 1/n^3, n = 1, 2, \ldots ?$

Problem 5.6.4 (Fa02) *Let k be an integer larger than 1. Find all entire functions f that satisfy $f(z^k) = (f(z))^k$.*

Problem 5.6.5 (Fa01) *Let F be a polynomial over $\mathbb{C}$ of positive degree d, and let S be its zero set. Prove that every rational function R whose finite poles lie in S can be written uniquely as*

$$ R = \sum_{k=m}^{n} a_k F^k , $$

where m and n are integers, $m \le n$, and the a_k are polynomials whose degrees are less than d.

Problem 5.6.6 (Fa99) *Let $A = \{0\} \cup \{1/n \mid n \in \mathbb{Z}, n > 1\}$, and let $\mathbb{D}$ be the open unit disc in the complex plane. Prove that every bounded holomorphic function on $\mathbb{D} \setminus A$ extends to a holomorphic function on $\mathbb{D}$.*

Problem 5.6.7 (Fa00) *Let U be a connected and simply connected open subset of the complex plane, and let f be a holomorphic function on U. Suppose a is a point of U such that the Taylor series of f at a converges on an open disc D that intersects the complement of U. Does it follow that f extends to a holomorphic function on $U \cup D$? Give a proof or a counterexample.*

Problem 5.6.8 (Su78)   1. *Suppose f is analytic on a connected open set $U \subset \mathbb{C}$ and f takes only real values. Prove that f is constant.*

2. *Suppose $W \subset \mathbb{C}$ is open, g is analytic on W, and $g'(z) \ne 0$ for all $z \in W$. Show that*
$$ \{\Re g(z) + \Im g(z) \mid z \in W\} \subset \mathbb{R} $$
is an open subset of $\mathbb{R}$.

Problem 5.6.9 (Sp78) *Let $f : \mathbb{C} \to \mathbb{C}$ be a nonconstant entire function. Prove that $f(\mathbb{C})$ is dense in $\mathbb{C}$.*

Problem 5.6.10 (Fa03) *Let L be a line in $\mathbb{C}$, and let f be an entire function such that $f(\mathbb{C}) \cap L = \emptyset$. Prove that f is constant. (Do not use the theorem of Picard that the image of a nonconstant entire function omits at most one complex number.)*

Problem 5.6.11 (Sp01)   1. *Prove that an entire function with a positive real part is constant.*

2. *Prove the analogous result for 2×2 matrix functions: If $F(z) = (f_{jk}(z))$ is a matrix function in the complex plane, each entry f_{jk} being entire, and if $F(z) + F(z)^*$ is positive definite for each z, then F is constant. (Here, $F(z)^*$ is the conjugate transpose of $F(z)$.)*

Problem 5.6.12 (Su82) *Let $s(y)$ and $t(y)$ be real differentiable functions of y, $-\infty < y < \infty$, such that the complex function*

$$f(x + iy) = e^x (s(y) + it(y))$$

is complex analytic with $s(0) = 1$ and $t(0) = 0$. Determine $s(y)$ and $t(y)$.

Problem 5.6.13 (Sp83) *Determine all the complex analytic functions f defined on the unit disc $\mathbb{D}$ which satisfy*

$$f''\left(\frac{1}{n}\right) + f\left(\frac{1}{n}\right) = 0$$

for $n = 2, 3, 4, \ldots$.

Problem 5.6.14 (Fa00) *Assume the nonconstant entire function f takes real values on two intersecting lines in the complex plane. Prove that the measure of either angle formed by the lines is a rational multiple of π.*

Problem 5.6.15 (Su83) *Let Ω be an open subset of $\mathbb{R}^2$, and let $f : \Omega \rightarrow \mathbb{R}^2$ be a smooth map. Assume that f preserves orientation and maps any pair of orthogonal curves to a pair of orthogonal curves. Show that f is holomorphic. Note: Here we identify $\mathbb{R}^2$ with $\mathbb{C}$.*

Problem 5.6.16 (Sp87) *Let f be a complex valued function in the open unit disc, $\mathbb{D}$, of the complex plane such that the functions $g = f^2$ and $h = f^3$ are both analytic. Prove that f is analytic in $\mathbb{D}$.*

Problem 5.6.17 (Fa84) *Prove or supply a counterexample: If f is a continuous complex valued function defined on a connected open subset of the complex plane and if f^2 is analytic, then f is analytic.*

Problem 5.6.18 (Sp88)   1. *Let G be an open connected subset of the complex plane, f an analytic function in G, not identically 0, and n a positive integer. Assume that f has an analytic n^{th} root in G; that is, there is an analytic function g in G such that $g^n = f$. Prove that f has exactly n analytic n^{th} roots in G.*

2. *Give an example of a continuous real valued function on $[0, 1]$ that has more than two continuous square roots on $[0, 1]$.*

Problem 5.6.19 (Sp99)   1. *Prove that if f is holomorphic on the unit disc $\mathbb{D}$ and $f(z) \neq 0$ for all $z \in \mathbb{D}$, then there is a holomorphic function g on $\mathbb{D}$ such that $f(z) = e^{g(z)}$ for all $z \in \mathbb{D}$.*

2. *Does the conclusion of Part 1 remain true if $\mathbb{D}$ is replaced by an arbitrary connected open set in $\mathbb{C}$?*

Problem 5.6.20 (Fa92) *Let the function f be analytic in the region $|z| > 1$ of the complex plane. Prove that if f is real valued on the interval $(1, \infty)$ of the real axis, then f is also real valued on the interval $(-\infty, -1)$.*

Problem 5.6.21 (Fa94) *Let the function f be analytic in the complex plane, real on the real axis, 0 at the origin, and not identically 0. Prove that if f maps the imaginary axis into a straight line, then that straight line must be either the real axis or the imaginary axis.*

Problem 5.6.22 (Fa01) *Let the entire function f be real valued on the lines $\Im z = 0$ and $\Im z = \pi$. Prove that f has $2\pi i$ as a period: $f(z + 2\pi i) = f(z)$ for all z.*

Problem 5.6.23 (Fa87) *Let $f(z)$ be analytic for $z \neq 0$, and suppose that $f(1/z) = f(z)$. Suppose also that $f(z)$ is real for all z on the unit circle $|z| = 1$. Prove that $f(z)$ is real for all real $z \neq 0$.*

Problem 5.6.24 (Fa91) *Let p be a nonconstant complex polynomial whose zeros are all in the half-plane $\Im z > 0$.*

1. *Prove that $\Im(p'/p) > 0$ on the real axis.*

2. *Find a relation between $\deg p$ and*

$$\int_{-\infty}^{\infty} \Im \frac{p'(x)}{p(x)} \, dx.$$

Problem 5.6.25 (Sp92) *Let f be an analytic function in the connected open subset G of the complex plane. Assume that for each point z in G, there is a positive integer n such that the nth derivative of f vanishes at z. Prove that f is a polynomial.*

Problem 5.6.26 (Sp92) *Let the function f be analytic in the entire complex plane, real valued on the real axis, and of positive imaginary part in the upper half-plane. Prove $f'(x) > 0$ for x real.*

Problem 5.6.27 (Sp94)   1. *Let U and V be open connected subsets of the complex plane, and let f be an analytic function in U such that $f(U) \subset V$. Assume $f^{-1}(K)$ is compact whenever K is a compact subset of V. Prove that $f(U) = V$.*

2. *Prove that the last equality can fail if analytic is replaced by continuous in the preceding statement.*

Problem 5.6.28 (Sp94) *Let $f = u + iv$ and $g = p + iq$ be analytic functions defined in a neighborhood of the origin in the complex plane. Assume $|g'(0)| < |f'(0)|$. Prove that there is a neighborhood of the origin in which the function $h = f + \bar{g}$ is one-to-one.*

Problem 5.6.29 (Sp87) *Prove or disprove: If the function f is analytic in the entire complex plane, and if f maps every unbounded sequence to an unbounded sequence, then f is a polynomial.*

Problem 5.6.30 (Fa88, Sp97) *Determine the group $\text{Aut}(\mathbb{C})$ of all one-to-one analytic maps of $\mathbb{C}$ onto $\mathbb{C}$.*

Problem 5.6.31 (Sp77) *Let $f(z)$ be a nonconstant meromorphic function. A complex number w is called a period of f if $f(z + w) = f(z)$ for all z.*

1. *Show that if w_1 and w_2 are periods, so are $n_1 w_1 + n_2 w_2$ for all integers n_1 and n_2.*

2. *Show that there are, at most, a finite number of periods of f in any bounded region of the complex plane.*

Problem 5.6.32 (Sp91) *Let the function f be analytic in the punctured disc $0 < |z| < r_0$, with Laurent series*

$$f(z) = \sum_{-\infty}^{\infty} c_n z^n.$$

Assume there is a positive number M such that

$$r^4 \int_0^{2\pi} |f(re^{i\theta})|^2 \, d\theta < M, \qquad 0 < r < r_0.$$

Prove that $c_n = 0$ for $n < -2$.

Problem 5.6.33 (Sp98) *Let $a > 0$. Show that the complex function*

$$f(z) = \frac{1 + z + az^2}{1 - z + az^2}$$

satisfies $|f(z)| < 1$ for all z in the open left half-plane $\Re z < 0$.

Problem 5.6.34 (Fa93) *Let f be a continuous real valued function on $[0, 1]$, and let the function h in the complex plane be defined by*

$$h(z) = \int_0^1 f(t) \cos zt \, dt.$$

1. *Prove that h is analytic in the entire plane.*

2. *Prove that h is the zero function only if f is the zero function.*

Problem 5.6.35 (Su79, Sp82, Sp91, Sp96) *Let f be a continuous complex valued function on* [0, 1], *and define the function g by*

$$g(z) = \int_0^1 f(t)e^{tz}\, dt \qquad (z \in \mathbb{C}).$$

Prove that g is analytic in the entire complex plane.

Problem 5.6.36 (Fa84, Fa95) *Let f and g be analytic functions in the open unit disc, and let C_r denote the circle with center 0 and radius r, oriented counterclockwise.*

1. *Prove that the integral*

$$\frac{1}{2\pi i} \int_{C_r} \frac{1}{w} f(w) g\left(\frac{z}{w}\right) dw$$

   is independent of r as long as $|z| < r < 1$ and that it defines an analytic function $h(z)$, $|z| < 1$.

2. *Prove or supply a counterexample: If $f \not\equiv 0$ and $g \not\equiv 0$, then $h \not\equiv 0$.*

Problem 5.6.37 (Sp84) *Let F be a continuous complex valued function on the interval* [0, 1]. *Let*

$$f(z) = \int_0^1 \frac{F(t)}{t - z}\, dt,$$

for z a complex number not in [0, 1].

1. *Prove that f is an analytic function.*

2. *Express the coefficients of the Laurent series of f about ∞ in terms of F. Use the result to show that F is uniquely determined by f.*

Problem 5.6.38 (Fa03) *Let $f(z)$ be a meromorphic function on the complex plane. Suppose that for every polynomial $p(z) \in \mathbb{C}[z]$ and every closed contour Γ avoiding the poles of f, we have*

$$\int_\Gamma p(z)^2 f(z)\, dz = 0.$$

Prove that $f(z)$ is entire.

5.7   Cauchy's Theorem

Problem 5.7.1 (Fa85) *Evaluate*

$$\int_0^{2\pi} e^{e^{i\theta}}\, d\theta.$$

Problem 5.7.2 (Su78) *Evaluate*

$$\int_0^{2\pi} e^{(e^{i\theta} - i\theta)}\, d\theta .$$

Problem 5.7.3 (Sp98) *Let a be a complex number with $|a| < 1$. Evaluate the integral*

$$\int_{|z|=1} \frac{|dz|}{|z - a|^2}$$

Problem 5.7.4 (Fa99) *For $0 < a < b$, evaluate the integral*

$$I = \frac{1}{2\pi} \int_0^{2\pi} \frac{1}{|ae^{i\theta} - b|^4}\, d\theta .$$

Problem 5.7.5 (Sp77, Sp82) *Prove the Fundamental Theorem of Algebra: Every nonconstant polynomial with complex coefficients has a complex root.*

Problem 5.7.6 (Su77) *Let f be continuous on $\mathbb{C}$ and analytic on $\{z \mid \Im z \neq 0\}$. Prove that f must be analytic on $\mathbb{C}$.*

Problem 5.7.7 (Fa00) *Let $f(z)$ be the rational function $p(z)/q(z)$, where $p(z)$ and $q(z)$ are nonzero polynomials with complex coefficients, such that the degree of $p(z)$ is less than the degree of $q(z)$, and such that $q(z)$ has no complex zeros with nonnegative imaginary part. Prove that if z_0 is a complex number with positive imaginary part, then*

$$f(z_0) = \frac{1}{2\pi i} \int_{-\infty}^{+\infty} \frac{f(t)}{t - z_0}\, dt.$$

Problem 5.7.8 (Fa78, Su79) *Let $f(z) = a_0 + a_1 z + \cdots + a_n z^n$ be a complex polynomial of degree $n > 0$. Prove*

$$\frac{1}{2\pi i} \int_{|z|=R} z^{n-1} |f(z)|^2\, dz = a_0 \bar{a}_n R^{2n} .$$

Problem 5.7.9 (Sp01) *Let f be an entire function such that*

$$\int_0^{2\pi} |f(re^{i\theta})|^2\, d\theta \leqslant Ar^{2k} \qquad (0 < r < \infty),$$

where k is a positive integer and A is a positive constant. Prove that f is a constant multiple of the function z^k.

Problem 5.7.10 (Fa95) *Let $f(z) = u(z) + iv(z)$ be holomorphic on $|z| < 1$, u and v real. Show that*

$$\int_0^{2\pi} u(re^{i\theta})^2 d\theta = \int_0^{2\pi} v(re^{i\theta})^2 d\theta$$

for $0 < r < 1$ if $u(0)^2 = v(0)^2$.

Problem 5.7.11 (Su83) *Let* $f : \mathbb{C} \to \mathbb{C}$ *be an analytic function such that*

$$\left(1 + |z|^k\right)^{-1} \frac{d^m f}{dz^m}$$

is bounded for some k and m. Prove that $d^n f/dz^n$ *is identically zero for sufficiently large n. How large must n be, in terms of k and m?*

Problem 5.7.12 (Su83) *Suppose* Ω *is a bounded domain in* $\mathbb{C}$ *with a boundary consisting of a smooth Jordan curve* γ. *Let f be holomorphic on a neighborhood of the closure of* Ω, *and suppose that* $f(z) \neq 0$ *for* $z \in \gamma$. *Let* $z_1, \ldots, z_k$ *be the zeros of f in* Ω, *and let* n_j *be the order of the zero of f at* z_j *(for* $j = 1, \ldots, k$).

1. *Use Cauchy's integral formula to show that*

$$\frac{1}{2\pi i} \int_\gamma \frac{f'(z)}{f(z)} \, dz = \sum_{j=1}^k n_j.$$

2. *Suppose that f has only one zero* z_1 *in* Ω *with multiplicity* $n_1 = 1$. *Find a boundary integral involving f whose value is the point* z_1.

Problem 5.7.13 (Fa88) *Let f be an analytic function on a disc D whose center is the point* z_0. *Assume that* $|f'(z) - f'(z_0)| < |f'(z_0)|$ *on D. Prove that f is one-to-one on D.*

Problem 5.7.14 (Fa89) *Let* $f(z)$ *be analytic in the annulus* $\Omega = \{1 < |z| < 2\}$. *Assume that f has no zeros in* Ω. *Show that there exists an integer n and an analytic function g in* Ω *such that, for all* $z \in \Omega$, $f(z) = z^n e^{g(z)}$.

Problem 5.7.15 (Sp02) *Define the function f on* $\mathbb{C} \setminus [0, 1]$ *by*

$$f(z) = \int_0^1 \frac{\sqrt{t}}{t - z} \, dt \, .$$

Prove that f is analytic, and find its Laurent series about ∞.

Problem 5.7.16 (Sp90) *Let the function f be analytic and bounded in the complex half-plane* $\Re z > 0$. *Prove that for any positive real number c, the function f is uniformly continuous in the half-plane* $\Re z > c$.

5.8   Zeros and Singularities

Problem 5.8.1 (Fa77, Fa96) *Let* $\mathbb{C}^3$ *denote the set of ordered triples of complex numbers. Define a map* $F : \mathbb{C}^3 \to \mathbb{C}^3$ *by*

$$F(u, v, w) = (u + v + w, uv + vw + wu, uvw).$$

Prove that F is onto but not one-to-one.

Problem 5.8.2 (Fa79, Fa89) *Prove that the polynomial*

$$p(z) = z^{47} - z^{23} + 2z^{11} - z^5 + 4z^2 + 1$$

has at least one root in the disc $|z| < 1$.

Problem 5.8.3 (Fa80) *Suppose that f is analytic inside and on the unit circle $|z| = 1$ and satisfies $|f(z)| < 1$ for $|z| = 1$. Show that the equation $f(z) = z^3$ has exactly three solutions (counting multiplicities) inside the unit circle.*

Problem 5.8.4 (Fa81)   *1. How many zeros does the function $f(z) = 3z^{100} - e^z$ have inside the unit circle (counting multiplicities)?*

2. Are the zeros distinct?

Problem 5.8.5 (Fa92)   *1. How many roots does the polynomial defined by $p(z) = 2z^5 + 4z^2 + 1$ have in the disc $|z| < 1$?*

2. How many roots does the same polynomial have on the real axis?

Problem 5.8.6 (Su80) *How many zeros does the complex polynomial*

$$3z^9 + 8z^6 + z^5 + 2z^3 + 1$$

have in the annulus $1 < |z| < 2$?

Problem 5.8.7 (Fa83) *Consider the polynomial*

$$p(z) = z^5 + z^3 + 5z^2 + 2.$$

How many zeros (counting multiplicities) does p have in the annular region $1 < |z| < 2$?

Problem 5.8.8 (Sp84, Fa87, Fa96) *Find the number of roots of*

$$z^7 - 4z^3 - 11 = 0$$

which lie between the two circles $|z| = 1$ and $|z| = 2$.

Problem 5.8.9 (Fa01) *Let ε be a positive number. Prove that the polynomial $p(z) = \varepsilon z^3 - z^2 - 1$ has exactly two roots in the half-plane $\Re z < 0$.*

Problem 5.8.10 (Sp96) *Let $r < 1 < R$. Show that for all sufficiently small $\varepsilon > 0$, the polynomial*

$$p(z) = \varepsilon z^7 + z^2 + 1$$

has exactly five roots (counted with their multiplicities) inside the annulus

$$r\varepsilon^{-1/5} < |z| < R\varepsilon^{-1/5}.$$

Problem 5.8.11 (Sp86) *Let the 3×3 matrix function A be defined on the complex plane by*

$$A(z) = \begin{pmatrix} 4z^2 & 1 & -1 \\ -1 & 2z^2 & 0 \\ 3 & 0 & 1 \end{pmatrix}.$$

How many distinct values of z are there such that $|z| < 1$ and $A(z)$ is not invertible?

Problem 5.8.12 (Fa85) *How many roots does the polynomial $z^4 + 3z^2 + z + 1$ have in the right half z-plane?*

Problem 5.8.13 (Sp87) *Prove that if the nonconstant polynomial $p(z)$, with complex coefficients, has all of its roots in the half-plane $\Re z > 0$, then all of the roots of its derivative are in the same half-plane.*

Problem 5.8.14 (Sp00) *Let f be a nonconstant entire function whose values on the real axis are real and nonnegative. Prove that all real zeros of f have even order.*

Problem 5.8.15 (Sp92) *Let p be a nonconstant polynomial with real coefficients and only real roots. Prove that for each real number r, the polynomial $p - rp'$ has only real roots.*

Problem 5.8.16 (Sp79, Su85, Sp89) *Prove that if $1 < \lambda < \infty$, the function*

$$f_\lambda(z) = z + \lambda - e^z$$

has only one zero in the half-plane $\Re z < 0$, and that this zero is real.

Problem 5.8.17 (Fa85) *Prove that for every $\lambda > 1$, the equation $ze^{\lambda-z} = 1$ has exactly one root in the disc $|z| < 1$ and that this root is real.*

Problem 5.8.18 (Sp85) *Prove that for any $a \in \mathbb{C}$ and any integer $n \geqslant 2$, the equation $1 + z + az^n = 0$ has at least one root in the disc $|z| \leqslant 2$.*

Problem 5.8.19 (Sp98) *Prove that the polynomial $z^4 + z^3 + 1$ has exactly one root in the quadrant $\{z = x + iy \mid x, y > 0\}$.*

Problem 5.8.20 (Sp98) *Let f be analytic in an open set containing the closed unit disc. Suppose that $|f(z)| > m$ for $|z| = 1$ and $|f(0)| < m$. Prove that $f(z)$ has at least one zero in the open unit disc $|z| < 1$.*

Problem 5.8.21 (Su82) *Let $0 < a_0 \leqslant a_1 \leqslant \cdots \leqslant a_n$. Prove that the equation*

$$a_0z^n + a_1z^{n-1} + \cdots + a_n = 0$$

has no roots in the disc $|z| < 1$.

Problem 5.8.22 (Fa86) *Show that the polynomial $p(z) = z^5 - 6z + 3$ has five distinct complex roots, of which exactly three (and not five) are real.*

Problem 5.8.23 (Sp90) *Let $c_0, c_1, \ldots, c_{n-1}$ be complex numbers. Prove that all the zeros of the polynomial*

$$z^n + c_{n-1}z^{n-1} + \cdots + c_1 z + c_0$$

lie in the open disc with center 0 and radius

$$\sqrt{1 + |c_{n-1}|^2 + \cdots + |c_1|^2 + |c_0|^2}\,.$$

Problem 5.8.24 (Sp95) *Let $P(x)$ be a polynomial with real coefficients and with leading coefficient 1. Suppose that $P(0) = -1$ and that $P(x)$ has no complex zeros inside the unit circle. Prove that $P(1) = 0$.*

Problem 5.8.25 (Su81) *Prove that the number of roots of the equation $z^{2n} + \alpha^2 z^{2n-1} + \beta^2 = 0$ (n a natural number, α and β real, nonzero) that have positive real part is*

1. *n if n is even, and*

2. *$n - 1$ if n is odd.*

Problem 5.8.26 (Su84) *Let $\rho > 0$. Show that for n large enough, all the zeros of*

$$f_n(z) = 1 + \frac{1}{z} + \frac{1}{2!z^2} + \cdots + \frac{1}{n!z^n}$$

lie in the circle $|z| < \rho$.

Problem 5.8.27 (Fa88) *Do the functions $f(z) = e^z + z$ and $g(z) = ze^z + 1$ have the same number of zeros in the strip $-\frac{\pi}{2} < \Im z < \frac{\pi}{2}$?*

Problem 5.8.28 (Sp93) *Let a be a complex number and ε a positive number. Prove that the function $f(z) = \sin z + \frac{1}{z-a}$ has infinitely many zeros in the strip $|\Im z| < \varepsilon$.*

Problem 5.8.29 (Sp99) *Suppose that f is holomorphic on some neighborhood of a in the complex plane. Prove that either f is constant on some neighborhood of a, or there exist an integer $n > 0$ and real numbers $\delta, \varepsilon > 0$ such that for each complex number b satisfying $0 < |b - f(a)| < \varepsilon$, the equation $f(z) = b$ has exactly n roots in $\{z \in \mathbb{C} \mid |z - a| < \delta\}$.*

Problem 5.8.30 (Su77, Sp81) *Let $\hat{a}_0 + \hat{a}_1 z + \cdots + \hat{a}_n z^n$ be a polynomial having $\hat{z}$ as a simple root. Show that there is a continuous function $r : U \to \mathbb{C}$, where U is a neighborhood of $(\hat{a}_0, \ldots, \hat{a}_n)$ in $\mathbb{C}^{n+1}$, such that $r(a_0, \ldots, a_n)$ is always a root of $a_0 + a_1 z + \cdots + a_n z^n$, and $r(\hat{a}_0, \ldots, \hat{a}_n) = \hat{z}$.*

Problem 5.8.31 (Su85) *Let*

$$f(z) = \sum_{n=0}^{\infty} a_n z^n$$

where all the a_n are nonnegative reals, and the series has radius of convergence 1. Prove that $f(z)$ cannot be analytically continued to a function analytic in a neighborhood of $z = 1$.

Problem 5.8.32 (Su80) *Let f be a meromorphic function on $\mathbb{C}$ which is analytic in a neighborhood of 0. Let its Maclaurin series be*

$$\sum_{k=0}^{\infty} a_k z^k$$

with all $a_k \geqslant 0$. Suppose there is a pole of modulus $r > 0$ and no pole has modulus $< r$. Prove there is a pole at $z = r$.

Problem 5.8.33 (Sp82) *Decide, without too much computation, whether a finite limit*

$$\lim_{z \to 0} \left((\tan z)^{-2} - z^{-2} \right)$$

exists, where z is a complex variable, and if yes, compute the limit.

Problem 5.8.34 (Sp89, Sp00) *Let f and g be entire functions such that $\lim_{z \to \infty} f(g(z)) = \infty$. Prove that f and g are polynomials.*

Problem 5.8.35 (Fa98) *Let $z_1, \ldots, z_n$ be distinct complex numbers, and let $a_1, \ldots, a_n$ be nonzero complex numbers such that $S_p = \sum_{j=1}^{n} a_j z_j^p = 0$ for $p = 0, 1, \ldots, m - 1$ but $S_m \neq 0$. Here $1 \leqslant m \leqslant n - 1$. How many zeros does the rational function $f(z) = \sum_{j=1}^{n} \frac{a_j}{z - z_j}$ have in $\mathbb{C}$? Why is $m \geqslant n$ impossible.*

5.9   Harmonic Functions

Problem 5.9.1 (Fa77, Fa81) *Let $u : \mathbb{R}^2 \to \mathbb{R}$ be the function defined by $u(x, y) = x^3 - 3xy^2$. Show that u is harmonic and find $v : \mathbb{R}^2 \to \mathbb{R}$ such that the function $f : \mathbb{C} \to \mathbb{C}$ defined by*

$$f(x + iy) = u(x, y) + iv(x, y)$$

is analytic.

Problem 5.9.2 (Fa80) *Let $f(z)$ be an analytic function defined for $|z| \leqslant 1$ and let*

$$u(x, y) = \Re f(z), \quad z = x + iy.$$

Prove that

$$\int_C \frac{\partial u}{\partial y} dx - \frac{\partial u}{\partial x} dy = 0$$

where C is the unit circle, $x^2 + y^2 = 1$.

Problem 5.9.3 (Fa83)   *1. Let f be a complex function which is analytic on an open set containing the disc $|z| \leqslant 1$, and which is real valued on the unit circle. Prove that f is constant.*

2. Find a nonconstant function which is analytic at every point of the complex plane except for a single point on the unit circle $|z| = 1$, and which is real valued at every other point of the unit circle.

Problem 5.9.4 (Fa92) *Let s be a real number, and let the function u be defined in $\mathbb{C} \backslash (-\infty, 0]$ by*

$$u(re^{i\theta}) = r^s \cos s\theta \qquad (r > 0, \quad -\pi < \theta < \pi).$$

Prove that u is a harmonic function.

Problem 5.9.5 (Fa87) *Let u be a positive harmonic function on $\mathbb{R}^2$; that is,*

$$\frac{\partial^2 u}{\partial x^2} + \frac{\partial^2 u}{\partial y^2} = 0.$$

Show that u is constant.

Problem 5.9.6 (Sp94) *Let u be a real valued harmonic function in the complex plane such that*

$$u(z) \leqslant a \, |\log |z|| + b$$

for all z, where a and b are positive constants. Prove that u is constant.

Problem 5.9.7 (Fa03) *Let $D = \{z \in \mathbb{C} : |z| \leq 1\} \backslash \{1, -1\}$. Find an explicit continuous function $f : D \to \mathbb{R}$ satisfying all the following conditions:*

- *f is harmonic on the interior of D (the open unit disk),*
- *$f(z) = 1$ when $|z| = 1$ and $\Im z > 0$, and*
- *$f(z) = -1$ when $|z| = 1$ and $\Im z < 0$.*

5.10   Residue Theory

Problem 5.10.1 (Fa83) *Let $r_1, r_2, \dots, r_n$ be distinct complex numbers. Show that a rational function of the form*

$$f(z) = \frac{b_0 + b_1 z + \cdots + b_{n-2} z^{n-2} + b_{n-1} z^{n-1}}{(z - r_1)(z - r_2) \cdots (z - r_n)}$$

can be written as a sum

$$f(z) = \frac{A_1}{z - r_1} + \frac{A_2}{z - r_2} + \cdots + \frac{A_n}{z - r_n}$$

for suitable constants $A_1, \dots, A_n$.

Problem 5.10.2 (Fa82) *Let*

$$\cot(\pi z) = \sum_{n=-\infty}^{\infty} a_n z^n$$

be the Laurent expansion for $\cot(\pi z)$ on the annulus $1 < |z| < 2$. Compute the a_n for $n < 0$.

Problem 5.10.3 (Sp78) *Show that there is a complex analytic function defined on the set $U = \{z \in \mathbb{C} \mid |z| > 4\}$ whose derivative is*

$$\frac{z}{(z-1)(z-2)(z-3)}.$$

Is there a complex analytic function on U whose derivative is

$$\frac{z^2}{(z-1)(z-2)(z-3)} \, ?$$

Problem 5.10.4 (Fa88) *Let n be a positive integer. Prove that the polynomial*

$$f(x) = \sum_{i=0}^{n} \frac{x^i}{i!} = 1 + x + \frac{x^2}{2} + \cdots + \frac{x^n}{n!}$$

in $\mathbb{R}[x]$ has n distinct complex zeros, $z_1, z_2, \ldots, z_n$, and that they satisfy

$$\sum_{i=1}^{n} z_i^{-j} = 0 \quad for \quad 2 \leqslant j \leqslant n.$$

Problem 5.10.5 (Sp79, Sp83) *Let P and Q be complex polynomials with the degree of Q at least two more than the degree of P. Prove there is an $r > 0$ such that if C is a closed curve outside $|z| = r$, then*

$$\int_C \frac{P(z)}{Q(z)} \, dz = 0.$$

Problem 5.10.6 (Sp93) *Prove that for any fixed complex number ζ,*

$$\frac{1}{2\pi} \int_0^{2\pi} e^{2\zeta \cos\theta} \, d\theta = \sum_{n=0}^{\infty} \left(\frac{\zeta^n}{n!} \right)^2.$$

Problem 5.10.7 (Fa99) *Evaluate the integral*

$$I = \frac{1}{2\pi i} \int_{|z|=1} \frac{(z+2)^2}{z^2(2z-1)} \, dz,$$

where the direction of integration is counterclockwise.

Problem 5.10.8 (Sp80) *Let $a > 0$ be a constant $\neq 2$. Let C_a denote the positively oriented circle of radius a centered at the origin. Evaluate*

$$\int_{C_a} \frac{z^2 + e^z}{z^2(z-2)}\, dz\,.$$

Problem 5.10.9 (Sp02) *Let r be a positive number and a a complex number such that $|a| \neq r$.*
Evaluate the integral

$$I(a, r) = \int_{|z|=r} \frac{1}{a - \bar{z}}\, dz$$

where the circle $|z| = r$ has the counterclockwise orientation.

Problem 5.10.10 (Su80) *Let C denote the positively oriented circle $|z| = 2$, $z \in \mathbb{C}$. Evaluate the integral*

$$\int_C \sqrt{z^2 - 1}\, dz$$

where the branch of the square root is chosen so that $\sqrt{2^2 - 1} > 0$.

Problem 5.10.11 (Fa00) *Evaluate the integral*

$$I = \frac{1}{2\pi i} \int_{|z|=1} \frac{dz}{\sin 4z}$$

where the direction of integration is counterclockwise.

Problem 5.10.12 (Fa02) *Evaluate the integrals*

$$I_n = \int_{C_n} \frac{1}{z^3 \sin z}\, dz\,, \qquad n = 0, 1, \ldots,$$

where C_n is the circle $|z| = \left(n + \frac{1}{2}\right)\pi$, with the counterclockwise orientation.

Problem 5.10.13 (Su81) *Compute*

$$\frac{1}{2\pi i} \int_C \frac{dz}{\sin \frac{1}{z}},$$

where C is the circle $|z| = \dfrac{1}{5}$, positively oriented.

Problem 5.10.14 (Su84)   *1. Show that there is a unique analytic branch out-side the unit circle of the function $f(z) = \sqrt{z^2 + z + 1}$ such that $f(t)$ is positive when $t > 1$.*

2. *Using the branch determined in Part 1, calculate the integral*

$$\frac{1}{2\pi i} \int_{C_r} \frac{dz}{\sqrt{z^2 + z + 1}}$$

where C_r is the positively oriented circle $|z| = r$ and $r > 1$.

Problem 5.10.15 (Sp86) *Let C be a simple closed contour enclosing the points $0, 1, 2, \ldots, k$ in the complex plane, with positive orientation. Evaluate the integrals*

$$I_k = \int_C \frac{dz}{z(z-1)\cdots(z-k)}, \qquad k = 0, 1, \ldots,$$

$$J_k = \int_C \frac{(z-1)\cdots(z-k)}{z} \, dz, \qquad k = 0, 1, \ldots.$$

Problem 5.10.16 (Sp86) *Evaluate*

$$\int_{|z|=1} (e^{2\pi z} + 1)^{-2} \, dz$$

where the integral is taken in counterclockwise direction.

Problem 5.10.17 (Fa86) *Evaluate*

$$\frac{1}{2\pi i} \int_{|z|=1} \frac{z^{11}}{12z^{12} - 4z^9 + 2z^6 - 4z^3 + 1} \, dz$$

where the direction of integration is counterclockwise.

Problem 5.10.18 (Sp89) *Evaluate*

$$\int_C (2z - 1)e^{z/(z-1)} \, dz$$

where C is the circle $|z| = 2$ with counterclockwise orientation.

Problem 5.10.19 (Fa90) *Evaluate the integral*

$$I = \frac{1}{2\pi i} \int_C \frac{dz}{(z-2)(1 + 2z)^2(1 - 3z)^3}$$

where C is the circle $|z| = 1$ with counterclockwise orientation.

Problem 5.10.20 (Fa91) *Evaluate the integral*

$$I = \frac{1}{2\pi i} \int_C \frac{z^{n-1}}{3z^n - 1} \, dz,$$

where n is a positive integer, and C is the circle $|z| = 1$, with counterclockwise orientation.

Problem 5.10.21 (Fa92) *Evaluate*

$$\int_C \frac{e^z}{z(2z+1)^2}\, dz,$$

where C is the unit circle with counterclockwise orientation.

Problem 5.10.22 (Fa01) *Let* $a = \dfrac{1+i}{2}$, $b = \dfrac{-1+i}{2}$. *Let* Γ *be the polygonal path in the complex plane with successive vertices* -1, $-1+i$, 1, $1+i$, -1. *Evaluate the integral*

$$I = \int_\Gamma \frac{1}{(z-a)^2(z-b)^3}\, dz .$$

Problem 5.10.23 (Fa93) *Evaluate the integral* $\frac{1}{2\pi i}\int_\gamma f(z)\, dz$ *for the function* $f(z) = z^{-2}(1-z^2)^{-1}e^z$ *and the curve* γ *depicted by*

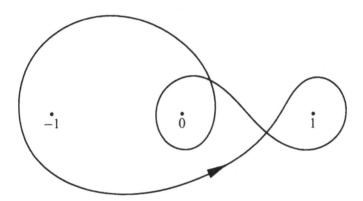

Problem 5.10.24 (Sp81) *Evaluate*

$$\int_C \frac{e^z - 1}{z^2(z-1)}\, dz$$

where C is the closed curve shown below:

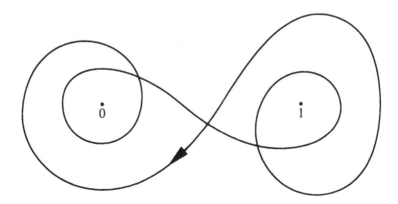

Problem 5.10.25 (Sp95) *Let n be a positive integer and $0 < \theta < \pi$. Prove that*

$$\frac{1}{2\pi i} \int_{|z|=2} \frac{z^n}{1 - 2z \cos \theta + z^2} dz = \frac{\sin n\theta}{\sin \theta}$$

where the circle $|z| = 2$ is oriented counterclockwise.

Problem 5.10.26 (Su77, Fa84, Sp94, Sp96) *Use the Residue Theorem to evaluate the integral*

$$I(a) = \int_0^{2\pi} \frac{d\theta}{a + \cos \theta}$$

where a is real and a > 1. Why the formula obtained for $I(a)$ is also valid for certain complex (nonreal) values of a ?

Problem 5.10.27 (Fa78) *Evaluate*

$$\int_0^{2\pi} \frac{d\theta}{1 - 2r \cos \theta + r^2}$$

where $r^2 \neq 1$.

Problem 5.10.28 (Sp87) *Evaluate*

$$I = \int_0^\pi \frac{\cos 4\theta}{1 + \cos^2 \theta} d\theta .$$

Problem 5.10.29 (Fa87) *Evaluate the integral*

$$I = \int_0^{2\pi} \frac{\cos^2 3\theta}{5 - 4 \cos 2\theta} d\theta .$$

Problem 5.10.30 (Sp00) *Evaluate*

$$I = \int_{|z|=1} \frac{\cos^3 z}{z^3} dz ,$$

where the direction of integration is counterclockwise.

Problem 5.10.31 (Sp95) *Let n be a positive integer. Compute*

$$\int_0^{2\pi} \frac{1 - \cos n\theta}{1 - \cos \theta} d\theta .$$

Problem 5.10.32 (Fa94) *Evaluate the integrals*

$$\int_{-\pi}^\pi \frac{\sin n\theta}{\sin \theta} d\theta, \quad n = 1, 2, \ldots .$$

Problem 5.10.33 (Sp88) *For a > 1 and $n = 0, 1, 2, \ldots$, evaluate the integrals*

$$C_n(a) = \int_{-\pi}^\pi \frac{\cos n\theta}{a - \cos \theta} d\theta , \qquad S_n(a) = \int_{-\pi}^\pi \frac{\sin n\theta}{a - \cos \theta} d\theta .$$

5.11   Integrals Along the Real Axis

Problem 5.11.1 (Sp86) *Let the complex valued functions f_n, $n \in \mathbb{Z}$, be defined on $\mathbb{R}$ by*

$$f_n(x) = \frac{(x-i)^n}{\sqrt{\pi}(x+i)^{n+1}} \cdot$$

Prove that these functions are orthonormal; *that is,*

$$\int_{-\infty}^{\infty} f_m(x)\overline{f_n(x)}\, dx = \begin{cases} 1 & \text{if} \quad m = n \\ 0 & \text{if} \quad m \neq n. \end{cases}$$

Problem 5.11.2 (Fa85) *Evaluate the integral*

$$\int_0^{\infty} \frac{1 - \cos ax}{x^2}\, dx$$

for $a \in \mathbb{R}$.

Problem 5.11.3 (Sp00) *Evaluate the integrals*

$$I(t) = \int_{-\infty}^{\infty} \frac{e^{itx}}{(x+i)^2}\, dx, \qquad -\infty < t < \infty.$$

Problem 5.11.4 (Fa01) *Evaluate the integrals* $\quad F(t) = \int_{-\infty}^{\infty} \frac{e^{-itx}}{(x+i)^3}\, dx$,

$-\infty < t < \infty.$

Problem 5.11.5 (Sp78, Sp83, Sp97) *Evaluate*

$$\int_{-\infty}^{\infty} \frac{\sin^2 x}{x^2}\, dx.$$

Problem 5.11.6 (Fa82, Sp92) *Evaluate*

$$\int_{-\infty}^{\infty} \frac{\sin^3 x}{x^3}\, dx.$$

Problem 5.11.7 (Sp93) *Evaluate*

$$\int_{-\infty}^{\infty} \frac{x^3 \sin x}{(1+x^2)^2}\, dx.$$

Problem 5.11.8 (Sp81) *Evaluate*

$$\int_{-\infty}^{\infty} \frac{x \sin x}{(1+x^2)^2}\, dx.$$

Problem 5.11.9 (Sp90, Fa92) *Let a be a positive real number. Evaluate the improper integral*

$$\int_0^\infty \frac{\sin x}{x(x^2 + a^2)}\, dx \,.$$

Problem 5.11.10 (Sp91) *Prove that*

$$\lim_{R \to \infty} \int_{-R}^R \frac{\sin x}{x - 3i}\, dx$$

exists and find its value.

Problem 5.11.11 (Sp83) *Evaluate*

$$\int_{-\infty}^\infty \frac{\sin x}{x(x - \pi)}\, dx \,.$$

Problem 5.11.12 (Fa02) *Evaluate*

$$\int_{-\infty}^\infty \frac{\cos x}{(1 + x^2)^3}\, dx \,.$$

Problem 5.11.13 (Fa97) *Evaluate the integral*

$$\int_{-\infty}^\infty \frac{\cos kx}{1 + x + x^2}\, dx$$

where $k \geqslant 0$.

Problem 5.11.14 (Fa82) *Evaluate*

$$\int_{-\infty}^\infty \frac{\cos \pi x}{4x^2 - 1}\, dx \,.$$

Problem 5.11.15 (Sp77, Fa81, Sp82) *Evaluate*

$$\int_{-\infty}^\infty \frac{\cos nx}{x^4 + 1}\, dx \,.$$

Problem 5.11.16 (Sp79) *Evaluate*

$$\int_0^\infty \frac{x^2 + 1}{x^4 + 1}\, dx \,.$$

Problem 5.11.17 (Sp02) *Evaluate   the   integrals* $I(a) = \int_{-\infty}^\infty \frac{\cos^3 x}{a^2 + x^2}\, dx$
for $a > 0$.

Problem 5.11.18 (Su84) *Evaluate*

$$\int_{-\infty}^\infty \frac{x \sin x}{x^2 + 4x + 20}\, dx \,.$$

Problem 5.11.19 (Fa84) *Evaluate*

$$\int_0^\infty \frac{x - \sin x}{x^3} \, dx.$$

Problem 5.11.20 (Fa84) *Evaluate*

$$\int_{-\infty}^\infty \frac{dx}{(1 + x + x^2)^2}.$$

Problem 5.11.21 (Fa79, Fa80, Sp85, Su85, Fa98, Sp99) *Prove that*

$$\int_0^\infty \frac{x^{\alpha-1}}{1 + x} \, dx = \frac{\pi}{\sin \pi \alpha}.$$

What restrictions must be placed on α?

Problem 5.11.22 (Fa96) *Evaluate the integral*

$$I = \int_0^\infty \frac{\sqrt{x}}{1 + x^2} \, dx.$$

Problem 5.11.23 (Sp01) *Evaluate*

$$\int_0^\infty \frac{1}{1 + x^5} dx.$$

Problem 5.11.24 (Fa77, Su82, Fa97) *Evaluate*

$$\int_{-\infty}^\infty \frac{dx}{1 + x^{2n}}$$

where n is a positive integer.

Problem 5.11.25 (Fa03) *Evaluate* $\displaystyle\int_{-\infty}^\infty \frac{x^2}{x^n + 1} \, dx$, *where $n \geq 4$ is an even integer.*

Problem 5.11.26 (Fa88) *Prove that*

$$\int_0^\infty \frac{x}{e^x - e^{-x}} \, dx = \frac{\pi^2}{8}.$$

Problem 5.11.27 (Fa93) *Evaluate*

$$\int_{-\infty}^\infty \frac{e^{-ix}}{x^2 - 2x + 4} \, dx.$$

Problem 5.11.28 (Fa86) *Evaluate*

$$\int_0^\infty \frac{\log x}{(x^2 + 1)(x^2 + 4)} \, dx.$$

Problem 5.11.29 (Fa94) *Evaluate*

$$\int_0^\infty \frac{(\log x)^2}{x^2 + 1}\, dx\,.$$

Problem 5.11.30 (Fa83) *Evaluate*

$$\int_0^\infty (\operatorname{sech} x)^2 \cos \lambda x\, dx$$

where λ is a real constant and

$$\operatorname{sech} x = \frac{2}{e^x + e^{-x}}\,.$$

Problem 5.11.31 (Sp85) *Prove that*

$$\int_0^\infty e^{-x^2} \cos(2bx)\, dx = \frac{1}{2}\sqrt{\pi}\, e^{-b^2}\,.$$

What restrictions, if any, need be placed on b?

Problem 5.11.32 (Sp03) *Evaluate*

$$\int_0^\infty e^{-x^2} \cos x^2\, dx\,.$$

Problem 5.11.33 (Sp97) *Prove that*

$$\int_{-\infty}^\infty \frac{e^{-(t-i\gamma)^2/2}}{\sqrt{2\pi}}\, dt$$

is independent of the real parameter γ.

6
Algebra

6.1 Examples of Groups and General Theory

Problem 6.1.1 (Sp77) *Let G be the collection of 2×2 real matrices with nonzero determinant. Define the product of two elements in G as the usual matrix product.*

1. *Show that G is a group.*

2. *Find the center Z of G; that is, the set of all elements z of G such that $az = za$ for all $a \in G$.*

3. *Show that the set O of real orthogonal matrices is a subgroup of G (a matrix is orthogonal if $AA^t = I$, where A^t denotes the transpose of A). Show by example that O is not a normal subgroup.*

4. *Find a nontrivial homomorphism from G onto an abelian group.*

Problem 6.1.2 (Fa77) *Let G be the set of 3×3 real matrices with zeros below the diagonal and ones on the diagonal.*

1. *Prove G is a group under matrix multiplication.*

2. *Determine the center of G.*

Problem 6.1.3 (Su78) *For each of the following either give an example or else prove that no such example is possible.*

1. *A nonabelian group.*

2. *A finite abelian group that is not cyclic.*

3. *An infinite group with a subgroup of index 5.*

4. *Two finite groups that have the same order but are not isomorphic.*

5. *A group G with a subgroup H that is not normal.*

6. *A nonabelian group with no normal subgroups except the whole group and the unit element.*

7. *A group G with a normal subgroup H such that the factor group G/H is not isomorphic to any subgroup of G.*

8. *A group G with a subgroup H which has index 2 but is not normal.*

Problem 6.1.4 (Fa80) *Let R be a ring with multiplicative identity 1. Call $x \in R$ a unit if $xy = yx = 1$ for some $y \in R$. Let $G(R)$ denote the set of units.*

1. *Prove $G(R)$ is a multiplicative group.*

2. *Let R be the ring of complex numbers $a + bi$, where a and b are integers. Prove $G(R)$ is isomorphic to $\mathbb{Z}_4$ (the additive group of integers modulo 4).*

Problem 6.1.5 (Sp83) *In the triangular network in $\mathbb{R}^2$ depicted below, the points P_0, P_1, P_2, and P_3 are respectively $(0, 0)$, $(1, 0)$, $(0, 1)$, and $(1, 1)$. Describe the structure of the group of all Euclidean transformations of $\mathbb{R}^2$ which leave this network invariant.*

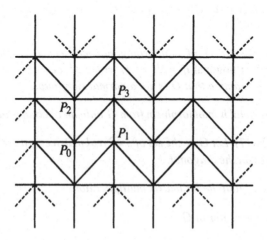

Problem 6.1.6 (Fa90) *Does the set $G = \{a \in \mathbb{R} \mid a > 0, a \neq 1\}$ form a group with the operation $a * b = a^{\log b}$?*

Problem 6.1.7 (Sp81) *Let G be a finite group. A* conjugacy class *is a set of the form*

$$C(a) = \{bab^{-1} \mid b \in G\}$$

for some a ∈ G.

1. *Prove that the number of elements in a conjugacy class divides the order of G.*

2. *Do all conjugacy classes have the same number of elements?*

3. *If G has only two conjugacy classes, prove G has order 2.*

Problem 6.1.8 (Sp91) *Let G be a finite nontrivial group with the property that for any two elements a and b in G different from the identity, there is an element c in G such that $b = c^{-1}ac$. Prove that G has order 2.*

Problem 6.1.9 (Sp99) *Let G be a finite group, with identity e. Suppose that for every $a, b \in G$ distinct from e, there is an automorphism σ of G such that $\sigma(a) = b$. Prove that G is abelian.*

Problem 6.1.10 (Sp84) *For a p-group of order p^4, assume the center of G has order p^2. Determine the number of conjugacy classes of G.*

Problem 6.1.11 (Sp85) *In a commutative group G, let the element a have order r, let b have order s $(r, s < \infty)$, and assume that the greatest common divisor of r and s is 1. Show that ab has order rs.*

Problem 6.1.12 (Fa85) *Let G be a group. For any subset X of G, define its* centralizer $C(X)$ *to be $\{y \in G \mid xy = yx, \text{ for all } x \in X\}$. Prove the following:*

1. *If $X \subset Y$, then $C(Y) \subset C(X)$.*

2. *$X \subset C(C(X))$.*

3. *$C(X) = C(C(C(X)))$.*

Problem 6.1.13 (Sp88) *Let D be a group of order 2n, where n is odd, with a subgroup H of order n satisfying $xhx^{-1} = h^{-1}$ for all h in H and all x in $D \setminus H$. Prove that H is commutative and that every element of $D \setminus H$ is of order 2.*

6.2   Homomorphisms and Subgroups

Problem 6.2.1 (Fa78) *How many homomorphisms are there from the group $\mathbb{Z}_2 \times \mathbb{Z}_2$ to the symmetric group on three objects?*

Problem 6.2.2 (Fa03)   *1. Let G be a finite group and let X be the set of pairs of commuting elements of G:*

$$X = \{(g, h) \in G \times G : gh = hg\}.$$

Prove that $|X| = c|G|$ where c is the number of conjugacy classes in G.

2. Compute the number of pairs of commuting permutations on five letters.

Problem 6.2.3 (Fa03) *The set of 5×5 complex matrices A satisfying $A^3 = A^2$ is a union of conjugacy classes. How many conjugacy classes?*

Problem 6.2.4 (Sp90) *Let $\mathbb{C}^*$ be the multiplicative group of nonzero complex numbers. Suppose that H is a subgroup of finite index of $\mathbb{C}^*$. Prove that $H = \mathbb{C}^*$.*

Problem 6.2.5 (Su80) *Let G be a finite group and $H \subset G$ a subgroup.*

1. *Show that the number of subgroups of G of the form xHx^{-1} for some $x \in G$ is $\leqslant$ the index of H in G.*

2. *Prove that some element of G is not in any subgroup of the form xHx^{-1}, $x \in G$.*

Problem 6.2.6 (Su79) *Prove that the group of automorphisms of a cyclic group of prime order p is cyclic and find its order.*

Problem 6.2.7 (Su81) *Let G be a finite group, and let φ be an automorphism of G which leaves fixed only the identity element of G.*

1. *Show that every element of G may be written in the form $g^{-1}\varphi(g)$.*

2. *If φ has order 2 (i.e., $\varphi \cdot \varphi = \mathrm{id}$) show that φ is given by the formula $g \mapsto g^{-1}$ and that G is an abelian group whose order is odd.*

Problem 6.2.8 (Fa79, Sp88, Fa91) *Prove that every finite group of order > 2 has a nontrivial automorphism.*

Problem 6.2.9 (Fa90) *Let A be an additively written abelian group, and $u, v : A \to A$ homomorphisms. Define the group homomorphisms $f, g : A \to A$ by*

$$f(a) = a - v(u(a)), \qquad g(a) = a - u(v(a)) \qquad (a \in A).$$

Prove that the kernel of f is isomorphic to the kernel of g.

Problem 6.2.10 (Su81) *Let G be an additive group, and $u, v : G \to G$ homomorphisms. Show that the map $f : G \to G$, $f(x) = x - v(u(x))$ is surjective if the map $h : G \to G$, $h(x) = x - u(v(x))$ is surjective.*

Problem 6.2.11 (Sp83) *Let H be the group of integers mod p, under addition, where p is a prime number. Suppose that n is an integer satisfying $1 \leqslant n \leqslant p$, and let G be the group $H \times H \times \cdots \times H$ (n factors). Show that G has no automorphism of order p^2.*

Problem 6.2.12 (Sp03)   *1. Suppose that H_1 and H_2 are subgroups of a group G such that $H_1 \cup H_2$ is a subgroup of G. Prove that either $H_1 \subseteq H_2$ or $H_2 \subseteq H_1$.*

2. Show that for each integer $n \geq 3$, there exists a group G with subgroups $H_1, H_2, \ldots, H_n$, such that no H_i is contained in any other, and such that $H_1 \cup H_2 \cup \ldots \cup H_n$ is a subgroup of G.

Problem 6.2.13 (Fa84) *Let G be a group and H a subgroup of index $n < \infty$. Prove or disprove the following statements:*

1. If $a \in G$, then $a^n \in H$.

2. If $a \in G$, then for some k, $1 \leq k \leq n$, we have $a^k \in H$.

Problem 6.2.14 (Fa78) *Find all automorphisms of the additive group of rational numbers.*

Problem 6.2.15 (Sp01) *Let α and β be real numbers such that the subgroup Γ of $\mathbb{R}$ generated by α and β is closed. Prove that α and β are linearly dependent over $\mathbb{Q}$.*

Problem 6.2.16 (Sp00) *Prove that the group $G = \mathbb{Q}/\mathbb{Z}$ has no proper subgroup of finite index.*

Problem 6.2.17 (Fa87, Fa93) *Let A be the group of rational numbers under addition, and let M be the group of positive rational numbers under multiplication. Determine all homomorphisms $\varphi : A \to M$.*

Problem 6.2.18 (Sp00) *Suppose that H_1 and H_2 are distinct subgroups of a group G such that $[G : H_1] = [G : H_2] = 3$. What are the possible values of $[G : H_1 \cap H_2]$?*

Problem 6.2.19 (Fa92) *Let G be a group and H and K subgroups such that H has a finite index in G. Prove that $K \cap H$ has a finite index in K.*

Problem 6.2.20 (Fa94) *Suppose the group G has a nontrivial subgroup H which is contained in every nontrivial subgroup of G. Prove that H is contained in the center of G.*

Problem 6.2.21 (Fa00) *Show that for each positive integer k there exists a positive integer N such that there are at least k nonisomorphic groups of order N.*

Problem 6.2.22 (Fa95) *Let G be a group generated by n elements. Find an upper bound $N(n, k)$ for the number of subgroups H of G with the index $[G : H] = k$.*

6.3   Cyclic Groups

Problem 6.3.1 (Su77, Sp92)   *1. Prove that every finitely generated subgroup of $\mathbb{Q}$, the additive group of rational numbers, is cyclic.*

2. Does the same conclusion hold for finitely generated subgroups of $\mathbb{Q}/\mathbb{Z}$, where $\mathbb{Z}$ is the group of integers?
Note: See also Problems 6.6.3 and 6.7.2.

Problem 6.3.2 (Sp98) *Let G be the group $\mathbb{Q}/\mathbb{Z}$. Show that for every positive integer t, G has a unique cyclic subgroup of order t.*

Problem 6.3.3 (Fa99) *Show that a group G is isomorphic to a subgroup of the additive group of the rationals if and only if G is countable and every finite subset of G is contained in an infinite cyclic subgroup of G.*

Problem 6.3.4 (Su85)   *1. Let G be a cyclic group, and let $a, b \in G$ be elements which are not squares. Prove that ab is a square.*

2. Give an example to show that this result is false if the group is not cyclic.

Problem 6.3.5 (Sp82) *Prove that any group of order 77 is cyclic.*

Problem 6.3.6 (Fa91) *Let G be a group of order $2p$, where p is an odd prime. Assume that G has a normal subgroup of order 2. Prove that G is cyclic.*

Problem 6.3.7 (Fa97) *A finite abelian group G has the property that for each positive integer n the set $\{x \in G \mid x^n = 1\}$ has at most n elements. Prove that G is cyclic, and deduce that every finite field has cyclic multiplicative group.*

Problem 6.3.8 (Fa00) *Let G be a finite group of order n with the property that for each divisor d of n there is at most one subgroup in G of order d. Show G is cyclic.*

6.4   Normality, Quotients, and Homomorphisms

Problem 6.4.1 (Fa78) *Let H be a subgroup of a finite group G.*

1. Show that H has the same number of left cosets as right cosets.

2. Let G be the group of symmetries of the square. Find a subgroup H such that $xH \neq Hx$ for some x.

Problem 6.4.2 (Fa80) *Let G be the group of orthogonal transformations of $\mathbb{R}^3$ to $\mathbb{R}^3$ with determinant 1. Let $v \in \mathbb{R}^3$, $|v| = 1$, and let $H_v = \{T \in G \mid Tv = v\}$.*

1. Show that H_v is a subgroup of G.

2. Let $S_v = \{T \in G \mid T$ is a rotation of $180°$ about a line orthogonal to $v\}$. Show that S_v is a coset of H_v in G.

Problem 6.4.3 (Su84) *Show that if a subgroup H of a group G has just one left coset different from itself, then it is a normal subgroup of G.*

Problem 6.4.4 (Su85) *Let G be a group of order 120, let H be a subgroup of order 24, and assume that there is at least one left coset of H (other than H itself) which is equal to some right coset of H. Prove that H is a normal subgroup of G.*

Problem 6.4.5 (Fa02) *Let G be a group of order 112. Prove that G has a non-trivial normal subgroup.*

Problem 6.4.6 (Fa02) *List all groups of order $\leqslant 6$ up to isomorphism. Find a group of order 120 that contains as a subgroup an isomorphic copy of each of them. Prove that no group of order < 120 has the preceding property.*

Problem 6.4.7 (Sp89) *For G a group and H a subgroup, let $C(G, H)$ denote the collection of left cosets of H in G. Prove that if H and K are two subgroups of G of infinite index, then G is not a finite union of cosets from $C(G, H) \cup C(G, K)$.*

Problem 6.4.8 (Fa82, Fa92) *Let*

$$G = \left\{ \begin{pmatrix} a & b \\ 0 & a^{-1} \end{pmatrix} \mid a, b \in \mathbb{R}, \ a > 0 \right\}$$

$$N = \left\{ \begin{pmatrix} 1 & b \\ 0 & 1 \end{pmatrix} \mid b \in \mathbb{R} \right\}.$$

1. *Show that N is a normal subgroup of G and prove that G/N is isomorphic to $\mathbb{R}$.*

2. *Find a normal subgroup N' of G satisfying $N \subset N' \subset G$ (where the inclusions are proper), or prove that there is no such subgroup.*

Problem 6.4.9 (Sp86) *Let $\mathbb{Z}^2$ be the group of lattice points in the plane (ordered pairs of integers, with coordinatewise addition as the group operation). Let H_1 be the subgroup generated by the two elements $(1, 2)$ and $(4, 1)$, and H_2 the subgroup generated by the two elements $(3, 2)$ and $(1, 3)$. Are the quotient groups $G_1 = \mathbb{Z}^2/H_1$ and $G_2 = \mathbb{Z}^2/H_2$ isomorphic?*

Problem 6.4.10 (Sp78, Fa81) *Let G be a group of order 10 which has a normal subgroup of order 2. Prove that G is abelian.*

Problem 6.4.11 (Sp79, Fa81) *Let G be a group with three normal subgroups N_1, N_2, and N_3. Suppose $N_i \cap N_j = \{e\}$ and $N_i N_j = G$ for all i, j with $i \neq j$. Show that G is abelian and N_i is isomorphic to N_j for all i, j.*

Problem 6.4.12 (Fa97) *Suppose H_i is a normal subgroup of a group G for $1 \leqslant i \leqslant k$, such that $H_i \cap H_j = \{1\}$ for $i \neq j$. Prove that G contains a subgroup isomorphic to $H_1 \times H_2 \times \cdots \times H_k$ if $k = 2$, but not necessarily if $k \geqslant 3$.*

Problem 6.4.13 (Sp80) *G is a group of order n, H a proper subgroup of order m, and $(n/m)! < 2n$. Prove G has a proper normal subgroup different from the identity.*

Problem 6.4.14 (Sp82, Sp93) *Prove that if G is a group containing no subgroup of index 2, then any subgroup of index 3 is normal.*

Problem 6.4.15 (Sp03) *Suppose G is a nonabelian simple group, and A is its automorphism group. Show that A contains a normal subgroup isomorphic to G.*

Problem 6.4.16 (Sp89) *Let G be a group whose order is twice an odd number. For g in G, let λ_g denote the permutation of G given by $\lambda_g(x) = gx$ for $x \in G$.*

1. *Let g be in G. Prove that the permutation λ_g is even if and only if the order of g is odd.*

2. *Let $N = \{g \in G \mid \operatorname{order}(g) \text{ is odd}\}$. Prove that N is a normal subgroup of G of index 2.*

Problem 6.4.17 (Fa89) *Let G be a group, G' its commutator subgroup, and N a normal subgroup of G. Suppose that N is cyclic. Prove that $gn = ng$ for all $g \in G'$ and all $n \in N$.*

Problem 6.4.18 (Fa90) *Let G be a group and N be a normal subgroup of G with $N \neq G$. Suppose that there does not exist a subgroup H of G satisfying $N \subset H \subset G$ and $N \neq H \neq G$. Prove that the index of N in G is finite and equal to a prime number.*

Problem 6.4.19 (Sp94) *Let G be a group having a subgroup A of finite index. Prove that there is a normal subgroup N of G contained in A such that N is of finite index in G.*

Problem 6.4.20 (Sp97) *Let H be the quotient of an abelian group G by a subgroup K. Prove or disprove each of the following statements:*

1. *If H is finite cyclic then G is isomorphic to the direct product of H and K.*

2. *If H is a direct product of infinite cyclic groups then G is isomorphic to the direct product of H and K.*

6.5 S_n, A_n, D_n, ...

Problem 6.5.1 (Fa80) *Let $\mathbf{F}_2 = \{0, 1\}$ be the field with two elements. Let G be the group of invertible 2×2 matrices with entries in $\mathbf{F}_2$. Show that G is isomorphic to S_3, the group of permutations of three objects.*

Problem 6.5.2 (Su84) *Let S_n denote the group of permutations of n objects. Find four different subgroups of S_4 isomorphic to S_3 and nine isomorphic to S_2.*

Problem 6.5.3 (Fa86) *Let G be a subgroup of S_5, the group of all permutations of five objects Prove that if G contains a 5-cycle and a 2-cycle, then $G = S_5$.*

Problem 6.5.4 (Fa85) *Let G be a subgroup of the symmetric group on six objects, S_6. Assume that G has an element of order 6. Prove that G has a normal subgroup H of index 2.*

Problem 6.5.5 (Sp79) *Let S_7 be the group of permutations of a set of seven objects. Find all n such that some element of S_7 has order n.*

Problem 6.5.6 (Sp80) *S_9 is the group of permutations of 9 objects.*

1. Exhibit an element of S_9 of order 20.

2. Prove that no element of S_9 has order 18.

Problem 6.5.7 (Sp88, Sp01) *Let S_9 denote the group of permutations of nine objects and let A_9 be the subgroup consisting of all even permutations. Denote by $1 \in S_9$ the identity permutation. Determine the minimum of all positive integers m such that every $\sigma \in S_9$ satisfies $\sigma^m = 1$. Determine also the minimum of all positive integers m such that every $\sigma \in A_9$ satisfies $\sigma^m = 1$.*

Problem 6.5.8 (Sp92) *Let S_{999} denote the group of permutations of 999 objects, and let $G \subset S_{999}$ be an abelian subgroup of order 1111. Prove that there exists $i \in \{1, \ldots, 999\}$ such that for all $\sigma \in G$, one has $\sigma(i) = i$.*

Problem 6.5.9 (Sp02) *Prove that S_n, the group of permutations of $\{1, 2, \ldots, n\}$, is isomorphic to a subgroup of A_{n+2}, the alternating subgroup of S_{n+2}.*

Problem 6.5.10 (Fa81, Sp95) *Let S_n be the group of all permutations of n objects and let G be a subgroup of S_n of order p^k, where p is a prime not dividing n. Show that G has a fixed point; that is, one of the objects is left fixed by every element of G.*

Problem 6.5.11 (Sp80) *Let G be a subgroup of S_n the group of permutations of n objects. Assume G is transitive; that is, for any x and y in S, there is some $\sigma \in G$ with $\sigma(x) = y$.*

1. Prove that n divides the order of G.

2. Suppose $n = 4$. For which integers $k \geqslant 1$ can such a G have order 4k?

Problem 6.5.12 (Su83) *Let G be a transitive subgroup of the group S_n of permutations of n objects $\{1, \ldots, n\}$. Suppose that G is a simple group and that $\sim$ is an equivalence relation on $\{1, \ldots, n\}$ such that $i \sim j$ implies that $\sigma(i) \sim \sigma(j)$ for all $\sigma \in G$. What can one conclude about the relation $\sim$?*

Problem 6.5.13 (Sp89) *Let D_n be the dihedral group, the group of rigid motions of a regular n-gon ($n \geqslant 3$). (It is a noncommutative group of order 2n.) Determine its center $Z = \{c \in D_n \mid cx = xc \text{ for all } x \in D_n\}$.*

Problem 6.5.14 (Fa92) *How many Sylow 2-subgroups does the dihedral group D_n of order $2n$ have, when n is odd?*

6.6 Direct Products

Problem 6.6.1 (Fa83) *Let G be a finite group and suppose that $G \times G$ has exactly four normal subgroups. Show that G is simple and nonabelian.*

Problem 6.6.2 (Sp99) *Let G be a finite simple group of order n. Determine the number of normal subgroups in the direct product $G \times G$.*

Problem 6.6.3 (Sp91) *Prove that $\mathbb{Q}$, the additive group of rational numbers, cannot be written as the direct sum of two nontrivial subgroups.*
Note: See also Problems 6.3.1 and 6.7.2.

Problem 6.6.4 (Su79, Fa93) *Let A, B, and C be finite abelian groups such that $A \times B$ and $A \times C$ are isomorphic. Prove that B and C are isomorphic.*

Problem 6.6.5 (Su83) *Let G_1, G_2, and G_3 be finite groups, each of which is generated by its commutators (elements of the form $xyx^{-1}y^{-1}$). Let A be a subgroup of $G_1 \times G_2 \times G_3$, which maps surjectively, by the natural projection map, to the partial products $G_1 \times G_2$, $G_1 \times G_3$ and $G_2 \times G_3$. Show that A is equal to $G_1 \times G_2 \times G_3$.*

Problem 6.6.6 (Fa82) *Let A be a subgroup of an abelian group B. Assume that A is a direct summand of B, i.e., there exists a subgroup X of B such that $A \cap X = 0$ and such that $B = X + A$. Suppose that C is a subgroup of B and satisfying $A \subset C \subset B$. Is A necessarily a direct summand of C?*

Problem 6.6.7 (Fa87, Sp96) *Let G and H be finite groups of relatively prime order. Show that $\mathrm{Aut}(G \times H)$, the group of automorphisms of $G \times H$, is isomorphic to the direct product of $\mathrm{Aut}(G)$ and $\mathrm{Aut}(H)$.*

Problem 6.6.8 (Fa03) *Let p be a prime, and let G be the group $\mathbb{Z}_{p^2} \times \mathbb{Z}_p$. How many automorphisms does G have?*

6.7 Free Groups, Generators, and Relations

Problem 6.7.1 (Sp77) *Let $\mathbb{Q}_+$ be the multiplicative group of positive rational numbers.*

1. *Is $\mathbb{Q}_+$ torsion free?*

2. *Is $\mathbb{Q}_+$ free?*

Problem 6.7.2 (Sp86) *Prove that the additive group of* $\mathbb{Q}$, *the rational number field, is not finitely generated.*
Note: See also Problems 6.3.1 and 6.6.3.

Problem 6.7.3 (Fa79, Fa82) *Let G be the abelian group defined by generators x, y, and z, and relations*

$$15x + 3y = 0$$
$$3x + 7y + 4z = 0$$
$$18x + 14y + 8z = 0.$$

1. *Express G as a direct product of two cyclic groups.*

2. *Express G as a direct product of cyclic groups of prime power order.*

3. *How many elements of G have order 2?*

Problem 6.7.4 (Sp82, Sp93) *Suppose that the group G is generated by elements x and y that satisfy* $x^5 y^3 = x^8 y^5 = 1$. *Is G the trivial group?*

Problem 6.7.5 (Su82) *Let G be a group with generators a and b satisfying*

$$a^{-1}b^2 a = b^3, \qquad b^{-1}a^2 b = a^3.$$

Is G trivial?

Problem 6.7.6 (Fa88, Fa97) *Let the group G be generated by two elements, a and b, both of order 2. Prove that G has a subgroup of index 2.*

Problem 6.7.7 (Fa89) *Let G_n be the free group on n generators. Show that G_2 and G_3 are not isomorphic.*

Problem 6.7.8 (Sp83) *Let G be an abelian group which is generated by, at most, n elements. Show that each subgroup of G is again generated by, at most, n elements.*

Problem 6.7.9 (Sp84) *Determine all finitely generated abelian groups G which have only finitely many automorphisms.*

Problem 6.7.10 (Fa89) *Let A be a finite abelian group, and m the maximum of the orders of the elements of A. Put $S = \{a \in A \mid |a| = m\}$. Prove that A is generated by S.*

6.8   Finite Groups

Problem 6.8.1 (Sp91) *List, to within isomorphism, all the finite groups whose orders do not exceed 5.*

Problem 6.8.2 (Fa84) *Show that all groups of order $\leqslant 5$ are commutative. Give an example of a noncommutative group of order 6.*

Problem 6.8.3 (Fa80) *Prove that any group of order 6 is isomorphic to either $\mathbb{Z}_6$ or S_3 (the group of permutations of three objects).*

Problem 6.8.4 (Sp87)   *1. Show that, to within isomorphism, there is just one noncyclic group G of order 4.*

2. Show that the group of automorphisms of G is isomorphic to the permutation group S_3.

Problem 6.8.5 (Fa88) *Find all abelian groups of order 8, up to isomorphism. Then identify which type occurs in each of*

1. $(\mathbb{Z}_{15})^*$,

2. $(\mathbb{Z}_{17})^* /(\pm 1)$,

3. the roots of $z^8 - 1$ in $\mathbb{C}$,

4. $\mathbf{F}_8^+$,

5. $(\mathbb{Z}_{16})^*$.

$\mathbf{F}_8$ *is the field of eight elements, and* $\mathbf{F}_8^+$ *is its underlying additive group;* R^* *is the group of invertible elements in the ring R, under multiplication.*

Problem 6.8.6 (Sp90, Fa93, Sp94) *Show that there are at least two nonisomorphic nonabelian groups of order 24, of order 30 and order 40.*

Problem 6.8.7 (Fa03) *List eight groups of order 36 and prove that they are not isomorphic.*

Problem 6.8.8 (Fa97) *Prove that if p is prime then every group of order p^2 is abelian.*

Problem 6.8.9 (Sp93) *Classify up to isomorphism all groups of order 45.*

Problem 6.8.10 (Sp02) *Let G be a group of order 56 having at least 7 elements of order 7. Prove that G has only one Sylow 2-subgroup P, and that all nonidentity elements of P have order 2.*

Problem 6.8.11 (Sp79, Sp97) *Classify all abelian groups of order 80 up to isomorphism.*

Problem 6.8.12 (Fa88) *Find (up to isomorphism) all groups of order $2p$, where p is a prime ($p \geqslant 2$).*

Problem 6.8.13 (Sp87) *Prove that any finite group of order n is isomorphic to a subgroup of $\mathbb{O}(n)$, the group of $n \times n$ orthogonal real matrices.*

Problem 6.8.14 (Fa98) *Suppose that G is a finite group such that every Sylow subgroup is normal and abelian. Show that G is abelian.*

Problem 6.8.15 (Sp01) *If G is a finite group, must $S = \{g^2 \mid g \in G\}$ be a subgroup? Provide a proof or a counterexample.*

Problem 6.8.16 (Fa99) *Let G be a finite group acting transitively on a set X of size at least 2. Prove that some element g of G acts without fixed points.*

Problem 6.8.17 (Fa02) *Let G be a finite non-Abelian group of order n. Show that there exists an integer d satisfying $2 \leqslant d \leqslant n/2$ and a set P of cardinality d, such that G acts transitively on P.*

Problem 6.8.18 (Su80, Fa96) *Prove that every finite group is isomorphic to*

1. *A group of permutations;*

2. *A group of even permutations.*

Problem 6.8.19 (Sp00) *Let G be a finite group and p a prime number. Suppose a and b are elements of G of order p such that b is not in the subgroup generated by a. Prove that G contains at least $p^2 - 1$ elements of order p.*

Problem 6.8.20 (Fa01) *Find all finite abelian groups G (up to isomorphism) such that the group of automorphisms of G has odd order.*

Problem 6.8.21 (Fa03) *Let n be a positive integer. Let $\phi(n)$ be the Euler phi function, so $\phi(n) = \#(\mathbb{Z}_n)^*$. Prove that if $\gcd(n, \phi(n)) > 1$, then there exists a noncyclic group of order n.*

6.9   Rings and Their Homomorphisms

Problem 6.9.1 (Fa80) *Let $M_{2\times2}$ be the ring of real 2×2 matrices and $S \subset M_{2\times2}$ the subring of matrices of the form*

$$\begin{pmatrix} a & -b \\ b & a \end{pmatrix}.$$

1. *Exhibit an isomorphism between S and $\mathbb{C}$.*

2. *Prove that*

$$A = \begin{pmatrix} 0 & 3 \\ -4 & 1 \end{pmatrix}$$

   lies in a subring isomorphic to S.

3. *Prove that there is an $X \in M_{2\times2}$ such that $X^4 + 13X = A$.*

Problem 6.9.2 (Sp03) *Let $M_{2\times2}(\mathbb{Q})$ denote the ring of 2×2 matrices with entries in $\mathbb{Q}$. Let R be the set of matrices in $M_{2\times2}(\mathbb{Q})$ that commute with $\left(\begin{smallmatrix} 1 & 1 \\ 0 & 1 \end{smallmatrix}\right)$.*

1. *Prove that R is a subring of $M_{2\times2}(\mathbb{Q})$.*

2. *Prove that R is isomorphic to the ring $\mathbb{Q}[x]/\langle x^2 \rangle$..*

Problem 6.9.3 (Sp86) *Prove that there exists only one automorphism of the field of real numbers; namely the identity automorphism.*

Problem 6.9.4 (Sp86) *Suppose addition and multiplication are defined on $\mathbb{C}^n$, complex n-space, coordinatewise, making $\mathbb{C}^n$ into a ring. Find all ring homomorphisms of $\mathbb{C}^n$ onto $\mathbb{C}$.*

Problem 6.9.5 (Fa88) *Let R be a finite ring. Prove that there are positive integers m and n with $m > n$ such that $x^m = x^n$ for every x in R.*

Problem 6.9.6 (Sp89) *Let R be a ring with at least two elements. Suppose that for each nonzero a in R there is a unique b in R (depending on a) with $aba = a$. Show that R is a division ring.*

Problem 6.9.7 (Sp91) *Let p be a prime number and R a ring with identity containing p^2 elements. Prove that R is commutative.*

Problem 6.9.8 (Fa00) *Let R be a ring with identity, having fewer than eight elements. Prove that R is commutative.*

Problem 6.9.9 (Sp03) *Let R be the set of complex numbers of the form*

$$a + 3bi, \quad a, b \in \mathbb{Z}.$$

Prove that R is a subring of $\mathbb{C}$, and that R is an integral domain but not a unique factorization domain.

Problem 6.9.10 (Fa93) *Let R be a commutative ring with identity. Let G be a finite subgroup of R^*, the group of units of R. Prove that if R is an integral domain, then G is cyclic.*

Problem 6.9.11 (Fa98) *Let R be a finite ring with identity. Let a be an element of R which is not a zero divisor. Show that a is invertible.*

Problem 6.9.12 (Fa94) *Let R be a ring with identity, and let u be an element of R with a right inverse. Prove that the following conditions on u are equivalent:*

1. *u has more than one right inverse;*

2. *u is a zero divisor;*

3. *u is not a unit.*

Problem 6.9.13 (Su81, Sp93) *Show that no commutative ring with identity has additive group isomorphic to $\mathbb{Q}/\mathbb{Z}$.*

Problem 6.9.14 (Sp81) *Let D be an ordered integral domain and a ∈ D. Prove that*

$$a^2 - a + 1 > 0.$$

Problem 6.9.15 (Fa95) *Prove that $\mathbb{Q}[x, y]/\langle x^2 + y^2 - 1\rangle$ is an integral domain and that its field of fractions is isomorphic to the field of rational functions $\mathbb{Q}(t)$.*

Problem 6.9.16 (Sp00) *Find the cardinality of the set of all subrings of $\mathbb{Q}$, the field of rational numbers.*

6.10   Ideals

Problem 6.10.1 (Sp98) *Let A be the ring of real 2×2 matrices of the form $\left(\begin{smallmatrix} a & b \\ 0 & c \end{smallmatrix}\right)$. What are the 2-sided ideals in A?*

Problem 6.10.2 (Fa79, Fa87) *Let $M_{n \times n}(\mathbf{F})$ be the ring of $n \times n$ matrices over a field $\mathbf{F}$. Prove that it has no 2-sided ideals except $M_{n \times n}(\mathbf{F})$ and $\{0\}$.*

Problem 6.10.3 (Fa83, Su85) *Let $M_{n \times n}(\mathbf{F})$ be the ring of $n \times n$ matrices over a field $\mathbf{F}$. For $n \geqslant 1$ does there exist a ring homomorphism from $M_{(n+1) \times (n+1)}(\mathbf{F})$ onto $M_{n \times n}(\mathbf{F})$?*

Problem 6.10.4 (Sp97) *Let R be the ring of $n \times n$ matrices over a field. Suppose S is a ring and $h : R \to S$ is a homomorphism. Show that h is either injective or zero.*

Problem 6.10.5 (Sp84) *Let $\mathbf{F}$ be a field and let X be a finite set. Let $R(X, \mathbf{F})$ be the ring of all functions from X to $\mathbf{F}$, endowed with the pointwise operations. What are the maximal ideals of $R(X, \mathbf{F})$?*

Problem 6.10.6 (Fa00) *Suppose V is a vector space over a field $\mathbf{K}$. If U and W are subspaces, let $E(U, W)$ be the set of linear endomorphisms F of V over $\mathbf{K}$ with the property that the image of FU in V/W is finite dimensional. Show that $E(U, U)$ is a subring of the ring of endomorphisms of V with two-sided ideals $E(V, U)$ and $E(U, 0)$.*

Problem 6.10.7 (Sp88) *Let R be a commutative ring with identity element and $a \in R$. Let n and m be positive integers, and write $d = \gcd\{n, m\}$. Prove that the ideal of R generated by $a^n - 1$ and $a^m - 1$ is the same as the ideal generated by $a^d - 1$.*

Problem 6.10.8 (Sp89)   *1. Let R be a commutative ring with identity containing an element a with $a^3 = a + 1$. Further, let $\mathfrak{J}$ be an ideal of R of index < 5 in R. Prove that $\mathfrak{J} = R$.*

2. Show that there exists a commutative ring with identity that has an element a with $a^3 = a + 1$ and that contains an ideal of index 5.

Note: The term index *is used here exactly as in group theory; namely the index of* $\Im$ *in R means the order of* $R/\Im$.

Problem 6.10.9 (Sp90) *Let R be a commutative ring with 1, and R^* be its group of units. Suppose that the additive group of R is generated by $\{u^2 \mid u \in R^*\}$. Prove that R has, at most, one ideal $\Im$ for which $R/\Im$ has cardinality 3.*

Problem 6.10.10 (Sp01) *Find all commutative rings R with identity such that R has a unique maximal ideal and such that the group of units of R is trivial.*

Problem 6.10.11 (Fa90) *Let R be a ring with identity, and let $\Im$ be the left ideal of R generated by $\{ab - ba \mid a, b \in R\}$. Prove that $\Im$ is a two-sided ideal.*

Problem 6.10.12 (Fa99) *Let R be a ring with identity element. Suppose that $\Im_1$, $\Im_2, \ldots, \Im_n$ are left ideals in R such that $R = \Im_1 \oplus \Im_2 \oplus \cdots \oplus \Im_n$ (as additive groups). Prove that there are elements $u_i \in \Im_i$ such that for any elements $a_i \in \Im_i$, $a_i u_i = a_i$ and $a_i u_j = 0$ if $j \neq i$.*

Problem 6.10.13 (Sp95) *Suppose that R is a subring of a commutative ring S and that R is of finite index n in S. Let m be an integer that is relatively prime to n. Prove that the natural map $R/mR \to S/mS$ is a ring isomorphism.*

Problem 6.10.14 (Sp81) *Let $\mathbf{M}$ be one of the following fields: $\mathbb{R}$, $\mathbb{C}$, $\mathbb{Q}$, and $\mathbf{F}_9$ (the field with nine elements). Let $\Im \subset \mathbf{M}[x]$ be the ideal generated by $x^4 + 2x - 2$. For which choices of $\mathbf{M}$ is the ring $\mathbf{M}[x]/\Im$ a field?*

Problem 6.10.15 (Sp84) *Let R be a principal ideal domain and let $\Im$ and $\Im$ be nonzero ideals in R. Show that $\Im\Im = \Im \cap \Im$ if and only if $\Im + \Im = R$.*

6.11 Polynomials

Problem 6.11.1 (Fa77) *Suppose the nonzero complex number α is a root of a polynomial of degree n with rational coefficients. Prove that $1/\alpha$ is also a root of a polynomial of degree n with rational coefficients.*

Problem 6.11.2 (Su85) *By the Fundamental Theorem of Algebra, the polynomial $x^3 + 2x^2 + 7x + 1$ has three complex roots, α_1, α_2, and α_3. Compute $\alpha_1^3 + \alpha_2^3 + \alpha_3^3$.*

Problem 6.11.3 (Sp85) *Let $\zeta = e^{\frac{2\pi i}{7}}$ be a primitive 7th root of unity. Find a cubic polynomial with integer coefficients having $\alpha = \zeta + \zeta^{-1}$ as a root.*

Problem 6.11.4 (Sp92, Su77, Fa81)   *1. Prove that $\alpha = \sqrt{5} + \sqrt{7}$ is algebraic over $\mathbb{Q}$, by explicitly finding a polynomial $f(x)$ in $\mathbb{Q}[x]$ of degree 4 having α as a root.*

   2. Prove that $f(x)$ is irreducible over $\mathbb{Q}$.

Problem 6.11.5 (Fa90) *Prove that $\sqrt{2} + \sqrt[3]{3}$ is irrational.*

Problem 6.11.6 (Su85) *Let $P(z)$ be a polynomial of degree $< k$ with complex coefficients. Let $\omega_1, \ldots, \omega_k$ be the kth roots of unity in $\mathbb{C}$. Prove that*

$$\frac{1}{k} \sum_{i=1}^{k} P(\omega_i) = P(0).$$

Problem 6.11.7 (Fa95) *Let $f(x) \in \mathbb{Q}[x]$ be a polynomial with rational coefficients. Show that there is a $g(x) \in \mathbb{Q}[x]$, $g \neq 0$, such that $f(x)g(x) = a_2x^2 + a_3x^3 + a_5x^5 + \cdots + a_px^p$ is a polynomial in which only prime exponents appear.*

Problem 6.11.8 (Fa91) *Let $\mathfrak{J}$ be the ideal in the ring $\mathbb{Z}[x]$ generated by $x - 7$ and 15. Prove that the quotient ring $\mathbb{Z}[x]/\mathfrak{J}$ is isomorphic to $\mathbb{Z}_{15}$.*

Problem 6.11.9 (Fa92) *Let $\mathfrak{J}$ denote the ideal in $\mathbb{Z}[x]$, the ring of polynomials with coefficients in $\mathbb{Z}$, generated by $x^3 + x + 1$ and 5. Is $\mathfrak{J}$ a prime ideal?*

Problem 6.11.10 (Su77) *In the ring $\mathbb{Z}[x]$ of polynomials in one variable over the integers, show that the ideal $\mathfrak{J}$ generated by 5 and $x^2 + 2$ is a maximal ideal.*

Problem 6.11.11 (Sp78) *let $\mathbb{Z}_n$ denote the ring of integers modulo n. Let $\mathbb{Z}_n[x]$ be the ring of polynomials with coefficients in $\mathbb{Z}_n$. Let $\mathfrak{J}$ denote the ideal in $\mathbb{Z}_n[x]$ generated by $x^2 + x + 1$.*

1. *For which values of n, $1 \leqslant n \leqslant 10$, is the quotient ring $\mathbb{Z}_n[x]/\mathfrak{J}$ a field?*

2. *Give the multiplication table for $\mathbb{Z}_2/\mathfrak{J}$.*

Problem 6.11.12 (Sp86) *Let $\mathbb{Z}$ be the ring of integers, p a prime, and $\mathbf{F}_p = \mathbb{Z}/p\mathbb{Z}$ the field of p elements. Let x be an indeterminate, and set $R_1 = \mathbf{F}_p[x]/\langle x^2 - 2\rangle$, $R_2 = \mathbf{F}_p[x]/\langle x^2 - 3\rangle$. Determine whether the rings R_1 and R_2 are isomorphic in each of the cases $p = 2, 5, 11$.*

Problem 6.11.13 (Fa79, Su80, Fa82) *Consider the polynomial ring $\mathbb{Z}[x]$ and the ideal $\mathfrak{J}$ generated by 7 and $x - 3$.*

1. *Show that for each $r \in \mathbb{Z}[x]$, there is an integer α satisfying $0 \leqslant \alpha \leqslant 6$ such that $r - \alpha \in \mathfrak{J}$.*

2. *Find α in the special case $r = x^{250} + 15x^{14} + x^2 + 5$.*

Problem 6.11.14 (Fa96) *Let $\mathbb{Z}[x]$ be the ring of polynomials in the indeterminate x with coefficients in the ring $\mathbb{Z}$ of integers. Let $\mathfrak{J} \subset \mathbb{Z}[x]$ be the ideal generated by 13 and $x - 4$. Find an integer m such that $0 \leqslant m \leqslant 12$ and*

$$(x^{26} + x + 1)^{73} - m \in \mathfrak{J}.$$

Problem 6.11.15 (Fa03) *Give an example, with proof, of a nonconstant irreducible polynomial $f(x)$ over $\mathbb{Q}$ with the property that $f(x)$ does not factor into linear factors over the field $K = \mathbb{Q}[x]/(f(x))$.*

Problem 6.11.16 (Sp77)   *1. In $\mathbb{R}[x]$, consider the set of polynomials $f(x)$ for which $f(2) = f'(2) = f''(2) = 0$. Prove that this set forms an ideal and find its monic generator.*

2. Do the polynomials such that $f(2) = 0$ and $f'(3) = 0$ form an ideal?

Problem 6.11.17 (Sp94) *Find all automorphisms of $\mathbb{Z}[x]$, the ring of polynomials over $\mathbb{Z}$.*

Problem 6.11.18 (Su78) *Let R denote the ring of polynomials over a field $\mathbf{F}$. Let $p_1, \ldots, p_n$ be elements of R. Prove that the greatest common divisor of $p_1, \ldots, p_n$ is 1 if and only if there is an $n \times n$ matrix over R of determinant 1 whose first row is $(p_1, \ldots, p_n)$.*

Problem 6.11.19 (Sp79) *Let $f(x)$ be a polynomial over $\mathbb{Z}_p$, the field of integers mod p. Let $g(x) = x^p - x$. Show that the greatest common divisor of $f(x)$ and $g(x)$ is the product of the distinct linear factors of $f(x)$.*

Problem 6.11.20 (Su79) *Let $\mathbf{F}$ be a subfield of a field $\mathbf{K}$. Let p and q be polynomials over $\mathbf{F}$. Prove that their greatest common divisor in the ring of polynomials over $\mathbf{F}$ is the same as their gcd in the ring of polynomials over $\mathbf{K}$.*

Problem 6.11.21 (Su81, Su82) *Show that $x^{10}+x^9+x^8+\cdots+x+1$ is irreducible over $\mathbb{Q}$. How about $x^{11} + x^{10} + \cdots + x + 1$?*

Problem 6.11.22 (Su84) *Let $\mathbb{Z}$ be the ring of integers and $\mathbb{Z}[x]$ the polynomial ring over $\mathbb{Z}$. Show that*

$$x^6 + 539x^5 - 511x + 847$$

is irreducible in $\mathbb{Z}[x]$.

Problem 6.11.23 (Sp82) *Prove that the polynomial $x^4 + x + 1$ is irreducible over $\mathbb{Q}$.*

Problem 6.11.24 (Fa83, Fa86) *Prove that if p is a prime number, then the polynomial*

$$f(x) = x^{p-1} + x^{p-2} + \cdots + 1$$

is irreducible in $\mathbb{Q}[x]$.

Problem 6.11.25 (Sp96) *Prove that $f(x) = x^4 + x^3 + x^2 + 6x + 1$ is irreducible over $\mathbb{Q}$.*

Problem 6.11.26 (Sp01) *Prove that the polynomial $f(x) = 16x^5 - 125x^4 + 50x^3 - 100x^2 + 75x + 25$ is irreducible over the rationals.*

Problem 6.11.27 (Su84) *Let* $\mathbb{Z}_3$ *be the field of integers* mod 3 *and* $\mathbb{Z}_3[x]$ *the corresponding polynomial ring. Decompose* $x^3 + x + 2$ *into irreducible factors in* $\mathbb{Z}_3[x]$.

Problem 6.11.28 (Sp85) *Factor* $x^4 + x^3 + x + 3$ *completely in* $\mathbb{Z}_5[x]$.

Problem 6.11.29 (Fa85)   *1. How many different monic irreducible polynomials of degree 2 are there over the field* $\mathbb{Z}_5$?

2. *How many different monic irreducible polynomials of degree 3 are there over the field* $\mathbb{Z}_5$?

Problem 6.11.30 (Sp78) *Is* $x^4 + 1$ *irreducible over the field of real numbers? The field of rational numbers? A field with 16 elements?*

Problem 6.11.31 (Sp81) *Decompose* $x^4 - 4$ *and* $x^3 - 2$ *into irreducibles over* $\mathbb{R}$, *over* $\mathbb{Z}$, *and over* $\mathbb{Z}_3$ *(the integers modulo 3).*

Problem 6.11.32 (Fa84) *Let* a *be an element in a field* $\mathbf{F}$ *and let* p *be a prime. Assume* a *is not a* p^{th} *power. Show that the polynomial* $x^p - a$ *is irreducible in* $\mathbf{F}[x]$.

Problem 6.11.33 (Sp92) *Let* p *be a prime integer,* $p \equiv 3$ (mod 4), *and let* $\mathbf{F}_p = \mathbb{Z}/p\mathbb{Z}$. *If* $x^4 + 1$ *factors into a product* $g(x)h(x)$ *of two quadratic polynomials in* $\mathbf{F}_p[x]$, *prove that* $g(x)$ *and* $h(x)$ *are both irreducible over* $\mathbf{F}_p$.

Problem 6.11.34 (Fa88) *Let* n *be a positive integer and let* f *be a polynomial in* $\mathbb{R}[x]$ *of degree* n. *Prove that there are real numbers* $a_0, a_1, \ldots, a_n$, *not all equal to zero, such that the polynomial*

$$\sum_{i=0}^{n} a_i x^{2^i}$$

is divisible by f.

Problem 6.11.35 (Fa89) *Let* $\mathbf{F}$ *be a field,* $\mathbf{F}[x]$ *the polynomial ring in one variable over* $\mathbf{F}$, *and* R *a subring of* $\mathbf{F}[x]$ *with* $\mathbf{F} \subset R$. *Prove that there exists a finite set* $\{f_1, f_2, \ldots, f_n\}$ *of elements of* $\mathbf{F}[x]$ *such that* $R = \mathbf{F}[f_1, f_2, \ldots, f_n]$.

Problem 6.11.36 (Sp87) *Let* $\mathbf{F}$ *be a finite field with* q *elements and let* x *be an indeterminate. For* f *a polynomial in* $\mathbf{F}[x]$, *let* φ_f *denote the corresponding function of* $\mathbf{F}$ *into* $\mathbf{F}$, *defined by* $\varphi_f(a) = f(a)$, $(a \in \mathbf{F})$. *Prove that if* φ *is any function of* $\mathbf{F}$ *into* $\mathbf{F}$, *then there is an* f *in* $\mathbf{F}[x]$ *such that* $\varphi = \varphi_f$. *Prove that* f *is uniquely determined by* φ *to within addition of a multiple of* $x^q - x$.

6.12   Fields and Their Extensions

Problem 6.12.1 (Su78, Fa87, Sp93) *Let R be the set of 2×2 matrices of the form*

$$\begin{pmatrix} a & -b \\ b & a \end{pmatrix}$$

where a, b are elements of a given field **F**. *Show that with the usual matrix operations, R is a commutative ring with identity. For which of the following fields* **F** *is R a field:* $\mathbf{F} = \mathbb{Q}, \mathbb{C}, \mathbb{Z}_5, \mathbb{Z}_7$?

Problem 6.12.2 (Fa83) *Prove that every finite integral domain is a field.*

Problem 6.12.3 (Sp77, Sp78) *Let* $\mathbf{F} \subset \mathbf{K}$ *be fields, and a and b elements of* **K** *which are algebraic over* **F**. *Show that a + b is algebraic over* **F**.

Problem 6.12.4 (Fa78, Fa85) *Prove that every finite multiplicative group of complex numbers is cyclic.*

Problem 6.12.5 (Sp87, Fa95) *Let* **F** *be a field. Prove that every finite subgroup of the multiplicative group of nonzero elements of* **F** *is cyclic.*

Problem 6.12.6 (Fa02) *Let* **K** *be a field such that the additive group of* **K** *is finitely generated as a group. Prove that* **K** *is finite.*

Problem 6.12.7 (Sp85) *Let* $\mathbf{F} = \{ a + b\sqrt[3]{2} + c\sqrt[3]{4} \mid a, b, c \in \mathbb{Q} \}$. *Prove that* **F** *is a field and each element in* **F** *has a unique representation as* $a + b\sqrt[3]{2} + c\sqrt[3]{4}$ *with* $a, b, c \in \mathbb{Q}$. *Find* $(1 - \sqrt[3]{2})^{-1}$ *in* **F**.

Problem 6.12.8 (Sp85) *Let* **F** *be a finite field. Give a complete proof of the fact that the number of elements of* **F** *is of the form* p^r, *where* $p \geq 2$ *is a prime number and r is an integer* ≥ 1.

Problem 6.12.9 (Fa02) *Let* **K** *be a field and* $\mathbf{L} \subset \mathbf{K}$ *a subfield containing* $\{ a^2 \mid a \in \mathbf{K} \}$. *Prove that if the characteristic of* **K** *is not 2 then* $\mathbf{L} = \mathbf{K}$ *and that the same conclusion holds for finite fields* **K** *of characteristic 2, but not in general for fields of characteristic 2.*

Problem 6.12.10 (Su85) *Let* **F** *be a field of characteristic* $p > 0$, $p \neq 3$. *If* α *is a zero of the polynomial* $f(x) = x^p - x + 3$ *in an extension field of* **F**, *show that* $f(x)$ *has p distinct zeros in the field* $\mathbf{F}(\alpha)$.

Problem 6.12.11 (Fa99) *Let* **K** *be the field* $\mathbb{Q}(\sqrt[10]{2})$. *Prove that* **K** *has degree 10 over* $\mathbb{Q}$, *and that the group of automorphisms of* **K** *has order 2.*

Problem 6.12.12 (Fa85) *Let* $f(x) = x^5 - 8x^3 + 9x - 3$ *and* $g(x) = x^4 - 5x^2 - 6x + 3$. *Prove that there is an integer d such that the polynomials* $f(x)$ *and* $g(x)$ *have a common root in the field* $\mathbb{Q}(\sqrt{d})$. *What is d?*

Problem 6.12.13 (Fa86) *Let* $\mathbf{F}$ *be a field containing* $\mathbb{Q}$ *such that* $[\mathbf{F} : \mathbb{Q}] = 2$. *Prove that there exists a unique integer* m *such that* m *has no multiple prime factors and* $\mathbf{F}$ *is isomorphic to* $\mathbb{Q}(\sqrt{m})$.

Problem 6.12.14 (Sp96) *Exhibit infinitely many pairwise nonisomorphic quadratic extensions of* $\mathbb{Q}$ *and show they are pairwise nonisomorphic.*

Problem 6.12.15 (Fa94) *Let* $\mathbb{Q}$ *be the field of rational numbers. For* θ *a real number, let* $\mathbf{F}_\theta = \mathbb{Q}(\sin \theta)$ *and* $\mathbf{E}_\theta = \mathbb{Q}\left(\sin \frac{\theta}{3}\right)$. *Show that* $\mathbf{E}_\theta$ *is an extension field of* $\mathbf{F}_\theta$, *and determine all possibilities for* $\dim_{\mathbf{F}_\theta} \mathbf{E}_\theta$.

Problem 6.12.16 (Fa98) *Show that the field* $\mathbb{Q}(t_1, \ldots, t_n)$ *of rational functions in n variables over the rational numbers is isomorphic to a subfield of* $\mathbb{R}$.

Problem 6.12.17 (Sp99) *Let* $f(x) \in \mathbb{Q}[x]$ *be an irreducible polynomial of degree* $n \geqslant 3$. *Let* L *be the splitting field of* f, *and let* $\alpha \in L$ *be a zero of* f. *Given that* $[L : \mathbb{Q}] = n!$, *prove that* $\mathbb{Q}(\alpha^4) = \mathbb{Q}(\alpha)$.

Problem 6.12.18 (Fa01) *Let* $\mathbf{K}$ *be a field. For what pairs of positive integers* (a, b) *is the subring* $\mathbf{K}[t^a, t^b]$ *of* $\mathbf{K}[t]$ *a unique factorization domain?*

Problem 6.12.19 (Sp95) *Let* $\mathbf{F}$ *be a finite field of cardinality* p^n, *with p prime and* $n > 0$, *and let* G *be the group of invertible* 2×2 *matrices with coefficients in* $\mathbf{F}$.

1. *Prove that* G *has order* $(p^{2n} - 1)(p^{2n} - p^n)$.

2. *Show that any p-Sylow subgroup of* G *is isomorphic to the additive group of* $\mathbf{F}$.

Problem 6.12.20 (Sp02) *Let* p *be a prime and* k, n *positive integers. Prove that the group* $GL_n(\mathbf{F}_p)$ *of invertible* $n \times n$ *matrices over* $\mathbf{F}_p$ *(the field of p elements) contains an element of order* p^k *if and only if* $n > p^{k-1}$.

Problem 6.12.21 (Fa01) *Let* $\mathbf{F}$ *be a field, and let* G *be the group of* 2×2 *upper-triangular matrices over* $\mathbf{F}$ *of determinant* 1.

1. *Determine the commutator subgroup of* G.

2. *Suppose* $\mathbf{F}$ *is the finite field of order* p^k. *Determine for this case the minimum number of generators for the subgroup found in (i).*

Problem 6.12.22 (Fa94) *Let* p *be an odd prime and* $\mathbf{F}_p$ *the field of p elements. How many elements of* $\mathbf{F}_p$ *have square roots in* $\mathbf{F}_p$? *How many have cube roots in* $\mathbf{F}_p$?

Problem 6.12.23 (Sp94) *Let* $\mathbf{F}$ *be a finite field with q elements. Say that a function* $f : \mathbf{F} \to \mathbf{F}$ *is a polynomial function if there are elements* $a_0, a_1, \ldots, a_n$ *of* $\mathbf{F}$ *such that* $f(x) = a_0 + a_1 x + \cdots + a_n x^n$ *for all* $x \in \mathbf{F}$. *How many polynomial functions are there?*

Problem 6.12.24 (Sp95) *Let* **F** *be a finite field, and suppose that the subfield of* **F** *generated by* $\{x^3 \mid x \in \mathbf{F}\}$ *is different from* **F**. *Show that* **F** *has cardinality 4.*

Problem 6.12.25 (Sp97) *Suppose that A is a commutative algebra with identity over* $\mathbb{C}$ *(i.e., A is a commutative ring containing* $\mathbb{C}$ *as a subring with identity). Suppose further that* $a^2 \neq 0$ *for all nonzero elements* $a \in A$. *Show that if the dimension of A as a vector space over* $\mathbb{C}$ *is finite and at least two, then the equations* $a^2 = a$ *is satisfied by at least three distinct elements* $a \in A$.

Problem 6.12.26 (Sp02) *Let A be a commutative ring with 1. Prove that the group* A^* *of units of A does not have exactly 5 elements.*

Problem 6.12.27 (Fa00) *Suppose* **K** *is a field and R is a nonzero* **K***-algebra generated by two elements a and b which satisfy* $a^2 = b^2 = 0$ *and* $(a + b)^2 = 1$. *Show R is isomorphic to* $M_2(\mathbf{K})$ *(the algebra of* 2×2 *matrices over K).*

6.13   Elementary Number Theory

Problem 6.13.1 (Fa86) *Prove that if six people are riding together in an Evans Hall elevator, there is either a three-person subset of mutual friends (each knows the other two) or a three-person subset of mutual strangers (each knows neither of the other two).*

Problem 6.13.2 (Fa03) *Let* $u_{m,n}$ *be an array of numbers for* $1 \leq m \leq N$ *and* $1 \leq n \leq N$. *Suppose that* $u_{m,n} = 0$ *when m is 1 or N, or when n is 1 or N. Suppose also that*

$$u_{m,n} = \frac{1}{4}\left(u_{m-1,n} + u_{m+1,n} + u_{m,n-1} + u_{m,n+1}\right)$$

whenever $1 < m < N$ *and* $1 < n < N$. *Show that all the* $u_{m,n}$ *are zero.*

Problem 6.13.3 (Sp02) *How many functions* $f : \{1, 2, 3, 4, 5\} \rightarrow \{1, 2, 3, 4, 5\}$ *have a range of size exactly 3 ?*

Problem 6.13.4 (Sp98) *Let* $m \geqslant 0$ *be an integer. Let* $a_1, a_2, \ldots, a_m$ *be integers and let*

$$f(x) = \sum_{i=1}^{m} \frac{a_i x^i}{i!}$$

Show that if $d \geqslant 0$ *is an integer then* $f(x)^d / d!$ *can be expressed in the form*

$$\sum_{i=0}^{md} \frac{b_i x^i}{i!}.$$

where the b_i *are integers.*

Problem 6.13.5 (Fa03) *Let* $f : [0, 1] \rightarrow [0, 1]$ *be an increasing (not strictly increasing) function such that*

$$f\left(\sum_{j=1}^{\infty} a_j 3^{-j}\right) = \sum_{j=1}^{\infty} \frac{a_j}{2} 2^{-j}$$

whenever the a_j *are 0 or 2. Prove that there is a constant* C_0 *such that*

$$|f(x) - f(y)| \le C_0 |x - y|^{\log 2/\log 3}$$

for all $x, y \in [0, 1]$.

Problem 6.13.6 (Sp77) *Let* p *be an odd prime. Let* $Q(p)$ *be the set of integers* a, $0 \le a \le p - 1$, *for which the congruence*

$$x^2 \equiv a \pmod{p}$$

has a solution. Show that $Q(p)$ *has cardinality* $(p + 1)/2$.

Problem 6.13.7 (Su77) *Let* p *be an odd prime. If the congruence* $x^2 \equiv -1$ (mod p) *has a solution, show that* $p \equiv 1 \pmod{4}$.

Problem 6.13.8 (Sp80) *Let* $n \ge 2$ *be an integer such that* $2^n + n^2$ *is prime. Prove that*

$$n \equiv 3 \pmod{6}.$$

Problem 6.13.9 (Fa77)   *1. Show that the set of all units in a ring with unity form a group under multiplication. (A unit is an element having a two-sided multiplicative inverse.)*

2. In the ring $\mathbb{Z}_n$ *of integers* mod n, *show that* k *is a unit if and only if* k *and* n *are relatively prime.*

3. Suppose $n = pq$, *where* p *and* q *are primes. Prove that the number of units in* $\mathbb{Z}_n$ *is* $(p - 1)(q - 1)$.

Problem 6.13.10 (Su79) *Which rational numbers* t *are such that*

$$3t^3 + 10t^2 - 3t$$

is an integer?

Problem 6.13.11 (Fa96) *Show the denominator of* $\binom{1/2}{n}$ *is a power of 2 for all integers* n.

Problem 6.13.12 (Su82) *Let* n *be a positive integer.*

1. *Show that the binomial coefficient*

$$c_n = \binom{2n}{n}$$

   is even.

2. *Prove that c_n is divisible by 4 if and only if n is not a power of 2.*

Problem 6.13.13 (Sp83) *Suppose that $n > 1$ is an integer. Prove that the sum*

$$1 + \frac{1}{2} + \cdots + \frac{1}{n}$$

is not an integer.

Problem 6.13.14 (Fa84, Fa96) *Let gcd abbreviate greatest common divisor and lcm abbreviate least common multiple. For three nonzero integers a, b, c, show that*

$$\gcd\{a, \operatorname{lcm}\{b, c\}\} = \operatorname{lcm}\{\gcd\{a, b\}, \gcd\{a, c\}\}.$$

Problem 6.13.15 (Sp92) *Let $a_1, a_2, \ldots, a_{10}$ be integers with $1 \leqslant a_i \leqslant 25$, for $1 \leqslant i \leqslant 10$. Prove that there exist integers $n_1, n_2, \ldots, n_{10}$, not all zero, such that*

$$\prod_{i=1}^{10} a_i^{n_i} = 1.$$

Problem 6.13.16 (Su83) *The number 21982145917308330487013369 is the thirteenth power of a positive integer. Which positive integer?*

Problem 6.13.17 (Sp96) *Determine the rightmost decimal digit of*

$$A = 17^{17^{17}}.$$

Problem 6.13.18 (Sp88) *Determine the last digit of*

$$23^{23^{23^{23}}}$$

in the decimal system.

Problem 6.13.19 (Sp88) *Show that one can represent the set of nonnegative integers, $\mathbb{Z}_+$, as the union of two disjoint subsets N_1 and N_2 ($N_1 \cap N_2 = \emptyset$, $N_1 \cup N_2 = \mathbb{Z}_+$) such that neither N_1 nor N_2 contains an infinite arithmetic progression.*

Problem 6.13.20 (Fa89) *Let φ be Euler's totient function; so if n is a positive integer, then $\varphi(n)$ is the number of integers m for which $1 \leqslant m \leqslant n$ and $\gcd\{n, m\} = 1$. Let a and k be two integers, with $a > 1$, $k > 0$. Prove that k divides $\varphi(a^k - 1)$.*

Problem 6.13.21 (Sp90) *Determine the greatest common divisor of the elements of the set* $\{n^{13} - n \mid n \in \mathbb{Z}\}$.

Problem 6.13.22 (Sp91) *For n a positive integer, let $d(n)$ denote the number of positive integers that divide n. Prove that $d(n)$ is odd if and only if n is a perfect square.*

Problem 6.13.23 (Fa01) *Which of the numbers $0,1,2,3,4,5,6,7,8,9$ occur as the last digit of n^n for infinitely many positive integers n?*

Problem 6.13.24 (Sp03) *Let $N = 30030$, which is the product of the first six primes. How many nonnegative integers x less than N have the property that N divides $x^3 - 1$?*

7
Linear Algebra

7.1 Vector Spaces

Problem 7.1.1 (Sp99) *Let p, q, r and s be polynomials of degree at most 3. Which, if any, of the following two conditions is sufficient for the conclusion that the polynomials are linearly dependent?*

1. At 1 each of the polynomials has the value 0.

2. At 0 each of the polynomials has the value 1.

Problem 7.1.2 (Sp03) *For an analytic function h on $\mathbb{C}$, let $h^{(i)}$ denote its i-th derivative, with $h^{(0)} = h$. Suppose that f and g are analytic functions on $\mathbb{C}$ satisfying*

$$f^{(n)} + a_{n-1} f^{(n-1)} + \cdots + a_0 f^{(0)} = 0$$
$$g^{(m)} + b_{m-1} g^{(m-1)} + \cdots + b_0 g = 0$$

for some constants $a_0, \ldots, a_{n-1}, b_0, \ldots, b_{m-1} \in \mathbb{C}$. Show that the product function $F = fg$ satisfies

$$c_{mn} F^{(mn)} + c_{mn-1} F^{(mn-1)} + \cdots + c_0 F = 0$$

for some constants $c_0, \ldots, c_{mn} \in \mathbb{C}$ not all zero.

Problem 7.1.3 (Su79, Sp82, Sp83, Su84, Fa91, Fa98) *Let $\mathbf{F}$ be a finite field with q elements and let V be an n-dimensional vector space over $\mathbf{F}$.*

1. *Determine the number of elements in V.*

2. *Let $GL_n(\mathbf{F})$ denote the group of all $n \times n$ nonsingular matrices over $\mathbf{F}$. Determine the order of $GL_n(\mathbf{F})$.*

3. *Let $SL_n(\mathbf{F})$ denote the subgroup of $GL_n(\mathbf{F})$ consisting of matrices with determinant 1. Find the order of $SL_n(\mathbf{F})$.*

Problem 7.1.4 (Sp01) *Let $\mathbf{F}$ be a finite field of order q, and let V be a two dimensional vector space over $\mathbf{F}$. Find the number of endomorphisms of V that fix at least one nonzero vector.*

Problem 7.1.5 (Sp97) *Let $GL_2(\mathbb{Z}_m)$ denote the multiplicative group of invertible 2×2 matrices over the ring of integers modulo m. Find the order of $GL_2(\mathbb{Z}_{p^n})$ for each prime p and positive integer n.*

Problem 7.1.6 (Fa00) *Let $\mathbf{F}_p$ denote the field of p elements (p prime). Let n be a positive integer. Prove that there is a transformation $A \in GL_n(\mathbf{F}_p)$ (the group of invertible linear transformations from $(\mathbf{F}_p)^n$ into itself) which, as a permutation of the nonzero vectors of $(\mathbf{F}_p)^n$, acts as a single cycle of length $p^n - 1$.*

Problem 7.1.7 (Sp96) *Let G be the group of 2×2 matrices with determinant 1 over the four-element field $\mathbf{F}$. Let S be the set of lines through the origin in $\mathbf{F}^2$. Show that G acts faithfully on S. (The action is faithful if the only element of G which fixes every element of S is the identity.)*

Problem 7.1.8 (Su77) *Prove the following statements about the polynomial ring $\mathbf{F}[x]$, where $\mathbf{F}$ is any field.*

1. *$\mathbf{F}[x]$ is a vector space over $\mathbf{F}$.*

2. *The subset $\mathbf{F}_n[x]$ of polynomials of degree $\leqslant n$ is a subspace of dimension $n + 1$ in $\mathbf{F}[x]$.*

3. *The polynomials $1, x - a, \ldots, (x - a)^n$ form a basis of $\mathbf{F}_n[x]$ for any $a \in \mathbf{F}$.*

Problem 7.1.9 (Sp02) *Let U, V, W be finite-dimensional subspaces of a vector space. Prove that $\dim(U) + \dim(V) + \dim(W) - \dim(U + V + W) \geq \max\{\dim(U \cap V), \dim(U \cap W), \dim(V \cap W)\}$.*

Problem 7.1.10 (Su84) *Suppose V is an n-dimensional vector space over the field $\mathbf{F}$. Let $W \subset V$ be a subspace of dimension $r < n$. Show that*

$$W = \bigcap \{U \mid U \text{ is an } (n-1) - \text{dimensional subspace of } V \text{ and } W \subset U\}.$$

Problem 7.1.11 (Sp80, Fa89) *Show that a vector space over an infinite field cannot be the union of a finite number of proper subspaces.*

Problem 7.1.12 (Fa88) *Let A be a complex $n \times n$ matrix, and let $C(A)$ be the commutant of A; that is, the set of complex $n \times n$ matrices B such that $AB = BA$. (It is obviously a subspace of $M_{n \times n}$, the vector space of all complex $n \times n$ matrices.) Prove that $\dim C(A) \geqslant n$.*

Problem 7.1.13 (Sp89, Fa97) *Let S be the subspace of $M_{n \times n}$ (the vector space of all real $n \times n$ matrices) generated by all matrices of the form $AB - BA$ with A and B in $M_{n \times n}$. Prove that $\dim(S) = n^2 - 1$.*

Problem 7.1.14 (Sp90) *Let A and B be subspaces of a finite-dimensional vector space V such that $A + B = V$. Write $n = \dim V$, $a = \dim A$, and $b = \dim B$. Let S be the set of those endomorphisms f of V for which $f(A) \subset A$ and $f(B) \subset B$. Prove that S is a subspace of the set of all endomorphisms of V, and express the dimension of S in terms of n, a, and b.*

Problem 7.1.15 (Sp81) *Let T be a linear transformation of a vector space V into itself. Suppose $x \in V$ is such that $T^m x = 0$, $T^{m-1} x \neq 0$ for some positive integer m. Show that $x, Tx, \ldots, T^{m-1}x$ are linearly independent.*

Problem 7.1.16 (Fa97) *Let $\alpha_1, \alpha_2, \ldots, \alpha_n$ be distinct real numbers. Show that the n exponential functions $e^{\alpha_1 t}, e^{\alpha_2 t}, \ldots, e^{\alpha_n t}$ are linearly independent over the real numbers.*

Problem 7.1.17 (Su83) *Let V be a real vector space of dimension n with a positive definite inner product. We say that two bases (a_i) and (b_i) have the same orientation if the matrix of the change of basis from (a_i) to (b_i) has a positive determinant. Suppose now that (a_i) and (b_i) are orthonormal bases with the same orientation. Show that $(a_i + 2b_i)$ is again a basis of V with the same orientation as (a_i).*

7.2   Rank and Determinants

Problem 7.2.1 (Sp78, Fa82, Fa86) *Let M be a matrix with entries in a field $\mathbf{F}$. The* row rank *of M over $\mathbf{F}$ is the maximal number of rows which are linearly independent (as vectors) over $\mathbf{F}$. The* column rank *is similarly defined using columns instead of rows.*

  1. Prove row rank = column rank.

  2. Find a maximal linearly independent set of columns of

$$\begin{pmatrix} 1 & 0 & 3 & -2 \\ 2 & 1 & 2 & 0 \\ 0 & 1 & -4 & 4 \\ 1 & 1 & 1 & 2 \\ 1 & 0 & 1 & 2 \end{pmatrix}$$

  taking $\mathbf{F} = \mathbb{R}$.

3. If **F** is a subfield of **K**, and M has entries in **F**, how is the row rank of M over **F** related to the row rank of M over **K**?

Problem 7.2.2 (Fa02) Let the $n \times n$ real matrix A be diagonalizable and have a one-dimensional null space. Prove that a nonzero left null vector of A cannot be orthogonal to a nonzero right null vector of A.

Problem 7.2.3 (Su85, Fa89) Let A be an $n \times n$ real matrix and A^t its transpose. Show that $A^t A$ and A^t have the same range.

Problem 7.2.4 (Sp97) Suppose that P and Q are $n \times n$ matrices such that $P^2 = P$, $Q^2 = Q$, and $1 - (P + Q)$ is invertible. Show that P and Q have the same rank.

Problem 7.2.5 (Fa03) Let $A(m, n)$ be the $m \times n$ matrix with entries

$$a_{ij} = j^i \quad (0 \le i \le m - 1, \ 0 \le j \le n - 1),$$

where $0^0 = 1$ by definition. Regarding the entries of $A(m, n)$ as representing congruence classes (mod p), determine the rank of $A(m, n)$ over the finite field $\mathbb{F}_p = \mathbb{Z}_p$ for all $m, n \ge 1$ and all primes p.

Problem 7.2.6 (Sp01) Let A be an $n \times n$ matrix over a field K. Prove that

$$\text{rank } A^2 - \text{rank } A^3 \le \text{rank } A - \text{rank } A^2$$

Problem 7.2.7 (Sp91) Let T be a real, symmetric, $n \times n$, tridiagonal matrix:

$$T = \begin{pmatrix} a_1 & b_1 & 0 & 0 & \cdots & 0 & 0 \\ b_1 & a_2 & b_2 & 0 & \cdots & 0 & 0 \\ 0 & b_2 & a_3 & b_3 & \cdots & 0 & 0 \\ \vdots & \vdots & \vdots & \vdots & \ddots & \vdots & \vdots \\ 0 & 0 & 0 & 0 & \cdots & a_{n-1} & b_{n-1} \\ 0 & 0 & 0 & 0 & \cdots & b_{n-1} & a_n \end{pmatrix}$$

(All entries not on the main diagonal or the diagonals just above and below the main one are zero.) Assume $b_j \neq 0$ for all j.
Prove:

1. rank $T \ge n - 1$.

2. T has n distinct eigenvalues.

Problem 7.2.8 (Sp83) Let $A = (a_{ij})$ be an $n \times n$ real matrix satisfying the conditions:

$$a_{ii} > 0 \quad (1 \le i \le n),$$
$$a_{ij} \le 0 \quad (i \neq j, \ 1 \le i, j \le n),$$
$$\sum_{i=1}^{n} a_{ij} > 0 \quad (1 \le j \le n).$$

Show that $\det(A) > 0$.

Problem 7.2.9 (Sp91) *Let* $A = (a_{ij})^r_{i,j=1}$ *be a square matrix with integer entries.*

1. *Prove that if an integer n is an eigenvalue of A, then n is a divisor of* det A, *the determinant of A.*

2. *Suppose that n is an integer and that each row of A has sum n:*

$$\sum_{j=1}^{r} a_{ij} = n, \qquad 1 \leqslant i \leqslant r.$$

Prove that n is a divisor of det A.

Problem 7.2.10 (Fa01) *Let A be a symmetric $n \times n$ matrix over $\mathbb{R}$ of rank $n - 1$. Prove there is a k in $\{1, 2, \ldots, n\}$ such that the matrix resulting from deletion of the kth row and k^{th} column from A has rank $n - 1$.*

Problem 7.2.11 (Fa84) *Let $\mathbb{R}[x_1, \ldots, x_n]$ be the polynomial ring over the real field $\mathbb{R}$ in the n variables $x_1, \ldots, x_n$. Let the matrix A be the $n \times n$ matrix whose i^{th} row is $(1, x_i, x_i^2, \ldots, x_i^{n-1})$, $i = 1, \ldots, n$. Show that*

$$\det A = \prod_{i>j}(x_i - x_j).$$

Problem 7.2.12 (Sp77) *A matrix of the form*

$$\begin{pmatrix} 1 & a_0 & a_0^2 & \cdots & a_0^n \\ 1 & a_1 & a_1^2 & \cdots & a_1^n \\ \vdots & \vdots & \vdots & \ddots & \vdots \\ 1 & a_n & a_n^2 & \cdots & a_n^n \end{pmatrix}$$

where the a_i are complex numbers, is called a Vandermonde matrix.

1. *Prove that the Vandermonde matrix is invertible if $a_0, a_1, \ldots, a_n$ are all different.*

2. *If $a_0, a_1, \ldots, a_n$ are all different, and $b_0, b_1, \ldots, b_n$ are complex numbers, prove that there is a unique polynomial f of degree n with complex coefficients such that $f(a_0) = b_0$, $f(a_1) = b_1, \ldots, f(a_n) = b_n$.*

Problem 7.2.13 (Fa03) *Let A and B be $n \times n$ complex unitary matrices. Prove that $|\det(A + B)| \leq 2^n$.*

Problem 7.2.14 (Sp90) *Give an example of a continuous function $v : \mathbb{R} \to \mathbb{R}^3$ with the property that $v(t_1)$, $v(t_2)$, and $v(t_3)$ form a basis for $\mathbb{R}^3$ whenever t_1, t_2, and t_3 are distinct points of $\mathbb{R}$.*

Problem 7.2.15 (Fa95) *Let* $f_1, f_2, \ldots, f_n$ *be continuous real valued functions on* $[a, b]$. *Show that the set* $\{f_1, \ldots, f_n\}$ *is linearly dependent on* $[a, b]$ *if and only if*

$$\det \left(\int_a^b f_i(x) f_j(x) dx \right) = 0.$$

Problem 7.2.16 (Fa81) *Let* $M_{2 \times 2}$ *be the vector space of all real* 2×2 *matrices. Let*

$$A = \begin{pmatrix} 1 & 2 \\ -1 & 3 \end{pmatrix} \qquad B = \begin{pmatrix} 2 & 1 \\ 0 & 4 \end{pmatrix}$$

and define a linear transformation $L : M_{2 \times 2} \to M_{2 \times 2}$ *by* $L(X) = AXB$. *Compute the trace and the determinant of* L.

Problem 7.2.17 (Su82) *Let* V *be the vector space of all real* 3×3 *matrices and let* A *be the diagonal matrix*

$$\begin{pmatrix} 1 & 0 & 0 \\ 0 & 2 & 0 \\ 0 & 0 & 1 \end{pmatrix}.$$

Calculate the determinant of the linear transformation T *on* V *defined by* $T(X) = \frac{1}{2}(AX + XA)$.

Problem 7.2.18 (Sp80) *Let* $M_{3 \times 3}$ *denote the vector space of real* 3×3 *matrices. For any matrix* $A \in M_{3 \times 3}$, *define the linear operator* $L_A : M_{3 \times 3} \to M_{3 \times 3}$, $L_A(B) = AB$. *Suppose that the determinant of* A *is* 32 *and the minimal polynomial is* $(t - 4)(t - 2)$. *What is the trace of* L_A?

Problem 7.2.19 (Su81) *Let* S *denote the vector space of real* $n \times n$ *skew-symmetric matrices. For a nonsingular matrix* A, *compute the determinant of the linear map* $T_A : S \to S$, $T_A(X) = AXA^t$.

Problem 7.2.20 (Fa94) *Let* $M_{7 \times 7}$ *denote the vector space of real* 7×7 *matrices. Let* A *be a diagonal matrix in* $M_{7 \times 7}$ *that has* $+1$ *in four diagonal positions and* -1 *in three diagonal positions. Define the linear transformation* T *on* $M_{7 \times 7}$ *by* $T(X) = AX - XA$. *What is the dimension of the range of* T?

Problem 7.2.21 (Fa93) *Let* $\mathbf{F}$ *be a field. For* m *and* n *positive integers, let* $M_{m \times n}$ *be the vector space of* $m \times n$ *matrices over* $\mathbf{F}$. *Fix* m *and* n, *and fix matrices* A *and* B *in* $M_{m \times n}$. *Define the linear transformation* T *from* $M_{n \times m}$ *to* $M_{m \times n}$ *by*

$$T(X) = AXB.$$

Prove that if $m \neq n$, *then* T *is not invertible.*

7.3   Systems of Equations

Problem 7.3.1 (Su77) *Determine all solutions to the following infinite system of linear equations in the infinitely many unknowns $x_1, x_2, \ldots$:*

$$
\begin{aligned}
x_1 + x_3 + x_5 &= 0 \\
x_2 + x_4 + x_6 &= 0 \\
x_3 + x_5 + x_7 &= 0 \\
\vdots \quad \vdots \quad \vdots \quad &\ \ \vdots
\end{aligned}
$$

How many free parameters are required?

Problem 7.3.2 (Fa77, Su78)   *1. Using only the axioms for a field* **F**, *prove that a system of m homogeneous linear equations in n unknowns with $m < n$ and coefficients in* **F** *has a nonzero solution.*

2. Use Part 1 to show that if V is a vector space over **F** *which is spanned by a finite number of elements, then every maximal linearly independent subset of V has the same number of elements.*

Problem 7.3.3 (Sp88, Sp96) *If a finite homogeneous system of linear equations with rational coefficients has a nontrivial complex solution, need it have a nontrivial rational solution? Give a proof or a counterexample.*

Problem 7.3.4 (Sp84, Sp87) *Let A be a real $m \times n$ matrix with rational entries and let b be an m-tuple of rational numbers. Assume that the system of equations $Ax = b$ has a solution x in complex n-space $\mathbb{C}^n$. Show that the equation has a solution vector with rational components, or give a counterexample.*

7.4   Linear Transformations

Problem 7.4.1 (Fa77) *Let E and F be vector spaces (not assumed to be finite-dimensional). Let $S : E \to F$ be a linear transformation.*

1. Prove $S(E)$ is a vector space.

2. Show S has a kernel $\{0\}$ if and only if S is injective (i.e., one-to-one).

3. Assume S is injective; prove $S^{-1} : S(E) \to E$ is linear.

Problem 7.4.2 (Sp82) *Let $T : V \to W$ be a linear transformation between finite-dimensional vector spaces. Prove that*

$$\dim(\ker T) + \dim(\operatorname{range} T) = \dim V .$$

Problem 7.4.3 (Fa99) *Let V and W be finite dimensional vector spaces, let X be a subspace of W, and let $T : V \to W$ be a linear map. Prove that the dimension of $T^{-1}(X)$ is at least $\dim V - \dim W + \dim X$.*

Problem 7.4.4 (Fa98) *Let A and B be linear transformations on a finite dimensional vector space V. Prove that* $\dim \ker(AB) \leqslant \dim \ker A + \dim \ker B$.

Problem 7.4.5 (Fa02) *Let V and W be vector spaces over a field K. Assume* $A : V \to W$ *and* $B : V \to W$ *are linear transformations such that A has rank at least 2, and for every vector v in V the vectors Av and Bv are linearly dependent. Prove that the linear transformations A and B are linearly dependent.*

Problem 7.4.6 (Sp95) *Suppose that* $W \subset V$ *are finite-dimensional vector spaces over a field, and let* $L : V \to V$ *be a linear transformation with* $L(V) \subset W$. *Denote the restriction of L to W by* L_W. *Prove that* $\det(1 - tL) = \det(1 - tL_W)$.

Problem 7.4.7 (Fa00) *Let V be a finite-dimensional vector space, and let* $f : V \to V$ *be a linear transformation. Let W denote the image of f. Prove that the restriction of f to W, considered as an endomorphism of W, has the same trace as* $f : V \to V$.

Problem 7.4.8 (Fa99) *Let* $T : V \to V$ *be a linear operator on an n dimensional vector space V over a field* $\mathbf{F}$. *Prove that T has an invariant subspace W other than* $\{0\}$ *and V if and only if the characteristic polynomial of T has a factor* $f \in \mathbf{F}[t]$ *with* $0 < \deg f < n$.

Problem 7.4.9 (Fa00) *Let* $T : \mathbb{R}^n \to \mathbb{R}^n$ *be a linear transformation, where* $n > 1$. *Prove that there is a 2-dimensional subspace* $M \subseteq \mathbb{R}^n$ *such that* $T(M) \subseteq M$.

Problem 7.4.10 (Sp95) *Let V be a finite-dimensional vector space over a field* $\mathbf{F}$, *and let* $L : V \to V$ *be a linear transformation. Suppose that the characteristic polynomial* χ *of L is written as* $\chi = \chi_1 \chi_2$, *where* χ_1 *and* χ_2 *are two relatively prime polynomials with coefficients in* $\mathbf{F}$. *Show that V can be written as the direct sum of two subspaces* V_1 *and* V_2 *with the property that* $\chi_i(L)V_i = 0$ *for* $i = 1, 2$.

Problem 7.4.11 (Su79) *Let E be a three-dimensional vector space over* $\mathbb{Q}$. *Suppose* $T : E \to E$ *is a linear transformation and* $Tx = y$, $Ty = z$, $Tz = x + y$, *for certain* $x, y, z \in E$, $x \neq 0$. *Prove that x, y, and z are linearly independent.*

Problem 7.4.12 (Su80) *Let* $T : V \to V$ *be an invertible linear transformation of a vector space V. Denote by G the group of all maps* $f_{k,a} : V \to V$ *where* $k \in \mathbb{Z}$, $a \in V$, *and for* $x \in V$,

$$f_{k,a}(x) = T^k x + a \quad (x \in V).$$

Prove that the commutator subgroup G' *of G is isomorphic to the additive group of the vector space* $(T - I)V$, *the image of* $T - I$. *(G' is generated by all* $ghg^{-1}h^{-1}$, *g and h in G.)*

Problem 7.4.13 (Sp86) *Let V be a finite-dimensional vector space and A and B two linear transformations of V into itself such that* $A^2 = B^2 = 0$ *and* $AB + BA = I$.

1. Prove that if N_A and N_B are the respective null spaces of A and B, then $N_A = AN_B$, $N_B = BN_A$, and $V = N_A \oplus N_B$.

2. Prove that the dimension of V is even.

3. Prove that if the dimension of V is 2, then V has a basis with respect to which A and B are represented by the matrices

$$\begin{pmatrix} 0 & 1 \\ 0 & 0 \end{pmatrix} \quad and \quad \begin{pmatrix} 0 & 0 \\ 1 & 0 \end{pmatrix}.$$

Problem 7.4.14 (Su84) Let $f : \mathbb{R}^m \to \mathbb{R}^n$, $n \geqslant 2$, be a linear transformation of rank $n - 1$. Let $f(v) = (f_1(v), f_2(v), \ldots, f_n(v))$ for $v \in \mathbb{R}^m$. Show that a necessary and sufficient condition for the system of inequalities $f_i(v) > 0$, $i = 1, \ldots, n$, to have no solution is that there exist real numbers $\lambda_i \geqslant 0$, not all zero, such that

$$\sum_{i=1}^{n} \lambda_i f_i = 0.$$

Problem 7.4.15 (Sp95) Let n be a positive integer, and let $S \subset \mathbb{R}^n$ a finite subset with $0 \in S$. Suppose that $\varphi : S \to S$ is a map satisfying

$$\varphi(0) = 0,$$
$$d(\varphi(s), \varphi(t)) = d(s, t) \qquad for\ all \quad s, t \in S,$$

where $d(\ ,\ )$ denotes Euclidean metric. Prove that there is a linear map $f : \mathbb{R}^n \to \mathbb{R}^n$ whose restriction to S is φ.

Problem 7.4.16 (Sp86) Consider $\mathbb{R}^2$ be equipped with the Euclidean metric $d(x, y) = \|x - y\|$. Let T be an isometry of $\mathbb{R}^2$ into itself. Prove that T can be represented as $T(x) = a + U(x)$, where a is a vector in $\mathbb{R}^2$ and U is an orthogonal linear transformation.

Problem 7.4.17 (Sp88) Let X be a set and V a real vector space of real valued functions on X of dimension n, $0 < n < \infty$. Prove that there are n points $x_1, x_2, \ldots, x_n$ in X such that the map $f \mapsto (f(x_1), \ldots, f(x_n))$ of V to $\mathbb{R}^n$ is an isomorphism.

Problem 7.4.18 (Sp97) Suppose that X is a topological space and V is a finite-dimensional subspace of the vector space of continuous real valued functions on X. Prove that there exist a basis $\{f_1, \ldots, f_n\}$ for V and points $x_1, \ldots, x_n$ in X such that $f_i(x_j) = \delta_{ij}$.

Problem 7.4.19 (Fa90) Let n be a positive integer and let P_{2n+1} be the vector space of real polynomials whose degrees are, at most, $2n + 1$. Prove that there exist unique real numbers $c_1, \ldots, c_n$ such that, for all $p \in P_{2n+1}$.

$$\int_{-1}^{1} p(x)\, dx = 2p(0) + \sum_{k=1}^{n} c_k(p(k) + p(-k) - 2p(0))$$

Problem 7.4.20 (Sp94) *Let $T : \mathbb{R}^n \to \mathbb{R}^n$ be a diagonalizable linear transformation. Prove that there is an orthonormal basis for $\mathbb{R}^n$ with respect to which T has an upper-triangular matrix.*

Problem 7.4.21 (Fa77) *Let P be a linear operator on a finite-dimensional vector space over a finite field. Show that if P is invertible, then $P^n = I$ for some positive integer n.*

Problem 7.4.22 (Fa82) *Let A be an $n \times n$ complex matrix, and let B be the Hermitian transpose of A (i.e., $b_{ij} = \bar{a}_{ji}$). Suppose that A and B commute with each other. Consider the linear transformations α and β on $\mathbb{C}^n$ defined by A and B. Prove that α and β have the same image and the same kernel.*

Problem 7.4.23 (Su79, Fa96) *Prove that a linear transformation $T : \mathbb{R}^3 \to \mathbb{R}^3$ has*

  1. *a one-dimensional invariant subspace, and*

  2. *a two-dimensional invariant subspace.*

Problem 7.4.24 (Fa83) *Let A be a linear transformation on $\mathbb{R}^3$ whose matrix (relative to the usual basis for $\mathbb{R}^3$) is both symmetric and orthogonal. Prove that A is either plus or minus the identity, or a rotation by $180°$ about some axis in $\mathbb{R}^3$, or a reflection about some two-dimensional subspace of $\mathbb{R}^3$.*

Problem 7.4.25 (Fa84) *Let θ and φ be fixed, $0 \leqslant \theta \leqslant 2\pi, 0 \leqslant \varphi \leqslant 2\pi$ and let R be the linear transformation from $\mathbb{R}^3$ to $\mathbb{R}^3$ whose matrix in the standard basis $\vec{i}, \vec{j},$ and $\vec{k}$ is*

$$\begin{pmatrix} 1 & 0 & 0 \\ 0 & \cos\theta & \sin\theta \\ 0 & -\sin\theta & \cos\theta \end{pmatrix}.$$

Let S be the linear transformation of $\mathbb{R}^3$ to $\mathbb{R}^3$ whose matrix with respect to the basis

$$\left\{ \frac{1}{\sqrt{2}}(\vec{i} + \vec{k}), \vec{j}, \frac{1}{\sqrt{2}}(\vec{i} - \vec{k}) \right\}$$

is

$$\begin{pmatrix} \cos\varphi & \sin\varphi & 0 \\ -\sin\varphi & \cos\varphi & 0 \\ 0 & 0 & 1 \end{pmatrix}.$$

Prove that $T = R \circ S$ leaves a line invariant.

Problem 7.4.26 (Sp86) *Let $e = (a, b, c)$ be a unit vector in $\mathbb{R}^3$ and let T be the linear transformation on $\mathbb{R}^3$ of rotation by $180°$ about e. Find the matrix for T with respect to the standard basis.*

Problem 7.4.27 (Su80) *Exhibit a real 3×3 matrix having minimal polynomial* $(t^2+1)(t-10)$, *which, as a linear transformation of* $\mathbb{R}^3$, *leaves invariant the line L through* $(0,0,0)$ *and* $(1,1,1)$ *and the plane through* $(0,0,0)$ *perpendicular to L.*

Problem 7.4.28 (Su77) *Show that every rotation of* $\mathbb{R}^3$ *has an* axis; *that is, given a 3×3 real matrix A such that* $A^t = A^{-1}$ *and* $\det A > 0$, *prove that there is a nonzero vector v such that* $Av = v$.

Problem 7.4.29 (Sp93) *Let P be the vector space of polynomials over* $\mathbb{R}$. *Let the linear transformation* $E : P \to P$ *be defined by* $Ef = f + f'$, *where* f' *is the derivative of* f. *Prove that E is invertible.*

Problem 7.4.30 (Fa84) *Let* P_n *be the vector space of all real polynomials with degrees at most n. Let* $D : P_n \to P_n$ *be given by differentiation:* $D(p) = p'$. *Let* π *be a real polynomial. What is the minimal polynomial of the transformation* $\pi(D)$?

Problem 7.4.31 (Su77) *Let V be the vector space of all polynomials of degree* $\leqslant$ 10, *and let D be the differentiation operator on V (i.e.,* $Dp(x) = p'(x)$*).*

1. *Show that* tr $D = 0$.

2. *Find all eigenvectors of D and* e^D.

Problem 7.4.32 (Sp02) *Let* $x_0, x_1, \ldots, x_n$ *be distinct points of* $\mathbb{R}$. *Prove that there are unique real numbers* $a_0, a_1, \ldots, a_n$ *such that*

$$\int_0^1 p(t)dt = \sum_{j=0}^n a_j\, p(x_j)$$

for all polynomials of degree n or less.

Problem 7.4.33 (Sp00) *Let* $I_1, \ldots, I_n$ *be disjoint closed nonempty subintervals of* $\mathbb{R}$.

1. *Prove that if p is a real polynomial of degree less than n such that*

$$\int_{I_j} p(x)dx = 0, \qquad for\ j = 1, \ldots, n$$

then $p = 0$.

2. *Prove that there is a nonzero real polynomial p of degree n that satisfies all the above equations.*

7.5 Eigenvalues and Eigenvectors

Problem 7.5.1 (Fa77) *Let M be a real 3×3 matrix such that $M^3 = I$, $M \neq I$.*

1. What are the eigenvalues of M?

2. Give an example of such a matrix.

Problem 7.5.2 (Fa79) *Let N be a linear operator on an n-dimensional vector space, $n > 1$, such that $N^n = 0$, $N^{n-1} \neq 0$. Prove there is no operator X with $X^2 = N$.*

Problem 7.5.3 (Sp89) *Let F be a field, n and m positive integers, and A an $n \times n$ matrix with entries in F such that $A^m = 0$. Prove that $A^n = 0$.*

Problem 7.5.4 (Fa98) *Let B be a 3×3 matrix whose null space is 2-dimensional, and let $\chi(\lambda)$ be the characteristic polynomial of B. For each assertion below, provide either a proof or a counterexample.*

1. λ^2 is a factor of $\chi(\lambda)$.

2. The trace of B is an eigenvalue of B.

3. B is diagonalizable.

Problem 7.5.5 (Sp99) *Suppose that the minimal polynomial of a linear operator T on a seven-dimensional vector space is x^2. What are the possible values of the dimension of the kernel of T?*

Problem 7.5.6 (Sp03) *Let L be a real symmetric $n \times n$ matrix with 0 as a simple eigenvalue, and let $v \in \mathbb{R}^n$.*

1. Show that for sufficiently small positive real ε, the equation $Lx + \varepsilon x = v$ has a unique solution $x = x(\varepsilon) \in \mathbb{R}^n$.

2. Evaluate $\lim_{\varepsilon \to 0+} \varepsilon x(\varepsilon)$ in terms of v, the eigenvectors of L, and the inner product $\langle\,,\,\rangle$ on $\mathbb{R}^n$.

Problem 7.5.7 (Su81, Su82) *Let V be a finite-dimensional vector space over the rationals $\mathbb{Q}$ and let M be an automorphism of V such that M fixes no nonzero vector in V. Suppose that M^p is the identity map on V, where p is a prime number. Show that the dimension of V is divisible by $p - 1$.*

Problem 7.5.8 (Fa92) *Let F be a field, V a finite-dimensional vector space over F, and T a linear transformation of V into V whose minimum polynomial, μ, is irreducible over F.*

1. Let v be a nonzero vector in V and let V_1 be the subspace spanned by v and its images under the positive powers of T. Prove that $\dim V_1 = \deg \mu$.

2. Prove that $\deg \mu$ divides $\dim V$.

Problem 7.5.9 (Fa02) *Suppose A and M are n × n matrices over* $\mathbb{C}$, *A is invertible and* $AMA^{-1} = M^2$. *Prove the nonzero eigenvalues of M are roots of unity.*

Problem 7.5.10 (Su79, Fa93) *Prove that the matrix*

$$\begin{pmatrix} 0 & 5 & 1 & 0 \\ 5 & 0 & 5 & 0 \\ 1 & 5 & 0 & 5 \\ 0 & 0 & 5 & 0 \end{pmatrix}$$

has two positive and two negative eigenvalues (counting multiplicities).

Problem 7.5.11 (Fa94) *Prove that the matrix*

$$\begin{pmatrix} 1 & 1.00001 & 1 \\ 1.00001 & 1 & 1.00001 \\ 1 & 1.00001 & 1 \end{pmatrix}$$

has one positive eigenvalue and one negative eigenvalue.

Problem 7.5.12 (Sp85) *For arbitrary elements a, b, and c in a field* **F**, *compute the minimal polynomial of the matrix*

$$\begin{pmatrix} 0 & 0 & a \\ 1 & 0 & b \\ 0 & 1 & c \end{pmatrix}.$$

Problem 7.5.13 (Fa01) *Let A be an n × n complex matrix such that the three matrices* $A + I$, $A^2 + I$, $A^3 + I$ *are all unitary. Prove that A is the zero matrix.*

Problem 7.5.14 (Sp03) *Let k be a field, and let* $n \geq 1$. *Prove that the following properties of an n × n matrix A with entries in k are equivalent:*

- *A is a scalar multiple of the identity matrix.*

- *Every nonzero vector* $v \in k^n$ *is an eigenvector of A.*

Problem 7.5.15 (Fa85, Sp97, Fa98) *Suppose that A and B are endomorphisms of a finite-dimensional vector space V over a field* **F**. *Prove or disprove the following statements:*

1. *Every eigenvector of AB is also an eigenvector of BA.*

2. *Every eigenvalue of AB is also an eigenvalue of BA.*

Problem 7.5.16 (Sp78, Sp98) *Let A and B denote real n×n symmetric matrices such that* $AB = BA$. *Prove that A and B have a common eigenvector in* $\mathbb{R}^n$.

Problem 7.5.17 (Sp86) *Let S be a nonempty commuting set of n × n complex matrices* ($n \geq 1$). *Prove that the members of S have a common eigenvector.*

Problem 7.5.18 (Sp84) *Let A and B be complex $n \times n$ matrices such that $AB = BA^2$, and assume A has no eigenvalues of absolute value 1. Prove that A and B have a common (nonzero) eigenvector.*

Problem 7.5.19 (Su78) *Let V be a finite-dimensional vector space over an algebraically closed field. A linear operator $T : V \to V$ is called completely reducible if whenever a linear subspace $E \subset V$ is invariant under T, that is $T(E) \subset E$, there is a linear subspace $F \subset V$ which is invariant under T and such that $V = E \oplus F$. Prove that T is completely reducible if and only if V has a basis of eigenvectors.*

Problem 7.5.20 (Fa79, Su81) *Let V be the vector space of sequences (a_n) of complex numbers. The shift operator $S : V \to V$ is defined by*

$$S((a_1, a_2, a_3, \ldots)) = (a_2, a_3, a_4, \ldots).$$

1. *Find the eigenvectors of S.*

2. *Show that the subspace W consisting of the sequences (x_n) with $x_{n+2} = x_{n+1} + x_n$ is a two-dimensional, S-invariant subspace of V and exhibit an explicit basis for W.*

3. *Find an explicit formula for the nth Fibonacci number f_n, where $f_2 = f_1 = 1$, $f_{n+2} = f_{n+1} + f_n$ for $n \geqslant 1$.*

Note: See also Problem 1.3.12.

Problem 7.5.21 (Fa82) *Let T be a linear transformation on a finite-dimensional $\mathbb{C}$-vector space V, and let f be a polynomial with coefficients in $\mathbb{C}$. If λ is an eigenvalue of T, show that $f(\lambda)$ is an eigenvalue of $f(T)$. Is every eigenvalue of $f(T)$ necessarily obtained in this way?*

Problem 7.5.22 (Fa83, Sp96) *Let A be the $n \times n$ matrix which has zeros on the main diagonal and ones everywhere else. Find the eigenvalues and eigenspaces of A and compute $\det(A)$.*

Problem 7.5.23 (Sp85) *Let A and B be two $n \times n$ self-adjoint (i.e., Hermitian) matrices over $\mathbb{C}$ and assume A is positive definite. Prove that all eigenvalues of AB are real.*

Problem 7.5.24 (Fa84) *Let $a, b, c,$ and d be real numbers, not all zero. Find the eigenvalues of the following 4×4 matrix and describe the eigenspace decomposition of $\mathbb{R}^4$:*

$$\begin{pmatrix} aa & ab & ac & ad \\ ba & bb & bc & bd \\ ca & cb & cc & cd \\ da & db & dc & dd \end{pmatrix}.$$

Problem 7.5.25 (Sp81) *Show that the following three conditions are all equivalent for a real 3×3 symmetric matrix A, whose eigenvalues are λ_1, λ_2, and λ_3:*

1. tr A *is not an eigenvalue of A.*

2. $(a + b)(b + c)(a + c) \neq 0$.

3. *The map $L : S \to S$ is an isomorphism, where S is the space of 3×3 real skew-symmetric matrices and $L(W) = AW + WA$.*

Problem 7.5.26 (Su84) *Let*

$$A = \begin{pmatrix} a & b \\ c & d \end{pmatrix}$$

be a real matrix with $a, b, c, d > 0$. Show that A has an eigenvector

$$\begin{pmatrix} x \\ y \end{pmatrix} \in \mathbb{R}^2$$

with $x, y > 0$.

Problem 7.5.27 (Sp90) *Let n be a positive integer, and let $A = (a_{ij})_{i,j=1}^{n}$ be the $n \times n$ matrix with $a_{ii} = 2$, $a_{i\,i\pm1} = -1$, and $a_{ij} = 0$ otherwise; that is,*

$$A = \begin{pmatrix} 2 & -1 & 0 & 0 & \cdots & 0 & 0 & 0 \\ -1 & 2 & -1 & 0 & \cdots & 0 & 0 & 0 \\ 0 & -1 & 2 & -1 & \cdots & 0 & 0 & 0 \\ 0 & 0 & -1 & 2 & \cdots & 0 & 0 & 0 \\ \vdots & \vdots & \vdots & \vdots & \ddots & \vdots & \vdots & \vdots \\ 0 & 0 & 0 & 0 & \cdots & 2 & -1 & 0 \\ 0 & 0 & 0 & 0 & \cdots & -1 & 2 & -1 \\ 0 & 0 & 0 & 0 & \cdots & 0 & -1 & 2 \end{pmatrix}.$$

Prove that every eigenvalue of A is a positive real number.

Problem 7.5.28 (Sp92) *Let A be a real symmetric $n \times n$ matrix with nonnegative entries. Prove that A has an eigenvector with nonnegative entries.*

Problem 7.5.29 (Sp00) *Let A_n be the $n \times n$ matrix whose entries a_{jk} are given by*

$$a_{jk} = \begin{cases} 1 & \text{if } \ |j - k| = 1 \\ 0 & \text{otherwise} . \end{cases}$$

Prove that the eigenvalues of A are symmetric with respect to the origin.

Problem 7.5.30 (Fa91) *Let $A = (a_{ij})_{i,j=1}^{n}$ be a real $n \times n$ matrix with nonnegative entries such that*

$$\sum_{j=1}^{n} a_{ij} = 1 \qquad (1 \leqslant i \leqslant n).$$

Prove that no eigenvalue of A has absolute value greater than 1.

Problem 7.5.31 (Sp01) *Let S be a* special orthogonal *$n \times n$ matrix, a real $n \times n$ matrix satisfying $S^t S = I$ and $det(S) = 1$.*

1. *Prove that if n is odd then 1 is an eigenvalue of S.*

2. *Prove that if n is even then 1 need not be an eigenvalue of S.*

Problem 7.5.32 (Sp85, Fa88) *Let A and B be two $n \times n$ self-adjoint (i.e., Hermitian) matrices over $\mathbb{C}$ such that all eigenvalues of A lie in $[a, a']$ and all eigenvalues of B lie in $[b, b']$. Show that all eigenvalues of $A + B$ lie in $[a+b, a'+b']$.*

Problem 7.5.33 (Fa85) *Let k be real, n an integer ≥ 2, and let $A = (a_{ij})$ be the $n \times n$ matrix such that all diagonal entries $a_{ii} = k$, all entries $a_{i\,i\pm1}$ immediately above or below the diagonal equal 1, and all other entries equal 0. For example, if $n = 5$,*

$$A = \begin{pmatrix} k & 1 & 0 & 0 & 0 \\ 1 & k & 1 & 0 & 0 \\ 0 & 1 & k & 1 & 0 \\ 0 & 0 & 1 & k & 1 \\ 0 & 0 & 0 & 1 & k \end{pmatrix}.$$

Let λ_{min} and λ_{max} denote the smallest and largest eigenvalues of A, respectively. Show that $\lambda_{min} \leqslant k - 1$ and $\lambda_{max} \geqslant k + 1$.

Problem 7.5.34 (Fa87) *Let A and B be real $n \times n$ symmetric matrices with B positive definite. Consider the function defined for $x \neq 0$ by*

$$G(x) = \frac{\langle Ax, x \rangle}{\langle Bx, x \rangle}.$$

1. *Show that G attains its maximum value.*

2. *Show that any maximum point U for G is an eigenvector for a certain matrix related to A and B and show which matrix.*

Problem 7.5.35 (Fa90) *Let A be a real symmetric $n \times n$ matrix that is positive definite. Let $y \in \mathbb{R}^n$, $y \neq 0$. Prove that the limit*

$$\lim_{m \to \infty} \frac{y^t A^{m+1} y}{y^t A^m y}$$

exists and is an eigenvalue of A.

7.6   Canonical Forms

Problem 7.6.1 (Sp90, Fa93) *Let A be a complex $n \times n$ matrix that has* finite order; *that is, $A^k = I$ for some positive integer k. Prove that A is diagonalizable.*

Problem 7.6.2 (Sp84) *Prove, or supply a counterexample: If A is an invertible n × n complex matrix and some power of A is diagonal, then A can be diagonalized.*

Problem 7.6.3 (Fa78) *Let*

$$A = \begin{pmatrix} 1 & 2 \\ 1 & -1 \end{pmatrix}.$$

Express A^{-1} as a polynomial in A with real coefficients.

Problem 7.6.4 (Sp81) *For $x \in \mathbb{R}$, let*

$$A_x = \begin{pmatrix} x & 1 & 1 & 1 \\ 1 & x & 1 & 1 \\ 1 & 1 & x & 1 \\ 1 & 1 & 1 & x \end{pmatrix}.$$

1. *Prove that $\det(A_x) = (x - 1)^3(x + 3)$.*

2. *Prove that if $x \neq 1, -3$, then $A_x^{-1} = -(x - 1)^{-1}(x + 3)^{-1} A_{-x-2}$.*

Problem 7.6.5 (Sp88) *Compute A^{10} for the matrix*

$$A = \begin{pmatrix} 3 & 1 & 1 \\ 2 & 4 & 2 \\ -1 & -1 & 1 \end{pmatrix}.$$

Problem 7.6.6 (Fa87) *Calculate A^{100} and A^{-7}, where*

$$A = \begin{pmatrix} 3/2 & 1/2 \\ -1/2 & 1/2 \end{pmatrix}.$$

Problem 7.6.7 (Sp96) *Prove or disprove: For any 2×2 matrix A over $\mathbb{C}$, there is a 2×2 matrix B such that $A = B^2$.*

Problem 7.6.8 (Su85)   1. *Show that a real 2×2 matrix A satisfies $A^2 = -I$ if and only if*

$$A = \begin{pmatrix} \pm\sqrt{pq - 1} & -p \\ q & \mp\sqrt{pq - 1} \end{pmatrix}$$

where p and q are real numbers such that $pq \geq 1$ and both upper or both lower signs should be chosen in the double signs.

2. *Show that there is no real 2×2 matrix A such that*

$$A^2 = \begin{pmatrix} -1 & 0 \\ 0 & -1 - \varepsilon \end{pmatrix}$$

with $\varepsilon > 0$.

Problem 7.6.9 (Fa96) *Is there a real 2×2 matrix A such that*

$$A^{20} = \begin{pmatrix} -1 & 0 \\ 0 & -1-\varepsilon \end{pmatrix} ?$$

Exhibit such an A or prove there is none.

Problem 7.6.10 (Sp88) *For which positive integers n is there a 2×2 matrix*

$$A = \begin{pmatrix} a & b \\ c & d \end{pmatrix}$$

with integer entries and order n; that is, $A^n = I$ but $A^k \neq I$ for $0 < k < n$? Note: See also Problem 7.7.8.

Problem 7.6.11 (Sp92) *Find a square root of the matrix*

$$\begin{pmatrix} 1 & 3 & -3 \\ 0 & 4 & 5 \\ 0 & 0 & 9 \end{pmatrix}.$$

How many square roots does this matrix have?

Problem 7.6.12 (Sp92) *Let A denote the matrix*

$$\begin{pmatrix} 0 & 0 & 0 & 1 \\ 0 & 0 & 0 & 0 \\ 0 & 0 & 0 & 0 \\ 0 & 0 & 0 & 0 \end{pmatrix}.$$

For which positive integers n is there a complex 4×4 matrix X such that $X^n = A$?

Problem 7.6.13 (Sp88) *Prove or disprove: There is a real $n \times n$ matrix A such that*

$$A^2 + 2A + 5I = 0$$

if and only if n is even.

Problem 7.6.14 (Su83) *Let A be an $n \times n$ Hermitian matrix satisfying the condition*

$$A^5 + A^3 + A = 3I.$$

Show that $A = I$.

Problem 7.6.15 (Su80) *Which of the following matrix equations have a real matrix solution X? (It is not necessary to exhibit solutions.)*

1.

$$X^3 = \begin{pmatrix} 0 & 0 & 0 \\ 1 & 0 & 0 \\ 2 & 3 & 0 \end{pmatrix},$$

2.

$$2X^5 + X = \begin{pmatrix} 3 & 5 & 0 \\ 5 & 1 & 9 \\ 0 & 9 & 0 \end{pmatrix},$$

3.

$$X^6 + 2X^4 + 10X = \begin{pmatrix} 0 & -1 \\ 1 & 0 \end{pmatrix},$$

4.

$$X^4 = \begin{pmatrix} 3 & 4 & 0 \\ 0 & 3 & 0 \\ 0 & 0 & -3 \end{pmatrix}.$$

Problem 7.6.16 (Sp80) *Find a real matrix B such that*

$$B^4 = \begin{pmatrix} 2 & 0 & 0 \\ 0 & 2 & 0 \\ 0 & -1 & 1 \end{pmatrix}.$$

Problem 7.6.17 (Fa87) *Let V be a finite-dimensional vector space and T:V → V a diagonalizable linear transformation. Let W ⊂ V be a linear subspace which is mapped into itself by T. Show that the restriction of T to W is diagonalizable.*

Problem 7.6.18 (Fa89) *Let A and B be diagonalizable linear transformations of $\mathbb{R}^n$ into itself such that $AB = BA$. Let E be an eigenspace of A. Prove that the restriction of B to E is diagonalizable.*

Problem 7.6.19 (Fa83, Sp87, Fa99) *Let V be a finite-dimensional complex vector space and let A and B be linear operators on V such that $AB = BA$. Prove that if A and B can each be diagonalized, then there is a basis for V which simultaneously diagonalizes A and B.*

Problem 7.6.20 (Sp80) *Let A and B be n × n complex matrices. Prove or disprove each of the following statements:*

1. If A and B are diagonalizable, so is $A + B$.

2. If A and B are diagonalizable, so is AB.

3. If $A^2 = A$, then A is diagonalizable.

4. If A is invertible and A^2 is diagonalizable, then A is diagonalizable.

Problem 7.6.21 (Fa77) *Let*

$$A = \begin{pmatrix} 7 & 15 \\ -2 & -4 \end{pmatrix}.$$

Find a real matrix B such that $B^{-1}AB$ is diagonal.

Problem 7.6.22 (Su77) *Let $A : \mathbb{R}^6 \to \mathbb{R}^6$ be a linear transformation such that $A^{26} = I$. Show that $\mathbb{R}^6 = V_1 \oplus V_2 \oplus V_3$, where V_1, V_2, and V_3 are two-dimensional invariant subspaces for A.*

Problem 7.6.23 (Sp78, Sp82, Su82, Fa90) *Determine the Jordan Canonical Form of the matrix*

$$A = \begin{pmatrix} 1 & 2 & 3 \\ 0 & 4 & 5 \\ 0 & 0 & 4 \end{pmatrix}.$$

Problem 7.6.24 (Su83) *Find the eigenvalues, eigenvectors, and the Jordan Canonical Form of*

$$A = \begin{pmatrix} 2 & 1 & 1 \\ 1 & 2 & 1 \\ 1 & 1 & 2 \end{pmatrix},$$

considered as a matrix with entries in $\mathbb{F}_3 = \mathbb{Z}/3\mathbb{Z}$.

Problem 7.6.25 (Su83) *Let A be an $n \times n$ complex matrix, and let χ and μ be the characteristic and minimal polynomials of A. Suppose that*

$$\chi(x) = \mu(x)(x - i),$$

$$\mu(x)^2 = \chi(x)(x^2 + 1).$$

Determine the Jordan Canonical Form of A.

Problem 7.6.26 (Fa78, Fa84) *Let M be the $n \times n$ matrix over a field $\mathbf{F}$, all of whose entries are equal to 1.*

1. *Find the characteristic polynomial of M.*

2. *Is M diagonalizable?*

3. *Find the Jordan Canonical Form of M and discuss the extent to which the Jordan form depends on the characteristic of the field $\mathbf{F}$.*

Problem 7.6.27 (Fa86) *Let $M_{2\times 2}$ denote the vector space of complex 2×2 matrices. Let*

$$A = \begin{pmatrix} 0 & 1 \\ 0 & 0 \end{pmatrix}$$

and let the linear transformation $T : M_{2\times 2} \to M_{2\times 2}$ be defined by $T(X) = XA - AX$. Find the Jordan Canonical Form for T.

Problem 7.6.28 (Sp01) *Let M_n be the vector space of $n \times n$ complex matrices. For A in M_n define the linear transformation of T_A on M_n by $T_A(X) = AX - XA$. Prove that the rank of T_A is at most $n^2 - n$.*

Problem 7.6.29 (Fa01) *Let A be an $n \times n$ matrix with real entries, let $\xi(t)$ denote its characteristic polynomial, and let $g(t) \in \mathbb{R}[t]$ be a polynomial of degree $n - 1$ dividing $\xi(t)$. What are the possibilities for the rank of $g(A)$?*

Problem 7.6.30 (Fa88) *Find the Jordan Canonical Form of the matrix*

$$\begin{pmatrix} 1 & 0 & 0 & 0 & 0 & 0 \\ 1 & 1 & 0 & 0 & 0 & 0 \\ 1 & 0 & 1 & 0 & 0 & 0 \\ 1 & 0 & 0 & 1 & 0 & 0 \\ 1 & 0 & 0 & 0 & 1 & 0 \\ 1 & 1 & 1 & 1 & 1 & 1 \end{pmatrix}.$$

Problem 7.6.31 (Sp02) *Let A and B be complex matrices of sizes 3×5 and 5×3, respectively, such that*

$$AB = \begin{pmatrix} 1 & 1 & 0 \\ 0 & 1 & 0 \\ 0 & 0 & -1 \end{pmatrix}$$

Find the Jordan canonical form of BA.

Problem 7.6.32 (Sp99) *Let $A = (a_{ij})$ be a $n \times n$ complex matrix such that $a_{ij} \neq 0$ if $i = j + 1$ but $a_{ij} = 0$ if $i \geq j + 2$. Prove that A cannot have more than one Jordan block for any eigenvalue.*

Problem 7.6.33 (Fa89) *Let A be a real, upper-triangular, $n \times n$ matrix that commutes with its transpose. Prove that A is diagonal.*

Problem 7.6.34 (Su78) *1. Prove that a linear operator $T : \mathbb{C}^n \to \mathbb{C}^n$ is diagonalizable if for all $\lambda \in \mathbb{C}$, $\ker(T - \lambda I)^n = \ker(T - \lambda I)$, where I is the $n \times n$ identity matrix.*

2. Show that T is diagonalizable if T commutes with its conjugate transpose T^* (i.e., $(T^*)_{jk} = \overline{T_{kj}}$).

Problem 7.6.35 (Fa79) *Let A be an $n\times n$ complex matrix. Prove there is a unitary matrix U such that $B = UAU^{-1}$ is upper-triangular: $B_{jk} = 0$ for $j > k$.*

Problem 7.6.36 (Sp81) *Let b be a real nonzero $n \times 1$ matrix (a column vector). Set $M = bb^t$ (an $n \times n$ matrix) where b^t denotes the transpose of b.*

1. Prove that there is an orthogonal matrix Q such that $QMQ^{-1} = D$ is diagonal, and find D.

2. Describe geometrically the linear transformation $M : \mathbb{R}^n \to \mathbb{R}^n$.

Problem 7.6.37 (Sp83) *Let M be an invertible real $n \times n$ matrix. Show that there is a decomposition $M = UT$ in which U is an $n \times n$ real orthogonal matrix and T is upper-triangular with positive diagonal entries. Is this decomposition unique?*

Problem 7.6.38 (Su85) *Let A be a nonsingular real $n\times n$ matrix. Prove that there exists a unique orthogonal matrix Q and a unique positive definite symmetric matrix B such that $A = QB$.*

Problem 7.6.39 (Fa03) *Let A be a 2 × 2 matrix with complex entries. Prove that the series $I + A + A^2 + \dots$ converges if and only if every eigenvalue of A has absolute value less than 1.*

Problem 7.6.40 (Sp95) *Let A be the 3×3 matrix*

$$\begin{pmatrix} 1 & -1 & 0 \\ -1 & 2 & -1 \\ 0 & -1 & 1 \end{pmatrix}$$

Determine all real numbers a for which the limit $\lim_{n\to\infty} a^n A^n$ exists and is nonzero (as a matrix).

Problem 7.6.41 (Fa96) *Suppose p is a prime. Show that every element of $GL_2(\mathbf{F}_p)$ has order dividing either $p^2 - 1$ or $p(p - 1)$.*

7.7   Similarity

Problem 7.7.1 (Fa80, Fa92) *Are the matrices give below similar ?*

$$A = \begin{pmatrix} 1 & 0 & 0 \\ -1 & 1 & 1 \\ -1 & 0 & 2 \end{pmatrix} \quad and \quad B = \begin{pmatrix} 1 & 1 & 0 \\ 0 & 1 & 0 \\ 0 & 0 & 2 \end{pmatrix}$$

Problem 7.7.2 (Fa78, Sp79) *Which of the following matrices are similar as matrices over $\mathbb{R}$?*

$$(a) \begin{pmatrix} 1 & 0 & 0 \\ 0 & 1 & 0 \\ 0 & 0 & 1 \end{pmatrix}, \quad (b) \begin{pmatrix} 0 & 0 & 1 \\ 0 & 1 & 0 \\ 1 & 0 & 0 \end{pmatrix}, \quad (c) \begin{pmatrix} 1 & 0 & 0 \\ 1 & 1 & 0 \\ 0 & 0 & 1 \end{pmatrix},$$

$$(d) \begin{pmatrix} 1 & 0 & 0 \\ 1 & 1 & 0 \\ 0 & 1 & 1 \end{pmatrix}, \quad (e) \begin{pmatrix} 1 & 1 & 0 \\ 0 & 1 & 0 \\ 0 & 0 & 1 \end{pmatrix}, \quad (f) \begin{pmatrix} 0 & 1 & 1 \\ 0 & 1 & 0 \\ 1 & 0 & 0 \end{pmatrix}.$$

Problem 7.7.3 (Sp00) *Are the 4×4 matrices*

$$A = \begin{pmatrix} 1 & 0 & 0 & 0 \\ 0 & -1 & 0 & 0 \\ 0 & 0 & 0 & 1 \\ 0 & 0 & 0 & 0 \end{pmatrix} \quad and \quad B = \begin{pmatrix} -1 & 0 & 0 & 0 \\ -1 & 1 & 1 & -1 \\ -1 & 0 & 0 & 0 \\ -1 & 0 & 1 & 0 \end{pmatrix}$$

similar?

Problem 7.7.4 (Sp79) *Let M be an n × n complex matrix. Let G_M be the set of complex numbers λ such that the matrix λM is similar to M.*

1. *What is G_M if*

$$M = \begin{pmatrix} 0 & 0 & 4 \\ 0 & 0 & 0 \\ 0 & 0 & 0 \end{pmatrix}?$$

2. *Assume M is not nilpotent. Prove G_M is finite.*

Problem 7.7.5 (Su80, Fa96) *Let A and B be real 2×2 matrices such that $A^2 = B^2 = I$ and $AB + BA = 0$. Prove there exists a real nonsingular matrix T with*

$$TAT^{-1} = \begin{pmatrix} 1 & 0 \\ 0 & -1 \end{pmatrix} \quad TBT^{-1} = \begin{pmatrix} 0 & 1 \\ 1 & 0 \end{pmatrix}.$$

Problem 7.7.6 (Su79, Fa82) *Let A and B be $n \times n$ matrices over a field **F** such that $A^2 = A$ and $B^2 = B$. Suppose that A and B have the same rank. Prove that A and B are similar.*

Problem 7.7.7 (Fa97) *Prove that if A is a 2×2 matrix over the integers such that $A^n = I$ for some strictly positive integer n, then $A^{12} = I$.*

Problem 7.7.8 (Su78) *Let G be a finite multiplicative group of 2×2 integer matrices.*

1. *Let $A \in G$. What can one prove about*

   (i) *det A?*

   (ii) *the (real or complex) eigenvalues of A?*

   (iii) *the Jordan or Rational Canonical Form of A?*

   (iv) *the order of A?*

2. *Find all such groups up to isomorphism.*

Note: See also Problem 7.6.10.

Problem 7.7.9 (Fa80) *Exhibit a set of 2×2 real matrices with the following property: A matrix A is similar to exactly one matrix in S provided A is a 2×2 invertible matrix of integers with all the roots of its characteristic polynomial on the unit circle.*

Problem 7.7.10 (Fa81, Su81, Sp84, Fa87, Fa95) *Let A and B be two real $n \times n$ matrices. Suppose there is a complex invertible $n \times n$ matrix U such that $A = UBU^{-1}$. Show that there is a real invertible $n \times n$ matrix V such that $A = VBV^{-1}$. (In other words, if two real matrices are similar over $\mathbb{C}$, then they are similar over $\mathbb{R}$.)*

Problem 7.7.11 (Sp91) *Let A be a linear transformation on an n-dimensional vector space over $\mathbb{C}$ with characteristic polynomial $(x - 1)^n$. Prove that A is similar to A^{-1}.*

Problem 7.7.12 (Sp94) *Prove or disprove: A square complex matrix, A, is similar to its transpose, A^t.*

Problem 7.7.13 (Sp79) *Let M be a real nonsingular 3×3 matrix. Prove there are real matrices S and U such that $M = SU = US$, all the eigenvalues of U equal 1, and S is diagonalizable over $\mathbb{C}$.*

Problem 7.7.14 (Sp77, Sp93, Fa94) *Find a list of real matrices, as long as possible, such that*

- *the characteristic polynomial of each matrix is $(x - 1)^5(x + 1)$,*

- *the minimal polynomial of each matrix is $(x - 1)^2(x + 1)$,*

- *no two matrices in the list are similar to each other.*

Problem 7.7.15 (Fa95) *Let A and B be nonsimilar $n \times n$ complex matrices with the same minimal and the same characteristic polynomial. Show that $n \geqslant 4$ and the minimal polynomial is not equal to the characteristic polynomial.*

Problem 7.7.16 (Sp98) *Let A be an $n \times n$ complex matrix with $\mathrm{tr}\, A = 0$. Show that A is similar to a matrix with all 0's along the main diagonal.*

Problem 7.7.17 (Sp99) *Let M be a 3×3 matrix with entries in the polynomial ring $\mathbb{R}[t]$ such that $M^3 = \begin{pmatrix} t & 0 & 0 \\ 0 & t & 0 \\ 0 & 0 & t \end{pmatrix}$. Let N be the matrix with real entries obtained by substituting $t = 0$ in M. Prove that N is similar to $\begin{pmatrix} 0 & 1 & 0 \\ 0 & 0 & 1 \\ 0 & 0 & 0 \end{pmatrix}$.*

Problem 7.7.18 (Sp99) *Let M be a square complex matrix, and let $S = \{XMX^{-1} \mid X \text{ is non-singular}\}$ be the set of all matrices similar to M. Show that M is a nonzero multiple of the identity matrix if and only if no matrix in S has a zero anywhere on its diagonal.*

Problem 7.7.19 (Sp00) *Let A be a complex $n \times n$ matrix such that the sequence $(A^n)_{n=1}^{\infty}$ converges to a matrix B. Prove that B is similar to a diagonal matrix with zeros and ones along the main diagonal.*

7.8 Bilinear, Quadratic Forms, and Inner Product Spaces

Problem 7.8.1 (Fa98) *Let $f : \mathbb{R}^n \to \mathbb{R}$ be a function such that*

1. *the function g defined by $g(x, y) = f(x + y) - f(x) - f(y)$ is bilinear,*

2. *for all $x \in \mathbb{R}^n$ and $t \in \mathbb{R}$, $f(tx) = t^2 f(x)$.*

Show that there is a linear transformation $A : \mathbb{R}^n \rightarrow \mathbb{R}^n$ such that $f(x) = \langle x, Ax \rangle$ where $\langle \cdot, \cdot \rangle$ is the usual inner product on $\mathbb{R}^n$ (in other words, f is a quadratic form).

Problem 7.8.2 (Sp98) *Let $A, B, \dots, F$ be real coefficients. Show that the quadratic form*

$$Ax^2 + 2Bxy + Cy^2 + 2Dxz + 2Eyz + Fz^2$$

is positive definite if and only if

$$A > 0, \qquad \begin{vmatrix} A & B \\ B & C \end{vmatrix} > 0, \qquad \begin{vmatrix} A & B & D \\ B & C & E \\ D & E & F \end{vmatrix} > 0.$$

Problem 7.8.3 (Fa00) *Find all real numbers t for which the quadratic form Q_t on $\mathbb{R}^3$, defined by*

$$Q_t(x_1, x_2, x_3) = 2x_1^2 + x_2^2 + 3x_3^2 + 2tx_1x_2 + 2x_1x_3,$$

is positive definite.

Problem 7.8.4 (Fa90) *Let $\mathbb{R}^3$ be 3-space with the usual inner product, and $(a, b, c) \in \mathbb{R}^3$ a vector of length 1. Let W be the plane defined by $ax + by + cz = 0$. Find, in the standard basis, the matrix representing the orthogonal projection of $\mathbb{R}^3$ onto W.*

Problem 7.8.5 (Fa93) *Let w be a positive continuous function on $[0, 1]$, n a positive integer, and P_n the vector space of real polynomials whose degrees are at most n, equipped with the inner product*

$$\langle p, q \rangle = \int_0^1 p(t)q(t)w(t)\, dt.$$

1. *Prove that P_n has an orthonormal basis $p_0, p_1, \dots, p_n$ (i.e., $\langle p_j, p_k \rangle = 1$ for $j = k$ and 0 for $j \neq k$) such that $\deg p_k = k$ for each k.*

2. *Prove that $\langle p_k, p_k' \rangle = 0$ for each k.*

Problem 7.8.6 (Sp98) *For continuous real valued functions f, g on the interval $[-1, 1]$ define the inner product $\langle f, g \rangle = \int_{-1}^1 f(x)g(x)dx$. Find that polynomial of the form $p(x) = a + bx^2 - x^4$ which is orthogonal on $[-1, 1]$ to all lower order polynomials.*

Problem 7.8.7 (Su80, Fa92) *Let E be a finite-dimensional vector space over a field $\mathbf{F}$. Suppose $B : E \times E \rightarrow \mathbf{F}$ is a bilinear map (not necessarily symmetric). Define subspaces*

$$E_1 = \{x \in E \mid B(x, y) = 0 \ for \ all \ y \in E\},$$

$$E_2 = \{y \in E \mid B(x, y) = 0 \ for \ all \ x \in E\}$$

Prove that $\dim E_1 = \dim E_2$.

Problem 7.8.8 (Su82) *Let A be a real n × n matrix such that $\langle Ax, x \rangle \geqslant 0$ for every real n-vector x. Show that $Au = 0$ if and only if $A^t u = 0$.*

Problem 7.8.9 (Fa85) *An n×n real matrix T is* positive definite *if T is symmetric and $\langle Tx, x \rangle > 0$ for all nonzero vectors $x \in \mathbb{R}^n$, where $\langle u, v \rangle$ is the standard inner product. Suppose that A and B are two positive definite real matrices.*

1. *Show that there is a basis $\{v_1, v_2, \ldots, v_n\}$ of $\mathbb{R}^n$ and real numbers $\lambda_1, \lambda_2, \ldots, \lambda_n$ such that, for $1 \leqslant i, j \leqslant n$:*

$$\langle Av_i, v_j \rangle = \begin{cases} 1 & i = j \\ 0 & i \neq j \end{cases}$$

    and

$$\langle Bv_i, v_j \rangle = \begin{cases} \lambda_i & i = j \\ 0 & i \neq j \end{cases}$$

2. *Deduce from Part 1 that there is an invertible real matrix U such that $U^t AU$ is the identity matrix and $U^t BU$ is diagonal.*

Problem 7.8.10 (Sp83) *Let V be a real vector space of dimension n, and let $S : V \times V \to \mathbb{R}$ be a nondegenerate bilinear form. Suppose that W is a linear subspace of V such that the restriction of S to $W \times W$ is identically 0. Show that $\dim W \leqslant n/2$.*

Problem 7.8.11 (Fa85) *Let A be the symmetric matrix*

$$\frac{1}{6} \begin{pmatrix} 13 & -5 & -2 \\ -5 & 13 & -2 \\ -2 & -2 & 10 \end{pmatrix}.$$

Denote by v the column vector

$$\begin{pmatrix} x \\ y \\ z \end{pmatrix}$$

in $\mathbb{R}^3$, and by x^t its transpose (x, y, z). Let $\|v\|$ denote the length of the vector v. As v ranges over the set of vectors for which $v^t Av = 1$, show that $\|v\|$ is bounded, and determine its least upper bound.

Problem 7.8.12 (Fa97) *Define the* index *of a real symmetric matrix A to be the number of strictly positive eigenvalues of A minus the number of strictly negative eigenvalues. Suppose A, and B are real symmetric n × n matrices such that $x^t Ax \leqslant x^t Bx$ for all n × 1 matrices x. Prove the the index of A is less than or equal to the index of B.*

Problem 7.8.13 (Fa78) *For $x, y \in \mathbb{C}^n$, let $\langle x, y \rangle$ be the Hermitian inner product $\sum_j x_j \bar{y}_j$. Let T be a linear operator on $\mathbb{C}^n$ such that $\langle Tx, Ty \rangle = 0$ if $\langle x, y \rangle = 0$. Prove that $T = kS$ for some scalar k and some operator S which is unitary: $\langle Sx, Sy \rangle = \langle x, y \rangle$ for all x and y.*

Problem 7.8.14 (Sp79) *Let E denote a finite-dimensional complex vector space with a Hermitian inner product $\langle x, y \rangle$.*

1. *Prove that E has an orthonormal basis.*

2. *Let $f : E \to \mathbb{C}$ be such that $f(x, y)$ is linear in x and conjugate linear in y. Show there is a linear map $A : E \to E$ such that $f(x, y) = \langle Ax, y \rangle$.*

Problem 7.8.15 (Fa86) *Let a and b be real numbers. Prove that there are two orthogonal unit vectors u and v in $\mathbb{R}^3$ such that $u = (u_1, u_2, a)$ and $v = (v_1, v_2, b)$ if and only if $a^2 + b^2 \leqslant 1$.*

7.9   General Theory of Matrices

Problem 7.9.1 (Fa01) *Prove that a commutative $\mathbb{C}$–algebra of 2×2 complex matrices has dimension at most 2 over $\mathbb{C}$.*

Problem 7.9.2 (Fa81) *Prove the following three statements about real $n \times n$ matrices.*

1. *If A is an orthogonal matrix whose eigenvalues are all different from -1, then $I + A$ is nonsingular and $S = (I - A)(I + A)^{-1}$ is skew-symmetric.*

2. *If S is a skew-symmetric matrix, then $A = (I - S)(I + S)^{-1}$ is an orthogonal matrix with no eigenvalue equal to -1.*

3. *The correspondence $A \leftrightarrow S$ from Parts 1 and 2 is one-to-one.*

Problem 7.9.3 (Fa79) *Let B denote the matrix*

$$\begin{pmatrix} a & 0 & 0 \\ 0 & b & 0 \\ 0 & 0 & c \end{pmatrix}$$

where a, b, and c are real and $|a|$, $|b|$, and $|c|$ are distinct. Show that there are exactly four symmetric matrices of the form BQ, where Q is a real orthogonal matrix of determinant 1.

Problem 7.9.4 (Sp79) *Let P be a $n \times n$ real matrix such that $x^t Py = -y^t Px$ for all column vectors x, y in $\mathbb{R}^n$. Prove that P is skew-symmetric.*

Problem 7.9.5 (Fa79) *Let A be a real skew-symmetric matrix $(A_{ij} = -A_{ji})$. Prove that A has even rank.*

Problem 7.9.6 (Fa98) *A real symmetric $n \times n$ matrix A is called positive semi-definite if $x^t Ax \geqslant 0$ for all $x \in \mathbb{R}^n$. Prove that A is positive semi-definite if and only if $\operatorname{tr} AB \geqslant 0$ for every real symmetric positive semi-definite $n \times n$ matrix B.*

Problem 7.9.7 (Fa80, Sp96) *Suppose that A and B are real matrices such that $A^t = A$,*

$$v^t A v \geqslant 0$$

for all $v \in \mathbb{R}^n$ and

$$AB + BA = 0.$$

Show that $AB = BA = 0$ and give an example where neither A nor B is zero.

Problem 7.9.8 (Fa02) *Let A and B be $n \times n$ matrices over $\mathbb{R}$ such that $A + B$ is invertible. Prove that*

$$A(A + B)^{-1} B = B(A + B)^{-1} A .$$

Problem 7.9.9 (Sp78) *Suppose A is a real $n \times n$ matrix.*

1. *Is it true that A must commute with its transpose?*

2. *Suppose the columns of A (considered as vectors) form an orthonormal set; is it true that the rows of A must also form an orthonormal set?*

Problem 7.9.10 (Sp98) *Let $M_1 = \left(\begin{smallmatrix} 3 & 2 \\ 1 & 4 \end{smallmatrix}\right)$, $M_2 = \left(\begin{smallmatrix} 5 & 7 \\ -3 & -4 \end{smallmatrix}\right)$, $M_3 = \left(\begin{smallmatrix} 5 & 6.9 \\ -3 & -4 \end{smallmatrix}\right)$. For which (if any) i, $1 \leqslant i \leqslant 3$, is the sequence (M_i^n) bounded away from ∞? For which i is the sequence bounded away from 0?*

Problem 7.9.11 (Su83) *Let A be an $n \times n$ complex matrix, all of whose eigenvalues are equal to 1. Suppose that the set $\{A^n \mid n = 1, 2, \ldots\}$ is bounded. Show that A is the identity matrix.*

Problem 7.9.12 (Fa81) *Consider the complex 3×3 matrix*

$$A = \begin{pmatrix} a_0 & a_1 & a_2 \\ a_2 & a_0 & a_1 \\ a_1 & a_2 & a_0 \end{pmatrix},$$

where $a_0, a_1, a_2 \in \mathbb{C}$.

1. *Show that $A = a_0 I_3 + a_1 E + a_2 E^2$, where*

$$E = \begin{pmatrix} 0 & 1 & 0 \\ 0 & 0 & 1 \\ 1 & 0 & 0 \end{pmatrix}.$$

2. *Use Part 1 to find the complex eigenvalues of A.*

3. *Generalize Parts 1 and 2 to $n \times n$ matrices.*

Problem 7.9.13 (Su78) *Let A be a $n \times n$ real matrix.*

1. *If the sum of each column element of A is 1 prove that there is a nonzero column vector x such that $Ax = x$.*

2. *Suppose that $n = 2$ and all entries in A are positive. Prove there is a nonzero column vector y and a number $\lambda > 0$ such that $Ay = \lambda y$.*

Problem 7.9.14 (Sp89) *Let the real $2n \times 2n$ matrix X have the form*

$$\begin{pmatrix} A & B \\ C & D \end{pmatrix}$$

where A, B, C, and D are $n \times n$ matrices that commute with one another. Prove that X is invertible if and only if $AD - BC$ is invertible.

Problem 7.9.15 (Sp03) *Let $GL_2(\mathbb{C})$ denote the group of invertible 2×2 matrices with coefficients in the field of complex numbers. Let $PGL_2(\mathbb{C})$ denote the quotient of $GL_2(\mathbb{C})$ by the normal subgroup $\left\{ \begin{pmatrix} \lambda & 0 \\ 0 & \lambda \end{pmatrix} \mid \lambda \in \mathbb{C}^* \right\}$. Let n be a positive integer, and suppose that a, b are elements of $PGL_2(\mathbb{C})$ of order exactly n. Prove that there exists $c \in PGL_2(\mathbb{C})$ such that cac^{-1} is a power of b.*

Problem 7.9.16 (Sp89) *Let $B = (b_{ij})_{i,j=1}^{20}$ be a real 20×20 matrix such that*

$$b_{ii} = 0 \quad for \quad 1 \leqslant i \leqslant 20,$$

$$b_{ij} \in \{1, -1\} \quad for \quad 1 \leqslant i, j \leqslant 20, \quad i \neq j.$$

Prove that B is nonsingular.

Problem 7.9.17 (Sp80) *Let*

$$A = \begin{pmatrix} 1 & 2 \\ 3 & 4 \end{pmatrix}.$$

Show that every real matrix B such that $AB = BA$ has the form $sI + tA$, where $s, t \in \mathbb{R}$.

Problem 7.9.18 (Su84) *Let A be a 2×2 matrix over $\mathbb{C}$ which is not a scalar multiple of the identity matrix I. Show that any 2×2 matrix X over $\mathbb{C}$ commuting with A has the form $X = \alpha I + \beta A$, where $\alpha, \beta \in \mathbb{C}$.*

Problem 7.9.19 (Sp02)   *1. Determine the commutant of the $n \times n$ Jordan matrix*

$$AB = \begin{pmatrix} \lambda & 1 & 0 & \cdots & 0 & 0 \\ 0 & \lambda & 1 & \cdots & 0 & 0 \\ 0 & 0 & \lambda & \cdots & 0 & 0 \\ \vdots & \vdots & \vdots & \ddots & \vdots & \vdots \\ 0 & 0 & 0 & \cdots & \lambda & 1 \\ 0 & 0 & 0 & \cdots & 0 & \lambda \end{pmatrix}$$

In particular, determine the dimension of the commutant as a complex vector space.

2. *What is the dimension of the commutant of the $2n \times 2n$ matrix*

$$J \oplus J = AB = \begin{pmatrix} J & 0 \\ 0 & J \end{pmatrix} ?$$

Problem 7.9.20 (Fa96) *Let*

$$A = \begin{pmatrix} 2 & -1 & 0 \\ -1 & 2 & -1 \\ 0 & -1 & 2 \end{pmatrix}.$$

Show that every real matrix B such that $AB = BA$ has the form

$$B = aI + bA + cA^2$$

for some real numbers a, b, and c.

Problem 7.9.21 (Sp77, Su82) *A square matrix A is nilpotent if $A^k = 0$ for some positive integer k.*

1. *If A and B are nilpotent, is $A + B$ nilpotent?*

2. *Prove: If A and B are nilpotent matrices and $AB = BA$, then $A + B$ is nilpotent.*

3. *Prove: If A is nilpotent then $I + A$ and $I - A$ are invertible.*

Problem 7.9.22 (Sp77) *Consider the family of square matrices $A(\theta)$ defined by the solution of the matrix differential equation*

$$\frac{dA(\theta)}{d\theta} = BA(\theta)$$

with the initial condition $A(0) = I$, where B is a constant square matrix.

1. *Find a property of B which is necessary and sufficient for $A(\theta)$ to be orthogonal for all θ; that is, $A(\theta)^t = A(\theta)^{-1}$, where $A(\theta)^t$ denotes the transpose of $A(\theta)$.*

2. *Find the matrices $A(\theta)$ corresponding to*

$$B = \begin{pmatrix} 0 & 1 \\ -1 & 0 \end{pmatrix}$$

and give a geometric interpretation.

Problem 7.9.23 (Su77) *Let A be an $r \times r$ matrix of real numbers. Prove that the infinite sum*

$$e^A = I + A + \frac{A^2}{2} + \cdots + \frac{A^n}{n!} + \cdots$$

of matrices converges (i.e., for each i, j, the sum of $(i, j)^{th}$ entries converges), and hence that e^A is a well-defined matrix.

Problem 7.9.24 (Sp97) *Show that*

$$\det(\exp(M)) = e^{\text{tr}(M)}$$

for any complex $n \times n$ matrix M, where $\exp(M)$ is defined as in Problem 7.9.23.

Problem 7.9.25 (Fa77) *Let T be an $n \times n$ complex matrix. Show that*

$$\lim_{k \to \infty} T^k = 0$$

if and only if all the eigenvalues of T have absolute value less than 1.

Problem 7.9.26 (Sp82) *Let A and B be $n \times n$ complex matrices. Prove that*

$$|\text{tr}(AB^*)|^2 \leqslant \text{tr}(AA^*)\text{tr}(BB^*).$$

Problem 7.9.27 (Fa84) *Let A and B be $n \times n$ real matrices, and k a positive integer. Find*

1.

$$\lim_{t \to 0} \frac{1}{t}\left((A + tB)^k - A^k\right).$$

2.

$$\frac{d}{dt}\text{tr}\,(A + tB)^k\bigg|_{t=0}.$$

Problem 7.9.28 (Fa91)   *1. Prove that any real $n \times n$ matrix M can be written as $M = A + S + cI$, where A is antisymmetric, S is symmetric, c is a scalar, I is the identity matrix, and $\text{tr}\,S = 0$.*

2. Prove that with the above notation,

$$\text{tr}(M^2) = \text{tr}(A^2) + \text{tr}(S^2) + \frac{1}{n}(\text{tr}\,M)^2.$$

Problem 7.9.29 (Fa99) *Let A be an $n \times n$ complex matrix such that $\text{tr}\,A^k = 0$ for $k = 1, \ldots, n$. Prove that A is nilpotent.*

Problem 7.9.30 (Sp98) *Let N be a nilpotent complex matrix. Let r be a positive integer. Show that there is a $n \times n$ complex matrix A with*

$$A^r = I + N.$$

Problem 7.9.31 (Fa94) *Let $A = (a_{ij})_{i,j=1}^n$ be a real $n \times n$ matrix such that $a_{ii} \geqslant 1$ for all i, and*

$$\sum_{i \neq j} a_{ij}^2 < 1.$$

Prove that A is invertible.

Problem 7.9.32 (Fa95) *Show that an $n \times n$ matrix of complex numbers A satisfying*

$$|a_{ii}| > \sum_{j \neq i} |a_{ij}|$$

for $1 \leqslant i \leqslant n$ must be invertible.

Problem 7.9.33 (Sp93) *Let $A = (a_{ij})$ be an $n \times n$ matrix such that $\sum_{j=1}^{n} |a_{ij}| < 1$ for each i. Prove that $I - A$ is invertible.*

Problem 7.9.34 (Sp94) *Let A be a real $n \times n$ matrix. Let M denote the maximum of the absolute values of the eigenvalues of A.*

1. *Prove that if A is symmetric, then $\|Ax\| \leqslant M\|x\|$ for all x in $\mathbb{R}^n$. (Here, $\|\cdot\|$ denotes the Euclidean norm.)*

2. *Prove that the preceding inequality can fail if A is not symmetric.*

Problem 7.9.35 (Sp00) *Let A be an $n \times n$ matrix over $\mathbb{C}$ whose minimal polynomial μ has degree k.*

1. *Prove that, if the point λ of $\mathbb{C}$ is not an eigenvalue of A, then there is a polynomial p_λ of degree $k - 1$ such that $p_\lambda(A) = (A - \lambda I)^{-1}$.*

2. *Let $\lambda_1, \ldots, \lambda_k$ be distinct points of $\mathbb{C}$ that are not eigenvalues of A. Prove that there are complex numbers $c_1, \ldots, c_k$ such that*

$$\sum_{j=1}^{k} c_j (A - \lambda_j I)^{-1} = I.$$

Problem 7.9.36 (Sp99) *Let $\|x\|$ denote the Euclidean norm of a vector x. Show that for any real $m \times n$ matrix M there is a unique non-negative scalar σ, and (possibly non-unique) unit vectors $u \in \mathbb{R}^n$ and $v \in \mathbb{R}^m$ such that*

1. *$\|Mx\| \leqslant \sigma\|x\|$ for all $x \in \mathbb{R}^n$,*

2. *$Mu = \sigma v$,*

3. *$M^T v = \sigma u$ (where M^T is the transpose of M).*

Part II

Solutions

1

Real Analysis

1.1 Elementary Calculus

Solution to 1.1.1: Let $f(\theta) = \cos p\theta - (\cos \theta)^p$. We have $f(0) = 0$ and, for $0 < \theta < \pi/2$,

$$f'(\theta) = -p \sin p\theta + p \cos^{p-1} \theta \sin \theta$$
$$= p\left(-\sin p\theta + \frac{\sin \theta}{\cos^{1-p} \theta}\right)$$
$$> 0$$

since sin is an increasing function on $[0, \pi/2]$ and $\cos^{1-p} \theta \in (0, 1)$. We conclude that $f(\theta) \geq 0$ for $0 \leq \theta \leq \pi/2$, which is equivalent to the inequality we wanted to establish.

Solution to 1.1.2: Let $x \in [0, 1]$. Using the fact that $f(0) = 0$ and the Cauchy–Schwarz Inequality [MH93, p. 69] we have,

$$|f(x)| = \left|\int_0^x f'(t)dt\right|$$
$$\leq \sqrt{\int_0^x |f'(t)|^2 dt} \sqrt{\int_0^x 1^2 dt}$$
$$\leq \sqrt{\int_0^x |f'(t)|^2 dt}$$

and the conclusion follows.

Solution to 1.1.3: As f' is positive, f is an increasing function, so we have, for $t > 1$, $f(t) > f(1) = 1$. Therefore, for $t > 1$,

$$f'(t) = \frac{1}{t^2 + f^2(t)} < \frac{1}{t^2 + 1},$$

so

$$f(x) = 1 + \int_1^x f'(t)dt$$

$$< 1 + \int_1^x \frac{1}{t^2 + 1}dt$$

$$< 1 + \int_1^\infty \frac{1}{t^2 + 1}dt$$

$$= 1 + \frac{\pi}{4};$$

hence, $\lim_{x \to \infty} f(x)$ exists and is, at most, $1 + \frac{\pi}{4}$. The strict inequality holds because

$$\lim_{x \to \infty} f(x) = 1 + \int_1^\infty f'(t)dt < 1 + \int_1^\infty \frac{1}{t^2 + 1}dt = 1 + \frac{\pi}{4}.$$

Solution to 1.1.4: Denote the common supremum of f and g by M. Since f and g are continuous and $[0, 1]$ is compact, there exist α, $\beta \in [0, 1]$ with $f(\alpha) = g(\beta) = M$. The function h defined by $h(x) = f(x) - g(x)$ satisfies $h(\alpha) = M - g(\alpha) \geq 0$, $h(\beta) = f(\beta) - M \leq 0$. Since h is continuous, it has a zero $t \in [\alpha, \beta]$. We have $f(t) = g(t)$, so $f(t)^2 + 3f(t) = g(t)^2 + 3g(t)$.

Solution to 1.1.5: Call a function of the desired form a periodic polynomial, and call its degree the largest k such that x^k occurs with a nonzero coefficient.

If a is 1-periodic, then $\Delta(af) = a\Delta f$ for any function f, so, by the Induction Principle [MH93, p. 7], $\Delta^n(af) = a\Delta^n f$ for all n.

We will use Complete Induction [MH93, p. 32]. For $n = 1$, the result holds: $\Delta f = 0$ if and only if f is 1-periodic. Assume it is true for $1, \ldots, n - 1$. If

$$f = a_0 + a_1 x + \cdots + a_{n-1} x^{n-1}$$

is a periodic polynomial of degree, at most, $n - 1$, then

$$\Delta^n f = a_1 \Delta^n x + \cdots + a_{n-1} \Delta x^{n-1}$$

and the induction hypothesis implies that all the terms vanish except, maybe, the last. We have $\Delta^n(x^{n-1}) = \Delta^{n-1}\Delta(x^{n-1})$, a polynomial of degree $n - 2$ by the Binomial Theorem [BML97, p. 15]. So the induction hypothesis also implies $\Delta^n(x^{n-1}) = 0$ and the first half of the statement is established.

For the other half, assume $\Delta^n f = 0$. By the induction hypothesis, Δf is a periodic polynomial of degree, at most, $n - 2$. Suppose we can find a periodic polynomial g, of degree, at most, $n - 1$, such that $\Delta g = \Delta f$. Then, as $\Delta(f - g) = 0$, the function $f - g$ will be 1-periodic, implying that f is a periodic polynomial of degree, at most, $n - 1$, as desired. Thus, it is enough to prove the following claim: *If h is a periodic polynomial of degree n ($n = 0, 1, \ldots$), then there is a periodic polynomial g of degree $n + 1$ such that $\Delta g = h$.*

If $n = 0$, we can take $g = hx$. Assume h has degree $n > 0$ and, as an induction hypothesis, that the claim is true for lower degrees than n. We can then, without loss of generality, assume $h = ax^n$, where a is 1-periodic. By the Binomial Theorem,

$$h - \Delta\left(\frac{ax^{n+1}}{n + 1}\right)$$

is a periodic polynomial of degree $n - 1$, so it equals Δg_1, for some periodic polynomial g_1 of degree n, and we have $h = \Delta g$, where

$$g = \frac{ax^{n+1}}{n + 1} + g_1$$

as desired.

Solution to 1.1.6: Since f is increasing and nonconstant, there exist $r_0 < s_0$ such that $f(r_0) < f(s_0)$. Increasing s_0 if necessary, we may assume $\delta = s_0 - r_0 > 1$. Let $\Delta = f(s_0) - f(r_0)$.

Choose $[r_1, s_1]$ to be either $[r_0, (r_0 + s_0)/2]$ or $[(r_0 + s_0)/2, s_0]$, whichever makes $f(s_1) - f(r_1)$ larger. (Choose either if they are equal.) Then $f(s_1) - f(r_1) \geqslant \Delta/2$. Similarly choose $[r_2, s_2]$ as the left or right half of $[r_1, s_1]$, to maximize $f(s_2) - f(r_2)$, and so on, so that $s_i - r_i = 2^{-i}\delta$, and $f(s_i) - f(r_i) \geqslant 2^{-i}\Delta$.

Let $a = \lim r_i$, and set $c = \Delta/(2\delta)$. Given $x \in (0, 1]$, choose the first integer $i \geqslant 0$ such that $2^{-i}\delta \leqslant x$. Then $2^{-i}\delta \geqslant x/2$. Also $[r_i, s_i]$ contains a and is of length $\leqslant x$, so $[r_i, s_i] \subseteq [a - x, a + x]$. Since f is increasing,

$$f(a + x) - f(a - x) \geqslant f(s_i) - f(r_i) \geqslant 2^{-i}\Delta = 2c2^{-i}\delta \geqslant cx.$$

Finally, the inequality $f(a + x) - f(a - x) \geqslant cx$ is immediate for $x = 0$.

Solution to 1.1.7:
1. This is false. For

$$f(t) = g(t) = \begin{cases} 0 & \text{for } t \neq 0 \\ 1 & \text{for } t = 0 \end{cases}$$

we have $\lim_{t\to 0} g(t) = \lim_{t\to 0} f(t) = 0$ but $\lim_{t\to 0} f(g(t)) = 1$.
2. This is false. $f(t) = t^2$ maps the open interval $(-1, 1)$ onto $[0, 1)$, which is not open.

3. This is true. Let $x, x_0 \in (-1, 1)$. By Taylor's Theorem [Rud87, p. 110], there is $\xi \in (-1, 1)$ such that

$$f(x) = \sum_{k=0}^{n-1} \frac{f^{(k)}(x_0)}{k!}(x - x_0)^k + \frac{f^{(n)}(\xi)}{n!}(x - x_0)^n \qquad (n \in \mathbb{N}).$$

We have

$$\lim_{n \to \infty} \left| \frac{f^{(n)}(\xi)}{n!}(x - x_0)^n \right| \leqslant \lim_{n \to \infty} \frac{|x - x_0|^n}{n!} = 0,$$

so

$$f(x) = \sum_{k=0}^{\infty} \frac{f^{(k)}(x_0)}{k!}(x - x_0)^k$$

for any $x_0 \in (-1, 1)$ and f is real analytic.

Solution to 1.1.8: Let $f(x) = \sin x - x + x^3/6$; then $f(0) = f'(0) = f''(0) = 0$ and $f'''(x) = 1 - \cos x$, so $f'''(x) \geqslant 0$ for all x, and $f'''(x) > 0$ for $0 < x < 2\pi$. Hence, for $x > 0$, $f''(x) = \int_0^x f'''(t)\, dt > 0$; similarly $f'(x) = \int_0^x f''(t)\, dt > 0$, and finally $f(x) = \int_0^x f'(t)\, dt > 0$.

Solution 2. The function $f(x) = \sin x - x + \frac{x^3}{6}$ has derivatives

$$f'(x) = \cos x - 1 + \frac{x^2}{2}$$
$$f''(x) = -\sin x + x$$
$$f'''(x) = -\cos x + 1.$$

Now $f'''(x) \geqslant 0$ with equality at the discrete set of points $2\pi n$, $n \in \mathbb{Z}$. Therefore f'' is strictly increasing on $\mathbb{R}$, and since $f''(0) = 0$, we obtain $f''(x) > 0$ for $x > 0$. Therefore f' is strictly increasing on $\mathbb{R}^+$; since $f'(0) = 0$, we obtain $f'(x) > 0$ for $x > 0$. Therefore f is strictly increasing on $\mathbb{R}^+$; since $f(0) = 0$ we obtain $f(x) > 0$ for $x > 0$, as required.

Solution 3. Consider the Taylor expansion of $\sin x$ around 0, of order 5, with the Lagrange remainder. For $x > 0$ we have

$$\sin x = x - \frac{x^3}{6} + \frac{x^5}{5!} + \frac{\cos^6 \xi}{6!}$$

for some $0 < \xi < x$. Since $\dfrac{x^5}{5!} > 0$ and $\dfrac{\cos^6 \xi}{6!} \geqslant 0$ we conclude that the result follows for positive x.

Solution to 1.1.9: The Maclaurin expansion of $\sin x$ of order 3 is

$$\sin x = x - \frac{x^3}{6} + O(x^5), \qquad (x \to 0),$$

therefore

$$\sin^2 x = x^2 + O(x^4) \qquad (x \to 0).$$

So, for h near 0, we have $y(h) = 1 - 8\pi^2 h^2 + O(h^4)$. (Alternatively, observe that $y(h) = \cos 4\pi ih$ and use the expansion for cosine.) Thus,

$$f(y(h)) = \frac{2}{1 + \sqrt{8\pi^2 h^2 + O(h^4)}} = \frac{2}{1 + 2\sqrt{2}\pi |h| + O(h^2)} \qquad (h \to 0).$$

Using the Maclaurin expansion $\dfrac{2}{1+x} = 2 - 2x + 2x^2 + O(x^2)$, we get

$$f(y(h)) = 2 - 4\sqrt{2}\pi |h| + O(h^2), \qquad (h \to 0).$$

Solution to 1.1.10: 1. Suppose $f : [0, 1] \to (0, 1)$ is a continuous surjection. Consider the sequence (x_n) such that $x_n \in f^{-1}((0, 1/n))$. By the Bolzano–Weierstrass Theorem [Rud87, p. 40], [MH93, p. 153], we may assume that (x_n) converges, to $x \in [0, 1]$, say. By continuity, we have $f(x) = 0$, which is absurd. Therefore, no such a function can exist.

2. $g(x) = |\sin 2\pi x|$.

3. Suppose $g : (0, 1) \to [0, 1]$ is a continuous bijection. Let $x_0 = g^{-1}(0)$ and $x_1 = g^{-1}(1)$. Without loss of generality, assume $x_0 < x_1$ (otherwise consider $1 - g$). By the Intermediate Value Theorem [Rud87, p. 93], we have $g([x_0, x_1]) = [0, 1]$. As $x_0, x_1 \in (0, 1)$, g is not an injection, which contradicts our assumption.

Solution to 1.1.11: Let $A(c)$ denote the area the problem refers. The condition on f'' implies the convexity of f, so the graph of f is always above any tangent to it, and we have

$$A(c) = \int_a^b \left(f(x) - f(c) - f'(c)(x - c) \right) \, dx \, .$$

The derivative of A is given by

$$A'(c) = -\int_a^b f''(c)(x - c) \, dx$$

$$= -f''(c)\frac{b^2 - a^2}{2} + (b - a)cf'(c)$$

$$= f''(c)(b - a)\left(c - \frac{a+b}{2} \right)$$

so the minimum occurs at $c = (a + b)/2$. As A' is an increasing function, A is convex, so its minimum in $[a, b]$ corresponds to the only critical point in (a, b).

Solution to 1.1.12: Using the parameterization

$$x = a \cos t, \quad y = b \sin t,$$

a triple of points on the ellipse is given by

$$(a \cos t_i, b \sin t_i), \quad i = 1, 2, 3.$$

So the area of an inscribed triangle is given by

$$\frac{1}{2} \begin{vmatrix} 1 & a \cos t_1 & b \sin t_1 \\ 1 & a \cos t_2 & b \sin t_2 \\ 1 & a \cos t_3 & b \sin t_3 \end{vmatrix} = \frac{ab}{2} \begin{vmatrix} 1 & \cos t_1 & \sin t_1 \\ 1 & \cos t_2 & \sin t_2 \\ 1 & \cos t_3 & \sin t_3 \end{vmatrix}$$

which is ab times the area of a triangle inscribed in the unit circle. In the case of the circle, among all inscribed triangles with a given base $2w$ ($0 < w \leqslant 1$), the one of maximum area is an isosceles triangle whose area equals

$$g(w) = w(1 + \sqrt{1 - w^2}).$$

Using elementary calculus one finds that the maximum of g on the interval $0 \leqslant w \leqslant 1$ occurs at $w = \sqrt{3}/2$, corresponding to an equilateral triangle, and equals $3\sqrt{3}/4$. Alternatively, fixing one side of the triangle as the basis, we easily see that among all the inscribed triangles the one with the greatest area is isosceles because of the maximum height, showing that the angle at the basis is the same. Fixing another side we see that the triangle is indeed equilateral. Hence, the area is maximal when

$$t_2 = t_1 + \frac{2\pi}{3} \quad \text{and} \quad t_3 = t_2 + \frac{2\pi}{3}$$

that is, when the corresponding triangle inscribed in the unit circle is regular.

For the ellipse with semiaxes a, b, this corresponds to an inscribed triangle with maximum area equals $3ab\sqrt{3}/4$.

Solution 2. Let $f : \mathbb{R}^2 \to \mathbb{R}^2$ be the stretch function $f(x, y) = (ax, by)$. By a well know lemma in the proof of the Change of Variable for Integration [MH93, p. 524, Lemma 1], or the theorem itself when proven in its whole generality,

$$\text{vol}(f(T)) = |\det f| \cdot \text{vol } A$$

since the determinant of f is constant, following on the steps of the previous proof, the maximum for the area is achieved over the image of an equilateral triangle and it is equal to

$$\text{vol}(f(T)) = ab \cdot \text{vol}(T) = 3ab\frac{\sqrt{3}}{4}.$$

Solution to 1.1.13: Assume that a and b are in A and that $a < b$. Suppose $a < c < b$. Let (x_n) and (y_n) be sequences in $[0, \infty)$ tending to $+\infty$ such that $a = \lim_{n \to \infty} f(x_n)$ and $b = \lim_{n \to \infty} f(y_n)$. Deleting finitely many terms from each sequence, if necessary, we can assume $f(x_n) < c$ and $f(y_n) > c$ for every n. Then, by the Intermediate Value Theorem [Rud87, p. 93], there is for each n a point z_n between x_n and y_n such that $f(z_n) = c$. Since obviously $\lim_{n \to \infty} z_n = +\infty$, it follows that c is in A, as desired.

Solution to 1.1.14: For each $x > 0$, the equation $x(1 + \log(1/\varepsilon \sqrt{x})) = 1$ may be solved for ε to obtain $\varepsilon = e/(x^{1/2}e^{1/x})$ that we will call $f(x)$.

For $x > 0$ define $g(x) = x^{1/2}e^{1/x} = e/f(x)$. Then $g'(x) = \frac{1}{2}x^{-3/2}e^{1/x}(x-2)$. Thus g is strictly decreasing on $(0, 2]$ and strictly increasing on $[2, \infty)$. Moreover $\lim_{x \to 0+} g(x) = \infty = \lim_{x \to \infty} g(x)$, and $g(2) > 0$. Thus $\lim_{x \to 0+} f(x) = 0 = \lim_{x \to \infty} f(x)$, f is strictly increasing on $(0, 2]$ and strictly decreasing on $[2, \infty)$, f is continuous on $(0, \infty)$, and $f(2) = \sqrt{e}/2$. Denote the restrictions of f to $(0, 2)$ and $(2, \infty)$ by f_1, f_2 respectively. Then for each $0 < \varepsilon < \sqrt{e}/2$, the given equation has exactly two solutions, namely $x = f_1^{-1}(\varepsilon)$, $f_2^{-1}(\varepsilon)$. The smaller solution is then $x = f_1^{-1}(\varepsilon)$. As f_1 is strictly increasing and continuous, with $f_1(0+) = 0$, we deduce that $x(\varepsilon) \to 0$ as $\varepsilon \to 0+$.

Fix $s > 0$. Now $\varepsilon^{-s}x(\varepsilon) = (e/x^{1/2}e^{1/x})^{-s}x = e^{-s}x^{1+s/2}e^{s/x}$, where $x = x(\varepsilon) = f_1^{-1}(\varepsilon)$. As $\varepsilon \to 0+$, $x \to 0+$. The function $x \mapsto e^{-s}x^{1+(s/2)}e^{s/x}$ has limit ∞ as $x \to 0+$. This proves the second part.

Solution to 1.1.15: Let g be a polynomial,

$$g(x) = a_0 + a_1(x - a) + a_2(x - a)^2 + \cdots + a_n(x - a)^n.$$

If we take

$$a_0 = \frac{1}{f(a)}, \qquad a_1 = -\frac{f'(a)g(a)}{f(a)}, \qquad a_2 = -\frac{f''(a)g(a) + f'(a)g'(a)}{f(a)},$$

a calculation shows that the requirements on g are met.

Solution to 1.1.16: Suppose that all the roots of p are real and let $\deg p = n$. We have $p(z) = (z - r_1)^{n_1}(z - r_2)^{n_2} \cdots (z - r_k)^{n_k}$, where $r_1 < r_2 < \cdots < r_k$ and $\sum n_i = n$. By differentiating this expression, we see that the r_i's are roots of p' of order $n_i - 1$ when $n_i > 1$. Summing these orders, we see that we have accounted for $n - k$ of the possible $n - 1$ roots of p'. Now by Rolle's Theorem [MH93, p. 200], for each i, $1 \leqslant i \leqslant k - 1$, there is a point s_i, $r_i < s_i < r_{i+1}$, such that $p'(s_i) = 0$. Thus, we have found the remaining $k - 1$ roots of p', and they are distinct. Now we know that a is a root of p' but not of p, so $a \neq r_i$ for all i. But a is a root of p'', so a is a multiple root of p'; hence, $a \neq s_i$ for all i. Therefore, a is not a root of p', a contradiction.

Solution to 1.1.17: Let $x \in \mathbb{R}$ and $h > 0$. By the Taylor's Theorem [Rud87, pag. 110], there is a $w \in (x, x + 2h)$ such that

$$f(x + 2h) = f(x) + 2hf'(x) + 2h^2 f''(w),$$

or rewriting

$$f'(x) = \frac{f(x + 2h) - f(x)}{2h} - hf''(w).$$

Taking absolute values and applying our hypotheses, we get

$$|f'(x)| \leqslant \frac{A}{h} + hB.$$

Now making $h = \sqrt{\frac{A}{B}}$ the conclusion follows:

$$|f'(x)| \leqslant 2\sqrt{AB},$$

Solution to 1.1.18: Consider the function $f(x) = \log x / x$. We have $a^b = b^a$ iff $f(a) = f(b)$. Now $f'(x) = (1 - \log x)/x^2$, so f is increasing for $x < e$ and decreasing for $x > e$. For the above equality to hold, we must have $0 < a < e$, so a is either 1 or 2, and $b > e$. For $a = 1$, clearly there are no solutions, and for $a = 2$ and $b = 4$ works; since f is decreasing for $x > e$, this is the only solution.

Solution 2. Clearly, a and b have the same prime factors. Let $b = a + t$, for some positive integer t. Then $a^a a^t = b^a$ which implies that $a^a | b^a$, therefore $a | b$. We have then $b = ka$, for some integer $k > 1$. Now $b^a = (ka)^a = a^b$ implies that k is a power of a, so $b = a^m$ for some $m > 1$. Now $b^a = a^{ma} = a^{a^m}$ exactly when $ma = a^m$, which can easily be seen to have the unique solution $a = m = 2$. So $a = 2$ and $b = 2^2 = 4$.

Solution 3. Let $b = a(1 + t)$, for some positive t. Then the equation $a^b = b^a$ is equivalent to any of the following

$$a^{a(1+t)} = (a(1 + t))^a$$
$$\left(a^a\right)^{1+t} = a^a(1 + t)^a$$
$$\left(a^a\right)^t = (1 + t)^a$$
$$a^t = 1 + t.$$

We have, by the power series expansion of the exponential function, that $e^t > 1 + t$ for positive t, so $a < e$. As $a = 1$ is impossible, we conclude $a = 2$. The original equation now becomes

$$2^b = b^2$$

which, considering the prime decomposition of b, clearly implies $b = 4$.

Solution to 1.1.19: The equation can be rewritten as $a^{x^b} = x$, or

$$\frac{\log x}{x^b} = \log a.$$

There is thus a solution for x if and only if $\log a$ is in the range of $x \mapsto (\log x)/x^b$. Using elementary calculus, we get that the range of this function is $(-\infty, 1/be]$. We conclude then that the original equation has a positive solution for x if and only if $\log a \leqslant 1/be$, that is, if and only if $1 < a < e^{1/be}$.

Solution to 1.1.20: Let $f(x) = 3^x x^{-3}$ for $x > 0$. We have

$$f'(x) = \frac{3^x(x \log 3 - 3)}{x^4} > 0 \quad \text{for} \quad x > \frac{3}{\log 3}.$$

As $3/\log 3 < 3 < \pi$, we have $f(3) = 1 < f(\pi) = 3^\pi/\pi^3$, that is, $\pi^3 < 3^\pi$.

Solution 2. The same kind of analysis can be performed to the function $f(x) = \ln(x)/x$, which is decreasing for $x > e$, as well as others like $g(x) = x^3 - 3^x$ and $h(x) = (3 + x)^{(\pi - x)}$.

Solution to 1.1.21: Fix a in $(1, \infty)$, and consider the function $f(x) = a^x x^{-a}$ on $(1, \infty)$, which we try to minimize. Since $\log f(x) = x \log a - a \log x$, we have

$$\frac{f'(x)}{f(x)} = \log a - \frac{a}{x},$$

showing that $f'(x)$ is negative on $(1, \frac{a}{\log a})$ and positive on $(\frac{a}{\log a}, \infty)$. Hence, f attains its minimum on $(1, \infty)$ at the point $x_a = \frac{a}{\log a}$, and

$$\log f(x_a) = a - a \log\left(\frac{a}{\log a}\right) = a \log\left(\frac{e \log a}{a}\right).$$

The number a thus has the required property if and only if $\frac{e \log a}{a} \geq 1$. To see which numbers a in $(1, \infty)$ satisfy this condition, we consider the function $g(y) = \frac{\log y}{y}$ on $(1, \infty)$. We have

$$g'(y) = \frac{1 - \log y}{y^2},$$

from which we conclude that g attains its maximum on $(1, \infty)$ at $y = e$, the maximum value being $g(e) = \frac{1}{e}$. Since $g(y) < \frac{1}{e}$ on $(1, \infty)\setminus\{e\}$, we conclude that $\frac{e \log a}{a} < 1$ for a in $(1, \infty)$, except for $a = e$. The number $a = e$ is thus the only number in $(1, \infty)$ with the required property.

Solution to 1.1.22: Let $g(x) = e^x/x^t$ for $x > 0$. Since $g(x) \to \infty$ as $x \to 0$ and as $x \to \infty$, there must be a minimum value in between. At the minimum,

$$g'(x) = e^x x^{-t}(1 - t/x) = 0,$$

so the minimum must occur at $x = t$, where

$$g(x) = g(t) = e^t/t^t = (e/t)^t.$$

Thus,

$$e^x \geq \left(\frac{xe}{t}\right)^t$$

and the right-hand side is strictly larger than x^t if and only if $t < e$.

Solution to 1.1.23: f can be written as its second degree Maclaurin polynomial [PMJ85, p. 127] on this interval:

$$f(x) = f(0) + f'(0)x + \frac{f''(0)}{2}x^2 + \frac{f^{(3)}(\xi)}{6}x^3$$

where ξ is between 0 and x. Letting $x = \pm\frac{1}{n}$ in this formula and combining the results, we get, for $n \geqslant 1$,

$$n\left(f\left(\frac{1}{n}\right) - f\left(-\frac{1}{n}\right)\right) - 2f'(0) = n\left(\frac{2f'(0)}{n} + \frac{f^{(3)}(\alpha_n)}{6n^3} + \frac{f^{(3)}(\beta_n)}{6n^3}\right) - 2f'(0)$$

$$= \frac{f^{(3)}(\alpha_n) + f^{(3)}(\beta_n)}{6n^2}$$

for some $\alpha_n, \beta_n \in [-1, 1]$. As f''' is continuous, there is some $M > 0$ such that $|f'''(x)| < M$ for all $x \in [-1, 1]$. Hence,

$$\left|\sum_{n=1}^{\infty} n\left(f\left(\frac{1}{n}\right) - f\left(-\frac{1}{n}\right)\right) - 2f'(0)\right| \leqslant \frac{M}{3}\sum_{n=1}^{\infty}\frac{1}{n^2} < \infty.$$

Solution to 1.1.24: Using Taylor's Theorem [Rud87, p. 110],

$$f(x + h) - f(x) = f'(x)h + \frac{f''(z)}{2}h^2 \quad \text{for some} \quad z \in (x, x + h)$$

and similarly

$$f(x - h) - f(x) = -f'(x)h + \frac{f''(w)}{2}h^2 \quad \text{for some} \quad w \in (x - h, x).$$

The result follows by adding the two expressions, dividing by h^2, and taking the limit when $h \to 0$.

Solution to 1.1.25: 1. The geometric construction

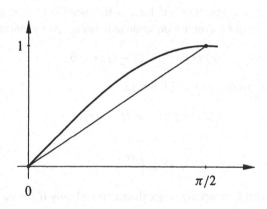

shows that

$$\sin\theta \geqslant \frac{2}{\pi}\theta \quad \text{for} \quad 0 \leqslant \theta \leqslant \frac{\pi}{2}.$$

An analytic proof can be written down from the fact that the sin function is concave down (second derivative negative) in the interval $0 \leqslant \theta \leqslant \frac{\pi}{2}$.

It can also be seen from the following geometric construction due to József Sándor [Sán88] and later rediscovered by Feng Yuefeng [Yue96]:

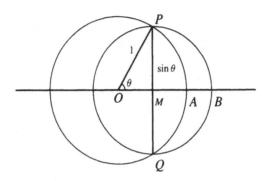

$$OB = OM + MP \geqslant OA \implies \overset{\frown}{PBQ} \geqslant \overset{\frown}{PAQ}$$
$$\implies \pi \sin\theta \geqslant 2\theta$$
$$\implies \sin\theta \geqslant \frac{2\theta}{\pi}.$$

2. The integral inequality

$$\begin{aligned}
J &= \int_0^{\pi/2} e^{-R\sin\theta} R\, d\theta \\
&\leqslant \int_0^{\pi/2} e^{-2R\theta/\pi} R\, d\theta \\
&= -\pi e^{-2R\theta/\pi} \Big|_0^{\pi/2} \\
&< \pi
\end{aligned}$$

is called Jordan's Lemma [MH87, p. 301]. Our limit is then

$$\lim_{R\to\infty} R^\lambda \int_0^{\pi/2} e^{-R\sin\theta}\, d\theta = \lim_{R\to\infty} R^{\lambda-1} \int_0^{\pi/2} e^{-R\sin\theta} R\, d\theta$$
$$< \lim_{R\to\infty} R^{\lambda-1}\pi = 0.$$

Solution 2. We have

$$R^\lambda \int_0^{\frac{\pi}{2}} e^{-R\sin\theta}\, d\theta = R^\lambda \int_0^{\frac{\pi}{3}} e^{-R\sin\theta}\, d\theta + R^\lambda \int_{\frac{\pi}{3}}^{\frac{\pi}{2}} e^{-R\sin\theta}\, d\theta.$$

As $\cos\theta \geqslant 1/2$ for $0 \leqslant \theta \leqslant \pi/3$, and $\sin\theta$ is an increasing function on $[0, \pi/2]$, we have

$$R^\lambda \int_0^{\frac{\pi}{2}} e^{-R\sin\theta}\, d\theta \leqslant 2R^\lambda \int_0^{\frac{\pi}{3}} e^{-R\sin\theta} \cos\theta\, d\theta + R^\lambda \int_{\frac{\pi}{3}}^{\frac{\pi}{2}} e^{-R\sin(\pi/3)}\, d\theta$$

$$= 2R^{\lambda-1}\left(1 - e^{-R\sin(\pi/3)}\right) + \frac{R^\lambda \pi}{6} e^{-R\sin(\pi/3)}$$

$$= o(1) \quad (R \to \infty)$$

Solution to 1.1.26: Let $T = \mathbb{R} \setminus S$, T is dense in $\mathbb{R}$ because each nonempty interval contains uncountably many numbers.

Fix $p \in T$ and define $F : \mathbb{R} \to \mathbb{R}$ by

$$F(x) = \int_p^x f(t)dt.$$

F vanishes on T, so, as it is continuous, F vanishes on $\mathbb{R}$. Therefore, we have $F' = f \equiv 0$.

Solution to 1.1.27: $f'(c) = 0$ for some $c \in (0, 1)$, by Rolle's Theorem [MH93, p. 200]. The concavity of f shows that f is increasing on $(0, c)$ and decreasing on $(c, 1)$. The arc length of the graph of f on $[0, c]$ is

$$L_{(0,c)} = \int_0^c \sqrt{1 + f'(x)^2}\, dx = \lim_{n\to\infty} \frac{c}{n} \sum_{k=0}^{n-1} \sqrt{1 + f'(\xi_k)^2}$$

where $\xi_k \in (kc/n, (k+1)c/n)$. By the Mean Value Theorem [Rud87, p. 108] we can assume the ξ_k's satisfy

$$f'(\xi_k) = \frac{f\left((k+1)c/n\right) - f(kc/n)}{c/n}.$$

We get

$$L_{(0,c)} = \lim_{n\to\infty} \sum_{k=0}^{n-1} \sqrt{(c/n)^2 + (f\left((k+1)c/n\right) - f(kc/n))^2}$$

$$\leqslant \lim_{n\to\infty} \sum_{k=0}^{n-1} (c/n) + (f\left((k+1)c/n\right) - f(kc/n))$$

$$= c + f(c)$$

since f is increasing. A similar reasoning shows that $L_{(c,1)} \leqslant 1 - c + f(c)$. So $L_{[0,1]} \leqslant c + f(c) + 1 - c + f(c) \leqslant 3$.

Solution to 1.1.28: The convergence of $\int_1^\infty |f'(x)|dx$ implies the convergence of $\int_1^\infty f'(x)dx$, which implies that $\lim_{x\to\infty} f(x)$ exists. If that limit is not 0, then $\sum_{n=1}^{\infty} f(n)$ and $\int_{1}^{\infty} f(x)dx$ both diverge. We may therefore, assume that $\lim_{x \to \infty} f(x) = 0$. Then $\int_{\lfloor r \rfloor}^{r} f(x)dx \to 0$ as $r \to \infty$ (where $\lfloor r \rfloor$ is the greatest integer $\leqslant$ r), implying that $\int_{1}^{\infty} f(x)dx$ converges if and only if $\lim_{n \to \infty} \int_{1}^{n} f(x)dx$ exists (where n here tends to ∞ through integer values). In other words, the convergence of $\int_{1}^{\infty} f(x)dx$ is equivalent to the convergence of $\sum_{n=1}^{\infty} \int_{n}^{n+1} f(x)dx$. It will therefore, suffice to prove that

$$\sum_{n=1}^{\infty} \left| \int_{n}^{n+1} f(x)dx - f(n) \right| < \infty.$$

We have

$$\left| \int_{n}^{n+1} f(x)dx - f(n) \right| = \left| \int_{n}^{n+1} (f(x) - f(n))dx \right|$$

$$= \left| \int_{n}^{n+1} \int_{n}^{x} f'(t)dt\,dx \right| \leqslant \int_{n}^{n+1} \int_{n}^{n+1} |f'(t)|dt\,dx$$

$$= \int_{n}^{n+1} |f'(t)|dt .$$

Hence, $\sum_{n=1}^{\infty} \left| \int_{n}^{n+1} f(x)dx - f(n) \right| \leqslant \int_{1}^{\infty} |f'(t)|dt < \infty$, as desired.

Solution to 1.1.29: We have

$$|u(x)| = |u(x) - u(0)| \leqslant |x|$$

and

$$|u^2(x) - u(x)| = |u(x)||u(x) - 1| \leqslant |x|(|x| + 1)$$

so

$$|\varphi(u)| \leqslant \int_{0}^{1} |u(x)^2 - u(x)| \leqslant \int_{0}^{1} x(x + 1)dx = \frac{5}{6}.$$

Equality can be achieved if $|u(x)| = x$ and $|u(x) - 1| = x + 1$. This is the case for $u(x) = -x$ which is in E.

Solution to 1.1.31: Let

$$u(t) = 1 + 2 \int_{0}^{t} f(s)ds .$$

We have

$$u'(t) = 2f(t) \leqslant 2\sqrt{u(t)},$$

so

$$\sqrt{u(t)} - 1 = \int_{0}^{t} \frac{u'(s)}{2\sqrt{u(s)}} ds \leqslant \int_{0}^{t} ds = t;$$

therefore,

$$f(t) \leqslant \sqrt{u(t)} \leqslant 1 + t.$$

Solution to 1.1.32: We will show b must be zero. By subtracting and multiplying by constants, we can assume $a = 0 \leqslant b$. Given $\varepsilon > 0$, choose $R \geqslant 1$ such that

$$|\varphi(x)| \leqslant \varepsilon$$

and

$$\varphi'(x) \geqslant b/2 \geqslant 0$$

for all $x \geqslant R$. By the Fundamental Theorem of Calculus [MH93, p. 209],

$$\varphi(x) = \varphi(R) + \int_R^x \varphi'(x)dx,$$

so

$$2\varepsilon \geqslant \varphi(x) - \varphi(R) \geqslant \int_R^x \frac{b}{2}dx = (x - R)b/2.$$

For $x = 5R$, we get

$$b \leqslant \varepsilon/R \leqslant \varepsilon.$$

Since $\varepsilon > 0$ was arbitrary, we must have $b = 0$.

Solution to 1.1.33: Let $0 \leqslant k_1 < k_2 < 1$, then for all $x \in (0, \pi/2)$,

$$-k_1 \cos^2 x > -k_2 \cos^2 x$$
$$\sqrt{1 - k_1 \cos^2 x} > \sqrt{1 - k_2 \cos^2 x}$$
$$\frac{1}{\sqrt{1 - k_1 \cos^2 x}} < \frac{1}{\sqrt{1 - k_2 \cos^2 x}}$$
$$\int_0^{\pi/2} \frac{1}{\sqrt{1 - k_1 \cos^2 x}} dx < \int_0^{\pi/2} \frac{1}{\sqrt{1 - k_2 \cos^2 x}} dx.$$

Solution to 1.1.34: With the change of variables $y = x\sqrt{t}$, we have

$$f(t) = \int_{-\infty}^{\infty} e^{-tx^2} dx = \int_{-\infty}^{\infty} e^{-y^2} \frac{dy}{\sqrt{t}} = \frac{1}{\sqrt{t}} \int_{-\infty}^{\infty} e^{-y^2} dy = \sqrt{\frac{\pi}{t}},$$

so

$$f'(t) = -\frac{\sqrt{\pi}}{2} t^{-3/2}.$$

Solution to 1.1.35: Let

$$G(u, v, x) = \int_v^u e^{t^2 + xt} dt.$$

Then $F(x) = G(\cos x, \sin x, x)$, so

$$F'(x) = \frac{\partial G}{\partial u}\frac{\partial u}{\partial x} + \frac{\partial G}{\partial v}\frac{\partial v}{\partial x} + \frac{\partial G}{\partial x}$$

$$= e^{u^2+xu}(-\sin x) - e^{(v^2+xv)}\cos x + \int_v^u te^{t^2+xt}\,dt$$

and

$$F'(0) = -1 + \int_0^1 te^{t^2}\,dt = \frac{1}{2}(e - 3).$$

Solution to 1.1.36: 1. Let $f(z) \neq 0$. Then

$$f(x)f(z) = f\left(\sqrt{x^2 + z^2}\right) = f(-x)f(z),$$

so $f(x) = f(-x)$ and f is even.

Also, $f(0)f(z) = f(z)$, so $f(0) = 1$.

2. We will show now that $f(\sqrt{n}x) = (f(x))^n$ for real x and natural n, using the Induction Principle [MH93, p. 7]. The result is clear for $n = 1$. Assume it holds for $n = k$. We have

$$f\left(\sqrt{k+1}x\right) = f\left(\sqrt{\left(\sqrt{k}x\right)^2 + x^2}\right)$$

$$= f\left(\sqrt{k}x\right)f(x)$$

$$= (f(x))^k f(x)$$

$$= (f(x))^{k+1}.$$

If $p, q \in \mathbb{N}$, then

$$f(p) = f\left(p^2 \cdot 1\right) = (f(1))^{p^2}$$

and

$$f(|p|) = f\left(\sqrt{p^2}\left|\frac{p}{q}\right|\right) = \left(f\left(\left|\frac{p}{q}\right|\right)\right)^{q^2}$$

from which follows

$$\left(f\left(\frac{p}{q}\right)\right)^{q^2} = (f(1))^{p^2}.$$

- If $f(1) > 0$, we have

$$f\left(\frac{p}{q}\right) = (f(1))^{\frac{p^2}{q^2}},$$

so, by continuity on $\mathbb{R}$,

$$f(x) = (f(1))^{x^2}.$$

- If $f(1) = 0$, then f vanishes on a dense set, so it vanishes everywhere, contradicting the hypothesis.

- To see that $f(1) < 0$ cannot happen, consider p even and q odd. We get $f(p/q) > 0$, so f is positive on a dense set, and $f(1) \geqslant 0$.

Note that we used only the continuity of f and its functional equation.

Differentiating, we easily check that f satisfies the differential equation. The most general function satisfying all the conditions is then

$$c^{x^2}$$

with $0 < c < 1$.

Solution 2. 1. Let $x = y = 0$. Then $f(0)^2 = f(0)$, so $f(0) = 0$ or 1. If $f(0) = 0$, then $0 = f(\sqrt{x^2})$ for any x, so, in fact, $f(x) = 0$ for all $x > 0$. If $f(y) \neq 0$ for any y, then $f(x)f(y) = 0$ implies $f(x) = 0$ for all x, so $f(x) = 0$ for all x if $f(0) = 0$. Since we assume f is nonzero, we must have $f(0) = 1$. Then evaluating at $y = 0$ gives $f(x) = f(\sqrt{x^2}) = f(-x)$, so f is an even function.
2. Differentiate with respect to y to get

$$f(x)f'(y) = f'(r)r_y$$

where $r = \sqrt{x^2 + y^2}$ and r_y denotes the partial derivative of r with respect to y. Differentiate again to get

$$f(x)f''(y) = f''(r)r_y^2 + f'(r)r_{yy} \ .$$

Since $r_y = y/r$ and $r_{yy} = x^2/r^3$, we get

$$f'(x) = f''(0)xf(x)$$

for $y = 0$. The solution of this differential equation is

$$f(x) = e^{f''(0)x^2/2}$$

and since f vanishes at infinity, we must have $f''(0)/2 = -\gamma < 0$. Thus, $f(x) = e^{-\gamma x^2}$ for some positive constant γ.

Solution to 1.1.37: Let C be the set of all sequences $\varepsilon = (\varepsilon_n)_1^\infty$ such that ε_n equals 1 or -1; it is an uncountable set. For ε in C let f_ε be the function on $[0,1]$ such that

- $f_\varepsilon(0) = 0$;

- $f_\varepsilon(1/n) = \varepsilon_n/n$ for n a positive integer;

- on each interval $[1/(n+1), 1/n]$, f_ε is the linear function whose values at the endpoints agree with those given by (2).

Each function f_ε is continuous: continuity at points of $(0,1]$ is obvious, and continuity at 0 follows because $|f(x)| \leqslant x$. Each f_ε takes rational values at rational points: if x is a rational point between $a = 1/(n+1)$ and $b = 1/n$, then $x = (1-t)a+tb$ with t a rational point of $[0,1]$, so $f_\varepsilon(x) = (1-t)f_\varepsilon(a)+tf_\varepsilon(b)$ is rational because $f_\varepsilon(a)$ and $f_\varepsilon(b)$ are. The functions f_ε thus form an uncountable subset of S, showing that S is uncountable.

1.2 Limits and Continuity

Solution to 1.2.1: Fix x_0. We have

$$|g(x) - g(x_0)| \leqslant \int_{-\infty}^{\infty} \frac{|f(x,t) - f(x_0,t)|}{1+t^2}\,dx = \int_{-\infty}^{-R} + \int_{-R}^{R} + \int_{R}^{\infty},$$

where R is any positive number. Fix $\varepsilon > 0$. Because of the boundedness of f and the convergence of $\displaystyle\int_{-\infty}^{\infty} \frac{1}{1+t^2}\,dt$, we can choose R so that the first and last integrals on the right are each less than $\varepsilon/3$. Since f is continuous on $\mathbb{R}^2$, it is uniformly continuous on the compact set $[x_0 - 1, x_0 + 1] \times [-R, R]$. Hence, there is a δ in $(0,1)$ such that $|f(x,t) - f(x_0,t)| < \varepsilon/6R$ for all t in $[-R, R]$ whenever $|x - x_0| < \delta$. Hence, for $|x - x_0| < \delta$, the middle integral is also less than $\varepsilon/3$, implying that $|g(x) - g(x_0)| < \varepsilon$. This establishes the continuity of g.

Solution to 1.2.2: Consider $f(x) = \sin x$. The Mean Value Theorem [Rud87, p. 108] implies that

$$f(x) - f(y) = f'(\xi)(x - y) = (\cos \xi)(x - y) \quad \text{for some } \xi \in (0, 1),$$

and since $|\cos \xi| < 1$, this implies

$$|f(x) - f(y)| < |x - y| \qquad \text{whenever } x \neq y.$$

However, if $M < 1$ were such that

$$|f(x) - f(y)| < M|x - y| \qquad \text{for all } x, y \in I,$$

then, putting $x = 0$ and letting $y \to 0$, we would get $|f'(0)| \leqslant M < 1$, which contradicts the fact that $f'(0) = 1$.

Solution to 1.2.3: Suppose f is not continuous at $\xi \in [0, 1]$. Then, for some $\varepsilon > 0$, there is a sequence (x_n) converging to ξ with $|f(x_n) - f(\xi)| > \varepsilon$ for all n. By the first condition, there is a sequence (y_n) such that y_n lies between ξ and x_n and $|f(y_n) - f(\xi)| = \varepsilon$. Then

$$y_n \in f^{-1}(f(\xi) + \varepsilon) \cup f^{-1}(f(\xi) - \varepsilon) \quad \xi \notin f^{-1}(f(\xi) + \varepsilon) \cup f^{-1}(f(\xi) - \varepsilon)$$

which contradicts the second condition.

Solution to 1.2.4: There exists $\delta > 0$ such that $|x - y| \leqslant \delta$ implies $|f(x) - f(y)| \leqslant 1$. Let $B = 1/\delta$. Take any $x > 0$ and let $n_x = \lfloor x/\delta \rfloor$, the greatest integer not exceeding x/δ. Then

$$|f(x)| = |f(x) - f(0)| \leqslant |f(x) - f(n_x\delta)| + \sum_{j=1}^{n_x} |f(j) - f((j-1)\delta)|$$

$$\leqslant 1 + n_x \leqslant 1 + Bx .$$

The proof for $x < 0$ is similar.

Solution to 1.2.5: 1. Let f_1 be the restriction of f to $[0, 2]$. The ranges of f and f_1 are the same, by periodicity, so f attains its extrema.
2. Let $\delta > 0$. f_1 is uniformly continuous, being a continuous function defined on a compact set, so there is $\varepsilon > 0$ such that

$$|f_1(a) - f_1(b)| < \delta \quad \text{for} \quad a, b \in [0, 2], |a - b| < \varepsilon.$$

Let $x, y \in \mathbb{R}$ with $|x - y| < \varepsilon$. Then, there are $x_1, x_2 = x_1 + 1, y_1, y_2 = y_1 + 1 \in [0, 2]$ with $f(x_1) = f(x_2) = f(x)$, $f(y_1) = f(y_2) = f(y)$, and $|x_i - y_j| < \varepsilon$ for some choice of $i, j \in \{1, 2\}$, and the conclusion follows.
3. Let f attain its maximum and minimum at ξ_1 and ξ_2, respectively. Then

$$f(\xi_1 + \pi) - f(\xi_1) \leqslant 0 \quad \text{and} \quad f(\xi_2 + \pi) - f(\xi_2) \geqslant 0;$$

as f is continuous, the conclusion follows from the Intermediate Value Theorem [Rud87, p. 93].

Solution to 1.2.6: Let (x_n) be a sequence of numbers in $[0, 1)$ converging to zero. As h is uniformly continuous, given $\delta > 0$ we can find $\varepsilon > 0$ such that $|h(x) - h(y)| < \delta$ if $|x - y| < \varepsilon$; therefore, we have

$$|h(x_n) - h(x_m)| < \delta$$

for n and m large enough. $(f(x_n))$ is a Cauchy sequence then, so it converges, to ξ, say. If (y_n) is another sequence with limit zero, a similar argument applied to $f(x_1), f(y_1), \ldots$ shows that $\lim f(y_n) = \xi$. The function $g : [0, 1] \to \mathbb{R}$ given by

$$g(x) = \begin{cases} h(x) & \text{for} \quad x \in [0, 1) \\ \xi & \text{for} \quad x = 0 \end{cases}$$

is clearly the unique extension of h to $[0, 1]$.

Solution to 1.2.7: Let $\lim_{x\to\infty} f(x) = a$. Given $\varepsilon > 0$, let $K > 0$ satisfy $|f(x) - a| < \varepsilon/2$ for $x \geqslant K$. If $x, y \geqslant K$ we have then

$$|f(x) - f(y)| \leqslant |f(x) - a| + |a - f(y)| \leqslant \varepsilon .$$

The interval $[0, K]$ is compact, so f is uniformly continuous there, that is, there exists $\delta > 0$ such that, if $x, y \in [0, K]$ and $|x - y| < \delta$, then $|f(x) - f(y)| < \varepsilon/2$. Finally, if $x \leqslant K$, $y > K$ verify $|x - y| < \delta$, we have

$$|f(x) - f(y)| \leqslant |f(x) - f(K)| + |f(K) - f(y)| < \varepsilon,$$

therefore, if $x, y \geqslant 0$ satisfy $|x - y| < \delta$, we have $|f(x) - f(y)| < \varepsilon$, as desired.

Solution to 1.2.8: Let E be the set of discontinuities of f. We have $E = E_1 \cup E_2 \cup E_3 \cup E_4$, where

$$E_1 = \{x \in E \mid f(x-) = f(x+) < f(x)\}, \quad E_2 = \{x \in E \mid f(x-) > f(x+)\}$$

$$E_3 = \{x \in E \mid f(x-) = f(x+) > f(x)\}, \quad E_4 = \{x \in E \mid f(x-) < f(x+)\}.$$

For $x \in E_1$, let $a_x \in \mathbb{Q}$ be such that $f(x-) < a_x < f(x+)$. Now let's take $b_x, c_x \in \mathbb{Q}$ in such a way that $b_x < x < c_x$ and

$$b_x < t < c_x, \; x \neq t \quad \text{implies} \quad f(t) < a_x.$$

This map $\varphi : E_1 \to \mathbb{Q}^3$ given by $x \mapsto (a_x, b_x, c_x)$ is injective since $(a_x, b_x, c_x) = (a_y, b_y, c_y)$ implies $f(y) < a_x < f(y)$ for $x \neq y$. So E_1 is, at most, countable.

For $x \in E_2$, take $a_x \in \mathbb{Q}$ with $f(x-) > a_x > f(x+)$ and choose $b_x, c_x \in \mathbb{Q}$ such that $b_x < x < c_x$ and

$$b_x < t < x \quad \text{implies} \quad f(t) > a_x$$

and

$$t < c_x \quad \text{implies} \quad f(t) < a_x;$$

this map is an injection $E_2 \to \mathbb{Q}^3$, so E_2 is, at most, countable.

Similar methods lead to analogous results for E_3 and E_4. As the union of countable sets is countable, the result follows.

Solution 2. Define the function $\sigma : \mathbb{R} \to \mathbb{R}$ by

$$\sigma(x) = \max\{|f(x) - f(x+)|, |f(x) - f(x-)|\};$$

observe that $\sigma(x) > 0$ if and only if x is a discontinuity of f.

For each $n \in \mathbb{N}$, let the set D_n be given by

$$D_n = \{x \in \mathbb{R} \mid \sigma(x) \geqslant \frac{1}{n}\}.$$

It is clear that the set of discontinuities of f is $D = \bigcup_{n=1}^{\infty} D_n$. We shall prove that each D_n has no accumulation points, so, it is countable. If $a \in D_n$, using the fact that $f(a+) = \lim_{x \to a+} f(x)$, we can find $\delta > 0$ such that, for all x, $a < x < a + \delta$, we have

$$f(a+) - \frac{1}{4n} < f(x) < f(a+) + \frac{1}{4n},$$

that is, for every point in this interval, $\sigma(x) \leqslant 1/2n$. In the same fashion, we can find an open set $a - \delta < x < a$ such that no point is in D_n, showing that D_n is made up of isolated points so it is countable, and so is D.

Solution to 1.2.9: By Problem 1.2.8, it is enough to show that f has lateral limits at all points. We have, for any $x \in \mathbb{R}$,

$$-\infty < \sup_{y<x}\{f(y)\} = f(x-) \leqslant f(x+) = \inf_{y<x}\{f(y)\} < \infty$$

since f is an increasing function.

Solution to 1.2.10: Fix $\varepsilon > 0$. For each $x \in [0, 1]$, let δ_x be as in the hypothesis and $I_x = (x - \delta_x, x + \delta_x)$. The open intervals $\{I_x\}$ cover $[0, 1]$ so, by compactness and the Heine–Borel Theorem [Rud87, p. 30], we can choose a finite subcover

$$[0, 1] \subset I_{x_1} \cup I_{x_2} \cup \cdots \cup I_{x_n}.$$

Let $M = \max\{f(x_i) + \varepsilon\}$. If $x \in [0, 1]$ then $f(x) < M$ and f is bounded from above.

Let N be the least upper bound of f on $[0, 1]$. Then there is a sequence of points (x_n) such that $(f(x_n))$ tends to N from below. Since $[0, 1]$ is compact, by the Bolzano–Weierstrass Theorem [Rud87, p. 40], [MH93, p. 153], (x_n) has a convergent subsequence, so (by passing to a subsequence) we may assume that (x_n) converges to some $p \in [0, 1]$. By the upper semicontinuity of f and the convergence of $(f(x_n))$, we have, for n sufficiently large, $f(x_n) < f(p) + \varepsilon$ and $N < f(x_n) + \varepsilon$. Combining these, we get $f(p) \leqslant N < f(p) + 2\varepsilon$. Since this holds for all $\varepsilon > 0$, $f(p) = N$.

Solution to 1.2.11: Suppose $f : \mathbb{R} \to \mathbb{R}$ is continuous, maps open sets to open sets but is not monotonic. Without loss of generality assume there are three real numbers $a < b < c$ such that $f(a) < f(b) > f(c)$. By Weierstrass Theorem [MH93, p. 189], f has a maximum, M, in $[a, c]$, which cannot occur at a or b. Then $f((a, c))$ cannot be open, since it contains M but does not contain $M + \varepsilon$ for any positive ε. We conclude then that f must be monotonic.

Solution to 1.2.12: The inequality given implies that f is one-to-one, so f is strictly monotone and maps open intervals onto open intervals, so $f(\mathbb{R})$ is open.

Let $z_n = f(x_n)$ be a sequence in $f(\mathbb{R})$ converging to $z \in \mathbb{R}$. Then z_n is Cauchy, and, by the stated inequality, so is x_n. Let $x = \lim x_n$. By continuity we have $f(x) = f(\lim x_n) = \lim f(x_n) = z$ so $f(\mathbb{R})$ is also closed. Thus, $f(\mathbb{R}) = \mathbb{R}$.

Solution to 1.2.13: 1. For $\varepsilon > 0$ let

$$L = \max_{x \in [0,1]} (|f(x)| + 1) \quad \text{and} \quad 0 < \delta < \min\left\{\frac{\varepsilon}{2L}, 1\right\}.$$

We have

$$\left| \int_{1-\delta}^{1} x^n f(x)dx \right| \leqslant \int_{1-\delta}^{1} x^n |f(x)|dx \leqslant L\delta \leqslant \frac{\varepsilon}{2}$$

and

$$\left| \int_0^{1-\delta} x^n f(x) dx \right| \leq \int_0^{1-\delta} (1-\delta)^n |f(x)| dx \leq L\delta^{n+1},$$

so

$$\lim_{n \to \infty} \int_0^1 x^n f(x) dx = 0.$$

2. We will show that

$$\lim_{n \to \infty} n \int_0^1 x^n (f(x) - f(1)) dx = 0.$$

For $\varepsilon > 0$ let δ be such that $|f(x) - f(1)| < \varepsilon/2$ if $x \in [1-\delta, 1]$. We have

$$\left| n \int_{1-\delta}^n x^n (f(x) - f(1)) dx \right| \leq n \int_{1-\delta}^1 x^n |f(x) - f(1)| dx \leq n \int_{1-\delta}^1 x^n \frac{\varepsilon}{2} dx \leq \frac{\varepsilon}{2},$$

and, letting $L = \sup_{x \in [0,1]} |f(x) - f(1)|$,

$$\left| n \int_0^{1-\delta} x^n (f(x) - f(1)) dx \right| \leq n \int_0^{1-\delta} x^n L dx = n \frac{(1-\delta)^{n+1}}{n+1}$$

and the result follows.

Now it suffices to notice that

$$n \int_0^1 x^n f(x) dx = n \int_0^1 x^n (f(x) - f(1)) dx + n \int_0^1 f(1) x^n dx.$$

Solution to 1.2.14: Suppose that f is not continuous. Then there exist $\varepsilon > 0$, $x \in [0, 1]$, and a sequence (x_n) tending to x such that $|f(x) - f(x_n)| \geq \varepsilon$ for all n. Consider the sequence $((x_n, f(x_n)))$ in G_f. Since the unit square is compact, by Bolzano–Weierstrass Theorem [Rud87, p. 40], [MH93, p. 153], this sequence has a convergent subsequence; using this subsequence, we may assume that $((x_n, f(x_n)))$ converges to some point (y, z). Then we must have the sequence (x_n) converging to y; so, by the uniqueness of limits, $x = y$. Since G_f is closed, we must have $z = f(x)$. Hence, $(f(x_n))$ converges to $f(x)$, contradicting our assumption.

The converse of the implication is also true, see the Solution to Problem 2.1.2.

Solution to 1.2.15: For each $y \in [0, 1]$, consider the function $g_y(x) = f(x, y)$. Then $g(x) = \sup g_y(x)$. The family $\{g_y\}$ is equicontinuous because f is uniformly continuous. It suffices then to show that the pointwise supremum of an equicontinuous family of functions is continuous. Let $\varepsilon > 0$, $x_0 \in [0, 1]$. There is y_0 such that

$$g_{y_0}(x_0) \leq g(x_0) < g_{y_0}(x_0) + \varepsilon.$$

Let δ be positive such that if $|r - s| < \delta$, then $|g_y(r) - g_y(s)| < \varepsilon$ for all y, and $|x_0 - x_1| < \delta$. For some y_1, we have that

$$g_{y_1}(x_1) \leqslant g(x_1) < g_{y_1}(x_1) + \varepsilon.$$

Further, by equicontinuity of $\{g_y\}$, we have the two inequalities $|g_{y_0}(x_0) - g_{y_0}(x_1)| < \varepsilon$ and $|g_{y_1}(x_0) - g_{y_1}(x_1)| < \varepsilon$. By combining them we get

$$g_{y_0}(x_0) < g_{y_0}(x_1) + \varepsilon < g(x_1) + \varepsilon < g_{y_1}(x_1) + 2\varepsilon$$

and

$$g_{y_1}(x_1) < g_{y_1}(x_0) + \varepsilon < g(x_0) + \varepsilon < g_{y_0}(x_0) + 2\varepsilon.$$

These two inequalities imply $|g_{y_1}(x_1) - g_{y_0}(x_0)| < 2\varepsilon$. This, combined with the first two inequalities, shows that $|g(x_0) - g(x_1)| < 3\varepsilon$. Since this holds for all ε and x_0 and all x_1 close to x_0, g is continuous.

Solution to 1.2.16: We need only show that $f^{-1}(F)$ is closed whenever F is closed. Let F be a closed subset of $\mathbb{R}$. Let $(x_n)_1^\infty$ be a convergent sequence in $f^{-1}(F)$ with limit x_0. We need only show that x_0 is in $f^{-1}(F)$. For $n > 0$ let $y_n = f(x_n)$. The sequence $(y_n)_1^\infty$ is in F and it is bounded (since the convergent sequence $(x_n)_1^\infty$ is). Passing to a subsequence, we can assume the sequence $(y_n)_1^\infty$ converges, say to y_0, which lies in F because F is closed. The set $K = \{y_0, y_1, y_2, \ldots\}$ is then compact, so $f^{-1}(K)$ is closed. Hence $f^{-1}(K)$ contains its limit points. But the sequence $(x_n)_1^\infty$ lies in $f^{-1}(K)$ and converges to x_0. Therefore x_0 is in $f^{-1}(K)$, and so also in $f^{-1}(F)$, as desired.

1.3   Sequences, Series, and Products

Solution to 1.3.1: $A_1^n \leqslant A_1^n + \cdots + A_k^n \leqslant kA_1^n$, so we have

$$A_1 = \lim_{n\to\infty} \left(A_1^n\right)^{1/n} \leqslant \lim_{n\to\infty} \left(A_1^n + \cdots + A_k^n\right)^{1/n} \leqslant \lim_{n\to\infty} \left(kA_1^n\right)^{1/n} = A_1.$$

showing that the limit equals A_1.

Solution to 1.3.2: Let $p_1 = 1$, $p_2 = (2/1)^2$, $p_3 = (3/2)^3, \ldots, p_n = (n/(n-1))^n$. Then

$$\frac{p_1 p_2 \cdots p_n}{n} = \frac{n^n}{n!},$$

and since $p_n \to e$, we have $\lim(n^n/n!)^{1/n} = e$ as well (using the fact that $\lim n^{1/n} = 1$).

Solution 2. As the exponential is a continuous function, $L = \exp(\lim_{n\to\infty} L_n)$ where

$$L_n = \log n - \frac{1}{n}(\log 1 + \log 2 + \cdots + \log n).$$

Since

$$\log 1 + \log 2 + \cdots + \log(n-1) \leqslant \int_1^n \log x \, dx = n \log n - n + 1,$$

we have

$$L_n \geqslant (1 - 1/n) \log n - \log n + 1 - 1/n = 1 - (1 + \log n)/n \to 1 \quad \text{as} \quad n \to \infty.$$

On the other hand,

$$\log 1 + \log 2 + \cdots + \log n \geqslant \int_1^n \log x \, dx = n \log n - n + 1,$$

so

$$L_n \leqslant \log n - (n \log n - n + 1)/n = 1 - 1/n.$$

Hence,

$$1 - (1 + \log n)/n \leqslant L_n \leqslant 1 - 1/n,$$

so $L_n \to 1$ and $L = \exp(1) = e$.

Solution to 1.3.3: Obviously, $x_n \geqslant 1$ for all n; so, if the limit exists, it is $\geqslant 1$, and we can pass to the limit in the recurrence relation to get

$$x_\infty = \frac{3 + 2x_\infty}{3 + x_\infty};$$

in other words, $x_\infty^2 + x_\infty - 3 = 0$. So x_∞ is the positive solution of this quadratic equation, that is, $x_\infty = \frac{1}{2}(-1 + \sqrt{13})$.

To prove that the limit exists, we use the recurrence relation to get

$$\begin{aligned} x_{n+1} - x_n &= \frac{3 + 2x_n}{3 + x_n} - \frac{3 + 2x_{n-1}}{3 + x_{n-1}} \\ &= \frac{3(x_n - x_{n-1})}{(3 + x_n)(3 + x_{n+1})}. \end{aligned}$$

Hence, $|x_{n+1} - x_n| \leqslant \frac{1}{3}|x_n - x_{n-1}|$. Iteration gives

$$|x_{n+1} - x_n| \leqslant 3^{-n}|x_1 - x_0| = \frac{1}{3^n \cdot 4}.$$

The series $\sum_{n=1}^\infty (x_{n+1} - x_n)$, of positive terms, is dominated by the convergent series $\frac{1}{4} \sum_{n=1}^\infty 3^{-n}$ and so converges. We have $\sum_{n=1}^\infty (x_{n+1} - x_n) = \lim x_n - x_1$ and we are done.

Solution 2. To prove the existence of the limit it is enough to notice that if g is defined by

$$g(x) = \frac{3 + 2x}{3 + x}$$

we have

$$|g'(x)| \leqslant \frac{3}{16} \leqslant 1 \quad \text{for} \quad x \geqslant 1$$

and apply the Fixed Point Theorem [Rud87, p. 220].

Solution to 1.3.4: We prove, by induction, that $0 < x_n < 1/2$ for $n \geqslant 1$. First, $0 < x_1 = 1/3 < 1/2$. Suppose for some $n \geqslant 1$, that $0 < x_n < 1/2$. Then $2/5 < x_{n+1} = 1/(2 + x_n) < 1/2$. This completes the induction.

Let $f(x) = 1/(2 + x)$. The equation $f(x) = x$ has a unique solution in the interval $0 < x < 1/2$ given by $p = \sqrt{2} - 1$. Moreover, $|f'(x)| = 1/(2 + x)^2 < 1/4$ for $0 < x < 1/2$. Thus, for $n \geqslant 1$,

$$|x_{n+1} - p| = |f(x_n) - f(p)|$$
$$\leqslant \tfrac{1}{4} |x_n - p|,$$

by the Mean Value Theorem.

Iterating, we obtain

$$|x_{n+1} - p| \leqslant (\tfrac{1}{4})^2 |x_{n-1} - p|$$
$$\leqslant \cdots$$
$$\leqslant (\tfrac{1}{4})^n |x_1 - p|.$$

Hence the sequence x_n tends to the limit $\sqrt{2} - 1$.

Solution 2. Define $f(x) = 1/(2 + x)$ for $0 \leqslant x \leqslant 1$. Then f maps the closed interval $[0, 1]$ to $[1/3, 1/2] \subset [0, 1]$. Also $|f'(x)| = \left| \dfrac{1}{(2 + x)^2} \right| \leqslant \dfrac{1}{4}$ for $x \in [0, 1]$. Therefore $f : [0, 1] \to [0, 1]$ is a contraction map with Lipschitz constant $1/4 < 1$, so f has a unique fixed point y, and the sequence $x_{n+1} = f(x_n)$ defined above converges to y. We have $y = 1/(2 + y)$ or $y^2 + 2y = 1$, whence $y = \sqrt{2} - 1$ (since $0 \leqslant y \leqslant 1$).

Solution to 1.3.5: By the given relation $x_{n+1} - x_n = (\alpha - 1)(x_n - x_{n-1})$. Therefore, by the Induction Principle [MH93, p. 7], we have $x_n - x_{n-1} = (\alpha - 1)^{n-1}(x_1 - x_0)$, showing that the sequence is Cauchy and then converges. Hence,

$$x_n - x_0 = \sum_{k=1}^{n} (x_k - x_{k-1}) = (x_1 - x_0) \sum_{k=1}^{n} (\alpha - 1)^{k-1}.$$

Taking limits, we get

$$\lim_{n \to \infty} x_n = \frac{(1 - \alpha)x_0 + x_1}{2 - \alpha}.$$

Solution 2. The recurrence relation can be expressed in matrix form as

$$\begin{pmatrix} x_{n+1} \\ x_n \end{pmatrix} = A \begin{pmatrix} x_n \\ x_{n-1} \end{pmatrix}, \quad \text{where} \quad A = \begin{pmatrix} \alpha & 1-\alpha \\ 1 & 0 \end{pmatrix}.$$

Thus,

$$\begin{pmatrix} x_{n+1} \\ x_n \end{pmatrix} = A^n \begin{pmatrix} x_1 \\ x_0 \end{pmatrix}.$$

A calculation shows that the eigenvalues of A are 1 and $\alpha - 1$, with corresponding eigenvectors $v_1 = (1, 1)^t$ and $v_2 = (\alpha - 1, 1)^t$. A further calculation shows that

$$\begin{pmatrix} x_1 \\ x_0 \end{pmatrix} = \left(\frac{(1 - \alpha)x_0 + x_1}{2 - \alpha} \right) v_1 + \left(\frac{x_0 - x_1}{2 - \alpha} \right) v_2 .$$

Hence,

$$\begin{pmatrix} x_{n+1} \\ x_n \end{pmatrix} = A^n \begin{pmatrix} x_1 \\ x_0 \end{pmatrix} = \frac{(1 - \alpha)x_0 + x_1}{2 - \alpha} v_1 + (\alpha - 1)^n \frac{x_0 - x_1}{2 - \alpha} v_2 .$$

Since $|\alpha - 1| < 1$ we have $\lim_{n \to \infty} (\alpha - 1)^n = 0$, and we can conclude that

$$\lim_{n \to \infty} x_n = \frac{(1 - \alpha)x_0 + x_1}{2 - \alpha}.$$

Solution to 1.3.6: The given relation can be written in matrix form as $\begin{pmatrix} x_{n+1} \\ x_n \end{pmatrix} = A \begin{pmatrix} x_{n-1} \\ x_n \end{pmatrix}$, where $A = \begin{pmatrix} 2c & -1 \\ 1 & 0 \end{pmatrix}$. The required periodicity holds if and only if $A^k = \begin{pmatrix} 1 & 0 \\ 0 & 1 \end{pmatrix}$. The characteristic polynomial of A is $\lambda^2 - 2c\lambda + 1$, so the eigenvalues of A are $c \pm \sqrt{c^2 - 1}$. A necessary condition for $A^k = \begin{pmatrix} 1 & 0 \\ 0 & 1 \end{pmatrix}$ is that the eigenvalues of A be k^{th} roots of unity, which implies that $c = \cos\left(\frac{2\pi j}{k}\right)$, $j = 0, 1, \ldots, \lfloor \frac{k}{2} \rfloor$. If c has the preceding form and $0 < j < \frac{k}{2}$ (i.e., $-1 < c < 1$), then the eigenvalues of A are distinct (i.e., A is diagonalizable), and the equality $A^k = \begin{pmatrix} 1 & 0 \\ 0 & 1 \end{pmatrix}$ holds. If $c = 1$ or -1, then the eigenvalues of A are not distinct, and A has the Jordan Canonical Form [HK61, p. 247] $\begin{pmatrix} 1 & 1 \\ 0 & 1 \end{pmatrix}$ or $\begin{pmatrix} -1 & 1 \\ 0 & -1 \end{pmatrix}$, respectively, in which case $A^k \neq \begin{pmatrix} 1 & 0 \\ 0 & 1 \end{pmatrix}$. Hence, the desired periodicity holds if and only if $c = \cos\left(\frac{2\pi j}{k}\right)$, where j is an integer, and $0 < j < k/2$.

Solution to 1.3.7: If $\lim x_n = x_\infty \in \mathbb{R}$, we have $x_\infty = a + x_\infty^2$; so

$$x_\infty = \frac{1 \pm \sqrt{1 - 4a}}{2}$$

and we must have $a \leqslant 1/4$.

Conversely, assume $0 < a \leqslant 1/4$. As $x_{n+1} - x_n = x_n^2 - x_{n-1}^2$, we conclude, by the Induction Principle [MH93, p. 7], that the given sequence is nondecreasing. Also,

$$x_{n+1} = a + x_n^2 < \frac{1}{4} + \frac{1}{4} = \frac{1}{2}$$

if $x_n < 1/2$, which shows that the sequence is bounded. It follows that the sequence converges when $0 < a \leqslant 1/4$.

Solution to 1.3.8: First we show, by induction, that the sequence (x_n) is bounded. For that, choose M large so that $\max |x_1|, |2x_{n+1} - x_n| \leq M$, for all n.

$$|x_{n+1}| = \left| \frac{x_n + (2x_{n+1} - x_n)}{2} \right| \leq \frac{1}{2}(|x_n| + |2x_{n+1} - x_n|) \leq M$$

showing that (x_n) is bounded. Now to compute the limit write

$$x_{n+1} = \frac{x_n + (2x_{n+1} - x_n)}{2}$$

and taking the lim sup, we have

$$\limsup x_n \leq \frac{\limsup x_n + x}{2}$$

showing that $\limsup x_n \leq x$. In the same way we obtain $\liminf x_n \geq x$, showing that $\lim x_n = x$.

Solution to 1.3.9: Let $y_n = x_n/\sqrt{a}$. It will be shown that $y_n \to 1$. The sequence (y_n) satisfies the recurrence relation

$$y_n = \frac{1}{2}\left(y_{n-1} + \frac{1}{y_{n-1}} \right).$$

We have $y_n \geqslant 1$ for all n because for any positive b

$$\frac{1}{2}\left(b + \frac{1}{b} \right) - 1 = \frac{1}{2}\left(\sqrt{b} - \frac{1}{\sqrt{b}} \right)^2 \geqslant 0.$$

Hence, for every n,

$$y_{n-1} - y_n = \frac{1}{2}\left(y_{n-1} - \frac{1}{y_{n-1}} \right) \geqslant 0,$$

so the sequence (y_n) is nonincreasing. As it is also bounded below, it converges. Let $y_n \to y$. The recurrence relation gives

$$y = \frac{1}{2}\left(y + \frac{1}{y} \right),$$

i.e., $y = 1/y$, from which $y = 1$ follows, since $y \geqslant 1$.

Solution to 1.3.10: Clearly, $0 \leqslant x_{n+1} = x_n(1 - x_n^n) \leqslant x_n \leqslant \cdots \leqslant x_1$ for all n. Thus,

$$x_{n+1} = x_n(1 - x_n^n) \geqslant x_n(1 - x_1^n),$$

and therefore

$$x_n \geq x_1 \prod_1^n (1 - x_1^k) = x_1 \exp\left(\sum_{k=1}^n \log(1 - x_1^k)\right).$$

Since $\log(1 - x_1^k) = O(x_1^k)$ as $k \to \infty$, the sum converges to a finite value L as $n \to \infty$ and we get

$$\liminf_{n\to\infty} x_n \geq x_1 \exp(L) > 0.$$

Solution to 1.3.11: 1. We have

$$f(x) = \frac{1}{2} - \left(x - \frac{1}{2}\right)^2$$

so x_n is bounded by $1/2$ and, by the Induction Principle [MH93, p. 7], nondecreasing. Let λ be its limit. Then

$$\lambda = \frac{1}{2} - \left(\lambda - \frac{1}{2}\right)^2$$

and, as the sequence takes only positive values,

$$\lambda = \frac{1}{2}.$$

2. It is clear, from the expression for f above, that

$$f(x) \leq x \quad \text{for} \quad x \leq -\frac{1}{2}$$

and

$$f(x) \leq -\frac{1}{2} \quad \text{for} \quad x \geq \frac{3}{2}$$

therefore, the sequence diverges for such initial values.
On the other hand, if $|x - 1/2| < 1$, we get

$$\left|f(x) - \frac{1}{2}\right| < \left|x - \frac{1}{2}\right|$$

so, for these initial values, we get

$$\left|x_{n+1} - \frac{1}{2}\right| < \left|x - \frac{1}{2}\right|^n = o(1).$$

Solution to 1.3.12: Suppose that $\lim f_{n+1}/f_n = a < \infty$. $a \geq 1$ since the the sequence f_n is increasing. We have

$$\frac{f_{n+1}}{f_n} = 1 + \frac{f_{n-1}}{f_n}.$$

Taking the limit as n tends to infinity, we get (since $a \neq 0$)

$$a = 1 + \frac{1}{a}$$

or

$$a^2 - a - 1 = 0.$$

This quadratic equation has one positive root,

$$\varphi = \frac{1 + \sqrt{5}}{2}.$$

We show now that the sequence (f_{n+1}/f_n) is a Cauchy sequence. Applying the definition of the f_n's, we get

$$\left| \frac{f_{n+1}}{f_n} - \frac{f_n}{f_{n-1}} \right| = \left| \frac{f_{n-1}^2 - f_n f_{n-2}}{f_{n-1}^2 + f_{n-1} f_{n-2}} \right|.$$

Since f_n is an increasing sequence,

$$f_{n-1}(f_{n-1} - f_{n-2}) \geq 0$$

or

$$f_{n-1}^2 + f_{n-1} f_{n-2} \geq 2 f_{n-1} f_{n-2}.$$

By substituting this in and simplifying, we get

$$\left| \frac{f_{n+1}}{f_n} - \frac{f_n}{f_{n-1}} \right| \leq \frac{1}{2} \left| \frac{f_n}{f_{n-1}} - \frac{f_{n-1}}{f_{n-2}} \right|.$$

By the Induction Principle [MH93, p. 7], we get

$$\left| \frac{f_{n+1}}{f_n} - \frac{f_n}{f_{n-1}} \right| \leq \frac{1}{2^{n-2}} \left| \frac{f_3}{f_2} - \frac{f_2}{f_1} \right|.$$

Therefore, by the Triangle Inequality [MH87, p. 20], for all $m > n$,

$$\left| \frac{f_{m+1}}{f_m} - \frac{f_{n+1}}{f_n} \right| \leq \left| \frac{f_3}{f_2} - \frac{f_2}{f_1} \right| \sum_{k=n}^{m-1} \frac{1}{2^{k-2}}.$$

Since the series $\sum 2^{-n}$ converges, the right-hand side tends to 0 as m and n tend to infinity. Hence, the sequence (f_{n+1}/f_n) is a Cauchy sequence, and we are done.

Solution to 1.3.13: We have

$$\frac{1}{n+1} + \cdots + \frac{1}{2n} = \sum_{k=1}^{n} \frac{1}{1 + \frac{k}{n}} \cdot \frac{1}{n}$$

which is a Riemann sum for $\int_0^1 (1+x)^{-1} dx$ corresponding to the partition of the interval $[0, 1]$ in n subintervals of equal length. Therefore, we get

$$\lim_{n \to \infty} \left(\frac{1}{n+1} + \cdots + \frac{1}{2n} \right) = \int_0^1 \frac{1}{1+x} dx = \log 2.$$

Solution 2. Using the inequalities

$$\left(1 + \frac{1}{k} \right)^k < e < \left(1 + \frac{1}{k-1} \right)^k \qquad (k \geqslant 2),$$

we get

$$\log 2 = \log \left(\prod_{k=n+1}^{2n} \frac{k}{k-1} \right) = \sum_{k=n+1}^{2n} \frac{1}{k} \log \left(\frac{k}{k-1} \right)^k$$

$$> \sum_{k=n+1}^{2n} \frac{1}{k} > \sum_{k=n+1}^{2n} \frac{1}{k} \log \left(\frac{k+1}{k} \right)^k = \log \left(\prod_{k=n+1}^{2n} \frac{k+1}{k} \right)$$

$$= \log \left(\frac{2n+1}{n+1} \right);$$

therefore, we have

$$\log 2 \geqslant \lim_{n \to \infty} \sum_{k=n+1}^{2n} \frac{1}{k} \geqslant \log 2$$

and the result follows.

Solution 3. We have

$$\frac{1}{n+1} + \cdots + \frac{1}{2n} = 1 + \frac{1}{2} + \cdots + \frac{1}{2n} - \left(1 + \frac{1}{2} + \cdots + \frac{1}{n} \right)$$

$$= 1 + \frac{1}{2} + \cdots + \frac{1}{2n} - 2 \left(\frac{1}{2} + \cdots + \frac{1}{2n} \right)$$

$$= 1 - \frac{1}{2} + \cdots + \frac{1}{2n-1} - \frac{1}{2n}$$

and the result now follows from the Maclaurin expansion [PMJ85, p. 127] of $\log(1+x)$.

Solution to 1.3.14: We have

$$H_{nk} - H_n = \frac{1}{n+1} + \frac{1}{n+2} + \ldots + \frac{1}{nk}$$

$$= \frac{1}{n} \left(\frac{1}{1 + \frac{1}{n}} + \frac{1}{1 + \frac{2}{n}} + \ldots + \frac{1}{k} \right).$$

The right side is the lower Riemann sum of the integral $\int_1^k \frac{1}{x}\,dx$ over the partition of $[1, k]$ into $(k-1)n$ intervals each of length $\frac{1}{n}$. Because the lower Riemann sum is less than the integral, whose value is $\log k$, we get $H_{nk} - H_n < \log k$.

To get the other inequality note that the change in $\frac{1}{x}$ on the interval $[1, 1+\frac{1}{n}]$ of the partition is $1 - \dfrac{1}{1+\frac{1}{n}} = \dfrac{1}{n+1}$, and it is less than that on every other interval of the partition. Hence

$$\int_1^k \frac{1}{x}\,dx - (H_{nk} - H_n) < (k-1)n \cdot \frac{1}{n} \cdot \frac{1}{n+1} = \frac{k-1}{n+1},$$

since each term in the Riemann sum contributes an error of less than $\dfrac{1}{n} \cdot \dfrac{1}{n+1}$, and there are $(k-1)n$ terms. So we can take $C = k-1$.

One can deduce a more precise error bound as follows. For each subinterval of the partition, translate the region between the curve $y = \frac{1}{x}$ and the approximating rectangle horizontally so as to put it above the interval $[1, 1 + \frac{1}{n}]$. The translated regions are then nonoverlapping. As they lie within a rectangle of width $\frac{1}{n}$ and height 1, the sum of their areas, which bounds the difference $\log k - (H_{nk} - H_n)$, is less than $\frac{1}{n}$. We can thus take $C = 1$. In fact, one easily sees that the translated regions fill up less than half of the preceding rectangle, so $C = \frac{1}{2}$ works.

Solution to 1.3.15: Let $n > 0$. For $m \geqslant n$, we have

$$x_m \leqslant x_n + \sum_{k=n}^{m-1} \frac{1}{k^2} \leqslant x_n + \xi_n$$

where

$$\xi_n = \sum_{k=n}^{\infty} \frac{1}{k^2}.$$

Taking the lim sup, with respect to m, we have

$$x_n \geqslant \limsup_{m\to\infty} x_m - \xi_n.$$

The series $\sum k^{-2}$ converges, so $\lim_{n\to\infty} \xi_n = 0$. Considering the lim inf with respect to n, we get

$$\liminf_{n\to\infty} x_n \geqslant \limsup_{m\to\infty} x_m - \liminf_{n\to\infty} \xi_n \geqslant \limsup_{m\to\infty} x_m.$$

The reverse inequality also holds, so $\lim x_n$ exists.

Solution to 1.3.16: Fix $\delta > 0$, and choose n_0 such that $\varepsilon_n < \delta$ for all $n \geqslant n_0$. Then

$$a_{n_0+1} \leqslant ka_{n_0} + \varepsilon_{n_0} < ka_{n_0} + \delta$$

$$a_{n_0+2} < k^2 a_{n_0} + k\delta + \varepsilon_{n_0+1} < k^2 a_{n_0} + (1+k)\delta$$
$$a_{n_0+3} < k^3 a_{n_0} + (k+k^2)\delta + \varepsilon_{n_0+2} < k^3 a_{n_0} + (1+k+k^2)\delta$$

and, by the Induction Principle [MH93, p. 7],

$$a_{n_0+m} < k^m a_{n_0} + (1 + k + \cdots + k^{m-1})\delta < k^m a_{n_0} + \frac{\delta}{1-k}.$$

Letting $m \to \infty$, we find that

$$\limsup_{n\to\infty} a_n \leqslant \frac{\delta}{1-k}.$$

Since δ is arbitrary, we have $\limsup_{n\to\infty} a_n \leqslant 0$, and thus (since $a_n > 0$ for all n) $\lim_{n\to\infty} a_n = 0$.

Solution to 1.3.17: If (x_n) is unbounded, then, without loss of generality, it has no finite upper bound. Take $x_{n_1} = x_1$ and, for each $k \in \mathbb{N}$, x_{n_k} such that $x_{n_k} > \max\{k, x_{n_{k-1}}\}$. This is clearly an increasing subsequence of x_n.

If x_n is bounded, it has a convergent subsequence: $\lim y_n = \xi$, say. y_n contains a subsequence converging to $\xi+$ or one converging to $\xi-$. Suppose (z_n) is a subsequence of (y_n) converging to $\xi+$. Let $z_{n_1} = z_1$ and, for each $k \geqslant 1$, let $\xi \leqslant z_{n_k} < z_{n_{k-1}}$. This is a monotone subsequence of (x_n).

Solution to 1.3.18: Suppose that there are $x > 1$, $\varepsilon > 0$ such that $|b_m/b_n - x| \geqslant \varepsilon$ for all $1 \leqslant n < m$. Since $\lim(b_n/b_{n+1}) = 1$, for all k sufficiently large there exists an integer $n_k > k$ such that $b_m/b_k < x$ if $m < n_k$ and $b_m/b_k > x$ if $m > n_k$. In particular, for each k,

$$\frac{b_{n_k+1}}{b_k} - \frac{b_{n_k}}{b_k} \geqslant 2\varepsilon$$

or

$$\frac{b_{n_k+1}}{b_{n_k}} - 1 \geqslant 2\varepsilon \frac{b_k}{b_{n_k}} > \frac{2\varepsilon}{x} > 0.$$

As n_k tends to infinity as k does, the left-hand side should tend to 0 as k tends to infinity, a contradiction.

Solution to 1.3.19: 1. Using the Ratio Test [Rud87, p. 66], we have

$$\frac{\frac{(2n)!(3n)!}{n!(4n)!}}{\frac{(2n+2)!(3n+3)!}{(n+1)!(4n+4)!}} = \frac{n!(4n)!(2n+2)(2n+1)(2n)!(3n+3)(3n+2)(3n+1)(3n)!}{(2n)!(3n)!(n+1)n!(4n+4)(4n+3)(4n+2)(4n+1)(4n)!}$$

$$= \frac{(2n+2)(2n+1)(3n+3)(3n+2)(3n+1)}{(n+1)(4n+4)(4n+3)(4n+2)(4n+1)}$$

$$\to \frac{27}{64} < 1$$

so the series converges.

2. Comparing with the series $\sum 1/(n \log n)$, which can be seen to diverge using the Integral Test [Rud87, p. 139],

$$\lim \frac{1/n^{1+1/n}}{1/n \log n} = \lim \frac{\log n}{n^{1/n}} = \infty$$

we conclude that the given series diverges.

Solution to 1.3.20: 1. Assume that $\sum a_n < \infty$. As $\left(\sqrt{a_{n+1}} - \sqrt{a_n}\right)^2 = a_{n+1} + a_n - 2\sqrt{a_n a_{n+1}}$, we have

$$\sum_{n=1}^{\infty} \sqrt{a_n a_{n+1}} \leqslant \frac{1}{2} \sum_{n=1}^{\infty} (a_n + a_{n+1}) = \frac{1}{2} a_1 + \sum_{n=2}^{\infty} a_n < \infty.$$

2. Since $\sum (a_n + a_{n+1}) = 2 \sum \sqrt{a_n a_{n+1}} + \sum \left(\sqrt{a_{n+1}} - \sqrt{a_n}\right)^2$, we require a sequence $a_n = b_n^2$, $b_n > 0$, such that $\sum b_n b_{n+1} < \infty$ but $\sum (b_{n+1} - b_n)^2 = \infty$. One such example is

$$b_n = \begin{cases} \frac{1}{\sqrt{n}} & \text{if } n \text{ is odd} \\ \frac{1}{n} & \text{if } n \text{ is even.} \end{cases}$$

Solution to 1.3.21: As

$$\lim \left| \frac{a^{n+1}}{(n+1)^b (\log n + 1)^c} \frac{n^b (\log n)^c}{a^n} \right| = |a|$$

the series converges absolutely for $|a| < 1$ and diverges for $|a| > 1$.

- $a = 1$.

    (i) $b > 1$. Let $b = 1 + 2\varepsilon$; we have

    $$\frac{1}{n^{1+2\varepsilon} (\log n)^c} = o\left(\frac{1}{n^{1+\varepsilon}}\right) \quad (n \to \infty)$$

    and, as the series $\sum n^{-(1+\varepsilon)}$ converges, the given series converges absolutely for $b > 1$.

    (ii) $b = 1$. The series converges (absolutely) only if $c > 1$ and diverges if $c \leqslant 1$, by the Integral Test [Rud87, p. 139].

    (iii) $b < 1$. Comparing with the harmonic series, we conclude that the series diverges.

- $a = -1$. By Leibniz Criterion [Rud87, p. 71], the series converges exactly when

    $$\lim \frac{1}{n^b (\log n)^c} = 0$$

    which is equivalent to $b > 0$ or $b = 0$, $c > 0$.

Solution to 1.3.22: Note that

$$\frac{\sqrt{n+1}-\sqrt{n}}{n^x} \sim \frac{1}{n^{x+1/2}} \quad (n \to \infty)$$

that is,

$$\lim_{n \to \infty} \frac{\sqrt{n+1}-\sqrt{n}/n^x}{1/n^{x+1/2}} = 1$$

so the given series and

$$\sum_{n=1}^{\infty} \frac{1}{n^{x+1/2}}$$

converge or diverge together. They converge when $x > 1/2$.

Solution to 1.3.23: If $a \leqslant 0$, the general term does not go to zero, so the series diverges. If $a > 0$, we have, using the Maclaurin series [PMJ85, p. 127] for $\sin x$,

$$\frac{1}{n} - \sin\frac{1}{n} = \frac{1}{6n^3} + o(n^{-3}) \quad (n \to \infty)$$

and, therefore,

$$\left(\frac{1}{n} - \sin\frac{1}{n}\right)^a = \frac{1}{6^a n^{3a}} + o(n^{-3a}) \quad (n \to \infty).$$

Thus, the series converges if and only if $3a > 1$, that is, $a > 1/3$.

Solution to 1.3.24: For $n = 1, 2, \ldots$ the number of terms in A that are less than 10^n is $9^n - 1$, so we have

$$\sum_{a \in A} \frac{1}{a} = \sum_{n \geqslant 1} \sum_{\substack{10^{n-1} \leqslant a < 10^n \\ a \in A}} \frac{1}{a} \leqslant \sum_{n \geqslant 1} \frac{9^n}{10^{n-1}} = 10 \sum_{n \geqslant 1} \left(\frac{9}{10}\right)^n < \infty.$$

Solution to 1.3.25: Let S be the sum of the given series. Let $N_0 = 0$. By convergence, for each $k > 0$ there exists an $N_k > N_{k-1}$ such that

$$\sum_{N_k+1}^{\infty} a_n \leqslant \frac{S}{4^k}.$$

For $N_k + 1 \leqslant n \leqslant N_{k+1}$ let $c_n = 2^k$. We have $\lim c_n = \infty$. As the terms are all positive, we may rearrange the sum and get

$$\sum_{n=1}^{\infty} c_n a_n = \sum_{k=0}^{\infty} \sum_{N_k+1}^{N_{k+1}} c_n a_n$$

$$\leqslant \sum_{k=0}^{\infty} 2^k \sum_{N_k+1}^{\infty} a_n$$

$$= \sum_{k=0}^{\infty} 2^{-k} S$$

$$= 2S.$$

Solution 2. The convergence of the given series shows that there is an increasing sequence of positive integers (N_k) with $\sum_{n=N_k}^{\infty} a_n < 1/k^{-3}$ for each k. Let

$$c_n = \begin{cases} 1 & \text{if } n < N_1 \\ k & \text{if } N_k \leqslant n < N_{k+1}. \end{cases}$$

Then $c_n \to \infty$, and

$$\sum_{n=1}^{\infty} c_n a_n \leqslant \sum_{n=1}^{N_1-1} a_n + \sum_{k=1}^{\infty} k \sum_{n=N_k}^{\infty} a_k$$

$$\leqslant \sum_{n=1}^{\infty} a_n + \sum_{k=1}^{\infty} \frac{k}{k^3}$$

$$\leqslant \sum_{n=1}^{\infty} a_n + \sum_{k=1}^{\infty} \frac{1}{k^2}$$

$$< \infty.$$

Solution 3. Let $\sum a_n$ converge to S. For positive n let $R_n = \sum_{k=1}^{n} a_k$, and $R_0 = 0$, so we have $a_n = R_n - R_{n-1}$ for all n. Consider the sequence c_n defined by $c_n = 1/(\sqrt{R_n} + \sqrt{R_{n-1}})$. As $R_n \to S$ we get $c_n \to \infty$.
   We have

$$\sum_{n=1}^{\infty} c_n a_n = \sum_{n=1}^{\infty} \frac{R_n - R_{n-1}}{\sqrt{R_n} + \sqrt{R_{n-1}}}$$

$$= \sum_{n=1}^{\infty} \sqrt{R_n} - \sqrt{R_{n-1}}$$

$$= \sqrt{S}.$$

Solution to 1.3.26: Using the formula $\sin 2x = 2 \sin x \cos x$ and the Induction Principle [MH93, p. 7], starting with $\sin \frac{\pi}{2} = 1$, we see that

$$\cos \frac{\pi}{2^2} \cos \frac{\pi}{2^3} \cdots \cos \frac{\pi}{2^n} = \frac{1}{2^{n-1} \sin \frac{\pi}{2^n}}.$$

So we have

$$\frac{1}{2^{n-1}\sin\frac{\pi}{2^n}} = \frac{2\,\frac{\pi}{2^n}}{\pi\,\sin\frac{\pi}{2^n}} \sim \frac{2}{\pi} \qquad (n\to\infty)$$

since $\sin x \sim x\ (x\to 0)$.

1.4 Differential Calculus

Solution to 1.4.1: Lemma 1: *If (x_n) is an infinite sequence in the finite interval $[a, b]$, then it has a convergent subsequence.*

Consider the sequence $y_k = \sup\{x_n \mid n \geq k\}$. By the least upper bound property, we know that y_k exists and is in $[a, b]$ for all k. By the definition of supremum, it is clear that the y_k's form a nonincreasing sequence. Let y be the infimum of this sequence. From the definition of infimum, we know that the y_k's converge to y. Again, by the definition of supremum, we know that we can find x_n's arbitrarily close to each y_k, so we can choose a subsequence of the original sequence which converges to y.

Lemma 2: *A continuous function f on $[a, b]$ is bounded.*

Suppose f is not bounded. Then, for each n, there is a point $x_n \in [a, b]$ such that $|f(x_n)| > n$. By passing to a subsequence, we may assume that the x_n's converge to a point $x \in [a, b]$. (This is possible by Lemma 1.) Then, by the continuity of f at x, we must have that $|f(x) - f(x_n)| < 1$ for n sufficiently large, or $|f(x_n)| < |f(x)| + 1$, contradicting our choice of the x_n's.

Lemma 3: *A continuous function f on $[a, b]$ achieves its extrema.*

It will suffice to show that f attains its maximum, the other case is proved in exactly the same way. Let $M = \sup f$ and suppose f never attains this value. Define $g(x) = M - f(x)$. Then $g(x) > 0$ on $[a, b]$, so $1/g$ is continuous. Therefore, by Lemma 2, $1/g$ is bounded by, say, N. Hence, $M - f(x) > 1/N$, or $f(x) < M - 1/N$, contradicting the definition of M.

Lemma 4: *If a differentiable function f on (a, b) has a relative extremum at a point $c \in (a, b)$, then $f'(c) = 0$.*

Define the function g by

$$g(x) = \begin{cases} \frac{f(x)-f(c)}{x-c} & \text{for}\quad x \neq c \\ 0 & \text{for}\quad x = c \end{cases}$$

and suppose $g(c) > 0$. By continuity, we can find an interval J around c such that $g(x) > 0$ if $x \in J$. Therefore, $f(x) - f(c)$ and $x - c$ always have the same sign in J, so $f(x) < c$ if $x < c$ and $f(x) > f(c)$ if $x > c$. This contradicts the fact that f has a relative extremum at c. A similar argument shows that the assumption that $g(c) < 0$ yields a contradiction, so we must have that $g(c) = 0$.

Lemma 5 (Rolle's Theorem [MH93, p. 200]): *Let f be continuous on $[a, b]$ and differentiable on (a, b) with $f(a) = f(b)$. There is a point $c \in (a, b)$ such that $f'(c) = 0$.*

Suppose $f'(c) \neq 0$ for all $c \in (a, b)$. By Lemma 3, f attains its extrema on $[a, b]$, but by Lemma 4 it cannot do so in the interior since otherwise the derivative at that point would be zero. Hence, it attains its maximum and minimum at the endpoints. Since $f(a) = f(b)$, it follows that f is constant, and so $f'(c) = 0$ for all $c \in (a, b)$, a contradiction.

Lemma 6 (Mean Value Theorem [Rud87, p. 108]): *If f is a continuous function on $[a, b]$, differentiable on (a, b), then there is $c \in (a, b)$ such that $f(b) - f(a) = f'(c)(b - a)$.*

Define the function $h(x) = f(x)(b - a) - x(f(b) - f(a))$. h is continuous on $[a, b]$, differentiable on (a, b), and $h(a) = h(b)$. By Lemma 5, there is $c \in (a, b)$ such that $h'(c) = 0$. Differentiating the expression for h yields the desired result.

There is a point c such that $f(b) - f(a) = f'(c)(b - a)$, but, by assumption, the right-hand side is 0 for all c. Hence, $f(b) = f(a)$.

Solution to 1.4.2: 1. We have $\exp(f(x)) = (1 + 1/x)^x$, which is an increasing function. As the exponential is also increasing, so is f.

2. We have

$$\lim_{x \to 0} f(x) = \lim_{x \to 0} \frac{\log(x + 1) - \log x}{1/x} = \lim_{x \to 0} \frac{1/(x + 1) - 1/x}{-1/x^2} = 0.$$

On the other hand,

$$\lim_{x \to \infty} \left(1 + \frac{1}{x}\right)^x = e$$

so $\lim_{x \to \infty} f(x) = 1$.

Solution to 1.4.3: Let $0 < s < t$. Then by the Mean Value Theorem, there exist u in the interval $(0, s)$ such that $f(s) = f(s) - f(0) = sf'(u)$ and v in the interval (s, t) such that $f(t) - f(s) = (t - s)f'(v)$. Then

$$\frac{f(s)}{s} = f'(u) \leqslant f'(v) = \frac{f(t) - f(s)}{t - s}.$$

This inequality reduces to $f(s)/s \leqslant f(t)/t$. Hence g is an increasing function on the interval $(0, \infty)$. Now $g(0) = f'(0) = \lim_{x \to 0} f(x)/x = \lim_{x \to 0} g(x)$. Fix $x > 0$. For $0 < t < x$, $g(t) \leqslant g(x)$. Taking limits as $t \to 0$ we conclude that $g(0) \leqslant g(x)$. Thus g is increasing on $[0, \infty)$.

For $x > 0$, $g(x)$ is the slope of the chord of the graph of f joining the points $(0, f(0))$ and $(x, f(x))$.

Solution 2. Since $f(0) = 0$, $f(x) = \int_0^x f'(t)dt$ and using the fact that $f'(x)$ is increasing, we get

$$f(x) = \int_0^x f'(t)dt < \int_0^x f'(x)dt = xf'(x),$$

so $g'(x) = \dfrac{xf'(x) - f(x)}{x^2}$ is positive and g an increasing function of x.

Solution to 1.4.4: 1. The function $f(x)$ given by

$$f(x) = x^2 \sin \frac{1}{x}$$

has a derivative that is not continuous at zero:

$$f'(x) = \begin{cases} 2x \sin \frac{1}{x} - \cos \frac{1}{x} & \text{for} \quad x \neq 0 \\ 0 & \text{for} \quad x = 0 \end{cases}$$

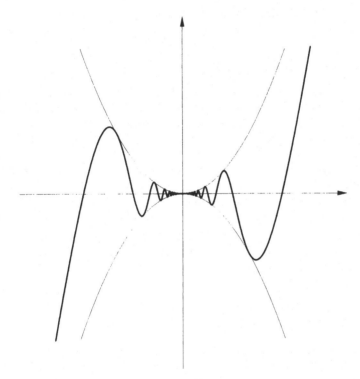

2. Consider the function g given by $g(x) = f(x) - 2x$. We then have $g'(0) < 0 < g'(1)$. Therefore, $g(x) < g(0)$ for x close to 0, and $g(x) < g(1)$ for x close to 1. Then the minimum of g in $[0, 1]$ occurs at an interior point $c \in (0, 1)$, at which we must have $g'(c) = 0$, which gives $f'(c) = 2$.

Solution to 1.4.5: Suppose y assumes a positive maximum at ξ. Then $y(\xi) > 0$, $y'(\xi) = 0$, and $y''(\xi) \leqslant 0$, contradicting the differential equation. Hence, the maximum of y is 0. Similarly, y cannot assume a negative minimum, so y is identically 0.

Solution to 1.4.6: 1. Suppose u has a local maximum at x_0 with $u(x_0) > 0$. Then $u''(x_0) \leqslant 0$, but $u''(x_0) = e^{x_0} u(x_0) > 0$ and we have a contradiction. So u cannot have a positive local maximum. Similarly, if u has a local minimum at x_0, then $u''(x_0) \geqslant 0$, so we must have $u(x_0) \geqslant 0$ and u cannot have a negative local minimum.

2. Suppose $u(0) = u(1) = 0$. If $u(x_0) \neq 0$ for some $x_0 \in (0, 1)$, then, as u is continuous, u attains a positive local maximum or a negative local minimum, which contradicts Part 1.

Solution to 1.4.7: From the given inequality we get

$$0 \geq e^{-Kt} y'(t) - K e^{-Kt} y(t) = \frac{d}{dt} \left(e^{-Kt} y(t) \right) \qquad (t \geq 0).$$

Integrating from 0 to t we find that, for $t \geq 0$,

$$e^{-Kt} y(t) - y(0) \leq 0,$$

from which the desired inequality follows.

Solution 2. Let $z(t) = \log y(t)$. Then $z'(t) = y'(t)/y(t) \leq K$. Fix $t > 0$. By the Mean Value Theorem, there exist u in the interval $(0, t)$ such that $z(t) - z(0) = z'(u)t \leq Kt$. Hence $z(t) \leq Kt + \log y(0)$. Since the exponential function is strictly increasing, we obtain $y(t) \leq e^{Kt} y(0)$. This inequality also holds for $t = 0$.

Solution to 1.4.8: Without loss of generality assume $x_0 = 0$. As f is continuous and $f(0) = 0$, we have $f'(x) > f(x) > 0$ in some interval $[0, \varepsilon)$. Suppose that $f(x)$ is not positive for all positive values of x. Let $c = \inf\{x > 0 \mid f(x) \leq 0\}$. Since f is continuous and positive in a neighborhood of the origin, we have $c > 0$ and $f(c) = 0$. By Rolle's Theorem, [MH93, p. 200] there is a point d with $0 < d < c$ and $f'(d) = 0$. However, by the definition of c, we have $f'(d) > f(d) > 0$, a contradiction.

Solution 2. Let $g(x) = e^{-x} f(x)$. Then $g'(x) = e^{-x}(f'(x) - f(x)) > 0$. As g is an increasing function, we have $g(x) = e^{-x} f(x) > g(x_0) = 0$ for $x > x_0$, and the conclusion follows.

Solution to 1.4.9: Suppose that $f(x) \leq 0$ for some positive value of x. Then $a = \inf\{x > 0 \mid f(x) \leq 0\}$ is positive. Since f is continuous, $f(a) = 0$. Let $b \in (0, a)$, then $f(b) > 0$. By the Mean Value Theorem [Rud87, p. 108], there exists $c \in (b, a)$ such that $f'(c) = (f(a) - f(b))/(a - b) < 0$. Applying the same theorem to f' on the interval $(0, c)$ we get a number $d \in (0, c)$ such that $f''(d) = (f'(c) - f'(0))/c < 0$. Since $0 < d < a$, $f(d) > 0$, contradicting the assumption that $f''(x) \geq f(x)$ for all $x \in (0, a)$.

Solution to 1.4.10: Let $f : \mathbb{R} \to \mathbb{R}$ be defined by $f(x) = ae^x - 1 - x - \dfrac{x^2}{2}$. We have

$$\lim_{x \to -\infty} f(x) = -\infty \quad \text{and} \quad \lim_{x \to \infty} f(x) = \infty$$

so f has at least one real root, x_0, say. We have

$$f'(x) = ae^x - 1 - x > ae^x - 1 - x - \frac{x^2}{2} = f(x) \quad \text{for all} \quad x \in \mathbb{R};$$

therefore, by Problem 1.4.8, f has no other root.

Solution to 1.4.11: Let $g : [0, 1] \to \mathbb{R}$ be defined by $g(x) = e^{-2x} f(x)$. We have

$$g'(x) = e^{-2x}(f'(x) - 2f(x)) \leqslant 0$$

so g is a decreasing function. As $g(0) = 0$ and g is nonnegative, we get $g \equiv 0$, so the same is true for f.

Solution to 1.4.12: Consider the function g defined on $[0, 1]$ by $g(x) = e^x f(x)$. We have

$$g''(x) = e^x \left(f''(x) + 2f'(x) + f(x) \right) \geqslant 0$$

so g is concave upward; that is, the point $(x, g(x))$ must lie below the chord joining $(0, g(0))$ and $(1, g(1)) = (1, 0)$ for $x \in (0, 1)$. Then $g(x) \leqslant 0$ and the conclusion follows.

Solution to 1.4.13: As φ_1 and φ_2 satisfy the given differential equations, we have $\varphi_1'(t) = v_1 (\varphi_1(t))$ and $\varphi_2'(t) = v_2 (\varphi_2(t))$. Since $\varphi_1(t_0) = \varphi_2(t_0)$, it follows from our hypotheses that $\varphi_1'(t_0) < \varphi_2'(t_0)$. Hence, there exists a point $s_0 > t_0$ such that $\varphi_1(t) \leqslant \varphi_2(t)$ for $t_0 \leqslant t \leqslant s_0$. Suppose there existed a point $s_0 < t < b$ such that $\varphi_1(t) > \varphi_2(t)$. Let $t_1 \geqslant s_0$ be the infimum of all such points t. By continuity, we must have that $\varphi_1(t_1) = \varphi_2(t_1)$. Hence, repeating the above argument, we see that there must be a point $s_1 > t_1$ such that $\varphi_1(t) \leqslant \varphi_2(t)$ if $t_1 < t < s_1$, contradicting our definition of t_1.

Solution to 1.4.14: Let $f(x) = x^2 \sin \dfrac{1}{x}$ and $g(x) = x$. Then we have $\lim_{x \to 0} f(x) = \lim_{x \to 0} g(x) = 0$ and

$$\lim_{x \to 0} \frac{f(x)}{g(x)} = \lim_{x \to 0} x \sin \frac{1}{x} = 0.$$

As $g'(x) = 1$ and $f'(x) = 2x \sin \dfrac{1}{x} - \cos \dfrac{1}{x}$, we have

$$\lim_{x \to 0} \frac{f'(x)}{g'(x)} = \lim_{x \to 0} 2x \sin \frac{1}{x} - \cos \frac{1}{x}$$

which does not exists.

Solution to 1.4.15: The given conditions imply that $f'(0) = 0$, so the question is whether $\lim_{x \to 0} f'(x)/x$ must exist. The answer is no and one counterexample is $f(x) = x^3 \sin \frac{1}{x}$. A calculation gives

$$\frac{f'(x)}{x} = 3x \sin \frac{1}{x} - \cos \frac{1}{x} \cdot$$

Solution to 1.4.16: By Rolle's Theorem [MH93, p. 200], $f'(x_1) = 0$ for some $x_1 \in (0, 1)$. Then, since $f'(0) = 0$, $f''(x_2) = 0$ for some $x_2 \in (0, x_1)$. Repeated applications of Rolle's Theorem give $f^{(n)}(x_n) = 0$ for some $x_n \in (0, x_{n-1})$, and therefore, $f^{(n+1)}(x) = 0$ for some $x \in (0, x_n) \subset (0, 1)$.

Solution to 1.4.17: Since $f' \leqslant 0$ and $f \geqslant 0$, we deduce that f is monotone and $\lim_{t \to \infty} f(t) = \xi$ exists. Hence, for every $\delta > 0$, we have

$$\lim_{t \to \infty} \frac{f(t + \delta) - f(t)}{\delta} = 0.$$

On the other hand

$$\frac{f(t + \delta) - f(t)}{\delta} = f'(t) + \frac{1}{2}\delta f''(\theta)$$

for some θ. Therefore,

$$\limsup_{t \to \infty} |f'(t)| \leqslant \frac{1}{2}\delta \sup_\theta |f''(\theta)|.$$

Letting $\delta \to 0$ we get $\lim_{t \to \infty} f'(t) = 0$.

Solution to 1.4.18: Let $x > 0$ and $\delta > 0$. Since f is positive and $\log$ is continuous,

$$\begin{aligned}
\log \lim_{\delta \to 0} \left(\frac{f(x + \delta x)}{f(x)} \right)^{1/\delta} &= \lim_{\delta \to 0} \log \left(\frac{f(x + \delta x)}{f(x)} \right)^{1/\delta} \\
&= \lim_{\delta \to 0} \frac{\log f(x + \delta x) - \log f(x)}{\delta} \\
&= \lim_{\delta \to 0} \frac{x \left(\log f(x + \delta x) - \log f(x) \right)}{\delta x} \\
&= x \left(\log f(x) \right)' \\
&= \frac{x f'(x)}{f(x)}
\end{aligned}$$

and the result follows, by exponentiating both sides.

Solution to 1.4.19: It is enough to show that

$$\lim_{h \to 0^+} \frac{f(h) - f(0)}{h} \quad \text{and} \quad \lim_{h \to 0^-} \frac{f(h) - f(0)}{h}$$

both exist and are equal. By L'Hôpital's Rule [Rud87, p. 109]

$$\lim_{h \to 0^+} \frac{f(h) - f(0)}{h} = \lim_{h \to 0^+} \frac{f'(h)}{1} = \lim_{h \to 0} f'(h).$$

The other lateral limit can be treated similarly.

Solution to 1.4.20: We have

$$\begin{aligned}
p_t(x) &= (1 + t^2)x^3 - 3t^3x + t^4 \\
p'_t(x) &= 3(1 + t^2)x^2 - 3t^3 \\
p''_t(x) &= 6(1 + t^2)x
\end{aligned}$$

- $t < 0$. In this case, $p'_t > 0$ and $p_t(x) < 0$ for x sufficiently negative, and $p_t(x) > 0$ for x sufficiently positive. Hence, by the Intermediate Value Theorem [Rud87, p. 93], p_t has exactly one root, of multiplicity 1, since the derivative is positive.

- $t = 0$. Now $p_t(x) = x^3$, which has a single zero of multiplicity 3.

- $t > 0$. We have

$$p'_t\left(\pm\sqrt{\frac{t^3}{1+t^2}}\right) = 0$$

and $p''_t(x) < 0$ for negative x; $p''_t(x) > 0$ for positive x. So

$$p_t\left(\sqrt{\frac{t^3}{1+t^2}}\right)$$

is a local minimum, and

$$p_t\left(-\sqrt{\frac{t^3}{1+t^2}}\right)$$

is a local maximum of p_t.

We will study the values of p_t at these critical points. As $p_t(0) > 0$ and $p'_t(0) < 0$, the relative maximum must be positive.

We have

$$p_t\left(\sqrt{\frac{t^3}{1+t^2}}\right) = t^4\left(1 - \sqrt{\frac{t}{1+t^2}}\right) = A_t$$

say. We get

(i) $0 < t < 2 - \sqrt{3}$. In this case, we have $A_t > 0$, so p_t has one single root.

(ii) $2 - \sqrt{3} < t < 2 + \sqrt{3}$. Now $A_t < 0$ and p_t has three roots.

(iii) $t > 2 + \sqrt{3}$. We have $A_t > 0$ and p_t has one root.

Solution to 1.4.21: Let

$$h(x) = \frac{f(x) - f(a)}{x - a} - f'(a)$$

so that $\lim_{x \to a} h(x) = 0$ and $f(x) = f(a) + \left(f'(a) + h(x)\right)(x - a)$. Then

$$\frac{f(y_n) - f(x_n)}{y_n - x_n} = \frac{f'(a)(y_n - x_n) + h(x)(y_n - a) - h(x_n)(x_n - a)}{y_n - x_n}$$

so that

$$\left| \frac{f(y_n) - f(x_n)}{y_n - x_n} - f'(a) \right| \leqslant |h(y_n)| \left(\frac{y_n - a}{y_n - x_n} \right) + |h(x_n)| \left(\frac{a - x_n}{y_n - x_n} \right)$$
$$\leqslant |h(y_n)| + |h(x_n)|$$
$$= o(1) \quad (n \to \infty).$$

Solution to 1.4.22: By changing variables, it is enough to show that $f(1) \geqslant f(0)$. Without loss of generality assume $f(0) = 0$. Consider the function g defined by

$$g(x) = f(x) - f(1)x.$$

As g is continuous, it attains a maximum at some point $\xi \in [0, 1]$. We can assume $\xi < 1$, because $g(1) = g(0) = 0$. As $g(\xi) \geqslant g(x)$ for $\xi < x < 1$, we have

$$0 \geqslant \limsup_{x \to \xi +} \frac{g(x) - g(\xi)}{x - \xi} = -f(1) + \frac{f(x) - f(\xi)}{x - \xi}.$$

As the rightmost term is nonnegative, we have $f(1) \geqslant 0$, as desired.

Solution to 1.4.24: We have

$$1 = f(z) \left(\frac{e^z - 1}{z} \right) = \left(\xi_0 + \xi_1 z + \xi_2 z^2 + \cdots \right) \left(1 + \frac{z}{2!} + \frac{z^2}{3!} + \cdots \right).$$

Multiplying this out, we get $\xi_0 = 1$ and

$$\sum_{k=0}^{n} \frac{\xi_{n-k}}{(k+1)!} = 0.$$

From this, it can easily be seen by the Induction Principle [MH93, p. 7] that all the ξ_i's are rational.

Solution to 1.4.25: The function f given by

$$f(x) = \begin{cases} (-1/3)e^{3-1/x+2/(2x-3)} & \text{for} \quad 0 < x < 3/2 \\ 0 & \text{for} \quad x \leqslant 0 \text{ or } x \geqslant 3/2 \end{cases}$$

is such a function.

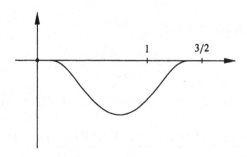

This is based on the example of a nonconstant function having derivatives of all orders, vanishing for negative x:

$$\begin{cases} e^{-1/x} & \text{for } x > 0 \\ 0 & \text{for } x \leqslant 0. \end{cases}$$

Solution to 1.4.26: Let $g(x) = \sin^n x$; it follows from the Taylor series of $\sin x$ that the first nonzero derivative of g at 0 is $g^{(n)}(0) = n!$. For any positive number λ, set $c = \alpha/(n!\lambda^n)$; then $f(x) = c \sin^n(\lambda x)$ satisfies $f^{(n)}(0) = c\lambda^n n! = \alpha$.

There is a real number M such that $|g^{(k)}(x)| \leqslant M$ for $x \in [0, 2\pi]$ and $k = 0, 1, \ldots n - 1$. Since g and its derivatives are periodic, with period 2π, $|g^{(k)}(x)| \leqslant M$ for all real x and $k < n$. Therefore $|f^{(k)}(x)| = |c|\lambda^k |g^{(k)}(x)| \leqslant |c|\lambda^k M = C\lambda^{k-n}$ for all real x, where $C = |\alpha|/n!$. Choosing $\lambda \geqslant \max(1, C/\varepsilon)$ ensures that $|f^{(k)}(x)| \leqslant \varepsilon$ for all $x \in \mathbb{R}$ and $k = 0, 1, \ldots n - 1$.

Solution to 1.4.28: We claim there is an $\varepsilon > 0$ such that $f(t) \neq 0$ for all $t \in (0, \varepsilon)$. Suppose, on the contrary, that there is a sequence $x_n \to 0$ such that $f(x_n) = 0$. Considering the real function $\Re f(x)$, to each subinterval $[x_{n+1}, x_n]$, we find a sequence $t_n \to 0$, $t_n \in [x_{n+1}, x_n]$, such that $\Re f'(t_n) = 0$ for all n, but since $\lim_{t \to 0+} f'(t) = C$, this would imply $\Re C = 0$. In the same fashion, using the imaginary part of $f(x)$, we see that $\Im C = 0$, which is a contradiction.

Since $f(t)$ is nonzero on a small interval starting at 0, the composition with the C^∞-function absolute value

$$| \quad | : \mathbb{C} \setminus \{0\} \to \mathbb{R}_+$$

will give a C^1-function $g(t) = |f(t)|$, on a small neighborhood of zero.

Solution to 1.4.29: By Taylor's Theorem [Rud87, p. 110], there is a constant C such that

$$|f(x) - f(0) - f'(0)x| \leqslant Cx^2$$

when $|x| < 1$. Since $f'(0) = 0$, we actually have $|f(x) - f(0) - f'(0)| \leqslant Cx^2$ and, consequently, by the triangle inequality, $f(x) \leqslant f(0) + Cx^2$ when $|x| < 1$. We conclude:

If (x, y) lies on or below the graph of f and $|x| < 1$, then $y \leqslant f(0) + Cx^2$.

Now consider the disc D centered at $(0, f(0) + b)$ with radius b, where $0 < b < 1$ will be chosen at the end. Clearly, $(0, f(0))$ is on the boundary of D. On the other hand, if $(x, y) \in D$, then $|x| < b < 1$ and

$$x^2 + (y - f(0) - b)^2 < b^2$$

$$|y - f(0) - b| < \sqrt{b^2 - x^2}$$

$$y > f(0) + b - \sqrt{b^2 - x^2} = f(0) + b - b\sqrt{1 - x^2/b^2} \geqslant f(0) + b - b(1 - x^2/2b^2)$$

since $\sqrt{1-x} \leqslant 1 - x/2$ when $0 \leqslant x \leqslant 1$. Thus,

$$y > f(0) + x^2/2b.$$

If $1/2b \geqslant c$, then it follows that (x, y) must be above the graph of f. So we are done if we take $b = \min\{1/2, 1/2c\}$.

1.5   Integral Calculus

Solution to 1.5.1: Let $a = x_0 < x_1 < \cdots < x_n = b$ be any partition of $[a, b]$. Since f' is Riemann integrable, on each interval $[x_{i-1}, x_i]$ it has a finite supremum M_i and infimum m_i.

By the Mean Value Theorem [Rud87, p. 108],

$$\sum_{i=1}^{n} m_i(x_i - x_{i-1}) \leqslant \sum_{i=1}^{n}(f(x_i) - f(x_{i-1})) \leqslant \sum_{i=1}^{n} M_i(x_i - x_{i-1}).$$

The middle sum simplifies to $f(b) - f(a)$, which therefore lies between the lower and upper sums for the partition. As these sums both converge to the integral of f' on $[a, b]$, we must have $\int_a^b f'(x)\,dx = f(b) - f(a)$.

Solution to 1.5.2: Assume f does not vanish identically. Then, for some $\delta > 0$, the set $\{x \in (0, 1) \mid f(x) > \delta\}$ is nonempty and, as f is continuous, open; therefore it contains a closed interval I of length $L > 0$. Let g be the function that equals δ on I and 0 off I. Then g is Riemann integrable with $\int_0^1 g(x)\,dx = \delta L$. Since $f \geqslant g$ we have

$$\int_0^1 f(x)\,dx \geqslant \int_0^1 g(x)\,dx = \delta L > 0,$$

as desired.

Solution 2. The properties of the Riemann integral we use are

- If f, g are (Riemann) integrable on $[a, b]$ and $f(x) \geqslant g(x)$ for all x then $\int_a^b f(x)\,dx \geqslant \int_a^b g(x)\,dx$.

- If f is integrable on $[a, b]$ and $a < c < b$ then f is integrable on $[a, c]$ and on $[c, b]$, and $\int_a^b f(x)\,dx = \int_a^c f(x)\,dx + \int_c^b f(x)\,dx$.

Suppose that $f(c) > 0$ for some c. We consider $0 < c < 1$. Since f is continuous at c, there is an interval $[a, b]$ containing c in which $f(x) > f(c)/2$. Then

$$\int_a^b f(x)\,dx \geqslant f(c)(b-a)/2 > 0,$$

and hence

$$\int_0^1 f(x)\,dx = \int_0^a f(x)\,dx + \int_a^b f(x)\,dx + \int_b^1 f(x)\,dx > 0,$$

a contradiction. A similar argument applies when c is an endpoint of $[0, 1]$.

Solution to 1.5.3: Since f is continuous, it attains its minimum and maximum at x_0 and y_0, respectively, in $[0, 1]$. So we have

$$f(x_0) \int_0^1 x^2\,dx \leqslant \int_0^1 x^2 f(x)\,dx \leqslant f(y_0) \int_0^1 x^2\,dx,$$

or

$$f(x_0) \leqslant 3 \int_0^1 x^2 f(x)\,dx \leqslant f(y_0).$$

Therefore, by the Intermediate Value Theorem [Rud87, p. 93], there is a point $\xi \in [0, 1]$ with

$$f(\xi) = 3 \int_0^1 x^2 f(x)\,dx.$$

Solution to 1.5.4: Since the discontinuities are only of the first type (the limit exists), they do not have any accumulation point (for a detailed proof of this, see the Solution to Problem 1.2.8), and form a finite set. Let $d_1 < d_2 < \cdots < d_n$ be the set of discontinuities of f. Then f is continuous in every interval $[x, y]$ with $d_n < x < y < d_{n+1}$; using the Solution to Problem 1.5.5 (on both endpoints of the interval), f is integrable on each interval of the type $[d_n, d_{n+1}]$, so f is integrable on $[a, b]$.

Solution to 1.5.5: 1. Let $|f(x)| \leqslant M$ for $x \in [0, 1]$. If (b_n) is a decreasing vanishing sequence, then $\int_{b_n}^1 |f| \leqslant M$ is a bounded, increasing sequence, so it must converge. We conclude that $|f|$ is Riemann integrable over $[0, 1]$, and so is f.
2. The function $f(x) = 1/x$ is integrable over any interval $[b, 1]$ for positive b, but is not integrable over $[0, 1]$.

Solution to 1.5.6: Using change of variables,

$$\int_{-\infty}^{\infty} |f(x)|dx = \int_0^{\infty} (|f(x)| + |f(-x)|)dx.$$

If, for $x > 1$, we had $x(|f(x)| + |f(-x)|) > 1$, we would have the convergence of the integral $\int_1^{\infty} \frac{1}{x}dx$. So, there is x_1 such that $x_1(|f(x_1)| + |f(-x_1)|) \leqslant 1$. If, for all $x > \max(2, x_1)$, we have $x(|f(x)| + |f(-x)|) > 1/2$ we similarly conclude the convergence of $\int_{\max(2,x_1)}^{\infty}(1/x)\,dx$, thus, there is $x_2 > x_1$ with $x_2(|f(x_2)| + |f(-x_2)|) \leqslant 1/2$. Recursively we can then define a sequence (x_n)

such that $x_{n+1} > \max(n + 1, x_n)$ and $x_n(|f(x_n)| + |f(-x_n)|) \leqslant 1/n$. It follows that $x_n \to \infty$ and

$$\lim_{n \to \infty} x_n |f(x_n)| = \lim_{n \to \infty} x_n |f(-x_n)| = 0.$$

Solution to 1.5.7: 1. Letting $t = x + s$, we get

$$f(x) = e^{x^2/2} \int_0^\infty e^{-(x+s)^2/2}\, ds = \int_0^\infty e^{-sx - s^2/2}\, ds.$$

Since $s > 0$, $e^{-s^2/2} < 1$, so $e^{-sx - s^2/2} < e^{-sx}$ for all positive x; then

$$0 < f(x) < \int_0^\infty e^{-sx}\, ds = \frac{1}{x}.$$

2. Let $0 < x_1 < x_2$. For $s > 0$, $e^{-sx_1 - s^2/2} > e^{-sx_2 - s^2/2}$, so

$$f(x_1) = \int_0^\infty e^{-sx_1 - s^2/2}\, ds > \int_0^\infty e^{-sx_2 - s^2/2}\, ds = f(x_2)$$

Solution 2.

1. The function f is clearly positive. Using integration by parts, we get

$$\int_x^\infty e^{-t^2/2}\, dt = \int_x^\infty \frac{1}{t}\left(-e^{-t^2/2}\right)'\, dt = \frac{e^{-x^2/2}}{x} - \int_x^\infty \frac{e^{-t^2/2}}{t^2}\, dt.$$

Therefore,

$$f(x) = \frac{1}{x} - e^{x^2/2} \int_x^\infty \frac{e^{-t^2/2}}{t^2}\, dt < \frac{1}{x}.$$

2. We have

$$f'(x) = \left(\frac{1}{x} - e^{x^2/2} \int_x^\infty \frac{e^{-t^2/2}}{t^2}\, dt\right)'$$

$$= -\frac{1}{x^2} + xe^{-x^2/2} \int_x^\infty \frac{e^{-t^2/2}}{t^2}\, dt + \frac{1}{x^2}$$

$$= xe^{-x^2/2} \int_x^\infty \frac{e^{-t^2/2}}{t^2}\, dt > 0,$$

so f is an increasing function.

Solution to 1.5.8: Integrating by parts and noting that φ vanishes at 1 and 2, we get

$$\int_1^2 e^{i\lambda x} \varphi(x)\, dx = \frac{e^{i\lambda x}}{i\lambda} \varphi(x)\Big|_1^2 - \frac{1}{i\lambda} \int_1^2 e^{i\lambda x} \varphi'(x)\, dx = -\frac{1}{i\lambda} \int_1^2 e^{i\lambda x} \varphi'(x)\, dx,$$

applying integration by parts a second time and using the fact that φ' also vanishes at the endpoints, we get

$$\int_1^2 e^{i\lambda x}\varphi(x)\,dx = -\frac{1}{\lambda^2}\int_1^2 e^{i\lambda x}\varphi''(x)\,dx\,.$$

Taking absolute values gives

$$\left|\int_1^2 e^{i\lambda x}\varphi(x)\,dx\right| \leqslant \frac{1}{\lambda^2}\int_1^2 |\varphi''(x)|\,dx\,.$$

Since $\varphi \in C^2$, the integral on the right-hand side is finite, and we are done.

Solution to 1.5.9: Suppose that f is such a function. Cauchy–Schwarz Inequality [MH93, p. 69] gives

$$a = \int_0^1 x f(x)\,dx$$

$$\leqslant \left(\int_0^1 x^2 f(x)\,dx \int_0^1 f(x)\,dx\right)^{1/2}$$

$$\leqslant a.$$

So we must have a chain of equalities. For equality to hold in the Cauchy–Schwarz Inequality, we must have $x\sqrt{f(x)} = k\sqrt{f(x)}$ for some constant k, so $\sqrt{f(x)} \equiv 0$, which contradicts

$$\int_0^1 f(x)\,dx = 1.$$

Thus, no such function f can exist.

Solution 2. Multiplying the given identities by α^2, -2α, and 1, respectively, we get

$$\int_0^1 f(x)(\alpha - x)^2 dx = 0$$

but the integral above is clearly positive for every positive continuous function, so no such function can exist.

Solution to 1.5.10: Dividing the integral in n pieces, we have

$$\left|\sum_{j=0}^{n-1} \frac{f(j/n)}{n} - \int_0^1 f(x)\,dx\right| = \left|\sum_{j=0}^{n-1}\left(\frac{f(j/n)}{n} - \int_{j/n}^{(j+1)/n} f(x)\,dx\right)\right|$$

$$\leqslant \sum_{j=0}^{n-1} \int_{j/n}^{(j+1)/n} |f(j/n) - f(x)|\,dx.$$

For every $x \in (j/n, (j+1)/n)$, applying the Mean Value Theorem [Rud87, p. 108], there is $c \in (j/n, x)$ with

$$f'(c) = \frac{f(x) - f(j/n)}{x - j/n}.$$

As the derivative of f is uniformly bounded by M, this gives us the inequality

$$|f(x) - f(j/n)| \leqslant M(x - j/n).$$

Therefore,

$$\left| \sum_{j=0}^{n-1} \frac{f(j/n)}{n} - \int_0^1 f(x)\,dx \right| \leqslant \sum_{j=0}^{n-1} \int_{j/n}^{(j+1)/n} M(x - j/n)\,dx$$

$$= M \sum_{j=0}^{n-1} \left(\frac{(j+1)^2}{2n^2} - \frac{j^2}{2n^2} \right) - \frac{j}{n^2}$$

$$= M \sum_{j=0}^{n-1} \frac{1}{2n^2}$$

$$= \frac{M}{2n}.$$

Solution to 1.5.11: Suppose not. Then, for some $\delta > 0$, there is a sequence of real numbers, (x_n), such that $x_n \to \infty$ and $|f(x_n)| \geqslant \delta$. Without loss of generality, we can assume $f(x_n) \geqslant \delta$.

Let $\varepsilon > 0$ verify

$$|f(x) - f(y)| < \frac{\delta}{2} \quad \text{for} \quad |x - y| < \varepsilon,$$

then

$$\sum_{n \geqslant 1} \int_{x_n - \varepsilon}^{x_n + \varepsilon} f(x)\,dx \geqslant \sum_{n \geqslant 1} 2\varepsilon \frac{\delta}{2} = \infty$$

contradicting the convergence of $\int_0^\infty f(x)\,dx$.

Solution to 1.5.12: Let

$$g(x) = f(x) + \int_0^x f(t)\,dt.$$

The result follows from the following claims.

Claim 1: $\liminf_{x \to \infty} f(x) \leqslant 0$.

If not, there are ε, $x_0 > 0$ such that $f(x) > \varepsilon$ for $x > x_0$. Then, we have

$$g(x) = f(x) + \int_0^{x_0} f(t)dt + \int_{x_0}^{x} f(t)dt$$

$$\geq \varepsilon + \int_0^{x_0} f(t)dt + \varepsilon(x - x_0).$$

This is a contradiction since the right side tends to ∞ with x.

Claim 2: $\limsup\limits_{x\to\infty} f(x) \geq 0$.

This follows from Claim 1 applied to the function $-f$.

Claim 3: $\limsup\limits_{x\to\infty} f(x) \leq 0$.

Assume not. Then, for some $\varepsilon > 0$ there is a sequence $x_1, x_2, \ldots$ tending to ∞ such that $f(x_n) > \varepsilon$ for all n. By Claim 1, the function f assumes values $\leq \varepsilon/2$ for arbitrarily large values of its argument. Thus, after possibly deleting finitely many of the x_n's, we can find another sequence $y_1, y_2, \ldots$ tending to ∞ such that $y_n < x_n$ for all n and $f(y_n) \leq \varepsilon/2$ for all n. Let z_n be the largest number in $[y_n, x_n]$ where f takes the value $\varepsilon/2$ (it exists by the Intermediate Value Theorem [Rud87, p. 93]). Then

$$g(x_n) - g(z_n) = f(x_n) - f(z_n) + \int_{z_n}^{x_n} f(t)dt$$

$$> \varepsilon - \frac{\varepsilon}{2} + \int_{z_n}^{x_n} \frac{\varepsilon}{2} dt$$

$$\geq \frac{\varepsilon}{2}$$

which contradicts the existence of $\lim\limits_{x\to\infty} g(x)$.

Claim 4: $\liminf\limits_{x\to\infty} f(x) \geq 0$.

Apply Claim 3 to the function $-f$.

Solution to 1.5.13: Suppose that for some $\varepsilon > 0$, there is a sequence $x_n \to \infty$ with $x_n f(x_n) \geq \varepsilon$. Then, as f is monotone decreasing, we have $f(x) \geq \varepsilon/x$ for x large enough, which contradicts the convergence of $\int_0^\infty f(x)dx$, and the result follows.

Solution to 1.5.14: Let $0 < \varepsilon < 1$. As

$$\int_0^\infty f(x)\,dx < \infty,$$

there is an $N > 0$ such that for $n > N$,

$$\int_n^\infty f(x)\,dx < \varepsilon.$$

Therefore, for n large enough, that is, such that $n\varepsilon > N$, we have

$$\int_0^n \left(\frac{x}{n}\right) f(x)\,dx = \int_0^{n\varepsilon} \left(\frac{x}{n}\right) f(x)\,dx + \int_{n\varepsilon}^{n} \left(\frac{x}{n}\right) f(x)\,dx$$

$$< \varepsilon \int_0^{n\varepsilon} f(x)\,dx + \int_{n\varepsilon}^n f(x)\,dx$$

$$< \varepsilon \int_0^{n\varepsilon} f(x)\,dx + \varepsilon$$

$$< \varepsilon \left(\int_0^\infty f(x)\,dx + 1 \right).$$

Since this inequality holds for all $\varepsilon > 0$ and for all n sufficiently large, it follows that

$$\lim_{n\to\infty} \frac{1}{n} \int_0^n xf(x)\,dx = 0.$$

Solution to 1.5.15: Using the Maclaurin expansion [PMJ85, p. 127] of $\sin x$, we get

$$\frac{\sin x}{x} = \sum_0^\infty (-1)^n \frac{x^{2n}}{(2n+1)!}.$$

The series above is alternating for every value of x, so we have

$$\left| \frac{\sin x}{x} - \sum_0^k (-1)^n \frac{x^{2n+1}}{(2n+1)!} \right| \le \frac{x^{2k+2}}{(2k+3)!}.$$

Taking $k = 2$, we have

$$\left| I - \int_0^{1/2} \left(1 - \frac{x^2}{3!} \right) dx \right| \le \int_0^{1/2} \frac{x^4}{5!}\,dx$$

which gives an approximate value of $71/144$ with an error bounded by 0.00013.

Solution to 1.5.16: Let

$$I(t) = \int_0^1 \frac{dx}{(x^4 + t^4)^{1/4}} + \log t.$$

It suffices to show that for $t > 0$, the function $I(t)$ is bounded below and monotonically increasing. For $x, t \ge 0$, we have $(x+t)^4 \ge x^4 + t^4$, so

$$I(t) \ge \int_0^1 \frac{dx}{x+t} + \log t = \int_t^{1+t} \frac{du}{u} + \log t = \log(1+t) \ge 0.$$

We now show that $I'(t) \ge 0$ for $t > 0$. We have

$$I(t) = \int_0^t \frac{dx}{t\left((x/t)^4 + 1\right)^{1/4}} + \int_t^1 \frac{dx}{t\left((x/t)^4 + 1\right)^{1/4}} + \log t,$$

letting $y = x/t$, we get

$$I(t) = \int_0^1 \frac{dy}{(y^4 + 1)^{1/4}} + \int_1^{1/t} \frac{dy}{(y^4 + 1)^{1/4}} + \log t,$$

so

$$I'(t) = \frac{-1}{t^2 (1/t^4 + 1)^{1/4}} + \frac{1}{t} \geq 0.$$

Solution to 1.5.17: Integrate by parts to get

$$\int_0^\infty f(x)^2 dx = \int_0^\infty \left(\frac{d}{dx} x\right) f(x)^2 dx = -\int_0^\infty x \cdot 2f(x) f'(x) dx .$$

The boundary terms vanish because $xf(x)^2 = 0$ at $x = 0$ and ∞. By the Cauchy–Schwarz Inequality [MH93, p. 69],

$$\left| \int_0^\infty x f(x) f'(x) dx \right| \leq \sqrt{\int_0^\infty x^2 f(x)^2 dx} \sqrt{\int_0^\infty f'(x)^2 dx} .$$

Solution to 1.5.18: Consider the figure

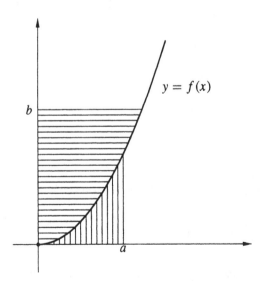

The left side of the desired inequality is the sum of the areas of the two shaded regions. Those regions together contain a rectangle of sides a, and b, from which the inequality follows. The condition for equality is $b = f(a)$, the condition that the two regions fill the rectangle.

Solution 2. Without loss of generality, assume $f(a) \leqslant b$. We have

$$ab = \int_0^a f(x)\,dx + \int_0^a (b - f(x))\,dx.$$

The second integral is

$$\lim_{n\to\infty} \frac{a}{n} \sum_{k=0}^{n-1} \left(b - f\left(\frac{(k+1)a}{n} \right) \right).$$

For $0 \leqslant k \leqslant n - 1$,

$$\frac{a}{n} = \frac{(k+1)a}{n} - \frac{ka}{n} = g \circ f\left(\frac{(k+1)a}{n} \right) - g \circ f\left(\frac{ka}{n} \right).$$

Substituting in the limit above,

$$\lim_{n\to\infty} \sum_{k=0}^{n-1} \left(b - f\left(\frac{(k+1)a}{n} \right) \right) \left(g \circ f\left(\frac{(k+1)a}{n} \right) - g \circ f\left(\frac{ka}{n} \right) \right).$$

Multiplying out each term in the sum and rearranging them, and noting that $f(0) = g(0) = 0$, we get

$$\lim_{n\to\infty} \sum_{k=0}^{n-1} g \circ f\left(\frac{ka}{n} \right) \left(f\left(\frac{(k+1)a}{n} \right) - f\left(\frac{ka}{n} \right) \right) + ab - af(a).$$

Since g is continuous, this equals

$$\int_0^{f(a)} g(y)\,dy + a\,(b - f(a)).$$

As $g(y) \geqslant a$ for $y \in (f(a), b)$, we have

$$a\,(b - f(a)) \leqslant \int_{f(a)}^b g(y)\,dy.$$

This gives the desired inequality. Also, we see that equality holds iff $f(a) = b$.

Solution to 1.5.19: Given $\varepsilon > 0$, choose R so that $\int_{|x| \geqslant R} |f(x)|dx < \varepsilon/4$. Then $\int_{|x| \geqslant R} |f(x) \cos xy|dx < \varepsilon/4$ for all y. So

$$|g(z) - g(y)| = \int_{|x| \geqslant R} f(x)\,(\cos xz - \cos xy)\,dx$$

$$+ \int_{|x| \geqslant R} f(x)\,(\cos xz - \cos xy)\,dx$$

$$\leqslant \frac{\varepsilon}{2} + \int_{|x| \geqslant R} |f(x)||\cos xz - \cos xy|dx.$$

The latter integral approaches 0 as $z \to y$ by uniform convergence of $\cos(xz)$ to $\cos(xy)$ on the compact interval $-R \leqslant x \leqslant R$. Hence, for $|z - y|$ sufficiently small,

$$|g(z) - g(y)| < \frac{\varepsilon}{2} + \frac{\varepsilon}{2} = \varepsilon$$

and g is continuous.

Solution to 1.5.20: Let M be an upper bound of $|f|$, and let $K = \int_{-\infty}^{\infty} |g(x)| dx$. Fix $\varepsilon > 0$. Since $\lim_{|x| \to \infty} f(x) = 0$, there exists $R_1 > 0$ such that $|f(x)| < \varepsilon$ for $|x| > R_1$. As K is finite, there is $R_2 > 0$ such that $\int_{|x| > R_2} |g(x)| dx < \varepsilon$. For $|x| > R_1 + R_2$ we have then

$$|h(x)| \leqslant \int_{-R_2}^{R_2} |f(x - y)||g(y)| dy + \int_{|y| > R_2} |f(x - y)||g(y)| dy$$

$$\leqslant \varepsilon \int_{-R_2}^{R_2} |g(y)| dy + M \int_{|y| > R_2} |g(y)| dy$$

$$\leqslant \varepsilon(K + M).$$

As ε is arbitrary, the desired conclusion follows.

Solution to 1.5.21: We will do the proof of the sine integral only. For $n \geqslant 0$, let

$$S_n = \int_{\sqrt{n\pi}}^{\sqrt{(n+1)\pi}} \sin x^2 \, dx.$$

We show that the series $\sum S_n$ converges and use this to show that the integral converges.

By the choice of the domains of integration, the S_n's alternate in sign. Also, setting $u = x^2$, we get

$$2|S_n| = \left| \int_{n\pi}^{(n+1)\pi} \frac{\sin u}{\sqrt{u}} \, du \right|$$

$$> \left| \int_{n\pi}^{(n+1)\pi} \frac{\sin u}{\sqrt{u + \pi}} \, du \right|$$

$$= \left| \int_{(n+1)\pi}^{(n+2)\pi} \frac{\sin u}{\sqrt{u}} \, du \right|$$

$$= 2|S_{n+1}|.$$

Finally, the S_n's tend to 0:

$$2|S_n| = \left| \int_{n\pi}^{(n+1)\pi} \frac{\sin u}{\sqrt{u}} \, du \right| < \frac{1}{\sqrt{n\pi}}$$

and the right-hand side gets arbitrarily small as n tends to infinity. Therefore, by Leibniz Criterion [Rud87, p. 71], the series $\sum S_n$ converges. Let $a > 0$ and n be such that $\sqrt{n\pi} \leqslant a < \sqrt{(n+1)\pi}$. Then

$$\int_0^a \sin x^2 \, dx - \sum_{k=0}^{\infty} S_k = \int_a^{\sqrt{(n+1)\pi}} \sin x^2 \, dx - \sum_{k=n+1}^{\infty} S_k.$$

The second term tends to zero as n tends to infinity. By estimates almost identical to those above,

$$\left| \int_a^{\sqrt{(n+1)\pi}} \sin x^2 \, dx \right| \leqslant \frac{|(n+1)\pi - a^2|}{2a} \leqslant \frac{\pi}{2\sqrt{n\pi}},$$

so the first term does as well. Therefore, we have

$$\int_0^{\infty} \sin x^2 \, dx = \sum_{n=0}^{\infty} S_n < \infty.$$

Solution to 1.5.22: Let $p(x) = \sum_{j=0}^{k} a_j x^j$ be a polynomial. We have

$$\lim_{n \to \infty} (n+1) \int_0^1 x^n p(x) \, dx = \lim_{n \to \infty} \sum_{j=0}^{k} \frac{n+1}{n+j+1} a_j = p(1).$$

So the result holds for polynomials. Now let f be a continuous function and $\varepsilon > 0$. By the Stone–Weierstrass Approximation Theorem [MH93, p. 284], there is a polynomial p with $\|f - p\|_\infty < \varepsilon$. So

$$\left| (n+1) \int_0^1 x^n f(x) \, dx - f(1) \right| \leqslant \left((n+1) \int_0^1 x^n |f(x) - p(x)| \, dx \right)$$

$$+ \left| (n+1) \int_0^1 x^n p(x) \, dx - f(1) \right|$$

$$\leqslant \varepsilon + \left| (n+1) \int_0^1 x^n p(x) \, dx - f(1) \right|$$

$$\to \varepsilon + |p(1) - f(1)|$$

$$< 2\varepsilon.$$

Since ε is arbitrary, the desired limit holds.

Solution to 1.5.23: $\log x$ is integrable near zero, and near infinity it is dominated by $\sqrt{x}$, so the given integral exists finitely. Making the change of variables $x = a/t$, it becomes

$$\int_0^\infty \frac{\log x}{x^2 + a^2}\,dx = \frac{\log a}{a}\int_0^\infty \frac{dt}{1+t^2} - \frac{1}{a}\int_0^\infty \frac{\log t}{1+t^2}\,dt$$

$$= \frac{\log a}{a}\,\arctan t\,\Big|_0^\infty - J$$

$$= \frac{\pi \log a}{2a} - J.$$

If we treat J in a similar way, we get $J = -J$, so $J = 0$ and the given integral equals

$$\frac{\pi \log a}{2a}.$$

Solution 2. We split the integral in two and use the substitution $x = a^2/y$.

$$\int_0^\infty \frac{\log x}{x^2 + a^2}\,dx = \int_0^a \frac{\log x}{x^2 + a^2}\,dx + \int_a^\infty \frac{\log x}{x^2 + a^2}\,dx$$

$$= \int_0^a \frac{\log x}{x^2 + a^2}\,dx + \int_a^0 \frac{\log(a^2/y)}{a^2 + (a^2/y)^2}\left(-\frac{a^2}{y^2}\right)dy$$

$$= \int_0^a \frac{\log x}{x^2 + a^2}\,dx + \int_0^a \frac{2\log a - \log y}{a^2 + y^2}\,dy$$

$$= \int_0^a \frac{2\log a}{a^2 + y^2}\,dy$$

$$= 2\,\frac{\log a}{a}\,\arctan \frac{y}{a}\,\Big|_0^a$$

$$= \frac{\pi \log a}{2a}.$$

Solution 3. Using residues, and arguing as in the solution to Problem 5.11.28, we get, for $f(z) = \dfrac{\log z}{z^2 + a^2}$,

$$\int_{-\infty}^0 f(z)\,dz + \int_0^\infty f(z)\,dz = 2\pi i\,\mathrm{Res}\,(f, ia) = \frac{\pi}{a}\left(\log a + \frac{i\pi}{2}\right).$$

On the negative real axis, we have

$$\int_{-\infty}^0 f(z)\,dz = \int_{-\infty}^0 \frac{\log(-x) + \pi i}{x^2 + a^2}$$

$$= \int_0^\infty \frac{\log x}{x^2 + a^2}\,dx + \int_0^\infty \frac{\pi i}{x^2 + a^2}\,dx$$

$$= \int_0^\infty \frac{\log x}{x^2 + a^2}\,dx + \frac{\pi^2 i}{2a},$$

therefore,

$$2 \int_0^\infty \frac{\log x}{x^2 + a^2} dx + \frac{\pi^2 i}{2a} = \frac{\pi}{a} \left(\log a + \frac{i\pi}{2} \right)$$

and

$$\int_0^\infty \frac{\log x}{x^2 + a^2} dx = \frac{\pi \log a}{2a} .$$

Solution to 1.5.24: As $\sin x \leqslant 1$, to show that I converges, it is enough to show that $I > -\infty$. By the symmetry of $\sin x$ around $\pi/2$, we have

$$\begin{aligned} I &= \int_0^{\pi/2} \log(\sin x) \, dx + \int_{\pi/2}^\pi \log(\sin x) \, dx \\ &= 2 \int_0^{\pi/2} \log(\sin x) \, dx \\ &\geqslant 2 \int_0^{\pi/2} \log(2x/\pi) \, dx \\ &> -\infty. \end{aligned}$$

The first inequality holds since on $[0, \pi/2]$, $\sin x \geqslant 2x/\pi$; see Problem 1.1.25. Letting $x = 2u$, we get

$$\begin{aligned} I &= 2 \int_0^{\pi/2} \log(\sin 2u) \, du \\ &= 2 \left(\int_0^{\pi/2} \log 2 \, du + \int_0^{\pi/2} \log(\sin u) \, du + \int_0^{\pi/2} \log(\cos u) \, du \right). \end{aligned}$$

The first integral equals $(\pi/2) \log 2$. As $\cos u = \sin(\pi/2 - u)$, the last integral is

$$\int_0^{\pi/2} \log\left(\sin(\pi/2 - u)\right) \, du = \int_0^{\pi/2} \log(\sin u) \, du = \int_{\pi/2}^\pi \log(\sin u) \, du.$$

The above equation becomes $I = \pi \log 2 + 2I$, so $I = -\pi \log 2$.

Solution to 1.5.25: The integrand is continuous at $x = 0$, so that no trouble arises there. For $r > 0$ let

$$I(r) = \int_0^r \frac{\sin x}{\sqrt{x}} \, dx .$$

For n a positive integer,

$$I(n\pi) = \sum_{k=1}^n (-1)^{k-1} a_k ,$$

where

$$a_k = \int_{(k-1)\pi}^{k\pi} \frac{|\sin x|}{\sqrt{x}} \, dx .$$

From the π-periodicity of $|\sin x|$ it follows that $a_k > a_{k+1}$ for all k, while $a_k < 1/\sqrt{(k-1)\pi} \to 0$. Hence $\lim\limits_{n\to\infty} I(n\pi)$ exists finitely by the alternating series test. For $r > 0$ let n_r be the smallest integer such that $r < n\pi$. Then

$$I(r) - I(n_r\pi) = O\left(\frac{1}{\sqrt{r}}\right) \qquad (r \to \infty),$$

so $\lim\limits_{n\to\infty} I(r)$ exists finitely, which means $\displaystyle\int_0^\infty \frac{\sin x}{\sqrt{x}} \, dx$ converges. On the other hand,

$$\int_0^\infty \frac{|\sin x|}{\sqrt{x}} \, dx \geqslant \sum_{k=0}^\infty \int_{k\pi+\frac{\pi}{6}}^{k\pi+\frac{5\pi}{6}} \frac{|\sin x|}{\sqrt{x}} \, dx$$

$$\geqslant \sum_{k=0}^\infty \frac{1}{2} \frac{1}{\sqrt{(k+1)\pi}} \frac{2\pi}{3}$$

$$= \frac{\sqrt{\pi}}{3} \sum_{k=0}^\infty \frac{1}{\sqrt{k+1}} = \infty.$$

An alternative proof of the convergence of the integral uses integration by parts:

$$\int_1^r \frac{\sin x}{\sqrt{x}} \, dx = -\int_1^r \frac{1}{\sqrt{x}} \, d(\cos x)$$

$$= -\left. \frac{\cos x}{\sqrt{x}} \right|_1^r + \frac{1}{2} \int_1^r \frac{\cos x}{x^{3/2}} \, dx$$

$$= \cos 1 - \frac{\cos r}{\sqrt{r}} + \frac{1}{2} \int_1^r \frac{\cos x}{x^{3/2}} \, dx \, .$$

The right side converges as $r \to \infty$ because $\displaystyle\int_1^\infty x^{-3/2} dx < \infty$.

1.6   Sequences of Functions

Solution to 1.6.1: Let B be the set of function that are the pointwise limit of continuous functions defined on $[0, 1]$. The characteristic functions of intervals, χ_I, are in B. Notice also that as f is monotone, the inverse image of an interval is an interval, and that linear combinations of elements of B are in B. Without loss of generality, assume $f(0) = 0$ and $f(1) = 1$. For $n \in \mathbb{N}$, let the functions g_n be defined by

$$g_n(x) = \sum_{k=0}^{n-2} \frac{k}{n} \chi_{f^{-1}\left(\left[\frac{k}{n}, \frac{k+1}{n}\right)\right)}(x) + \frac{n-1}{n} \chi_{f^{-1}\left(\left[\frac{n-1}{n}, 1\right]\right)}(x).$$

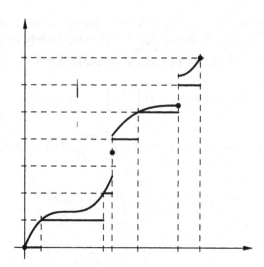

From the construction we can easily see that $\max\limits_{x\in[0,1]} |g_n(x) - f(x)| \leqslant \dfrac{1}{n}$, consider now the following result:

Lemma: Let $\{h_n\} \subset B$ with $\max\limits_{x\in[0,1]} |h_n(x)| \leqslant A_n$ and $\sum\limits_{n=1}^{\infty} A_n < \infty$. Then

$\sum\limits_{n=1}^{\infty} h_n \in B$. As $|g_{2^{k+1}} - g_{2^k}| \leqslant |g_{2^{k+1}} - f| + |g_{2^k} - f| \leqslant \dfrac{1}{2^{k-1}}$ and $\sum \dfrac{1}{2^{k-1}} < \infty$, we get

$$\sum_{k=1}^{\infty} (g_{2^{k+1}} - g_{2^k}) = f - g_2 \in B$$

so $f - g_2 + g_2 = f \in B$.

Proof of the Lemma: For each n let h_n be the pointwise limit of $\{\varphi_k^n\} \subset B$ such that $|h_n(x)| \leqslant A_n$ on $[0, 1]$. Consider the functions $\Phi_k = \sum_{n=1}^{k} \varphi_k^n$. Given $\varepsilon > 0$, take m such that $\sum_{n=m+1}^{\infty} A_n < \varepsilon/3$. Then the sum $\sum_{n=m+1}^{\infty} |h_n(x)| < \varepsilon/3$ and $\sum_{n=m+1}^{\infty} |\varphi_k^n(x)| < \varepsilon/3$.

For $x \in [0, 1]$, take K so that

$$|h_n(x) - \varphi_K^n(x)| < \frac{\varepsilon}{3m} \quad \text{for} \quad n = 1, \ldots, m.$$

For $k > K$ we then have

$$\left| \sum_{n=1}^{\infty} h_n(x) - \Phi_k(x) \right| \leqslant \sum_{k=1}^{m} |h_n(x) - \varphi_k^n(x)| + \sum_{n=m+1}^{\infty} |h_n(x)| + \sum_{n=m+1}^{k} |\varphi_k^n(x)| < \varepsilon$$

so $\sum\limits_{n=1}^{\infty} h_n \in B$.

Solution to 1.6.2: Let $a < b$ be real numbers and $\varepsilon > 0$. Take n large enough so

$$|f_n(a) - g(a)| < \varepsilon \quad \text{and} \quad |f_n(b) - g(b)| < \varepsilon.$$

Then, using the Mean Value Theorem [Rud87, p. 108],

$$|g(a)-g(b)| \leqslant |g(a)-f_n(a)|+|f_n(a)-f_n(b)|+|f_n(b)-g(b)| < 2\varepsilon+|f_n'(\xi)||b-a|$$

where $a < \xi < b$. As the inequality holds for any $\varepsilon > 0$,

$$|g(a) - g(b)| \leqslant |b - a|$$

and the continuity of g follows.

Solution 2. Let $N > 0$. We will show that g is continuous in $[-N, N]$. We have, for any $n \in \mathbb{N}$, by the Mean Value Theorem,

$$|f_n(x) - f_n(y)| = |f_n'(\xi)(x - y)| \leqslant 2N \quad \text{for } x, y \in [-N, N].$$

So the sequence $\{f_n\}$ is bounded. The relation

$$|f_n(x) - f_n(y)| = |f_n'(\xi)(x - y)| \leqslant |x - y|$$

shows that it is also equicontinuous. By the Arzelà–Ascoli Theorem [MH93, p. 273] we may assume that $\{f_n\}$ converges uniformly on $[-N, N]$. The function g, being the uniform limit of continuous functions, is continuous as well. As N is arbitrary, we are done.

Solution to 1.6.3: Let $\varepsilon > 0$. By uniform continuity, for some $\delta > 0$ we have $|f(x) - f(y)| < \varepsilon$ for $|x - y| < \delta$. Take N satisfying $1/N < \delta$. For $0 \leqslant k \leqslant N$, let $\xi_k = k/N$, and divide $[0, 1]$ into the intervals $[\xi_{k-1}, \xi_k]$, $1 \leqslant k \leqslant N$. Since f_n tends to f pointwise, by taking the maximum over the finite set $\{\xi_k\}$ we know that there exists $M > 0$ such that if $n \geqslant M$, then $|f_n(\xi_k) - f(\xi_k)| < \varepsilon$ for $0 \leqslant k \leqslant N$. Each of the f_n's is nondecreasing, so we have, for $x \in [\xi_{k-1}, \xi_k]$,

$$f(\xi_{k-1}) - \varepsilon < f_n(x) < f(\xi_{k-1}) + 2\varepsilon,$$

or

$$|f_n(x) - f(\xi_{k-1})| < 2\varepsilon.$$

Therefore,

$$|f_n(x) - f(x)| \leqslant |f_n(x) - f(\xi_{k-1})| + |f(\xi_{k-1}) - f(x)| < 3\varepsilon.$$

Since this bound does not depend on x, the convergence is uniform.

Solution to 1.6.4: Assume the contrary. Then, for some real x and positive ε, given any $\delta > 0$, we can find $y \in (x - \delta, x + \delta)$ such that $|f(x) - f(y)| \geqslant \varepsilon$.

For each positive integer m let $y_m \in (x - \frac{1}{m}, x + \frac{1}{m})$ with $|f(x) - f(y_m)| \geq \varepsilon$. We know that $\lim_{n \to \infty} f_n(y_m) = f(y_m)$, therefore, given $\varepsilon > 0$, we can find, for each m, an integer n_m such that

$$|f_{n_m}(y_m) - f(y_m)| < \varepsilon/2,$$

and choose them so $n_1 < n_2 < \cdots$.

Consider the sequence (x_n) defined by $x_n = y_m$ for $n_{m-1} < n \leq n_m$. Then $x_n \to x$, but $f_n(x_n) \not\to f(x)$ since, for each m,

$$\begin{aligned}
|f_{n_m}(x_{n_m}) - f(x)| &= |f_{n_m}(y_m) - f(x)| \\
&\geq |f(x) - f(y_m)| - |f(y_m) - f_{n_m}(y_m)| \\
&> \varepsilon - \varepsilon/2 = \varepsilon/2.
\end{aligned}$$

Solution to 1.6.6: 1. For $k \in \mathbb{N}$, consider the continuous functions g_k given by

$$g_k(x) = \begin{cases} 4k - 16k^2 |x - \frac{3}{4}k| & \text{if } x \in [\frac{1}{2k}, \frac{1}{k}] \\ \\ 0 & \text{if } x \notin [\frac{1}{2k}, \frac{1}{k}] \end{cases}$$

Define $f_0 \equiv 0$, and, for $k > 0$, $f_k(x) = \int_0^x g_k(t)dt$. We have $f_k \in C^1(\mathbb{R}_+)$, $f_k(0) = 0$, and $f_k'(x) = g_k(x) \to 0 = f_0'(x)$ for all $x \in \mathbb{R}_+$. However,

$$\lim_{n \to \infty} f_k(x) = \lim_{n \to \infty} \int_0^x g_k(t)dt = \int_0^\infty g_k(t)dt = 1 \neq f_0(x).$$

2. $f_k'(x) \to f_0'(x)$ uniformly on $\mathbb{R}$.

Solution to 1.6.7: As f is a homeomorphism of $[0,1]$ onto itself, we may assume without loss of generality (by replacing f by $1 - f$) that f is strictly increasing, with $f(0) = 0$ and $f(1) = 1$. We first treat the case where f' is a continuous function. By the Stone–Weierstrass Approximation Theorem [MH93, p. 284], there is a sequence of polynomials $\{P_n\}$ which converge to f uniformly. Since $f' > 0$, we may assume (by adding a small constant) that each of the P_n is positive. Further, since the P_n's converge uniformly,

$$\int_0^1 P_n(t)\,dt \to \int_0^1 f'(t)\,dt = f(1) = 1.$$

Defining a_n by

$$a_n^{-1} = \int_0^1 P_n(t)\,dt$$

we can replace each P_n by $a_n P_n$, so we may assume that

$$\int_0^1 P_n(t)\,dt = 1.$$

Now consider the polynomials

$$Q_n(x) = \int_0^x P_n(t)\, dt.$$

$Q_n(0) = 0$, $Q_n(1) = 1$, and $Q_n'(x) = P_n(x) > 0$ for all x and n. Hence, each Q_n is a homeomorphism of the unit interval onto itself, and by their definition, the Q_n's converge to f uniformly.

It is enough now to show that any increasing homeomorphism of the unit interval onto itself can be uniformly approximated by C^1 homeomorphisms. Let $r > 0$ and

$$f_r(x) = \begin{cases} e^{1-x^{-r}} & \text{for} \quad 0 < x \leqslant 1 \\ 0 & \text{for} \quad x = 0 \end{cases}$$

A calculation shows that f_r is C^1 on $[0, 1]$, $f_r(1) = 1$, $f_r'(0) = 0$, and $f_r'(1) = r$. For $r, s > 0$, let

$$g_{rs}(x) = \begin{cases} f_r(x) & \text{for} \quad x \in [0, 1] \\ -f_s(-x) & \text{for} \quad x \in [-1, 0]. \end{cases}$$

Each g_{rs} is a C^1-function such that $g_{rs}(0) = 0$, $g_{rs}(1) = 1$, $g_{rs}(-1) = -1$, $g_{rs}'(-1) = s$, and $g_{rs}'(1) = r$. By scaling and translating, we can find a C^1 function on any interval such that its values and the values of its derivative at both endpoints are any given positive values desired.

We can now approximate any continuous homeomorphism f as follows: Given $\varepsilon > 0$, choose $n > 0$ such that if $|x - y| < 1/n$, for these values $|f(x) - f(y)| < \varepsilon$. (This is possible since $[0, 1]$ is compact, so f is uniformly continuous there.) Partition $[0, 1]$ into $2n$ intervals of equal length. On the intervals $[2k/2n, (2k + 1)/2n]$, $0 \leqslant k \leqslant n - 1$, approximate f by the line segment joining $f(2k/2n)$ and $f((2k + 1)/2n)$. On the other intervals, join the line segments by suitable functions as defined above to make the approximating function C^1. Since f is an increasing function, this approximating function will always lie within ε of it.

Solution to 1.6.8: 1. Let $S_\varepsilon = \{x \in [0, 1] \mid f(x) \geqslant M - \varepsilon\}$ for positive ε. $f(x) \geqslant M - \varepsilon$ if and only if for each n, $f_n(x) \geqslant M - \varepsilon$, so $S_\varepsilon = \bigcap_{n \geqslant 1} f_n^{-1}([M - \varepsilon, \infty))$. So each S_ε is closed. By definition of supremum, each set S_ε is nonempty. Also, if ε_i are finitely many positive numbers, $\bigcap_i S_{\varepsilon_i} = S_{\min \varepsilon_i} \neq \emptyset$. As $[0, 1]$ is compact, the intersection of all sets S_ε is nonempty. Let t belong to this intersection. Then $M \geqslant f(t) \geqslant M - \varepsilon$ for arbitrary $\varepsilon > 0$, so $f(t) = M$.
2. Take $f_n(x) = \min\{nx, 1 - x\}$.

Solution to 1.6.9: Fix $\varepsilon > 0$. For each n, let $G_n = \{x \mid f_n(x) < \varepsilon\}$. Then

- G_n is open, since f_n is continuous.

- $G_n \subset G_{n+1}$ since $f_n \geqslant f_{n+1}$.

- $[0, 1] = \cup_{n=1}^{\infty} G_n$, since $f_n(x) \to 0$, for each x.

Since $[0, 1]$ is compact, a finite number of the G_n cover $[0, 1]$, and by the second condition above, there is N such that $G_n = [0, 1]$ for all $n \geqslant N$. By the definition of G_n, for all $n \geqslant N$, we have $0 \leqslant f_n(x) < \epsilon$ for all $x \in [0, 1]$. This proves that the sequence f_n converges uniformly to 0 on $[0, 1]$.

Solution to 1.6.10: If $f_n(x) \to f(x)$ pointwise on E, then f_n converges to f uniformly on E if and only if $\sup_E |f_n(x) - f(x)|$ tends to 0 as n tends to ∞. We note that if $A \subset B$ then $\sup_A |f_n(x) - f(x)| \leqslant \sup_B |f_n(x) - f(x)|$.

$k = 0$. In this case $|f_n(x)| \leqslant 1/n$ so $\sup_{\mathbb{R}} |f_n| \to 0$ as $n \to \infty$. Hence f_n converges uniformly on $\mathbb{R}$, and also on every bounded subset of $\mathbb{R}$.

$k = 1$. $f_n(x) = \dfrac{x}{x^2 + n} \to 0$ pointwise, and the derivative is $f_n'(x) = \dfrac{n - x^2}{(x^2 + n)^2}$, so $|f_n|$ attains its maximum at $x = \sqrt{n}$, where

$$\sup |f_n| = f_n(\sqrt{n}) = \frac{1}{2\sqrt{n}}$$

so that $\sup |f_n| \to 0$ as $n \to \infty$, hence the convergence is uniform on $\mathbb{R}$ and also on bounded sets.

$k = 2$. $f_n(x) = \dfrac{x^2}{x^2 + n}$ and $\sup |f_n| = 1$, thus the convergence is not uniform on $\mathbb{R}$. However, since the function is even and increasing on $\mathbb{R}_+$, the supremum on $[-a, a]$ is $\sup |f_n| = f_n(a) = \dfrac{a^2}{a^2 + n}$ so that $\sup |f_n| \to 0$ as $n \to \infty$ and the convergence is uniform on bounded subsets.

$k \geqslant 3$. Again $f_n(x) \to 0$ pointwise, but $\sup |f_n| = \infty$, thus the convergence is not uniform on $\mathbb{R}$. In the same way as above, $|f_n|$ is even, and increasing on $\mathbb{R}_+$, this can be seen directly or by computing the derivative

$$f_n'(x) = \frac{x^{k-1}((k-2)x^2 + kn)}{(x^2 + n)^2}$$

so $\sup |f_n| = f_n(a) \to 0$ as $n \to \infty$. Thus the convergence is uniform on bounded sets.

Hence the convergence is uniform on $\mathbb{R}$ when $k = 0$ and 1 only, and uniform on bounded sets for all k.

Solution to 1.6.11: Let $\mathcal{P}_k$ be the set of polynomials of degree $\leqslant k$, and let $L = d(f, \mathcal{P}_k) = \inf\{\|f - P\| \mid P \in \mathcal{P}_k\}$, where the norm is that of $\mathcal{C}_{[0,1]}$. Then, there exists a sequence $\{P_n\}$ contained in $\mathcal{P}_k$ such that $\|f - P_n\| \to L$. Then, for some $M > 0$, $\|P_n\| \leqslant \|P_n - f\| + \|f\| \leqslant M$, for all $n \in \mathbb{N}$. Let $x_1, ..., x_k \in [0, 1]$ be

k distinct points. Thus, $|P_n(x_i)| \leq \|P_n\| \leq M$, for all $i = 1, \ldots, k$ and $n \in \mathbb{N}$. The sequence $(P_n(x_1))$ is bounded, so, passing to a subsequence if necessary, we may assume that it converges. We can repeat this argument and suppose that $\lim_{l \to \infty} P_n(x_i) = y_i$ for any $i = 1, \ldots, k$, for some constants y_i. Let $P(x) = \sum_{i=1}^{k} y_i \omega_i(x)$ where

$$\omega_i(x) = \frac{(x - x_1) \cdots (x - x_{i-1})(x - x_{i+1}) \cdots (x - x_k)}{(x_i - x_1) \cdots (x_i - x_{i-1})(x_i - x_{i+1}) \cdots (x_i - x_k)}$$

i.e., P is Lagrange's interpolation polynomial. Clearly, there are positive constants L_i such that $|\omega_i(x)| \leq L_i$, for all $i = 1, \ldots, k$, and all $x \in [0, 1]$. Therefore, for all $x \in [0, 1]$, we have $|P_n(x) - P(x)| \leq \sum_{i=1}^{k} |P_n(x_i) - y_i| |\omega_i(x)| \leq \sum_{i=1}^{k} L_i |P_n(x_i) - y_i| \to 0$. This means that for this polynomial P in $\mathcal{P}_k$, we have $P_n \to P$; but $\|P_n - f\| \to L$, whence $L = \|P - f\|$, i.e., $d(f, \mathcal{P}_k) = \|f - P\|$, $P \in \mathcal{P}_k$, therefore P is a polynomial of degree $\leq k$, which minimizes $d(f, \mathcal{P}_k)$.

Solution to 1.6.12: Let $a_0, \ldots, a_D$ be $D + 1$ distinct points in $[0, 1]$. The polynomials f_m, defined by

$$f_m(x) = \prod_{\substack{i=0 \\ i \neq m}}^{D} \frac{x - a_i}{a_m - a_i} \quad \text{for} \quad m = 0, \ldots, D$$

satisfy $f_m(a_i) = 0$ for $i \neq m$, $f_m(a_m) = 1$. Any polynomial of degree, at most, D can be written

$$P(x) = \sum_{m=0}^{D} P(a_m) f_m(x)$$

since the right-hand side is a polynomial of degree, at most, D which agrees with P in $D + 1$ points.

Let M be an upper bound of $|f_m(x)|$ for $x \in [0, 1]$, $m = 0, \ldots, D$. Given $\varepsilon > 0$ let $N \in \mathbb{N}$ be such that $|P_n(a_m)| \leq \frac{\varepsilon}{(D+1)M}$ for $n \geq N$. Then we have

$$|P_n(x)| \leq \sum_{m=0}^{D} |P_n(a_m)| |f_m(x)| < \varepsilon$$

therefore the convergence is uniform.

Solution to 1.6.14: Suppose that $f_{n_j} \to f$ uniformly. Then f is continuous, and $f(0) = \lim_{j \to \infty} \cos 0 = 1$. So there is $\varepsilon > 0$ with $f(x) > 1/2$ for $|x| < \varepsilon$. If j is large enough, we have, by uniform convergence,

$$|f(x) - f_{n_j}(x)| < \frac{1}{2} \quad \text{for all } x, \qquad \frac{\pi}{2n_j} < \varepsilon.$$

For one such j, and $x = \dfrac{\pi}{2n_j}$, we get

$$\frac{1}{2} < f(x) \leqslant |f(x) - f_{n_j}(x)| + |f_{n_j}(x)| < \frac{1}{2} + f_{n_j}(x) = \frac{1}{2} + |\cos\frac{\pi}{2}| = \frac{1}{2}$$

a contradiction.

Solution to 1.6.15: 1. In an obvious way we see that $d(f, f) = 0$ and $d(f, g) = d(g, f)$, so we only need to verify that $d(f, g) > 0$ for $f \neq g$ and the Triangle Inequality [MH87, p. 20].

If $f \neq g$ then $|f - g|$ is positive in a small interval and so is the integrand and the integral is nonzero. Now the function $a \mapsto \frac{a}{1+a}$ is increasing in $[0, \infty)$. Hence, for $a = |f - g|, b = |g - h|, c = |f - h|$, we have $c \leqslant a + b$ and

$$\frac{c}{1+c} \leqslant \frac{a+b}{1+a+b} = \frac{a}{1+a+b} + \frac{b}{1+a+b} \leqslant \frac{a}{1+a} + \frac{b}{1+b}.$$

which implies the Triangle Inequality.

2. For natural n define f_n by

$$f_n(x) = \begin{cases} n^2 x, & 0 \leqslant x \leqslant 1/n \\ 1/x, & 1/n \leqslant x \leqslant 1. \end{cases}$$

It is easy to verify that the sequence $\{f_n\}$ is Cauchy, since for any n, m we have,

$$\begin{aligned} d(f_m, f_n) &= \int_0^{\max\{1/m, 1/n\}} \frac{|f_m(x) - f_n(x)|}{1 + |f_m(x) - f_n(x)|} dx \\ &\leqslant \int_0^{\max\{1/m, 1/n\}} 1\, dx \\ &= \max\{1/m, 1/n\}. \end{aligned}$$

Suppose that $(C_{[0,1]}, d)$ is complete and take f as the limit of the sequence $\{f_n\}$. If $f(a) \neq 1/a$ for some $a \in (0, 1]$, then, by continuity, there exists $\varepsilon > 0$ such that $|1/x - f(x)| \leqslant \varepsilon$ for $x \in (a - \varepsilon, a]$. Therefore,

$$d(f_n, f) \leqslant \int_{a-\varepsilon}^a \frac{\varepsilon}{1+\varepsilon} dx$$

for sufficiently large n. But the right hand side is a positive constant independent of n, which contradicts the convergence $f_n \to f$. Thus $f(a) = 1/a$ for all $a \in (0, 1]$. This contradicts the continuity of f on $[0, 1]$. We conclude then that $(C_{[0,1]}, d)$ is not complete.

Solution to 1.6.16: (a). Let $f_n : [0, 1] \to \mathbb{R}$ be defined by $f_n(x) = x^n$. $[0, 1]$ is compact, $\|f_n\| = 1$, but the sequence f_n is not equicontinuous.

(b). Let $\Omega = [0, 1]$, and $g_n(x) = n$. This sequence is clearly equicontinuous, Ω is compact, but no subsequence of g_n can converge.

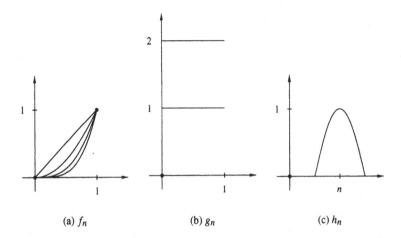

(a) f_n (b) g_n (c) h_n

(c). Consider now $h_n : \mathbb{R} \to \mathbb{R}$, $h_n(x) = \chi_{[n-1,n+1]}(x) \cos((x - n)\pi/2)$, where $\chi_{[a,b]}$ is the characteristic function for the interval $[a, b]$. $\|h_n\| \leqslant 1$ and the sequence is equicontinuous, however no subsequence can converge.

Solution to 1.6.17: For each n, let g_n be the function that equals f_n at the points k/n, $k = 0, 1, \ldots, n$, and that on each interval $[(k - 1)/n, \, k/n]$ interpolates its endpoint values linearly. By the assumption on f_n, the slope of g_n on each of the preceding intervals is at most 1, so g_n is a Lipschitz function with Lipschitz constant at most 1. Hence the sequence $\{g_n\}_{n=1}^{\infty}$ is uniformly equicontinuous, so by the Arzelà–Ascoli Theorem [MH93, p. 273], it has a uniformly convergent subsequence $\{g_{n_j}\}_{j=1}^{\infty}$. Let g be the limit function. Fix a point x in $[0, 1]$. For each j, let x_j be a point in $[0,1]$ of the form k/n_j $(k = 0, \ldots, n_j)$ such that $1/n_j \leqslant |x - x_j| < 2/n_j$. We have

$$|g(x) - f_{n_j}(x)| \leqslant |g(x) - g_{n_j}(x)| + |g_{n_j}(x) - g_{n_j}(x_j)| + |f_{n_j}(x_j) - f_{n_j}(x)|$$
$$\leqslant |g(x) - g_{n_j}(x)| + \frac{2}{n_j} + \frac{2}{n_j}.$$

As $j \to \infty$, the first summand on the right tends to 0 uniformly, showing that $f_{n_j} \to g$ uniformly.

Solution to 1.6.18: By the Arzelà–Ascoli Theorem [MH93, p. 273], it will suffice to prove that the sequence $\{f_n\}$ is equicontinuous and uniformly bounded. *Equicontinuity.* For $0 \leqslant x < y \leqslant 1$ and any n,

$$|f_n(y) - f_n(x)| = \left| \int_x^y f_n'(t)dt \right| \leqslant \int_x^y t^{-\frac{1}{2}}dt = 2\sqrt{y} - 2\sqrt{x}.$$

The function $F(x) = 2\sqrt{x}$ is continuous on $[0, 1]$, hence uniformly continuous. Therefore, given $\varepsilon > 0$, there is a $\delta > 0$ such that $|F(y) - F(x)| < \varepsilon$ whenever

x and y are in $[0, 1]$ and $|y - x| < \delta$. By the inequality above, we then have $|f_n(y) - f_n(x)| < \varepsilon$ for all n when $|y - x| < \delta$, establishing the equicontinuity of the sequence.

Uniform boundedness. Since $\int_0^1 f_n(x)dx = 0$, the function f_n cannot be always positive or always negative, so there is a point x_n on $[0, 1]$ such that $f_n(x_n) = 0$. Then, by the estimate found above, for all x:

$$|f_n(x)| \leqslant 2\left|\sqrt{x} - \sqrt{x_n}\right| \leqslant 2.$$

Solution to 1.6.19: We claim that a subset A of M is compact iff A is closed, bounded and $\{f' \mid f \in A\}$ is equicontinuous. If A satisfies all conditions and $\{f_n\}$ is a subsequence in A then $\{f_n\}$ and $\{f_n'\}$ are bounded and equicontinuous and by the Theorem of Arzelà–Ascoli [MH93, p. 273], there is a subsequence $\{f_{n_j}\}$ such that $\{f_{n_j}\}$ and $\{f_{n_j}'\}$ are uniformly convergent and therefore, sequences of Cauchy. Since M is complete and A is closed, $\{f_{n_j}\}$ converges to $f \in A$ in M, and A is compact.

On the other hand, if A is compact, consider the spaces:

$$\tilde{M} = \{(f, f') \mid f \in M\} \quad \tilde{A} = \{(f, f') \mid f \in A\}$$

$\tilde{A}$ is compact in $\tilde{M}$, and so are each of the projections, and, by Arzelà–Ascoli Theorem, $\{f' \mid f \in A\}$ is equicontinuous.

Solution to 1.6.20: We consider three cases.

- (a_n) has a vanishing subsequence. Then, the corresponding subsequence of $\{f_n\}$ converges to $x + \cos x$, a continuous function.

- (a_n) has a subsequence with limit $a \neq 0$. Then, the corresponding subsequence of $\{f_n\}$ converges to

$$\frac{1}{a}\sin(ax) + \cos(x + a),$$

  which is continuous.

- $|a_n| \to \infty$. In this case, $\frac{1}{a_n}\sin a_n x \to 0$, and $\cos(x + a_n)$ depends on $a_n \pmod{2\pi}$. Let (b_n) be the sequence defined by $b_n \equiv a_n \pmod{2\pi}$, $b_n \in [0, 2\pi]$. As $[0, 2\pi]$ is compact, (b_n) has a convergent subsequence, to b say. Then the corresponding subsequence of $\{f_n\}$ converges to $\cos(x + b)$, a continuous function.

Solution to 1.6.21: The answer is no. Consider the sequence of functions $f_n : [0, 1] \to \mathbb{R}$ whose graphs are given by the straight lines through the points $(0, 0)$, $(1/2n, n)$, $(1/n, 0)$ to $(1, 0)$.

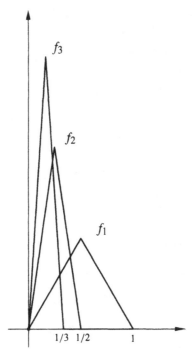

The sequence approximates the zero-function pointwise, but $\int_0^1 f_n(x)dx = \dfrac{1}{2}$ for all n.

Solution to 1.6.22: We have

$$g_n(x) = g_n(0) + g_n'(0)x + \frac{g_n''(\xi)}{2}x^2 = \frac{g_n''(\xi)}{2}x^2 \quad \text{for some} \quad \xi \in (0, 1)$$

so

$$|g_n(x)| \leqslant \frac{1}{2} \quad \text{for} \quad x \in [0, 1].$$

Also,

$$|g_n'(x)| = |g_n'(x) - g_n'(0)| \leqslant |g_n''(\xi)(x - 0)| \leqslant 1 \quad \text{for} \quad x \in [0, 1].$$

Therefore,

$$|g_n(x) - g_n(y)| \leqslant |x - y| \quad \text{for} \quad x, y \in [0, 1].$$

The sequence $\{g_n\}$ is then equicontinuous and uniformly bounded; so, by the Arzelà–Ascoli's Theorem [MH93, p. 273], it has a uniformly convergent subsequence.

Solution to 1.6.23: Fix $\varepsilon > 0$. Since K is continuous on a compact set, it is uniformly continuous. So, there is a $\delta > 0$ such that $|K(x_1, y_1) - K(x_2, y_2)| < \varepsilon$ whenever $\sqrt{(x_1 - x_2)^2 + (y_1 - y_2)^2} < \delta$. Let f and g be as above, and suppose

x_1 and x_2 are in $[0,1]$ and satisfy $|x_1 - x_2| < \delta$. Then

$$|f(x_1) - f(x_2)| = \left| \int_0^1 g(y)\,(K(x_1, y) - K(x_2, y))\,dy \right|$$

$$\leqslant \int_0^1 |g(y)|\,|K(x_1, y) - K(x_2, y)|\,dy$$

$$\leqslant \int_0^1 1 \cdot \varepsilon\,dy = \varepsilon\,.$$

As the estimate holds for all f in F, the family F is equicontinuous.

Solution to 1.6.25: 1. By the Cauchy–Schwarz Inequality [MH93, p. 69], we have

$$|g_n(x)| \leqslant \sqrt{\int_0^1 (x + y)\,dy} \sqrt{\int_0^1 (f_n(y))^2\,dy} \leqslant \sqrt{\int_0^1 (1 + y)\,dy}\,\sqrt{5} = \sqrt{\frac{15}{2}}$$

2. Since $\sqrt{x + y}$ is a continuous function of x and y on the compact unit square, it is uniformly continuous there. Hence, given any $\varepsilon > 0$, there exists $\delta > 0$ such that

$$\left| \sqrt{x_1 + y_1} - \sqrt{x_2 + y_2} \right| < \varepsilon$$

whenever $|x_1 - x_2| + |y_1 - y_2| < \delta$. In particular, $\left| \sqrt{x_1 + y} - \sqrt{x_2 + y} \right| < \varepsilon$ whenever $|x_1 - x_2| < \delta$, therefore,

$$|g_n(x_1) - g_n(x_2)| = \left| \int_0^1 \left(\sqrt{x_1 + y} - \sqrt{x_2 + y} \right) f_n(y)\,dy \right|$$

$$\leqslant \varepsilon \int_0^1 |f_n(y)|\,dy$$

$$\leqslant 5\varepsilon$$

whenever $|x_1 - x_2| < \delta$. Since the same value of δ works for all values of n simultaneously, the family $\{g_n\}$ is equicontinuous. Using the uniform bound established above the conclusion follows from the Arzelà–Ascoli's Theorem.

Solution to 1.6.26: We will first show that $\{g_n\}$ is a Cauchy sequence in sup-norm. Using the Cauchy–Schwarz Inequality [MH93, p. 69], we have

$$|g_n(x) - g_m(x)| \leqslant \int_0^1 |K(x, y)|(f_n(y) - f_m(y))dy$$

$$\leqslant \sqrt{\int_0^1 |K(x, y)|^2 dy} \sqrt{\int_0^1 |f_n(y) - f_m(y)|^2 dy}$$

hence,

$$\sup_{x \in [0,1]} |g_n(x) - g_n(x)| \leqslant \sup_{x \in [0,1]} \sqrt{\int_0^1 |K(x, y)|^2 dy} \sqrt{\int_0^1 |f_n(y) - f_m(y)|^2 dy}$$

Since K is continuous, it is integrable, and taking $M = \sup_{x,y \in [0,1]} |K(x,y)|$, we have

$$\|g_n(x) - g_m(y)\| \leq M \sqrt{\int_0^1 |f_n(y) - f_m(y)|^2 dy} \to 0$$

showing that the sequence $\{g_n\}$ is a Cauchy sequence in the sup-norm; as $C[0, 1]$ is complete on this norm, the sequence converges uniformly.

Solution to 1.6.27: We'll use the Stone–Weierstrass Approximation Theorem [MH93, p. 284] in the space $C_{[0,1]}$ equipped with the sup norm $\|f - g\| = \sup\{|f(x) - g(x)| \mid x \in [0, 1]\}$. Let $f \in C_{[0,1]}$ and let $\{p_n\}$ be a sequence of polynomials converging to f in the sup norm. Using our hypothesis with $k = 0$ we conclude that, for some positive M, we have $|\int_0^1 \varphi_n dx| \leq M$ for all n. Given $\varepsilon > 0$ there exists an integer $k(\varepsilon)$ such that $\|f - p_{k(\varepsilon)}\| \leq \frac{\varepsilon}{3M}$ for $k \geq k(\varepsilon)$. As $\int_0^1 p_{k(\varepsilon)} \varphi_n$ converges, it is a Cauchy sequence, therefore there is an integer $n(\varepsilon)$ such that

$$\left| \int_0^1 p_{k(\varepsilon)} \varphi_n - \int_0^1 p_{k(\varepsilon)} \varphi_m \right| < \frac{\varepsilon}{3} \quad \text{for } m, n \geq n(\varepsilon),$$

for $m, n \geq n(\varepsilon)$, we have

$$\left| \int_0^1 f \varphi_n - \int_0^1 f \varphi_m \right| \leq \left| \int_0^1 f \varphi_n - \int_0^1 p_{k(\varepsilon)} \varphi_m \right|$$

$$+ \left| \int_0^1 p_{k(\varepsilon)} \varphi_n - \int_0^1 p_{k(\varepsilon)} \varphi_m \right|$$

$$+ \left| \int_0^1 p_{k(\varepsilon)} \varphi_m - \int_0^1 f \varphi_m \right|$$

$$\leq \int_0^1 |f - p_{k(\varepsilon)} \varphi_n| + \frac{\varepsilon}{3} + \int_0^1 |f - p_{k(\varepsilon)} \varphi_m|$$

$$\leq \|f - p_{k(\varepsilon)}\| \int_0^1 \varphi_n + \frac{\varepsilon}{3} + \|f - p_{k(\varepsilon)}\| \int_0^1 \varphi_m$$

$$\leq \varepsilon$$

therefore the sequence $\int_0^1 f \varphi_n$, being Cauchy, converges.

Solution to 1.6.28: As

$$\left| \frac{e^{i\lambda_n x}}{n^2} \right| \leq \frac{1}{n^2} \quad \text{and} \quad \sum_{n=1}^{\infty} \frac{1}{n^2} < \infty$$

by the Weierstrass *M-test* [Rud87, p. 148], the given series converges uniformly on $\mathbb{R}$ to a continuous function f. We have, since the convergence is uniform,

$$\frac{1}{2T}\int_{-T}^{T}\sum_{n=1}^{\infty}\frac{e^{i\lambda_n x}}{n^2}dx = \frac{1}{2T}\sum_{n=1}^{\infty}\int_{-T}^{T}\frac{e^{i\lambda_n x}}{n^2}dx = \sum_{n=1}^{\infty}\frac{\sin\lambda_n T}{n^2\lambda_n T}.$$

As

$$\left|\frac{\sin\lambda_n T}{n^2\lambda_n T}\right| \leqslant \frac{1}{n^2}$$

we have, again by the Weierstrass *M-test*, that

$$\sum_{n=1}^{\infty}\frac{\sin\lambda_n T}{n^2\lambda_n T}$$

converges uniformly in T. Therefore, we get

$$\lim_{T\to\infty}\frac{1}{2T}\int_{-T}^{T}f(x)dx = \lim_{T\to\infty}\sum_{n=1}^{\infty}\frac{\sin\lambda_n T}{n^2\lambda_n T} = \sum_{n=1}^{\infty}\lim_{T\to\infty}\frac{\sin\lambda_n T}{n^2\lambda_n T} = \sum_{\lambda_n=0}\frac{1}{n^2}.$$

Solution to 1.6.29: Let $\sigma > 1$. It suffices to show that $\zeta(x)$ is defined and has continuous derivatives for $x \geqslant \sigma$. The series $\sum n^{-x}$ converges for such x. As $n^{-x} \leqslant n^{-\sigma}$, it follows from the Weierstrass *M-test* [Rud87, p. 148] that the series converges uniformly, so ζ is a continuous function. To see that it has continuous derivatives of all orders, we formally differentiate the series k times, getting

$$\sum_{n=2}^{\infty}\frac{(-\log n)^k}{n^x}.$$

It is enough to show that this series converges uniformly on k. Since

$$\left|\frac{(-\log n)^k}{n^x}\right| \leqslant \frac{(\log n)^k}{n^\sigma},$$

by the Weierstrass *M-test*, it will suffice to show that the series

$$\sum_{n=2}^{\infty}\frac{(\log n)^k}{n^\sigma}$$

converges. But

$$\frac{(\log n)^k}{n^\sigma} = o\left(\frac{1}{n^{\sigma-\varepsilon}}\right) \quad (n \to \infty),$$

for any positive ε. As

$$\sum_{n=1}^{\infty}\left(\frac{1}{n^{\sigma-\varepsilon}}\right)$$

converges for $\sigma - \varepsilon > 1$, we are done.

Solution to 1.6.30: Fix an interval $[a, b]$ and $\varepsilon > 0$. Since f is continuous, it is uniformly continuous on the interval $[a, b+1]$, then there exists an $N > 0$ such that if $n \geqslant N$ and $|x - y| < 1/n$ we have $|f(x) - f(y)| < \varepsilon$. We will show that $f_n(x)$ converges uniformly to

$$\int_x^{x+1} f(y)\,dy$$

for all x in the given interval. Fix x and $n \geqslant N$. We have

$$\left| \int_x^{x+1} f(y)\,dy - f_n(x) \right| = \left| \sum_{k=0}^{n-1} \int_{x+k/n}^{x+(k+1)/n} f(y)\,dy - f_n(x) \right|.$$

By the Mean Value Theorem for Integrals [MH93, p. 457], for each k there is $a_k \in (x + k/n, x + (k+1)/n)$ such that

$$\int_{x+k/n}^{x+(k+1)/n} f(y)\,dy = f(a_k)/n.$$

Substituting this in the above, expanding using the definition of f_n, and using uniform continuity, we get

$$\left| \int_x^{x+1} f(y)\,dy - f_n(x) \right| \leqslant \frac{1}{n} \sum_{k=0}^{n-1} |f(a_k) - f(x + k/n)| < \varepsilon.$$

Since this holds for any x, we are done.

Solution to 1.6.31: Let $\alpha > 0$, then for $|n| > 4\alpha$, the bound on f gives, for $x \in [-\alpha, \alpha]$,

$$|f(x + n)| \leqslant \frac{C}{1 + n^2/2} = M_n.$$

As the series

$$\sum_{-\infty}^{\infty} M_n$$

converges, by the Weierstrass M-test [Rud87, p. 148], the series

$$\sum_{|n|>4\alpha} f(x + n)$$

converges uniformly. So the series for $F(x)$ converges uniformly on $[-\alpha, \alpha]$ and F is continuous there. As α is arbitrary, F is continuous on $\mathbb{R}$.

We have

$$F(x+1) - F(x) = \lim_{\alpha \to \infty} \sum_{-\alpha}^{\alpha} (f(x+1+n) - f(x+n))$$

$$= \lim_{\alpha \to \infty} (f(x+1+\alpha) - f(x-\alpha))$$

$$= 0$$

the last equality holding by our assumption on f.

If G is continuous and periodic with period 1, then, since the series for F converges uniformly,

$$\int_0^1 F(x)G(x)\,dx = \sum_{-\infty}^{\infty} \int_0^1 f(x+n)G(x)\,dx.$$

In each integral on the right-hand side, let $y = x + n$, and get, since $G(y - n) = G(y)$,

$$\sum_{-\infty}^{\infty} \int_n^{n+1} f(y)G(y)\,dy = \int_{-\infty}^{\infty} f(y)G(y)\,dy.$$

Solution to 1.6.32: Given f and ε, let $h : [0, 1] \to \mathbb{R}$ be defined by $h(x) = f\left(\sqrt[4]{x}\right)$. By the Stone–Weierstrass Approximation Theorem [MH93, p. 284], there is a polynomial P such that $|P(x) - h(x)| < \varepsilon/2$ for $x \in [0, 1]$, from which it follows that

$$\left|P\left(x^4\right) - f(x)\right| = \left|P\left(x^4\right) - h\left(x^4\right)\right| < \varepsilon/2 \quad \text{for} \quad x \in [0, 1].$$

If $P = \sum_{k=0}^{n} a_k x^k$, take $C_0, \ldots, C_n \in \mathbb{Q}$ such that $\sum_{k=0}^{n} |a_k - C_k| < \varepsilon/2$. Then we have

$$\left|\sum_{k=0}^{n} C_k x^{4k} - f(x)\right| \leqslant \left|\sum_{k=0}^{n} C_k x^{4k} - \sum_{k=0}^{n} a_k x^{4k}\right| + \left|\sum_{k=0}^{n} a_k x^{4k} - f(x)\right| < \varepsilon.$$

1.7 Fourier Series

Solution to 1.7.1: 1. We have, for $n \in \mathbb{N}$,

$$\frac{1}{\pi} \int_{-\pi}^{\pi} f(x) \cos nx \, dx = 0$$

because the integrand is an odd function. Also

$$\frac{1}{\pi} \int_{-\pi}^{\pi} f(x) \sin nx \, dx = \frac{2}{\pi} \int_0^{\pi} f(x) \sin nx \, dx$$

$$= \frac{2}{\pi} \left(-\frac{x \cos nx}{n} \Big|_0^\pi + \int_0^\pi \frac{\cos nx}{n} \right)$$

$$= \frac{2}{\pi} \frac{(-1)^{n+1}}{n},$$

so the Fourier series of f is

$$\sum_{n=1}^{\infty} \frac{(-1)^{n+1}2}{n} \sin nx.$$

2. If the series converged uniformly the function f would be continuous, which is not the case.

3. As f and f' are sectionally continuous, we have

$$\sum_{n=1}^{\infty} \frac{(-1)^{n+1}2 \sin nx}{n} = \frac{f(x-) + f(x+)}{2} = \begin{cases} f(x) & \text{if } x \neq (2n+1)\pi \\ 0 & \text{if } x = (2n+1)\pi \end{cases}$$

where $n \in \mathbb{Z}$.

Solution to 1.7.2: 1. Since $f(x)$ is an odd function, the integrals

$$\frac{1}{\pi} \int_{-\pi}^{\pi} f(x) \cos nx \, dx$$

vanish for $n \in \mathbb{N}$. The Fourier series has only terms in $\sin nx$ given by

$$b_n = \frac{1}{\pi} \int_{-\pi}^{\pi} x^3 \sin nx \, dx.$$

2. As f and f' are sectionally continuous, we have

$$\sum_{n=1}^{\infty} b_n \sin nx = \frac{f(x-) + f(x+)}{2} = \begin{cases} f(x) & \text{if } x \neq (2n+1)\pi \\ 0 & \text{if } x = (2n+1)\pi \end{cases}$$

where $n \in \mathbb{Z}$.

3. Using Parseval's Identity [MH93, p. 577]

$$\frac{1}{2} a_0^2 + \sum_{n=1}^{\infty} (a_n^2 + b_n^2) = \frac{1}{\pi} \int_{-\pi}^{\pi} f^2(x) \, dx$$

and the fact that all $a_n = 0$,

$$\sum_{n=1}^{\infty} b_n^2 = \frac{1}{\pi} \int_{-\pi}^{\pi} x^6 \, dx = \frac{2}{7} \pi^6.$$

Solution to 1.7.3: The answer is no; $f(x) = 1$ satisfies the above equation and is not identically zero.

Solution to 1.7.4: As f is $\sqrt{2}$–periodic, we have

$$\hat{f}(n) = \int_0^1 f(x)e^{-2n\pi ix}\,dx = \int_0^1 f(x+\sqrt{2})e^{-2n\pi ix}\,dx.$$

Letting $y = x + \sqrt{2}$, we get

$$\hat{f}(n) = e^{2n\pi i\sqrt{2}} \int_{\sqrt{2}}^{1+\sqrt{2}} f(y)e^{-2n\pi iy}\,dy.$$

Since f is also 1-periodic, we have

$$\hat{f}(n) = e^{2n\pi i\sqrt{2}} \int_0^1 f(y)e^{-2n\pi iy}\,dy = e^{2n\pi i\sqrt{2}}\hat{f}(n).$$

$e^{2n\pi i\sqrt{2}} \neq 1$ for $n \neq 0$, so $\hat{f}(n) = 0$ for $n \neq 0$ and f is constant.

Solution to 1.7.5: Suppose that such an f exists. As the power series for e^x converges uniformly, we have, for $n > 0$,

$$\hat{f}(n) = \int_0^1 f(x)e^{-2\pi inx}\,dx = \sum_{k=0}^\infty \frac{(-2\pi ni)^k}{k!} \int_0^1 f(x)x^k\,dx = -2\pi ni.$$

This equality contradicts the Riemann–Lebesgue Lemma [MH93, p. 628], which says that $\lim_{n\to\infty} \hat{f}(n) = 0$, so no such a function can exist.

Solution 2. From the assumptions the function $x^2 f(x)$ is orthogonal to all polynomials, so by the Stone–Weierstrass Approximation Theorem [MH93, p. 284] it is identically zero, and so it is $f(x)$, which is a contradiction with the first integral.

Solution to 1.7.6: The Fourier series of f converges to f because f'' exists. Let the Fourier series of f be

$$\frac{\alpha_0}{2} + \sum_{n=1}^\infty (\alpha_n \cos nx + \beta_n \sin nx).$$

As $f'' = g - kf$ is continuous, we obtain its Fourier series by termwise differentiating the series for f, and get

$$\frac{k\alpha_0}{2} + \sum_{n=1}^\infty \left(\left(k - n^2\right)\alpha_n \cos nx + \left(k - n^2\right)\beta_n \sin nx \right) =$$

$$\frac{a_0}{2} + \sum_{n=1}^\infty (a_n \cos nx + b_n \sin nx).$$

So we have

$$\alpha_0 = \frac{a_0}{k}, \quad \alpha_n = \frac{a_n}{k - n^2}, \quad \beta_n = \frac{b_n}{k - n^2} \quad \text{for} \quad n \geqslant 1.$$

Solution to 1.7.7: Consider the Fourier series of f,

$$f(x) = \frac{a_0}{2} + \sum_{n=1}^{\infty}(a_n \cos nx + b_n \sin nx).$$

We have $a_0 = 0$, and, by Parseval's Identity [MH93, p. 577],

$$\int_0^{2\pi} f^2(x)dx = \pi \sum_{n=1}^{\infty} \left(a_n^2 + b_n^2\right) \leqslant \pi \sum_{n=1}^{\infty} \left(n^2 a_n^2 + n^2 b_n^2\right) = \int_0^{2\pi} (f'(x))^2 dx.$$

Solution to 1.7.8: The Riemann–Lebesgue Lemma [MH93, p. 628] states that the result is valid for all functions $g(x)$ of the type

$$\cos k\pi x \quad \text{and} \quad \sin k\pi x$$

using linearity of the integral the result extends to all finite trigonometric polynomials

$$p(x) = \sum_{k=0}^{n} a_k \cos k\pi x + b_n \sin k\pi x .$$

We will now use the fact that the set of trigonometric polynomials is dense in the space of continuous functions with the *sup* norm (Stone–Weierstrass Approximation Theorem [MH93, p. 284]) to extend it to all continuous functions. Given any $\varepsilon > 0$, there exists a $p_\varepsilon(x)$ as above such that

$$|g(x) - p_\varepsilon(x)| < \varepsilon$$

then

$$\left| \lim_{n\to\infty} \int_0^1 f(x)g(nx)dx - \int_0^1 f(x)dx \int_0^1 g(x)dx \right.$$

$$\left. - \left(\lim_{n\to\infty} \int_0^1 f(x)p_\varepsilon(nx)dx - \int_0^1 f(x)dx \int_0^1 p_\varepsilon(x)dx \right) \right| \leqslant$$

$$\leqslant \left| \lim_{n\to\infty} \int_0^1 f(x)(g(nx) - p_\varepsilon(nx))dx \right| + \left| \int_0^1 f(x)dx \int_0^1 (g(x) - p_\varepsilon(x))dx \right|$$

$$\leqslant \lim_{n\to\infty} \int_0^1 |f(x)| \, |g(nx) - p_\varepsilon(nx)| \, dx + \int_0^1 |f(x)| \, dx \int_0^1 |g(x) - p_\varepsilon(x)| \, dx$$

$$\leqslant 2\varepsilon \int_0^1 |f(x)| \, dx$$

1.8   Convex Functions

Solution to 1.8.1: Let $M = \max_{x \in [0,1]} |f(x)|$. Consider a sequence (x_n) such that $x_0 = 1$, and $0 < x_n < 3^{-1}x_{n-1}$ satisfies $|f(x)| < M/2^n$ for $0 < x < x_n$. Define $g : [0, 1] \to [0, 1]$ by $g(0) = 0$ and, using the fact that $x_n \to 0$,

$$g(x) = t\frac{M}{2^n} + (1 - t)\frac{M}{2^{n-1}}$$

for $0 < x = tx_{n+1} + (1 - t)x_n, t \in [0, 1]$, $n = 0, 1, \ldots$. We have $g \geqslant f$ and

$$\frac{g(x_n) - g(x_{n+1})}{x_n - x_{n+1}} = \frac{M/2^n}{x_n - x_{n+1}} > \frac{M/2^{n-1}}{x_n - x_{n+1}} = \frac{g(x_{n-1}) - g(x_n)}{x_{n-1} - x_n}$$

so g is concave.

Solution to 1.8.2: For $x \neq y$ and $t \in [0, 1]$, let

$$c = (\log f(y) - \log f(x))/(x - y).$$

By hypothesis, we have

$$e^{c(tx+(1-t)y)} f(tx + (1 - t)y) \leqslant te^{cx} f(x) + (1 - t)e^{cy} f(y)$$

so

$$\begin{aligned}
f(tx + (1 - t)y) &\leqslant te^{c(x-y)(1-t)} f(x) + (1 - t)e^{-c(x-y)t} f(y) \\
&= te^{(\log f(y) - \log f(x))(1-t)} f(x) + (1 - t)e^{(\log f(x) - \log f(y))t} f(y) \\
&= t\left(\frac{f(x)}{f(y)}\right)^{t-1} f(x) + (1 - t)\left(\frac{f(x)}{f(y)}\right)^{t} f(y) \\
&= f(x)^t f(y)^{1-t}
\end{aligned}$$

taking logarithms, we get that $\log f$ is convex.

Solution to 1.8.3: Consider an interval $[a, b]$ and suppose that the maximum of f does not occur at one of its endpoints. Then, by Weierstrass Theorem [MH93, p. 189], there is $c \in (a, b)$ maximizing f. By the continuity of f, there are intervals $A = [a, a_0]$ and $B = [b_0, b]$ in $[a, b]$ with $f(x) < f(c)$ if x lies in A or B. By the Mean Value Inequality for Integrals [MH93, p. 457], we have

$$\begin{aligned}
f(c) &\leqslant \frac{1}{2h}\int_A f(y)\,dy + \frac{1}{2h}\int_{[a_0,b_0]} f(y)\,dy + \frac{1}{2h}\int_B f(y)\,dy \\
&< \frac{a_0 - a}{2h}f(c) + \frac{b_0 - a_0}{2h}f(c) + \frac{b - b_0}{2h}f(c) \\
&= f(c).
\end{aligned}$$

This contradiction shows that f must attain its maximum at a or b.

If $L(x)$ is any linear function, a straightforward calculation shows that L is convex and satisfies the *mean value inequality* above, and that both of these inequalities are, in fact, equalities. Now let L be given by

$$L(x) = \frac{(x-a)f(b) - (x-b)f(a)}{b-a}$$

and consider $G(x) = f(x) - L(x)$. By the linearity of the integral, since f and L satisfy the Mean Value Inequality, G does as well. Therefore, G takes its maximum value at a or b. A calculation shows that $G(a) = G(b) = 0$. Therefore, we must have that $f(x) \leqslant L(x)$ for all $x \in [a, b]$. For any $t \in [0, 1]$, $(1-t)a + tb \in [a, b]$. Substituting this into the inequality gives that f is convex.

2
Multivariable Calculus

2.1 Limits and Continuity

Solution to 2.1.1: Let $x \in \mathbb{R}^n$, $\varepsilon > 0$, and let B denote the open ball with center $f(x)$ and radius ε. For $n = 1, 2, \ldots$, let K_n be the closed ball with center x and radius $1/n$. By (ii) we have $\bigcap_1^\infty f(K_n) = \{f(x)\}$. By (i) the sets $(\mathbb{R}^n - B) \cap f(K_n)$ are compact for $n = 1, 2, \ldots$. They form a decreasing sequence, and their intersection is empty, by the preceding equality. Hence, there is an n_0 such that $(\mathbb{R}^n - B) \cap f(K_{n_0}) = \emptyset$, which means that $|f(y) - f(x)| < \varepsilon$ whenever $|y - x| < 1/n_0$. So f is continuous at x.

Solution to 2.1.2: Let $(x, y) \notin G(g)$. Then $y \neq g(x)$, and there exist disjoint neighborhoods V of y and V' of $g(x)$ in $\mathbb{R}^n$. By hypothesis, we can find a neighborhood U of x such that $g(U) \subset V'$. Then $U \times V$ is a neighborhood of (x, y) disjoint from $G(g)$, and this proves that $G(g)$ is closed.

The converse is false. Take $n = 1$, and let $g : \mathbb{R} \to \mathbb{R}$ be defined by $g(x) = 1/x$, for $x \neq 0$; $g(0) = 0$. Then the graph of g is closed in $\mathbb{R} \times \mathbb{R}$, and g is discontinuous at 0.

Solution 2. Let $\rho : \mathbb{R}^n \times \mathbb{R}^n \to \mathbb{R}^n$ defined by $\rho(x, y) = y - f(x)$. Then ρ is continuous and $G_f = \rho^{-1}(0)$ is closed being the inverse image of a closed set.

Solution to 2.1.3: If $U \neq \mathbb{R}^n$ let x be in the boundary of U. Therefore there is a sequence (x_n) of elements of U converging to x. $(h(x_n))$ is a Cauchy sequence, so it is convergent, to y say. Let $y = h(z)$. Then $z \neq x$ since $z \in U$ and $x \notin U$.

Then,

$$h(x_n) \to h(z) \quad \text{and} \quad x_n \not\to z,$$

which contradicts the fact that h is a homeomorphism.

Solution to 2.1.4: We show that f is continuous at $(0,0)$; for the general case, use a change of variables. By adding a constant, if necessary, we may assume $f(0,0) = 0$. Suppose f is not continuous at the origin. Then, for any $\varepsilon > 0$, there is a sequence $((x_n, y_n))$ tending to the origin with $|f(x_n, y_n)| \geqslant \varepsilon$ for each n. Since f is continuous in the first variable, there exists a $\delta > 0$ such that if $|x| < \delta$, then $|f(x,0)| < \varepsilon/2$. Applying this to our sequence, we see that there is an $N > 0$, such that if $n \geqslant N$ then $|x_n| < \delta$, so $|f(x_n, 0)| < \varepsilon/2$. However, for each such n, $f(x_n, y)$ is continuous in the second variable, so by the Intermediate Value Theorem [Rud87, p. 93], there exists y'_n, $0 < y'_n < y_n$, such that $|f(x_n, y'_n)| = n\varepsilon/(n+1)$. Since the y_n's tend to 0 as n tends to infinity; the y'_n's do so as well. Hence, the set $E = \{(x_n, y'_n) \mid n \geqslant N\} \cup \{(0,0)\}$ is compact. Then by our hypothesis, the set $f(E)$ is compact. But $f(E) = \{n\varepsilon/(n+1) \mid n \geqslant N\} \cup \{0\}$, and ε is a limit point of this set which is not contained in it, a contradiction. Hence, f is continuous at the origin and we are done.

Solution to 2.1.5: Continuity implies $f(0) = 0$, so if any x_k is 0, then so are all subsequent ones, and the desired conclusion holds. Assume therefore, that $x_k \neq 0$ for all k. The sequence $(\|x_k\|)$ is then a decreasing sequence of positive numbers, so it has a nonnegative limit, say c. Suppose $c > 0$. The sequence (x_k), being bounded, has a convergent subsequence, say (x_{k_j}), with limit α. Then $\|\alpha\| = \lim_{j \to \infty} \|x_{k_j}\| = c$. Hence, $\|f(\alpha)\| < c$. But, by the continuity of f,

$$f(\alpha) = \lim_{j \to \infty} f(x_{k_j}) = \lim_{j \to \infty} x_{k_j+1},$$

and $\|x_{k_j+1}\| \geqslant c$ for all j, so we have a contradiction, and the desired conclusion follows.

Solution to 2.1.6: 1. If $\{e_1, \ldots, e_n\}$ denotes the standard basis of $\mathbb{R}^n$ and $N(e_i)$ the norm of each base vector, then any vector v in the unit sphere can be written as

$$v = c_1 e_1 + \cdots + c_n e_n \quad \text{where} \quad 0 \leqslant c_i \leqslant 1$$

so $N(v) \leqslant |c_1| N(v_1) + \cdots + |c_n| N(e_n) \leqslant n \max\{N(e_1), \ldots N(e_n)\}$ showing that N is bounded on the unit sphere.

2. Let B be the supremum of N in the unit sphere, then for any vector $x \neq y$

$$y = x + \|y - x\| \frac{y - x}{\|y - x\|}$$

so $N(y) \leqslant N(x) + \|y - x\| N\left(\dfrac{y - x}{\|y - x\|}\right)$, that is

$$N(y) - N(x) \leqslant B\|y - x\|$$

changing the places of x and y we then get $N(x) - N(y) \leqslant B\|x - y\|$, that is

$$|N(x) - N(y)| \leqslant B\|x - y\|$$

and N is continuous.

3. Let $A > 0$ be the minimum of the continuous function N on the compact unit sphere, where $B > 0$ is already the maximum, then

$$N(v) = \|v\|N\left(\frac{v}{\|v\|}\right)$$

so taking the maximum and minimum over all such v,

$$A\|v\| \leqslant N(v) \leqslant B\|v\| .$$

Solution to 2.1.7: 1. Let A be a closed subset of $\mathbb{R}^m$, and suppose $b \in \mathbb{R}^n \setminus f(A)$. For $r > 0$ let B_r be the closed ball in $\mathbb{R}^n$ with center b and radius r. Since B_r is compact and f is proper, $f^{-1}(B_r)$ is compact. As A is closed, $A \cap f^{-1}(B_r)$ is compact. Since $\bigcap_{r>0}(A \cap f^{-1}(B_r)) = A \cap f^{-1}(b) = \emptyset$, $A \cap f^{-1}(B_s) = \emptyset$ for some $s > 0$, so $f(A)$ is disjoint from the neighborhood B_s of b, so $f(A)$ is closed.

2. Suppose B is a compact subset of $\mathbb{R}^n$. Let $A_R = \{x \in \mathbb{R}^m \mid \|x\| \geqslant R\}$. Since f is closed, $f(A_R)$ is closed, so $f(A_R) \cap B$ is compact. As f is 1-1, $\bigcap_{R>0}(f(A_R) \cap B) = \emptyset$. Therefore $f(A_S) \cap B = \emptyset$ for some $S > 0$, so $f^{-1}(B)$ is contained in the compact set $\mathbb{R}^n \setminus A_S$. Since f is continuous, $f^{-1}(B)$ is closed. Therefore $f^{-1}(B)$ is compact.

2.2 Differential Calculus

Solution to 2.2.1: We maximize the function $f(x, y) = (x^2 + y^2)e^{-x-y}$ in the first quadrant, $x \geqslant 0$ and $y \geqslant 0$. The function attains a maximum there because it is nonnegative and tends to 0 as (x, y) tends to infinity. We have

$$\frac{\partial f}{\partial x} = (2x - x^2 - y^2)e^{-x-y} , \qquad \frac{\partial f}{\partial y} = (2y - x^2 - y^2)e^{-x-y} .$$

The critical points of f are thus the points (x, y) that satisfy

$$2x - x^2 - y^2 = 0 = 2y - x^2 - y^2 .$$

These equalities imply $x = y$ and $2x^2 - 2x = 0$. Hence, the only critical point in the open quadrant is $(1, 1)$.

On the x–axis, we have $f(x, 0) = x^2 e^{-x}$ and $\frac{df(0,x)}{dx} = (x^2 - 2x)e^{-x}$, so $\frac{df(0,x)}{dx} = 0$ only for $x = 0$ and $x = 2$. The point $(2,0)$ is thus another candidate for the point at which f attains its maximum. By the same reasoning, the point $(0,2)$ is another such candidate. The points $(2,0)$ and $(0,2)$ are the only candidates on the boundary of the quadrant.

We have

$$f(1, 1) = 2e^{-2}, \qquad f(2, 0) = 4e^{-2} = f(0, 2).$$

Hence, the maximum value of f in the quadrant is $4e^{-2}$, that is,

$$(x^2 + y^2)e^{-x-y} \leqslant 4e^{-2}$$
$$\frac{x^2 + y^2}{4} \leqslant e^{x+y-2}$$

for $x \geqslant 0$, $y \geqslant 0$.

Solution 2. Letting $u = t + 1$ in the inequality $e^t \geqslant t + 1$, which holds for any real t, we get $e^{u-1} \geqslant u$. For $u = x/2 + y/2$ we obtain

$$e^{\frac{x+y}{2}-1} \geqslant \frac{x + y}{2}.$$

As $\dfrac{x + y}{2} \geqslant \sqrt{\dfrac{x^2 + y^2}{4}}$, squaring we get

$$e^{x+y-2} \geqslant \left(\frac{x + y}{2}\right)^2 \geqslant \frac{x^2 + y^2}{4}.$$

Solution 3. On each level set $x + y = k$ the maximum for the left-hand side occurs at the extremes, where one of the variables is equal to 0. To see this observe that $x^2 + y^2$ is the square of the distance of (x, y) to the origin, which is maximal at the extremes of the segment $x + y = k$ in the first quadrant. So we only need to prove that

$$\frac{x^2}{4} \leqslant e^{x-2},$$

but this can be easily seem by the same argument as in the last solution, $u \leqslant e^{u-1}$, substituting $u = x/2$ and squaring both sides. Alternatively, we can also differentiate the function $f(x) = e^{-x}x^2$ to see that the maximum for positive values of x happens at 2 and then obtain the inequality again.

Solution 4. On each quart-of-circle $0 \leqslant \vartheta \pi/2$ of radius r the maximum of the left-hand side is $r^2/4$ and the value of the right-hand side is

$$e^{r(\cos\theta+\sin\theta)-2}$$

which is a one variable function assuming the minimum at the extremes 0 and $\pi/2$, where the value is e^{r-2}. Using the same argument as before we complete the proof.

Solution to 2.2.2: Let the arc length of the circle between the points of contact of the $(i-1)$th and ith sides be $2\varphi_i$. The area of the corresponding part of the hexagon is $\tan\varphi_i$, the total area is $A = \sum_{i=1}^{6} \tan\varphi_i$. This is the function to be minimized on $(0, \pi/2)^6$ with the restriction $\sum_{i=1}^{6} \varphi_i = \pi$. Since A is a convex function on $(0, \pi/2)^6 \subset \mathbb{R}^6$, any critical point is an absolute minimum.

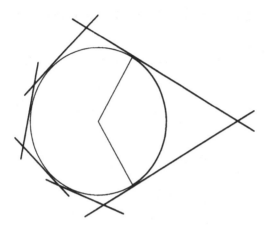

We use the method of Lagrange Multipliers [MH93, p. 414]. Lagrange's equations for a critical point are $\sec^2\varphi_i = \lambda$, which are satisfied by $\varphi_i = \pi/6$, corresponding to a regular hexagon. The minimal value of A is $2\sqrt{3}$.

Solution 2. With the same notation as above, the function tan is convex in the interval $(0, \pi/2)$, so, by Jensen's Inequality [Str81, p. 201], the minimum for the area will happen when all angles are the same, that is, when the hexagon is regular.

Solution to 2.2.3: 1. $|f(x, y)| \leq 2\sqrt{x^2 + y^2}$, and when (x, y) approaches the origin the limit is 0, and so is the limit of $f(x, y)$.
2. Computing the the directional derivative of f in the direction of (x, y) with $y \neq 0$, we have

$$\lim_{h \to 0} \frac{f(hx, hy)}{h} = \lim_{h \to 0} \left(1 - \cos\frac{h^2 x^2}{hy}\right)\sqrt{x^2 + y^2} = 0$$

and in the direction of the x-axis the limit is trivialy zero, because the function is identically zero.
3. From the previous calculation, we know that , if f were differentiable, the derivative would be zero, and then the quotient

$$\frac{f(x, y)}{\sqrt{x^2 + y^2}} \to 0$$

as $(x, y) \to (0, 0)$, but is false. To see it, approach the origin following the curve $y = x^2/\pi$, the limit of the quotient is then $1 - \cos\pi = 2$.

Solution to 2.2.4: The function f is differentiable at the point $z = (x_0, y_0) \in U$ if there is a linear transformation $f'(z) \in L\left(\mathbb{R}^2, \mathbb{R}^1\right)$ such that

$$\lim_{h \to 0} \frac{|f(z + h) - f(z) - f'(z)h|}{\|h\|} = 0.$$

Continuity of the partial derivatives is a sufficient condition for differentiability. A calculation gives

$$\frac{\partial f}{\partial x}(x, y) = \begin{cases} (4/3)x^{1/3} \sin y/x - yx^{-2/3} \cos y/x & \text{if } x \neq 0 \\ 0 & \text{if } x = 0 \end{cases}$$

$$\frac{\partial f}{\partial y}(x, y) = \begin{cases} x^{1/3} \cos(y/x) & \text{if } x \neq 0 \\ 0 & \text{if } x = 0 \end{cases}$$

which are continuous on $\mathbb{R}^2 \setminus \{(0, y) \mid y \in \mathbb{R}\}$. Thus, f is differentiable there. At any point $(0, y)$, we have

$$\frac{f(h, k) - f(0, y)}{\|(h, k)\|} = O(|h|^{1/3}) = o(1) \quad (h \to 0)$$

so f is differentiable at these points also.

Solution to 2.2.5: Let $A = (a_{jk})$ be the Jacobian matrix of F at the origin. According to the definition of differentiability we have $F(x) = Ax + R(x)$ with $\lim_{x \to 0} \|R(x)\|/\|x\| = 0$. Moreover,

$$\|Ax\|^2 = \sum_{j=1}^n \left(\sum_{k=1}^n a_{jk}x_k\right)^2$$

$$\leq \sum_{j=1}^n \left(\sum_{k=1}^n a_{jk}^2 \sum_{k=1}^n x_k^2\right)$$

$$= \left(\sum_{j,k=1}^n a_{jk}^2\right)\left(\sum_{k=1}^n x_k^2\right)$$

$$= c\|x\|^2,$$

where we used Cauchy–Schwarz's inequality.

Choose ε positive such that $\|R(x)\|/\|x\| < 1 - \sqrt{c}$ for $\|x\| < \varepsilon$, for these values of x, we have

$$\|F(x)\| \leq \|Ax\| + \|R(x)\| \leq \sqrt{c}\|x\| + \left(1 - \sqrt{c}\right)\|x\| = \|x\|,$$

hence F maps the ball centered at 0 with radius ε into itself.

Solution to 2.2.6: The Jacobian determinant of f is given by

$$Jf(x, y) = \det \begin{pmatrix} p'(x + y) & p'(x + y) \\ p'(x - y) & -p'(x - y) \end{pmatrix} = -2p'(x + y)p'(x - y) .$$

The derivative $Df(x, y)$ is invertible exactly when $Jf(x, y) \neq 0$, thus if and only if $p'(x + y)p'(x - y) \neq 0$.

Since p has positive degree, its derivative p' is not identically zero, so it has only finitely many zeros, say $\lambda_1, \ldots, \lambda_n$. The set $\{Jf = 0\}$ is the union of finitely many lines, the lines $x + y = \lambda_k$, $k = 1, \ldots, n$, and the lines $x - y = \lambda_k$, $k = 1, \ldots, n$. This is a closed nowhere-dense set, so $\{Jf \neq 0\}$ is an open dense set.

Solution to 2.2.7: We have

$$\nabla h(x, y) = (g'(x)g(y), \, g'(y)g(x)) .$$

For the desired condition to hold, $\nabla h(x, y)$ must be a scalar multiple of (x, y), at least for $(x, y) \neq (0, 0)$. Thus,

$$\frac{g'(x)g(y)}{x} = \frac{g'(y)g(x)}{y},$$

assuming $xy \neq 0$. This can be written as

$$\frac{g'(x)}{xg(x)} = \frac{g'(y)}{yg(y)},$$

which implies that $\dfrac{g'(x)}{xg(x)} = A = \text{constant}$. The preceding differential equation for g has the general solution

$$g(x) = B \, e^{Ax^2/2}$$

with B a constant, positive because g is positive by assumption. Every such g obviously has the required property.

Solution to 2.2.8: Since f is continuous and $\mathbb{R}^n$ is connected, $f(\mathbb{R}^n)$ is connected. We will prove that $f(\mathbb{R}^n)$ is both open and closed in $\mathbb{R}^n$. This will imply that $f(\mathbb{R}^n) = \mathbb{R}^n$, because $f(\mathbb{R}^n) \neq \emptyset$.

Let $y = f(x) \in f(\mathbb{R}^n)$. As the rank of $(\partial f_i/\partial x_j)$ is n, by the Inverse Function Theorem [Rud87, p. 221], there are open neighborhoods V_x and V_y of x and y such that $f|_{V_x} : V_x \rightarrow V_y$ is a diffeomorphism; therefore, V_y is an open neighborhood of y, and $f(\mathbb{R}^n)$ is open.

Let (y_n) be a sequence in $f(\mathbb{R}^n)$ converging to $y \in \mathbb{R}^n$, $f(x_n) = y_n$, say. The set $K = \{y_n \mid n \in \mathbb{N}\} \cup \{y\}$ is compact; therefore, $f^{-1}(K)$ is also compact. But $\{x_n \mid n \in \mathbb{N}\} \subset f^{-1}(K)$; therefore, it contains a convergent subsequence, say (x_{n_j}) with $x_{n_j} \rightarrow x \in \mathbb{R}^n$. Since f is continuous, $f(x_{n_j}) \rightarrow f(x)$. But $\lim_{j \rightarrow \infty} y_{n_j} = y$; therefore, $f(x) = y$, and $f(\mathbb{R}^n)$ is closed.

Solution to 2.2.9: We have $\mathbb{R}^2 = f(S) \cup \left(\mathbb{R}^2 \setminus f(S)\right)$ where S is the set of singularities of f. It suffices to show that f maps $R = \mathbb{R}^2 \setminus f^{-1}(f(S))$ onto $\mathbb{R}^2 \setminus f(S)$. $f(S)$ is finite, so $\mathbb{R}^2 \setminus f(S)$ is connected. As $f(S)$ is closed, R is open. It suffices to show that $f(R)$ is open and closed in $\mathbb{R}^2 \setminus f(S)$.

As $R \cap S = \emptyset$, by the Inverse Function Theorem [Rud87, p. 221], f is invertible in a neighborhood of each point of R. Hence, locally, f is an open map, and as the union of open sets is open, $f(R)$ is open.

Let ξ be a limit point of $f(R)$ in $\mathbb{R}^2 \setminus f(S)$, and (ξ_n) be a sequence in $f(R)$ converging to ξ. The set (ξ_n) is bounded, therefore, by hypothesis, $(f^{-1}(\xi_n))$ is bounded. Let (x_n) be a sequence such that $f(x_n) = \xi_n$. This sequence is bounded, so it must have a limit point, x, say. As f is continuous, we have $f(x) = \xi$. Since $\xi \notin f(S)$, $x \notin f^{-1}(f(S))$, so $x \in R$. Therefore, $\xi \in f(R)$, so $f(R)$ is closed in $\mathbb{R}^2 \setminus f(S)$.

Solution to 2.2.10: Consider the scalar field $G : \mathbb{R}^n \to \mathbb{R}$ given by $G(y) = \|\nabla f(y)\|^2$. We have

$$DG(y) = \left(\frac{\partial^2 f}{\partial x_i \partial x_j}(y)\right).$$

By the hypothesis, we have $G(x) = 0$, G is C^1, and $G'(x) \neq 0$. Therefore, by the Inverse Function Theorem [Rud87, p. 221], G is locally a diffeomorphism onto a neighborhood of 0 in $\mathbb{R}$. In particular, it is injective in some neighborhood of x, so it has no other zeros there.

Solution to 2.2.11: 1. Since f is C^2, we can expand f in a Taylor series [MH93, p. 359] around a and obtain

$$f(a+h) = f(a) + f'(a) \cdot h + \frac{1}{2} f''(a) \cdot h^2 + O(|h|^3) \quad (h \to 0)$$

where $f''(a) \cdot h^2 = h^t H h = \langle h, Hh \rangle$, $H = \left(\frac{\partial^2 f}{\partial x_i \partial x_j}\right)$, and the big Oh notation means that for h in some neighborhood of 0, $h \in V_0$, we have $|O(|h|^3)| < K|h|^3$ for some $K > 0$.

The hypothesis that a is a critical point implies that $f'(a) \cdot h = 0$ for all $h \in \mathbb{R}^n$, and the hypothesis that H is positive definite implies that $\langle h, Hh \rangle \geqslant 0$ for all $h \in \mathbb{R}^n$ and zero only at $h = 0$. Therefore, all the eigenvalues of H are positive and there exists some $c > 0$ such that $\langle h, Hh \rangle \geqslant c|h|^2$ (namely c is the minimum of all the eigenvalues, which are all real because H is symmetric, by the Schwarz Theorem [HK61, p. 340]). Let $W_a = \{a + h \mid h \in V_0, |h| < c/k\}$. We have

$$\begin{aligned} f(a+h) - f(a) &= \langle h, Hh \rangle + O(|h|^3) \geqslant c|h|^2 + O(|h|^3) \\ &\geqslant c|h|^2 - K|h|^3 = |h|^2(c - K|h|) > 0 \end{aligned}$$

which shows that $f(a+h) > f(a)$ for $h \neq 0$ and $h \in W_a$, and, therefore, a is a local minimum.

2. Assume f has two critical points, p_1 and p_2. Since the Hessian matrix is positive definite, p_1 and p_2 are local minima. Let $tp_1 + (1 - t)p_2, t \in \mathbb{R}$, be the line containing p_1 and p_2. Consider the real function g given by $g(t) = f(tp_1 + (1 - t)p_2)$. g has local minima at $t = 0$ and at $t = 1$. Therefore, g has a local maximum at some $t_0 \in (0, 1)$. We have $g''(t_0) \leqslant 0$, but

$$g''(t_0) = f''(tp_1 + (1 - t_0)p_2)(p_1 - p_2)^2 = \langle p_1 - p_2, H(p_1 - p_2) \rangle$$

and our assumptions on H imply $\langle p_1 - p_2, H(p_1 - p_2) \rangle > 0$, a contradiction.

Solution to 2.2.12: As the Laplacian of f is positive, the Hessian of f has positive trace everywhere. However, since $f \in C^3$, for f to have a relative maximum its Hessian must have negative eigenvalues and so its trace must be negative.

Solution to 2.2.13: With respect to polar coordinates r, θ, we can write u as $u(r, \theta) = r^d g(\theta)$. We need the expression for the Laplacian in polar coordinates, which one can derive on the basis of the easily obtained identities

$$\frac{\partial r}{\partial x} = \frac{x}{r}, \quad \frac{\partial r}{\partial y} = \frac{y}{r}, \quad \frac{\partial \theta}{\partial x} = -\frac{\sin \theta}{r}, \quad \frac{\partial y}{\partial r} = \frac{\cos \theta}{r}.$$

After some computations one finds that

$$\Delta = \frac{\partial^2}{\partial x^2} + \frac{\partial^2}{\partial y^2} = \frac{\partial^2}{\partial r^2} + \frac{1}{r}\frac{\partial}{\partial r} + \frac{1}{r^2}\frac{\partial^2}{\partial \theta^2}.$$

Hence

$$0 = \Delta(r^d g(\theta)) = d(d-1)r^{d-2}g(\theta) + dr^{d-2}g(\theta) + r^{d-2}g''(\theta)$$
$$= r^{d-2}(d^2 g(\theta) + g''(\theta)) .$$

Thus g must be a solution of the ordinary differential equation $g'' + d^2 g = 0$. This equation has the general solution

$$g(\theta) = a \cos d\theta + b \sin d\theta .$$

Since g must be 2π-periodic, d is an integer.

Solution to 2.2.15: The derivative of T is given by

$$DT = \begin{pmatrix} 1 & 2u \\ 1 & 2v \end{pmatrix}$$

which is always nonsingular since $\det(DT) = 2v - 2u$ is never 0. By the Inverse Function Theorem [Rud87, p. 221], this means that T is locally one-to-one.
2. Considering the function $f(u, v) = u + v$ restricted to $u^2 + v^2 = y$, we conclude that $-\sqrt{2y} \leqslant u + v \leqslant \sqrt{2y}$; therefore, the range of T is

$$\left\{ (x, y) \mid y > 0, -\sqrt{2y} \leqslant x \leqslant \sqrt{2y} \right\}.$$

Let $(x, y) \in \mathrm{range}(T)$. $u+v = x$ is the equation of a straight line with slope -1 in the u, v-plane which intersects the circle $u^2 + v^2 = y$ centered at the origin with radius $\sqrt{y}$. These two lines intersect exactly at one point in U, so T is globally injective.

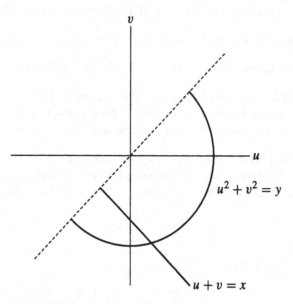

Solution to 2.2.16: Letting f_1 and f_2 be the components of f, we have

$$\frac{d}{dt} \|f(t)\|^2 = 2f_1 f_1'(t) + 2f_2 f_2'(t).$$

Assume $t > 0$ and use the Mean Value Theorem [Rud87, p. 108] to rewrite the right side as

$$2t \left(f_1'(\xi_1) f_1'(t) + f_2'(\xi_2) f_2'(t) \right)$$

where $0 < \xi_1 = \xi_1(t) < t$ and $0 < \xi_2 = \xi_2(t) < t$. As $t \searrow 0$, the continuity of f' gives

$$f_1'(\xi_1) f_1'(t) + f_2'(\xi_2) f_2'(t) \longrightarrow \|f'(0)\|^2 > 0.$$

Hence, there is an $\varepsilon > 0$ such that $\dfrac{d}{dt} \|f'(t)\|^2 > 0$ for $0 < t < \varepsilon$, and the desired conclusion follows.

Solution to 2.2.17: For $X \in \Sigma$, we have

$$\begin{aligned}
\|A - X\|^2 &= (1 - x)^2 + y^2 + z^2 + (2 - t)^2 \\
&= y^2 + z^2 + 1 - 2x + x^2 + (2 - t)^2 \\
&\geqslant \pm 2yz + 1 - 2x + 2|x|(2 - t) \\
&= 4|x| - 2x + 2(\pm yz - |x|t) + 1.
\end{aligned}$$

We can choose the sign, so $\pm yz - |x|t = 0$ because $\det X = 0$. As $4|x| - 2x \geqslant 0$, we have $\|A - X\| \geqslant 1$ with equality when $4|x| - 2x = 0$, $|x| = 2 - t$, $y = \pm z$, and $\det X = xt - yz = 0$, which give $S = \left(\begin{smallmatrix} 0 & 0 \\ 0 & 2 \end{smallmatrix}\right)$.

Solution 2. We can use the Method of Lagrange Multipliers [MH93, p. 414] to find the minimum of the function

$$d(x, y, z, t) = (x - 1)^2 + y^2 + z^2 + (t - 2)^2$$

over the surface defined by $\Sigma = det^{-1}(0)$. Setting up the equations

$$x - 1 = \lambda t$$
$$y = -\lambda z$$
$$z = -\lambda y$$
$$t - 2 = \lambda x$$

substituting the third equation into the second we see that if $y \neq 0$ then $\lambda = \pm 1$, and substituting back into the first and last equations we get a contradiction, so indeed $y = 0$, and then so is $z = 0$. Now solving the first and last equations together we find

$$x = \frac{2\lambda + 1}{1 - \lambda^2}$$

$$t = \frac{\lambda + 2}{1 - \lambda^2}.$$

To have determinant zero then either $\lambda = -1/2$ or $\lambda = -2$, and the two matrices are

$$\begin{pmatrix} 0 & 0 \\ 0 & 2 \end{pmatrix} \quad \text{and} \quad \begin{pmatrix} 1 & 0 \\ 0 & 0 \end{pmatrix}$$

and we can easily see that $\left(\begin{smallmatrix} 0 & 0 \\ 0 & 2 \end{smallmatrix}\right)$ is the closest one.

Solution to 2.2.18: Let d denote the Euclidean metric in $\mathbb{R}^3$. We first prove that there exist $p \in S$ and $q \in T$ such that $d(p, q) = d(S, T)$, where

$$d(s, T) = \inf\{d(s, t) \mid t \in T\}$$

and

$$d(S, T) = \inf\{d(s, t) \mid s \in S, t \in T\}$$
$$= \inf\{d(s, T) \mid s \in S\}.$$

The function $S \to \mathbb{R}$ defined by $x \mapsto d(x, T)$ is continuous and attains its infimum on the compact set S, i.e., there exists $p \in S$ such that $d(p, T) = d(S, T)$. The function $\varphi : T \to \mathbb{R}$ given by $x \mapsto d(p, x)$ is continuous, and $\inf_T \varphi = \inf_{T'} \varphi$ where T' is the compact set

$$T' = \{x \in T \mid d(x, p) \leqslant d(S, T) + 1\}.$$

Hence there exists $q \in T' \subset T$ such that $d(p, q) = d(p, T) = d(S, T)$.

If $(x, y, z) \in S$ then $z \geqslant 9$, whereas if $(x, y, z) \in T$ then $z \leqslant 1$. Therefore $p \neq q$. Now S, T are level sets of

$$g_1(x, y, z) = 2x^2 + (y - 1)^2 + (z - 10)^2$$
$$g_2(x, y, z) = z(x^2 + y^2 + 1)$$

respectively.

Define, on $\mathbb{R}^3 \times \mathbb{R}^3$,

$$F(x, y, z, u, v, w) = (x - u)^2 + (y - v)^2 + (z - w)^2$$
$$f_1(x, y, z, u, v, w) = g_1(x, y, z)$$
$$f_2(x, y, z, u, v, w) = g_2(u, v, w).$$

Then $S \times T = \{(\xi, \eta) \in \mathbb{R}^3 \times \mathbb{R}^3 \mid f_1(\xi, \eta) = 1, \ f_2(\xi, \eta) = 1\}$. We showed above that there exists $(p, q) \in S \times T$ which minimizes F on $S \times T$. For $(\xi, \eta) \in S \times T$, the two vectors $\nabla f_1(\xi, \eta) = (\nabla g_1(\xi), 0)$, $\nabla f_2(\xi, \eta) = (0, \nabla g_2(\eta))$ are linearly independent.

By Lagrange's Multiplier Theorem [MH93, p. 414], there exist $\lambda, \mu \in \mathbb{R}$ such that

$$\nabla F(p, q) = \lambda \nabla f_1(p, q) + \mu \nabla f_2(p, q).$$

Now

$$\nabla F(p, q) = (2(p - q), -2(p - q)),$$

and

$$\lambda \nabla f_1(p, q) + \mu \nabla f_2(p, q) = \lambda (\nabla g_1(p), 0) + \mu (0, \nabla g_2(q))$$
$$= (\lambda \nabla g_1(p), \mu \nabla g_2(q)).$$

It follows from the equations above that $\frac{1}{2}\lambda \nabla g_1(p) = p - q = -\frac{1}{2}\mu \nabla g_2(q)$. None of $p - q$, $\nabla g_1(p)$ and $\nabla g_2(q)$ are zero. Thus the vector $p - q$ is parallel to both $\nabla g_1(p)$ and $\nabla g_2(q)$, from which the result follows.

Solution to 2.2.19: Each element of P_2 has the form $ax^2 + bx + c$ for $(a, b, c) \in \mathbb{R}^3$, so we can identify P_2 with $\mathbb{R}^3$ and J becomes a scalar field on $\mathbb{R}^3$:

$$J(a, b, c) = \int_0^1 (ax^2 + bx + c)^2 \, dx = \frac{a^2}{5} + \frac{ab}{2} + \frac{2ac}{3} + \frac{b^2}{3} + bc + c^2.$$

To Q corresponds the set $\{(a, b, c) \mid a+b+c = 1\}$. If J achieves a minimum value on Q, then, by the Method of Lagrange Multipliers [MH93, p. 414], we know that there is a constant λ with $\nabla J = \lambda \nabla g$, where $g(a, b, c) = a + b + c - 1$. We have

$$\nabla J = \left(\frac{2a}{5} + \frac{b}{2} + \frac{2c}{3}, \ \frac{a}{2} + \frac{2b}{3} + c, \ \frac{2a}{3} + b + 2c \right)$$

and $\nabla g = (1, 1, 1)$. These and the constraint equation $g(a, b, c) = 0$ form the system

$$\begin{pmatrix} 2/5 & 1/2 & 2/3 & -1 \\ 1/2 & 2/3 & 1 & -1 \\ 2/3 & 1 & 2 & -1 \\ 1 & 1 & 1 & 0 \end{pmatrix} \begin{pmatrix} a \\ b \\ c \\ \lambda \end{pmatrix} = \begin{pmatrix} 0 \\ 0 \\ 0 \\ 1 \end{pmatrix}$$

which has the unique solution $\lambda = 2/9$, $(a, b, c) = (10/3, -8/3, 1/3)$. Therefore, if J attains a minimum, it must do so at this point. To see that J does attain a minimum, parameterize the plane Q with the xy coordinates and consider the quadratic surface with a linear z term defined by $z = J(x, y, 1 - x - y)$ in $\mathbb{R}^3$. The surface is the graph of the map $J : P_2 \to \mathbb{R}$. Rotating around the z–axis will eliminate the xy cross-terms in the equation, reducing it to the standard equation of either an elliptic paraboloid or a hyperbolic paraboloid. However, J is always nonnegative, so the surface must be an elliptic paraboloid and, as such, has a minimum.

Solution to 2.2.20: First observe that the group of orthogonal matrices is transitive in $\mathbb{R}^n$, that is, given any two vectors u and v there is a orthogonal matrix A, such that $Au = v$. Now using the Cauchy–Schwarz Inequality [MH93, p. 69] we see that the maximum and minimum of the inner product of any two vectors of the same size is $\pm\|v\|^2$ and they are both achieved when the vectors are the same and have opposite directions, respectively. So the maximum of the function f is achieved on the matrices that leave v_0 invariant and the minimum on the ones that reverse it.

Solution to 2.2.21: Let $e_1, \ldots, e_n$ be the unit vectors in $\mathbb{R}^n$. Let $a, b \in W$ be such that $b = a + he_j$ for some $h > 0$ and some $1 \le j \le n$, and that the line segment $[a, b] = \{a + te_j \mid 0 \le t \le h\}$ lies in W. The function $t \mapsto f(a + te_j)$ has derivative $D_j f = 0$, by hypothesis. By the mean value theorem, $f(b) = f(a)$.

In order to prove that f is constant, it is therefore sufficient to note that any two points in W can be joined by a sequence of line segments in W each of which is parallel to some axis of $\mathbb{R}^n$.

Solution to 2.2.22: Let

$$g(x, y) = \frac{-y}{x^2 + y^2} \quad \text{and} \quad f(x, y) = \frac{x}{x^2 + y^2}.$$

Then

$$\frac{\partial g}{\partial y} = \frac{y^2 - x^2}{(x^2 + y^2)^2} = \frac{\partial f}{\partial x}.$$

Let γ be the curve in $\mathcal{A}$ given by

$$(x, y) = (2\cos t, 2\sin t), \qquad 0 \le t \le 2\pi.$$

Then

$$\int_\gamma g(x, y)\, dx + f(x, y)\, dy = \int_0^{2\pi} \left(\frac{-2\sin t}{4}(-2\sin t) + \frac{2\cos t}{4}(2\cos t) \right) dt$$

$$= \int_0^{2\pi} dt$$

$$= 2\pi .$$

Suppose there is a function h such that $(g, f) = \nabla h = (\frac{\partial h}{\partial x}, \frac{\partial h}{\partial y})$. Then, since γ is closed, by the fundamental theorem of vector calculus, we obtain

$$\int_\gamma g \, dx + f \, dy = 0.$$

which contradicts the calculation above.

Solution to 2.2.24: Let $(x, t) \in \mathbb{R}^2$. By the Mean Value Theorem, [Rud87, p. 108] and the hypothesis, we have, for some (ξ, η) in the segment connecting (x, y) to $(x + y, 0)$,

$$f(x, t) - f(x + t, 0) = Df(\xi, \eta) \cdot ((x, t) - (x + t, 0))$$

$$= \left(\frac{\partial f}{\partial x}(\xi, \eta), \frac{\partial f}{\partial t}(\xi, \eta) \right) \cdot (-t, t)$$

$$= t \left(\frac{\partial f}{\partial x}(\xi, \eta) - \frac{\partial f}{\partial t}(\xi, \eta) \right)$$

$$= 0$$

so $f(x, t) = f(x + t, 0) > 0$.

Solution to 2.2.25: Given two points x and $y \in \mathbb{R}^n$ one can build a polygonal path from x to y with n segments all parallel to the axis (adjusting one coordinate at a time). Applying the Mean Value Theorem [Rud87, p. 108] to each of the segments of the path, we have

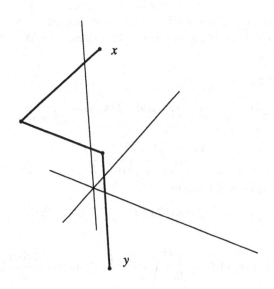

$$|f(x_1, \ldots, x_{i-1}, x_i, y_{i+1}, \ldots, y_n) - f(x_1, \ldots, x_{i-1}, y_i, y_{i+1}, \ldots, y_n)| \leqslant K|x_i - y_i|$$

and then

$$
\begin{aligned}
|f(x) - f(y)| &\leqslant \sum_{i=1}^{n} |f(x_1, \ldots, x_{i-1}, x_i, y_{i+1}, \ldots, y_n) \\
&\quad - f(x_1, \ldots, x_{i-1}, y_i, y_{i+1}, \ldots, y_n)| \\
&\leqslant K \sum_{i=1}^{n} |x_i - y_i|.
\end{aligned}
$$

Now applying the Cauchy–Schwarz Inequality [MH93, p. 69] to the vectors $(1, 1, \ldots, 1)$ and $x - y$, we get

$$
\begin{aligned}
|f(x) - f(y)| &\leqslant K \sqrt{\sum_{i=1}^{n} 1} \sqrt{\sum_{i=1}^{n} |x_i - y_i|^2} \\
&= \sqrt{n}\, K\, \|x - y\|.
\end{aligned}
$$

Solution to 2.2.26: Let $((x_1, \ldots, x_n))_k$ be a sequence in $\mathbb{R}^n$ converging to $0 \in \mathbb{R}^n$. This sequence is Cauchy, so there is an $N > 0$ such that if $k, l > N$, then for each of the coordinates we have $|x_{ik} - x_{il}| < \varepsilon/2nM$. Then we draw a polygonal path, as in the Solution to Problem 2.2.25, from $(x_{1k}, \ldots, x_{nk})$ to $(x_{1l}, \ldots, x_{nl})$, parallel to the axes.

If this path does not goes through the origin, then as before

$$|f(x_{1k}, \ldots, x_{nk}) - f(x_{1l}, \ldots, x_{nl})| < M \sum_{i=1}^{n} |x_{ik} - x_{il}| < \varepsilon$$

and if the origin is in one of the segments of the polygonal path, we can perturb it a bit, by traversing in the same direction but $\varepsilon/4M$ away from the origin. On this altered path

$$|f(x_{1k}, \ldots, x_{nk}) - f(x_{1l}, \ldots, x_{nl})| < M \sum_{i=1}^{n} |x_{ik} - x_{il}| + 2M \frac{\varepsilon}{4M} \leqslant \varepsilon$$

in both case the sequence $(f(x_1, \ldots, x_n))_k$ is Cauchy and, thus, it converges. Given any other sequence $((y_1, \ldots, y_n))_k$, an identical argument shows that $|f(x_{in}, \ldots, x_{in}) - f(y_{in}, \ldots, y_{in})|$ tends to 0, so all sequences must converge to the same value, which can be defined as the continuous extension of f to the origin. For $n = 1$, consider the function $f(x) = 1$ if $x < 0$ and $f(x) = 0$ if $x > 0$.

Solution to 2.2.27: 1. The answer is no, and a counterexample is the function

$$f(x, y) = \frac{xy}{x^2 + y^2}, \qquad \text{for } (x, y) \neq (0, 0)$$

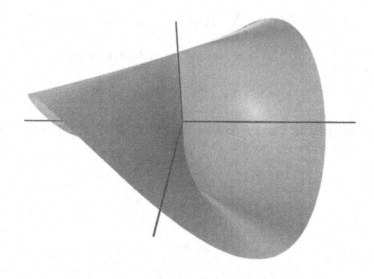

f is differentiable everywhere, but cannot be extended continuously to the origin, because it is constant equal to $k/(1 + k^2)$ on each line $y = kx$ passing through the origin.

2. The answer is again no, with the counterexample a variant of the previous one, the function

$$g(x, y) = \frac{xy^2}{x^2 + y^2} \qquad \text{and} \qquad g(0, 0) = 0$$

g is now continuous everywhere, but not differentiable because the directional derivative does not depend linearly on the vector. Let $(u, v) \neq (0, 0)$. We have

$$\lim_{t \to 0} \frac{g((0, 0) + t(u, v)) - g(0, 0)}{t} = \lim_{t \to 0} t \frac{uv^2}{u^2 + v^2} = 0.$$

So the directional derivatives at the origin exist in all directions. If g were differentiable at $(0, 0)$, as all the directional derivatives vanish there, we would have $Dg(0, 0) = 0$ (the zero linear map). Then, by definition of differentiability, we would have

$$g(x, y) = o(\|(x, y)\|) \qquad ((x, y) \to (0, 0))$$

which is absurd, since

$$\lim_{\substack{(x,y) \to (0,0) \\ x=y}} \frac{g(x, y)}{\|(x, y)\|} = \lim_{x \to 0} \frac{x^3}{2x^2\sqrt{2x^2}} \neq 0.$$

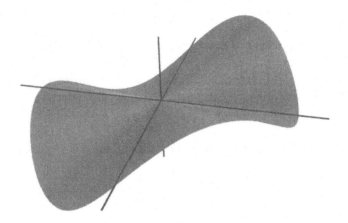

Both examples are from [Lim82].

Solution to 2.2.28: A simple counterexample is

$$f(x, y) = \begin{cases} x & \text{if } y = 0 \\ 0 & \text{if } y \neq 0 \end{cases}$$

and not even continuity at the origin and C^1 on the rest of the plane is enough to guarantee differentiability, as shown in the counterexample of Problem 2.2.27, Part 2.

Solution to 2.2.29: Let $\varepsilon > 0$. By the hypothesis, there is δ such that $\|Df(w)\| < \varepsilon$ if $\|w\| < \delta$. For $\|x\| < \delta$, by the Mean Value Theorem [Rud87, p. 108], applied to the line segment joining 0 and x, we have

$$\|f(x) - f(0)\| \leqslant \|Df(\xi x)\| \|x - 0\| < \varepsilon \|x\| \quad \text{for some } \quad 0 < \xi < 1,$$

which implies differentiability at the origin.

Solution to 2.2.31: Define $f : \mathbb{R}^2 \to \mathbb{R}$ by

$$f(x, y) = \int_{(0,0)}^{(x,y)} u\, dx + v\, dy,$$

the line integral taken along the straight line segment from $(0, 0)$ to (x, y). Fix $a = (a_1, a_2) \in \mathbb{R}^2$, and $t \neq 0$. Using Green's Theorem [Rud87, p. 253], on the triangle with vertices $(0, 0)$, (a_1, a_2) and $(a_1 + t, a_2)$, the parametrization $x = a_1 + \tau, y = a_2, 0 \leqslant \tau \leqslant t$, and the mean value theorem for integrals, we obtain

$$\frac{f(a_1 + t, a_2) - f(a_1, a_2)}{t} = \frac{1}{t} \int_{(a_1, a_2)}^{(a_1 + t, a_2)} u\, dx + v\, dy$$

$$= \frac{1}{t} \int_0^t u(a_1 + \tau, a_2) \, d\tau$$

$$= u(a_1 + \theta t, a_2) \quad \text{for} \quad \text{some} \quad 0 < \theta < 1.$$

Taking the limit when $t \to 0$; we obtain $\dfrac{\partial f}{\partial x} = u$ and similarly, $\dfrac{\partial f}{\partial y} = v$.

2. Let $U = \mathbb{R}^2 \setminus \{0\}$. If such a function $f : U \to \mathbb{R}$ existed then, for any closed curve C in U,

$$\int_C \frac{\partial f}{\partial x} \, dx + \frac{\partial f}{\partial y} \, dy = 0.$$

However, for the unit circle with the parametrization $x = \cos t$, $y = \sin t$, $0 \leqslant t \leqslant 2\pi$,

$$\int_C \frac{-y}{x^2 + y^2} \, dx + \frac{x}{x^2 + y^2} \, dy = 2\pi.$$

which is a contradiction.

Solution to 2.2.32: The answer is no; to see it, consider the function $f(x, y, z) = 1 - x^2 - y^2 - z^2$.

Solution to 2.2.33: Fix a point $x \in \mathbb{R}^n$. By the Chain Rule [Rud87, p. 214],

$$D(g \circ f)(x) = ((Dg) (f(x))) ((Df)(x)) = 0.$$

The transformation $(Dg) (f(x)) : \mathbb{R}^n \to \mathbb{R}$ is nonzero because g has no critical points. The preceding equality therefore implies that the transformation $(Df)(x) : \mathbb{R}^n \to \mathbb{R}^n$ is noninvertible, so its determinant vanishes. That determinant is the Jacobian determinant of f at x.

Solution to 2.2.34: 1. Let $F : \mathbb{R}^2 \to \mathbb{R}$ be defined by $F(x, t) = f(x) - tg(x)$. Then F is a smooth scalar field with $F(0, 0) = 0$ and

$$\frac{\partial F}{\partial x}(0, 0) = f'(0) - 0g'(0) \neq 0.$$

Therefore, by the Implicit Function Theorem [Rud87, p. 224], there exists a positive δ such that, for $|t| < \delta$, x is a smooth function of t, with $x(0) = 0$.
2. Differentiating both sides of $f(x(t)) = tg(x(t))$ with respect to t, we have, for $|t| < \delta$,

$$x'(t) = \frac{g(t)}{f'(t)}.$$

As $x(0) = 0$, the desired expansion of $x(t)$ is

$$\frac{g(0)}{f'(0)} t.$$

Solution to 2.2.35: Subtracting the second and third equations from the first we get $u^4 - 3u = 0$, so that $u = 0$ or $u = \sqrt[3]{3}$.

If $u = 0$, then

$$3x + y - z = 0$$
$$x - y + 2z = 0$$
$$2x + 2y - 3z = 0$$

whose solution is given by $x = -z/4$, $y = 7z/4$.
If $u = \sqrt[3]{3}$, then

$$3x + y - z = -3\sqrt[3]{3}$$
$$x - y + 2z = -\sqrt[3]{3}$$
$$2x + 2y - 3z = -2\sqrt[3]{3}$$

whose solution is given by $x = -(z/4) - \sqrt[3]{3}$, $y = 7z/4$.

The solution set of the system of equations is a disjoint union of two parallel *lines* in $\mathbb{R}^4$:

$$(x, y, z, u) = z\left(-\frac{1}{4}, \frac{7}{4}, 1, 0\right)$$

$$(x, y, z, u) = z\left(-\frac{1}{4}, \frac{7}{4}, 1, 0\right) + \left(-\sqrt[3]{3}, 0, 0, \sqrt[3]{3}\right).$$

Thus, for any $\epsilon > 0$, the system can be solved for (x, y, z) as a function of $z \in [-\epsilon, \epsilon]$, with $x(0) = y(0) = u(0) = 0$. Namely, let A, B be disjoint sets with $A \cup B = [-\epsilon, \epsilon] \setminus \{0\}$, and define

$$x(z) = \begin{cases} -z/4 & z \in A \\ -(z/4) - \sqrt[3]{3} & z \in B \\ 0 & z = 0 \end{cases} \quad y(z) = 7z/4 \quad u(z) = \begin{cases} 0 & z \in A \\ \sqrt[3]{3} & z \in B \\ 0 & z = 0 \end{cases}$$

The function $y(z)$ is unique, continuous, and differentiable. Neither functions $x(z)$ and $u(z)$ are unique. The function $x(z)$ is continuous, and differentiable, if and only if $B = \varnothing$. The function $u(z)$ is continuous, and differentiable, if and only if $B = \varnothing$.

2. We have seen above that the system has a solution if and only if $u = 0$ or $u = \sqrt[3]{3}$. So if $0 < u < \sqrt[3]{3}$, the system has no solutions. Thus, the system cannot be solved for (x, y, z) as a function of $u \in [-\delta, \delta]$, for any $\delta > 0$.

Solution to 2.2.36: Parametrizing the circle by the angle θ, the function becomes

$$f_{a,b}(\theta) = a \sin^2 \theta + b \cos \theta$$

and the derivative is

$$f'_{a,b}(\theta) = 2a \sin \theta \cos \theta - b \sin \theta$$
$$= \sin \theta (2a \cos \theta - b)$$

So $f_{a,b}$ always has at least two critical points at $\theta = \pm\pi/2$ and maybe two additional ones when $\cos\theta = b/2a$. For this to happen $|b/2a| \leqslant 1$, that is, the (a, b) region is the region in between the lines $b = \pm 2a$, containing the x-axis, including the lines itself and excluding the origin.

Solution 2. Using the method of Lagrange Multipliers [MH93, p. 414] for $f_{a,b}$ and $g(x, y) = x^2 + y^2$, we end up with the system of equations:

$$b = \lambda x$$
$$2ay = \lambda y$$
$$x^2 + y^2 = 1.$$

From this we can easily see that the pair $(x, y) = (\pm 1, 0)$ is always a critical point, and if we suppose $y \neq 0$ then $\lambda = 2a$, which implies that the pair of points

$$\left(\frac{b}{2a}, \pm\frac{\sqrt{4a^2 - b^2}}{2a} \right)$$

might be two additional critical points. They will occur only when $a \neq 0$ and $4a^2 - b^2 \geqslant 0$, that is $|b| \leqslant 2|a|$, that is, the region in between the lines $b = \pm 2a$, containing the x-axis, including the lines itself and excluding the origin.

Solution to 2.2.37: Consider the function $G : \mathbb{R}^2 \to \mathbb{R}^2$ given by

$$G(x, y) = \begin{pmatrix} u(x, y) \\ v(x, y) \end{pmatrix}.$$

Since ∇u and ∇v are linearly dependent and ∇u is never 0, G' has rank 1 everywhere. Therefore, by the Rank Theorem [Bar76, p. 391], given a point $p_0 \in \mathbb{R}^2$, there is a neighborhood V of p_0, an open set $W \subset \mathbb{R}^2$, a diffeomorphism $h : W \to V$, and a C^1–function $g = (g_1, g_2) : \mathbb{R} \to \mathbb{R}^2$ such that $G(h(x, y)) = g(x)$ on W. So $g_1'(x)$ is never 0. Therefore, by the Inverse Function Theorem [Rud87, p. 221], g_1 is locally invertible. By shrinking the set W (and so the set V), we may assume that it is invertible. Therefore, $g_1^{-1}(u(h(x, y))) = x$ or $g_2 \circ g_1^{-1}(u(h(x, y))) = v(h(x, y))$ for all $(x, y) \in W$. Since h is a diffeomorphism of W onto V, it follows that $g_2 \circ g_1^{-1}(u(x, y)) = v(x, y)$ for all $(x, y) \in V$. $F = g_2 \circ g_1^{-1}$ satisfies the required condition.

Solution to 2.2.38: The conclusion is trivial if f is constant, so we assume f is not a constant. There is $(x_0, y_0) \in \mathbb{R}^2$ such that $Df(x_0, y_0) \neq 0$. After performing a rotation of the coordinates, if necessary, we assume $f_x(x_0, y_0) \neq 0$. Let $a = f(x_0, y_0)$, and consider the function $F : \mathbb{R}^2 \to \mathbb{R}^2$ given by

$$F(x, y) = (f(x, y), y).$$

The Jacobian of F is nonzero at (x_0, y_0), so, by the Inverse Function Theorem [Rud87, p. 221], the function F has a local inverse, G, defined in a neighborhood of (a, y_0). Thus, $F(G(a, y_0)) = (a, y)$ for all y in some closed interval I

containing y_0. Let γ be any one-to-one map of $[0, 1]$ onto I. The function

$$g(t) = G(a, \gamma(t)) \quad (t \in [0, 1])$$

has the desired properties.

Solution to 2.2.39: Consider $F : \mathbb{R}^2 \to \mathbb{R}^2$ given by

$$F(x, y) = (f(x), -y + xf(x)).$$

A calculation gives that the Jacobian of F at (x_0, y_0) is $-f'(x_0) \neq 0$. So, by the Inverse Function Theorem [Rud87, p. 221], F is invertible in a neighborhood of (x_0, y_0). Similarly, f has a local inverse, g, close to x_0. In a sufficiently small neighborhood of (x_0, y_0) we can then solve for each component of F^{-1} explicitly and get

$$g(u) = g(f(x)) = x$$

and

$$y = -v + xf(x) = -v + g(u)f(g(u)) = -v + ug(u).$$

Solution to 2.2.40: f is clearly a C^∞ function, so

$$f(X + hY) - f(X) = (X + hY)(X + hY) - X^2$$
$$= hXY + hYX + h^2Y^2$$

therefore

$$f'(X) \cdot Y = XY + YX$$

Solution to 2.2.41: 1. Let $F : \mathbb{R}^4 \to \mathbb{R}^4$ be defined by

$$F(x, y, z, w) = \left(x^2 + yz, y(x + w), z(x + w), zy + w^2\right).$$

This map is associated with the given map because

$$\begin{pmatrix} x & y \\ z & w \end{pmatrix}^2 = \begin{pmatrix} x^2 + yz & y(x + w) \\ z(x + w) & zy + w^2 \end{pmatrix}.$$

The Jacobian of F at $(1, 0, 0, 1)$ is 2^4; therefore, F is locally invertible near that point.
2. We have $F(1, \varepsilon, \varepsilon, -1) = (1, 0, 0, 1)$ for any ε, so F is not invertible near $(1, 0, 0, -1)$.

Solution to 2.2.42: Identify the matrix $X = \begin{pmatrix} x & y \\ z & w \end{pmatrix}$ with the element of $(x, y, z, w) \in \mathbb{R}^4$ in the usual way. Let F be defined by $F(X) = X^2 + X^t$. We have

$$DF(X)(x, y, z, w) = \begin{pmatrix} 2x + 1 & z & y & 0 \\ y & x + w & 1 & y \\ z & 1 & x + w & z \\ 0 & z & y & 2w + 1 \end{pmatrix}$$

so

$$DF(X)(0,0,0,0) = \begin{pmatrix} 1 & 0 & 0 & 0 \\ 0 & 0 & 1 & 0 \\ 0 & 1 & 0 & 0 \\ 0 & 0 & 0 & 1 \end{pmatrix}$$

is invertible; therefore, by the Inverse Function Theorem [Rud87, p. 221], there is such an ε.

Global unicity fails for $X = \left(\begin{smallmatrix} -1 & 0 \\ 0 & 0 \end{smallmatrix}\right)$ since $X^2 + X^t = 0 = 0^2 + 0^t$.

Solution to 2.2.43: Since $F(0) = 0$ and F is clearly a C^∞–function, the Inverse Function Theorem [Rud87, p. 221] will yield the result if we can prove that $DF(0)$ is invertible. We have

$$F(X+hY) - F(X) = X + hY + (X+hY)^2 - X - X^2 = hY + hXY + hYX + h^2Y^2$$

therefore,

$$DF(X)Y = Y + XY + YX.$$

In particular, $DF(0)$ is the identity operator which is invertible.

Solution to 2.2.44: Define the function $F : M_{n \times n} \to M_{n \times n}$ by $F(X) = X^4$ F is clearly a C^∞–function and $F(I) = I$, so the Inverse Function Theorem [Rud87, p. 221] will yield the result once we prove that $DF(I)$ is invertible, but the computation of the derivative, as above, is

$$DF(X) \cdot Y = X^3Y + X^2YX + XYX^2 + YX^3.$$

In particular, $DF(I) = 4I$ which is invertible, showing that F is a diffeomorphism in a neighbourhood of the Identity matrix.

Solution to 2.2.45: 1. Using the method of Laplace Expansions [HK61, p. 179] we can see that finding the determinant involves only sums and multiplications of the entries of a matrix, therefore, it is a C^∞–function.
2. For $i, j = 1, \ldots, n$, let x_{ij} denote the $(i, j)^{th}$ entry of X, and let X_{ij} denote the cofactor of x_{ij}, so that

$$\det(X) = \sum_{j=1}^n x_{ij} X_{ij} \qquad (i = 1, \ldots, n).$$

Since $\dfrac{\partial X_{ik}}{\partial x_{ij}} = 0$ for each i, j, k, it follows from the preceding expression that

$$\frac{\partial \det}{\partial x_{ij}} = X_{ij}.$$

Thus, X is a critical point of $\det$ if and only if $X_{ij} = 0$ for every i and j or, what is equivalent, if and only if the rank of X does not exceed $n - 2$.

Solution to 2.2.46: $u(t)$ is differentiable as the composition of differentiable functions, and the derivative can be computed using the chain rule.

$$u'(t) = \text{tr}(F^2(t)F'(t) + F(t)F'(t)F(t) + F'(t)F^2(t))$$
$$= \text{tr}(F^2(t)F'(t)) + \text{tr}(F(t)F'(t)F(t)) + \text{tr}(F'(t)F^2(t))$$
$$= 3\text{tr}(F^2(t)F'(t)).$$

Solution to 2.2.47: Let A_j be the 1–column matrix

$$A_j = \begin{pmatrix} a_{1j} \\ \vdots \\ a_{nj} \end{pmatrix}$$

so A is the matrix whose column j is A_j, that is, $A = (A_1, \ldots, A_n)$. Since det is an n–linear function of $(\mathbb{R}^n)^n$ into $\mathbb{R}$, the derivative is given by

$$\frac{d}{dt} \det(A) = \sum_{j=1}^{n} \det\left(A_1, \ldots, \frac{d}{dt}A_j, \ldots, A_n\right).$$

Let $A(i, j)$ denote the cofactor of a_{ij}, that is,

$$A(i, j) = (-1)^{i+j} \begin{vmatrix} a_{11} & \cdots & a_{1j-1} & a_{1j+1} & \cdots & a_{1n} \\ \vdots & & \vdots & \vdots & & \vdots \\ a_{i-11} & \cdots & a_{i-1j-1} & a_{i-1j+1} & \cdots & a_{i-1n} \\ a_{i+11} & \cdots & a_{i+1j-1} & a_{i+1j+1} & \cdots & a_{i+1n} \\ \vdots & & \vdots & \vdots & & \vdots \\ a_{n1} & \cdots & a_{nj-1} & a_{nj+1} & \cdots & a_{nn} \end{vmatrix}.$$

Using Laplace's Expansion Method to evaluate the determinant [HK61, p. 179]

$$\det\left(A_1, \ldots, \frac{d}{dt}A_j, \ldots, A_n\right)$$

of each component of the derivative, by developing the jth column we get

$$\det\left(A_1, \ldots, \frac{d}{dt}A_j, \ldots, A_n\right) = \sum_{i=1}^{n} \frac{d}{dt}a_{ij} A(i, j)$$

and

$$\frac{d}{dt} \det(A) = \sum_{j=1}^{n}\sum_{i=1}^{n} \frac{d}{dt}a_{ij} A(i, j) = \sum_{j=1}^{n}\sum_{i=1}^{n} \frac{d}{dt}a_{ij} \det(A)b_{ji}$$

where the last equality follows from the fact that the inverse matrix is given by $b_{ij} = \frac{1}{\det(A)} \cdot A(j, i)$. Therefore, we have

$$\frac{d}{dt}\log(\det(A)) = \frac{1}{\det(A)} \cdot \frac{d}{dt}\det(A)$$

$$= \frac{1}{\det(A)} \cdot \sum_{j=1}^{n}\sum_{i=1}^{n}\frac{d}{dt}a_{ij} \cdot \det(A) \cdot b_{ji}$$

$$= \sum_{j=1}^{n}\sum_{i=1}^{n}\frac{d}{dt}a_{ij}b_{ji}.$$

Solution to 2.2.48: 1. The measure preserving condition is that the derivative, f', has absolute value 1 at every point. Since the function is continuous it must be either $f' = -1$ or $f' = 1$ on the entire domain. Therefore, $f(x) = -x + c$ or $f(x) = x + c$ where c is a constant. Thus, f cannot map $\mathbb{R}$ into $\mathbb{R}_{+}$.
2. Take $f(x, y) = (e^{-y}x, e^{y})$.

2.3   Integral Calculus

Solution to 2.3.1: Let $f : \mathbb{R}^3 \to \mathbb{R}$, $f(x, y, z) = (ax, by, cz)$. The volume given is the image, under f, of the unit ball of $\mathbb{R}^3$, $\mathcal{B}$. As the Jacobian of f is abc everywhere, we have

$$\text{vol}\,(f(\mathcal{B})) = \iiint_{f(\mathcal{B})} dxdydz = \iiint_{\mathcal{B}} abc\,dxdydz = \frac{4}{3}\pi abc.$$

Solution to 2.3.2: Using polar coordinates, we have

$$\iint_{A} e^{-x^2-y^2}\,dxdy = \int_{0}^{2\pi}\int_{0}^{1} \rho e^{-\rho^2}\,d\rho d\theta$$

$$= -\frac{1}{2}\int_{0}^{2\pi}\int_{0}^{1} -2\rho e^{-\rho^2}\,d\rho d\theta$$

$$= -\frac{1}{2}\int_{0}^{2\pi} (e^{-1} - 1)$$

$$= \pi(e^{-1} - 1).$$

Solution to 2.3.3: We write

$$I = \int_{-\infty}^{\infty} e^{-3y^2/2}\left(\int_{-\infty}^{\infty} e^{-2x^2+2xy-y^2/2}\,dx\right)dy$$

$$= \int_{-\infty}^{\infty} e^{-3y^2/2}\left(\int_{-\infty}^{\infty} e^{-2(x-\frac{y}{2})^2}\,dx\right)dy$$

making the substitution $t = \sqrt{2}\left(x - \dfrac{y}{2}\right)$, $dt = \sqrt{2}\,dx$ on the inner integral, we get

$$I = \frac{1}{\sqrt{2}} \int_{-\infty}^{\infty} e^{-3y^2/2} \left(\int_{-\infty}^{\infty} e^{-t^2} dt \right) dy$$

$$= \sqrt{\frac{\pi}{2}} \int_{-\infty}^{\infty} e^{-3y^2/2} \, dy$$

now making the substitution: $s = \sqrt{\frac{3}{2}}\, y$, $ds = \sqrt{\frac{3}{2}}\, dy$ we obtain

$$I = \sqrt{\frac{\pi}{3}} \int_{-\infty}^{\infty} e^{-s^2} \, ds$$

$$= \frac{\pi}{\sqrt{3}}\,.$$

Solution 2. As the integrand is positive, the iterated integral equals the double integral:

$$I = \int_{\mathbb{R}^2} e^{-(x^2+(y-x)^2+y^2)} \, d(x,\,y)\,.$$

Let $u = x + y$, $v = -x + y$; this map is a differentiable bijection $\mathbb{R}^2 \to \mathbb{R}^2$. Moreover

$$x^2 + (y-x)^2 + y^2 = \frac{1}{2}(u^2 + 3v^2) \quad \text{and} \quad \frac{\partial(u,\,v)}{\partial(x,\,y)} = 2\,.$$

By the Change of Variable Formula for multiple integrals [MH93, p. 505],

$$\int_{\mathbb{R}^2} e^{-(x^2+(y-x)^2+y^2)} \, d(x,\,y) = \int_{\mathbb{R}^2} e^{-\frac{1}{2}(u^2+3v^2)} \left| \frac{\partial(x,\,y)}{\partial(u,\,v)} \right| d(u,\,v)$$

$$= \frac{1}{2} \int_{\mathbb{R}^2} e^{-\frac{1}{2}u^2} e^{-\frac{3}{2}v^2} \, d(u,\,v)$$

$$= \frac{1}{2} \int_{\mathbb{R}} e^{-\frac{1}{2}u^2} \, du \int_{\mathbb{R}} e^{-\frac{3}{2}v^2} \, dv$$

$$= \frac{1}{2} \left(\sqrt{2\pi} \right) \left(\sqrt{2\pi/3} \right)$$

$$= \frac{\pi}{\sqrt{3}}\,.$$

Solution to 2.3.4: Let $f(x,\,y) = x^3 - 3xy^2$. Derivating twice (see Problem 5.9.1), we can see that $\Delta f = 0$, so f is harmonic. Let $\mathcal{B}_1 = \{(x,\,y) \mid (x+1)^2 + y^2 \leqslant 9\}$ and $\mathcal{B}_2 = \{(x,\,y) \mid (x-1)^2 + y^2 \leqslant 1\}$. We have

$$\iint_{\mathcal{R}} f = \iint_{\mathcal{B}_1} f - \iint_{\mathcal{B}_2} f$$

and

$$\iint_{B_1} f(x, y)\, dx dy = \int_0^3 \left(\int_0^{2\pi} f(\xi + 3e^{i\theta})\, d\theta \right) r\, dr,$$

where $\xi = (-1, 0)$ is the center of B_1. As f is harmonic, the inner integral is equal to $2\pi f(\xi) = -2\pi$ and the integral of f over B_1 is -9π. An identical calculation shows that the integral over B_2 is π, so the integral of f over $\mathcal{R}$ is -10π.

Solution 2. The integrand is harmonic, being the real part of the holomorphic function z^3. Its integral over any disc is therefore the value at the center of the disc times the area of the disc. Hence, if we call $\mathcal{D}_1$ and $\mathcal{D}_2$ the external and internal discs that make up the boundary of $\mathcal{R}$, respectively, we have

$$\iint_{\mathcal{R}} (x^3 - 3xy^2)\, dx dy = \iint_{\mathcal{D}_1} (x^3 - 3xy^2)\, dx dy - \iint_{\mathcal{D}_2} (x^3 - 3xy^2)\, dx dy$$

$$= -9\pi - \pi$$

$$= -10\pi .$$

Solution to 2.3.5: Consider the annular region $\mathcal{A}$ between the circles of radius r and R, then by the Green's Theorem

$$\int_R e^x(-\varphi_y\, dx + \varphi_x\, dy) - \int_r e^y(-\varphi_y\, dx + \varphi_x\, dy) =$$

$$= \int_{\partial\mathcal{A}} e^x(-\varphi_x\, dx + \varphi_x\, dy)$$

$$= \iint_{\mathcal{A}} d(e^x(-\varphi_y\, dx + \varphi_x\, dy))$$

$$= \iint_{\mathcal{A}} -e^x\varphi_{yy}\, dy \wedge dx + (e^x\varphi_x + e^x\varphi_{xx})\, dx \wedge dy$$

$$= \iint_{\mathcal{A}} e^x(\varphi_{xx} + e^x\varphi_{xx} + \varphi_x)\, dx \wedge dy = 0$$

showing that the integral does not depend on the radius r. Now, parametrizing the circle of radius r

$$\int_{C_r} e^x(-\varphi_y\, dx + \varphi_x\, dy) =$$

$$= \int_0^{2\pi} e^{r\cos\theta}(\varphi_y(r\cos\theta, r\sin\theta)\sin\theta + \varphi_x(r\cos\theta, r\sin\theta)\cos\theta)r\, d\theta$$

but when $r \to 0$ the integrand converges uniformly to

$$\frac{\sin\theta}{2\pi} \cdot \sin\theta + \frac{\cos\theta}{2\pi} \cdot \cos\theta = \frac{1}{2\pi}$$

so the integral approaches 1 when $r \to 0$ and that is the value of the integral.

Solution to 2.3.6: Using the change of variables

$$\begin{cases} x = \sin \varphi \cos \theta & 0 < \theta < 2\pi \\ y = \sin \varphi \sin \theta & 0 < \varphi < \pi \\ z = \cos \varphi \end{cases}$$

we have

$$dA = \sin \varphi \, d\theta \, d\varphi$$

and

$$\iint_S (x^2 + y + z) dA = \int_0^\pi \int_0^{2\pi} (\sin^2 \varphi \cos^2 \theta + \sin \varphi \sin \theta + \cos \varphi) \sin \varphi \, d\theta \, d\varphi.$$

Breaking the integral in three terms, we get

$$\int_0^\pi \int_0^{2\pi} \sin \varphi \cos \varphi \, d\theta \, d\varphi = 2\pi \cdot \frac{1}{2} \int_0^\pi \sin 2\varphi \, d\varphi = 0$$

$$\int_0^\pi \int_0^{2\pi} \sin^2 \varphi \sin \theta \, d\theta \, d\varphi = \left(\int_0^\pi \sin^2 \varphi \, d\varphi \right) \int_0^{2\pi} \sin \theta \, d\theta = 0$$

$$\int_0^\pi \int_0^{2\pi} \sin^3 \varphi \cos^2 \theta \, d\theta \, d\varphi = \left(\int_0^\pi \sin^3 \varphi \, d\varphi \right) \left(\int_0^{2\pi} \cos^2 \theta \, d\theta \right)$$

$$= \int_0^\pi \frac{1}{4} (3 \sin \varphi - \sin^3 \varphi) d\varphi \int_0^{2\pi} \cos^2 \theta \, d\theta$$

$$= \frac{1}{4} \left(-3 \cos \varphi + \frac{1}{3} \cos^3 \varphi \Big|_0^\pi \right) \left(\int_0^{2\pi} \frac{1 + \cos 2\theta}{2} d\theta \right)$$

$$= \frac{1}{4} \left(-3(-2) + \frac{1}{3}(-2) \right) \pi$$

$$= \frac{1}{4} \left(6 - \frac{2}{3} \right) \pi = \frac{4}{3}\pi.$$

Therefore,

$$\iint_S (x^2 + y + z) dA = \frac{4}{3}\pi.$$

Solution 2. The outward unit normal to S is $\vec{n} = (x, y, z)$. Consider the vector field $\vec{F}(x, y, z) = (x, 1, 1)$. Then $\vec{F} \cdot \vec{n} = x^2 + y + z$, and $\operatorname{div} \vec{F} = 1$. Using the Divergence Theorem,

$$\iint_S (x^2 + y + z) dA = \iint_S \vec{F} \cdot \vec{n} \, dA$$

$$= \iiint_{x^2+y^2+z^2 \leqslant 1} \operatorname{div} \vec{F} \, dV$$

$$= \iiint dV$$

$$= \frac{4}{3}\pi.$$

Solution to 2.3.7: 1. We have

$$\begin{vmatrix} \vec{\imath} & \vec{\jmath} & \vec{k} \\ \partial/\partial x & \partial/\partial y & \partial/\partial z \\ x^2 + y - 4 & 3xy & 2xz + z^2 \end{vmatrix} = -2z\vec{\jmath} + (3y - 1)\vec{k}.$$

2. Let $\mathcal{H} = \{(x, y, z) \in \mathbb{R}^3 \mid x^2 + y^2 + z^2 = 16, \ z \geqslant 0\}$, and consider the set $\mathcal{D}$ given by $\mathcal{D} = \{(x, y, 0) \in \mathbb{R}^3 \mid x^2 + y^2 \leqslant 16\}$. $\mathcal{H}$ and $\mathcal{D}$ have the same boundary, so, by Stokes' Theorem [Rud87, p. 253]

$$\int_{\mathcal{H}} (\nabla \times F) \cdot d\vec{S} = \int_{\partial \mathcal{H}} F \cdot d\vec{l} = \int_{\partial \mathcal{D}} F \cdot d\vec{l}$$

$$= \int_{\partial \mathcal{D}} (\nabla F \times F) \cdot dS = \iint_{\mathcal{D}} (-2z\vec{\jmath} + (3y - 1)\vec{k}) \cdot \vec{k} \, dxdy$$

$$= \iint_{\mathcal{D}} (3y - 1)dxdy = -16\pi.$$

Solution to 2.3.8: Let C be a smooth closed path in $\mathbb{R}^3$ which does not contain the origin, and let L be any polygonal line from the origin to infinity that does not intersect C. $V = \mathbb{R}^3 \setminus L$ is simply connected; so to show that

$$\int_C F \cdot ds = 0$$

it suffices to show that $\nabla \times F = 0$. Let $r = (x, y, z)$ and $F = (P, Q, R)$. We have

$$F(r) = (g(\|r\|) \, x, \ g(\|r\|) \, y, \ g(\|r\|) \, z).$$

By the Chain Rule [Rud87, p. 214],

$$\frac{\partial R}{\partial y} - \frac{\partial Q}{\partial z} = g'(\|r\|) \frac{\partial \|r\|}{\partial y} z + g'(\|r\|) \frac{\partial z}{\partial y} - g'(\|r\|) \frac{\partial \|r\|}{\partial z} y - g(\|r\|) \frac{\partial y}{\partial z}$$

$$= g'(\|r\|) \frac{yz}{\|r\|} - g'(\|r\|) \frac{yz}{\|r\|}$$

$$= 0.$$

Similarly,

$$\frac{\partial P}{\partial z} - \frac{\partial R}{\partial z} = \frac{\partial Q}{\partial x} - \frac{\partial P}{\partial y} = 0$$

and we are done.

Solution to 2.3.9: 1. From the identity

$$\nabla \cdot \left(f \vec{J} \right) = (\nabla f) \cdot \vec{J} + f \nabla \cdot \vec{J} = \nabla f \cdot \vec{J}$$

and Gauss Theorem [Rud87, p. 272], we obtain

$$\iiint_{\mathcal{B}} (\nabla f) \cdot \vec{J} \, dx \, dy \, dz = \iiint_{\mathcal{B}} \nabla \cdot \left(f \vec{J} \right) \, dx \, dy \, dz$$
$$= \iint_{\partial \mathcal{B}} f \vec{J} \cdot \vec{n} \, dA$$
$$= 0.$$

2. Apply Part 1 with $f(x, y, z) = x$.

Solution to 2.3.10: By Gauss Theorem [Rud87, p. 272],

$$\iint_{\mathcal{D}} \operatorname{div} (u \operatorname{grad} u) dx dy = \int_{\partial \mathcal{D}} (u \operatorname{grad} u) \cdot \vec{n} \, ds \,,$$

where $\vec{n}$ is the unit outward normal, and ds is the differential of arc length. The right-hand side vanishes because $u = 0$ on $\partial \mathcal{D}$. The left side equals the left side of $(*)$ because

$$\operatorname{div} (u \operatorname{grad} u) = \frac{\partial}{\partial x} \left(u \frac{\partial u}{\partial x} \right) + \frac{\partial}{\partial y} \left(u \frac{\partial u}{\partial y} \right)$$
$$= \left(\frac{\partial u}{\partial x} \right)^2 + \left(\frac{\partial u}{\partial y} \right)^2 + u \left(\frac{\partial^2 u}{\partial x^2} + \frac{\partial^2 u}{\partial y^2} \right)$$
$$= |\operatorname{grad} u|^2 + \lambda u^2.$$

Solution to 2.3.11:

$$\iint_{\mathcal{D}} u \left(\frac{\partial^2 u}{\partial x^2} + \frac{\partial^2 u}{\partial y^2} - \lambda u \right) = \iint_{\mathcal{D}} 0 = 0$$

substituting

$$u \left(\frac{\partial^2 u}{\partial x^2} + \frac{\partial^2 u}{\partial y^2} \right) = \nabla \cdot (u \nabla u) - \|\nabla u\|^2$$

the equality becomes:

$$\iint_{\mathcal{D}} \nabla \cdot (u \nabla u) - \iint_{\mathcal{D}} \|\nabla u\|^2 - \iint_{\mathcal{D}} \lambda u^2 = 0$$

Now applying Green's Theorem [Rud87, p. 253], to the first term of the equation, we get:

$$\int_{\partial D} u D_n u - \iint_D \|\nabla u\|^2 - \iint_D \lambda u^2 = 0$$

and since $D_n u = au$ on ∂D, we get

$$a \int_{\partial D} u^2 - \iint_D \|\nabla u\|^2 - \lambda \iint_D u^2 = 0$$

since u is not identically zero inside the disk, the last integral is positive. Now if u were constant on $\mathcal{D}$, the second equation $D_n u = au$ on ∂D would force $u = 0$, so ∇u is not identically zero on $\mathcal{D}$ and the integral $\iint_D \|\nabla u\|^2$ is also positive, showing that $a \int_{\partial D} u^2 \geq 0$ and consequently that $\lambda < 0$.

Solution to 2.3.12: By the Change of Variable Formula for multiple integrals [MH93, p. 505],

$$\text{vol } f\left(Q_r(x_0)\right) = \int_{Q_r(x_0)} |J(x)| \, dx.$$

Hence, if M_r is the maximum and m_r the minimum of $|J(x)|$ for $x \in Q_r(x_0)$, we have

$$m_r \leqslant r^{-3} \text{vol } f\left(Q_r(x_0)\right) \leqslant M_r.$$

By the continuity of J, we have $m_r \to |J(x_0)|$ and $M_r \to |J(x_0)|$ as $r \to 0$, from which the desired equality follows.

To establish the inequality, we note that the same reasoning gives

$$(*) \qquad |J(x_0)| = \lim_{r \to 0} \frac{\text{vol } f\left(B_r(x_0)\right)}{\frac{4}{3}\pi r^3}$$

where $B_r(x_0)$ denotes the ball of radius r and center x_0. Let

$$K = \limsup_{x \to x_0} \frac{\|f(x) - f(x_0)\|}{\|x - x_0\|}.$$

Then, given $\varepsilon > 0$, there is an $r_\varepsilon > 0$ such that $\|f(x) - f(x_0)\| \leqslant (K+\varepsilon)\|x - x_0\|$ for $\|x - x_0\| \leqslant r_\varepsilon$. The latter means that, for $r \leqslant r_\varepsilon$,

$$f\left(B_r(x_0)\right) \subset B_{(K+\varepsilon)r}\left(f(x_0)\right)$$

so that

$$\frac{\text{vol } f\left(B_r(x_0)\right)}{\frac{4}{3}\pi r^3} \leqslant \frac{\frac{4}{3}(K+\varepsilon)^3 r^3}{\frac{4}{3}\pi r^3} = (K+\varepsilon)^3.$$

In view of the equation $(*)$, this gives $|J(x_0)| \leqslant (K+\varepsilon)^3$. Since ε is arbitrary, we get $|J(x_0)| \leqslant K^3$, the desired inequality.

3
Differential Equations

3.1 First Order Equations

Solution to 3.1.1: For (x_0, y_0) to be the midpoint of $L(x_0, y_0)$, the y intercept of L must be $2y_0$ and the x intercept must be $2x_0$. Hence, the slope of the tangent line is $-y_0/x_0$. Let the curve have the equation $y = f(x)$. We get the differential equation

$$f'(x) = -\frac{f(x)}{x},$$

or

$$-\frac{1}{x} = \frac{f'(x)}{f(x)} = \frac{1}{y}\frac{dy}{dx}.$$

By separation of variables, we get

$$\log y = -\log x + C.$$

Hence,

$$f(x) = y = \frac{D}{x}$$

for some constant D. Solving for the initial condition $f(3) = 2$, we get $f(x) = 6/x$.

Solution to 3.1.2: We have $g'(x)/g(x) = 2$, so $g(x) = Ke^{2x}$ where K is a constant. The initial condition $g(0) = a$ gives $g(x) = ae^{2x}$.

Solution to 3.1.3: Suppose y is such a function. Then

$$y'(x) = y(x)^n \, ,$$

or

$$\frac{dy}{y^n} = dx$$

$$d\left(\frac{y^{-n+1}}{1-n}\right) = dx$$

$$y^{-n+1} = (1-n)x + c \, .$$

Moreover, $c = 1/y(0)^{n-1} > 0$. We thus have

$$y(x) = \frac{1}{(c - (n-1)x)^{1/(n-1)}} \, .$$

This function solves the initial value problem $y' = y^n$, $y(0) = c^{-1/(n-1)}$ in the interval $[0, \frac{c}{n-1})$, and, by the Picard's Theorem [San79, p. 8], it is the only solution. Since the function tends to ∞ as $x \to \frac{c}{n-1}$, there is no function meeting the original requirements.

Solution to 3.1.4: 1. The zero function $x(t) \equiv 0$ is a solution for all $\alpha > 0$.
2. If $\alpha > 1$, then $f(x) = x^\alpha$ is lipschitzian in a neighborhood of zero and by Picard's Theorem this is the only solution.

If $0 < \alpha < 1$, then, as in the solution to Problem 3.1.3, we get

$$x(t)^{1-\alpha} = (1-\alpha)t + c \, ,$$

and the initial condition implies $c = 0$. Therefore

$$x(t) = ((1-\alpha)t)^{\frac{1}{1-\alpha}}$$

is a solution defined on $\mathbb{R}$.

Solution to 3.1.6: Separating the variables, we obtain

$$\frac{dy}{dx} = x^2(y - 3)$$

$$\int \frac{dy}{y-3} = \int x^2 \, dx$$

$$\log(y-3) = \tfrac{1}{3}x^3 + C$$

$$y - 3 = Ce^{x^3/3}$$

$$y = 3 + Ce^{x^3/3} \, .$$

From initial conditions, $C = -2$. The unique maximal solution is given by

$$y = 3 - 2e^{x^3/3}.$$

Solution to 3.1.7: Picard's Theorem [San79, p. 8] applies because the function $x^2 - x^6$ is Lipschitzian on finite subintervals of the x-axis. Thus, two distinct solution curves are non intersecting. The constant functions $x(t) \equiv 0$ and $x(t) \equiv 1$ are solutions. Hence, if the solution $x(t)$ satisfies $x(0) > 1$, then $x(t) > 1$ for all t, and if it satisfies $0 < x(0) < 1$, then $0 < x(t) < 1$ for all t.

Since

$$x^2 - x^6 = x^2(1 - x^4) = (1 - x)x^2(1 + x + x^2 + x^3),$$

we have

$$(*) \qquad \frac{d}{dt}(x - 1) = -(x - 1)x^2(1 + x + x^2 + x^3).$$

We see from this (or directly from the original equation) that if $x(0) > 1$, then $x - 1$ decreases as t increases, and if $0 < x(0) < 1$, then $1 - x$ decreases as x increases.

Case 1: $x(0) > 1$. In this case, $(*)$ implies

$$\frac{d}{dt}(x - 1) \leqslant -(x - 1)$$

(since $x(t) > 1$ for all t), so that

$$\frac{d}{dt}(e^t(x - 1)) \leqslant 0.$$

Hence, $e^t(x(t) - 1) \leqslant x(0) - 1$, that is, $x(t) - 1 \leqslant e^{-t}(x(0) - 1)$, from which the desired conclusion follows.

Case 2: $0 < x(0) < 1$. In this case, $(*)$ implies

$$\frac{d}{dt}(1 - x) \leqslant -x(0)^2(1 - x)$$

(since $x(t) \geqslant x(0)$ for all t), so that

$$\frac{d}{dt}(e^{x(0)^2 t}(1 - x)) \leqslant 0.$$

Therefore, $e^{x(0)^2 t}(1 - x(t)) \leqslant 1 - x(0)$, that is, $1 - x(t) \leqslant e^{-x(0)^2 t}(1 - x(0))$, and the desired conclusion follows.

Solution 2. The constant solutions are given by the roots of $x^2 - x^6 = 0$, and are $x(t) \equiv 0, 1, -1$. The phase portrait for this equation is

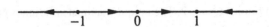

so 1 is a stable singularity and the only one in the positive axis, so it is the limit of any orbit starting on the positive real axis.

Solution to 3.1.8: From the equation, we get $x' = 0$ iff $x^3 - x = 0$, so the constant solutions are $x \equiv -1$, $x \equiv 0$, and $x \equiv 1$.
2. Considering the sign of x', we get the phase portrait

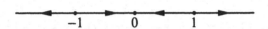

so 0 is a stable singularity, and 1 a unstable one. There are no other singularities in $[0, 1]$, so the limit of the orbit of the solution $x(t)$ that verifies $x(0) = 1/2$ is 0.

Solution 2. The function $\dfrac{x^3 - x}{1 + e^x}$ is lipschitzian on the finite subintervals of the x-axis, so by Picard's Theorem the solutions by each point are unique. For any $0 < x < 1$, $dx/dt < 0$ and since $x(t) \equiv 0$ is one of the solutions of the equation, the solution by $x(0) = 1/2$ is strictly decreasing and in between 0 and $1/2$ for all $t > 0$. Now for these values of x ($0 < x < 1/2$) we can easily estimate

$$\frac{1}{2} > \frac{1 - x^2}{1 + e^x} > \frac{1}{4}$$

so $\dfrac{dx}{dt} = -x\dfrac{1 - x^2}{1 + e^x}$ can be estimated as

$$\frac{dx}{dt} < -\frac{1}{4}x \qquad \text{for} \quad 0 < x < \frac{1}{2}$$

and

$$\frac{d}{dt}\left(e^{t/4}x\right) = e^{t/4}\left(\frac{x}{4} + \frac{dx}{dt}\right) < 0.$$

Hence $e^{t/4}x < x(0)$, that is, $x < \dfrac{1}{2}e^{-t/4}$, from which the conclusion follow.

Solution to 3.1.9: The given equation satisfies the hypotheses of Picard's Theorem [San79, p. 8], so a solution $x(t)$ exists in a neighborhood of the origin. Since $x'(0) = 231 + 85 \cos 77 \neq 0$, by the Inverse Function Theorem [Rud87, p. 221], x is locally invertible. Its inverse satisfies the initial value problem:

$$\frac{dt}{dx} = \frac{1}{3x + 85 \cos x}, \qquad t(77) = 0.$$

So

$$t(x) = \int_{77}^{x} \frac{1}{3\xi + 85 \cos \xi} d\xi$$

in some neighborhood of 77. Let α be the first zero of $3x + 85 \cos x$ to the left of 77, since there are no zeroes to the right of it, $t(x)$ above is defined in the entire interval (α, ∞), our maximal solution. The series expansion around α is

$$3x + 85 \cos x = (3 - 85 \sin \alpha)(x - \alpha) + O((x - \alpha)^2) \quad (x \to \alpha).$$

We cannot have $3 - 85 \sin \alpha = 0$, since this, together with $3\alpha + 85 \cos \alpha = 0$ imply $9 + 9\alpha^2 = 0$. Therefore

$$\lim_{x \to \alpha} \frac{3x + 85 \cos x}{x - \alpha} \neq 0.$$

It is clear that

$$\lim_{x \to \infty} \frac{3x + 85 \cos x}{x - \alpha} = 3.$$

Thus,

$$\lim_{x \to \alpha} t(x) = -\infty \quad \text{and} \quad \lim_{x \to \infty} t(x) = \infty.$$

We may take the inverse of $t : (\alpha, \infty) \to \mathbb{R}$ and get a function $x(t)$ that solves our initial value problem and is defined in all of $\mathbb{R}$.

Solution 2. Because the function $3x + 85 \cos x$ is continuous in $|x - b| \leq K$ and $|t - a| \leq T$, for any positive K and T, the differential equation has a unique solution with $x(a) = b$ defined for

$$|t - a| \leq \min\left(T, \frac{K}{3K + 3|b| + 85}\right)$$

For $T = 1/4$ and $K = 3|b| + 85$, the solution is defined for $|t - a| \leq 1/4$, since a and b are arbitrary, the solution can then be continued in patches of size $1/2$ to the entire real line.

Solution to 3.1.11: Suppose $y(t) > 0$ for $t \in (t_0, t_1)$, $y(t_0) = 0$. Integrating the equation

$$\frac{y'}{\sqrt{y}} = 1$$

we get the solution $y(t) = (t + c)^2/4$ where c is a constant. Each such solution can be extended to a differentiable function on $\mathbb{R}$:

$$y(t) = \begin{cases} 0 & \text{if } t \leq t_0 \\ (t - t_0)^2/4 & \text{if } t \geq t_0 \end{cases}$$

We must have $t_0 \geq 0$ for y to satisfy the given initial condition. $y \equiv 0$ is also a solution.

Solution to 3.1.12: The two functions $y \equiv 0$ and $y \equiv 2$ are obviously solutions to the differential equation, and the first one satisfies the given initial condition. We will show that this is the only one. Suppose there is a solution $y(x)$ that satisfies the initial condition and it is not identically zero, then there is a point x_0 where $y(x_0) \neq 0$ and let (a, b) denote a maximal interval where $y(x) \neq 0$ containing x_0. Ate least one of the two extremes of the interval exists, let's call it a, the other might be infinite and for that extreme $y(a) = 0$.

Inside this interval the separable differential equation can be written as

$$\frac{dy}{\sqrt{y(y-2)}} = dx$$

and with the substitution $w = x - 1$ the first integral becomes

$$\int \frac{dw}{\sqrt{w^2 - 1}} = \ln\left(w + \sqrt{w^2 - 1}\right) + C$$

so integrating both sides we get

$$\ln\left(y - 1 + \sqrt{y(y-2)}\right) = x + C$$

which is the same as

$$y - 1 + \sqrt{y(y-2)} = ke^x \qquad \text{for some } k > 0$$

but taking the limit when $x \to a$, y is zero at the limit and we have a contradiction because the right side is positive. So no such functions exist, except for $y(x) \equiv 0$.

Solution to 3.1.13: The given equation is equivalent to $\dfrac{d}{dx}(xy) = x$. Integrating both sides we get $xy = \dfrac{x^2}{2} + k$, so $y = \dfrac{x}{2} + \dfrac{k}{x}$ and k has to be 0 for the function to be differentiable in the given interval.

Solution 2. The equation is equivalent to

$$\frac{dy}{dx} = 1 - \frac{y}{x}$$

which is homogeneous in the sense of $\dfrac{dy}{dx} = f(x, y)$ and $f(tx, ty) = f(x, y)$, so the substitution $y = xv(x)$ converts it into a separable equation, in this case

$$v + x\frac{dv}{dx} = 1 - v(x)$$

or, after the separation of variables,

$$\frac{dv}{1 - 2v} = \frac{dx}{x}$$

integrating both sides, we get,

$$\frac{\ln|1-2v|}{-2} = \ln|x| + C$$

$$\ln|1-2v| = \ln\frac{1}{k^2x^2}$$

$$|1-2v| = \frac{1}{k^2x^2}$$

solving both cases and renaming the constant we get $y = \frac{x}{2} + \frac{k}{x}$. We obtain the only one that is differentiable in the given domain making $k = 0$, that is, $f(x) = \frac{x}{2}$.

Solution to 3.1.15: 1. Let u be defined by $u(x) = \exp(-3x^2/2)y(x)$. The given equation becomes

$$\frac{du}{dx} = -3xe^{-\frac{3}{2}x^2}y(x) + e^{-\frac{3}{2}x^2}\left(3xy(x) + \frac{y(x)}{1+y^2}\right)$$

$$= \frac{u(x)}{1+e^{3x^2}u(x)^2}$$

$$= f(x, u).$$

f is clearly C^1, so it satisfies the Lipschitz condition on any compact convex domain. The initial value problem

$$\frac{du}{dx} = \frac{u}{1+e^{3x^2}u}, \qquad u(0) = \frac{1}{n}$$

then has a unique solution for any $n \in \mathbb{N}$.
2. $f \equiv 0$ is the unique solution of the initial value problem associated with the condition $u(0) = 0$. Therefore, f_n cannot have any zero, so $f_n(x) > 0$ for $x \in [0, 1]$. For $u(x) = \exp(-3x^2/2)f_n(x)$, we have

$$0 \leqslant \frac{u'}{u} \leqslant \frac{1}{1+e^{3x^2}u^2} \leqslant 1$$

so

$$u(0) = \frac{1}{n} \leqslant u(x) \leqslant \frac{1}{n}e^x$$

therefore,

$$\frac{1}{n}e^{\frac{3}{2}} \leqslant f_n(1) \leqslant \frac{1}{n}e^{\frac{5}{2}}$$

and $f_n(1) \to 0$ when $n \to \infty$.

Solution to 3.1.16: If $y(t) \leqslant 0$ for some t, then $y'(t) \geqslant 1$, so $y(t)$ is growing faster than $z(t) = t$ for all t where $y(t) \leqslant 0$. Hence, there is a t_0 with $y(t_0) > 0$.

For $y > 0$, $y' > 0$, so for $t \geqslant t_0$, y is positive. Further, for $y > 0$, $e^{-y} > e^{-3y}$, so $y' > e^{-5y}$. Now consider the equation $z' = e^{-5z}$. Solving this by separation of variables, we get $z(t) = \log(t/5 + C)/5$, and for some choice of C, we have $z(t_0) = y(t_0)$. For all $t \geqslant t_0$, $y' > z'$, so $z(t) \leqslant y(t)$ for $t > t_0$. Since $z(t)$ tends to infinity as t does, so does y.

Solution to 3.1.17: Since $\frac{d}{dt}(e^{\lambda t}) = \lambda e^{\lambda t}$, the function $y(t) = e^{\lambda t}$ is a solution if and only if $\lambda = -e^{\lambda t_0}$, in other words, if and only if λ is a zero of the entire function $f(z) = z + e^{-t_0 z}$.

Assume $0 < t_0 < \pi/2$. We must show that f has no zeros in the closed half-plane $\Re z \geqslant 0$. If $\Re z \geqslant 0$ and $|z| > 1$, then

$$|f(z)| \geqslant |z| - |e^{-t_0 z}| > 1 - e^{-t_0 \Re z} \geqslant 1 - 1 = 0,$$

so $f(z) \neq 0$. If $\Re z \geqslant 0$ and $|\Im z| \leqslant 1$, then

$$\Re f(z) = \Re z + \Re e^{-t_0 z} = \Re z + e^{-t_0 \Re z} \cos(t_0 \Im z) > 0$$

because $\cos \theta > 0$ for $-t_0 \leqslant \theta \leqslant t_0$, so $f(z) \neq 0$. Hence f has no zeros for $\Re z \geqslant 0$.

Solution 2. Suppose $e^{\lambda t}$ with $\lambda = a + ib$ is a solution. We must prove $a < 0$. Assume $a \geqslant 0$. Then, from $f(\lambda) = 0$, we get

$$a + ib = -e^{-(a+ib)t_0} = -e^{-a}(\cos bt_0 - i \sin bt_0).$$

Hence

$$a = -e^{-a} \cos bt, \qquad b = e^{-a} \sin bt_0.$$

Since $a \geqslant 0$, the second equality implies that $|b| \leqslant 1$. Then $\cos bt_0 > 0$ since $0 < t_0 < \pi/2$, whence $a < 0$ by the first equality, a contradiction.

Solution to 3.1.18: Multiplying the first equation by the integrating factor $\exp\left(\int_0^x q(t)dt\right)$, we get

$$\frac{d}{dx}\left(f(x)\exp\left(\int_0^x q(t)dt\right)\right) = 0.$$

The general solution is therefore,

$$f(x) = C \exp\left(-\int_0^x q(t)dt\right)$$

where C is a constant. The hypothesis is that $\lim_{x\to\infty}\int_0^x q(t)dt = +\infty$. Even if $|p| \geqslant |q|$, the corresponding property may fail for p. For example, for $p \equiv -1$ and $q \equiv 1$, the general solutions are respectively $C \exp(-x)$ and $C \exp(x)$.

Solution to 3.1.19: Consider the equation

$$0 = F(x, y, z) = (e^x \sin y)z^3 + (e^x \cos y)z + e^y \tan x.$$

$F(0, 0, 0) = 0$, and all the partial derivatives of F are continuous, with

$$\left.\frac{\partial F}{\partial z}\right|_{(0,0,0)} = \left.\left(3z^2(e^x \sin y) + e^x \cos y\right)\right|_{(0,0,0)} = 1.$$

By the Implicit Function Theorem [Rud87, p. 224], there is a real valued function f with continuous partial derivatives, such that, $F(x, y, f(x, y)) = 0$ in a neighborhood of $(0, 0, 0)$. Locally, then, the given differential equation is equivalent to $y' = f(x, y)$. Since f satisfies the hypotheses of Picard's Theorem [San79, p. 8], there is a unique solution y in a neighborhood of 0 with $y(0) = 0$.

Solution to 3.1.20: Since f and g are positive, the solutions of both problems are monotonically increasing. The first differential equation can be rewritten as $dx/f(x) = dt$, so its solution is given by $x = h^{-1}(t)$, where the function h is defined by

$$h(x) = \int_0^x \frac{d\xi}{f(\xi)}.$$

Because the solution is defined for all t, we must have

$$\int_0^\infty \frac{d\xi}{f(\xi)} = \infty, \qquad \int_0^{-\infty} \frac{d\xi}{f(\xi)} = -\infty.$$

Since $g \leqslant f$, it follows that

$$\int_0^\infty \frac{d\xi}{g(\xi)} = \infty, \qquad \int_0^{-\infty} \frac{d\xi}{g(\xi)} = -\infty.$$

Using a similar reasoning we can see that the solution of the second equation is given by $x = H^{-1}(t)$, where

$$H(x) = \int_0^x \frac{d\xi}{g(\xi)}.$$

The conditions $\int_0^\infty \frac{d\xi}{g(\xi)} = \infty$, and $\int_0^{-\infty} \frac{d\xi}{g(\xi)} = -\infty$ guarantee that H maps $\mathbb{R}$ onto $\mathbb{R}$, hence that H^{-1} is defined on all of $\mathbb{R}$. Thus, the solution of the second equation is defined on $\mathbb{R}$.

Solution to 3.1.21: Let y be a solution of the given differential equation. If y' never vanishes, then y' has constant sign, so y is monotone.

Suppose that $y'(x_1) = 0$ for some x_1. Then the constant function $y_1(x) = y(x_1)$ is a solution of $f(y_1) = 0$. Consider the function z, $z(x) = y_1$ for all x. Then the differential equation $y' = f(y)$ with initial condition $y(x_1) = y_1$ is satisfied by y and by z. f is continuously differentiable and by Picard's Theorem [San79, p. 8], $y = z$, so y is constant.

3.2 Second Order Equations

Solution to 3.2.1: By Picard's Theorem [San79, p. 8] there is, at most, one real valued function f on $[0, \infty)$ such that $f(0) = 1$, $f'(0) = 0$ and $f''(x) = (x^2 - 1)f(x)$. Since the function $e^{-x^2/2}$ satisfies these conditions, we must have $f(x) = e^{-x^2/2}$. We then have

$$\lim_{x \to \infty} f(x) = \lim_{x \to \infty} e^{-x^2/2} = 0.$$

Solution to 3.2.2: The characteristic polynomial of the given differential equation is $(r - 1)^2$ so the general solution is

$$\alpha e^t + \beta t e^t.$$

The initial conditions give $\alpha = 1$, and $\beta = 0$, so the solution is $y(t) = e^t$.

Solution to 3.2.3: The characteristic polynomial of the associated homogeneous equation is

$$r^2 - 2r + 1 = (r - 1)^2$$

so the general solution of the homogeneous equation

$$\frac{d^2x}{dt^2} - 2\frac{dx}{dt} + x = 0$$

is

$$Ae^t + Bte^t \qquad (A, B \in \mathbb{R}).$$

$(\cos t)/2$ is easily found to be a particular solution of the original equation, so the general solution is

$$Ae^t + Bte^t + \frac{\cos t}{2}.$$

The initial conditions give $A = -\frac{1}{2}$ and $B = \frac{1}{2}$, so the solution is

$$\frac{1}{2}(e^t - te^t + \cos t).$$

Solution to 3.2.4: The characteristic polynomial of the given equation is

$$5r^2 + 10r + 6$$

which has roots $-1 \pm i/\sqrt{5}$, so the general solution is given by

$$x(t) = c_1 e^{-t} \cos\left(\frac{t}{\sqrt{5}}\right) + c_2 e^{-t} \sin\left(\frac{t}{\sqrt{5}}\right)$$

where c_1 and c_2 are constants. We can assume $c_1 \neq 0$ or $c_2 \neq 0$. Using calculus, we can see that $u^2(1 + u^4)^{-1} \leqslant 1/2$ with equality when $u = \pm 1$. Then f attains a maximum of $1/2$ iff x attains one of the values ± 1. We have $\lim_{t \to \infty} x(t) = 0$. Suppose $c_1 \neq 0$. Then, if k is a large enough integer, we have

$$\left| x\left(-\sqrt{5}k\pi\right) \right| = |c_1| e^{\sqrt{5}k\pi} > 1$$

so, by the Intermediate Value Theorem [Rud87, p. 93], x attains one of the values ± 1. If $c_2 \neq 0$, a similar argument gives the same conclusion.

Solution to 3.2.5: We first solve the homogeneous equation $x'' + 8x' + 25 = 0$. The general solution is $x_0(t) = c_1 e^{ir_1 t} + c_2 e^{ir_2 t}$, where c_1 and c_2 are constants and $r_k = -4 \pm 3i$, $k = 1, 2$, are the roots of the characteristic equation $r^2 + 8r + 25 = 0$.

All the solutions of the differential equation $x'' + 8x' + 25x = 2\cos t$ are of the form $x(t) = x_0(t) + s(t)$, where $s(t)$ is any particular solution. We solve for an $s(t)$ by the Method of Undetermined Coefficients [BD65, p. 115]. Consider $s(t) = A \cos t + B \sin t$. Differentiating this expression twice, we get

$$2\cos t = s'' + 8s' + 25s = (24A + 8B)\cos t + (24B - 8A)\sin t.$$

Solving the two linear equations gives $A = 3/40$ and $B = 1/40$. Therefore, the desired solution $x(t)$ is given by $x_0(t) + s(t)$, where c_1 and c_2 are chosen to give the correct initial conditions. $x_0(t)$ tends to 0 as t tends to infinity; therefore, to finish the problem, we need to find constants α and δ with $\alpha \cos(t - \delta) = A \cos t + B \sin t$. We have $\alpha \cos(t - \delta) = \alpha \cos t \cos \delta + \alpha \sin t \sin \delta$, so the problem reduces to solving $\alpha \cos \delta = 3/40$ and $\alpha \sin \delta = 1/40$. These equations imply $\tan \delta = 1/3$ or $\delta = \arctan(1/3)$. Hence, by elementary trigonometry, $\cos \delta = 3/\sqrt{10}$, so $\alpha = \sqrt{10}/40$.

Solution to 3.2.6: 1. The differential equation is equivalent to $y' = z$ and $z' = -|y|$. We have

$$\|(z_1, |y_1|) - (z_2, |y_2|)\| = \sqrt{(z_1 - z_2)^2 + (|y_2| - |y_1|)^2}$$
$$\leqslant \sqrt{(z_1 - z_2)^2 + (y_2 - y_1)^2}$$
$$= \|(z_1, y_1) - (z_2, y_2)\|$$

so the Lipschitz condition is verified and our initial value problem has, by Picard's Theorem [San79, p. 8], a unique solution. If y is such a solution, define the function z by $z(x) = y(-x)$. We have $z''(x) = y''(-x) = -|y(-x)| = -|z(x)|$, $z(0) = y(0) = 1$ and $z'(0) = -y'(0) = 0$, so $z = y$ and y is even.
2. We have

$$y'(x) = \int_0^x y''(t)dt = -\int_0^x |y(t)|dt < 0$$

so y is a decreasing function; therefore, it has, at most, one positive zero. If y is positive on $\mathbb{R}_+$, by continuity, y is positive in some interval of the form $(-\varepsilon, \infty)$

for some $\varepsilon > 0$. Together with $y(0) = 1$, $y'(0) = 0$ gives $y(x) = \cos x$, which is absurd. We conclude then that y has exactly one positive zero.

Solution to 3.2.7: Substituting $y(x) = x^a$ gives the quadratic equation $a(a-1)+1 = 0$. The two roots are

$$\frac{1}{2} \pm \frac{\sqrt{3}i}{2},$$

so the general solution is

$$y(x) = A\sqrt{x}\cos\left(\frac{\sqrt{3}}{2}\log x\right) + B\sqrt{x}\cos\left(\frac{\sqrt{3}}{2}\log x\right).$$

The boundary condition $y(1) = 0$ implies $A = 0$ and then the boundary condition $y(L) = 0$ can be satisfied for nonzero B only if

$$\sin\left(\frac{\sqrt{3}}{2}\log L\right) = 0.$$

Equivalently,

$$L = e^{2n\pi/\sqrt{3}}$$

where n is any positive integer.

Solution to 3.2.8: The general solution of the corresponding homogeneous equation is $Ae^{rx} + Be^{sx}$ where r, s are solutions of the characteristic equation $x^2 + 2p + x1 = 0$, i.e.,

$$r, s = \frac{-2p \pm \sqrt{4p^2 - 4}}{2}.$$

Since the given equation has a constant solution, $y \equiv 3$, it will admit solutions with infinitely many critical points exactly when r, s have a nonvanishing imaginary part, because in this case the homogeneous equation will admit solutions which are product of exponentials and sines or cosines. This happens when $4p^2 - 4 < 0$, i.e., for $|p| < 1$, $p \in \mathbb{R}$.

Solution to 3.2.9: Multiplying the first equation by $a(t)/p(t)$ where a is a differentiable function, we get

$$a(t)x''(t) + \frac{a(t)q(t)}{p(t)}x'(t) + \frac{a(t)r(t)}{p(t)}x(t) = 0.$$

Expanding the second given equation, we get

$$a(t)x''(t) + a'(t)x'(t) + b(t)x(t) = 0.$$

For the two equations to be equivalent, we must have $a'(t) = a(t)q(t)/p(t)$. Solving this by separation of variables, we get

$$a(t) = \exp\left(\int_0^t \frac{q(x)}{p(x)}\,dx\right).$$

Letting $b(t) = a(t)r(t)/p(t)$, we are done.

Solution to 3.2.10: The function $x'(1)x(t) - x'(0)x(t+1)$ satisfies the differential equation and vanishes along with its first derivative at $t = 0$. By Picard's Theorem [San79, p. 8] this function vanishes identically. Assuming x is not the zero function, we have $x'(0) \neq 0$ (again, by Picard's Theorem), so $x(t + 1) = cx(t)$, where $c = x'(1)/x'(0)$. It follows that the zero set of x is invariant under translation by one unit, which implies the desired conclusion.

Solution to 3.2.11: The characteristic equation is $l^2 - 2cl + 1 = 0$, which has the roots $c \pm \sqrt{c^2 - 1}$.
Case 1: $|c| > 1$. Let $\omega = \sqrt{c^2 - 1}$. Then the general solution is

$$x(t) = e^{ct}(A\cosh\omega t + B\sinh\omega t).$$

The condition $x(0) = 0$ implies $A = 0$, and then the condition $x(2\pi k) = 0$ implies $B = 0$, that is, $x(t) \equiv 0$. There are no nontrivial solutions in this case.
Case 2: $c = 1$. The general solution is

$$x(t) = Ae^t + Bte^t.$$

The condition $x(0) = 0$ implies $A = 0$, and then the condition $x(2\pi k) = 0$ implies $B = 0$. There are no nontrivial solutions in this case.
Case 3: $c = -1$. Similar reasoning shows that there are no nontrivial solutions in this case.
Case 4: $-1 < c < 1$. Let $\omega = \sqrt{1 - c^2}$. The general solution is then

$$x(t) = e^{ct}(A\cos\omega t + B\sin\omega t).$$

The condition $x(0) = 0$ implies $A = 0$. If $B \neq 0$, the condition $x(2\pi k) = 0$ then implies $2\pi k\omega = \pi n$ ($n \in \mathbb{Z}$), that is, $\omega = n/2k$, and

$$c^2 = 1 - \omega^2 = 1 - \frac{n^2}{4k^2}.$$

The right side is nonnegative and less than 1 only for $0 < |n| \leqslant 2k$. The required values of c are thus

$$c = \pm\sqrt{1 - \frac{n^2}{4k^2}}, \qquad n = 1, 2, \ldots, 2k.$$

Solution to 3.2.12: By the basic Existence–Uniqueness Theorem, the solution set is a two-dimensional vector space, V. Fix a real number s, and define the functions

p_s and q_s by $p_s(t) = p(t+s)$, $q_s(t) = q(t+s)$. Let f be in V. By the translation invariance of V, f is a solution of the equation

$$\frac{d^2x}{dt^2} + p_s\frac{dx}{dt} + q_s x = 0.$$

Hence it is also a solution of the equation

$$(p - p_s)\frac{dx}{dt} + (q - q_s)x = 0. \tag{*}$$

If $p - p_s$ does not vanish identically, there is an interval on which it is nowhere zero. But on such an interval the solutions of (*) form a one-dimensional vector space (by the Basic Uniqueness Theorem). Hence $p - p_s \equiv 0$, from which one sees that also $q - q_s \equiv 0$. Therefore p and q are constant.

3.3 Higher Order Equations

Solution to 3.3.1: 1. Since the differential equation is linear homogeneous with constant coefficients we can build its characteristic equation, which is,

$$\lambda^3 + \lambda^2 - 2 = 0.$$

By inspection we can easily see that 1 is one of the roots, factoring as

$$(\lambda - 1)(\lambda^2 + 2\lambda + 2) = 0$$

so the roots are 1 and $-1 \pm i$, and the solution to the differential equation has the form

$$f(x) = c_1 e^x + c_2 e^{-x}\cos x + c_3 e^{-x}\sin x$$

so the space of solutions has dimension 3.

2. E_0 is the two-dimensional subspace defined by $c_1 = 0$, that is, the function $g(t) = c_2 e^{-t}\cos t + c_3 e^{-t}\sin t$, and after the substitution for the initial condition we see that $c_2 = 0$ and $c_3 = 2$, so

$$g(t) = 2e^{-t}\sin t.$$

Solution to 3.3.2: The four complex fourth roots of -1 are $\dfrac{\pm 1 \pm i}{\sqrt{2}}$. Therefore the functions $e^{\pm x/\sqrt{2}}\cos(x/\sqrt{2})$ and $e^{\pm x/\sqrt{2}}\sin(x/\sqrt{2})$ satisfy the differential equation $y^{(4)} = -y$.

For $u(t) = e^{-x/\sqrt{2}}\sin(x/\sqrt{2})$ we have $u(0) = \lim_{x\to\infty} u(x) = \lim_{x\to\infty} u'(x) = 0$.

As $u'(0) = 1/\sqrt{2}$, $y(t) = \sqrt{2}\,e^{-x/\sqrt{2}}\sin(x/\sqrt{2})$ also satisfies $y'(0) = 1$.

Solution to 3.3.4: 1. The characteristic polynomial of the equation is

$$x^7 + \cdots + x + 1 = \frac{x^8 - 1}{x - 1}$$

which has roots -1, $\pm i$, and $(\pm 1 \pm i)/\sqrt{2}$. For each such root $z_k = u_k + i v_k$, $k = 1, \ldots, 7$, we have the corresponding solution

$$e^{z_k t} = e^{u_k t}(\cos v_k t + i \sin v_k t)$$

and these form a basis for the space of complex solutions. To get a basis for the real solutions, we take the real and imaginary parts, getting the basis

$$x_1(t) = e^{-t}, \quad x_2(t) = \cos t, \quad x_3(t) = \sin t, \quad x_4(t) = e^{\frac{1}{\sqrt{2}}t} \cos \frac{1}{\sqrt{2}}t,$$

$$x_5(t) = e^{\frac{1}{\sqrt{2}}t} \sin \frac{1}{\sqrt{2}}t, \quad x_6(t) = e^{\frac{-1}{\sqrt{2}}t} \cos \frac{1}{\sqrt{2}}t, \quad x_7(t) = e^{\frac{-1}{\sqrt{2}}t} \sin \frac{1}{\sqrt{2}}t.$$

2. A solution tends to 0 at ∞ iff it is a linear combination of solutions in the basis with the same property. Hence, the functions x_1, x_6, and x_7 form a basis for the space of solutions tending to 0 at ∞.

Solution to 3.3.5: The set of complex solutions of the equation forms a complex vector space which is invariant under differentiation. Hence, the functions $\cos t$ and $\cos 2t$ are also solutions, and, therefore, so are $e^{\pm it} = \cos t \pm i \sin t$ and $e^{\pm 2it} = \cos 2t \pm i \sin 2t$. It follows that the characteristic polynomial of the equation has at least the four roots $\pm i$, $\pm 2i$, so it is divisible by the polynomial $(\lambda^2 + 1)(\lambda^2 + 4)$. The differential equation is therefore, at least of order 4. The smallest possible order is, in fact, 4, because the given functions are both solutions of the equation

$$\left(\frac{d^2}{dt^2} + 1 \right) \left(\frac{d^2}{dt^2} + 4 \right) = 0,$$

that is,

$$\frac{d^4 x}{dt^4} + 5 \frac{d^2 x}{dt^2} + 4x = 0.$$

The preceding reasoning applies for both real and complex coefficients.

Solution to 3.3.6: Solving the characteristic equation $r^3 - 1 = 0$, we find that the general solution to $y''' - y = 0$ is given by

$$y(x) = c_1 e^x + c_2 e^{-x/2} \cos \frac{\sqrt{3}}{2}x + c_3 e^{-x/2} \sin \frac{\sqrt{3}}{2}x,$$

with c_1, c_2, and $c_3 \in \mathbb{R}$. $\lim_{x \to \infty} y(x) = 0$ when $c_1 = 0$. But the solution above with $c_1 = 0$, is the general solution of the differential equation with characteristic polynomial $(r^3 - 1)/(r - 1) = r^2 + r + 1$, that is,

$$y'' + y' + y = 0.$$

So $y''(0) + y'(0) + y(0) = 0$, and we can take $a = b = c = 1$ and $d = 0$.

3.4   Systems of Differential Equations

Solution to 3.4.2: Rewriting the system as a liner homogeneous system of equations we have

$$\begin{pmatrix} x'(t) \\ y'(t) \end{pmatrix} = \begin{pmatrix} 2 & -1 \\ 1 & 0 \end{pmatrix} \begin{pmatrix} x(t) \\ y(t) \end{pmatrix}$$

so the solutions are of the form

$$\begin{pmatrix} x(t) \\ y(t) \end{pmatrix} = e^{At} \begin{pmatrix} c_1 \\ c_2 \end{pmatrix}$$

By the Cayley–Hamilton Theorem [HK61, p. 194], the computation of the exponential of a matrix reduces to a finite sum of terms, in this case two,

$$e^{At} = \alpha_1 A t + \alpha_0 I$$

where the $\alpha_i(t)$ are functions of t depending on the matrix A. Furthermore, if $r(\lambda) = \alpha_1 \lambda + \alpha_0$, then for each eigenvalue λ_i of A, $r(\lambda_i) = e^{\lambda_i}$ and for each one with multiplicity two

$$e^{\lambda_i} = \left. \frac{d}{d\lambda} r(\lambda) \right|_{\lambda=\lambda_i}$$

which is a way to compute the values of α_0 and α_1, and then the exponential e^{At}. More generally, if e^{At} can be computed with the polynomial

$$e^{At} = \alpha_{n-1} A^{n-1} t^{n-1} + \alpha_{n-2} A^{n-2} t^{n-2} + \cdots + \alpha_1 A t + \alpha_0 I$$

then if we consider the polynomial

$$r(\lambda) = \alpha_{n-1} \lambda^{n-1} + \alpha_{n-2} \lambda^{n-2} + \cdots + \alpha_1 \lambda + \alpha_0$$

for each eigenvalue λ_i of At

$$e^{\lambda_i} = r(\lambda_i)$$

and furthermore, if the multiplicity of λ_i is $k > 1$, the each of the equations is also valid:

$$e^{\lambda_i} = \left. \frac{d^j}{d\lambda^j} r(\lambda) \right|_{\lambda=\lambda_i} \qquad \text{for} \qquad 1 \leqslant j \leqslant k - 1.$$

For more details on this and other interesting ways to compute the exponential of a matrix see [MVL78], specially Section 5, and the references cited there.

The characteristic polynomial of At is

$$\chi_{At}(\lambda) = \lambda^2 - (2t)\lambda + t^2$$

so the eigenvalues are $\lambda_1 = \lambda_2 = t$, and using the above formulas for the computation of α_i, we get

$$r(\lambda) = \alpha_1 \lambda + \alpha_0$$
$$\frac{dr(\lambda)}{d\lambda} = \alpha_1$$

and we end up with the system

$$e' = t\alpha_1 + \alpha_0$$
$$e' = \alpha_1 .$$

Therefore the solutions are $\alpha_1 = e'$ and $\alpha_0 = e'(1 - t)$ so the exponential is

$$e^{At} = e' \begin{pmatrix} 1+t & -t \\ t & 1-t \end{pmatrix} .$$

The solutions are then

$$\begin{pmatrix} x(t) \\ y(t) \end{pmatrix} = e' \begin{pmatrix} 1+t & -t \\ t & 1-t \end{pmatrix} \begin{pmatrix} c_1 \\ c_2 \end{pmatrix} ,$$

that is,

$$x(t) = e'(k_1 + k_2 t)$$
$$y(t) = e'(k_1 - k_2 + k_2 t) .$$

Solution 2. Alternatively, we can follow the previous solution closely, but compute the matrix e^{tA} by decomposing $A = S + N$, where S is complex diagonalizable and N nilpotent, for details we refer the reader to [HS74, Chap. 6]. This decomposition changes the computation of the exponential into a finite sum, in our case, since A has only one eigenvalue with multiplicity two

$$S = \begin{pmatrix} 1 & 0 \\ 0 & 1 \end{pmatrix} \qquad N = A - S = \begin{pmatrix} 1 & -1 \\ 1 & -1 \end{pmatrix}$$

obviously $N^2 = 0$, and the exponential can now be computed as

$$e^{tA} = e^{tS} e^{tN} = e'(I + tN) = e' \begin{pmatrix} 1+t & -t \\ t & 1-t \end{pmatrix}$$

and the solutions follow as before.

Solution 3. Computing the Jordan Canonical Form [HK61, p. 247] of the system we have

$$\begin{pmatrix} 0 & 1 \\ 1 & -1 \end{pmatrix} \begin{pmatrix} 2 & -1 \\ 1 & 0 \end{pmatrix} \begin{pmatrix} 1 & 1 \\ 1 & 0 \end{pmatrix} = \begin{pmatrix} 1 & 1 \\ 0 & 1 \end{pmatrix}$$

which we will write $T^{-1}AT = J_A$, so with the change of coordinates

$$\begin{pmatrix} x \\ y \end{pmatrix} = T \begin{pmatrix} z \\ w \end{pmatrix} ,$$

that is,

$$\begin{pmatrix} x \\ y \end{pmatrix} = \begin{pmatrix} 1 & 1 \\ 1 & 0 \end{pmatrix} \begin{pmatrix} z \\ w \end{pmatrix}$$

the system becomes, according to the form J_A,

$$z'(t) = z(t) + w(t)$$
$$w'(t) = w(t).$$

These equations can be readily solved in the reverse order, and we obtain,

$$w(t) = c_1 e^t \qquad z(t) = c_1 t e^t + c_2 e^t$$

and the solutions $x(t)$ and $y(t)$ can now be written as:

$$x(t) = e^t (c_1 t + c_1 + c_2)$$
$$y(t) = e^t (c_1 t + c_2).$$

Solution 4. Derivating the first equation and substituting the second we end up with a second order homogeneous differential equation

$$x''(t) - 2x'(t) + x(t) = 0$$

whose characteristic equation $\lambda^2 - 2\lambda + 1 = 0$ has one double root $\lambda = 1$, so the general solution will be

$$x(t) = c_1 e^t + c_2 t e^t$$

substituting back in the first equation we get

$$y(t) = (c_1 - c_2)e^t + c_2 t e^t.$$

Solution to 3.4.3: We have

$$\frac{d}{dt}\left(x^2 + y^2\right) = 2x\frac{dx}{dt} + 2y\frac{dy}{dt}$$
$$= 2x(-x + y) + 2y(\log(20 + x))$$
$$= -2x^2 + 2xy - 2y^2 + 2y\log(20 + x).$$

As, for any positive ε,

$$\log(20 + x) = o(x^\varepsilon) \quad (x \to +\infty)$$

and

$$-2x^2 + 2xy - 2y^2 \leqslant -2(x^2 + y^2 - |xy|)$$
$$\leqslant -2(x - y)^2$$
$$\leqslant 0$$

we conclude that

$$\frac{d}{dt}\left(x^2 + y^2\right) \leqslant 0$$

for $\|(x, y)\|$ large enough so the distance of (x, y) to the origin is bounded.

Solution to 3.4.4: 1. Using polar coordinates, $x = r \cos \theta$ and $y = r \sin \theta$, we get

$$\frac{dr}{dt} = \frac{x}{r}\frac{dx}{dt} + \frac{y}{r}\frac{dy}{dt} = r(1 - r^2)$$

$$\frac{d\theta}{dt} = \frac{x}{r^2}\frac{dy}{dt} - \frac{y}{r^2}\frac{dx}{dt} = -1$$

solving these, we get

$$r = \frac{c_1 e^t}{\sqrt{1 + c_1^2 e^{2t}}} \quad \text{and} \quad \theta = -t + c_2$$

where c_1, c_2 are constants.

For $(x_0, y_0) = (0, 0)$, we have $\frac{dx}{dt} = \frac{dy}{dt} = 0$; therefore, $x = y \equiv 0$. For $(x_0, y_0) \neq (0, 0)$ let $x_0 = r_0 \cos \theta_0$ and $y_0 = r_0 \sin \theta_0$. We have

$$c_1 = \frac{r_0}{\sqrt{1 - r_0^2}} \quad \text{and} \quad c_2 = \theta_0$$

so

$$x(t) = \frac{c_1 e^t}{\sqrt{1 + c_1 e^{2t}}} \cos(\theta_0 - t), \qquad y(t) = \frac{c_1 e^t}{\sqrt{1 + c_1 e^{2t}}} \sin(\theta_0 - t).$$

2. We have

$$\lim_{t \to \infty} r = \lim_{t \to \infty} \frac{c_1 e^t}{\sqrt{1 + c_1 e^{2t}}} = 1.$$

Solution to 3.4.5: We have

$$\frac{d}{dt}\begin{pmatrix} x \\ y \\ z \end{pmatrix} = \begin{pmatrix} 0 & 1 & 0 \\ 2 & 0 & 0 \\ 0 & 0 & 3 \end{pmatrix}\begin{pmatrix} x \\ y \\ z \end{pmatrix}$$

Let A denote the 3×3 matrix above. Its characteristic polynomial is

$$\chi(x) = -x^3 + 3x^2 + 2x - 6 = (x - 3)(-x + \sqrt{2})(x + \sqrt{2})$$

therefore $-\sqrt{2}$ is a simple eigenvalue of A. Using an eigenvector $v \in \mathbb{R}^3$ associated with this eigenvalue, we get a solution $e^{-\sqrt{2}t}v$ which tends to ∞ as $t \to -\infty$ and to the origin as $t \to +\infty$.

Solution to 3.4.6: We have

$$\frac{d}{dt}\|x(t)\|^2 = \frac{d}{dt}\langle x(t), x(t) \rangle$$

$$= \langle x'(t), x(t) \rangle + \langle x(t), x(t)' \rangle$$

$$= \langle Ax(t), x(t) \rangle + \langle x(t), Ax(t) \rangle$$

$$= \langle (A + A^*)x(t), x(t) \rangle$$

so it suffices to prove that $A + A^*$ is positive definite. We have

$$A + A^* = \begin{pmatrix} 2 & 2 & -2 \\ 2 & 8 & 2 \\ -2 & 2 & 16 \end{pmatrix}$$

and by the Solution to Problem 7.8.2 it is enough to check that the determinant of the principal minors are positive, which is a simple calculation.

Solution to 3.4.7: We have

$$\frac{dx}{dt}(0) = 1,$$

so, for small positive t, $(x(t), y(t))$ lies in the right half-plane. For $0 < x < 1$, we have

$$\frac{dx}{dt} = 1 + \frac{1}{2}x^2 \sin y \geq 1 - \frac{1}{2}x^2 \geq \frac{1}{2}.$$

Thus, the function $x(t)$ is increasing with slope at least $1/2$. Therefore, by the time $t = 2$, the curve $(x(t), y(t))$ will cross the line $x = 1$.

Solution to 3.4.8: 1. Let the function g be defined by $g(t) = f(x(t), x'(t))$. We have

$$(*) \qquad g'(t) = -2(x'(t)^2 + x(t)^4)$$

so g is a decreasing function.

2. It is enough to show that $\lim_{t \to \infty} g(t) = 0$.

Since g is a positive decreasing function, the limit exists and satisfies $\lim_{t \to \infty} g(t) = c \geq 0$. If $c > 0$, then, for some $\varepsilon > 0$, $T \in \mathbb{R}$ we have $x'(t)^2 + x(t)^4 > \varepsilon$ for $t \geq T$. Then, by $(*)$, we have

$$\frac{g(T) - c}{2} = \int_T^\infty (x'(t)^2 + x(t)^4)\, dt > \int_T^\infty \varepsilon\, dt = \infty$$

which is absurd. We must then have $c = 0$, as desired.

Solution to 3.4.9: 1. We have

$$\frac{\partial F}{\partial t} = \frac{\partial F}{\partial x}\frac{\partial x}{\partial t} + \frac{\partial F}{\partial y}\frac{\partial y}{\partial t} = y(x^3 + x^5) - y(ay + x^3 + x^5) = -ay^2.$$

Thus, $\partial F / \partial t \leq 0$, which implies that F decreases along any solution $(x(t), y(t))$ of the differential equation.

2. Let $\varepsilon > 0$. Since F is continuous, there exists a $\delta > 0$ such that $F(x, y) < \varepsilon^2/2$ if $\|(x, y)\| < \delta$. Further, by letting (x, y) vary over all points such that $\|(x, y)\| = \varepsilon$, elementary calculus shows that $F(x, y) \geq \varepsilon^2/2$.

Let the initial conditions $x(0)$ and $y(0)$ be such that $\|(x(0), y(0))\| < \delta$. By Picard's Theorem [San79, p. 8], there exist unique solutions $x(t)$ and $y(t)$ to the differential equation satisfying these initial conditions. Since $F(x, y)$ decreases along solutions, we must have that $F(x(t), y(t)) < \varepsilon^2/2$ for all $t > 0$

in the domain of the solution. Now suppose that for some $t > 0$ in this domain, $\|(x(t), y(t))\| = \varepsilon$. We would have $F(x(t), y(t)) \geq \varepsilon^2/2$, a contradiction. Therefore, $\|(x(t), y(t))\| < \varepsilon$ for all $t > 0$ in the domain of the solution. But this bound is independent of t, so the Extension Theorem [San79, p. 131] shows that this solution exists on all positive t.

Solution to 3.4.12: 1. We have

$$\frac{dH}{dt}(t) = \frac{1}{2}\frac{d}{dt}\langle x'(t), x'(t)\rangle + \langle \operatorname{grad} V(x(t)), x'(t)\rangle$$
$$= \langle x''(t), x'(t)\rangle + \langle \operatorname{grad} V(x(t)), x'(t)\rangle$$
$$= \sum_{i=1}^{n} f_i(x(t)) \cdot x_i'(t) + \sum_{i=1}^{n} \frac{\partial V}{\partial x_i}(x(t)) \cdot x_i'(t)$$
$$= 0$$

since $f_i = -\partial V/\partial x_i$. Therefore $H(t)$ is constant on (a, b).
2. Fix $t_0 \in (a, b)$. Since H is constant, $H(t) = H(t_0)$, for all $t \in (a, b)$. We have

$$H(t_0) = \frac{1}{2}\|x'(t)\|^2 + V(x(t)) \geq \frac{1}{2}\|x'(t)\|^2 + M$$

and then

$$0 \leq \|x'(t)\|^2 \leq 2(H(t_0) - M)$$

showing that $x'(t)$ is bounded. Noting that

$$|x(t_1) - x(t_2)| = \int_{t_1}^{t_2} x'(t)dt \leq \sqrt{2(H(t_0) - M)} \, |t_1 - t_2|$$

we conclude that x is bounded and has limits at the endpoints of the interval because it is Lipschitz. Using the continuity of f we get

$$\lim_{t \to b} x_i''(t) = \lim_{t \to b} f_i(x(t)) = f_i(\lim_{t \to b} x(t))$$

so $\lim_{t \to b} x''(t)$ exists, for $i = 1, \ldots, n$. The corresponding facts for $t \to a$ can be treated similarly. Therefore the closure of $x((a, b))$ is compact and x_i'' is bounded because its image is contained in the image of a compact under the continuous extension of f_i.

Using the maximum of $\|x''(t)\|$ on (a, b) we see that x' is a Lipschitz function, and that the limit $\lim_{t \to b} x'(t)$ exists, completing the proof.

Solution to 3.4.14: Let $f_1, \ldots, f_n$ be a basis for V. Since V is closed under differentiation, we have the system of equations

$$f_i' = \sum_{j=1}^{n} a_{ij} f_j \qquad 1 \leq i \leq n.$$

Let $A=(a_{ij})$. This system has solutions of the form $f_i(x) = C_i e^{\lambda_i x}$, where the C_i's are constants and the λ_i's are the (complex) eigenvalues of the matrix A. By the properties of the exponential function, we immediately have that $f_i(x + a) = C f_i(x)$ for some constant C depending on a, so $f_i(x + a) \in V$. Since the f_i's form a basis of V, V is closed under translation.

Solution to 3.4.15: Let V be such a space, and let D denote the operator of differentiation acting on V. If χ is the characteristic polynomial of D, the functions in V are solutions of the linear differential equation $\chi(D)f = 0$. The set of solutions of that equation is a three dimensional vector space, since χ has degree 3, so it equals V.

Case 1. χ has three distinct roots $\lambda_1, \lambda_2, \lambda_3$. The functions $e^{\lambda_1 t}, e^{\lambda_2 t}, e^{\lambda_3 t}$ form a basis for the set of solutions of $\chi(D)f = 0$ and so for V.

Case 2. χ has a root λ_1 of multiplicity 2 and a root λ_2 of multiplicity 1. The functions $e^{\lambda_1 t}, t e^{\lambda_1 t}, e^{\lambda_2 t}$ form a basis for V.

Case 3. χ has a single root of multiplicity 3. In this case the functions $e^{\lambda t}, t e^{\lambda t}, t^2 e^{\lambda t}$ form a basis for V.

Solution to 3.4.16: We will show that for $1 \leqslant k \leqslant n+1$, $f_k(t) = \sum_{j=k}^{n+1} \xi_j e^{-jt}$ where the ξ_j's are constants depending on k. From this, it follows that each $f_k(t)$ approaches 0 as t tends to infinity.

Solving the second equation, we get

$$f_{n+1}(t) = \xi_{n+1} e^{-(n+1)t}$$

for some $\xi_{n+1} \in \mathbb{R}$, which has the desired form. Assume that, for some k, the formula holds for f_{k+1}. Differentiating it and substituting it into the first equation gives

$$f_k' = \sum_{j=k+1}^{n+1} (k + 1 + j)\xi_j e^{-jt} - k f_k.$$

This is a first order linear differential equation which we can solve. Letting $\mu_j = (k + 1 + j)\xi_j$, we get

$$f_k = \left(\int_0^t e^{ks} \left(\sum_{j=k+1}^{n+1} \mu_j e^{-js} \right) ds + C \right) e^{-kt}$$

where C is a constant. Changing the order of summation and evaluating, we get

$$f_k = \left(\sum_{j=k+1}^{n+1} \frac{\mu_j}{k - j} e^{(k-j)x} \Bigg|_{x=0}^{t} + C \right) e^{-kt} = \sum_{j=k}^{n+1} \xi_j e^{-jt}$$

where the ξ_j's are some real constants, and we are done.

Solution to 3.4.17: We solve the case $n = 1$ in two different ways. *First method:* Let B be the indefinite integral of A vanishing at 0. One can then integrate the equation $\dfrac{dx}{dt} = Ax$ with the help of the integrating factor e^{-B}, namely

$$0 = e^{-B}\frac{dx}{dt} - e^{-B}A\frac{dx}{dt} = \frac{d}{dt}\left(e^{-B}x\right),$$

giving $x(t) = e^{B(t)}x(0)$. Since $A(t) \leqslant \beta$, we have $B(t) \leqslant \beta t$ for $t > 0$, so

$$|x(t)| = |x(0)|e^{B(t)} \leqslant |x(0)|e^{-\beta t},$$

as desired.

Second method ($n = 1$): Consider the derivative of $e^{-\beta t}x$:

$$\frac{d}{dt}\left(e^{-\beta t}x\right) = e^{-\beta t}\left(\frac{dx}{dt} - \beta x\right) = e^{-\beta t}\,(A - \beta)\,x.$$

By Picard's Theorem [San79, p. 8], x either has a constant sign or is identically 0. Hence, $e^{-\beta t}x$ is nonincreasing when x is positive and nondecreasing when x is negative, which gives the desired conclusion.

$n > 1$. We have for a solution $x(t)$,

$$\frac{d}{dt}\|x\|^2 = 2\langle\frac{dx}{dt}, x\rangle = 2\langle Ax, x\rangle \leqslant 2\beta\|x\|^2$$

which reduces the case $n > 1$ to the case $n = 1$.

Solution to 3.4.18: 1. We have

$$\begin{aligned}
\frac{d}{dt}\|X(t)\|^2 &= \frac{d}{dt}(X(t)\cdot X(t)) = 2X(t)\cdot\frac{dX(t)}{dt}\\
&= 2X(t)\cdot WX(t) = 2W^t X(t)\cdot X(t)\\
&= -2WX(t)\cdot X(t) = -2X(t)\cdot WX(t)\\
&= -\frac{d}{dt}\|X(t)\|^2,
\end{aligned}$$

from which it follows that $\dfrac{d}{dt}\|X(t)\|^2 = 0$, hence that $\|X(t)\|$ is constant.

2. We have

$$\frac{d}{dt}(X(t)\cdot v) = \frac{dX(t)}{dt}\cdot v = WX(t)\cdot v = X(t)\cdot W^t v = -X(t)\cdot Wv = 0.$$

3. It will suffice to show that the null space of W is nontrivial. For if v is a nonzero vector in that null space, then

$$\|X(t) - v\|^2 = \|X(t)\|^2 + \|v\|^2 - 2X(t)\cdot v,$$

which is constant by Part 1 and Part 2, implying that $X(t)$ lies on the intersection of two spheres, one with center 0 and one with center v.

The nontriviality of the null space of W follows from the antisymmetry of W:

$$\det W = \det W^t = \det(-W) = (-1)^3 \det W = -\det W .$$

Hence, $\det W = 0$, so W is singular.

Solution to 3.4.19: Consider the function u defined by $u(t) = \|x(t)\|^2$. We have, using Rayleigh's Theorem [ND88, p. 418],

$$u'(t) = 2\langle x(t), x'(t)\rangle = 2\langle x(t), P(t)x(t)\rangle \leqslant -2\langle x(t), x(t)\rangle = -2u(t)$$

which implies that $u(t) \leqslant u(0)\exp(-2t)$ for $t > 0$, so $\lim_{t \to \infty} u(t) = 0$, and the result follows.

Solution to 3.4.20: Expanding the matrix differential equation, we get the family of differential equations

$$\frac{df_{ij}}{dt} = f_{i-1\,j-1}, \qquad 1 \leqslant i, j \leqslant n,$$

where $f_{ij} \equiv 0$ if i or j equals 0. Solving these, we get

$$X(t) = \begin{pmatrix} \xi_{11} & \xi_{12} & \xi_{13} & \xi_{14} \\ \xi_{21} & \xi_{11}t + \xi_{22} & \xi_{12}t + \xi_{23} & \xi_{13}t + \xi_{24} \\ \xi_{31} & \xi_{21}t + \xi_{32} & \frac{1}{2}\xi_{11}t^2 + \xi_{22}t + \xi_{33} & \frac{1}{2}\xi_{12}t^2 + \xi_{23}t + \xi_{34} \\ \xi_{41} & \xi_{31}t + \xi_{42} & \frac{1}{2}\xi_{21}t^2 + \xi_{32}t + \xi_{43} & \frac{1}{6}\xi_{11}t^3 + \frac{1}{2}\xi_{22}t^2 + \xi_{33}t + \xi_{44} \end{pmatrix}$$

where the ξ_{ij}'s are constants.

Solution 2. We will use a power series. Assume $X(t) = \sum_{n=0}^{\infty} t^n C_n$. The given equation gives

$$\sum_{n=1}^{\infty} nt^{n-1}C_n = \sum_{n=0}^{\infty} At^n C_n B$$

which can be written as

$$\sum_{n=0}^{\infty} t^n \left((n+1)C_{n+1} - AC_n B\right) = 0$$

giving the recurrence relation

$$C_{n+1} = \frac{1}{n+1}AC_n B,$$

so we have

$$C_n = \frac{1}{n!}A^n C_0 B^n.$$

Since $A^4 = B^4 = 0$, the solution reduces to a polynomial of degree at most 3:

$$X(t) = C_0 + t\,AC_0 B + \frac{t^2}{2}A^2 C_0 B^2 + \frac{t^3}{6}A^3 C_0 B^3$$

where $C_0 = X(0)$ is the initial value of X.

4

Metric Spaces

4.1 Topology of $\mathbb{R}^n$

Solution to 4.1.1: Suppose x is a point of $S \setminus C$. Then there is an open interval containing x whose intersection with S is finite or countable and which thus is disjoint from C. By the density of $\mathbb{Q}$ in $\mathbb{R}$, there is such an interval with rational endpoints. There are countably many open intervals in $\mathbb{R}$ with rational endpoints. Hence $S \setminus C$ is a finite or countable union of finite or countable sets. Therefore, $S \setminus C$ is finite or countable.

Solution to 4.1.2: 1. For $n \in \mathbb{Z}$ let $A_n = A \cap [n, n+1)$. Then $A = \cup_{n \in \mathbb{Z}} A_n$. Since A is uncountable, at least one of these sets, A_k say, is uncountable. Then A_k contains a bounded sequence with distinct terms, which has a convergent subsequence. The limit of this subsequence is an accumulation point of A.
2. Let M be the set of accumulation points of A. Suppose M is at most countable. As M is closed, its complement $\mathbb{R} \setminus M$ is open. Then we can represent it as a countable union of closed sets, $\mathbb{R} \setminus M = \cup C_n$. If M is at most countable, then A must have uncountable many elements in $\mathbb{R} \setminus M$, therefore in some C_m. By part 1., $A \cap C_m$ has an accumulation point. C_m is closed, so this accumulation point is in C_m. This contradicts the fact that all accumulation points of A are in M.
Solution 2. Suppose there are no accumulation points, then for each point $r \in \mathbb{R}$ there is a open interval I_r around r with only finitely many points of A. Reducing the interval a bit further we may assume it contains at most one point of the set A (the possibility being r itself). This open set form an open covering of the compact set $[n, n+1]$, so it has a finite subcover and as such A only has finite number of

elements in the interval $[n, n+1]$, hence at most a countable number of elements, a contradiction, so A must be uncountable.

Solution to 4.1.3: For each i, the given condition guarantees that the sequence $(x(i, j))_{j=1}^\infty$ is Cauchy, hence convergent. Let a_i denote its limit. We have, by the same condition,

$$\rho(x(i, j), x(k, j)) \leqslant \max\left\{\frac{1}{i}, \frac{1}{k}\right\},$$

giving $\rho(a_i, a_k) \leqslant \max\{\frac{1}{i}, \frac{1}{k}\}$. Hence the sequence $(a_i)_{i=1}^\infty$ is Cauchy, therefore convergent. Let a denote its limit.

Each sequence $(x(i, j))_{i=1}^\infty$ is Cauchy, hence convergent, say to b_j, and the sequence $(b_j)_{j=1}^\infty$ is Cauchy, hence convergent to b.

Again, by the same condition as before,

$$\rho(x(i, i), x(i, j)) \leqslant \max\{\frac{1}{i}, \frac{1}{j}\}.$$

In the limit as $j \to \infty$ this gives $\rho(x(i, i), a_i) \leqslant \dfrac{1}{i}$. Hence $\lim_{i\to\infty} x(i, i) = a$. By the same reasoning, $\lim_{i\to\infty} x(i, i) = b$.

Solution to 4.1.4: The closures $\overline{T}_n$ form a nested sequence of compact sets whose diameters tend to 0. The intersection of the closures therefore consists of a single point. The question is whether that point belongs to every T_n.

It is asserted that the centroid $x_0 = \frac{1}{3}(A + B + C)$ of the vertices A, B, C is in every T_n. It is in T_0, being a convex combination of the vertices A, B, C with nonzero coefficients, in fact, $T_0 = \{\alpha A + \beta B + \gamma C \mid \alpha, \beta, \gamma > 0, \ \alpha + \beta + \gamma = 1\}$. Moreover the centroid of the vertices of T_1 is

$$\frac{1}{3}\left(\frac{A+B}{2} + \frac{B+C}{2} + \frac{C+A}{2}\right) = x_0.$$

The obvious induction argument shows that x_0 is the centroid of the vertices of T_n for every n. Hence $\bigcap_{n=0}^\infty T_n = \{x_0\}$.

Solution to 4.1.5: Suppose there is no such ε. Then there exists a sequence (x_n) in K such that none of the balls $B_{1/n}(x_n)$ is contained in any of the balls B_j. Since K is compact, this sequence has a limit point, by the Bolzano–Weierstrass Theorem [Rud87, p. 40], [MH93, p. 153], $x \in K$. Then, since the B_j's are an open cover of K, there is a j and an $\varepsilon > 0$ such that $B_\varepsilon(x) \subset B_j$. Let $1/N < \varepsilon/2$, and choose $n > N$ such that $|x - x_n| < \varepsilon/2$. Then $B_{1/n}(x_n) \subset B_\varepsilon(x) \subset B_j$, contradicting our choice of x_n's. Hence, the desired ε must exist.

Solution 2. Suppose the conclusion is false. Then, for each positive integer n, there are two points x_n and y_n in K such that $|x_n - y_n| < 1/n$, yet no B_j contains both x_n and y_n. Since K is compact, the sequence (x_n) has a convergent subsequence, (x_{n_k}) say, with limit $\rho \in K$. Then, obviously, $y_{n_k} \to \rho$. There is a B_j that contains ρ. Since B_j is open and $\rho = \lim x_{n_k} = \lim y_{n_k}$, both x_{n_k} and y_{n_k} must be in B_j for k sufficiently large, in contradiction to the way the points x_n and y_n were chosen.

Solution 3. By compactness, we can choose a finite subcover $\{B_j\}_{j=1}^N$ of K, [Rud87, p. 30]. For $x \in K$, define

$$f(x) = \max\{\text{dist}(x, \mathbb{R}^n \setminus B_j) \mid x \in B_j\}.$$

Then $f(x) > 0$ for each $x \in K$, because each B_j is open and there are only finitely many of them. Since K is compact and f is continuous and strictly positive on K, f has a positive minimum $\varepsilon > 0$. By definition of f, every ε–ball centered at a point of K is contained in some B_j.

Solution 4. For each positive integer k let $E_k = \{x \in M \mid B(x, 1/k) \subset U_\alpha$ for some $\alpha \in I\}$. For each $x \in E_k$ we clearly have $B(x, 1/k) \subset E_k$ so E_k is open. The sets E_k form an open cover of C, and are nested. Since C is compact, $C \subset E_K$ for some K, and we are done.

Solution to 4.1.6: Suppose U_n is an open set of real numbers for $n \in \mathbb{N}$, such that $\mathbb{Q} = \cap U_n$. Then each set $\mathbb{R} \setminus U_n$ is nowhere dense, since it is a closed set which contains only irrational numbers. We then have

$$\mathbb{R} = \bigcup_{n \in \mathbb{N}} U_n \bigcup_{q \in \mathbb{Q}} \{q\}$$

but $\mathbb{R}$ is not a countable union of nowhere dense sets, by Baire's Category Theorem [MH93, p. 175]. So $\mathbb{Q}$ cannot be a countable intersection of open sets.

Solution to 4.1.7: Suppose $x, y \in X$. Without loss of generality, assume $x < y$. Let z be such that $x < z < y$ (for instance, z irrational verifying the double inequality). Then

$$(-\infty, z) \cap X, \quad (z, \infty) \cap X$$

is a disconnection of X. We conclude then that X can have only one element.

Solution to 4.1.8: The Cantor set [Rud87, p. 41] is an example of a closed set having uncountably many connected components.

Let A be an open set and suppose C_α, $\alpha \in \Gamma$ are its connected components. Each C_α is an open set, so it contains a rational number. As the components are disjoint, we have an injection of Γ in $\mathbb{Q}$, so Γ is, at most, countable.

Solution to 4.1.9: We show S is bounded and closed.

To establish boundedness, choose n_0 such that $\|f - f_n\| \leqslant 1$ on K for $n \geqslant n_0$. The function f is continuous, being the uniform limit of continuous functions, so $f(K)$ is bounded, hence $f(K) \cup \bigcup_{n=n_0}^\infty f_n(K)$ is bounded. Each $f_n(K)$ is bounded. Therefore so is $\bigcup_{1 \leqslant n < n_0} f_n(K)$, whence S is bounded.

To prove that S is closed, let $\{y_j\}$ be a convergent sequence in S with limit y. It will suffice to show that y is in S. If infinitely many terms of the sequence lie in $f(K)$, or in $f_n(K)$ for some n, then y belongs to the same set, and hence to S. (Reason: $f(K)$ and all the $f_n(K)$ are compact.) In the contrary case we may assume, after passing to a subsequence, that y_j is in $f_{m_j}(K)$, where $m_1 < m_2 <$

..., say $y_j = f_{m_j}(x_j)$. Since K is compact we may assume, passing to a further subsequence, that the sequence (x_j) converges, say to x. Then

$$\|y - f(x)\| \leqslant \|y - f_{m_j}(x_j)\| + \|f_{m_j}(x_j) - f(x_j)\| + \|f(x_j) - f(x)\|,$$

and all three summands on the right tend to 0 as $j \to \infty$. Hence $y = f(x)$, so y is in S.

Solution to 4.1.10: Suppose we have

$$[0, 1] = \bigcup_{i \in \mathbb{N}} [a_i, b_i]$$

where the $[a_i, b_i]$'s are non empty pairwise disjoint intervals. Let X be the set of the corresponding endpoints:

$$X = \{a_1, a_2, \ldots\} \cup \{b_1, b_2, \ldots\}.$$

We will show that X is a perfect set, so it cannot be countable.

The complement of X in $[0, 1]$ is a union of open intervals, so it is open, and X is closed. By the assumption, there must be elements of X in $(a_i - \varepsilon, a_i)$ for each $\varepsilon > 0$, and each $i \in \mathbb{N}$, and similarly for the b_i's. Each element of X is then an accumulation point, and X is perfect.

Solution to 4.1.11: 1. Let $X = \{x\}$ and (y_n) be a sequence in Y such that $|x - y_n| < d(X, Y) + 1/n$. As (y_n) is bounded, passing to a subsequence, we may assume that it converges, to y, say. As Y is closed, $y \in Y$ and, by the continuity of the norm, $|x - y| = d(X, Y)$.
2. Let (x_n) be a sequence in X such that $d((x_n), Y) < d(X, Y) + 1/n$. As X is compact, by the Bolzano–Weierstrass Theorem [Rud87, p. 40], [MH93, p. 153], we may assume, passing to a subsequence, that (x_n) converges, to x, say. We then have $d(X, Y) = d(\{x\}, Y)$ and the result follows from Part 1.
3. Take $X = \{(x, 1/x) \mid x > 0\}$ and $Y = \{(x, 0) \mid x > 0\}$ in $\mathbb{R}^2$.

Solution to 4.1.12: Suppose that S contains no limit points. Then, for each $x \in S$, there is a $\delta_x > 0$ such that $B_{\delta_x} \cap S = \{x\}$. Let $\varepsilon_x = \delta_x/2$. The balls $B_{\varepsilon_x}(x)$ are disjoint, so we can choose a distinct point from each one with rational coordinates. Since the collection of points in $\mathbb{R}^n$ with rational coordinates is countable, the set S must be countable, a contradiction. Hence, S must contain one of its limit points.

Solution to 4.1.13: Let y be a limit point of Y and (y_n) a sequence in Y converging to y. Without loss of generality, we may suppose that $|y_n - y| < r$. By the definition of Y, there is a sequence (x_n) in X with $|x_n - y_n| = r$. Therefore, $|x_n - y| \leqslant |x_n - y_n| + |y_n - y| < 2r$, so the sequence (x_n) is bounded. Hence, it has a limit point $x \in X$. By passing to subsequences of (x_n) and (y_n), if necessary, we may assume that $\lim x_n = x$. Let $\varepsilon > 0$. For n large, we have

$$|x - y| \leqslant |x - x_n| + |x_n - y_n| + |y_n - y| \leqslant r + 2\varepsilon$$

and

$$r = |x_n - y_n| \leqslant |x_n - x| + |x - y| + |y - y_n| \leqslant |x - y| + 2\varepsilon.$$

Since ε is arbitrary, $|x - y| = r$. Hence, $y \in Y$ and Y is closed.

Solution to 4.1.14: Let A be an infinite closed subset of $\mathbb{R}^n$. For $k = 1, \ldots,$ let B_k be the family of open balls in $\mathbb{R}^n$ whose centers have rational coordinates and whose radii are $1/k$. Each family B_k is countable. For each ball $B \in B_k$ such that $B \cap A \neq \emptyset$, choose a point in $B \cap A$, and let A_k be the set of chosen points. Each A_k is a countable subset of A, so the set $A_\infty = \cup A_k$ is a countable subset of A. Since A is closed, the inclusion $\overline{A_\infty} \subset A$ is obvious. Suppose $a \in A$ and fix a positive integer k. Then a lies in some ball $B \in B_k$, and this ball B must then contain a point of A_k, hence of A_∞. Thus, some point of A_∞ lies within a distance of $2/k$ of a. Since k is arbitrary, it follows that $a \in \overline{A_\infty}$ and, thus, that $A \subset \overline{A_\infty}$.

Solution to 4.1.15: For $k = 1, 2, \ldots,$ let B_k be the closed ball in $\mathbb{R}^n$ with center at 0 and radius k. Each compact of $\mathbb{R}^n$ is contained in some B_k. As each B_k is compact, it is covered by finitely many U_j's, by the Heine–Borel Theorem, [Rud87, p. 30]. Let j_k be the smallest index such that B_k is covered by $U_1, \ldots, U_{j_k}$. Define V_j by setting $V_j = U_j \setminus B_k$ for $j_k + 1 \leqslant j \leqslant j_{k+1}$ (if the indices j_k are all equal from some point on, set $V_j = \emptyset$ for j larger than their ultimate value.) The sets $V_1, V_2, \ldots$ have the required property.

Solution to 4.1.16: If K is not bounded, then the function $x \mapsto \|x\|$ is not bounded on K. If K is bounded but not compact, then it is not closed, by the Heine–Borel Theorem, [Rud87, p. 40], [MH93, p. 155]; therefore, there exists $\xi \in \overline{K} \setminus K$. In this case, the function $x \mapsto \|x - \xi\|^{-1}$ is not bounded on K.

Solution to 4.1.17: 1. Suppose not. Then there is a positive number δ and a subsequence of (x_i), (y_n), such that

$$|y_n - x| \geqslant \delta.$$

As A is compact, by the Bolzano–Weierstrass Theorem [Rud87, p. 40], [MH93, p. 153], (y_n) has a convergent subsequence, which, by hypothesis, converges to x, contradicting the inequality.
2. Let $A = \mathbb{R}$ and consider

$$x_i = \begin{cases} i & \text{if } i \text{ is odd} \\ 1/i & \text{if } i \text{ is even.} \end{cases}$$

All the convergent subsequences converge to zero, but (x_i) diverges.

Solution to 4.1.18: Let $\varepsilon > 0$. As f is uniformly continuous on X, there is a $\delta > 0$ such that

$$|f(x) - f(y)| < \varepsilon \leqslant \varepsilon + M_1|x - y|$$

for $|x - y| < \delta$ and any $M_1 \geq 0$.

Assume $|x - y| \geq \delta$. As the function f is bounded, there is an $M_2 > 0$ with $|f(x) - f(y)| \leq M_2$ for all x and y. Let $M_1 = M_2/\delta$. We have

$$|f(x) - f(y)| \leq \delta M_1 \leq M_1|x - y| \leq M_1|x - y| + \varepsilon$$

for all $x, y \in X$.

Solution to 4.1.19: Let

$$S = \bigcup_\alpha S_\alpha = A \cup B$$

where A and B are open. The origin belongs to A or to B. Without loss of generality, assume $O \in A$. For every α, we have

$$S_\alpha = (S_\alpha \cap A) \cup (S_\alpha \cap B)$$

so, as S_α is connected and $O \in S_\alpha \cap A$, we get $S_\alpha \cap B = \emptyset$. Therefore,

$$S \cap B = \bigcup_\alpha (S_\alpha \cap B) = \emptyset$$

and S is connected.

Solution to 4.1.20: Proof of 1 $\Rightarrow$ 2: Suppose that $G(f)$ is disconnected. Let A, B be disjoint non-empty open subsets of $G(f)$ with $A \cup B = G(f)$. Let $A_0 = \{x \in \mathbb{R}^n \mid (x, f(x)) \in A\}$, $B_0 = \{x \in \mathbb{R}^n \mid (x, f(x)) \in B\}$. If $x \in A_0$ then $(x, f(x)) \in A$. Since A is open in $G(f)$ there exist neighborhoods U of x, and V of $f(x)$, with $(U \times V) \cap G(f) \subset A$. Since f is continuous at x, we can find a neighborhood U' of x such that $f(U') \subset V$. Then $U \cap U'$ is a neighborhood of x, and for any $z \in U \cap U'$, $(z, f(z)) \in (U \times V) \cap G(f) \subset A$, and hence A_0 contains $U \cap U'$. This proves that A_0 is open in $\mathbb{R}^n$. Similarly, B_0 is an open set. Clearly A_0 and B_0 are non-empty, $A_0 \cap B_0 = \emptyset$ and $A_0 \cup B_0 = \mathbb{R}^n$. This is a contradiction since $\mathbb{R}^n$ is connected.

Counterexample to 2 $\Rightarrow$ 1: Take $n = 1$, and let $S_1 = \{(x, \sin(1/x)) \mid x < 0\}$, $S_2 = \{(x, \sin(1/x)) \mid x > 0\}$. S_1 is the image of the continuous map $(-\infty, 0) \to \mathbb{R} \times \mathbb{R}, x \mapsto (x, \sin(1/x))$ and therefore it is a connected subset of $\mathbb{R} \times \mathbb{R}$. Similarly S_2 is a connected subset of $\mathbb{R} \times \mathbb{R}$. Since $(0, 0)$ belongs to the closure in $\mathbb{R} \times \mathbb{R}$ of both S_1 and S_2, the sets $S_1 \cup \{(0, 0)\}$ and $S_2 \cup \{(0, 0)\}$ are connected. Since these sets have a point in common, their union $(S_1 \cup \{(0, 0)\}) \cup (S_2 \cup \{(0, 0)\})$ is connected. This union is the graph of the function $f : \mathbb{R} \to \mathbb{R}$ defined by $f(x) = \sin(1/x), x \neq 0, f(0) = 0$, which is not continuous at $x = 0$.

Solution to 4.1.21: Let S be a countable dense subset of U, for example, the set of points in U with rational coordinates. Let f be the function that equals 0 off S and, at any point x in S, equals the distance from x to $\mathbb{R}^n \backslash U$. Then f has the required property.

Solution to 4.1.22: The assertion is true. Let S be such a set and a a point in S. Define the set $S_a = \{b \in S \mid a \text{ and } b \text{ are connected by a path in } S\}$. S_a is open (because S is locally path connected) as well as its complement in S, so these two sets make up a partition of S, which is connected, therefore, the partition is trivial and S_a is the whole S.

Solution to 4.1.24: As X is compact, there is a constant $B > 0$ such that $B \geqslant |a_{ij}|$ for all matrices $A = (a_{ij})$ in X. This bound, together with the equation $Ax = \lambda x$ give us the bound $|\lambda| \leqslant \sum_{i,j} |a_{ij}| \leqslant n^2 B$, so S is bounded.

Let (λ_i) be a sequence in S converging to μ. For each λ_i there is an element M_i of X such that λ_i is an eigenvalue of M_i. The sequence (M_i) has a convergent subsequence, by compacity, so we may assume it converges to $M \in X$. The set of singular matrices in $M_{n \times n}$ is closed, since it is the inverse image of $\{0\}$ under the continuous map $\det : M_{n \times n} \to \mathbb{R}$, therefore $\lim_{i \to \infty}(M_i - \lambda_i I)$ is singular. As this limit equals $M - \lambda I$ we conclude that $\lambda \in S$, so S is also closed, therefore it is compact.

Solution to 4.1.25: 1. P^2 is the quotient of the sphere S^2 by the equivalence relation that identifies two antipode points x and $-x$. If $\pi : S^2 \to P^2$ is the natural projection which associates each point $x \in S^2$ to its equivalence class $\pi(x) = \{x, -x\} \in P^2$, the natural topology is the quotient topology; that is, $A \subset P^2$ is open if and only if $\pi^{-1}(A) \subset S^2$ is open. With this topology, the projection π is a continuous function and $P^2 = \pi(S^2)$ is compact, being the image of a compact by a continuous function.

Another topology frequently referred to as the *usual* topology of P^2 is the one defined by the metric

$$d(x, y) = \min\{|x - y|, |x + y|\}.$$

It is a straightforward verification that the function d above satisfies all axioms of a metric. We will show now that it defines the same topology as the one above, on the space that we will call (P^2, d).

The application $\pi : S^2 \to P^2$ with the metric as above satisfies the inequality

$$d(\pi(x), \pi(y)) \leqslant |x - y|$$

so d is continuous. This defines a function $\overline{\pi}$ on the quotient which is

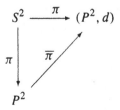

the identity and then continuous. Since P^2 is compact and (P^2, d) Hausdorff, $\overline{\pi}$ is a homeomorphism and the two topologies in P^2 are equivalent.

Now $S\mathbb{O}(3)$ is the group of orthogonal transformations of $\mathbb{R}^3$ with determinant 1, so every matrix in this set of satisfies

$$X \cdot X^t = \begin{pmatrix} 1 & 0 & 0 \\ 0 & 1 & 0 \\ 0 & 0 & 1 \end{pmatrix}$$

therefore, $\sum_{k=1}^{3} X_{ik}^2 = 1$, for $i = 1, 2, 3$, implying that $S\mathbb{O}(3)$ is bounded. Consider now the transformation

$$\begin{aligned} f : M_{3\times 3} \approx \mathbb{R}^9 & \rightarrow & M_{3\times 3} \times \mathbb{R} \\ X & \rightarrow & (X^t X, \det X). \end{aligned}$$

f is continuous and $S\mathbb{O}(3) = f^{-1}(I, 1)$, that is, the inverse image of a closed set, then itself a closed set, showing that $S\mathbb{O}(3)$ is compact. Another way to see this is to observe that the function

$$X \rightarrow \sqrt{\operatorname{tr}(X^t X)}$$

is a norm on the space of matrices $M_{n\times n} \equiv \mathbb{R}^{n^2}$ and that for matrices in the orthogonal group $\operatorname{tr}(X^t X) = n$, so $\mathbb{O}(n)$ and, consequentially, $S\mathbb{O}(n)$ are compact.
2. To see the homeomorphism between P^2 and Q, first define the application $\varphi : P^2 \rightarrow Q$ given by the following construction: For each line x through the origin, take $\varphi_x : \mathbb{R}^3 \rightarrow \mathbb{R}^3$ as the rotation of $180°$ around the x axis. This is well defined and continuous. To see that it is surjective, notice that every orthogonal matrix in dim 3 is equivalent to one of the form

$$\begin{pmatrix} 1 & 0 & 0 \\ 0 & \cos\theta & \sin\theta \\ 0 & -\sin\theta & \cos\theta \end{pmatrix}$$

and with the additional condition of symmetry $\theta = \pi$, which is a rotation of $180°$ around an axis. For more details see the Solution to Problem 7.4.24. Since φ is continuous and injective on a compact, it is an homeomorphism.

Solution to 4.1.27: Convergence in $M_{n\times n}$ is entrywise convergence. In other words, the sequence (A_k) in $M_{n\times n}$ converges to the matrix A if and only if, for each i and j, the $(i, j)^{th}$ entry of A_k converges to the $(i, j)^{th}$ entry of A. It follows that the operator of multiplication in $M_{n\times n}$ is continuous; in other words, if $A_k \rightarrow A$ and $B_k \rightarrow B$, then $A_k B_k \rightarrow AB$. Now suppose (A_k) is a sequence of nilpotent matrices in $M_{n\times n}$ and assume $A_k \rightarrow A$. Then $A_k^n \rightarrow A^n$ by the continuity of multiplication. But $A_k^n = 0$ for each k since A_k is nilpotent. Hence, $A^n = 0$, that is, A is nilpotent. As a subset of a metric space is closed exactly when it contains all its limit points, the conclusion follows.

Solution to 4.1.28: For real l let $T_l = \{t \in \mathbb{R} \mid S \cap (-\infty, t) \text{ is countable}\}$. If (t_n) is sequence in T_l with limit $t \in \mathbb{R}$, then

$$S \cap (-\infty, t) \subset \bigcup_{i=1}^{\infty} (S \cap (-\infty, t_i)).$$

As a subset of a countable union of countable sets is countable we conclude that $t \in T_l$, so T_l is closed. If $S \cap (-\infty, n)$ were countable for every integer n, then S would be countable, a contradiction. Therefore $T_l \neq \mathbb{R}$.

Similarly, $T_r = \{t \in \mathbb{R} \mid S \cap (r, \infty) \text{ is countable}\}$ is closed and $T_R \neq \mathbb{R}$. We have $T_l \cap T_r = \emptyset$ since if $t \in T_l \cap T_r$ then $S \subset (S \cap (-\infty, t)) \cup (S \cap (r, \infty)) \cup \{t\}$ would be countable. If $T_l \cup T_r = \mathbb{R}$, then we would have $\mathbb{R}$ as a union of disjoint proper subsets, contradicting the connectivity of $\mathbb{R}$. Hence there exists $t \in \mathbb{R} \setminus T_l \cup T_r$. For such a t $S \cap (-\infty, t)$ and $S \cap (r, \infty)$ are uncountable.

4.2   General Theory

Solution to 4.2.1: We'll prove that $\rho(g(y), g_n(y)) \to 0$ uniformly as $n \to \infty$. Since a uniformly continuous map preserves uniform convergence, and g is uniformly continuous, it will suffice to show that $\sigma(y, f(g_n(y)) \to 0$ uniformly as $n \to \infty$. But

$$\sigma(y, f(g_n(y))) = \sigma(f_n(g_n(y)), f(g_n(y))) ,$$

so the desired conclusion follows from the hypothesis that $f_n \to f$ uniformly as $n \to \infty$.

Solution to 4.2.2: As the E_n's are non empty we can form a sequence (x_n) satisfying $x_n \in E_n$ for all positive integers n. Given $\varepsilon > 0$, there is an integer $n(\varepsilon)$ such that diam $E_n < \varepsilon$ for $n \geq n(\varepsilon)$. Thus, $d(x_n, x_m) < \varepsilon$ for $n, m \geq n(\varepsilon)$. This sequence is Cauchy, so it converges, to x, say. As E_n is closed and contains the sequence (x_n) we must have $x \in E_n$ for all n, i.e., $x \in \bigcap_{n=1}^{\infty} E_n$.

Solution to 4.2.3: Suppose that p is in the closure of A. Define the function $f(x) = -d(x, p)$ on X. By assumption, the restriction $f|_A$ attains a maximum on A. But since p is in the closure of A, $f|_A$ is nonpositive while attaining values arbitrarily close to 0, so the maximum value can only be 0; hence $p \in A$. Thus A is a closed subset of a compact space, so A is compact.

Solution to 4.2.4: Let $\mathcal{U}$ be an open cover of C. Then there is a set U_0 in $\mathcal{U}$ that contains x_0. Since $\lim_{n \to \infty} x_n = x_0$, there is an n_0 such that x_n is in U_0 for all $n > n_0$. For each $n \leq n_0$ there is a set U_n in $\mathcal{U}$ that contains x_n. The subfamily $\{U_0, U_1, \ldots, U_{n_0}\}$ is then a finite subcover of C, proving, by the Heine–Borel Theorem [Rud87, p. 30], that C is compact.

Solution to 4.2.5: Let X be a compact metric space. For each $n \in \mathbb{N}$, consider a cover of X by balls with radius $1/n$, $\mathcal{B}(1/n) = \{B_\alpha(x_\alpha, 1/n) \mid x_\alpha \in X\}$. As X is compact, a finite subcollection of $\mathcal{B}(1/n)$, $\mathcal{B}'(1/n)$, covers X, by the Heine–Borel Theorem [Rud87, p. 30]. Let A be the set consisting of the centers of the balls in $\mathcal{B}'(1/n)$, $n \in \mathbb{N}$. A is a countable union of finite sets, so it is countable. It is also clearly dense in X.

Solution to 4.2.6: Suppose $x \notin f(X)$. As $f(X)$ is closed, there exists a positive number ε such that $d(x, f(X)) \geqslant \varepsilon$.

As X is compact, using the Bolzano–Weierstrass Theorem [Rud87, p. 40], [MH93, p. 153], the sequence of iterates $(f^n(x))$ has a convergent subsequence, $(f^{n_i}(x))$, say. For $i < j$, we have

$$d\left(f^{n_i}(x), f^{n_j}(x)\right) = d\left(x, f^{n_j - n_i}(x)\right) \geqslant \varepsilon$$

which contradicts the fact that every convergent sequence in X is a Cauchy sequence, and the conclusion follows.

Solution to 4.2.7: For $x \in C$ we clearly have $f(x) = 0$. Conversely, if $f(x) = 0$, then there is a sequence (y_n) in C with $d(x, y_n) \to 0$. As C is closed, we have $x \in C$.

Given $x, z \in M$ and $y \in C$, we have, by the Triangle Inequality [MH87, p. 20],

$$d(x, y) \leqslant d(x, z) + d(z, y).$$

Taking the infimum of both sides over $y \in C$, we get

$$f(x) \leqslant d(x, z) + f(z)$$

or

$$f(x) - f(z) \leqslant d(x, z),$$

and, by symmetry,

$$f(z) - f(x) \leqslant d(x, z).$$

Therefore,

$$|f(x) - f(z)| \leqslant d(x, z)$$

and f is continuous.

Solution to 4.2.8: $\|f\| \geqslant 0$ for all f in $C^{1/3}$ is clear. If $f \equiv 0$, it is obvious that $\|f\| = 0$. Conversely, suppose that $\|f\| = 0$. Then, for all $x \neq 0$, we have

$$0 \leqslant \frac{|f(x) - f(0)|}{|x - 0|} \leqslant \|f\| = 0.$$

Since $f(0) = 0$, this implies $f(x) = 0$ for all x. Let $f, g \in C^{1/3}$ and $\varepsilon > 0$. There exists $x \neq y$ such that

$$
\begin{aligned}
\|f + g\| &\leqslant \frac{|(f + g)(x) - (f + g)(y)|}{|x - y|^{1/3}} + \varepsilon \\
&\leqslant \frac{|f(x) - f(y)|}{|x - y|^{1/3}} + \frac{|g(x) - g(y)|}{|x - y|^{1/3}} + \varepsilon \\
&\leqslant \|f\| + \|g\| + \varepsilon.
\end{aligned}
$$

Since ε was arbitrary, the Triangle Inequality [MH87, p. 20] holds.

The property $\|cf\| = |c|\|f\|$ for $f \in C^{1/3}$ and $c \in \mathbb{R}$ is clear.

Let $\{f_n\}$ be a Cauchy sequence in $C^{1/3}$. By the definition of the norm, for all $x \in [0, 1]$ and any $\varepsilon > 0$ there is an $N > 0$ such that if $n, m > N$, we have

$$|(f_n - f_m)(x) - (f_n - f_m)(0)| \leqslant |x - 0|^{1/3}\varepsilon$$

or

$$|f_n(x) - f_m(x)| \leqslant \varepsilon.$$

Hence, the sequence $\{f_n\}$ is uniformly Cauchy. A similar calculation shows that functions in $C^{1/3}$ are continuous. Since the space of continuous functions on $[0, 1]$ is complete with respect to uniform convergence, there exists a continuous function f such that the f_n's converge to f uniformly. Suppose $f \notin C^{1/3}$. Then, for any $M > 0$, there exist $x \neq y$ such that

$$\frac{|f(x) - f(y)|}{|x - y|^{1/3}} > M.$$

So

$$\frac{|f(x) - f_n(x)|}{|x - y|^{1/3}} + \frac{|f_n(x) - f_n(y)|}{|x - y|^{1/3}} + \frac{|f(y) - f_n(y)|}{|x - y|^{1/3}} > M.$$

Since the f_n's converge to f uniformly, for fixed x and y we can make the first and third terms as small as desired. Hence, $\|f_n\| > M$ for all M and n sufficiently large, contradicting the fact that $f_n \in C^{1/3}$ and that, since the f_n's are Cauchy, their norms are uniformly bounded.

Suppose now that the sequence $\{f_n\}$ does not converge to f in $C^{1/3}$. Then there is an $\varepsilon > 0$ such that $\|f_n - f\| > \varepsilon$ for infinitely many n's. But then there exist $x \neq y$ with

$$\frac{|f_n(x) - f(x)|}{|x - y|^{1/3}} + \frac{|f_n(y) - f(y)|}{|x - y|^{1/3}} > \varepsilon$$

for those n's. But, as we have uniform convergence, we can make the left-hand side as small as desired for fixed x and y, a contradiction.

Solution to 4.2.9: We first show that $\{g_n\}$ is equicontinuous. Fix $\varepsilon > 0$. By the equicontinuity of the sequence $\{f_n\}$, there is a $\delta > 0$ such that

$$|f_n(x) - f_n(y)| < \varepsilon$$

for all n, whenever $d(x, y) < \varepsilon$. Fix n and fix x and y with $d(x, y) < \delta$. Let $j, k \leqslant n$ be such that $g_n(x) = f_j(x)$ and $g_n(y) = f_k(y)$. Then

$$g_n(x) - g_n(y) = f_j(x) - f_k(y) = f_j(x) - f_j(y) + f_i(y) - f_k(y)$$
$$\leqslant f_j(x) - f_j(y) < \varepsilon.$$

By the same reasoning, $g_n(y) - g_n(x) < \varepsilon$, from which follows the equicontinuity of the sequence $\{g_n\}$.

The sequence $\{g_n\}$ is clearly uniformly bounded, so, by the Arzelà-Ascoli Theorem, it has a uniformly convergent subsequence $\{g_{n_k}\}$, with limit g, say. Since the sequence $\{g_n\}$ is nondecreasing, for $n > n_k$, we have

$$g_{n_k} \leqslant g_n \leqslant g,$$

implying that $g_n \to g$ uniformly.

Solution to 4.2.11: Since $f(K) \subset f(K_n)$ for all n, the inclusion $f(K) \subset \bigcap_1^\infty f(K_n)$ is clear. Let y be a point in $\bigcap_1^\infty f(K_n)$. Then, for each n, the set $f^{-1}(\{y\}) \cap K_n$ is nonempty and compact (the latter because it is a closed subset of the compact set K_n). Also, $f^{-1}(\{y\}) \cap K_{n+1} \subset f^{-1}(\{y\}) \cap K_n$. Hence, by the Nested Set Property [MH93, p. 157], the set

$$\bigcap_1^\infty \left(f^{-1}(\{y\}) \cap K_n \right) = f^{-1}(\{y\}) \cap K$$

is nonempty; that is, $y \in f(K)$.

Solution to 4.2.12: 1. The completeness of X_1 implies the completeness of X_2. In fact, assume X_1 is complete, and let (y_n) be a Cauchy sequence in X_2. The conditions on f imply that it is one-to-one, so each y_n can be written uniquely as $f(x_n)$ with x_n in X_1. Then $d_1(x_m, x_n) \leqslant d_2(y_m, y_n)$, implying that (x_n) is a Cauchy sequence, hence convergent, say to x. Since f is continuous, we then have $\lim y_n = f(x)$, proving that X_2 is complete.
2. The completeness of X_2 does not imply the completeness of X_1. For an example, take $X_1 = (-\frac{\pi}{2}, \frac{\pi}{2})$, $X_2 = \mathbb{R}$, and $f(x) = \tan x$. Since $f'(x) = \sec^2 x \geqslant 1$ on X_1, the condition $|x - y| \leqslant |f(x) - f(y)|$ holds.

4.3   Fixed Point Theorem

Solution to 4.3.1: The map is the image, by a contraction, of a complete metric space (California!). The result is a consequence of the Fixed Point Theorem [Rud87, p. 220].

Solution to 4.3.2: Let $g(x) = (1 + x)^{-1}$. We have

$$g'(x) = \frac{-1}{(1 + x)^2}$$

therefore,

$$|g'(x)| \leqslant \frac{1}{(1 + x_0/2)^2} < 1 \quad \text{for} \quad x > x_0.$$

Then, by the Fixed Point Theorem [Rud87, p. 220], the sequence given by

$$x_0 > 0, \qquad x_{n+1} = g(x_n)$$

converges to the unique fixed point of g in $[x_0, \infty)$. Solving $g(x) = x$ in that domain gives us the limit

$$\frac{-1 + \sqrt{5}}{2}.$$

Solution to 4.3.3: Let $S = \{x \in [0, \infty) \,|\, x - f(x) \leqslant 0\}$. S is not empty because $0 \in S$; also, every element of S is less than 100, so S has a supremum, x_0, say. For any $\varepsilon > 0$, there exists an element of S, x, with

$$x \leqslant x_0 < x + \varepsilon$$

so

$$x_0 - f(x_0) \leqslant x_0 - f(x) < x + \varepsilon - f(x) \leqslant \varepsilon$$

and we conclude, since ε is arbitrary, that $x_0 \leqslant f(x_0)$.

Suppose $f(x_0) - x_0 = \delta > 0$. Then, for some $x \in S$, we have $x \leqslant x_0 < x_0 + \delta$; therefore,

$$x \leqslant x_0 < f(x_0) - x_0 + x + \delta$$

and we get

$$x \leqslant x_0 < f(x_0) - x_0 + x$$

from which follows, since f is an increasing function,

$$f(x_0) - x_0 + x \leqslant f(x_0) \leqslant f(f(x_0) - x_0 + x)$$

but then $x_0 < f(x_0) - x_0 + x \in S$, which contradicts the definition of x_0. We must then have $f(x_0) = x_0$.

Solution to 4.3.4: Consider $F : K \rightarrow \mathbb{R}$ defined by $F(x) = d(x, \varphi(x))$. φ, being a contraction, is continuous, and so is F. Since K is compact and nonempty, F attains its minimum ε at a point $m \in K$, $d(m, \varphi(m)) = \varepsilon$. From the minimality of ε, it follows that $d(\varphi(m), \varphi(\varphi(m))) = F(\varphi(m)) \geqslant \varepsilon = d(m, \varphi(m))$. The contractiveness assumption implies that $m = \varphi(m)$.

Suppose $n \in K$ also satisfies $n = \varphi(n)$. Then $d(\varphi(n), \varphi(m)) = d(n, m)$, which, by the contractiveness assumption, implies $n = m$.

Solution to 4.3.5: As the unit square is compact, $\max |K(x, y)| = M < 1$. Consider the map $T : C_{[0,1]} \rightarrow C_{[0,1]}$ defined by

$$T(f)(x) = e^{x^2} - \int_0^1 K(x, y) f(y) dy.$$

We have

$$|T(f)(x) - T(g)(x)| = \left| \int_0^1 K(x, y)(g(y) - f(y)) dy \right|$$

$$\leqslant \int_0^1 M|g(y) - f(y)|dy$$

$$\leqslant M \max_{0 \leqslant y \leqslant 1} |g(y) - f(y)|$$

$$= M\|g - f\|.$$

By the Contraction Mapping Principle [MH93, p. 275], $\hat{T}$ has a unique fixed point, $h \in \mathcal{C}_{[0,1]}$. We have

$$h(x) + \int_0^1 K(x, y)h(y)dy = e^{x^2}.$$

Any such a solution is a fixed point of T, so it must equal h.

Solution to 4.3.6: Consider the map $T : \mathcal{C}_{[0,1]} \rightarrow \mathcal{C}_{[0,1]}$ defined by

$$T(f) = g(x) + \int_0^x f(x - t)e^{-t^2} dt.$$

Given $f, h \in \mathcal{C}_{[0,1]}$, we have

$$\|T(f) - T(h)\|_\infty \leqslant \sup_{x \in [0,1]} \int_0^x |f(x - t) - h(x - t)|e^{-t^2} dt$$

$$\leqslant \|f - h\|_\infty \sup_{x \in [0,1]} \int_0^x e^{-t^2} dt$$

$$= \|f - h\|_\infty \int_0^1 e^{-t^2} dt < \|f - h\|_\infty$$

so T is a contraction. Since $\mathcal{C}_{[0,1]}$ is a complete metric space, by the Contraction Mapping Principle [MH93, p. 275] there is $f \in \mathcal{C}_{[0,1]}$ such that $T(f) = f$, as desired.

Solution to 4.3.7: Define the operator T on $\mathcal{C}_{[0,1]}$ by

$$T(f)(x) = \sin x + \int_0^1 \frac{f(y)}{e^{x+y+1}} dy.$$

Let $f, g \in \mathcal{C}_{[0,1]}$. We have

$$\|T(f) - T(g)\| \leqslant \sup_x \left\{ \int_0^1 \frac{|f(y) - g(y)|}{e^{x+y+1}} dy \right\}$$

$$\leqslant \sup_x \{|f(x) - g(x)|e^{-x}\} \int_0^1 \frac{dy}{e^{y+1}}$$

$$\leqslant \|f - g\| \left(\frac{1}{e} - \frac{1}{e^2} \right)$$

$$\leqslant \lambda \|f - g\|,$$

where $0 < \lambda < 1$ is a constant. Hence, T is a strict contraction. Therefore, by the Contraction Mapping Principle [MH93, p. 275], there is a unique $f \in C_{[0,1]}$ with $T(f) = f$.

Solution to 4.3.8: Since M is a complete metric space and S^2 is a strict contraction, by the Contraction Mapping Principle [MH93, p. 275] there is a unique point $x \in M$ such that $S^2(x) = x$. Let $S(x) = y$. Then $S^2(y) = S^3(x) = S(x) = y$. Hence, y is a fixed point of S^2, so $x = y$. Any fixed point of S is a fixed point S^2, so S has a unique fixed point.

5
Complex Analysis

5.1 Complex Numbers

Solution to 5.1.1: We have

$$1 = e^{2k\pi i} \quad \text{for} \quad k \in \mathbb{Z};$$

therefore,

$$
\begin{aligned}
1^{\frac{1}{3}+i} &= e^{\left(\frac{1}{3}+i\right)\log 1} = e^{\left(\frac{1}{3}+i\right)2k\pi i} \\
&= e^{-2k\pi + i\frac{2k\pi}{3}} \\
&= e^{-2k\pi}\left(\cos\frac{2k\pi}{3} + i\sin\frac{2k\pi}{3}\right) \quad (k \in \mathbb{Z}).
\end{aligned}
$$

Solution to 5.1.2: We have $i^i = e^{i\log i}$ and $\log i = \log|i| + i\arg i = i(\pi/2 + 2k\pi)$, $k \in \mathbb{Z}$. So the values of i^i are $\{e^{-(\pi/2+2k\pi)} \mid k \in \mathbb{Z}\}$.

Solution to 5.1.3: Let A, B, C be the points a, b, c in the Argand diagram. We suppose that a, b, c are noncollinear.

To prove necessity, suppose ABC is equilateral. To be definite suppose ABC has counterclockwise orientation. Then $c - a = (b - a)e^{i\pi/3}$ and $a - b = (c - b)e^{i\pi/3}$. Therefore $(c - a)(c - b) = (a - b)(b - a)$ which is equivalent to the condition $a^2 + b^2 + c^2 = bc + ca + ab$.

Conversely, suppose $a^2 + b^2 + c^2 = bc + ca + ab$. Then $(c - a)(c - b) = (a - b)(b - a)$, and

$$\frac{c - a}{b - a} = \frac{a - b}{c - b} = re^{i\theta} \text{ say.}$$

Hence

$$\frac{AC}{AB} = \frac{BA}{BC} = r$$

and

$$\angle BAC = \angle CBA = \pm\theta \quad (\text{mod } 2\pi).$$

From the above equations we get $AC = BC$. Therefore ABC is an equilateral triangle.

Now consider the case that a, b, c are collinear. The condition $a^2 + b^2 + c^2 = bc + ca + ab$ is equivalent to $(c - a)(c - b) = (a - b)(b - a)$. The latter condition is clearly invariant under translation and rotation about the origin. Thus we may assume that a, b, c are real and that $a = 0$. The latter condition then reduces to $c^2 + b^2 = bc$, which is satisfied only by the real numbers $c = b = 0$, as can be seen by using the quadratic formula to solve for c. Thus in this case the given condition is equivalent to a, b, c forming a degenerate equilateral triangle.

Solution to 5.1.4: Multiplying by a unimodular constant, if necessary, we can assume $c = 1$. Then $\Im a + \Im b = 0$. So $a = \bar{b}$. Their real part must be negative, since otherwise the real parts of a, b, and c would sum to a positive number. Therefore, there is θ such that $a = \cos\theta + i\sin\theta$ and $b = \cos\theta - i\sin\theta$, $\cos\theta = -1/2$. Then $\theta = 2\pi/3$ and we are done.

Solution 2. Since $b + c = -a$, we get $1 = \|b + c\|^2 = 2 + 2\Re b\bar{c}$, thus $2\Re b\bar{c} = -1$, hence $\|b - c\|^2 = 2 - 2\Re b\bar{c} = 3$. The same calculation gives $\|a - b\|^2 = 3$ and $\|a - c\|^2 = 3$.

Solution to 5.1.5: 1. We have

$$P_{n-1}(x) = \frac{x^n - 1}{x - 1}$$
$$= x^{n-1} + \cdots + 1$$

for $x \neq 1$, so $P_{n-1}(1) = n$.

2. Let

$$p_k = e^{\frac{2\pi i(k-1)}{n}} \quad \text{for } k = 1, \ldots, n$$

be the nth roots of 1. As $p_1 = 1$, we have

$$\prod_{i=2}^{n} (z - p_k) = P_{n-1}(z).$$

Letting $z = 1$, and using Part 1, we get the desired result.

Solution to 5.1.6: Let $q = e^{iz}$, so $2\cos z = q + q^{-1}$, and $2\cos nz = q^n + q^{-n}$. The problem reduces to find the polynomial such that $T_N(q + q^{-1}) = q^n + q^{-n}$. Now

$$(q + q^{-1})^n = \sum_{k=0}^{n} \binom{n}{k} q^{2k-n} = q^n + q^{-n} + \sum_{\substack{0 < j < n \\ n-j \, \text{even}}} \binom{n}{(n-j)/2} (q^j + q^{-j}) +$$

$$+ \begin{cases} \binom{n}{n/2} & \text{if n is even,} \\ 0 & \text{otherwise.} \end{cases}$$

Now assuming that we know T_j for $j < n$ and proceeding by induction,

$$T_n(x) = x^n - \sum_{\substack{0 < j < n \\ n-j \, \text{even}}} \binom{n}{(n-j)/2} (T_j(x)) - \begin{cases} \binom{n}{n/2} & \text{if n is even,} \\ 0 & \text{otherwise.} \end{cases}$$

is the sought polynomial.

Solution to 5.1.7: Consider the complex plane divided into four quadrants by the lines $\Re z = \pm \Im z$, and let Δ_i be the set of indices j such that z_j lies in the ith quadrant. The union of the four sets Δ_i is $\{1, 2, \ldots, n\}$, so there is an i such that $\Delta = \Delta_i$ satisfies

$$\sum_{j \in \Delta} |z_j| \geq \frac{1}{4} \sum_{j=1}^{n} |z_j|.$$

Since multiplying all of the z_j by a unimodular constant will not affect this sum, we may assume that Δ is the quadrant in the right half-plane, where $\Re z_j > 0$ and $|z_j| \leq \sqrt{2} \Re z_j$. So we have

$$\left| \sum_{j \in \Delta} z_j \right| \geq \sum_{j \in \Delta} \Re z_j \geq \frac{1}{\sqrt{2}} \sum_{j \in \Delta} |z_j|.$$

Combining this with the previous inequality, we get the desired result.

Solution to 5.1.8: The functions $1, e^{2\pi ix}, \ldots, e^{2\pi inx}$ are orthonormal on $[0,1]$. Hence,

$$\int_0^1 \left| 1 - \sum_{k=1}^{n} a_k e^{2\pi ikx} \right|^2 dx = 1 + \sum_{k=1}^{n} |a_s|^2 \geq 1.$$

Since the integrand is continuous and nonnegative, it must be ≥ 1 at some point.

Solution 2. Since $\int_0^1 e^{2\pi ikx} dx = 0$ for $k \neq 0$, we have

$$1 = \int_0^1 \left(1 - \sum_{k=1}^{n} a_k e^{ikx} \right) dx \leq \int_0^1 \left| 1 - \sum_{k=1}^{n} a_k e^{ikx} \right| dx.$$

Now argue as above.

Solution to 5.1.9: We have

$$e^b - e^a = \int_a^b e^z \, dz$$

for all complex numbers a and b, where the integral is taken over any path connecting them. Suppose that a and b lie in the left half-plane. Then we can take a path also in the same half-plane, and for any z on this line, $|e^z| \leq 1$. Therefore, integrating along this line, we get

$$|e^b - e^a| \leq \int_a^b |e^z| \, |dz| \leq |b - a|.$$

Solution to 5.1.10: The boundary of $N(A, r)$ consists of a finite set of circular arcs C_i, each centered at a point a_i in A. The sectors S_i with base C_i and vertex a_i are disjoint, and their total area is $Lr/2$, where L is the length of the boundary. Since everything lies in a disc of radius 2, the total area is at most 4π, so $L \leq 8\pi/r$.

Solution to 5.1.11: Without loss of generality, suppose that

$$|\alpha_1| \leq |\alpha_2| \leq \cdots \leq |\alpha_l| < |\alpha_{l+1}| = \cdots = |\alpha_k|$$

that is, exactly $k - l$ of the α's with maximum modulus (l may be zero.)

We will first show that $|\alpha_k| = \sup_j |\alpha_j|$ is an upper bound for the expression, and then prove that a subsequence gets arbitrarily close to this value. We have

$$\left| \sum_{j=1}^k \alpha_j^n \right|^{1/n} \leq \left(\sum_{j=1}^k |\alpha_j|^n \right)^{1/n} \leq \left(k \, |\alpha_k|^n \right)^{1/n} = k^{1/n} \, |\alpha_k|$$

the limit on the right exists and is $|\alpha_k| = \sup_j |\alpha_j|$, so

$$\limsup_n \left| \sum_{j=1}^k \alpha_j^n \right|^{1/n} \leq \sup_j |\alpha_j| \, .$$

Now dividing the whole expression by α_k^n we get

$$\sum_{j=1}^k \alpha_j^n = \alpha_k^n \sum_{j=1}^k \left(\frac{\alpha_j}{\alpha_k} \right)^n = \alpha_k^n \left(\sum_{j=1}^l \left(\frac{\alpha_j}{\alpha_k} \right)^n + \sum_{j=l+1}^k e^{in\theta_j} \right)$$

since the last $k - l$ terms all have absolute value 1.

It suffices to show that

$$(*) \qquad \limsup_{n} \left| \sum_{j=1}^{l} \left(\frac{\alpha_j}{\alpha_k} \right)^n + \sum_{j=l+1}^{k} e^{in\theta_j} \right| = k - l$$

since

$$\left| \sum_{j=1}^{k} \alpha_j^n \right|^{1/n} = |\alpha_k| \left| \sum_{j=1}^{l} \left(\frac{\alpha_j}{\alpha_k} \right)^n + \sum_{j=l+1}^{k} e^{in\theta_j} \right|^{1/n}.$$

$(*)$ is a consequence of the the fact that orbits of irrational rotation on the circle are dense in the unit circle. To see that, discard the first term $\sum_{j=1}^{l} \left(\frac{\alpha_j}{\alpha_k} \right)^n$ because its limit exists and equals zero, being a finite sum of terms that converge to zero, and distribute the rest in two sums, one containing all rational angles (p_i/q_i), and another one containing all irrational angles (s_i). Without loss of generality we are left to prove that

$$\limsup_{n} \left| \sum_{j=l+1}^{k'} e^{in\frac{p_j}{q_j}} + \sum_{j=k'+1}^{k} e^{ins_j} \right| = k - l.$$

If the sequence contains only rational angles choose $P = 2 \prod_j q_j$, twice the product of the denominators of the rational angles. Then the sequence nP where $n \in \mathbb{N}$ will land all angles at zero and the summation is equal to $k - l$.

Now if there is at least one irrational angle among them the set of points

$$\left(e^{ni\frac{p_{l+1}}{q_{l+1}}}, \ldots, e^{ni\frac{p_{k'}}{q_{k'}}}, e^{nis_{k'+1}}, \ldots, e^{nis_k} \right) \qquad n \in \mathbb{N}$$

in the tori $S^1 \times S^1 \times \cdots \times S^1$ is infinite (the last coordinates will never repeat) and so has an accumulation point, that is, for any $\varepsilon > 0$ there are two iterates $m > n$ such that

$$\left| \left(e^{mi\frac{p_{l+1}}{q_{l+1}}}, \ldots, e^{mi\frac{p_{k'}}{q_{k'}}}, \ldots, e^{mis_k} \right) - \left(e^{ni\frac{p_{l+1}}{q_{l+1}}}, \ldots, e^{ni\frac{p_{k'}}{q_{k'}}}, \ldots, e^{nis_k} \right) \right| < \varepsilon$$

and then

$$\left(e^{(m-n)i\frac{p_{l+1}}{q_{l+1}}}, \ldots, e^{(m-n)i\frac{p_{k'}}{q_{k'}}}, e^{(m-n)is_{k'+1}}, \ldots, e^{(m-n)is_k} \right)$$

is ε-close to $(1, \ldots, 1)$, and we are done.

5.2   Series and Sequences of Functions

Solution to 5.2.1: Multiplying the first $N + 1$ factors we get

$$\frac{1 - z^{10}}{1 - z} \frac{1 - z^{100}}{1 - z^{10}} \frac{1 - z^{1000}}{1 - z^{100}} \cdots \frac{1 - z^{10N}}{1 - z^{10N-1}} = \frac{1 - z^{10N}}{1 - z}$$

so the product converges to $1/(1-z)$ as $N \to \infty$.

Solution to 5.2.2: From the recurrence relation, we see that the coefficients a_n grow, at most, at an exponential rate, so the series has a positive radius of convergence. Let f be the function it represents in its disc of convergence, and consider the polynomial $p(z) = 3 + 4z - z^2$. We have

$$p(z)f(z) = (3 + 4z - z^2) \sum_{n=0}^{\infty} a_n z^n$$

$$= 3a_0 + (3a_1 + 4a_0)z + \sum_{n=0}^{\infty} (3a_n + 4a_{n-1} - a_{n-2})z^n$$

$$= 3 + z.$$

So

$$f(z) = \frac{3 + z}{3 + 4z - z^2}.$$

The radius of convergence of the series is the distance from 0 to the closest singularity of f, which is the closest root of p. The roots of p are $2 \pm \sqrt{7}$. Hence, the radius of convergence is $\sqrt{7} - 2$.

Solution to 5.2.3: Let

$$f(z) = \sum_{n=0}^{\infty} a_n z^n$$

be the Maclaurin series of f. Substituting the equation we get:

$$a_0 = 0, \quad a_1 = 1, \quad a_2 = 0, \quad a_k = \frac{a_{k-3}}{k(k-1)} \quad (k \geqslant 3).$$

and by induction for $k \geqslant 1$, we get,

$$a_{3k} = \prod_{j=1}^{k} \frac{1}{3j(3j-1)}, \quad a_{3k+1} = \prod_{j=1}^{k} \frac{1}{3j(3j+1)} \quad a_{3k+2} = 0.$$

So dividing the sum up in two components and analyzing the convergence of each piece we see that $\lim_{k \to \infty} \frac{a_{3k+3}}{a_{3k}} = 0$, so the series $\sum a_{3k} z^{3k}$ has infinite radius of convergence. The series relative to the indexes $3k1 + 1$ can be treated in a similar way. We conclude then that the series for f has infinite radius of convergence, which proves the unicity of the solution to the given problem.

Solution to 5.2.4: Let $f(z) = \exp(z/(z - 2))$. The series can then be rewritten as $\sum_{n=1}^{\infty} \frac{1}{n^2} (f(z))^n$, so, by the standard theory of power series, it converges if and only if $|f(z)| \leqslant 1$. The preceding inequality holds when $\Re \frac{z}{z-2} \leqslant 0$, so the problem reduces to that of finding the region sent into the closed left half-plane by the linear fractional map $z \mapsto \frac{z}{z-2}$. The inverse of the preceding map is the map g defined by $g(z) = \frac{2z}{z-1}$. Since $g(0) = 0$ and $g(\infty) = 2$, the image of the imaginary axis under g is a circle passing through the points 0 and 2. As g sends the real axis onto itself, that circle must be orthogonal to the real axis, so it is the circle $|z - 1| = 1$. Thus, g sends the open left half-plane either to the interior or to the exterior of that circle. Since $g(-1) = 1$, the first possibility occurs. We can conclude that $|f(z)| \leqslant 1$ if and only if $|z - 1| \leqslant 1$ and $z \neq 2$, which is the region of convergence of the original series.

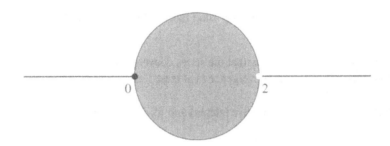

Solution to 5.2.5: The radius of convergence, R, of this power series is given by

$$\frac{1}{R} = \limsup_{n \to \infty} |a_n|^{1/n}.$$

For $|z| < 1$, we have

$$\sum_{n=1}^{\infty} n z^{n-1} = \left(\sum_{n=0}^{\infty} z^n \right)' = \left(\frac{1}{1-z} \right)' = \frac{1}{(1-z)^2}.$$

By the Identity Theorem [MH87, p. 397],

$$f(z) = \frac{1}{(1-z)^2}$$

where the right-hand side is analytic. Since this happens everywhere except at $z = 1$, the power series expansion of f centered at -2 will have a radius of convergence equal to the distance between -2 and 1. Hence, $R = 3$.

Solution to 5.2.6: As

$$1 - x^2 + x^4 - x^6 + \cdots = \frac{1}{1 + x^2}$$

which has singularities at $\pm i$, the radius of convergence of

$$\sum_{n=0}^{\infty} a_n (x - 3)^n$$

is the distance from 3 to $\pm i$, $|3 \mp i| = \sqrt{10}$. We then have

$$\limsup_{n \to \infty} \left(|a_n|^{\frac{1}{n}} \right) = \frac{1}{\sqrt{10}}.$$

Solution to 5.2.7: As $\lim_{n \to \infty} \sqrt[n]{n^2} = 1$, we have

$$\frac{1}{R} = \limsup_{n \to \infty} \sqrt[n]{|a_n|} = \limsup_{n \to \infty} \sqrt[n]{n^2 |a_n|}$$

so $\sum a_n z^n$ and $\sum n^2 a_n z^n$ have the same radius of convergence, and the conclusion follows.

Solution to 5.2.8: We'll show that the series converges for $|z| = 1, z \neq 1$, which implies that the radius of convergence is at least 1, therefore the series is convergent for $|z| < 1$.

Let $R_{n,p}$ (where p is a positive integer) and R_n be defined by

$$R_{n,p}(z) = \sum_{i=n+1}^{n+p} b_i z^i, \qquad R_n(z) = \sum_{i \geqslant n+1} b_i z^i.$$

We have,

$$(z-1)R_{n,p}(z) = -b_{n+1} z^{n+1} + \sum_{i=n+1}^{n+p-1} (b_i - b_{i+1}) z^{i+1} + b_{n+p} z^{n+p+1},$$

therefore,

$$|z-1||R_{n,p}(z)| \leqslant b_{n+1}|z|^{n+1} + \sum_{i=n+1}^{n+p-1} (b_i - b_{i+1})|z|^{i+1} + b_{n+p}|z|^{n+p+1}.$$

Letting $|z| = 1, z \neq 1$ in this inequality, we get,

$$|R_{n,p}(z)| \leqslant \frac{1}{|z-1|} \left(b_{n+1} + \sum_{i=n+1}^{n+p-1} (b_i - b_{i+1}) + b_{n+p} \right) = \frac{2b_{n+1}}{|z-1|}.$$

Since this holds for any positive p, we may let $p \to \infty$ and obtain,

$$|R_n(z)| = \left| \sum_{i \geqslant n+1} b_i z^i \right| \leqslant \frac{2b_{n+1}}{|z-1|} = o(1) \quad (n \to \infty),$$

as we wanted.

Nothing can be said about convergence at $z = 1$. The series $\sum \frac{z^i}{i}$ diverges at $z = 1$, while $\sum \frac{z^i}{i^2}$ converges at the same point.

Solution to 5.2.9: The series $\sum_{n=0}^{\infty} \frac{z^n}{n!}$ converges for all z.

If $z \neq 0$ then

$$\left| \frac{n^2}{z^n} \right|^{1/n} = \frac{(n^{1/n})^2}{|z|} \to \frac{1}{|z|}.$$

By Cauchy's Root Test, the series $\sum n^2/z^n$ converges for $|z| > 1$, and diverges for $|z| < 1$. Hence the series $\sum (z^n/n! + n^2/z^n)$ converges for $|z| > 1$, and diverges for $|z| < 1$; the series is undefined for $z = 0$.

Let $|z| = 1$. The sequence $z^n/n!$ tends to zero and the sequence n^2/z^n tends to ∞. Therefore the sequence $(z^n/n! + n^2/z^n)$ tends to ∞, which implies that the series $\sum(z^n/n! + n^2/z^n)$ is divergent for $|z| = 1$. Thus the given series converges exactly when $|z| > 1$.

Solution to 5.2.10: Fix $z \neq \pm i$. Then

$$\left| \frac{z}{(1 + z^2)^n} \right|^{1/n} = \frac{|z|^{1/n}}{|1 + z^2|} \to \frac{1}{|1 + z^2|}.$$

By Cauchy's Root Test the given series converges if $|1 + z^2| > 1$, that is, for points exterior to the lemniscate $|1 + z^2| = 1$.

Cauchy's test also shows that the given series is divergent for $|1 + z^2| < 1$. The series clearly converges for $z = 0$. If $|1 + z^2| = 1$, $z \neq 0$ then $|z|/|1 + z^2|^n = |z|$ which does not tend to 0 as $n \to \infty$. Thus the series diverges for points $z \neq 0$ on the lemniscate.

Thus the series is convergent precisely for points exterior to the lemniscate, together with the point $z = 0$ on the lemniscate, is undefined at $z = \pm i$, and diverges at all other points.

Solution to 5.2.11: Let R denote the radius of convergence of this power series.

$$R = \limsup_{n} |n^{\log n}|^{1/n} = \limsup_{n} e^{(\log n)^2/n} = e^0 = 1.$$

The series and all term by term derivatives converge absolutely on $|z| < 1$ and diverge for $|z| > 1$. Let $|z| = 1$. For $k \geq 0$ the kth derivative of the power series is

$$\sum_{n=k}^{\infty} n(n - 1) \cdots (n - k + 1) \frac{z^{n-k}}{n^{\log n}}.$$

To see that this converges absolutely, note that

$$\sum_{n=k}^{\infty} n(n - 1) \cdots (n - k + 1) \frac{1}{n^{\log n}} \leq \sum_{n=k}^{\infty} \frac{1}{n^{\log n - k}}.$$

Since, for n sufficiently large, $\log n - k > 2$, and $\sum 1/n^2$ converges, by the Comparison Test [Rud87, p. 60] it follows that the power series converges absolutely on the circle $|z| = 1$.

Solution to 5.2.12: We have

$$
\begin{aligned}
\limsup \left| \frac{a_n}{n!} \right|^{1/n} &= \limsup |a_n|^{1/n} \limsup \left| \frac{1}{n!} \right|^{1/n} \\
&= \frac{1}{R} \limsup \left| \frac{1}{n!} \right|^{1/n} \\
&= \frac{1}{R} \cdot 0 \\
&= 0
\end{aligned}
$$

so h is entire.

Let $0 < r < R$. Then $1/R < 1/r$, so there is an $N > 0$ such that $|a_n| \leqslant 2/r^n$ for $n > N$. Further, there exists a constant $M > 2$ such that $|a_n| \leqslant M/r^n$ for $1 \leqslant n \leqslant N$. Therefore, for all z,

$$
|h(z)| \leqslant \sum_{n=1}^{\infty} |a_n| \frac{|z|^n}{n!} \leqslant M \sum_{n=1}^{\infty} \frac{|z|^n}{r^n n!} = M e^{|z|/r} .
$$

Solution to 5.2.13: We have

$$
|f(z) - s_k(z)| = \left| \sum_{n=k+1}^{\infty} c_n z^n \right| \leqslant \sum_{n=k+1}^{\infty} |c_n| |z^n| .
$$

Therefore

$$
\begin{aligned}
\sum_{k=0}^{\infty} |f(z) - s_k(z)| &\leqslant \sum_{k=0}^{\infty} \sum_{n=k+1}^{\infty} |c_n| |z^n| \\
&= \sum_{n=1}^{\infty} \sum_{k=0}^{n-1} |c_n| |z^n| = \sum_{n=1}^{\infty} n |c_n| |z^n| .
\end{aligned}
$$

From the basic theory of power series, the series $\sum_{n=0}^{\infty} c_n z^n$ and its formal derivative, the series $\sum_{n=1}^{\infty} n c_n z^{n-1}$, have the same radius of convergence. Therefore the series $\sum_{n=1}^{\infty} n c_n z^n$ also has radius of convergence R, so it converges absolutely for $|z| < R$, and the desired conclusion follows.

Solution to 5.2.14: Let the residue of f at 1 be K. We have

$$
\sum_{n=0}^{\infty} a_n z^n = \frac{K}{1-z} + \sum_{n=0}^{\infty} b_n z^n \quad \text{with} \quad \limsup_{n \to \infty} |b_n|^{1/n} > 1.
$$

Therefore,

$$\sum_{n=0}^{\infty} a_n z^n = \sum_{n=0}^{\infty} (K + b_n) z^n$$

and $a_n = K + b_n$. As $\sum b_n < \infty$, we have $\lim b_n = 0$ and $\lim a_n = K$.

Solution to 5.2.15: The rational function

$$f(z) = \frac{1 - z^2}{1 - z^{12}}$$

has poles at all nonreal twelfth roots of unity (the singularities at $z^2 = 1$ are removable). Thus, the radius of convergence is the distance from 1 to the nearest singularity:

$$R = |\exp(\pi i/6) - 1| = \sqrt{(\cos(\pi/6) - 1)^2 + \sin^2(\pi/6)} = \sqrt{2 - \sqrt{3}}.$$

Solution to 5.2.16: Suppose that $\{f_n\}$ is a sequence of functions analytic in an open set G, and that f_n converges to the function f uniformly on G. Then f is continuous in G. Fix $\zeta \in G$. There is $R > 0$ with $D(\zeta, R) \subset G$. Let γ be a triangular contour in $D(\zeta, R)$. Then $\int_\gamma f_n(z)\, dz = 0$ for each n, by Cauchy's Theorem [MH87, p. 152]. As $f_n \to f$ uniformly on the image of γ, we deduce that $\int_\gamma f_n(z)\, dz \to \int_\gamma f(z)\, dz$; thus $\int_\gamma f(z)\, dz = 0$. By Morera's Theorem, [MH87, p. 173] f is analytic in $D(\zeta, R)$. This proves the result.

The analogous result for real analytic functions is false. Consider the function $f : (-1, 1) \to \mathbb{R}$ defined by $f(x) = |x|$. As f is continuous, it is the uniform limit of polynomials, by Stone–Weierstrass Approximation Theorem [MH93, p. 284]. However, f is not even continuous in all of its domain.

Solution to 5.2.17: By the Hurwitz Theorem [MH87, p. 423], each zero of g is the limit of a sequence of zeros of the g_n's, which are all real, so the limit will be real as well.

Solution to 5.2.18: Let $\varepsilon_k = \lim_{n\to\infty} g_n^{(k)}(0)$. Then, clearly, $|\varepsilon_k| \leqslant |f^{(k)}(0)|$ for all k. Since f is an entire function, its Maclaurin series [MH87, p. 234] converges absolutely for all z. Therefore, by the Comparison Test [Rud87, p. 60], the series

$$\sum_{k=0}^{\infty} \varepsilon_k z^k$$

converges for all z and defines an entire function $g(z)$. Let $R > 0$ and $\varepsilon > 0$. For $|z| \leqslant R$, we have

$$|g_n(z) - g(z)| \leqslant \sum_{k=0}^{N} |g_n^{(k)}(0) - \varepsilon_k| R^k$$

$$\leqslant \sum_{k=0}^{N} |g_n^{(k)}(0) - \varepsilon_k| R^k + \sum_{k=N+1}^{\infty} 2|f^{(k)}(0)| R^k$$

taking N sufficiently large, the second term is less than $\varepsilon/2$ (since the power series for f converges absolutely and uniformly on the disc $|z| \leqslant R$). Let n be so large that $|g_n^{(k)}(0) - \varepsilon_k| < \varepsilon/2M$ for $1 \leqslant k \leqslant N$, where

$$M = \sum_{k=0}^{N} R^k.$$

Thus, for such n, we have $|g_n(z) - g(z)| < \varepsilon$. Since this bound is independent of z, the convergence is uniform.

Solution to 5.2.19: We have

$$\log\left(\frac{z(2-z)}{1-z}\right) = \log(2-z) + \log\left(\frac{1}{\frac{1}{z}-1}\right)$$

$$= \log 2 + \log\left(1 - \frac{z}{2}\right) + \pi i + \log\left(\frac{1}{1-\frac{1}{z}}\right).$$

In the unit disc, the principal branch of $\log\left(\frac{1}{1-z}\right)$ is represented by the series $\sum_{n=1}^{\infty} \frac{z^n}{n}$, which one can obtain by termwise integration of the geometric series $\frac{1}{1-z} = \sum_{n=0}^{\infty} z^n$. Hence,

$$\log\left(1 - \frac{z}{2}\right) = -\sum_{n=1}^{\infty} \frac{z^n}{2^n n} \qquad (|z| < 2),$$

$$\log\left(\frac{1}{1-\frac{1}{z}}\right) = \sum_{n=1}^{\infty} \frac{z^{-n}}{n} \qquad (|z| > 1),$$

and

$$\log\left(\frac{z(2-z)}{1-z}\right) = -\sum_{n=-\infty}^{-1} \frac{z^n}{n} + \log 2 + \pi i - \sum_{n=1}^{\infty} \frac{z^n}{2^n n} \qquad \text{for} \quad 1 < |z| < 2.$$

5.3   Conformal Mappings

Solution to 5.3.1: We will show that the given transformations also map straight lines into circles or straight lines.

$z \mapsto z + b$ and $z \mapsto kz$ clearly map circles and straight lines into circles and straight lines.

Let $S = \{z \mid |z - \alpha| = r\}$, $\alpha = x_0 + iy_0$, and $f(z) = 1/z = w = u + iv$. The equation for S is

$$(z - \alpha)(\overline{z} - \overline{\alpha}) = r^2$$

or

$$\frac{1}{w\overline{w}} - \frac{\alpha}{\overline{w}} - \frac{\overline{\alpha}}{w} = r^2 - |\alpha|^2.$$

- If $r = |\alpha|$, that is, when S contains the origin, we get

$$1 - \alpha w - \overline{\alpha}\overline{w} = 0$$

or

$$\Re(\alpha w) = \frac{1}{2}.$$

This is equivalent to

$$u x_0 - v y_0 = \frac{1}{2}$$

which represents a straight line.

- If $r \neq |\alpha|$, we obtain

$$w\overline{w} - \left(\frac{\overline{\alpha}}{|\alpha|^2 - r^2}\right)\overline{w} - \left(\frac{\alpha}{|\alpha|^2 - r^2}\right)w = \frac{-1}{|\alpha|^2 - r^2}.$$

Letting

$$\beta = \frac{\overline{\alpha}}{|\alpha|^2 - r^2}$$

we get

$$w\overline{w} - \beta\overline{w} - \overline{\beta}w + |\beta|^2 = \frac{r^2}{\left(|\alpha|^2 - r^2\right)^2}$$

and

$$|w - \beta|^2 = \left(\frac{r}{|\alpha|^2 - r^2}\right)^2$$

which represents the circle centered at β with radius $r/(|\alpha|^2 - r^2)$.

If S is a straight line, then, for some real constants a, b, and c, we have, for $z = x + iy \in S$,

$$ax + by = c.$$

Letting $\alpha = a - ib$, we get

$$\Re(\alpha z) = c$$

or

$$\alpha z + \overline{\alpha}\overline{z} = 2c$$

and it follows, as above, that $f(S)$ is a straight line or a circle.

Finally, let

$$f(z) = \frac{az+b}{cz+d}.$$

If $c = 0$ f is linear, so it is the sum of two functions that map circles and lines into circles and lines, so f itself has that mapping property. If $c \neq 0$, we have

$$\frac{az+b}{cz+d} = \frac{1}{c}\left(a - \frac{ad-bc}{cz+d}\right)$$

so $f(z) = f_3(f_2(f_1(z)))$ where

$$f_1(z) = cz+d, \qquad f_2(z) = \frac{1}{z}, \qquad f_3(z) = \frac{a}{c} - \frac{ad-bc}{c}z,$$

each of which has the desired property, and so does f.

Solution to 5.3.2: 1. The locus $|z+1| = |z-1|$ is the perpendicular bisector of the line segment joining -1 to 1, that is, the imaginary axis. The set $|z+1| < |z-1|$ is then the set of points z closer to -1 than to 1, that is, the left half-plane $\Re z < 0$. Hence, $\Re z < 0$ iff $\left|\frac{z+1}{z-1}\right| < 1$. The map $f : z \mapsto w = \frac{z+1}{z-1}$ maps $\{z \mid \Re z < 0\}$ onto $\{w \mid |w| < 1\}$, and is conformal as $f' \neq 0$. The inverse map is easily seen to be $w \mapsto z = \frac{w+1}{w-1}$.
2. The map $f(z) = iz$ has the specified properties.
3. The map $z \mapsto w = z^3$ is conformal (when $z \neq 0$), and sends $\{z \mid 0 < \arg z < \frac{\pi}{2}\}$ onto $\{w \mid 0 < \arg w < \frac{3\pi}{2}\}$. The inverse is the map $w \mapsto z = w^{1/3} = \exp(\frac{1}{3}(\log|w| + i \arg w))$ where $\arg w$ is chosen in $0 < \arg w < \frac{3\pi}{2}$.

Solution to 5.3.3: Let $A = \{z \mid |z| < 1, \ |z - 1/4| > 1/4\}$, $B = \{z \mid r < |z| < 1\}$. Let $f(z) = (z - \alpha)/(\alpha z - 1)$ be a linear fractional transformation mapping A onto B, where $-1 < \alpha < 1$. We have

$$f(\{z \mid |z - 1/4| = 1/4\}) = \{z \mid |z| = r\}$$

so

$$\{f(0), f(1/2)\} = \{-r, r\}$$

and

$$0 = r - r = f(0) + f(1/2) = \alpha + \frac{1/2 - \alpha}{\alpha/2 - 1}$$

which implies $\alpha = 2 - \sqrt{3}$. Therefore, $r = |f(0)| = 2 - \sqrt{3}$.

Suppose now that g is a linear fractional transformation mapping $C = \{z \mid s < |z| < 1\}$ onto A. Then $g^{-1}(\mathbb{R})$ is a straight line through the origin, because the real line is orthogonal to the circles $\{z \mid |z - 1/4| = 1/4\}$ and $\{z \mid |z| = 1\}$. Multiplying by a unimodular constant, we may assume $g^{-1}(\mathbb{R}) = \mathbb{R}$. Then $f \circ g(C) = A$ and $f \circ g(\mathbb{R}) = \mathbb{R}$. Replacing, if necessary, $g(z)$ by $g(s/z)$, we may suppose $f \circ g(\{z \mid |z| \leqslant 1\}) = \{z \mid |z| \leqslant 1\}$, so

$$f \circ g(z) = \beta \frac{z - \alpha}{\bar{\alpha} z - 1} \quad \text{with} \quad |\alpha| < |\beta| = 1.$$

Using the relation $0 = f(s) + f(-s)$, we get $\alpha = 0$, so $f \circ g(z) = \beta z$ and $s = r = 2 - \sqrt{3}$.

Solution to 5.3.4: Suppose f is such a function. Let $g : A \rightarrow B$ be defined by $g(z) = f(z)^2/z$. Then, as on $C_1 \cup C_4$, the absolute value of g is 1, then g is a constant, c, say. Therefore, $f(z) = \sqrt{cz}$ which is not continuous on A. We conclude that no such function can exist.

Solution to 5.3.5: The map $z \mapsto iz$ maps the given region conformally onto $A = \mathbb{D} \cap \{z \mid \Im z > 0\}$. The map

$$w \mapsto \frac{1 + w}{1 - w}$$

maps A onto the first quadrant, Q. The square function takes Q onto $\{z \mid \Im z > 0\}$. Finally,

$$\xi \mapsto \frac{\xi - i}{\xi + i}$$

takes $\{\xi \mid \Im \xi > 0\}$ onto $\mathbb{D}$. Combining these, we get for the requested map:

$$z \mapsto \frac{(1 + iz)^2 - i(1 - iz)^2}{(1 + iz)^2 + i(1 - iz)^2}.$$

Solution to 5.3.6: The map $\varphi_1(z) = 2z - 1$ maps conformally the semidisc

$$\left\{z \mid \Im z > 0, \; \left|z - \frac{1}{2}\right| < \frac{1}{2}\right\}$$

onto the upper half of the unit disc. The map

$$\varphi_2(z) = \frac{1 + z}{1 - z}$$

maps the unit disc conformally onto the right half-plane. Letting $z = re^{i\theta}$, it becomes

$$\frac{1 + re^{i\theta}}{1 - re^{i\theta}} = \frac{1 - r^2 + 2ir \sin \theta}{|1 + re^{i\theta}|^2}.$$

Since $\sin \theta > 0$ for $0 < \theta < \pi$, φ_2 maps the upper half of $\mathbb{D}$ onto the upper-right quadrant. The map $\varphi_3(z) = z^2$ maps the upper-right quadrant conformally onto the upper half-plane. The composition of φ_1, φ_2, and φ_3 is the desired map, namely the function $z \mapsto \frac{z^2}{(1-z)^2}$.

Solution to 5.3.7: Let R be the given domain. The map

$$z \mapsto \frac{1}{z} = \xi$$

transforms R onto the vertical strip $S = \{\xi \in \mathbb{C} \mid \frac{1}{2} < \Re\xi < 1\}$. Now

$$\xi \mapsto 2\pi i(\xi - \frac{1}{2}) = \eta$$

maps S onto the horizontal strip $T = \{\eta \in \mathbb{C} \mid 0 < \Im\eta < \pi\}$. Finally

$$\eta \mapsto e^{\eta} = \rho$$

maps T onto the upper half-plane $U = \{\rho \in \mathbb{C} \mid \Im\rho > 0\}$. Putting everything together we get the conformal map

$$z \mapsto e^{2\pi i(\frac{1}{z} - \frac{1}{2})}$$

Solution 2. Let C_1, C_2 denote the circles $|z - 1| = 1$, $|z - \frac{1}{2}| = \frac{1}{2}$ respectively. Consider the map $z \mapsto \zeta = 1/z$. C_1 contains the points $0, 1 + i, 2$ which are sent to ∞, $\frac{1}{2} - \frac{1}{2}i$, $\frac{1}{2}$ respectively. Thus C_1 is sent to the line $\Re\zeta = \frac{1}{2}$, and the inside of C_1 is sent to $\Re\zeta > \frac{1}{2}$ (consider the image of 1). C_2 contains the points $0, \frac{1}{2} + \frac{1}{2}i, 1$ which are sent to ∞, $1 - i$, 1; thus C_2 is sent to the line $\Re\zeta = 1$ and the outside of C_2 is sent to $\Re\zeta < 1$ (consider the image of 2). Thus the *lune* is sent to the strip $\frac{1}{2} < \Re\zeta < 1$. We follow with the map $\zeta \mapsto w = \exp(-2\pi i\zeta)$ which sends this strip to the upper half-plane. Composing, we obtain the map

$$z \mapsto \exp\left(-\frac{2\pi i}{z}\right)$$

which is a conformal map from $\{z \in \mathbb{C} \mid |z - 1| \leqslant 1\} \cap \{z \in \mathbb{C} \mid |z - \frac{1}{2}| \leqslant \frac{1}{2}\}$ onto the upper half-plane.

Solution to 5.3.8: Suppose f is such a map. f is bounded, so the singularity at the origin is removable, $p = \lim_{z \to 0} f(z)$. Since f is continuous, p is in the closure of A.

Suppose that p is on the boundary of A. Then $f(G) = A \cup \{p\}$, which is not an open set, contradicting the Open Mapping Theorem, [MH87, p. 436].

Let $p \in A$ and $a \in G$ be such that $f(a) = p$. Take disjoint open neighborhoods U of 0 and V of a. By the Open Mapping Theorem, $f(U)$ and $f(V)$ are open sets containing p. Then $f(U) \cap f(V)$ is a nonempty open set. Take $x \in f(U) \cap f(V)$, $x \neq p$. Then $x = f(z)$ for some nonzero $z \in U$ and $x = f(w)$ for some $w \in V$. Then z and w are distinct elements of G with $f(z) = f(w)$, contradicting the injectivity of f.

Solution to 5.3.9: Let ω be a primitive n^{th} root of unity. The transformations $\psi_j(z) = \omega^j z$, $j = 0, \ldots, n - 1$, form a cyclic group of order n and fix the points $z = 0$ and $z = \infty$. Let $\chi(z) = \frac{z+1}{z-1}$. Then $\chi(1) = \infty$, $\chi(-1) = 0$, and $\chi^{-1} = \chi$. The transformations $\varphi_j = \chi \circ \psi_j \circ \chi$, $j = 0, \ldots, n - 1$, thus form a cyclic group of order n that fix the points 1 and -1. We have explicitly

$$\varphi_j(z) = \frac{\omega^j\left(\frac{z+1}{z-1}\right) + 1}{\omega^j\left(\frac{z+1}{z-1}\right) - 1} = \frac{(\omega^j + 1)z + \omega^j - 1}{(\omega^j - 1)z + \omega^j + 1} = \frac{z + \left(\frac{\omega^j-1}{\omega^j+1}\right)}{\left(\frac{\omega^j-1}{\omega^j+1}\right)z + 1}.$$

5.4   Functions on the Unit Disc

Solution to 5.4.1: The function f is continuous on S_α, and $\overline{S}_\alpha = S_\alpha \cup \{0\}$ is compact, so it will be enough to show that f has a continuous extension to $\overline{S}_\alpha$. It will be shown that $f_\alpha(z) \to 0$ as $z \to 0$ from within S_α. We have

$$|f_\alpha(r\,e^{i\theta})| = \left|\exp\left(\frac{-e^{-i\theta}}{r}\right)\right| = \left|\exp\left(\frac{-\cos\theta + i\sin\theta}{r}\right)\right| = \exp\left(\frac{-\cos\theta}{r}\right)$$

Inside S_α we have $\cos\theta \geqslant \cos\alpha$, so

$$|f(z)| \leqslant \exp\left(\frac{-\cos\alpha}{|z|}\right) \longrightarrow 0,$$

when $z \to 0$, as desired.

Solution to 5.4.2: Let $b/a = re^{i\beta}$ and consider the function g defined by $g(z) = f(z)a^{-1}e^{-i\beta/2}$. We have

$$g\left(e^{i\theta}\right) = e^{i(\theta - \beta/2)} + re^{i(\beta/2 - \theta)}$$
$$= (1+r)\cos(\theta - \beta/2) + i(1-r)\sin(\theta - \beta/2)$$

so the image of the unit circle under g is the ellipse in standard position with axes $1 + r$ and $|1 - r|$. As $f(z) = a\exp(i\beta/2)g(z)$, f maps the unit circle onto the ellipse of axes $|a|(1 + r)$ and $|a(1 - r)|$, rotated from the standard position $\arg a + \beta/2$.

Solution to 5.4.3: We have

$$L = \int_0^{2\pi} \left|f'(e^{i\theta})\right|\left|ie^{i\theta}\right|d\theta = \int_0^{2\pi} \left|f'(e^{i\theta})\right|d\theta$$
$$\geqslant \left|\int_0^{2\pi} f'(e^{i\theta})d\theta\right|$$
$$= 2\pi \left|f'(0)\right|$$

by the Mean Value Property [MH87, p. 185].

Solution to 5.4.4: As the Jacobian of the transformation is $|f'(z)|^2$, we have

$$A = \int_{\mathbb{D}} |f'(z)|^2 \, dx\, dy.$$

$f'(z)$ can be found by term by term differentiation:

$$f'(z) = \sum_{n=1}^{\infty} nc_n z^{n-1}$$

so

$$|f'(z)|^2 = \sum_{j,k=1}^{\infty} jk c_j \bar{c}_k z^{j-1} \bar{z}^{k-1}.$$

We then have

$$A = \iint_{\mathbb{D}} \sum_{j,k=1}^{\infty} jk c_j \bar{c}_k z^{j-1} \bar{z}^{k-1} \, dxdy.$$

Letting $z = re^{i\theta}$, we get

$$A = \sum_{j,k=1}^{\infty} jk c_j \bar{c}_k \int_0^1 \int_0^{2\pi} r^{j+k-1} e^{i(j-k)\theta} \, d\theta dr.$$

Since

$$\int_0^{2\pi} e^{in\theta} \, d\theta = 0$$

for $n \neq 0$, we have

$$A = 2\pi \sum_{n=1}^{\infty} n^2 |c_n|^2 \int_0^1 r^{2n-1} \, dr = \pi \sum_{n=1}^{\infty} n |c_n|^2.$$

Solution to 5.4.5: We have, for $z, w \in \mathbb{D}$,

$$f(w) = f(z) \quad \text{iff} \quad (w - z)\left(1 + \frac{w+z}{2}\right) = 0$$

so f in injective. Then the area of its image is given by

$$\begin{aligned}
\int_{\mathbb{D}} |f'(z)|^2 \, dxdy &= \int_{\mathbb{D}} \left(1 + 2\Re z + |z|^2\right) dz \\
&= \int_{\mathbb{D}} \left(1 + 2x + x^2 + y^2\right) dxdy \\
&= \int_0^{2\pi} \int_0^1 \left(1 + 2r\cos\theta + r^2\right) drd\theta \\
&= \frac{3\pi}{2}.
\end{aligned}$$

Solution to 5.4.6: Fix $z_0 \in \mathbb{D}$ and let h be defined by $h(z) = f(z) - f(z_0)$. As h has only isolated zeros in $\mathbb{D}$, we can find an increasing sequence $\rho_i \to 1$ with $h(z) \neq 0$ for $|z| = \rho_i$, $i = 1, \ldots$. Let g be the function given by $g(z) = a_1(z - z_0)$. For $|z| = \rho_i$, we have

$$|g(z) - h(z)| = \left| \sum_{n \geq 2} a_n z^n - \sum_{n \geq 2} a_n z_0^n \right|$$

$$\leqslant \max \left| \frac{d}{dz} \sum_{n \geqslant 2} a_n z^n \right|_{z=w} |z - z_0| \quad (w \text{ in the segment } [z, z_0])$$

$$\leqslant \sum_{n \geqslant 2} n a_n \rho_i^{n-1} |z - z_0|$$

$$\leqslant |a_1||z - z_0|$$

$$= |g(z)|.$$

By Rouché's Theorem [MH87, p. 421], h has a unique zero in the disc $\{z \mid |z| < \rho_i\}$, so f assumes the value $f(z_0)$ only once there. Letting $\rho_i \to 1$, we get that f is injective in $\mathbb{D}$.

Solution to 5.4.7: Let ψ be a linear fractional transformation which maps $\mathbb{D}$ onto itself with $\psi(0) = z_1$. Then the conjugate $g = \psi^{-1} \circ f \circ \psi$ is analytic in the unit disc f, and has two fixed points, namely $0 = \psi^{-1}(z_1)$ and $w = \psi^{-1}(z_2)$.

As $g(0) = 0$ we can define the analytic function h by $h(z) = g(z)/z$. For each $0 < \varepsilon < 1$, on the circle $|z| = 1 - \varepsilon$, we have $|h(z)| = |g(z)/|z| \leqslant 1/(1 - \varepsilon)$, so the maximum modulus principle implies $|h(z)| \leqslant 1/(1 - \varepsilon)$ on $|z| \leqslant 1 - \varepsilon$. As ε can be arbitrarily small, we have $|h(z)| \leqslant 1$ for any $z \in \mathbb{D}$.

We know that $h(w) = 1$, so h attains its maximum inside $\mathbb{D}$. By the Maximum Modulus Principle h must be constant, that is, h is identically 1, and in this case $g(z) = f(z) = z$.

Solution to 5.4.8: Let $f \in X_k$, $z \in \mathbb{D}$, and γ be the circle around z with radius $r = (1 - |z|)/2$. γ lies inside the unit disc, so, by Cauchy's Integral Formula for derivatives [MH87, p. 169], we have

$$|f'(z)| \leqslant \frac{1}{2\pi} \int_\gamma \frac{|f(w)|}{|z - w|^2} |dw| \leqslant \frac{C}{r(1 - |z|)^k} = \frac{2C}{(1 - |z|)^{k+1}},$$

where C is a constant, so $f' \in X_{k+1}$.

Let $f' \in X_{k+1}$ with $f(0) = 0$ (the general case follows easily from this). Letting $z = re^{i\theta}$, we have

$$|f(z)| \leqslant \int_0^z |f'(w)| \, |dw|$$

$$= \int_0^r \left| f'(te^{i\theta}) \right| dt$$

$$\leqslant \int_0^r \frac{C}{(1 - r)^{k+1}} dt$$

$$= \frac{kC}{(1 - r)^k}.$$

Hence, $f \in X_k$.

Solution to 5.4.9: Let $f(z) = \sum_{n \geqslant 0} a_n z^n$ for $z \in \mathbb{D} = \{z \mid |z| < 1\}$.

As $f(z)$ is analytic, so is $g : \mathbb{D} \to \mathbb{C}$ defined by $g(z) = f(z) - \overline{f(\overline{z})}$. We have, for real $z \in \mathbb{D}$,

$$
\begin{aligned}
g(z) &= \sum (a_n - \overline{a}_n) z^m \\
&= f(z) - \overline{f(\overline{z})} \\
&= f(z) - f(z) \\
&= 0
\end{aligned}
$$

therefore, $g(z) \equiv 0$ on $\mathbb{D}$, so the coefficients a_n are all real.

Put $z_0 = e^{i\pi\sqrt{2}}$, and let a_k be the nonzero coefficient of smallest index. We have

$$
\frac{f(tz_0) - a_0}{a_k t^k} = z_0^k + \frac{t}{a_k} \sum_{n \geq k+1} a_n z_0^n t^{n-k+1} .
$$

$tz_0 \in \mathbb{D}$ for $t \in [0, 1)$ and the left-hand side expression is a real number for all t, so

$$
\lim_{t \to 0} \frac{f(tz_0) - a_0}{a_k t^k} = z_0^k \in \mathbb{R}
$$

which implies $k = 0$, by the irrationality of $\sqrt{2}$, thus f is a constant.

Solution to 5.4.10: Let $\sum_{n=0}^{\infty} c_n z^n$ be the Maclaurin series for f. Then $f'(z) = \sum_{n=1}^{\infty} n c_n z^{n-1}$. The Cauchy Inequalities [MH87, p. 170] give

$$
\begin{aligned}
|c_n| &= \left| \frac{1}{2\pi i} \int_{|z|=r} \frac{f'(z)}{z^n} dz \right| \\
&= \left| \frac{1}{2\pi r^{n-1}} \int_0^{2\pi} f'(re^{i\theta}) e^{-i(n-1)\theta} d\theta \right| \\
&\leq \frac{M}{r^{n-1}}, \qquad 0 < r < 1.
\end{aligned}
$$

Letting $r \to 1$, we get $|c_n| \leq \dfrac{M}{n}$ $(n = 1, 2, \ldots)$. Hence,

$$
\begin{aligned}
\int_{[0,1)} |f(x)| dx &\leq \int_0^1 \left(\sum_{n=0}^{\infty} |c_n| x^n \right) dx \\
&\leq |c_0| + M \int_0^1 \left(\sum_{n=0}^{\infty} \frac{x^n}{n} \right) dx \\
&= |c_0| + M \int_0^1 \log \frac{1}{1-x} dx \\
&= |c_0| + M \lim_{r \to 1} \left(-(1-x) \log \frac{1}{1-x} + x \right) \Big|_0^r \\
&= |c_0| + M.
\end{aligned}
$$

Solution to 5.4.11: As $|h(0)| = 5$, by the Maximum Modulus Principle [MH87, p. 185], h is constant in the unit disc. Therefore, $h'(0) = 0$.

Solution to 5.4.12: For $r = \log 2$,

$$T(z) = \frac{z - r}{1 - rz}$$

maps the closed unit disc onto itself with $T(0) = -r$. Then $g(z) = f(T(z))$ is analytic on the closed unit disc and $g(0) = 0$. For z on the unit circle we have $|g(z)| \leqslant |e^{T(z)}|$. Therefore $h(z) = e^{-T(z)}g(z)$ is analytic in the closed unit disc, $h(0) = 0$, and $|h(z)| \leqslant 1$ for all z with $|z| = 1$. By Schwartz Lemma, [MH87, p. 190], $|h(z)| \leqslant |z|$ for all z with $|z| \leqslant 1$, hence $|f(T(z))| \leqslant |z||e^{T(z)}|$ on the closed unit disc. If $w = T(z)$ then $z = -T(-w)$, giving $|f(z)| \leqslant |e^z T(-z)|$. Setting $z = r = \log 2$, we get

$$|f(\log 2)| \leqslant e^r \frac{2r}{1 + r^2} = \frac{4\log 2}{1 + (\log 2)^2}.$$

This bound is attained by the function $\dfrac{z + \log 2}{1 + z\log 2} e^z$.

Solution to 5.4.13:

$$\varphi(z) = i\left(\frac{1 + z}{1 - z}\right)$$

maps the unit disc to the upper half-plane with $\varphi(0) = i$. Thus, $f \circ \varphi$ maps the unit disc into itself fixing 0. By the Schwarz Lemma [MH87, p. 190], $|f \circ \varphi(z)| \leqslant |z|$. Solving $\varphi(z) = 2i$, we get $z = 1/3$. Hence, $|f(2i)| \leqslant 1/3$. Letting $f = \varphi^{-1}$, we see that this bound is sharp.

Solution to 5.4.14: The function

$$\varphi(z) = \frac{1 + z}{1 - z}$$

maps the unit disc, $\mathbb{D}$, onto the right half-plane, with $\varphi(0) = 1$. Therefore, the function $f \circ \varphi$ maps $\mathbb{D}$ into itself, with $f \circ \varphi(0) = 0$. By the Schwarz Lemma [MH87, p. 190], we have $|(f \circ \varphi)'(0)| \leqslant 1$, which gives $|f'(\varphi(0))\varphi'(0)| \leqslant 1$ and $|f'(1)| \leqslant 1/2$. A calculation shows that equality happens for $f = \varphi^{-1}$.

Solution to 5.4.15: Suppose f has infinitely many zeros in $\mathbb{D}$. If they have a cluster point in $\mathbb{D}$, then $f \equiv 0$ and the result is trivial. Otherwise, since $\{z \in \mathbb{C} \mid |z| \leqslant 1\}$ is compact, there is a sequence of zeros converging to a point in the boundary of $\mathbb{D}$, and the conclusion follows.

Assume now that f has only finitely many zeros in $\mathbb{D}$, $w_1, \ldots, w_m$. Then f can be written as

$$f(z) = (z - w_1)^{\alpha_1} \cdots (z - w_m)^{\alpha_m} g(z)$$

where g is analytic and never zero on $\mathbb{D}$. Applying the Maximum Modulus Principle [MH87, p. 185], we get that $1/g$ attains a maximum in the disc $|z| \leqslant (1 - 1/n)$

at a point z_n with $|z_n| = 1 - 1/n$ $(n \geqslant 2)$. Then $|g(z_2)| \geqslant |g(z_3)| \geqslant \cdots$. The product $(z - w_1)^{\alpha_1} \cdots (z - w_m)^{\alpha_m}$ is clearly bounded, and so is $f(z_n)$.

Solution to 5.4.16: 1. We can assume f has only finitely many zeros. (Otherwise, assuming $f \not\equiv 0$, its zero sequence has the required property, since the zeros of a nonconstant analytic function in an open connected set can cluster only on the boundary of the set.) That done, we can, after replacing f by its quotient with a suitable polynomial, assume f has no zeros. Then $1/f$ is analytic in the disc. For $n = 1, 2, \ldots$, let M_n be the maximum of $|1/f(z)|$ for $|z| = 1 - \frac{1}{n}$. By the Maximum Modulus Principle [MH87, p. 185], $M_n \geqslant M_1$ for all n. Hence, for each n, there is a point a_n such that $|a_n| = 1 - \frac{1}{n}$ and $|f(a_n)| = 1/M_n \leqslant 1/M_1 = |f(0)|$. Then $(f(a_n))$ is a bounded sequence of complex numbers and so has a convergent subsequence, which gives the desired conclusion.

2. Let (z_n) be a sequence with the properties given in Part 1. Subtracting a constant from f, if needed, we can assume $\lim f(z_n) = 0$. We can suppose also that $|z_{n+1}| > |z_n| > 0$ for all n. For each n, let M_n be the maximum of $|f(z)|$ for $|z| = |z_n|$. The numbers M_n are positive (since f is nonconstant) and increase with n (by the Maximum Modulus Principle [MH87, p. 185]). Since $f(z_n) \to 0$, there is an n_0 such that $|f(z_n)| < M_1$ for $n \geqslant n_0$. For such n, the restriction of $|f|$ to the circle $|z| = |z_n|$ is a continuous function that takes values both larger than M_1 and smaller than M_1. By the Intermediate Value Theorem [Rud87, p. 93], there is for each $n \geqslant n_0$, a point b_n such that $|b_n| = |z_n|$ and $|f(b_n)| = M_1$. Then, for the desired sequence (w_n), we can take any subsequence of (b_n) along which f converges. (There will be such a subsequence by the boundedness of the sequence $(f(b_n))$.)

Solution to 5.4.17: Suppose $f(a) = a \in \mathbb{D}$, $f(b) = b \in \mathbb{D}$, and $a \neq b$. Let $\varphi : \mathbb{D} \to \mathbb{D}$ be the automorphism of the unit disc that maps 0 to a, that is, $\varphi(z) = (a - z)/(1 - \bar{a}z)$. Then the function $g = \varphi^{-1} \circ f \circ \varphi$ maps $\mathbb{D}$ into itself with $g(0) = 0$ and $g(\varphi^{-1}(b)) = \varphi^{-1}(b)$. Since φ is one-to-one and $a \neq b$, $\varphi^{-1}(b) \neq 0$. Hence, by the Schwarz Lemma [MH87, p. 190], there exists a unimodular constant λ such that $g(z) = \lambda z$, and letting $z = \varphi^{-1}(b)$, we see that $\lambda = 1$; that is, g is the identity map and so is f.

Solution to 5.4.18: Using Cauchy's Integral Formula [MH87, p. 167], for $0 < r < 1$, we have

$$\left| \frac{f^{(n)}(0)}{n!} \right| \leqslant \frac{1}{2\pi} \int_{|w|=r} \frac{|f(w)|}{|w|^{n+1}} |dw|$$

$$\leqslant \frac{1}{2\pi} \int_{|w|=r} \frac{1}{(1-r)r^{n+1}} |dw| = \frac{1}{(1-r)r^n}.$$

Letting $r = n/(n+1)$, we get $|f^{(n)}(0)/n!| \leqslant (n+1)(1 + 1/n)^n < (n+1)e$.

Solution to 5.4.19: By the Schwarz Lemma [MH87, p. 190],

$$|f(z)| \leqslant |z|.$$

If $f(z) = a_1z + a_2z^2 + \cdots$, let g be defined by

$$g(z) = \frac{f(z) + f(-z)}{2z} = a_2z + a_4z^3 + a_6z^5 + \cdots$$

g is analytic in $\mathbb{D}$, and since

$$|g(z)| \leqslant \frac{|f(z)| + |f(-z)|}{2|z|} \leqslant 1$$

g maps $\mathbb{D}$ into $\mathbb{D}$. Hence, by the Schwarz Lemma,

$$|g(z)| \leqslant |z|$$

or

$$|f(z) + f(-z)| \leqslant 2|z|^2.$$

Now suppose equality held for $z_0 \in \mathbb{D}$. We would have $|g(z_0)| = |z_0|$ so, by the Schwarz Lemma,

$$g(z) = \lambda z$$

for some unimodular λ, or

$$f(z) + f(-z) = 2\lambda z^2.$$

Plugging this back into the power series for $g(z)$, we get $a_2 = \lambda$ and $a_4 = a_6 = \cdots = 0$. Hence,

$$f(z) = \lambda z^2 + h(z)$$

where $h(z)$ is odd. We have

$$1 \geqslant |f(z)| = |\lambda z^2 + h(z)|$$

and

$$1 \geqslant |f(-z)| = |\lambda z^2 + h(-z)| = |\lambda z^2 - h(z)|.$$

Therefore,

$$\begin{aligned}(\lambda z^2 + h(z))(\overline{\lambda z^2} + \overline{h(z)}) &\leqslant 1 \\ (\lambda z^2 - h(z))(\overline{\lambda z^2} - \overline{h(z)}) &\leqslant 1.\end{aligned}$$

Expanding and adding, we get

$$\begin{aligned}|z|^4 + |h(z)|^2 &\leqslant 1 \\ |h(z)|^2 &\leqslant 1 - |z|^4\end{aligned}$$

which, by the Maximum Modulus Principle [MH87, p. 185], implies $h(z) \equiv 0$.

Solution to 5.4.20: Schwarz's Lemma [MH87, p. 190] implies that the function $f_1(z) = f(z)/z$ satisfies $|f_1(z)| \leqslant 1$. The linear fractional map $z \mapsto \frac{z-r}{1-rz}$ sends the unit disc onto itself. Applying Schwarz's Lemma to the function

$f_2(z) = f_1\left(\frac{z-r}{1-rz}\right)$, we conclude that the function $f_3(z) = f_1(z)/\left(\frac{z-r}{1-rz}\right)$ satisfies $|f_3(z)| \leqslant 1$. Similarly, the map $z \mapsto \frac{z+r}{1+rz}$ sends the unit disc onto itself, and Schwarz's Lemma applied to the function $f_4(z) = f_3(z)/\left(\frac{z+r}{1+rz}\right)$ implies that the function $f_5(z) = f_3\left(\frac{z+r}{1+rz}\right)$ satisfies $|f_5(z)| \leqslant 1$. All together, then,

$$|f(z)| \leqslant |z| \left|\frac{z-r}{1-rz}\right| \left|\frac{z+r}{1+rz}\right| \quad |f_5(z)| \leqslant |z| \left|\frac{z-r}{1-rz}\right| \left|\frac{z+r}{1+rz}\right|$$

which is the desired inequality.

Solution to 5.4.21: Let φ_{z_0} be the automorphism of the unit disc given by

$$\varphi_{z_0}(z) = \frac{z - z_0}{1 - \overline{z_0}z},$$

we have

$$\varphi_{z_0}'(z) = \frac{1 - |z_0|^2}{(1 - \overline{z_0}z)^2}.$$

Now consider the composition

$$g(z) = \varphi_{f \circ \varphi_{z_0}(0)} \circ f \circ \varphi_{z_0}(z) = \varphi_{f(-z_0)} \circ f \circ \varphi_{z_0}(z)$$

then $g(0) = 0$ and as composition of maps of the unit disc into itself, we can apply the Schwarz Lemma [MH87, p. 190] to obtain $|g'(0)| \leqslant 1$. Computing $g'(0)$ using the chain rule, we have

$$|g'(0)| = \left|\frac{1}{1 - |f(-z_0)|^2} \cdot f'(-z_0) \cdot (1 - |z_0|^2)\right| \leqslant 1$$

so we can conclude that

$$|f'(z)| \leqslant \frac{1 - |f(z)|^2}{1 - |z|^2} \leqslant \frac{1}{1 - |z|^2}.$$

The first inequality is known as Picks' Lemma and is the main ingredient in the proof that an analytic map of the disc that preserves the hyperbolic distance between any two points, preserves all distances, for more detail see [Car60, Vol. 2, §290] or [Kra90, p. 16].

Solution 2. Using the same notation as above,

$$|(f \circ \varphi_{z_0})'(0)| \leqslant \frac{1}{2\pi} \int_{|\omega|=r} \frac{|f \circ \varphi_{z_0}(w)|}{|\omega|^2} |d\omega|$$

that is

$$|f'(-z_0)|\left|1 - |z_0|^2\right| \leqslant \frac{1}{2\pi} \int_0^{2\pi} \frac{d\theta}{r} = \frac{1}{r}$$

which holds for any $|z_0| < r = |w| < 1$, so the conclusion follows.

Solution to 5.4.22: Fix $z \in \mathbb{D}$ and let γ be the circle centered at z parametrized by

$$w = z + \frac{1 - |z|}{2} e^{i\theta}$$

which is completely contained inside the unit disc. Using Cauchy's Integral Formula for derivatives [MH87, p. 169], we have

$$f'(z) = \frac{1}{2\pi i} \int_\gamma \frac{f(w)}{(w - z)^2} dw .$$

taking absolute values and applying the inequality we get

$$|f'(z)| \leqslant \frac{1}{2\pi} \int_\gamma \frac{C}{|1 - w||w - z|^2} |dw| .$$

From the parametrization we cann see that $|dw| = \dfrac{|1 - |z||}{2} d\theta$, and obtain

$$|f'(z)| \leqslant \frac{C}{\pi} \int_0^{2\pi} \frac{1 - |z|}{(1 - |w|)(1 - |z|)^2} d\theta .$$

Now $1 - |w|$ is minimum at the point of γ farthest from the origin, at which we have $1 - |w| = (1 - |z|)/2$, therefore,

$$|f'(z)| \leqslant \frac{C}{\pi} \int_0^{2\pi} \frac{2(1 - |z|)}{(1 - |z|)(1 - |z|)^2} d\theta = \frac{4C}{(1 - |z|)^2} .$$

5.5   Growth Conditions

Solution to 5.5.1: Consider the function $g(z) = e^{-f(z)}$, it is entire with $|g(z)| = e^{\Re f(z)} \leqslant e^{-2}$. Liouville's Theorem implies that g is constant, and since it is obviously a nonzero contant, f maps the connected set $\mathbb{C}$ into the discrete set of the logarithms, hence f itself is constant.

Solution to 5.5.2: Let $g(z) = f(z) - f(0)$. Then $g(0) = 0$, so $g(z)/z$ has a removable singularity at 0 and extends to an entire function. $g(z)/z$ tends to 0 as $|z|$ tends to infinity since $f(z)/z$ does. Let $\varepsilon > 0$. There is an $R > 0$ such that $|g(z)/z| < \varepsilon$ for $|z| \geqslant R$. By the Maximum Modulus Principle [MH87, p. 185], $|g(z)/z| < \varepsilon$ for all z. Since ε is arbitrary, $g(z)/z$ is identically 0. Hence, $g(z) = 0$ for all z and f is constant.

Solution to 5.5.3: Let $f = p/q$ where p, q are polynomials with q zero free in the upper half plane. For positive R consider the curve $\Gamma_R = [-R, R] \cup C_R$, where C_R is the upper semicircle centered at the origin, with radius R. By the Maximum Modulus Principle [MH87, p. 185], we have

$$\sup\{|f(z)| \mid \Im z \geqslant 0\} = \lim_{R \to \infty} \sup\{|f(z)| \mid z \in \Gamma_R\}.$$

Let's call this value M. If $\deg q > \deg p$, we have $\lim_{|z| \to \infty} |f(z)| = 0$ so the result is clear. If $\deg q \leqslant p$, then $\lim_{|z| \to \infty} |f(z)| = M$ and the result follows also.

Solution 2. Case 1. f has a pole at ∞. Then $\lim_{z \to \infty} f(z) = \infty$, so both the suprema equal ∞.

Case 2. f does not have a pole at ∞. Then f has a finite limit at ∞, so f extends continuously to the closure of $H = \{z \in \mathbb{C} \mid \Im z > 0\}$ in the extended plane. Since that closure is compact, $|f|$ attains a maximum there. By the Maximum Modulus Theorem, $|f|$ cannot attain a local maximum in H unless f is constant. Thus, the maximum of $|f|$ on the extended closure of H is attained at a point of the real axis or at ∞. In either case the desired equality holds.

Solution to 5.5.4: The function $g(z) = f(z)/\sin z$ is analytic in the open connected set $G = \mathbb{C} \setminus \pi\mathbb{Z}$. Write $g = u + iv$. By hypothesis, $u^2 + v^2 = 1$ in G. Taking partial derivatives, we have

$$uu_x + vv_x = 0, \quad uu_y + vv_y = 0.$$

Using the Cauchy–Riemann equations we get

$$uu_x - vu_y = 0, \quad uu_y + vu_x = 0.$$

From these we get $u_x(u^2 + v^2) = 0$ and $u_y(u^2 + v^2) = 0$. Thus $u_x = 0 = u_y$ on G. It follows that $v_x = 0 = v_y$ on G. By the Mean Value Theorem, both u and v are constant on every line segment lying in G parallel to the real or imaginary axis. G is step connected: that is, given any two points z_0, z_1 in G there is a path in G from z_0 to z_1, consisting of line segments parallel to the axes. Thus u and v are constant in G, whence $g = C$ in G, for some constant C, with $|C| = |g| = 1$. That is, $f(z) = C \sin z$ in G. By continuity this equality holds in $\mathbb{C}$.

Solution to 5.5.5: If $g \equiv 0$, the result is trivially true. Otherwise, the zeros of g are isolated points. $|f/g|$ is bounded by 1 in $\mathbb{C}$, so all the singularities of f/g are removable, and f/g can be extended to an entire function. Liouville's Theorem [MH87, p. 170] now guarantees that f/g must be a constant.

Solution to 5.5.6: Since p and q are of the same degree, the ratio p/q has a nonzero finite limit at ∞. The function $f(z) = p(1/z)/q(1/z)$ thus has a removable singularity at 0, as does $1/f$. Moreover f and $1/f$ are both analytic in a neighborhood of the disk $|z| \leqslant 1$ (since p and q are without zeros in $|z| \geqslant 1$) and f and $1/f$ have unit modulus on $|z| = 1$. By the Maximum Modulus Principle [MH87, p. 185], $|f(z)| \leqslant 1$ and $|1/f(z)| \leqslant 1$ for $|z| \leqslant 1$. So $|f(z)| = 1$ and for

$|z| \leqslant 1$, and another application of the Maximum Modulus Principle shows that $f = $ constant, the desired conclusion.

Solution to 5.5.7: Let g be defined on $\mathbb{C}$ by $g(z) = \dfrac{f(z) - f(0)}{z}$. g is an entire function such that, on the circle of radius R,

$$|g(z)| \leqslant \frac{|\log z| + |f(0)|}{R} \leqslant \frac{\sqrt{\log^2 R + \pi^2} + |f(0)|}{R},$$

where the inequality for $z = -R$ follows by continuity. Using the Maximum Modulus Principle [MH87, p. 185], we get, for $|z| \leqslant R$,

$$|g(z)| \leqslant \frac{\sqrt{\log^2 R + \pi^2} + |f(0)|}{R}.$$

Taking the limit $R \to \infty$ we get $g \equiv 0$ so f is a constant function. As $|f(1)| \leqslant |\log 1|$ we conclude that f vanishes identically.

Solution 2. Since f is entire, its Maclaurin series converges everywhere, $f(z) = \sum_{n=0}^{\infty} a_n z^n$. Using Cauchy's Integral Formula for derivatives [MH87, p. 169], we have

$$a_n = \frac{1}{2\pi i} \int_{C_R} \frac{f(\xi)}{\xi^{n+1}} d\xi$$

where C_R is the circle of radius R centered at the origin, positively oriented. Using the given inequality we get, for $n \geqslant 1$,

$$|a_n| \leqslant \frac{1}{2\pi} 2\pi R \frac{\log R}{R^{n+1}} = \frac{\log R}{R} = o(R) \quad (R \to \infty).$$

Therefore f equals a constant, a_0. As in the previous solution, we must have $a_0 = 0$.

Solution to 5.5.8: Let $h(z) = f(z) - kg(z)$. Then h is entire and $\Re h(z) \leqslant 0$. We then have

$$\left| e^{h(z)} \right| \leqslant 1 \quad \text{for all} \quad z \in \mathbb{C}$$

therefore, by Liouville's Theorem [MH87, p. 170], e^h is constant, and so is h.

Solution to 5.5.9: Suppose the maximum of φ over the compact set K is attained at a point z_0 in the interior of K. Let r be the distance of z_0 from the complement of K. The circle $|z - z_0| = r$ is then in K, so $\varphi(z) \leqslant \varphi(z_0)$ on that circle, giving

$$\varphi(z_0) \geq \frac{1}{2\pi} \int_0^{2\pi} \varphi(z_0 + re^{i\theta}) d\theta.$$

Hence equality holds by the sub-mean-value property. Thus

$$\int_0^{2\pi} (\varphi(z_0) - \varphi(z_0 + re^{i\theta})) d\theta = 0.$$

The integrand here is nonnegative, so, being continuous, it must vanish identically. Thus $\varphi(z) = \varphi(z_0)$ for $|z - z_0| = r$. Since r is the distance from z_0 to the complement of K, the circle $|z - z_0| = r$ contains at least one point of the boundary of K. Hence φ attains its maximum over K on the boundary of K.

Solution to 5.5.10: 1. Using Cauchy's Integral Formula for derivatives [MH87, p. 169], we get

$$\left| f^{(k)}(0) \right| \leqslant \frac{k!}{2\pi} \int_{|z|=R} \left| \frac{f(z)}{z^{k+1}} \right| |dz|$$

$$\leqslant \frac{k!}{2\pi R^{k+1}} \int_{|z|=R} \left| a\sqrt{|z|} + b \right| |dz|$$

$$= \frac{k! \left(a\sqrt{|z|} + b \right)}{R^k}$$

$$= o(1) \quad (R \to \infty)$$

so $f^{(k)}(0) = 0$ for $k \geqslant 1$, and f reduces to a constant, $f(0)$.

2. Using the same method as above for $k \geqslant 3$, we get

$$\left| f^{(k)}(0) \right| \leqslant \frac{k!}{2\pi} \int_{|z|=R} \left| \frac{f(z)}{z^{k+1}} \right| |dz|$$

$$\leqslant \frac{k!}{2\pi R^{k+1}} \int_{|z|=R} \left| a\sqrt{|z|^5} + b \right| |dz|$$

$$= \frac{k! \left(a\sqrt{|z|^5} + b \right)}{R^k} = o(1) \quad (R \to \infty)$$

so $f^{(k)}(0) = 0$ for $k \geqslant 3$ and f reduces to a polynomial of degree, at most, 2, $f(0) + f'(0)z + f''(0)z^2/2$.

Solution to 5.5.11: For $r > 0$, let $z = re^{i\theta}$ in Cauchy's Integral Formula for derivatives [MH87, p. 169] to get

$$f^{(n)}(0) = \frac{n!}{2\pi i} \int_0^{2\pi} \frac{f(re^{i\theta})}{r^n e^{in\theta}} \, d\theta.$$

Combining this with the inequality given yields

$$\frac{|f^{(n)}(0)|}{n!} \leqslant \frac{r^{17/3-n}}{2\pi}.$$

For $n > 5$, letting r tend to infinity, we get $f^{(n)}(0) = 0$. If $n \leqslant 5$, letting r tend to 0 gives the same result. Hence, the coefficients of the Maclaurin series [MH87, p. 234] of f are all 0, so $f \equiv 0$.

Solution to 5.5.12: If such a function f exists then $g = 1/f$ is also analytic on $\mathbb{C} \setminus \{0\}$, and satisfies $|g(z)| \leqslant \sqrt{|z|}$. As g is bounded on $\{z \mid 0 < |z| < 1\}$, g has a removable singularity at 0, and extends as an analytic function over the complex plane. Fix z, choose $R > |z|$, and let C_R be the circle with center 0 and radius R. Then

$$g'(z) = \frac{1}{2\pi i} \int_{C_R} \frac{g(w)}{(w-z)^2} dw$$

so

$$|g'(z)| \leqslant \frac{1}{2\pi} \cdot 2\pi R \cdot \frac{\sqrt{R}}{(R - |z|)^2} \to 0 \text{ as } R \to \infty.$$

Thus, $g' = 0$ everywhere, so g (and, hence, f) is constant. But this contradicts the hypothesis $|f(z)| \geqslant \frac{1}{\sqrt{|z|}}$ for small z, so no such function exists.

Solution to 5.5.13: By Liouville's Theorem [MH87, p. 170], it will be enough to prove that f is bounded. For $|\Re z| \geqslant 1/2$, we have $|f(z)| \leqslant \sqrt{2}$. Let z_0 be a point such that $|\Re z_0| < 1/2$. Let S be the square with vertices $i\Im z_0 \pm 1 \pm i$, oriented counterclockwise.

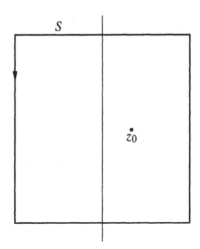

Then z_0 is in the interior of S, so Cauchy's Integral Formula [MH87, p. 167] gives

$$f(z_0) = \frac{1}{2\pi i} \int_S \frac{f(z)}{z - z_0} dz.$$

The absolute value of the integrand is, at most, $2|\Re z|^{-1/2}$. The contribution to the integral from each vertical edge is thus, at most, 4 in absolute value. The contribution from each horizontal edge is, at most, $2\int_{-1}^{1} |x|^{-1/2} dx = 8$ in absolute value. Hence,

$$|f(z_0)| \leqslant \frac{1}{2\pi} (4 + 4 + 8 + 8) = 12\pi$$

proving that f is bounded.

Solution to 5.5.14: Since f is nonconstant its zero at the origin is isolated. Hence there is a circle C with center at the origin on and within which f does not vanish, except at 0. Let ε be the minimum of $|f|$ on C. Then the set $\{z \mid |f(z)| < \varepsilon\}$ intersects the interior of C but is disjoint from C. Since it is connected it must be contained in the interior of C. It follows that the origin is the only zero of f, and that f is bounded away from 0 near ∞. The latter conclusion implies that ∞ is not an essential singularity of f (Casorati–Weierstrass Theorem [MH87, p. 256]), so ∞ must be a pole of f (it is not a removable singularity, by Liouville's Theorem [MH87, p. 170]). Therefore f is a polynomial. Since its only zero is at the origin, it is a constant times a positive power of z.

5.6 Analytic and Meromorphic Functions

Solution to 5.6.1: By the Cauchy–Riemann equations [MH87, p. 72],

$$u_x = v_y \qquad \text{and} \qquad u_y = -v_x.$$

Thus, $au + bv = c$ implies

$$au_x + bv_x = 0 = au_y + bv_y$$

and, therefore,

$$au_x - bu_y = 0 = au_y + bu_x.$$

In matrix form, this reads

$$\begin{pmatrix} a & -b \\ b & a \end{pmatrix} \begin{pmatrix} u_x \\ u_y \end{pmatrix} = \begin{pmatrix} 0 \\ 0 \end{pmatrix}.$$

Since the matrix has nonzero determinant $a^2 + b^2$, the homogeneous system has only the zero solution. Hence, $u_x = u_y = 0$. By the Cauchy–Riemann equations, $v_x = v_y = 0$. Since D is connected, f is constant.

Solution to 5.6.2: $f(z) = \sqrt{z}$ is a counterexample. Define it by making a cut on the negative real axis and choosing an associated branch of the logarithm:

$$f(z) = \begin{cases} e^{\frac{1}{2} \log z} & z \neq 0 \\ 0 & z = 0. \end{cases}$$

f is analytic in the right half-plane and so on the disc $|z - 1| < 1$. Since $\sqrt{z}$ tends to 0 as z tends to 0, f is continuous on the disc $|z - 1| \leq 1$. However, f cannot be analytic on any open disc of radius larger than 1. For if it were, f would be analytic at 0, so

$$f'(0) = \lim_{z \to 0} \frac{f(z)}{z}$$

would exist and be finite, which is absurd.

Solution to 5.6.3: 1. $f(z) = z^2$ is entire and satisfies

$$f(1/n) = f(-1/n) = 1/n^2.$$

2. By the Identity Theorem [MH87, p. 397], in a disc centered at the origin, g would have to be z^3 and $-z^3$, which is not possible; therefore, no such function g can exist.

Solution to 5.6.4: Let $\sum_{n=0}^{\infty} c_n z^n$ be the power series of such an f at the origin. Then

$$c_0 = f(0) = (f(0))^k = c_0^k,$$

so c_0 is either 0 or a $(k-1)^{th}$ root of unity. Assume $c_0 \neq 0$ but that f is not constant. Let j be the smallest positive integer such that $c_j \neq 0$. Then $f(z^k) = c_0 + c_j z^{jk} +$ higher order terms, so $(f(z))^k = c_0 + kc_j z^j +$ higher order terms, which gives a contradiction. Hence, if $f(0) \neq 0$, then f is constant and equal to a $(k-1)^{th}$ root of unity.

Assume $f(0) = 0$ but f is not identically 0. Let j, as above, be the smallest positive integer such that $c_j \neq 0$. We then have $f(z) = z^j g(z)$, where g is entire and $g(0) = c_j$. The function $z \mapsto z^j$ satisfies the condition imposed on f, so g also satisfies that condition. By the preceding argument, g is constant, equal to a $(k-1)^{th}$ root of unity.

Then f is either the zero function or $f(z) = cz^j$, where c is a $(k-1)^{th}$ root of unity and j is a nonnegative integer.

Solution to 5.6.5: We first prove existence. Multiplying the rational function R by a suitably high power of F, we can reduce the general case to the case where R is a polynomial. Assuming R is a polynomial, we can use the division algorithm to write $R = a_0 + R_1 F$ where a_0 and R_1 are polynomials and $\deg a_0 < d$. Necessarily, $\deg R_1 < \deg R$. If $\deg R_1 \geq d$ we can use the division algorithm again to write $R_1 = a_1 + R_2 F$ where $\deg a_1 < d$ and $\deg R_2 < \deg R_1$. The procedure can be iterated and with each iteration the degree of the polynomial multiplying F is reduced. After finitely many iterations we arrive at the desired representation.

To prove uniqueness it will suffice to show that if

$$\sum_{k=m}^{n} a_k F^k = 0 \tag{$*$}$$

then every a_k is 0. Assume not. After multiplying the expression by a suitable power of F, we can reduce to the case where $m = 0$ and $a_0 \neq 0$. Then the $\sum_{k=1}^{n} a_k F^k \neq 0$, and $(*)$ implies that F divides a_0, a contradiction because $\deg a_0 < d = \deg F$.

Solution to 5.6.6: Let f be such a function. For $n = 2, 3, \ldots$ the singularity at $1/n$ is removable because f is bounded in a punctured neighborhood of $1/n$.

Therefore f extends holomorphically to $\mathbb{D} \setminus \{0\}$. Then the origin becomes an isolated singularity of the extended f, so, by the same reasoning, f extends holomorphically to $\mathbb{D}$.

Solution to 5.6.7: No. Let $U = \mathbb{C} \setminus \{z \in \mathbb{R} \mid z \leqslant 0\}$, let f be the branch of $\sqrt{z}$ on U with $f(1) = 1$, and let $a = -1 + i$. Then the Taylor series for f at a has radius of convergence $\sqrt{2}$, since there is a holomorphic branch of $\sqrt{z}$ defined on $D = \{z \mid |z - a| < \sqrt{2}\}$ that agrees with f in a neighborhood of $a = -1 + i$.

But f cannot extend to a holomorphic function on $U \cup D$, because $-1 \in D$, and the limit of $f(z)$ as z approaches -1 along the unit circle does not exist:

$$\lim_{t \to \pi_-} f(e^{it}) = \lim_{t \to \pi_-} e^{it/2} = i \quad \text{and} \quad \lim_{t \to (-\pi)_+} f(e^{it}) = \lim_{t \to (-\pi)_+} e^{it/2} = -i \,.$$

Solution to 5.6.8: 1. Write $f = u + iv$. Then the Cauchy–Riemann equations hold in U. By hypothesis, $v = 0$ and hence $u_x = u_y = 0$. The open connected set U is step-path connected in that given z_1, z_2 in U there is a path consisting of line segments parallel to the axes contained in U from z_1 to z_2. The Mean Value Theorem applied to u then shows that $u(z_1) = u(z_2)$, and thus $f = u + iv$ is constant.

2. Write $A = \{\Re g(z) + \Im g(z) \mid z \in W\} \subset \mathbb{R}$, and let $h : \mathbb{C} \to \mathbb{R}$; $w \mapsto \Re w + \Im w$. Then $A = h(g(W))$. Fix $x = h(g(z)) \in A$. The hypothesis implies that g is nonconstant. By the Open Mapping Theorem [MH87, p. 436] $g(z)$ is an interior point of $g(W)$ and thus $g(W)$ contains some disc centered at $g(z)$ with radius $r > 0$. It is clear that $h(g(W))$ then contains the open interval $(x - r, x + r)$, proving that A is an open set.

Solution to 5.6.9: Suppose $f(\mathbb{C})$ is not dense. Then, for some $w \in \mathbb{C}$ and $\varepsilon > 0$, we have $|f(z) - w| \geqslant \varepsilon$ for all $z \in \mathbb{C}$. The function $1/(f(z) - w)$ is then entire and bounded in modulus by $1/\varepsilon$, so, by Liouville's Theorem [MH87, p. 170], is a constant, and so is f.

Solution to 5.6.10: We may assume that $0 \in L$, otherwise we use $f + c$ instead of f, where c is a constant. We may even assume that L is the imaginary axis, otherwise we use αf in the place of f, where α is a constant of modulus 1. Since $f(\mathbb{C})$ is connected, it is contained in one of the halfplanes $\Re z > 0$, $\Re z < 0$. We may assume that $f(\mathbb{C})$ is contained in the left halfplane (otherwise we use $-f$ instead of f). Then g is an entire bounded function, where $g(z) = e^{f(z)}$. By Liouville's Theorem, g is constant, $g(z) = k$, say. Therefore, $f(\mathbb{C})$ is contained in the solution set of $e^z = k$, which is discrete. As $f(\mathbb{C})$ is connected we conclude that it must contain only a point, so f is constant as well.

Solution 2. By Picard's Theorem the image of an nonconstant entire function omits at most one complex number, [Ahl79, Sec. 8.3] and [Car63b, Sec 4.4], so the result follows. See the previous problem.

Solution to 5.6.11: 1. If f is entire and $\Re f > 0$, then $1/(1 + f)$ is bounded and entire, hence constant by Liouville's Theorem [MH87, p. 170].

2. Assume $F(z) + F(z)^*$ is positive definite for each z. Since the diagonal entries of a positive definite matrix are positive, it follows that f_{11} and f_{22} have positive real parts. Hence f_{11} and f_{22} are constant by Part 1. Also, the determinant of a positive definite matrix is positive, so

$$f_{11} f_{22} - |f_{12} + \overline{f_{21}}|^2 \geqslant 0.$$

The function $f_{12} + \overline{f_{21}}$ is thus bounded. It follows that the real part of the entire function $f_{12} + f_{21}$ is bounded, hence $f_{12} + f_{21}$ is constant, by Part 1. Similarly, the real part of the entire function $i(f_{12} - f_{21})$ is bounded, so $f_{12} - f_{21}$ is constant, by Part 1, therefore f_{12} and f_{21} are constant.

Solution to 5.6.12: Let $f(x + iy) = u(x, y) + iv(x, y)$, where $u(x, y) = e^x s(y)$ and $v(x, y) = e^x t(y)$. From the Cauchy–Riemann equations [MH87, p. 72], we get $e^x s(y) = e^x t'(y)$, so $s(y) = t'(y)$. Similarly, $s'(y) = -t(y)$. This equation has the unique solution $s(y) = \cos y$ satisfying the initial conditions $s(0) = 1$ and $s'(0) = -t(0) = 0$, which, in turn, implies that $t(y) = -s'(y) = \sin y$.

Solution to 5.6.13: $f'' + f$ is analytic on $\mathbb{D}$ and vanishes on $X = \{1/n \mid n \geqslant 0\}$, so it vanishes identically. Using the Maclaurin expansion [MH87, p. 234] of f, we get

$$\sum_{k \geqslant 0} \frac{f^{(k)}(0)}{k!} z^k = -\sum_{k \geqslant 0} \frac{f^{(k+2)}(0)}{k!} z^k.$$

So we have

$$f(0) = -f''(0) = \cdots = (-1)^k f^{(2k)}(0) = \cdots$$

and

$$f'(0) = -f'''(0) = \cdots = (-1)^k f^{(2k+1)}(0) = \cdots.$$

Therefore,

$$f(z) = f(0) \sum_{k \geqslant 0} \frac{(-1)^k}{(2k)!} z^{2k} + f'(0) \sum_{k \geqslant 0} \frac{(-1)^k}{(2k+1)!} z^{2k+1}$$

$$= f(0) \cos z + f'(0) \sin z.$$

Conversely, any linear combination of $\cos z$ and $\sin z$ satisfies the given equation, so these are all such functions.

Solution to 5.6.14: Let a be the intersection of the lines. Replacing $f(z)$ by $f(z + a) - f(a)$, we may reduce to the case $a = f(a) = 0$. Then $f(z) = cz^n(1 + o(1))$ as $z \to 0$, for some nonzero $c \in \mathbb{C}$ and $n \geqslant 1$. In particular, $f(z)$ is nonvanishing in some punctured neighborhood of 0. If the lines are in the directions of $e^{i\alpha}$ and $e^{i\beta}$, then for real t, $f(te^{i\alpha})$ and $f(te^{i\beta})$ are real, and so is

$$\lim_{t \to 0} \frac{f(te^{i\alpha})}{f(te^{i\beta})} = \lim_{t \to 0} \frac{ce^{ni\alpha} t^n(1 + o(1))}{ce^{ni\beta} t^n(1 + o(1))} = e^{ni(\alpha - \beta)}.$$

This forces $ni(\alpha - \beta) \in \pi i \mathbb{Z}$, so $(\alpha - \beta)/\pi$ is rational, as desired.

Solution to 5.6.15: It is enough to show that for any $z \in \Omega$, the derivative in the sense of $\mathbb{R}^2$ has an associated matrix

$$Df(z) = \begin{pmatrix} a & c \\ b & d \end{pmatrix}$$

satisfying $a = d$ and $c = -b$. As $Df(z)(1, 0) \perp Df(z)(0, 1)$, we have $c = -kb$ and $d = ka$ for some k. As f preserves orientation, $\det Df(z) > 0$, so $k > 0$.

If $a^2 + b^2 = 0$, then $Df(z) = 0$ and there is nothing to show.

Assume $a^2 + b^2 \neq 0$. As, for $(x, y) \neq 0$, $Df(z)(x, y) \perp Df(z)(-y, x)$, we have $0 = (k^2 - 1)(a^2 + b^2)xy$. Therefore, $k = 1$ and the result follows.

Solution to 5.6.16: We have $f = f^3/f^2 = g/h$. It is clear that f^3 and f^2 have the same zero set. If z_0 is a common zero, there are analytic functions g_1 and h_1 which are not zero at z_0 such that $f^3(z) = (z - z_0)^k h_1(z)$ and $f^2(z) = (z - z_0)^j g_1(z)$. But $\left(f^3\right)^2 = f^6 = \left(f^2\right)^3$, so $(z - z_0)^{2k} h_1(z)^2 = (z - z_0)^{3j} g_1(z)^3$. Rearranging, we get $h_1(z)^2/g_1(z)^3 = (z - z_0)^{3j-2k}$. Neither h_1 nor g_1 are zero at z_0, so the left side is analytic and nonzero at z_0. Hence, we must have $3j - 2k = 0$, so $k > j$. Therefore, z_0 has a higher multiplicity as a zero of f^3 than it does as a zero of f^2. Thus, the function f^3/f^2 has a removable singularity at z_0. Since this holds for every zero, $f = f^3/f^2$ can be extended to an analytic function on $\mathbb{D}$.

Solution to 5.6.17: Let $g = f^2$ and let the common domain of f and g be G. We will show that $g(G)$ contains no path with winding number 1 about 0. Suppose that for some path $\gamma : [0, 1] \to G$, $g(\gamma)$ had winding number 1 about 0. Since $g(\gamma)$ is compact, there is a finite cover of it by n open, overlapping balls, none of which contain 0. In the first ball, define the function $h_1(z) = \sqrt{g(z)}$, where the branch of the square root is chosen so that $h_1(\gamma(0)) = f(\gamma(0))$. In each successive ball, define the function $h_k(z) = \sqrt{g(z)}$, with the branch chosen so that h_k is an analytic continuation of h_{k-1}. This implies that if $\gamma(t)$ is in the domain h_k, we must have $h_k(\gamma(t)) = f(\gamma(t))$. However, since these analytic extensions wrap around the origin, $h_n(\gamma(1)) = h_n(\gamma(0)) \neq h_1(\gamma(0))$, which contradicts the continuity of f.

Therefore, since $g(G)$ contains no path with winding number 1 about the origin, there exists a branch of the square root on it which is analytic and such that $f(z) = \sqrt{g(z)}$ for all z in G. Thus, f is an analytic function.

Solution to 5.6.18: 1. Let w be an nth primitive root of unity. Then the function $w^k g, 0 \leqslant k \leqslant n - 1$ are all n^{th} roots of f, analytic and distinct.

Let h be any analytic n^{th} root of f. Fix $z_0 \in G$, $\varepsilon > 0$ such that $f(z) \neq 0$ for $|z - z_0| < \varepsilon$. h/g is continuous in $|z - z_0| < \varepsilon$ and since $h^n/g^n = 1$, h/g has its range among the n points $1, w, \ldots, w^{n-1}$. Since it is continuous, it must be constant. Therefore, $h = w^k g$ for some k in this little neighborhood, so $h \equiv w^k g$.

2. The function $f : [0, 1] \to \mathbb{R}$ defined by $f(x) = (x - 1/2)^2$ has four continuous square roots, f_1, f_2, f_3, and f_4, given by

$$f_1(x) = x - 1/2 \qquad f_2(x) = -f_1(x)$$

$$f_3(x) = \begin{cases} x - 1/2 & \text{for } 0 \leqslant x \leqslant 1/2 \\ 1/2 - x & \text{for } 1/2 \leqslant x \leqslant 1 \end{cases} \qquad f_4(x) = -f_3(x).$$

Solution to 5.6.19: 1. Since $f(z) \neq 0$ for all $z \in \mathbb{D}$, $\dfrac{f'(z)}{f(z)}$ is a holomorphic function on $\mathbb{D}$. For $z \in \mathbb{D}$ set $h(z) = \displaystyle\int_{[0,z]} \dfrac{f'(z)}{f(z)} dz$, where $[0, z]$ stands for the oriented segment from the origin to z. Using the independence of path, which is a consequence of Cauchy's Theorem [MH87, p. 152], we get $h'(z) = \dfrac{f'(z)}{f(z)}$ for all $z \in \mathbb{D}$, so $e^{-h(z)} f(z)$ has derivative $e^{-h(z)}(f'(z) - h'(z)f(z)) = 0$, therefore $e^{-h(z)} f(z) = f(0)$ for all $z \in \mathbb{D}$. Since $f(0) \neq 0$ there exists $c \in \mathbb{C}$ such that $e^c = f(0)$. Set $g(z) = h(z) + c$; then $e^{g(z)} = f(z)$ for all $z \in \mathbb{D}$.

2. No. Consider $D = \mathbb{C} \setminus \{0\}$ and $f(z) = z$. Suppose $z = e^{g(z)}$ with g holomorphic on D. Then $g'(z) = 1/z$, so, for any closed curve in D, $\displaystyle\int_C \dfrac{dz}{z}$ would vanish, which is absurd since $\displaystyle\int_C \dfrac{dz}{z} = 2\pi i$ if C is the unit circle.

Solution to 5.6.20: The function $g(z) = \overline{f(\bar{z})}$ is analytic in the same region as f, and $f - g = 0$ on $(1, \infty)$. Since the zero set of $f - g$ has limit points in the region $|z| > 1$, the Identity Theorem [MH87, p. 397] implies that $f - g \equiv 0$. Hence, $f(z) \equiv \overline{f(\bar{z})}$. In particular, for x in $(-\infty, -1)$, $f(x) = \overline{f(x)}$.

Solution 2. Let $\sum_{-\infty}^{\infty} c_n z^n$ be the Laurent expansion [MH87, p. 246] of f about ∞. It will suffice to show that c_n is real for all n. The series $\sum_{-\infty}^{\infty} (\Re c_n) z^n$ converges everywhere the original series does (since its terms are dominated in absolute value by those of the original series); let $g(z) = \sum_{-\infty}^{\infty} (\Re c_n) z^n$. For x in $(1, \infty)$,

$$g(x) = \Re f(x) = f(x).$$

As above, the Identity Theorem [MH87, p. 397] implies $g = f$, so each c_n is real, as desired.

Solution to 5.6.21: The function $g(z) = \overline{f(\bar{z})}$ is analytic and coincides with f on the real axis; therefore, it equals f. The line in question is its own reflection with respect to the real axis. Since it also passes through the origin, it must be one of the axes.

Solution to 5.6.22: The function $\overline{f(\bar{z})}$ is also entire, and it agrees with f on the real axis. By the Identity Theorem [MH87, p. 397], it equals f everywhere. In particular, then, $f(x - \pi i) = f(x + \pi i)$ for all real x. On the line $\Im z = -\pi$

the function $f(z + 2\pi i) - f(z)$ vanishes identically, so it is identically 0, by the Identity Theorem.

Solution to 5.6.23: By the Schwarz Reflection Principle for circles [BN82, p. 85], we have

$$f(z) = \overline{f(1/\bar{z})}.$$

For x real, we get

$$f(x) = \overline{f(1/x)} = \overline{f(x)}$$

so $f(x)$ is real.

Solution to 5.6.24: 1. Let $z_1, z_2, \ldots, z_n$ be the zeros of p, ennumerated with multiplicities, so that

$$p(z) = c(z - z_1)(z - z_2) \cdots (z - z_n)$$

where c is a constant. Then

$$\frac{p'(z)}{p(z)} = \frac{1}{z - z_1} + \frac{1}{z - z_2} + \cdots + \frac{1}{z - z_n}$$

and, for x real,

$$\Im \frac{p'(z)}{p(z)} = \frac{\Im z_1}{|x - z_1|^2} + \frac{\Im z_2}{|x - z_2|^2} + \cdots + \frac{\Im z_n}{|x - z_n|^2}.$$

Part 1 is now obvious.

2. Write $z_j = x_j + y_j$, so that

$$\int_{-\infty}^{\infty} \frac{\Im z_j}{|x - z_j|^2} = \int_{-\infty}^{\infty} \frac{y_j}{(x - x_j)^2 + y_j^2} \, dx$$

$$= \int_{-\infty}^{\infty} \frac{y_j}{x^2 + y_j^2} \, dx$$

$$= \int_{-\infty}^{\infty} \frac{d}{dx} \left(\arctan \frac{x}{y_j} \right) dx$$

$$= \pi.$$

Hence,

$$\int_{-\infty}^{\infty} \Im \frac{p'(x)}{p(x)} \, dx = \pi n = \pi \deg p.$$

Solution to 5.6.25: Let D be an open disc with $\overline{D} \subset G$. It will suffice to show that there is an n such that $f^{(n)}$ has infinitely many zeros in D. For then, the zeros of $f^{(n)}$ will have a limit point in G, forcing $f^{(n)}$ to vanish identically in G by the Identity Theorem [MH87, p. 397], and it follows that f is a polynomial of degree, at most, $n - 1$.

By hypothesis, D is the union of the sets $Z_n = \{z \in D \mid f^{(n)}(z) = 0\}$ for $n = 1, 2, \ldots$. Since D is uncountable, at least one Z_n is, in fact, uncountable (because a countable union of finite sets is, at most, countable).

Solution to 5.6.26: Let $u = \Re f$ and $v = \Im f$. For x real, we have

$$f'(x) = \frac{\partial u}{\partial x}(x, 0) = \frac{\partial v}{\partial y}(x, 0),$$

where the first equality holds because $v = 0$ on the real axis and the second one follows from the Cauchy–Riemann equations [MH87, p. 72]. Since v is positive in the upper half-plane, $\frac{\partial v}{\partial y} \geq 0$ on the real axis. It remains to show that f' does not vanish on the real axis.

It suffices to show that $f'(0) \neq 0$. In the contrary case, since f is nonconstant, we have

$$f(z) = cz^k (1 + O(z)) \qquad (z \to 0)$$

where $c \neq 0$ is real and $k \geq 2$. For small z, the argument of the factor $1 + O(z)$ lies between $-\frac{\pi}{4}$ and $\frac{\pi}{4}$, say, whereas on any half-circle in the upper half-plane centered at 0, the factor cz^k assumes all possible arguments. On a sufficiently small such half-circle, therefore, the product will assume arguments between π and 2π, contrary to the assumption that $\Im f(z) > 0$ for $\Im z > 0$. This proves $f'(0) \neq 0$.

Solution to 5.6.27: 1. The function f is not constant, because if f took the constant value c, then $f^{-1}(\{c\})$ would equal U, a noncompact set. Since f is holomorphic and nonconstant, it is an open map, and $f(U)$ is open. Since V is connected, it only remains to show that $f(U)$ is closed relative to V. Let $a \in V \cap \overline{f(U)}$. There is a sequence (w_n) in $f(U)$ such that $a = \lim w_n$. For each n, there is a point z_n in U with $w_n = f(z_n)$. The set $K = \{a, w_1, w_2, \ldots\}$ is a compact subset of V, so $f^{-1}(K)$ is also compact. Since the sequence (z_n) lies in a compact subset of U, it has a subsequence, (z_{n_k}), converging to a point b of U. Then $f(b) = \lim f(z_{n_k}) = \lim w_{n_k} = a$, proving that a is in $f(U)$ and hence that $f(U)$ is closed relative to V.
2. Take $U = V = \mathbb{C}$ and $f(z) = |z|$.

Solution to 5.6.28: By the Inverse Function Theorem [Rud87, p. 221], it will suffice to prove that $Jh(0) \neq 0$, where Jh denotes the Jacobian of h:

$$Jh = \det \begin{pmatrix} \frac{\partial}{\partial x}(u + p) & \frac{\partial}{\partial x}(v - q) \\ \frac{\partial}{\partial y}(u + p) & \frac{\partial}{\partial y}(v - q) \end{pmatrix}$$

By the Cauchy–Riemann equations [MH87, p. 72],

$$\frac{\partial u}{\partial x} = \frac{\partial v}{\partial y} \qquad \frac{\partial u}{\partial y} = -\frac{\partial v}{\partial x} \qquad \frac{\partial p}{\partial x} = \frac{\partial q}{\partial y} \qquad \frac{\partial p}{\partial y} = -\frac{\partial q}{\partial x}.$$

Hence,

$$Jh = \det \begin{pmatrix} \frac{\partial u}{\partial x} + \frac{\partial p}{\partial x} & \frac{\partial v}{\partial x} - \frac{\partial q}{\partial x} \\ -\frac{\partial v}{\partial x} - \frac{\partial q}{\partial x} & \frac{\partial u}{\partial x} - \frac{\partial p}{\partial x} \end{pmatrix}$$

$$= \left(\frac{\partial u}{\partial x}\right)^2 - \left(\frac{\partial p}{\partial x}\right)^2 + \left(\frac{\partial v}{\partial x}\right)^2 - \left(\frac{\partial q}{\partial x}\right)^2$$

$$= |f'|^2 - |g'|^2 .$$

Since $|g'(0)| < |f'(0)|$, it follows that $(Jh)(0) \neq 0$, as desired.

Solution to 5.6.29: f does not have a removable singularity at ∞. If f had an essential singularity at infinity, for any $w \in \mathbb{C}$ there would exist a sequence $z_n \to \infty$ with $\lim f(z_n) = w$. Therefore, f has a pole at infinity and is a polynomial.

Solution to 5.6.30: Clearly, entire functions of the form $f(z) = az + b$, $a, b \in \mathbb{C}$ $a \neq 0$, are one-to-one maps of $\mathbb{C}$ onto $\mathbb{C}$. We will show that these are all such maps by considering the kind of singularity such a map f has at ∞. If it has a removable singularity, then it is a bounded entire function, and, by Liouville's Theorem [MH87, p. 170], a constant.

If it has an essential singularity, then, by the Casorati–Weierstrass Theorem [MH87, p. 256], it gets arbitrarily close to any complex number in any neighborhood of ∞. But if we look at, say, $f(0)$, we know that for some ε and δ, the disc $|z| < \delta$ is mapped onto $|f(0) - z| < \varepsilon$ by f. Hence, f is not injective.

Therefore, f has a pole at ∞, so is a polynomial. But all polynomials of degree 2 or more have more than one root, so are not injective.

Solution to 5.6.31: 1. If w is a period of f, an easy induction argument shows that all integer multiples of w are periods of f. It is also clear that any linear combination of periods of f, with integer coefficients, is a period of f.
2. If f had infinitely many periods in a bounded region, by the Identity Theorem [MH87, p. 397], f would be constant.

Solution to 5.6.32: For $0 < r < r_0$, by the formula for Laurent coefficients [MH87, p. 246], we have

$$|c_n| \leq \frac{1}{2\pi r^{n+1}} \int_{|z|=r} |f(z)| \, |dz|$$

$$= \frac{1}{2\pi r^n} \int_0^{2\pi} |f(re^{i\theta})| \, d\theta$$

$$\leq \sqrt{\frac{M}{2\pi}} r^{-(n+2)} .$$

If $n < -2$, as r gets arbitrarily close to zero, this upper bound gets arbitrarily small. Hence, for $n < -2$, $c_n = 0$.

Solution to 5.6.33: We think of the rational function f as a continuous map of $\overline{\mathbb{C}}$, the Riemann sphere, into itself. The roots of the denominator of f are $\frac{1 \pm \sqrt{1-4a}}{2a}$, whose real parts are positive. Hence f is holomorphic on the closed left half-plane. Let $M = \sup \{|f(z)| \mid \Re z < 0\}$. By the Maximum Modulus Principle [MH87, p. 185], there is no point z in the open left half-plane where $|f(z)| = M$. Take a sequence $(z_n)_{n=1}^{\infty}$ in the open left half-plane such that $|f(z_n)| \to M$. Passing to a subsequence if necessary, we can assume, by the compactness of $\overline{\mathbb{C}}$, that the sequence converges in $\overline{\mathbb{C}}$, to z_0, say. Then, as f is continuous, we have $|f(z_0)| = M$; therefore z_0 is not in the open left half-plane. Hence, either $z_0 = \infty$ or z_0 is on the imaginary axis. We have $f(\infty) = 1$. On the imaginary axis

$$|f(iy)| = \left| \frac{1 + iy - ay^2}{1 - iy - ay^2} \right| = 1.$$

Since numerator and denominator are complex conjugates of each other. Hence $M = |f(z_0)| = 1$, as desired.

Solution 2. The three facts below are verified by straightforward algebra.

- If $w = u + iv \neq -1$, then

$$\Re\left(\frac{w-1}{w+1}\right) = \frac{u^2 + v^2 - 1}{(u+1)^2 + v^2}.$$

Thus, $\Re\left(\frac{w-1}{w+1}\right) < 0 \Leftrightarrow |w| < 1$.

- If $w = \dfrac{1 + z + az^2}{1 - z + az^2}$, then

$$\frac{w-1}{w+1} = \frac{z}{1 + az^2}.$$

- If $z = x + iy$, then

$$\Re\left(\frac{z}{1 + az^2}\right) = \frac{x(1 + ax^2 + ay^2)}{|1 + az^2|^2}.$$

Thus, when $a > 0$, we have $\Re\left(\frac{z}{1+az^2}\right) < 0 \Leftrightarrow \Re z < 0$.

Now let $\Re z < 0$. From the last statement, we have $\Re\left(\frac{z}{1+az^2}\right) < 0$ and from the first two it follows that $|f(z)| < 1$.

Solution to 5.6.34: 1.   The Maclaurin series for $\cos z$ is $\displaystyle\sum_{n=0}^{\infty} \frac{(-1)^n z^{2n}}{(2n)!}$, and it converges uniformly on compact sets. Hence, for fixed z,

$$f(t) \cos zt = \sum_{n=0}^{\infty} \frac{(-1)^n f(t) t^{2n} z^{2n}}{(2n)!}$$

with the series converging uniformly on $[0,1]$. We can therefore, interchange the order of integration and summation to get

$$h(z) = \sum_{n=0}^{\infty} \frac{(-1)^n}{(2n)!} \left(\int_0^1 t^{2n} f(t) dt \right) z^n$$

in other words, h has the power series representation

$$h(z) = \sum_{n=0}^{\infty} c_{2n} z^{2n} \quad \text{with} \quad c_{2n} = \frac{(-1)^n}{(2n)!} \int_0^1 t^{2n} f(t) dt.$$

Since h is given by a convergent power series, it is analytic.

2. Suppose h is the zero function. Then, by Part 1, $\int_0^1 t^{2n} f(t) dt = 0$ for $n = 0, 1, 2, \ldots$. Hence, if p is any polynomial, then $\int_0^1 p(t^2) f(t) dt = 0$. By the Stone–Weierstrass Approximation Theorem [MH93, p. 284], there is a sequence $\{p_k\}$ of polynomials such that $p_k(t) \to f\left(\sqrt{t}\right)$ uniformly on $[0,1]$. Then $p_k(t^2) \to f(t)$ uniformly on $[0,1]$, so

$$\int_0^1 f(t^2) dt = \lim_{k \to \infty} \int_0^1 p_k(t^2) f(t) dt = 0$$

implying that $f \equiv 0$.

Solution to 5.6.35: Let $z \in \mathbb{C}$. We have

$$g(z) = \int_0^1 \sum_{n=0}^{\infty} f(t) \frac{t^n z^n}{n!} dt.$$

Since f is bounded, this series converges uniformly in t, so we can change the order of summation and get

$$g(z) = \sum_{n=0}^{\infty} z^n \left(\frac{1}{n!} \int_0^1 f(t) t^n \, dt \right) = \sum_{n=0}^{\infty} \xi_n z^n$$

where

$$\xi_n = \frac{1}{n!} \int_0^1 f(t) t^n \, dt.$$

We have

$$|\xi_n| \leq \frac{1}{n!} \int_0^1 |f(t)| \, dt$$

so the radius of convergence of the series of g is ∞.

Solution 2. Let $z_0 \in \mathbb{C}$. We have

$$\frac{g(z) - g(z_0)}{z - z_0} = \int_0^1 f(t) \frac{e^{tz} - e^{tz_0}}{z - z_0} dt.$$

From the power series expansion $e^{t(z-z_0)} = \sum_{n=0}^{\infty} \dfrac{t^n(z-z_0)^n}{n!}$, one gets

$$\frac{e^{tz} - e^{tz_0}}{z - z_0} = te^{tz_0} + O(z - z_0)$$

uniformly on $0 \leqslant t \leqslant 1$, when $z \to z_0$. Thus, as $z \to z_0$, the integrand in the integral above converges uniformly on $[0, 1]$ to $tf(t)e^{tz_0}$, and one can pass to the limit under the integral sign to get

$$\lim_{z \to z_0} \frac{g(z) - g(z_0)}{z - z_0} = \int_0^1 tf(t)e^{tz_0} dt,$$

proving that g is differentiable at z_0.

Solution 3. The integrand in the integral defining g is a continuous function of the pair of variables $(t, z) \in [0, 1] \times \mathbb{C}$, implying that g is continuous. If $R \subset \mathbb{C}$ is a rectangle, then by Fubini's Theorem [MH93, p. 500] and the analyticity of e^{tz} with respect to z,

$$\int_R g(z)dz = \int_R \int_0^1 f(t)e^{tz} dt\, dz = \int_0^1 \int_R f(t)e^{tz} dz\, dt = 0.$$

By Morera's Theorem [MH87, p. 173], g is analytic.

Solution to 5.6.36: 1. Let f and g have the Maclaurin expansions [MH87, p. 234]

$$f(z) = \sum_{k=0}^{\infty} a_n z^n, \qquad g(z) = \sum_{k=0}^{\infty} b_n z^n.$$

By Cauchy's Integral Formula [MH87, p. 167] and the uniform convergence of the series for g, we have

$$\frac{1}{2\pi i} \int_{C_r} \frac{1}{w} f(w) g\left(\frac{z}{w}\right) dw = \frac{1}{2\pi i} \int_{C_r} \sum_{k=0}^{\infty} b_n z^n \frac{f(w)}{w^{n+1}} dw$$

$$= \sum_{k=0}^{\infty} b_n z^n \left(\frac{1}{2\pi i} \int_{C_r} \frac{f(w)}{w^{n+1}} dw \right)$$

$$= \sum_{k=0}^{\infty} a_n b_n z^n.$$

As

$$\limsup_{n \to \infty} \sqrt[n]{|a_n|} \limsup_{n \to \infty} \sqrt[n]{|b_n|} \geqslant \limsup_{n \to \infty} \sqrt[n]{|a_n b_n|}$$

the radius of convergence of h is at least 1.

2. If we take $f(z) = \sin z$ and $g(z) = \cos z$, we have $a_{2n} = b_{2n-1} = 0$ for $n = 1, 2, \ldots$, so $h \equiv 0$.

Solution to 5.6.37: 1. The function f is defined on the open set $\Omega = \mathbb{C} \setminus [0, 1]$. Suppose $D(a, R) \subset \Omega$. For every $z \in D(a, R)$ and $t \in [0, 1]$,

$$\left| \frac{z-a}{t-a} \right| \leqslant \frac{|z-a|}{R} < 1.$$

Thus,

$$\sum_{n=0}^{\infty} \frac{F(t)(z-a)^n}{(t-a)^{n+1}} = \frac{F(t)}{(t-a)} \frac{1}{\left(1 - \frac{z-a}{t-a}\right)} = \frac{F(t)}{t-z}.$$

The convergence of the series, for a fixed $z \in D(a, R)$, is uniform in $t \in [0, 1]$. This follows from Weierstrass's M-test and the inequality

$$\left| \frac{F(t)(z-a)^n}{(t-a)^{n+1}} \right| \leqslant \frac{\|F\|}{R} \left(\frac{|z-a|}{R} \right)^n.$$

Thus we have the power series expansion, valid for $z \in D(a, R)$,

$$f(z) = \int_0^1 \frac{F(t)}{t-z} dt = \int_0^1 \sum_{n=0}^{\infty} \frac{F(t)(z-a)^n}{(t-a)^{n+1}} dt = \sum_{n=0}^{\infty} c_n(z-a)^n$$

where

$$c_n = \int_0^1 \frac{F(t)}{(t-a)^{n+1}} dt \quad (n = 0, 1, 2, \ldots).$$

This proves that f is analytic in Ω.

2. For $|z| > 1$ and $0 \leqslant t \leqslant 1$, we have,

$$\frac{F(t)}{t-z} = -\frac{F(t)}{z} \cdot \frac{1}{\left(1 - \frac{t}{z}\right)} = -\frac{F(t)}{z} \sum_{n=0}^{\infty} \frac{t^n}{z^n}$$

where the convergence, for a fixed z in $|z| > 1$, is uniform for $t \in [0, 1]$ since $|t/z| \leqslant 1/|z| < 1$.

We thus have the Laurent expansion, valid in $|z| > 1$,

$$f(z) = -\sum_{n=0}^{\infty} z^{-(n+1)} \int_0^1 t^n F(t) dt = \sum_{n=1}^{\infty} b_n z^{-n}$$

where

$$b_n = -\int_0^1 t^{n-1} F(t) dt \quad (n = 1, 2, \ldots).$$

The above is a transform $F \in C[0, 1] \mapsto f \in H(\Omega)$. We have seen above that the Laurent coefficients of f are the moments $\int_0^1 t^n F(t) dt$ $(n \geqslant 0)$ of F. Now the Laurent coefficients are determined by f. To show that F is determined by f it is sufficient to show that the moments of F determine F. This in turn is a consequence of the following result.

If $\varphi \in C[0, 1]$, and $\int_0^1 t^n \varphi(t)\, dt = 0$ for all $n \geqslant 0$, then $\varphi = 0$.

It suffices to consider real-valued φ. By hypothesis, $\int_0^1 p(t)\varphi(t)\, dt = 0$ for all polynomials p. The Weierstrass Approximation Theorem asserts that there is a sequence of polynomials p_k which converges to φ uniformly on $[0, 1]$. Thus $\int_0^1 p_k(t)\varphi(t)\, dt \to \int_0^1 \varphi^2(t)\, dt$ as $k \to \infty$, and so $\int_0^1 \varphi^2(t)\, dt = 0$. This implies that $\varphi = 0$ on $[0, 1]$.

Solution to 5.6.38: First we will show that $\int_\Gamma p(z) f(z) dz = 0$ for every polynomial p. To see this consider the polynomial $p(z)$ and $p(z) + 1$ apply the condition and subtract them, we find that

$$\int_\Gamma (2p(z) + 1) f(z) dz = 0.$$

and since every polynomial can be written in the form $2p(z) + 1$, we conclude that

$$\int_\Gamma p(z) f(z) dz = 0,$$

for every polynomial p.

Suppose that f has a pole of order n at $a \in \mathbb{C}$. Then $(z - a)^{n-1} f(z)$ has a nonzero residue at a, therefore,

$$\int_\Gamma (z - a)^{n-1} f(z) dz \neq 0,$$

for a sufficiently small loop around a. Thus, f cannot have any poles and must be entire.

5.7   Cauchy's Theorem

Solution to 5.7.1: By Cauchy's Integral Formula [MH87, p. 167], we have

$$e^0 = \frac{1}{2\pi i} \int_{|z|=1} \frac{e^z}{z} dz = \frac{1}{2\pi} \int_0^{2\pi} e^{e^{i\theta}}\, d\theta$$

therefore,

$$\int_0^{2\pi} e^{e^{i\theta}}\, d\theta = 2\pi.$$

Solution 2. We can also compute it using residues. Parametrize the unit circle by $\gamma(t) = e^{it}$ for $0 \leqslant t \leqslant 2\pi$. Then

$$\int_\gamma \frac{e^z}{z} dz = \int_0^{2\pi} \frac{e^{\gamma(t)}}{\gamma(t)} \gamma'(t)\, dt = \int_0^{2\pi} \frac{e^{e^{it}}}{e^{it}} i e^{it}\, dt = i \int_0^{2\pi} e^{e^{it}}\, dt.$$

Since e^z/z is a quotient of analytic functions, it is itself analytic everywhere except at $z = 0$, hence the only singularity lying inside γ is the origin, so

$$\int_0^{2\pi} e^{e^{it}}\, dt = \frac{1}{i}\int_\gamma \frac{e^z}{z}\, dz = 2\pi\,\mathrm{Res}\left(\frac{e^z}{z}, 0\right) = 2\pi \lim_{z\to 0} z\cdot\frac{e^z}{z} = 2\pi\,.$$

Solution to 5.7.2: By Cauchy's Integral Formula for derivatives [MH87, p. 169], we have

$$\frac{d}{dz}e^z\Big|_{z=0} = \frac{1}{2\pi i}\int_{|z|=1}\frac{e^z}{z^2}\, dz = \frac{1}{2\pi}\int_0^{2\pi} e^{e^{i\theta}-i\theta}\, d\theta$$

therefore,

$$\int_0^{2\pi} e^{e^{i\theta}-i\theta}\, d\theta = 2\pi.$$

Solution to 5.7.3: We'll apply Cauchy's Integral Formula [MH87, p. 167] to $f(z) = 1/(1-\bar{a}z)$, which is holomorphic on a neighborhood of $|z| \leqslant 1$. We have

$$\begin{aligned}
\int_{|z|=1}\frac{|dz|}{|z-a|^2} &= \int_{-\pi}^{\pi}\frac{1}{|e^{i\theta}-a|^2}\, d\theta \\
&= \int_{-\pi}^{\pi}\frac{1}{(e^{i\theta}-a)(e^{-i\theta}-\bar{a})}\, d\theta \\
&= \int_{-\pi}^{\pi}\frac{e^{i\theta}}{(e^{i\theta}-a)(1-\bar{a}e^{i\theta})}\, d\theta \\
&= \frac{1}{i}\int_{|z|=1}\frac{1}{(z-a)(1-\bar{a}z)}\, dz \\
&= \frac{2\pi}{1-|a|^2}\,.
\end{aligned}$$

Solution 2. The function $f(z) = 1/(\bar{a}z-1)(z-a)$ has a simple pole inside the circle $|z| = 1$ at $z = a$, with residue $1/(|a|^2 - 1)$. Therefore, we have, by the Residue Theorem [MH87, p. 280],

$$\begin{aligned}
\int_{|z|=1}\frac{|dz|}{|z-a|^2} &= i\int_{|z|=1}\frac{dz}{(\bar{a}z-1)(z-a)} \\
&= 2\pi\,\mathrm{Res}\left(\frac{1}{(\bar{a}z-1)(z-a)}, a\right) \\
&= \frac{2\pi}{1-|a|^2}\,.
\end{aligned}$$

Solution to 5.7.4: The integrand equals

$$\frac{1}{\left(ae^{i\theta}-b\right)^2\left(ae^{-i\theta}-b\right)^2} = \frac{e^{2i\theta}}{\left(ae^{i\theta}-b\right)^2\left(a-be^{i\theta}\right)^2}\,.$$

Thus, I can be written as a complex integral,

$$I = \frac{1}{2\pi i} \int_{|z|=1} \frac{z}{(az-b)^2(a-bz)^2}\, dz = \frac{1}{2\pi i b^2} \int_{|z|=1} \frac{z}{(az-b)^2(z-\frac{a}{b})^2}\, dz.$$

By Cauchy's Theorem for derivatives, we have

$$I = \frac{1}{b^2} \frac{d}{dz}\left(\frac{z}{(az-b)^2}\right)\Bigg|_{z=\frac{a}{b}} = \frac{a^2+b^2}{(b^2-a^2)^3}.$$

Solution to 5.7.5: Let $p(z) = a_n z^n + \cdots + a_0$. If p has no zeros then $1/p$ is entire. As $\lim_{|z|\to\infty} p(z) = \infty$, $1/p$ is bounded. By Liouville's Theorem [MH87, p. 170] $1/p$ is constant, and so is p.

Solution 2. Let $p(z) = a_n z^n + \cdots + a_0$, $n \geqslant 1$. If p has no zeros, then $1/p$ is entire. As $\lim_{|z|\to\infty} p(z) = \infty$, the Maximum Modulus Principle [MH87, p. 185] gives

$$\max_{z\in\mathbb{C}} \frac{1}{|p(z)|} = \lim_{R\to\infty} \max_{|z|\leqslant R} \frac{1}{|p(z)|} = \lim_{R\to\infty} \max_{|z|=R} \frac{1}{|p(z)|} = 0$$

which is a contradiction.

Solution 3. For $p(z) = a_n z^n + \cdots + a_0$ with $n \geqslant 1$ let the functions f and g be given by $f(z) = a_n z^n$, $g(z) = p(z) - f(z)$. For $R > 1$ consider the circle centered at the origin with radius R, C_R. For $z \in C_R$ we have

$$|f(z)| = |a_n|R^n \quad \text{and} \quad |g(z)| \leqslant (|a_0| + \cdots + |a_{n-1}|)\, R^{n-1}.$$

Therefore, on C_R,

$$|g| < |f| \quad \text{if} \quad R > \frac{|a_0| + \cdots + |a_{n-1}|}{|a_n|}$$

so, by Rouché's Theorem [MH87, p. 421], $f + g = p$ has n zeros in $\{z \in \mathbb{C} \mid |z| < R\}$.

Solution 4. Let $P(z)$ be a nonconstant polynomial. We may assume $P(z)$ is real for real z, otherwise we consider $P(z)\overline{P}(z)$. Suppose that P is never zero. Since $P(z)$ does not either vanish or change sign for real z, we have

$$(*) \qquad \int_0^{2\pi} \frac{d\theta}{P(2\cos\theta)} \neq 0.$$

But

$$\int_0^{2\pi} \frac{d\theta}{P(2\cos\theta)} = \frac{1}{i} \int_{|z|=1} \frac{dz}{z P(z + z^{-1})}$$

$$= \frac{1}{i} \int_{|z|=1} \frac{z^{n-1} dz}{Q(z)}$$

where $Q(z) = zP(z + z^{-1})$ is a polynomial. For $z \neq 0$, $Q(z) \neq 0$; in addition, if a_n is the leading coefficient of P, we have $Q(0) = a_n \neq 0$. Since $Q(z)$ is never zero, the last integrand is analytic and, hence, the integral is zero, by Cauchy's Theorem [MH87, p. 152], contradicting $(*)$.

This solution is an adaptation of [Boa64].

Solution 5. Let $p(z) = a_n z^n + \cdots + a_0$, $n \geqslant 1$. We know that $\lim_{|z| \to \infty} |p(z)| = \infty$, thus the preimage, by p, of any bounded set is bounded. Let w be in the closure of $p(\mathbb{C})$. There exists a sequence $\{w_n\} \subset p(\mathbb{C})$ with $\lim_n w_n = w$. The set $\{w_n \mid n \in \mathbb{N}\}$ is bounded so, by the previous observation, so is its preimage, $p^{-1}(\{w_n \mid n \in \mathbb{N}\}) = X$. X contains a convergent sequence, $z_n \to z_0$, say. By continuity we have $p(z_0) = w$, so $w \in p(\mathbb{C})$. We proved then that $p(\mathbb{C})$ is closed. As any analytic function is open we have that $p(\mathbb{C})$ is closed and open. As only the empty set and $\mathbb{C}$ itself are closed and open we get that $p(\mathbb{C}) = \mathbb{C}$ and p is onto. In fact, all we need here, in order to show that p is open, is that $p : \mathbb{R}^2 \to \mathbb{R}^2$ has isolated singularities, which guides us into another proof.

Solution 6. By the Solution to Problem 2.2.9 the map $p : \mathbb{R}^2 \to \mathbb{R}^2$ is onto guaranteeing a point where $p(x, y) = (0, 0)$.

Solution 7. Let $p(z) = a_n z^n + \cdots + a_0$, $n \geqslant 1$. Consider the polynomial q given by $q(z) = \overline{a_n} z^n + \cdots + \overline{a_0}$. Assume p has no zeros. As the conjugate of any root of q is a root of p, q is also zero free. Then the function $1/pq$ is entire. By Cauchy's Theorem [MH87, p. 152], we have

$$\int_\Gamma \frac{dz}{p(z)q(z)} = 0$$

where Γ is the segment from $-R$ to R in the horizontal axis together with the half-circle $C = \{z \in \mathbb{C} \mid |z| = R, \Im(z) > 0\}$. But we have

$$\int_\Gamma \frac{dz}{p(z)q(z)} = \int_C \frac{dz}{p(z)q(z)} + \int_{-R}^R \frac{dz}{|p(z)|^2}$$

$$= o(1) + \int_{-R}^R \frac{dz}{|p(z)|^2} \quad (R \to \infty)$$

this gives, for large R,

$$\int_{-R}^R \frac{dz}{|p(z)|^2} = 0$$

which is absurd since the integrand is a continuous positive function.

Solution 8. Let $p(z) = a_n z^n + \cdots + a_0$, $n \geqslant 1$. For R large enough, as $\lim_{|z| \to \infty} p(z) = \infty$, $|p|$ has a minimum in $\{z \in \mathbb{C} \mid |z| < R\}$, at z_0, say. Suppose $p(z_0) \neq 0$. Expanding p around z_0 we get

$$p(z) = p(z_0) + \sum_{j=k}^n b_j (z - z_0)^j \quad b_k \neq 0.$$

Let w be a k-root of $-p(z_0)/b_k$. We get, for $\varepsilon > 0$,

$$p(z_0 + w\varepsilon) = p(z_0) + b_k w^k \varepsilon^k + \sum_{j=k+1}^{n} b_j w^j \varepsilon^j$$

$$= p(z_0)(1 - \varepsilon^k) + \sum_{j=k+1}^{n} b_j w^j \varepsilon^j$$

therefore, for ε small enough, we have $|z_0 + w\varepsilon| < R$ and

$$|p(z_0 + w\varepsilon)| \leqslant |p(z_0)||1 - \varepsilon^k| - \sum_{j=k+1}^{n} |b_j w^j| \varepsilon^j$$

$$= |p(z_0)| - \left(|p(z_0)| - \sum_{j=k+1}^{n} |b_j w^j| \varepsilon^{j-k}\right) \varepsilon^k$$

$$< |p(z_0)|$$

which contradicts the definition of z_0. We conclude then that $p(z_0) = 0$.

Solution 9. Let $p(z) = a_n z^n + \cdots + a_0$, $n \geqslant 1$. We have

$$\mathrm{Res}\left(\frac{p'}{p}, \infty\right) = - \lim_{|z| \to \infty} z \frac{p'(z)}{p(z)} = -n.$$

As the singularities of $\dfrac{p'}{p}$ occur at the zeros of its denominator, the conclusion follows.

Solution to 5.7.6: By Morera's Theorem [MH87, p. 173], it suffices to show that

$$\int_{\gamma} f(z)\, dz = 0$$

for all rectangles γ in $\mathbb{C}$. Since f is analytic on $\{z \mid \Im z \neq 0\}$, which is simply connected, it is enough to consider rectangles which contain part of the real axis in their interiors.

Let γ be such a rectangle and l be the segment of $\mathbb{R}$ in its interior. For $\varepsilon > 0$ small enough, draw line segments l_1 and l_2 parallel to the real axis at distance ε above and below it, forming contours γ_1 and γ_2.

Since f is continuous, its integral depends continuously on the path. So, as ε tends to 0,

$$(*) \qquad \int_{\gamma_1} f(z)\, dz + \int_{\gamma_2} f(z)\, dz = \int_{\gamma_1 + \gamma_2} f(z)\, dz \longrightarrow \int_{\gamma} f(z)\, dz,$$

since the integrals along l_1 and l_2 have opposite orientation, in the limit, they cancel each other. By Cauchy's Theorem [MH87, p. 152], the left side of $(*)$ is always 0, so

$$\int_{\gamma} f(z)\, dz = 0.$$

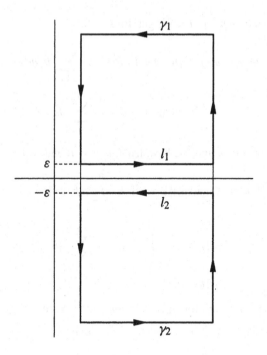

Solution to 5.7.7: Since $p(z)$ has lower degree than $q(z)$, $f(z) = O(1/z)$ as $z \to \infty$, so the integrand is $O(1/t^2)$ as $t \to \infty$, and the integral converges. For $R > |z_0|$, let γ_1 be the straight-line path from $-R$ to R, and let γ_2 be the counterclockwise semicircle $\gamma_2(t) = Re^{it}$, $t \in [0, \pi]$. Let γ be the closed contour consisting of γ_1 followed by γ_2. The function $f(z)$ is holomorphic on and inside γ. By Cauchy's formula,

$$f(z_0) = \frac{1}{2\pi i} \int_\gamma \frac{f(z)}{z - z_0} \, dz = \frac{1}{2\pi i} \int_{-R}^{R} \frac{f(z)}{z - z_0} \, dz + \frac{1}{2\pi i} \int_{\gamma_2} \frac{f(z)}{z - z_0} \, dz.$$

We will obtain the desired formula by taking the limit of both sides as $R \to \infty$. The first term on the right tends to

$$\frac{1}{2\pi i} \int_{-\infty}^{+\infty} \frac{f(t)}{t - z_0} \, dt \qquad \text{as} \quad R \to \infty.$$

The second term on the right tends to 0 as $R \to \infty$, since the integrand is $O(1/R^2)$ while the length of γ_2 is $O(R)$.

Solution to 5.7.8: We have, using the fact that the exponential is 2π-periodic,

$$\frac{1}{2\pi i} \int_{|z|=R} z^{n-1} |f(z)|^2 \, dz = \frac{1}{2\pi i} \int_{|z|=R} z^{n-1} \left| \sum_{i=0}^{n} a_i z^i \right|^2 \, dz$$

$$= \frac{1}{2\pi i} \int_{|z|=R} z^{n-1} \sum_{i=0}^{n} a_i z^i \sum_{i=0}^{n} \overline{a_i} \overline{z}^i \, dz$$

$$= \frac{1}{2\pi} \int_0^{2\pi} R^n e^{in\theta} \sum_{i,j=0}^{n} a_i \overline{a_j} R^{i+j} e^{(i-j)\theta} \, d\theta$$

$$= \frac{1}{2\pi} \int_0^{2\pi} \sum_{i,j=0}^{n} a_i \overline{a_j} R^{n+i+j} e^{(n+i-j)\theta} \, d\theta$$

$$= \frac{1}{2\pi} \int_0^{2\pi} a_0 \overline{a_n} R^{2n} \, d\theta$$

$$= a_0 \overline{a_n} R^{2n}.$$

Solution to 5.7.9: Let $\sum_{n=0}^{\infty} c_n z^n$ be the power series for f. Since the series converges uniformly on each circle $|z| = r$, we have

$$\int_0^{2\pi} |f(r \, e^{i\theta})|^2 d\theta = \sum_{n=0}^{\infty} \int_0^{2\pi} c_n r^n e^{in\theta} \, \overline{f(re^{i\theta})} d\theta$$

$$= \sum_{n=0}^{\infty} \sum_{m=0}^{\infty} \int_0^{2\pi} c_n \bar{c}_m r^{n+m} e^{i(n-m)\theta} d\theta$$

$$= 2\pi \sum_{n=0}^{\infty} |c_n|^2 r^{2n}$$

Thus, for each n we have $2\pi |c_n|^2 \leqslant A r^{2k-2n}$. If $n > k$ we let $r \to \infty$ to get $c_n = 0$. If $n < k$ we let $r \to 0$ to get $c_n = 0$. Hence $c_n = 0$ whenever $n \neq k$, and the desired conclusion follows.

Solution 2. Assume f is not the zero function, and let c_n be its lowest order nonzero power series coefficient at the origin, that is,

$$f(z) = c_n z^n + c_{n+1} z^{n+1} + \cdots$$

then $z^{-n} f(z) \to c_n$ as $z \to 0$, so

$$\lim_{r \to 0} \frac{1}{r^{2n}} \int_0^{2\pi} |f(re^{i\theta})|^2 d\theta = 2\pi |c_n|^2.$$

On the other hand,

$$\limsup_{r \to 0} \frac{1}{r^{2k}} \int_0^{2\pi} |f(re^{i\theta})|^2 d\theta \leqslant A,$$

so r^{2n}/r^{2k} stays bounded as $r \to 0$, implying that $n \geqslant k$. Hence the function $g(z) = f(z)/z^k$ is entire.

Fix z_0 in $\mathbb{C}$. For $r > |z_0|$, Cauchy's formula gives

$$g(z_0)^2 = \frac{1}{2\pi i} \int_{|z|=r} \frac{g(z)^2}{z - z_0} dz,$$

so

$$|g(z_0)|^2 \leqslant \frac{r}{2\pi(r - |z_0|)} \int_0^{2\pi} |g(re^{i\theta})|^2 \, d\theta$$

$$= \frac{r^{1-2k}}{2\pi(r - |z_0|)} \int_0^{2\pi} |f(re^{i\theta})|^2 \, d\theta$$

$$\leqslant \frac{rA}{2\pi(r - |z_0|)} \longrightarrow \frac{A}{2\pi}.$$

It follows that g is bounded, hence constant, as desired.

Solution to 5.7.10: By Cauchy's Theorem [MH87, p. 152],

$$\frac{1}{2\pi i} \int_{|z|=r} f(z)^2 \frac{dz}{z} = f(0)^2$$

for $r > 1$. Parameterizing the domain of integration by $z = re^{i\theta}$, we find

$$\frac{1}{2\pi i} \int_0^{2\pi} f(re^{i\theta})^2 rie^{i\theta} \frac{d\theta}{re^{i\theta}} = f(0)^2.$$

Simplifying and taking real parts gives

$$\int_0^{2\pi} \left(u(re^{i\theta})^2 - v(re^{i\theta})^2 \right) d\theta = 2\pi \left(u(0)^2 - v(0)^2 \right) = 0.$$

Solution to 5.7.11: We have

$$\left| \frac{d^m f}{dz^m} \right| \leqslant M(1 + |z|^k)$$

for all z. Dividing both sides by $|z|^k$ and taking the limit as $|z|$ tends to infinity, we see that $d^m f/dz^m$ has a pole at infinity of order at most, k so $d^m f/dz^m$ is a polynomial of degree, at most, k. Letting $n = m + k + 1$, we must have that $d^n f/dz^n = 0$ and that n is the best possible such bound.

Solution to 5.7.12: 1. We have

$$f(z) = (z - z_1)^{n_1} \cdots (z - z_k)^{n_k} g(z)$$

where g is an analytic function with no zeros in Ω. So

$$\frac{f'(z)}{f(z)} = \frac{n_1}{z - z_1} + \cdots + \frac{n_k}{z - z_k} + \frac{g'(z)}{g(z)}.$$

Since g is never 0 in Ω, g'/g is analytic there, and, by Cauchy's Theorem [MH87, p. 152], its integral around γ is 0. Therefore,

$$\frac{1}{2\pi i} \int_\gamma \frac{f'(z)}{f(z)} \, dz = \sum_{j=1}^k \frac{1}{2\pi i} \int_\gamma \frac{n_j}{z - z_j} \, dz$$

$$= \sum_{j=1}^{k} n_j$$

2. We have

$$\frac{zf'(z)}{f(z)} = \frac{z}{z - z_1} + \frac{g'(z)}{g(z)}$$

so

$$\frac{1}{2\pi i} \int_\gamma \frac{zf'(z)}{f(z)} \, dz = z_1.$$

Solution to 5.7.13: Suppose $f(z_1) = f(z_2)$ and let γ be the segment connecting these two points. We have $0 = \int_\gamma f'(z)dz$. Hence,

$$\int_\gamma (f'(z) - f'(z_0))dz = -f'(z_0)(z_2 - z_1).$$

Taking absolute values, we get

$$|f'(z_0)| \, |z_2 - z_1| \leqslant \int_\gamma |f'(z) - f'(z_0)| \, |dz| < \int_\gamma |f'(z_0)| \, |dz| = |f'(z_0)| \, |z_2 - z_1|,$$

an absurd. We conclude, then, that f is injective.

Solution to 5.7.14: It suffices to show that there exists an integer n such that the image of Ω under $h(z) = f(z)/z^n$ contains no curves with positive winding number about 0; because it implies the existence of an analytic branch of the logarithm in $h(\Omega)$. Each closed curve in $h(\Omega)$ is the image of a closed curve in Ω, so it is enough to show that the images of simple closed curves in Ω have winding number 0 about the origin. Consider two classes of simple closed curves in Ω:

- Γ_1, the curves with 0 in their interiors, and

- Γ_2, the curves with 0 in their exteriors.

Since f has no zeros in Ω, it is clear that if $\gamma \in \Gamma_2$, then $\mathrm{Ind}_{f(\gamma)}(0) = 0$. From the shape of Ω, it follows that all the curves in Γ_1 are homotopic. Let n be the winding number about 0 of $f(\gamma)$ for $\gamma \in \Gamma_1$. Since h has no zeros in Ω, we must have $\mathrm{Ind}_{h(\gamma)}(0) = 0$ for $\gamma \in \Gamma_2$. Fix $\gamma \in \Gamma_1$; then

$$\mathrm{Ind}_{h(\gamma)}(0) = \frac{1}{2\pi i} \int_\gamma \frac{h'(z)}{h(z)} \, dz = \frac{1}{2\pi i} \int_\gamma \left(\frac{f'(z)}{f(z)} - \frac{n}{z} \right) dz = \mathrm{Ind}_{f(\gamma)}(0) - n = 0$$

and we are done.

Solution to 5.7.15: The analyticity of f can be proved with the aid of Morera's Theorem [MH87, p. 173]. If γ is a rectangle contained with its interior in $\mathbb{C} \setminus [0, 1]$, then

$$\int_\gamma f(z)dz = \int_0^1 \left(\int_\gamma \frac{\sqrt{t}}{t - z} \, dz \right) dt = 0$$

(by Cauchy's Theorem [MH87, p. 152]). Morera's Theorem thus implies that f is analytic. Alternatively, one can argue directly: for z_0 in $\mathbb{C} \setminus [0, 1]$

$$\lim_{z \to z_0} \frac{f(z) - f(z_0)}{z - z_0} = \lim_{z \to z_0} \int_0^1 \frac{\sqrt{t}}{(t - z_0)(t - z)} \, dt = \int_0^1 \frac{\sqrt{t}}{(t - z_0)^2} \, dt \,,$$

where the passage to the limit in the integral is justified by the uniform convergence of the integrands.

To find the Laurent expansion [MH87, p. 246] about ∞ we assume $|z| > 1$ and write

$$f(z) = -\frac{1}{z} \int_0^1 \frac{\sqrt{t}}{1 - \frac{t}{z}} \, dt$$

$$= -\frac{1}{z} \int_0^\infty \sum_{n=0}^\infty \frac{t^{n+\frac{1}{2}}}{z^n} \, dt \,.$$

The series converges uniformly on $[0,1]$, so we can integrate term by term to get

$$f(z) = -\sum_{n=0}^\infty \left(\int_0^1 t^{n+\frac{1}{2}} \, dt \right) z^{-n-1} = -\sum_{n=0}^\infty \frac{z^{-n-1}}{n + \frac{3}{2}}.$$

Solution to 5.7.16: Let $c > 0$. It suffices to show that there is a constant M such that

$$|f(z_1) - f(z_2)| \leqslant M|z_1 - z_2| \quad \text{for all} \quad z_1, z_2 \in \{z \mid \Re z > c \,, |z_1 - z_2| < c\}.$$

Fix two such points and let γ be the circle of radius c whose center is the midpoint of the segment joining them. γ lies in the right half-plane, so, by Cauchy's Integral Formula [MH87, p. 167], we have

$$|f(z_1) - f(z_2)| \leqslant \frac{1}{2\pi} \int_\gamma \left| \frac{f(\zeta)}{z_1 - \zeta} - \frac{f(\zeta)}{z_2 - \zeta} \right| |d\zeta|$$

$$\leqslant \frac{N|z_1 - z_2|}{2\pi} \int_\gamma \frac{|d\zeta|}{|z_1 - \zeta||z_2 - \zeta|}$$

where N is the supremum of $|f|$ in the right half-plane. On γ, $|z_i - \zeta| \geqslant \frac{c}{2}$ for $i = 1, 2$, so

$$|f(z_1) - f(z_2)| \leqslant \frac{4N}{c}|z_1 - z_2|.$$

5.8   Zeros and Singularities

Solution to 5.8.1: F is a map from $\mathbb{C}^3$ to the space of monic polynomials of degree 3, that takes the roots of a monic cubic polynomial to its coefficients, because if α, β, and γ are the zeros of $z^3 - Az^2 + Bz - C$, we have

$$A = \alpha + \beta + \gamma, \qquad \alpha\beta + \alpha\gamma + \beta\gamma = B, \qquad \alpha\beta\gamma = C.$$

Thus, by the Fundamental Theorem of Algebra (for several different proofs see the Solution to Problem 5.7.5), it is clear that F is onto. $F(1, 1, 0) = F(1, 0, 1)$, so F is not injective, in fact, $F(u, v, w) = F(v, w, u) = F(w, u, v)$.

Solution to 5.8.2: Using Rouché's Theorem [MH87, p. 421], it is easy to conclude that $p(z)$ has two zeros inside the circle $|z| = 3/4$.

Solution 2. The constant term of p is 1, so the product of its roots is 1, in absolute value. They either all have absolute value 1, or at least one lies inside $|z| < 1$. The former is not possible, since the degree of p is odd, it has at least one real root, and a calculation shows that neither 1 nor -1 is a root. So p has a root in the unit disc.

Solution to 5.8.3: For $|z| = 1$, we have

$$\left| -z^3 \right| = 1 > |f(z)|$$

so, by Rouché's Theorem [MH87, p. 421], $f(z) - z^3$ and z^3 have the same number of zeros in the unit disc.

Solution to 5.8.4: Let f_1 and f_2 be defined by $f_1(z) = 3z^{100}$ and $f_2(z) = -e^z$. On the unit circle, we have

$$|f_1(z)| = 3 > 1 = |f_2(z)|.$$

By Rouché's Theorem [MH87, p. 421], we know that f and f_1 have the same number of zeros in the unit disc, namely 100.

Let ξ be a zero of f. Then

$$f'(\xi) = 300\xi^{99} - e^\xi = 300\xi^{99} - 3\xi^{100} = 3\xi^{99}(100 - 3\xi) \neq 0$$

so all the zeros of f are simple.

Solution to 5.8.5: 1. Let $f(z) = 4z^2$ and $g(z) = 2z^5 + 1$. For $|z| = 1$, we have

$$|f(z)| = 4 > 3 \geqslant |g(z)|.$$

By Rouché's Theorem [MH87, p. 421], f and $p = f + g$ have the same number of roots in $|z| < 1$. Since f has two roots in $|z| < 1$, so does p.

2. There is at least one real root, since p has odd degree. The derivative is $p'(z) = 10z^4 + 8z$, so p' has two real zeros, namely at 0 and $-\sqrt[3]{4/5}$. Moreover, on the real axis, p' is positive on $(-\infty, -\sqrt[3]{4/5})$ and $(0, \infty)$, and negative on $(-\sqrt[3]{4/5}, 0)$. Thus, p is increasing on the first two intervals and decreasing on the last one. Since $p(0) = 1 > 0$, also $p(-\sqrt[3]{4/5}) > 0$, so p has no root

in $[-\sqrt[3]{4/5}, \infty)$ and exactly one in $(-\infty, -\sqrt[3]{4/5})$. (The real root is actually in $(-2, -1)$, since $p(-1) > 0$ and $p(-2) < 0$.)

Solution to 5.8.6: Let $p(z) = 3z^9 + 8z^6 + z^5 + 2z^3 + 1$. For $|z| = 2$, we have

$$|p(z) - 3z^9| = |8z^6 + z^5 + 2z^3 + 1|$$
$$\leqslant 8|z|^6 + |z|^5 + 2|z|^3 + 1$$
$$= 561 < 1536 = |3z^9|$$

so, by Rouché's Theorem [MH87, p. 421], p has nine roots in $|z| < 9$. For $|z| = 1$, we have

$$|p(z) - 8z^6| = |3z^9 + z^5 + 2z^3 + 1|$$
$$\leqslant 3|z|^9 + |z|^5 + 2|z|^3 + 1$$
$$= 7 < 8 = |8z^6|$$

and we conclude that p has six roots in $|z| < 1$. Combining these results, we get that p has three roots in $1 < |z| < 2$.

Solution to 5.8.7: For z in the unit circle, we have

$$\left|5z^2\right| = 5 > 4 \geqslant \left|z^5 + z^3 + 2\right|$$

so, by Rouché's Theorem [MH87, p. 421], $p(z)$ has two zeros in the unit disc. For $|z| = 2$,

$$\left|z^5\right| = 32 > 30 \geqslant \left|z^3 + 5z^2 + 2\right|$$

so $p(z)$ has five zeros in $\{z \mid |z| < 2\}$. We conclude then that $p(z)$ has three zeros in $1 < |z| < 2$.

Solution to 5.8.8: Let $p(z) = z^7 - 4z^3 - 11$. For z in the unit circle, we have

$$|p(z) - 11| = \left|z^7 - 4z^3\right| \leqslant 5 < 11$$

so, by Rouché's Theorem [MH87, p. 421], the given polynomial has no zeros in the unit disc. For $|z| = 2$,

$$\left|p(z) - z^7\right| = \left|4z^3 + 11\right| \leqslant 43 < 128 = \left|z^7\right|$$

so there are seven zeros inside the disc $\{z \mid |z| < 2\}$ and they are all between the two given circles.

Solution to 5.8.9: Since $p(0) = -1$ and $p(x) \to +\infty$ as $x \to \infty$ on the positive real axis, p has at least one root on $(0, \infty)$. On the negative real axis p is negative, in particular nonzero.

We have $p'(z) = 3\varepsilon z^2 - 2z$, which has roots at 0 and $2/3\varepsilon$. The roots of p' lie in the convex hull of the roots of p, by the Gauss–Lucas Theorem [LR70, p.

94] On the imaginary axis, p has a nonzero imaginary part, except at the origin. Therefore p has no roots on the imaginary axis. Consequently p must have at least one root in $\Re z < 0$. Such a root is nonreal (since $p \neq 0$ on $(-\infty, 0)$), and nonreal roots of p occur in conjugate pairs. Therefore p has two roots in $\Re z < 0$.

Solution 2. From the expression above for p' one sees that p is decreasing on $(0, 2/3\varepsilon)$ and increasing on $(2/3\varepsilon, \infty)$. Therefore p has exactly one root on $(0, \infty)$. Let the root be λ. Since p is nonzero on $(-\infty, 0]$, the other two roots form a conjugate pair, say μ and $\bar{\mu}$. Since the linear term in p vanishes, we have

$$\lambda\mu + \lambda\bar{\mu} + \mu\bar{\mu} = 0,$$

which can be written as

$$-\lambda = \frac{2\Re\mu}{|\mu|^2},$$

implying that $\Re\mu < 0$.

Solution to 5.8.10: Rescale by setting $z = \varepsilon^{-1/5}w$. Then we need to show that exactly five roots of the rescaled polynomial

$$p_\varepsilon(w) = w^7 + w^2 + \delta,$$

with $\delta = \varepsilon^{2/5} \to 0$ as $\varepsilon \to 0$, converge to the unit circle as $\varepsilon \to 0$. We have $p_0(w) = w^2(w^7 + 1)$. Since two roots of p_0 are at $w = 0$ and the other five are on the unit circle, the result follows from the continuity of the roots of a polynomial as functions of the coefficients, see Problem 5.8.30.

Solution 2. Let $q(z) = z^2 + 1$, so

$$|p(z) - q(z)| = \varepsilon|z|^7 = r^7\varepsilon^{-2/5}$$

on the circle $|z| = r\varepsilon^{-1/5}$. Also,

$$|q(z)| = |z^2 + 1| > r^2\varepsilon^{-2/5} - 1$$

on $|z| = r\varepsilon^{-1/5}$. Since $r < 1$, $r^7 < r^2$, and $r^7\varepsilon^{-2/5} < r^2\varepsilon^{-2/5} - 1$ for ε sufficiently small. Then $|p(z) - q(z)| < |q(z)|$ on $|z| = r\varepsilon^{-1/5}$, and by Rouché's Theorem [MH87, p. 421], p and q have the same number of zeros inside $|z| = r\varepsilon^{-1/5}$, namely two. By the Fundamental Theorem of Algebra (for several different proofs see the Solution to Problem 5.7.5), the other five roots must lie in $|z| > r\varepsilon^{-1/5}$.

Now take $q(z) = \varepsilon z^7$, so

$$|p(z) - q(z)| = |z^2 + 1| \leqslant R^2\varepsilon^{-2/5} + 1$$

on $|z| = R\varepsilon^{-1/5}$, where

$$|q(z)| = R^7\varepsilon^{-2/5}.$$

Since $R > 1$, we have $R^7 > R^2$ and

$$R^2\varepsilon^{-2/5} + 1 < R^7\varepsilon^{-2/5}$$

for ε sufficiently small. Thus, $|p(z) - q(z)| < |q(z)|$ on $|z| = R\varepsilon^{-1/5}$, so p and q have the same number of zeros inside $|z| = R\varepsilon^{-1/5}$, namely seven. This leaves precisely five roots between the two circles.

Solution to 5.8.11: The determinant of $A(z)$ is $8z^4 + 6z^2 + 1$. For z in the unit circle, we have

$$\left| 8z^4 \right| = 8 > 7 \geqslant \left| 6z^2 + 1 \right|$$

so, by Rouché's Theorem [MH87, p. 421], det $A(z)$ has four zeros in the unit disc. Also,

$$\frac{d}{dz} (\det A(z)) = z(32z^3 + 12)$$

with roots

$$0, \pm i\sqrt{\frac{3}{8}}$$

which are not zeros of det $A(z)$. Thus, all the four zeros are simple, so they are distinct.

Solution to 5.8.12: Let $\gamma : [a, b] \to \mathbb{C}$ be a continuous function, with $0 \notin \gamma^*$, where we denote the image of γ by γ^*. There is a continuous branch of arg γ on $[a, b]$. If θ, ϕ are branches of arg γ on $[a, b]$, then $[\theta]_a^b = \theta(b) - \theta(a) = [\phi]_a^b$. We may therefore define $\omega(\gamma, 0) = (\theta(b) - \theta(a))/2\pi$, where $\theta : [a, b] \to \mathbb{R}$ is any branch of arg γ. Note we are not assuming that γ is closed. If γ lies in some open half-plane, then $|\omega(\gamma, 0)| < \frac{1}{2}$. If γ is closed then $\omega(\gamma, 0)$ is an integer. If $\gamma(t) = \gamma_1(t)\gamma_2(t)$ where γ_1, γ_2 are curves not containing 0, then $\omega(\gamma, 0) = \omega(\gamma_1, 0) + \omega(\gamma_2, 0)$; in particular, if $\gamma^n(t) = \gamma(t)^n$ then $\omega(\gamma^n, 0) = n\omega(\gamma, 0)$ for $n \geqslant 1$. If $\gamma = \gamma_1 + \cdots + \gamma_k$ is a join of curves not containing 0, then $\omega(\gamma, 0) = \sum_{j=0}^{k} \omega(\gamma_j, 0)$.

Choose R sufficiently large, so that the zeros of $f(z) = z^4 + 3z^2 + z + 1$ lie inside $|z| = R$. Let $\gamma = \gamma_1 + \gamma_2 + \gamma_3$, where $\gamma_1 = [iR, 0]$ (line segment from iR to 0 on imaginary axis), $\gamma_2 = [0, R]$ and $\gamma_3(t) = Re^{it}$, $0 \leqslant t \leqslant \frac{\pi}{2}$. Denote $\Gamma_i = f \circ \gamma_i$, $\Gamma = f \circ \gamma$; so $\Gamma = \Gamma_1 + \Gamma_2 + \Gamma_3$.

For $z = iy \in \gamma_1^*$, $f(z) = f(iy) = y^4 - 3y^2 + 1 + iy \neq 0$. As $0 \leqslant \Im f(z) = y < R$, Γ_1 lies in the upper half-plane and therefore $|\omega(\Gamma_1, 0)| \leqslant \frac{1}{2}$.

For $z = x \in \gamma_2^*$, $f(z) = x^4 + 3x^2 + x + 1 \neq 0$, this being a strictly increasing function in $x \geqslant 0$. Γ_2 is a subset of the positive real axis. Thus $\omega(\Gamma_2, 0) = 0$.

We have $f(z) = z^4(1 + g(z))$, where $g(z) = (3z^2 + z + 1)/z^4$. Now $g(z) \to 0$ as $z \to \infty$. For large R, $|g(Re^{it})| < 1$, and $1 + g(Re^{it})$ lies in the disc centered at 1 radius 1; that is $1 + g(Re^{it})$ lies in the right half-plane, and thus $|\omega(\sigma, 0)| < \frac{1}{2}$, where $\sigma(t) = 1 + g(\gamma_3(t))$. Thus $\omega(\Gamma_3, 0) = \omega(\gamma_3^4, 0) + \omega(\sigma, 0) = 4\omega(\gamma_3, 0) + \omega(\sigma, 0) = 1 + \omega(\sigma, 0)$.

Thus $|\omega(\Gamma, 0) - 1| = |\omega(\Gamma_1, 0) + \omega(\Gamma_2, 0) + \omega(\sigma, 0)| < \frac{1}{2} + \frac{1}{2} = 1$. As Γ is closed, $\omega(\Gamma, 0)$ is an integer, and so $\omega(\Gamma, 0) = 1$. The Argument principle asserts that $\omega(\Gamma, 0)$ equals the number of zeros of f inside γ. As R is arbitrarily large, we deduce that f has one zero in the first quadrant. The conjugate of this root is the only other zero in the right half-plane. Thus f has two zeros in the right half-plane.

Solution to 5.8.13: Let $z_1, \ldots, z_n$ be the zeros of p, and z a zero of p', $z \neq z_i$, $i = 1, \ldots, n$. We have

$$0 = \frac{p'}{p}(z) = \sum_{i=1}^{n} \frac{1}{z - z_i}$$

Using the fact that $1/\alpha = \bar{\alpha}/|\alpha|^2$ and conjugating we get

$$\sum_{i=1}^{n} \frac{z - z_i}{|z - z_i|^2} = 0$$

which is clearly impossible if $\Re z \leqslant 0$.

This result can be generalized to give the Gauss–Lucas Theorem [LR70, p. 94]: *The zeros of p' lie in the convex hull of the zeros of p. If $z_1, \ldots, z_n$ are the zeros of p, and z is a zero of p', $z \neq z_i$, $i = 1, \ldots, n$. We have, similar to the above,*

$$\sum_{i=1}^{n} \frac{z - z_i}{|z - z_i|^2} = 0$$

which is impossible if z is not in the convex hull of $z_1, \ldots, z_n$.

Solution to 5.8.14: Let a be a real zero of f. Then at a the function f restricted to $\mathbb{R}$ has a minimum, therefore $f'(a) = 0$. Hence the order of the zero of f at a is at least 2. The function $g(z) = f(z)/(z - a)^2$ thus satisfies the same hypotheses as f. If $g(a) = 0$ we can repeat the preceding argument. After finitely many repetitions we reach the desired conclusion.

Solution to 5.8.15: We may assume $r \neq 0$. Let $n = \deg p$ and $x_1 < x_2 < \cdots < x_k$ be the roots of p, with multiplicities $m_1, m_2, \ldots, m_k$, respectively. If any m_j exceeds 1, then $p - rp'$ has a root at x_j of multiplicity $m_j - 1$ (giving a total of $n - k$ roots all together). We have

$$p(x) = c(x - x_1)^{m_1} \cdots (x - x_k)^{m_k}.$$

The logarithmic derivative p'/p is given by

$$\frac{p'(x)}{p(x)} = \sum_{j=1}^{k} \frac{m_j}{x - x_j}.$$

Its range on the interval (x_j, x_{j+1}) $(j = 1, \ldots, k - 1)$ is all of $\mathbb{R}$, since it is continuous there and

$$\lim_{x \to x_j+} \frac{p'(x)}{p(x)} = +\infty, \qquad \lim_{x \to x_j-} \frac{p'(x)}{p(x)} = -\infty.$$

Hence, there is a point $x \in (x_j, x_{j+1})$ where $p'(x)/p(x) = 1/r$; in other words, where $p - rp'$ has a root. Thus, $p - rp'$ has at least $k - 1$ real roots other than the $n - k$ that are roots of p. Hence, $p - rp'$ has at least $n - 1$ real roots all together, and the nonreal ones come in conjugate pairs. Hence, it has only real roots.

Solution to 5.8.16: Let $R > \lambda + 1$ and consider the contour $C_R = \Gamma_R \cup [-Ri, Ri]$, where $\Gamma_R = \{z \mid |z| = R, \Re z \leqslant 0\}$ and $[-Ri, Ri] = \{z \mid \Re z = 0, -R \leqslant \Im z \leqslant R\}$.

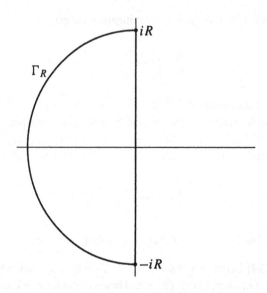

Let the functions f and g be defined by $f(z) = z + \lambda$, $g(z) = -e^z$. On Γ_R, we have

$$|f(z)| \geqslant |z| - \lambda > 1 \geqslant |g(z)|$$

and on $[-Ri, Ri]$,

$$|f(z)| \geqslant \lambda > 1 = |g(z)|.$$

By Rouché's Theorem [MH87, p. 421], f_λ and f have the same number of zeros inside the contour C_R, so f_λ has exactly one zero there. As this conclusion is valid for every $R > \lambda + 1$, we conclude that f_λ has one zero in the left half-plane. As f is real on the real axis and $f(x)f(0) < 0$ for x small enough, we get that the zero of f_λ is real.

Solution 2. We find the number of zeros of f_λ in the left half-plane by considering a Nyquist diagram [Boa87, p. 106] relative to the rectangle with corners iy, $-x+iy$, $-x-iy$, and $-iy$, x, $y > \lambda$. This will give the change in $(1/2\pi)$ arg $f_\lambda(z)$. Then we let $x, y \to \infty$.

On the right side of the rectangle, as t ranges from $-y$ to y, then $f_\lambda(it) = it + \lambda - \cos t - i \sin t$ has a positive real part, and its imaginary part changes sign from negative to positive. On the top of the rectangle, as s ranges from 0 to $-x$, $f_\lambda(s+iy) = s+iy+\lambda-e^s \cos y - ie^s \sin y$ has positive imaginary part, and its real part changes sign from positive to negative.

Similar reasoning shows that on the left side of the rectangle, $\Re f_\lambda < 0$ and $\Im f_\lambda$ changes sign from positive to negative. On the bottom of the rectangle, $\Im f_\lambda < 0$ and $\Re f_\lambda$ changes sign from negative to positive. Hence, f_λ is never 0 on this rectangle and the image of the rectangle winds around the origin exactly once. By the Argument Principle [MH87, p. 419], f_λ has exactly one zero in the interior of this rectangle. Letting x and y tend to infinity, we see that f_λ has exactly one zero in the left half-plane. As f_λ is real on the real axis and $f_\lambda(x)f_\lambda(0) < 0$ for x small enough, we get that the zero of f_λ is real.

Solution to 5.8.17: For $|z| = 1$, we have

$$\left|ze^{\lambda-z}\right| = e^{\Re(\lambda-z)} > e^0 = 1 = |-1|$$

so, by Rouché's Theorem [MH87, p. 421], the given equation has one solution in the unit disc. Let $f(z) = ze^{\lambda-z}$. As, for z real, f increases from $f(0) = 0$ to $f(1) = e^{\lambda-1} > 1$, by the Intermediate Value Theorem [Rud87, p. 93], $f(\xi) = 1$ for some $\xi \in (0, 1)$.

Solution to 5.8.18: By the Gauss–Lucas Theorem [LR70, p. 94] (see Solution to 5.8.13), if $p(z)$ is a polynomial, then all of the roots of $p'(z)$ lie in the convex hull of the roots of $p(z)$. Let $z = 1/w$. The given equation becomes, after multiplying by w^n, $w^n + w^{n-1} + a = 0$. The derivative of the polynomial on the left-hand side is $nw^{n-1} + (n-1)w^{n-2}$, which has roots 0 and $-(n-1)/n \geqslant 1/2$. For these two roots to lie in the convex hull of the roots of $w^n + w^{n-1} + a$, the latter must have at least one root in $|w| \geqslant 1/2$, which implies that $az^n + z + 1$ has at least one root in $|z| \leqslant 2$.

Solution 2. The product of the roots of $p(z) = az^n + z + 1$ is its constant term, namely 1, so all p's roots are unimodular or at least one is in $|z| < 1$.

Solution to 5.8.19: Let Γ_R be the closed curve in the first quadrant consisting of the interval $[0, R]$, the interval $[0, iR]$, and the circular arc centered at 0 joining the points R and iR. It suffices to show that, for R arbitrarily large, the function $g(z) = z^4 + z^3 + 1$ has exactly one zero in the interior of Γ_R. We'll compare g with $f(z) = z^4 + 1$. The function f has one simple zero in the interior of Γ_R, at $z = e^{\pi i/4}$. By Rouché's Theorem [MH87, p. 421], it remains to verify that $|f(z) - g(z)| < |f(z)|$ for z in Γ_R. We have

- $|f(x) - g(x)| = |x|^3 < x^4 + 1 = |f(x)|$ for x in $\mathbb{R}$ (since $|x|^3 \leqslant 1$ for x in $[-1, 1]$ and $|x|^3 < x^4$ for x in $\mathbb{R} \setminus [-1, 1]$)

- $|f(iy) - g(iy)| = |y|^3 < y^4 + 1 = |f(iy)|$ for y in $\mathbb{R}$

- $|f(z) - g(z)| = |z|^3 = R^3 < R^4 - 1 \leqslant |f(z)|$ for $|z| = R$ (with $R \geqslant 2$, so $R^4 - 1 > 2R^3 - R^3 = R^3$).

Solution to 5.8.20: Suppose that f is zero free in the disc $|z| < 1$, and hence also in the disc $|z| \leqslant 1$. Then f has no zero in an open set containing the latter disc, so the function $g = 1/f$ is holomorphic there. By the Maximum Modulus Principle [MH87, p. 185] and our hypotheses, we have

$$\frac{1}{m} < |g(0)| \leqslant \max\{|g(z)| \mid |z| = 1\} = \frac{1}{m},$$

a contradiction.

Solution 2. For $|z| \leqslant 1$ Let $g(z) = -f(0)$. Then on $|z| = 1$,

$$|g(z)| = |f(0)| < m < |f(z)|.$$

By Rouché's theorem [MH87, p. 421], f and $f + g$ have the same number of zeros in the unit disc. Since $f + g = f - f(0)$ has at least one zero, the result follows.

Solution to 5.8.21: Let $p(z)$ denote the polynomial and suppose $p(z_0) = 0$ for some $z_0 \in \mathbb{D}$. Then z_0 is also a root of $(z - 1)p(z)$. We then have

$$0 = a_0 z_0^{n+1} + (a_1 - a_0)z_0^n + \cdots + (a_n - a_{n-1})z_0 - a_n.$$

Since all the a_i's are positive and $|z_0| < 1$, we have, by the Triangle Inequality [MH87, p. 20],

$$\begin{aligned}
a_n &= |a_0 z_0^{n+1} + (a_1 - a_0)z_0^n + \cdots + (a_n - a_{n-1})z_0| \\
&< a_0 + (a_1 - a_0) + \cdots + (a_n - a_{n-1}) \\
&= a_n,
\end{aligned}$$

a contradiction.

Solution to 5.8.22: By Descartes' Rule of Signs [Caj69, p. 7], [Coh95, vol. 1, pag. 172], the polynomial $p(z)$ has zero or two positive real roots. As $p(0) = 3$ and $p(1) = -2$, by the Intermediate Value Theorem [Rud87, p. 93], $p(z)$ has one and so, two, positive real roots. Replacing z by $-z$, and again applying Descartes' Rule of Signs, we see that $p(-z)$ has one positive real root, so $p(z)$ has one negative real root. Applying Rouché's Theorem [MH87, p. 421] to the functions $f = p$ and $g = 6z$ on the unit circle, we see that p has exactly one zero in the unit disc, which is positive as seen above. Hence, the real roots are distinct. (The same conclusion would follow from noticing that p and p' have no common roots.) The imaginary roots are conjugate, so they are distinct as well.

Solution 2. Graphing the polynomial $y = x^5 - 6x + 3$ (for real x), we can see the result easily. First, $y' = 5x^4 - 6$ and the only two real roots are $x = \pm\sqrt[4]{6/5}$ and none of them are multiple. Now looking at the limits when $x \to -\infty$ and $x \to \infty$, we can conclude that the graph looks like

So there are three distinct real roots. There cannot be a forth, otherwise y' would have a third root. The other two roots are then complex and not real; since they are conjugate, they are distinct, making for five distinct roots, three of them real.

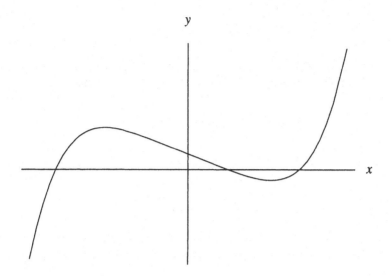

Solution to 5.8.23: Let z be a zero of the given polynomial with $|z| = r$. If $r \leqslant 1$, then z lies in the given disc. For $r > 1$, we have

$$r^{2n} = |-z^n|^2 = \left| \sum_{i=0}^{n-1} c_i z^i \right|^2.$$

By the Cauchy–Schwarz Inequality [MH93, p. 69], we get

$$r^{2n} \leqslant \sum_{i=0}^{n-1} |c_i|^2 \sum_{i=0}^{n-1} |z^i|^2.$$

The second sum is a finite geometric series which sums to $\dfrac{r^{2n} - 1}{r^2 - 1}$. Combining these, we have

$$r^{2n} < \left(\sum_{i=0}^{n-1} |c_i|^2 \right) \left(\frac{r^{2n}}{r^2 - 1} \right).$$

Multiplying both sides by $\dfrac{r^2 - 1}{r^{2n}}$, we get the result wanted.

Solution to 5.8.24: The product of all zeros of $P(x)$ (with multiplicities) equals ± 1, so, if there are no roots inside the unit circle, then there are no roots outside the unit circle either. Hence, all roots are on the unit circle. From $P(0) = -1 < 0$ and $\lim_{x \to \infty} P(x) = +\infty$, it follows that $P(x)$ has a real zero in the interval $(0, \infty)$. Since it lies on the unit circle, it must be 1, so $P(1) = 0$.

Solution to 5.8.25: Let $R > 0$. Consider the semicircle with diameter $[-Ri, Ri]$ containing R and its diameter. We will apply the Argument Principle [MH87, p. 419] to the given function on this curve.

Suppose n is even. Then

$$(iy)^{2n} + \alpha^2(iy)^{2n-1} + \beta^2 = y^{2n} + \beta^2 - i\alpha^2 y^{2n-1}$$

is always in the first quadrant for $y < 0$, so the change in the argument when we move from 0 to $-Ri$ is close to zero, for R large. On the semicircle,

$$z^{2n} + \alpha^2 z^{2n-1} + \beta^2 = z^{2n}\left(1 + \frac{\alpha^2}{z} + \frac{\beta^2}{z^{2n}}\right)$$

which is close to z^{2n} for R large. So the argument changes by $2\pi n$ when we go from $-Ri$ to Ri. From Ri to 0,

$$y^{2n} + \beta^2 - i\alpha^2 y^{2n-1}$$

is always in the fourth quadrant, so the change in the argument is close to zero, for R large. The total change is then $2\pi n$, so there are n roots with positive real part.

Now, suppose that n is odd. We have

$$(iy)^{2n} + \alpha^2(iy)^{2n-1} + \beta^2 = -y^{2n} + \beta^2 + i\alpha^2 y^{2n-1}$$

so when we go from the origin to $-Ri$, for R large, the argument change is close to $-\pi$. The variation on the semicircle is again about $2\pi n$. The change when y goes from R to 0 in the argument of

$$-y^{2n} + \beta^2 + i\alpha^2 y^{2n-1}$$

is about $-\pi$. Therefore, the number of zeros with, positive real part is now $(-\pi + 2\pi - \pi)/2\pi = n - 1$.

Solution to 5.8.26: Let $\rho > 0$ and consider the functions

$$g_n(z) = f_n(1/z) = 1 + z + \frac{z^2}{2!} + \cdots + \frac{z^n}{n!}.$$

Since $g_n(0) \neq 0$ for all n, $g_n(z)$ has a zero in $|z| \leqslant \rho$ if and only if $f_n(z)$ has a zero in $|z| \geqslant \rho$. $g_n(z)$ is a partial sum of the power series for e^z. Since this series converges locally uniformly and $\{z \mid |z| \leqslant \rho\}$ is compact, for any $\varepsilon > 0$ there is $N > 0$ such that if $n \geqslant N$, then $|g_n(z) - e^z| < \varepsilon$ for all z in this disc. e^z attains its minimum $m > 0$ in this disc. Taking N_0 such that if $n \geqslant N_0$, $|g_n(z) - e^z| < m/2$ for all z in the disc, we get that $g_n(z)$ is never zero for $|z| \leqslant \rho$. Therefore, $f_n(z)$ has no zeros outside this disc.

Solution to 5.8.27: If $z = x + iy$ then $e^z = e^x e^{iy}$, so $w = \exp z$ maps the horizontal line $\Im z = \lambda$ onto the ray $w = re^{i\lambda}$, $r > 0$. Thus if $0 < \Im z < \pi/2$

then e^z lies in the first quadrant. As $\Im(z)$ is also positive, we have $e^z + z \neq 0$. If $-\pi/2 < \Im z < 0$ then e^z lies in the fourth quadrant; whereas z lies in the lower half-plane. Again $e^z + z \neq 0$. Consider $z = x$ real. It is easy to see that $e^x = -x$ has one root. Thus $f(z) = e^z + z$ has precisely one zero in the strip $-\pi/2 < \Im z < \pi/2$.

We now consider the zeros of $g(z)$ in the same strip. As $\arg(-1/z) = \pi - \arg z$ we deduce that, (writing $z = x + iy$), if $x > 0$, $0 < y < \pi/2$, then $-1/z$ lies in the second quadrant, whereas e^z lies in the first quadrant; and if $x > 0$, $-\pi/2 < y < 0$ then $-1/z$ lies in the third quadrant, whereas e^z lies in the fourth quadrant; that is, $e^z \neq -1/z$. When $z = x > 0$ then $ze^z + 1 > 0$. Therefore $ze^z + 1 \neq 0$ for $x > 0$, $-\pi/2 < y < \pi/2$.

We'll use the notation of the solution to Problem 5.8.12. Consider the rectangular contour γ, which is the join of the line segments $\gamma_1 = [-i\pi/2, 0]$, $\gamma_2 = [0, i\pi/2]$, $\gamma_3 = [i\pi/2, -R + i\pi/2]$, $\gamma_4 = [-R + i\pi/2, -R - i\pi/2]$, $\gamma_5 = [-R - i\frac{\pi}{2}, -i\frac{\pi}{2}]$. Write $h(z) = ze^z$, $\Gamma_i = h \circ \gamma_i$. We will study each piece separately.

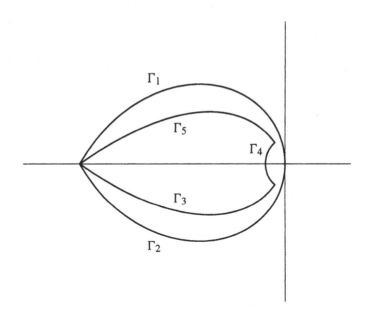

Γ_1: We see that $h(-i\pi/2) = -\pi/2$, $h(0) = 0$. If $0 < y < \pi/2$ then $h(-iy) = -iye^{-iy} = ye^{-i(\frac{\pi}{2}+y)}$ and $-\pi < \arg h(iy) = -(\frac{\pi}{2}+y) < -\pi/2$. Thus Γ_1 is a curve from $-\frac{\pi}{2}$ to 0 which, apart from the end points, lies in the third quadrant.

Γ_2: We find that $h(i\pi/2) = -\frac{\pi}{2}$, $h(iy) = iye^{iy} = ye^{i(\frac{\pi}{2}+y)}$. For $0 < y < \pi/2$, we have $\frac{\pi}{2} < \arg h(iy) = \frac{\pi}{2} + y < \pi$. Thus Γ_2 is a curve from 0 to $-\frac{\pi}{2}$ which, apart from the end points, lies in the second quadrant.

Γ_3: If $z = -x + i\frac{\pi}{2}$, $x > 0$ then $h(z) = -\frac{\pi}{2}e^{-x} - ixe^{-x}$; so $h(z)$ lies in the third quadrant. Thus Γ_3 is a curve from $-\frac{\pi}{2}$ to $\zeta_1 = h(-R + i\frac{\pi}{2}) = -e^{-R}(\frac{\pi}{2} + iR))$, which apart from the initial point, lies in the third quadrant.

Γ_4: We see that $\zeta_2 = h(-R - i\frac{\pi}{2}) = e^{-R}(-\frac{\pi}{2} + iR) = \overline{\zeta_1}$. If $z = -R + iy$, $-\frac{\pi}{2} < y < \frac{\pi}{2}$ then $|h(z)| = |-R + iy||e^{-R+iy}| \leqslant (R + \frac{\pi}{2})e^{-R}$; thus $h(z)$ lies in the unit disc, for large R. For such an R, Γ_4 is a curve in the unit disc from ζ_1 to ζ_2.

Γ_5: If $z = -x - i\frac{\pi}{2}$, $x > 0$ then $h(z) = -\frac{\pi}{2}e^{-x} + ixe^{-x}$ which is in the second quadrant. Thus Γ_5 is a curve from ζ_2 to $-\frac{\pi}{2}$ which lies, apart from the final point, in the second quadrant.

The curve $1 + \Gamma_1(t)$ lies in the lower half-plane; by choosing a branch of $\arg(1 + \Gamma_1(t))$, we see that $\omega(\Gamma_1, -1) = 1/2$. Likewise $\omega(\Gamma_2, -1) = 1/2$. If $\alpha = \arg(\zeta_2 + 1)$ with $0 < \alpha < \pi/2$, then $\omega(\Gamma_5, -1) = (\pi - \alpha)/2\pi = \omega(\Gamma_3, -1)$. The curve $1 + \Gamma_4(t)$ lies in the disc $D(1; 1)$, and thus in the right half-plane. Hence $\omega(\Gamma_4, -1) = \alpha/\pi$. Adding, we have $\omega(\Gamma, -1) = 2$. By the argument principle, $h(z) = ze^z$ takes the value -1 twice inside γ. As R is arbitrarily large, we conclude that $g(z) = ze^z + 1$ has two zeros in the specified strip. So f has one zero, whereas g has two zeros in the strip.

Solution to 5.8.28: The function $\sin z$ satisfies the identity $\sin(z + \pi) = -\sin z$, and vanishes at the points $n\pi$, $n \in \mathbb{Z}$, and only at those points. For m a positive integer, let R_m denote the closed rectangle with vertices $(m - \frac{1}{2})\pi + i\varepsilon$ and $(m + \frac{1}{2})\pi \pm i\varepsilon$. The function $\sin z$ has no zeros on the boundary of R_m, so its absolute value has a positive lower bound, say δ, there. (The number δ is independent of m because of the identity $\sin(z + \pi) = -\sin z$.) Suppose $(m - \frac{1}{2})\pi - |a| > \frac{1}{\delta}$. Then, for z in R_m, we have

$$\frac{1}{|z - a|} \leqslant \frac{1}{|z| - a} \leqslant \frac{1}{(m - \frac{1}{2})\pi - |a|} < \delta$$

implying that $\frac{1}{|z-a|} < |\sin z|$ on the boundary of R_m. By Rouché's Theorem [MH87, p. 421] then, the functions $\sin z$ and $f(z) = \sin z + \frac{1}{z-a}$ have the same number of zeros in the interior of R_m. Since $\sin z$ has one zero there, so does $f(z)$. As the condition on m holds for all sufficiently large m, the desired conclusion follows.

Solution to 5.8.30: We will prove that the simple zeros of a polynomial depend continuously on the coefficients of the polynomial, around a simple root.

Consider

$$p(z) = \hat{a}_0 + \hat{a}_1 z + \cdots + \hat{a}_n z^n = \hat{a}_n \prod_{j=1}^{s} (z - z_j)^{m_j} \quad (\hat{a}_n \neq 0).$$

For $(\xi_0, \ldots, \xi_{n-1}) \in \mathbb{C}^n$, let F be the polynomial given by

$$F(z) = \hat{a}_0 \xi_0 + (\hat{a}_1 + \xi_1)z + \cdots + (\hat{a}_{n-1} + \xi_{n-1})z^{n-1} + \hat{a}_n z^n$$

and, for each $1 \leqslant k \leqslant s$, let $0 < r_k < \min_{k \neq j} |z_k - z_j|$.

We will show that for some $\varepsilon > 0$, $|\xi_i| < \varepsilon$ for $i = 0, \ldots, n-1$ implies that F has m_j zeros inside the circle C_k centered at z_k with radius r_k.

Let ζ be the polynomial given by

$$\zeta(z) = \xi_0 + \xi_1 z + \cdots + \xi_{n-1} z^{n-1}.$$

On C_k, we have

$$|\zeta(z)| \leqslant \varepsilon M_k, \quad M_k = \sum_{j=1}^{n-1} (r_k + |z_k|)^j$$

and

$$|p(z)| \geqslant |\hat{a}_n| r_k^{m_k} \prod_{\substack{j=1 \\ j \neq k}}^{k} (|z_j - z_k| - r_k)^{m_j} = \delta_k > 0.$$

Taking $\varepsilon < \delta_k / M_k$, we get $|\zeta(z)| < |p(z)|$ on C_k; therefore, by Rouché's Theorem [MH87, p. 421], F has the same number of zeros in C_k as p. As in this domain p has a single zero with multiplicity m_j, we are done.

Solution to 5.8.31: From

$$f^{(k)}(z) = \sum_{n \geqslant k} n(n-1) \cdots (n-k+1) a_n z^{n-k},$$

we conclude that

$$\left| f^{(k)}(re^{i\theta}) \right| \leqslant \left| f^{(k)}(r) \right|$$

for $0 < r < 1$, and $0 < \theta \leqslant 2\pi$. Suppose f can be analytically continued in a neighborhood of $z = 1$; then its power series expansion around $z = 1/2$,

$$\sum_{n=0}^{\infty} \beta_n (z - 1/2)^n$$

has a radius of convergence $R > 1/2$. Let (γ_n) be the Taylor coefficients [MH87, p. 233] of the power series expansion of f around the point $(1/2)e^{i\theta}$. By the above inequality, $|\gamma_n| \leqslant |\beta_n|$. Therefore, the power series around the point $(1/2)e^{i\theta}$ has a radius of convergence of at least R. So f can be analytically continued in a neighborhood of every point of the unit circle. However, the Maclaurin series [MH87, p. 234] of f has radius of convergence 1, which implies that at least one point on the unit circle is a singularity of f, a contradiction.

Solution to 5.8.32: Without loss of generality, assume $r = 1$.

Suppose f is analytic at $z = 1$. Then f has a power expansion centered at 1 with positive radius of convergence. Therefore, f has a power series expansion centered at $z = 1/2$ with radius $1/2 + \varepsilon$ for some positive ε.

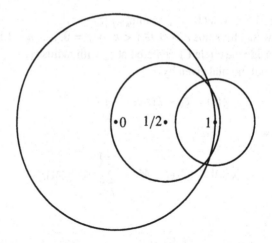

As

$$f^{(n)}\left(\frac{1}{2}\right) = \sum_{k \geq n} \frac{k!}{(k-n)!} a_k \left(\frac{1}{2}\right)^{k-n}$$

we have, for $1 < x < 1 + \varepsilon$,

$$f(x) = \sum_{n \geq 0} \frac{f^{(n)}(1/2)}{n!} \left(x - \frac{1}{2}\right)^n$$

$$= \sum_{n \geq 0} \frac{1}{n!} \left(x - \frac{1}{2}\right)^n \sum_{k \geq n} \frac{k!}{(k-n)!} a_k \left(\frac{1}{2}\right)^{k-n}$$

$$= \sum_{k \geq 0} a_k \sum_{n=0}^{k} \frac{k!}{n!(k-n)!} \left(x - \frac{1}{2}\right)^n \left(\frac{1}{2}\right)^{k-n}$$

$$= \sum_{k \geq 0} a_k x^k$$

which is absurd because we assumed the radius of convergence of $\sum a_n z^n$ to be 1. This contradiction shows that f cannot be analytic at $z = 1$.

Solution to 5.8.33: As

$$(\tan z)^{-2} - z^{-2} = \frac{z^2 - (\tan z)^2}{z^2(\tan z)^2}$$

the Maclaurin expansion [MH87, p. 234] of the numerator has no terms of degree up to 3, whereas the expansion of the denominator starts with z^4, therefore, the limit is finite.

As

$$\tan z = z + \frac{1}{3}z^3 + o(z^4) \qquad (z \to 0)$$

we have

$$(\tan z)^{-2} - z^{-2} = \frac{z^2 - z^2 - \frac{2}{3}z^4 + o(z^4)}{z^4 + o(z^4)} \qquad (z \to 0)$$

so the limit at 0 is $-2/3$.

Solution to 5.8.34: We use the fact that an entire function is a polynomial exactly when it has limit ∞ at ∞. (If the function tends to ∞ at ∞ then it has a pole at ∞. In this case, its Laurent series about ∞, which is its power series, has only finitely many nonzero terms with positive exponent. The other direction is clear.)

Suppose g is not a polynomial. Then $(g(z_n))$ is bounded for some sequence (z_n) tending to ∞. Therefore $(f(z_n))$ is bounded, against the hypothesis. Hence g is a polynomial.

Suppose f is not a polynomial. Then $(f(w_n))$ is bounded for some sequence $(w_n$ tending to ∞. As g is onto, there is, for each $n \in \mathbb{N}$, $w_n' \in \mathbb{C}$ such that $w_n = g(w_n')$. As $w_n \to \infty$ we have $w_n' \to \infty$, which is contrary to the hypothesis since $f(g(w_n'))$ stays bounded. Hence f is a polynomial.

Solution to 5.8.35: When we write f as a single fraction, with denominator $\prod_{i=1}^{n}(z - z_i)$ we see that the number of zeros of f in $\mathbb{C}$ is given by the degree of the numerator of f, since the z_j's are no zeros of f. The order of the zero at infinity of $f(1/t)$ gives the difference between the degree of the denominator and the degree of the numerator. We have

$$f\left(\frac{1}{t}\right) = \sum_{i=1}^{n} \frac{a_i t}{1 - z_i t} = \sum_{i=1}^{n} a_i t (1 + z_i t + z_i^2 t^2 + \cdots)$$

$$= \sum_{p=0}^{\infty} S_p t^{p+1} = \sum_{p=m}^{\infty} S_p t^{p+1}.$$

Thus $f(1/t)$ has a zero of order $m + 1$ at $t = 0$. Since the denominator of $f(z)$ has degree n, the numerator has degree $n - m - 1$, so f has $n - m - 1$ zeros in $\mathbb{C}$.

We cannot have $m \geqslant n$ because the Vandermonde determinant, see the Solution to Problem 7.2.11 or [HK61, p. 125]

$$\begin{vmatrix} 1 & z_1 & z_1^2 & \cdots & z_1^{n-1} \\ 1 & z_2 & z_2^2 & \cdots & z_2^{n-1} \\ \vdots & \vdots & \vdots & \ddots & \vdots \\ 1 & z_n & z_n^2 & \cdots & z_n^{n-1} \end{vmatrix}$$

is not zero, so the linear system $\sum_{i=1}^{k} a_i z_i^k = 0$ has only the trivial solution for $k = 0, \ldots, n - 1$.

5.9   Harmonic Functions

Solution to 5.9.1: Derivating twice, we can see that

$$\frac{\partial^2 u}{\partial x^2} = 6x = -\frac{\partial^2 u}{\partial y^2},$$

so $\Delta u = 0$. The function f is then given by (see [Car63b, pp. 126-127])

$$f(z) = 2u\left(\frac{z}{2}, \frac{z}{2i}\right) = 2\left(\frac{z^3}{8} - 3\frac{z}{2}\frac{z^2}{(-4)}\right) = z^3.$$

so, up to a constant, $v(x, y) = 3x^2y - y^3 + k$.

Solution 2. Using the Cauchy–Riemann equations [MH87, p. 72], we see that

$$\frac{\partial v}{\partial y} = 3x^2 - 3y^2$$

and integrating with respect to y we obtain

$$v(x, y) = 3x^2y - y^3 + \psi(x)$$

and using the Cauchy–Riemann equations again, this time $\dfrac{\partial v}{\partial x} = -\dfrac{\partial u}{\partial y}$ we see that $\psi'(x) = 0$, so $v(x, y) = 3x^2y - y^3 + k$.

Solution to 5.9.2: Since u is the real part of an analytic function, it is harmonic in the unit disc $\mathbb{D}$. By Green's Theorem [Rud87, p. 253],

$$\int_{\partial \mathbb{D}} \frac{\partial u}{\partial y}\, dx - \frac{\partial u}{\partial x}\, dy = -\int_{\mathbb{D}} \left(\frac{\partial^2 u}{\partial x^2} + \frac{\partial^2 u}{\partial y^2}\right) dxdy = 0.$$

Solution to 5.9.3: 1. Let $f = u + iv$. Then v is identically 0 on the unit circle. By the Maximum Modulus Principle [MH87, p. 185] for harmonic functions, v is identically zero on $\mathbb{D}$. By the Cauchy–Riemann equations [MH87, p. 72], the partial derivatives of u vanish; hence, u is constant also and so is f.
2. Consider

$$f(z) = i\frac{z+1}{z-1}.$$

f is analytic everywhere in $\mathbb{C}$ except at 1. We have

$$f\left(e^{i\theta}\right) = i\frac{e^{i\theta}+1}{e^{i\theta}-1} = i\frac{e^{i\theta/2}+e^{-i\theta/2}}{e^{i\theta/2}-e^{-i\theta/2}} = \cot\left(\frac{\theta}{2}\right) \in \mathbb{R}.$$

Solution to 5.9.4: We have $u = \Re z^s$, $z^s \equiv e^{s\log z}$, and $\log z =$ principal branch of $\log$ with $-\pi < \arg z < \pi$. z^s is analytic in the slit plane $\mathbb{C} \setminus (-\infty, 0]$ and $\frac{d}{dz}e^{s\log z} = sz^{s-1}$. Hence, $u = \Re z^s$ is harmonic in the same domain.

Solution to 5.9.5: Let v be the harmonic conjugate of u. Then, $f = u + iv$ is an entire function. Consider $h = e^{-f}$. Since $u \geq 0$, $|h| \leq 1$, h is a bounded entire function and, by Liouville's Theorem [MH87, p. 170], a constant. Therefore, u is constant as well.

Solution to 5.9.6: Let v be a harmonic conjugate of u, and let $f = e^{u+iv}$. Then f is an entire function and, for $|z| > 1$, we have

$$|f(z)| = e^{u(z)} \leq e^{a \log|z|+b} = e^b |z|^a .$$

Let n be a positive integer such that $n \geq a$. Then the function $z^{-n} f(z)$ has an isolated singularity at ∞ and, by the preceding inequality, is bounded in a neighborhood of ∞. Hence, ∞ is a removable singularity of $z^{-n} f(z)$ and, thus, is, at worst, a pole of f. That means f is an entire function with, at worst, a pole at ∞, and so f is a polynomial. Since nonconstant polynomials are surjective and f omits the value 0, f must be constant, and so is u, as desired.

Solution to 5.9.7: The linear fractional transformation $w = (1 + z)/(1 - z)$ maps the unit disc to the halfplane $\Re w > 0$, with the upper and lower boundary semicircles mapped to the halflines $i\mathbb{R}_+$ and $i\mathbb{R}_-$, respectively. A branch of $\log w$, defined on $\mathbb{C} \setminus \mathbb{R}_-$, has

$$\Im \log w = \begin{cases} \pi/2, & w \in i\mathbb{R}_+ \\ -\pi/2, & w \in \mathbb{R}_-, \end{cases}$$

so a solution is given by f defined by

$$f(z) = \frac{2}{\pi} \Im \log \frac{1+z}{1-z} .$$

5.10   Residue Theory

Solution to 5.10.1: Since the r_i's are distinct, f has a simple pole at each of these points. If $A_1, A_2, \ldots, A_n$ are the residues of f at each of these points, then

$$g(z) = f(z) - \frac{A_1}{z - r_1} - \frac{A_2}{z - r_2} - \cdots - \frac{A_n}{z - r_n}$$

is entire. Clearly, g tends to zero as z tends to infinity, so, by the Maximum Modulus Principle [MH87, p. 185], g must be identically zero and we are done.

Solution to 5.10.2: We have

$$a_{-1} = \mathrm{Res}\,(\cot \pi z, -1) + \mathrm{Res}\,(\cot \pi z, 0) + \mathrm{Res}\,(\cot \pi z, 1) = \frac{3}{\pi}.$$

For $n < -1$, the coefficients are given by

$$a_n = \frac{1}{2\pi} \int_{|z|=3/2} \frac{\cot \pi z}{z^{n+1}} dz$$

$$= \text{Res} \left(\frac{\cot \pi z}{z^{n+1}}, -1 \right) + \text{Res} \left(\frac{\cot \pi z}{z^{n+1}}, 1 \right)$$

$$= \lim_{z \to -1} (z+1) \frac{\cot \pi z}{z^{n+1}} + \lim_{z \to 1} (z-1) \frac{\cot \pi z}{z^{n+1}}$$

$$= \left((-1)^{-n-1} + 1 \right) \frac{1}{\pi}.$$

Solution to 5.10.3: The function $f(z) = z/(z-1)(z-2)(z-3)$ has simple poles at $z \in A = \{1, 2, 3\}$, with residues $\text{Res}(f, 1) = 1/2$, $\text{Res}(f, 2) = -2$, and $\text{Res}(f, 3) = 3/2$. For f to have a primitive in U, it is necessary and sufficient that $\int_\gamma f(z)\, dz = 0$ for all (piecewise C^1) closed paths γ in U.

Let γ be a closed path in U, with image γ^*. Then $U^c = \{|z| \leq 4\}$ is connected and lies in one connected component of $\mathbb{C} \setminus \gamma^*$. Therefore the winding numbers are the same, $\omega(\gamma, 1) = \omega(\gamma, 2) = \omega(\gamma, 3)$.

By the Residue Theorem, [MH87, p. 280],

$$\int_\gamma f(z)\, dz = 2\pi i \sum_{z \in A} \text{Res}(f, z)\, \omega(\gamma, z)$$

$$= 2\pi i \omega(\gamma, 1) \sum_{z \in A} \text{Res}(f, z)$$

$$= 0.$$

Therefore f has a primitive in U. On the other hand, if

$$g(z) = \frac{z^2}{(z-1)(z-2)(z-3)}$$

then $\text{Res}(g, 1) = 1/2$, $\text{Res}(g, 2) = -4$ and $\text{Res}(g, 3) = 9/2$. If γ is the circle $|z| = 5$, then

$$\int_\gamma g(z)\, dz = 2\pi i \sum_{z \in A} \text{Res}(g, z)\, \omega(\gamma, z) \neq 0,$$

and g has no primitive in U.

Solution to 5.10.4: If the roots of f are not distinct, then some x_0 satisfies $f(x_0) = f'(x_0) = 0$. But $f'(x) = 1 + x + \cdots + \frac{x^{m-1}}{(m-1)!}$, so

$$0 = f(x_0) = f'(x_0) = \frac{x_0^m}{m!}.$$

and $x_0 = 0$. However, 0 is clearly not a root of f. Hence, the roots of f are distinct and nonzero.

For $0 \leqslant k \leqslant n - 2$, consider the integral

$$I_k = \int_{C_r} \frac{z^k}{f(z)} dz$$

where C_r is a circle of radius r centered at the origin such that all the roots of f lie inside it. By Cauchy's Theorem [MH87, p. 152], I_k is independent of r, and as $r \to \infty$, the integral tends to 0. Hence, $I_k = 0$. By the Residue Theorem [MH87, p. 280],

$$0 = \sum \operatorname{Res} \frac{z^k}{f(z)} = \sum_{i=1}^{N} \frac{z_i^k}{f'(z_i)} = \sum_{i=1}^{n} \frac{z_i^k}{f(z_i) - \frac{z_i^n}{n!}} = n! \sum_{i=1}^{n} z_i^{k-n}.$$

Since $2 \leqslant m - k \leqslant n$, we get the desired result.

Solution to 5.10.5: Let the disc centered at the origin with radius r contain all the zeros of Q. Let C_R be a circle centered at the origin with radius $R > r$. Then, by the Deformation Theorem [MH87, p. 148],

$$\int_C \frac{P(z)}{Q(z)} dz = \int_{C_R} \frac{P(z)}{Q(z)} dz$$

where C is any closed curve outside $|z| = r$. As

$$\frac{P(z)}{Q(z)} = O\left(|z|^{-2}\right) \quad (|z| \to \infty)$$

we have

$$\int_{C_R} \frac{P(z)}{Q(z)} dz = O\left(|z|^{-2}\right) 2\pi R = o(1) \quad (R \to \infty)$$

and the result follows.

Solution to 5.10.6: Letting $z = i\theta$, we have

$$\cos \theta = \tfrac{1}{2}(z + z^{-1})$$

and $d\theta = dz/iz$, so that

$$\frac{1}{2\pi} \int_0^{2\pi} e^{2\zeta \cos\theta} d\theta = \frac{1}{2\pi} \int_\gamma e^{\zeta(z+z^{-1})} \frac{dz}{z}$$

where γ is the unit circle. Next,

$$\frac{1}{2\pi i} \int_\gamma e^{\zeta(z+z^{-1})} \frac{dz}{z} = \frac{1}{2\pi i} \int_\gamma \sum_{n=0}^{\infty} \frac{1}{n!} \left(\frac{\zeta}{z}\right)^n e^{\zeta z} \frac{dz}{z}$$

$$= \sum_{n=0}^{\infty} \frac{1}{2\pi i} \int_\gamma \frac{\zeta^n}{n!} \frac{e^{\zeta z}}{z^n} \frac{dz}{z}.$$

Now,

$$\frac{e^{\zeta z}}{z^{n+1}} = \frac{1}{z^{n+1}}\left(1 + \zeta z + \tfrac{1}{2}(\zeta z)^2 + \cdots + \frac{1}{k!}(\zeta z)^k + \cdots\right)$$

$$= \frac{1}{z^{n+1}} + \frac{\zeta}{z^n} + \cdots + \frac{1}{n!}\frac{\zeta^n}{z} + \frac{\zeta^{n+1}}{(n+1)!} + \cdots .$$

Thus, the residue at zero is $\dfrac{\zeta^n}{n!}$ and

$$\frac{1}{2\pi i}\int_\gamma \frac{e^{\zeta z}}{z^{n+1}}\, dz = \frac{\zeta^n}{n!}$$

hence

$$\left(\frac{\zeta^n}{n!}\right)^2 = \frac{1}{2\pi i}\int_\gamma \frac{\zeta^n e^{\zeta z}}{n! z^n}\,\frac{dz}{t}$$

and the result follows.

Solution to 5.10.7: As the singularities of the integrand, call it f, all lie inside the unit disc, we have [MH87, p. 286],

$$I = -\operatorname{Res}\,(f(z), \infty) = \operatorname{Res}\,\left(\frac{1}{z^2}f\left(\frac{1}{z}\right), 0\right).$$

We have $\dfrac{1}{z^2}f\left(\dfrac{1}{z}\right) = \dfrac{(1+2z)^2}{z(2-z)}$, which has residue $\dfrac{1}{2}$ at 0. Therefore $I = \dfrac{1}{2}$.

Solution 2. Let $f(z) = \dfrac{(z+2)^2}{z^2(2z-1)}$. Then f is meromorphic with a double pole at $z = 0$ and a simple pole at $z = 1/2$. Both singularities are inside the contour of integration. By the residue theorem we have

$$I = \operatorname{Res}\,(f(z), 0) + \operatorname{Res}\,\left(f(z), \frac{1}{2}\right).$$

We have, near the origin,

$$f(z) = -\frac{1}{z^2}\left(\frac{(z+2)^2}{1-2z}\right) = -\frac{1}{z^2}(4+4z+z^2)(1+2z+4z^2+\cdots) = -\frac{4}{z^2} - \frac{12}{z} + \cdots$$

so the residue at the origin is -12.

To find the residue at $\dfrac{1}{2}$ we write f as

$$f(z) = \frac{(z+2)^2}{2z^2(z-\tfrac{1}{2})},$$

and get

$$\text{Res}\left(f(z), \frac{1}{2}\right) = \left.\frac{(z+2)^2}{2z^2}\right|_{z=\frac{1}{2}} = \frac{25}{2}.$$

Therefore, $I = -12 + 12.5 = 0.5$.

Solution 3. Since f has no singularities outside the unit disc, we have, by Cauchy's Theorem [MH87, p. 152],

$$I = \frac{1}{2\pi i}\int_{|z|=r} f(z)dz$$

for any $r > 1$. Thus

$$I = \frac{1}{2\pi}\int_0^{2\pi} \frac{\left(re^{i\theta}+2\right)^2 re^{i\theta}}{r^2 e^{2i\theta}\left(2re^{i\theta}-1\right)}d\theta \qquad (r > 1).$$

It is easy to see that the integrand approaches $\dfrac{1}{2}$ uniformly when $r \to \infty$, so $I = \dfrac{1}{2}$.

Solution to 5.10.8: We have

$$\int_{C_a} \frac{z^2+e^z}{z^2(z-2)}dz = \int_{C_a} \frac{1}{(z-2)}dz + \int_{C_a} \frac{e^z}{z^2}\frac{1}{(z-2)}dz.$$

We will use the Residue Theorem [MH87, p. 280], to compute the integral, and for the first integral:

$$\text{Res}\left(\frac{1}{z-2},0\right) = 0 \qquad \text{and} \qquad \text{Res}\left(\frac{1}{z-2},2\right) = 1.$$

Let $f(z) = \dfrac{e^z}{z^2(z-2)}$. The following expansions hold:

$$\frac{1}{z-2} = -\frac{1}{2}\left(1+\frac{z}{2}+\frac{z^2}{4}+\frac{z^3}{8}+\cdots\right)$$

$$\frac{1}{z^2(z-2)} = -\frac{1}{2z^2}-\frac{1}{4z}-\frac{1}{8}-\cdots$$

$$\frac{e^z}{z^2(z-2)} = \left(1+z+\frac{z^2}{2}+\frac{z^3}{3!}+\cdots\right)\left(-\frac{1}{2z^2}-\frac{1}{4z}-\frac{1}{8}-\cdots\right)$$

$$= -\frac{1}{2z^2}-\frac{3}{4z}-\frac{5}{8}-\cdots$$

thus

$$\text{Res}\,(f(z),0) = -\frac{3}{4}.$$

Also,

$$\text{Res}\,(f(z),2) = \lim_{z \to 2} \frac{e^z}{z^2} = \frac{e^2}{4}.$$

Therefore, we have, for $a > 2$,

$$\int_{C_a} \frac{z^2 + e^z}{z^2(z-2)}\,dz = 2\pi i\,\left(\text{Res}\,\left(\frac{1}{z-2},2\right) + \text{Res}\,(f,0) + \text{Res}\,(f,2)\right)$$
$$= \frac{\pi i}{2}\left(e^2 + 1\right)$$

and for $a < 2$,

$$\int_{C_a} \frac{z^2 + e^z}{z^2(z-2)}\,dz = 2\pi i\,\text{Res}\,(f(z),0) = -\frac{3\pi i}{2}.$$

Solution to 5.10.9: For $|z| = r$ we have $\bar{z} = \dfrac{r^2}{z}$, so we can rewrite $I(a,r)$ as

$$I(a,r) = \int_{|z|=r} \frac{1}{a - \frac{r^2}{z}}\,dz = \int_{|z|=r} \frac{z}{az - r^2}\,dz$$
$$= \frac{1}{a}\int_{|z|=r} \frac{z}{z - \frac{r^2}{a}}\,dz \qquad (a \neq 0).$$

If $|a| < r$ the integrand is analytic in and on the circle $|z| = r$, so $I(a,r) = 0$ by Cauchy's Theorem [MH87, p. 152]. (Use the second-last expression for the case $a = 0$.)

If $|a| > r$ then Cauchy's Integral Formula [MH87, p. 167] gives

$$I(a,r) = \frac{2\pi i}{a}\left(\frac{r^2}{a}\right) = \frac{2\pi i r^2}{a^2}.$$

Solution to 5.10.10: Let $f(z) = \sqrt{z^2 - 1} = \displaystyle\sum_{k=0}^{\infty}\binom{1/2}{k}z^{2k}$. f is analytic for $|z| > 1$; therefore,

$$\int_C \sqrt{z^2 - 1}\,dz = -2\pi i\,\text{Res}\,(f(z),\infty).$$

We have

$$\frac{1}{z^2}f\left(\frac{1}{z}\right) = \frac{1}{z^3}\sqrt{1 - z^2}$$
$$= \frac{1}{z^3}\sum_{k \geqslant 0}\binom{1/2}{k}(-1)^k z^{2k}$$
$$= \sum_{k \geqslant 0}\binom{1/2}{k}(-1)^k z^{2k-3}$$

so

$$\text{Res}\,(f(z),\infty) = \text{Res}\left(\frac{1}{z^2}f\left(\frac{1}{z}\right),0\right) = \frac{1}{2}.$$

We then obtain

$$\int_C \sqrt{z^2-1}\,dz = -\pi i.$$

Solution to 5.10.11: The function $\sin z = (e^{iz} - e^{-iz})/(2i)$ vanishes if and only if $e^{2iz} = 1$, which happens if and only if z is an integer multiple of π. Since $2 < \pi < 4$, the zeros of $\sin 4z$ within the unit circle are $0, \pm\pi/4$.

The residue of $1/\sin 4z$ at $z = 0$ is $1/4$, because the numerator is nonvanishing, while the denominator $\sin 4z$ has a simple zero and its derivative at $z = 0$ is 4. Since $\sin(4(z + \pi/4)) = -\sin 4z$, the residue of $1/\sin 4z$ at $z = \pm\pi/4$ is $-1/4$. By the Residue Theorem [MH87, p. 280], $I = 1/4 + (-1/4) + (-1/4) = -1/4$.

Solution to 5.10.12: The integrand has poles at the integer multiples of π, all simple poles except for the one at the origin. By the Residue Theorem [MH87, p. 280],

$$I_n = 2\pi i \sum_{k=-n}^{n} \text{Res}\left(\frac{1}{z^3 \sin z}, k\pi\right).$$

Since the integrand is an even function, its Laurent series at the origin contains only even powers of z, implying that the residue at the origin is 0. Hence $I_0 = 0$. For $k \neq 0$,

$$\text{Res}\left(\frac{1/z^3}{\sin z}, k\pi\right) = \left.\frac{1/z^3}{\frac{d}{dz}(\sin z)}\right|_{z=k\pi} = \frac{1/k^3\pi^3}{\cos k\pi} = \frac{(-1)^k}{k^3\pi^3}.$$

So for $n > 0$, $I_n = \dfrac{4i}{\pi^2} \displaystyle\sum_{k=1}^{n} \frac{(-1)^k}{k^3}$.

Solution to 5.10.13: Using the change of variables $z = \dfrac{1}{w}$, we obtain

$$\frac{1}{2\pi i}\int_C \frac{dz}{\sin\frac{1}{z}} = \frac{1}{2\pi i}\int_{C'} \frac{-dw}{w^2 \sin w} = \frac{1}{2\pi i}\int_{C''} \frac{dw}{w^2 \sin w}$$

where the circles C' and C'' have radius 5 and are oriented negatively and positively, respectively. By the Residue Theorem [MH87, p. 280], we have

$$\frac{1}{2\pi i}\int_C \frac{-dw}{w^2 \sin w} = \text{Res}\left(\frac{1}{w^2 \sin w}, \pi\right) + \text{Res}\left(\frac{1}{w^2 \sin w}, -\pi\right)$$

$$+\,\text{Res}\left(\frac{1}{w^2 \sin w}, 0\right).$$

We have

$$\text{Res}\left(\frac{1}{w^2 \sin w}, \pm\pi\right) = \lim_{w\to\pm\pi} \frac{w \mp \pi}{w^2 \sin w} = -\frac{1}{\pi^2}$$

and

$$\frac{1}{w^2 \sin w} = \frac{1}{w^2 \left(w - w^3/3! + w^5/5! + \cdots\right)}$$

$$= \frac{1}{w^3} \frac{1}{1 - (w^2/3! - w^4/5! + \cdots)}$$

$$= \frac{1}{w^3}\left(1 + \left(w^2/3! - w^4/5! + \cdots\right)\right.$$

$$\left. + \left(w^2/3! - w^4/5! + \cdots\right)^2 + \cdots\right)$$

$$= \frac{1}{w^3} + \frac{1}{3!w} - \frac{w}{5!} + \cdots$$

Then,

$$\frac{1}{2\pi i}\int_C \frac{dw}{w^2 \sin w} = \frac{1}{6} - \frac{2}{\pi^2}.$$

Solution to 5.10.15: We assume C has counterclockwise orientation. First, $I_0 = J_0 = \int_C \frac{dz}{z} = 2\pi i$. Now consider $k \geqslant 1$. We have, for some constants $A_0, \ldots, A_k$,

$$f(z) = \frac{1}{z(z-1)\cdots(z-k)} = \frac{A_0}{z} + \frac{A_1}{z-1} + \cdots + \frac{A_k}{z-k}.$$

Then

$$1 = \sum_{m=0}^{k} A_m \prod_{\substack{n=0 \\ n\neq m}}^{k} (z-n).$$

Equating coefficients gives $0 = \sum_{m=0}^{k} A_m$. For $n = 0, 1, \ldots, k$

$$\int_C \frac{dz}{z-n} = 2\pi i\, \omega(C, n) = 2\pi i.$$

Hence

$$I_k = \int_C f(z)\, dz = \sum_{n=0}^{k} A_n \int_C \frac{dz}{z-n} = 2\pi i \sum_{n=0}^{k} A_n = 0.$$

On the other hand,

$$J_k = 2\pi i\, \text{Res}\left(\frac{(z-1)\ldots(z-k)}{z}, 0\right)$$

$$= (-1)^k k!\, 2\pi i.$$

Solution to 5.10.16: $(e^{2\pi z} + 1)^{-2}$ has a double pole at $\pm i/2$. By the Residue Theorem [MH87, p. 280], the value of this integral is $2\pi i$ times the sum of the residues of $(e^{2\pi z} + 1)^{-2}$ at these two points. We have

$$-e^{2\pi z} = e^{-\pi i}e^{2\pi z} = e^{2\pi(z-i/2)} = 1 + 2\pi(z - i/2) + 2\pi^2(z - i/2)^2 + \cdots$$

hence,

$$e^{2\pi z} + 1 = -2\pi(z - i/2) - 2\pi^2(z - i/2)^2 - \cdots$$

and so

$$(e^{2\pi z} + 1)^2 = 4\pi^2(z - i/2)^2 + 8\pi^3(z - i/2)^3 + \cdots$$

The residue at $\dfrac{i}{2}$ is

$$\frac{d}{dz}\left(\frac{(z - i/2)^2}{(e^{2\pi z} + 1)^2}\right)\bigg|_{z=i/2} = \frac{d}{dz}\left(\frac{1}{4\pi^2 + 8\pi^3(z - i/2) + O((z - i/2)^2)}\right)\bigg|_{z=i/2}$$

$$= \frac{-8\pi^3 + O(z - i/2)}{(4\pi^2 + O(z - i/2))^2}\bigg|_{z=i/2}$$

$$= -\frac{1}{2\pi}.$$

Using the fact that $-e^{2\pi z} = e^{\pi i}e^{2\pi z}$, an identical calculation shows that the other residue is also $-1/2\pi$, so the integral equals $-2i$.

Solution to 5.10.17: A standard application of the Rouché's Theorem [MH87, p. 421] shows that all the roots of the denominator lie in the open unit disc. By the Deformation Theorem [MH87, p. 148], therefore, the integral will not change if we replace the given contour by the circle centered at the origin with radius $R > 1$. Using polar coordinates on this circle, the integral becomes

$$\frac{1}{2\pi i}\int_0^{2\pi} \frac{Re^{11i\theta}\, iRe^{i\theta}\, d\theta}{12R^{12}e^{12i\theta} - 9R^9e^{9i\theta} + 2R^6e^{6i\theta} - 4R^3e^{3i\theta} + 1} =$$

$$\frac{1}{2\pi}\int_0^{2\pi} \frac{d\theta}{12 - 9R^{-3}e^{-3i\theta} + 2R^{-6}e^{-6i\theta} - 4R^{-9}e^{-9i\theta} + R^{-12}}$$

which has the limit $\dfrac{1}{12}$ as $R \to \infty$. The value of the given integral is then, $\dfrac{1}{12}$.

Solution to 5.10.18: We make the change of variables $u = z - 1$. The integral becomes

$$\int_{|u+1|=2} (2u + 1)e^{1+1/u}\, du.$$

Using the power series for the exponential function,

$$e^{z/(z-1)} = e^{1+1/(z-1)} = e \cdot e^{1/(z-1)} = e\sum_{n=0}^{\infty} \frac{1}{n!(z - 1)^n}$$

we get

$$(2u + 1)e^{1+1/u} = e\left(2u + 3 + \frac{2}{u} + \cdots\right).$$

The residue of this function at zero, which lies inside $|u + 1| = 2$, is $2e$, so the integral is $4e\pi i$.

Solution to 5.10.19: Denote the integrand by f. By the Residue Theorem [MH87, p. 280], I is equal to the sum of the residues of f at $-1/2$ and $1/3$, which lie in the interior of C. I is also the negative of the sum of the residues in the exterior of C, namely at 2 and ∞. We have

$$\text{Res}\,(f, 2) = \lim_{z \to 2}(z - 2)f(z) = \frac{-1}{5^5}.$$

As $\lim_{z \to \infty} f(z) = 0$,

$$\text{Res}\,(f, \infty) = -\lim_{z \to \infty} zf(z) = 0.$$

So $I = \dfrac{1}{5^5}$.

Solution to 5.10.20: The numerator in the integrand is $\frac{1}{3n}$ times the derivative of the denominator. Hence, I equals $\frac{1}{3n}$ times the number of zeros of the denominator inside C; that is, $I = \frac{1}{3}$.

Solution 2. For $r > 1$, let C_r be the circle $|z| = r$, oriented counterclockwise. By Cauchy's Theorem [MH87, p. 152], and using the parameterization $z = re^{i\theta}$,

$$I = \frac{1}{2\pi i} \int_{C_r} \frac{z^{n-1}}{3z^n - 1}\, dz$$

$$= \frac{1}{2\pi} \int_0^{2\pi} \frac{r^n e^{in\theta}}{3r^n e^{in\theta} - 1}\, d\theta.$$

As $r \to \infty$, the integrand converges uniformly to $\frac{1}{3}$, giving $I = \frac{1}{3}$.

Solution to 5.10.21: The integrand has two singularities inside C, a pole of order 1 at the origin and a pole of order 2 at $-1/2$. Hence,

$$\int_C \frac{e^z}{z(2z + 1)^2}\, dz = 2\pi i\left(\text{Res}\left(\frac{e^z}{z(2z + 1)^2}, 0\right) + \text{Res}\left(\frac{e^z}{z(2z + 1)^2}, -\frac{1}{2}\right)\right).$$

The residues can be evaluated by standard methods:

$$\text{Res}\left(\frac{e^z}{z(2z + 1)^2}, 0\right) = \frac{e^z}{z(2z + 1)^2}\bigg|_{z=0} = 1$$

$$\text{Res}\left(\frac{e^z}{z(2z + 1)^2}, -\frac{1}{2}\right) = \frac{1}{4}\frac{d}{dz}\left(\frac{e^z}{z}\right)\bigg|_{z=-\frac{1}{2}}$$

$$= \frac{1}{4} \left(\frac{e^z}{z} - \frac{e^z}{z^2} \right)\Big|_{z=-\frac{1}{2}} = -\frac{3e^{-\frac{1}{2}}}{2}.$$

Hence,

$$\int_C \frac{e^z}{z(2z+1)^2}\, dz = 2\pi i \left(1 - \frac{3}{2\sqrt{e}} \right).$$

Solution to 5.10.22: Let $f(z)$ denote the integrand. It has a pole of order 2 at a and a pole of order 3 at b, and no other singularities. The contour Γ has winding number 1 about a and -1 about b. By the Residue Theorem [MH87, p. 280],

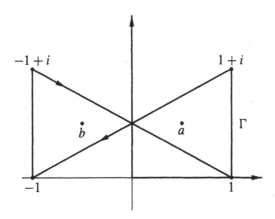

$$I = 2\pi i (\mathrm{Res}(f, a) - \mathrm{Res}(f, b)).$$

We have

$$\mathrm{Res}(f, a) = \frac{d}{dz}(z - b)^{-3}\Big|_{z=a} = \frac{-3}{(a - b)^4}$$

$$\mathrm{Res}(f, b) = \frac{1}{2}\frac{d^2}{dz^2}(z - a)^{-2}\Big|_{z=b} = \frac{3}{(b - a)^4}.$$

Hence

$$I = 2\pi i \left(\frac{-3}{(a - b)^4} - \frac{3}{(a - b)^4} \right) = \frac{-12\pi i}{(a - b)^4} = -12\pi i.$$

Solution to 5.10.23: The function f has poles of order 1 at $z = \pm 1$ and a pole of order 2 at $z = 0$. These are the only singularities of f. The winding numbers of γ around -1, 0, 1 are 1, 2, -1, respectively. By the Residue Theorem [MH87, p. 280],

$$\frac{1}{2\pi i}\int_\gamma f(z)\,dz = \mathrm{Res}(f, -1) + 2\mathrm{Res}(f, 0) - \mathrm{Res}(f, 1).$$

Since 1 and −1 are simple poles, we have

$$\text{Res}\,(f, 1) = \lim_{z \to 1}(z - 1)f(z) = \frac{-e^z}{z^2(z + 1)}\bigg|_{z=1} = \frac{-e}{2}$$

$$\text{Res}\,(f, -1) = \lim_{z \to -1}(z + 1)f(z) = \frac{-e^z}{z^2(z - 1)}\bigg|_{z=-1} = \frac{e^{-1}}{2}.$$

To find the residue at 0, we use power series:

$$\frac{e^z}{z^2(1 - z^2)} = \frac{1}{z^2}\left(1 + z + \frac{z^2}{2} + \cdots\right)\left(1 + z^2 + z^4 + \cdots\right)$$

$$= \frac{1}{z^2}(1 + z + \cdots) = \frac{1}{z^2} + \frac{1}{z} + \cdots$$

It follows that $\text{Res}(f, 0) = 1$. Hence,

$$\frac{1}{2\pi i}\int_\gamma f(z)dz = \frac{e^{-1}}{2} + 2 + \frac{e}{2} = 2 + \cosh 1.$$

Solution to 5.10.24: The function $f(z) = (e^z - 1)/z^2(z - 1)$ has a pole of order 2 at $z = 0$, and a simple pole at $z = 1$. We have

$$\text{Res}\,(f, 0) = \frac{d}{dz}\frac{e^z - 1}{z - 1}\bigg|_{z=0} = -1,$$

$$\text{Res}\,(f, 1) = \lim_{z \to 1}\frac{e^z - 1}{z^2} = e - 1,$$

$$\omega(C, 0) = -2,$$

$$\omega(C, 1) = 2.$$

By the Residue Theorem [MH87, p. 280], we have then

$$\int_C \frac{e^z - 1}{z^2(z - 1)}\,dz = 2\pi i\,(\text{Res}\,(f, 0)\,\omega(C, 0) + \text{Res}\,(f, 1)\,\omega(C, 1))$$

$$= 4\pi i e.$$

Solution to 5.10.25: The roots of $1 - 2z\cos\theta + z^2 = 0$ are $z = \cos\theta \pm i\sqrt{1 - \cos^2\theta} = e^{\pm i\theta}$. Using the Residue Theorem, [MH87, p. 280], we get

$$\frac{1}{2\pi i}\int_{|z|=2}\frac{z^n}{1 - 2z\cos\theta + z^2}\,dz = \frac{1}{2\pi i}\int_{|z|=2}\frac{z^n}{(z - e^{i\theta})(z - e^{-i\theta})}\,dz$$

$$= \text{Res}\left(\frac{z^n}{(z - e^{i\theta})(z - e^{-i\theta})}, e^{i\theta}\right)$$

$$+ \text{Res}\left(\frac{z^n}{(z - e^{i\theta})(z - e^{-i\theta})}, e^{-i\theta}\right)$$

$$= \frac{z^n}{z - e^{-i\theta}}\bigg|_{z=e^{i\theta}} + \frac{z^n}{z - e^{i\theta}}\bigg|_{z=e^{-i\theta}}$$

$$= \frac{e^{in\theta}}{e^{i\theta} - e^{-i\theta}} + \frac{e^{-in\theta}}{e^{-i\theta} - e^{i\theta}}$$

$$= \frac{e^{in\theta} - e^{-in\theta}}{e^{i\theta} - e^{-i\theta}}$$

$$= \frac{\sin n\theta}{\sin \theta}.$$

Solution to 5.10.26: Substituting $z = e^{i\theta}$, we have

$$I(a) = \int_0^{2\pi} \frac{d\theta}{a + \frac{e^{i\theta} + e^{-i\theta}}{2}} = 2 \int_0^{2\pi} \frac{e^{i\theta}\, d\theta}{2ae^{i\theta} + e^{2i\theta} + 1}$$

$$= \frac{2}{i} \int_{|z|=1} \frac{dz}{z^2 + 2az + 1}.$$

The roots of the polynomial in the denominator are $-a + \sqrt{a^2 - 1}$ and $-a - \sqrt{a^2 - 1}$, of which only the former is within the unit circle. By the Residue Theorem [MH87, p. 280],

$$I(a) = 4\pi \, \text{Res}\left(\frac{1}{z^2 + 2az + 1}, -a + \sqrt{a^2 - 1}\right).$$

Since the function in question has a single pole at $z = -a + \sqrt{a^2 - 1}$, the residue equals

$$\frac{1}{2z + 2a}\bigg|_{z=-a+\sqrt{a^2-1}} = \frac{1}{2\sqrt{a^2 - 1}}$$

giving $I(a) = \dfrac{2\pi}{\sqrt{a^2 - 1}}$.

Consider the function F defined for $\xi \notin [-1, 1]$ by

$$F(\xi) = \int_0^{2\pi} \frac{d\theta}{\xi + \cos \theta}.$$

As $F'(\xi)$ exists, F is analytic on its domain. Combining with the previous results, we have that the function

$$F(\xi) - \frac{2\pi}{\sqrt{\xi^2 - 1}}$$

is analytic and vanishes for $\xi > 1$; therefore, it must be identically zero. From this, we obtain that

$$\int_0^{2\pi} \frac{d\theta}{\xi + \cos \theta} = \frac{2\pi}{\sqrt{\xi^2 - 1}}$$

in the domain of F.

Solution to 5.10.27: Let f be the function defined by

$$f(z) = \frac{i}{(z - r)(rz - 1)}.$$

We have

$$\int_{|z|=1} f(z)dz = \int_0^{2\pi} \frac{d\theta}{1 - 2r\cos\theta + r^2}.$$

For $|r| < 1$,

$$\text{Res}(f(z), r) = \frac{i}{r^2 - 1}$$

and

$$\int_0^{2\pi} \frac{d\theta}{1 - 2r\cos\theta + r^2} = \frac{2\pi}{1 - r^2}.$$

For $|r| > 1$, we get

$$\text{Res}\left(f(z), \frac{1}{r}\right) = \frac{i}{1 - r^2}$$

and

$$\int_0^{2\pi} \frac{d\theta}{1 - 2r\cos\theta + r^2} = \frac{2\pi}{r^2 - 1}.$$

Solution 2. Suppose that $r \in \mathbb{R}$ and $0 < r < 1$. Let u_0 be defined on the unit circle by $u_0(z) = 1$. The solution of the corresponding Dirichlet Problem [MH93, p. 600], that is, the harmonic function on $\mathbb{D}$, u, that agrees with u_0 on $\partial\mathbb{D}$ is given by Poisson's Formula [MH87, p. 195]:

$$u(r) = \frac{1 - r^2}{2\pi} \int_0^{2\pi} \frac{d\theta}{1 - 2r\cos\theta + r^2}$$

but $u \equiv 1$ is clearly a solution of the same problem. Therefore, by unicity, we get

$$\int_0^{2\pi} \frac{d\theta}{1 - 2r\cos\theta + r^2} = \frac{2\pi}{1 - r^2}.$$

If $r > 1$, consider a similar Dirichlet problem with $u_0(z) = 1$ for $|z| = r$. We get

$$1 = u(1) = \frac{r^2 - 1}{2\pi} \int_0^{2\pi} \frac{d\theta}{1 - 2r\cos\theta + r^2}$$

so

$$\int_0^{2\pi} \frac{d\theta}{1 - 2r\cos\theta + r^2} = \frac{2\pi}{r^2 - 1}.$$

If $r < 0$, a similar argument applied to $u(-r)$, $u(-1)$ leads to the results above, noting that $\cos(\theta - \pi) = -\cos\theta$.

Solution to 5.10.28: Evaluating the integral using the Residue Theorem [MH87, p. 280], we have

$$\int_0^\pi \frac{\cos 4\theta}{1+\cos^2\theta}\,d\theta = \Re \int_0^\pi \frac{e^{4i\theta}}{1+\cos^2\theta}\,d\theta$$

$$= \Re \int_0^\pi \frac{e^{4i\theta}}{1+\frac{(e^{i\theta}+e^{-i\theta})^2}{4}}\,d\theta$$

$$= 4\Re \int_0^\pi \frac{e^{4i\theta}}{6+e^{2i\theta}+e^{-2i\theta}}\,d\theta$$

$$= 4\Re \int_{|z|=1} \frac{z^3}{z^2+6z+1}\frac{dz}{2iz}$$

$$= 2\Re \int_{|z|=1} \frac{z^2}{z^2+6z+1}\,dz$$

$$= 2\Re\, 2\pi \left(\mathrm{Res}\left(\frac{z^2}{z^2+6z+1}, -3+2\sqrt{2}\right)\right)$$

$$= 4\pi \left(-3+\frac{17}{4\sqrt{2}}\right) = -12\pi + \frac{17}{\sqrt{2}}\pi.$$

Solution to 5.10.29: As $2\cos^2 x = \cos 2x + 1$, we have

$$I = \frac{1}{2}\int_0^{2\pi} \frac{\cos 6\theta + 1}{5-4\cos 2\theta}\,d\theta$$

$$= \frac{1}{2}\Re \int_0^{2\pi} \frac{e^{6i\theta}+1}{5-4\cos 2\theta}\,d\theta$$

$$= \frac{-1}{2}\Re \int_0^{2\pi} \frac{e^{8i\theta}+e^{2i\theta}}{2e^{4i\theta}-5e^{2i\theta}+2}\,d\theta$$

$$= \frac{1}{2}\Re i \int_{|z|=1} \frac{z^3+1}{2z^2-5z+2}\,dz.$$

We evaluate this integral using the Residue Theorem [MH87, p. 280]. As the integrand has a simple pole at $z = 1/2$ inside the unit circle and no others, we have

$$I = \frac{1}{2}\Re \left(i2\pi i \left.\frac{z^3+1}{2(z-2)}\right|_{z=1/2}\right) = \frac{3\pi}{8}.$$

Solution to 5.10.30: The integrand is holomorphic except for a pole of order three at the origin. By the Residue Theorem, we have

$$I = 2\pi i\,\mathrm{Res}\left(\frac{\cos^3 z}{z^3}, 0\right).$$

Near $z = 0$ we have

$$\frac{\cos^3 z}{z^3} = \frac{\left(1 - \frac{z^2}{2} + O(z^4)\right)^3}{z^3}$$

$$= \frac{\left(1 - z^2 + O(z^4)\right)\left(1 - \frac{z^2}{2} + O(z^4)\right)}{z^3}$$

$$= \frac{1 - \frac{3}{2}z^2 + O(z^4)}{z^3},$$

showing that the residue in question is $-\dfrac{3}{2}$. Hence $I = -3\pi i$.

Solution to 5.10.31: Consider the contour in the figure

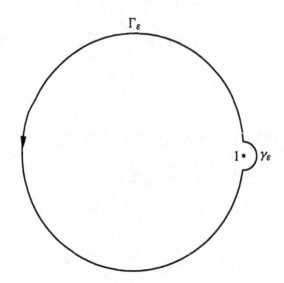

We have

$$\int_0^{2\pi} \frac{1 - \cos n\theta}{1 - \cos \theta}\,d\theta = \lim_{\varepsilon \to 0} \int_\varepsilon^{2\pi - \varepsilon} \frac{1 - \cos n\theta}{1 - \cos \theta}\,d\theta = \lim_{\varepsilon \to 0} \int_\varepsilon^{2\pi - \varepsilon} \frac{1 - e^{in\theta}}{1 - \cos \theta}\,d\theta,$$

the last equality because the integral of $f(\theta) = \dfrac{i \sin n\theta}{1 - \cos \theta}$ vanishes (since $f(2\pi - \theta) = -f(\theta)$). Next,

$$\int_\varepsilon^{2\pi - \varepsilon} \frac{1 - e^{in\theta}}{1 - \cos \theta}\,d\theta = \int_{\Gamma_\varepsilon} \frac{1 - z^n}{1 - (z + 1/z)/2}\,\frac{dz}{iz},$$

where Γ_ε is the almost-circle $\{e^{i\theta} \mid \varepsilon \leqslant \theta \leqslant 2\pi - \varepsilon\}$, traversed counterclockwise. The latter integral equals

$$\frac{2}{i} \int_{\Gamma_\varepsilon} \frac{z^n - 1}{(z - 1)^2}\,dz,$$

which, when $\varepsilon \to 0$, tends to

$$\pi i \frac{2}{i} \operatorname{Res}\left(\frac{z^n - 1}{(z-1)^2}, 1\right) = 2\pi n,$$

so the integral in the statement of the problem has the value $2\pi n$.

Solution to 5.10.32: Let $z = e^{i\theta}$. Then

$$
\begin{aligned}
\int_{-\pi}^{\pi} \frac{\sin n\theta}{\sin \theta} d\theta &= \int_{|z|=1} \frac{z^n - z^{-n}}{z - z^{-1}} \frac{dz}{iz} \\
&= \frac{1}{i} \int_{|z|=1} \frac{z^{2n} - 1}{z^2 - 1} \frac{1}{z^n} dz \\
&= \frac{1}{i} \int_{|z|=1} \left(1 + z^2 + \cdots + z^{2n-2}\right) \frac{1}{z^n} dz \\
&= \sum_{k=0}^{n-1} \int_{|z|=1} z^{2k-n} dz.
\end{aligned}
$$

If n is even, only even powers of z occur, and each term vanishes. For odd n, we have

$$\int_{-\pi}^{\pi} \frac{\sin n\theta}{\sin \theta} d\theta = \frac{1}{i} \int_{|z|=1} \frac{dz}{z} = 2\pi.$$

Solution to 5.10.33: We have

$$
\begin{aligned}
C_n(a) + iS_n(a) &= \int_{-\pi}^{\pi} \frac{e^{in\theta}}{a - \cos \theta} d\theta \\
&= -2 \int_{-\pi}^{\pi} \frac{e^{i(n+1)\theta}}{e^{2i\theta} - 2ae^{i\theta} + 1} d\theta \\
&= 2i \int_{|z|=1} \frac{z^n}{z^2 - 2az + 1} dz.
\end{aligned}
$$

Let $f(z)$ denote this last integrand. Its denominator has two zeros, $a \pm \sqrt{a^2 - 1}$, of which $a - \sqrt{a^2 - 1}$ is inside the unit circle. The residue is given by

$$\operatorname{Res}\left(f, a - \sqrt{a^2 - 1}\right) = \frac{\left(a - \sqrt{a^2 - 1}\right)^n}{\left(a - \sqrt{a^2 - 1} - a - \sqrt{a^2 - 1}\right)} = \frac{\left(a - \sqrt{a^2 - 1}\right)^n}{-2\sqrt{a^2 - 1}}.$$

Therefore, by the Residue Theorem [MH87, p. 280],

$$C_n(a) + iS_n(a) = \frac{2\pi \left(a - \sqrt{a^2 - 1}\right)^n}{\sqrt{a^2 - 1}}.$$

Since the right-hand side is real, this must be the value of $C_n(a)$, and $S_n(a) = 0$.

5.11   Integrals Along the Real Axis

Solution to 5.11.1: We have

$$\int_{-\infty}^{\infty} f_m(x)\overline{f_n(x)}\,dx = \frac{1}{\pi}\int_{-\infty}^{\infty} \frac{(x-i)^m}{(x+i)^{m+1}}\frac{(x+i)^n}{(x-i)^{n+1}}\,dx.$$

If $m = n$, we get

$$\frac{1}{\pi}\int_{-\infty}^{\infty}\frac{dx}{1+x^2} = \frac{1}{\pi}\arctan x\,\Big|_{-\infty}^{\infty} = 1.$$

If $m < n$, as $x^2 + 1 = (x-i)(x+i)$, we have

$$\frac{1}{\pi}\int_{-\infty}^{\infty}\frac{1}{x^2+1}\frac{(x+i)^{n-m}}{(x-i)^{n-m}}\,dx.$$

Since the numerator has degree 2 less than the denominator, the integral converges absolutely. We evaluate it using residue theory. Let $C_R = [-R, R] \cup \Gamma_R$ be the contour

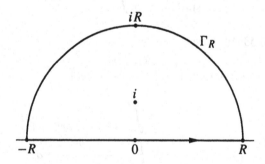

We evaluate the integral over C_R. For $R > 0$ sufficiently large, the integrand has a pole at $x = i$ inside the contour. Calculating the residue, we get

$$\frac{d^{n-m+1}}{dx}\left((x-i)^{n-m+1}\frac{(x+i)^{n-m-1}}{(x-i)^{n-m+1}}\right)\Bigg|_{x=i} = 0.$$

By the Residue Theorem [MH87, p. 280], the integral over C_R is 0 for all such R. Letting R tend to infinity, we see that the integral over the semicircle Γ_R tends to 0 since the numerator has degree 2 less than the denominator. So

$$\frac{1}{\pi}\int_{-\infty}^{\infty}\frac{1}{x^2+1}\frac{(x+i)^{n-m}}{(x-i)^{n-m}}\,dx = 0.$$

Solution to 5.11.2: Consider the following contour around the pole of the function

$$f(z) = \frac{1 - e^{i|a|z}}{z^2}$$

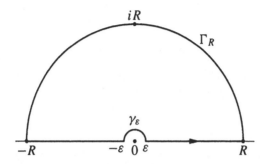

On the larger arc $z = R(\cos\theta + i\sin\theta)$, so

$$\left| \frac{1 - e^{i|a|z}}{z^2} \right| \leq \frac{1 + e^{-R|a|\sin\theta}}{R^2} = O\left(R^{-2}\right) \quad (R \to \infty).$$

Then

$$\int_0^\infty \frac{1 - \cos ax}{x^2} dx = \frac{1}{2}\Re\left(\pi i \operatorname{Res}\left(\frac{1 - e^{i|a|z}}{z^2}, 0\right)\right)$$

$$= \frac{1}{2}\Re\left(\pi i \, 2\frac{-i|a|e^{i|a|z}}{2}\bigg|_{z=0}\right)$$

$$= \frac{\pi|a|}{2}.$$

Solution to 5.11.3: Let $f_t(z) = e^{itz}/(z + i)^2$, and consider first the case $t > 0$. Then $|f_t(z)|$ is bounded in the upper half-plane by $|z + i|^{-2}$. For $R > 1$ let $C_R = \Gamma_R \cup [-R, R]$, where Γ_R is the semicircle centered at the origin joining R and $-R$, oriented counterclockwise. The function f_t is holomorphic on C_R and its interior, so, by Cauchy's Theorem, we have

$$0 = \int_{C_R} f_f(z)dz = \int_{-R}^R f_t(x)dx + \int_{\Gamma_R} f_t(z)dz.$$

The absolute value of the second summand on the right is at most $\pi R/R^2$, since $|f_t(z)| \leq R^{-2}$ on Γ_R. Taking the limit as $R \to \infty$ we obtain $I(t) = 0$ ($f \geq 0$).

Suppose now $t < 0$. Then $|f_t|$ is bounded in the lower half-plane by $|z + i|^{-2}$. Let C'_R be the reflection of C_R with respect to the real axis, oriented clockwise. By the Residue Theorem, we have

$$\int_{C'_R} f_t(z)dz = -2\pi i \operatorname{Res}\left(f_t(z), -i\right).$$

A calculation shows that the residue equals ite^t, so

$$\int_{C_R'} f_t(z)dz = 2\pi t e^t \,.$$

As $R \to \infty$, the contribution to the last integral from the semicircle tends to 0 since $|f_t| \leqslant (R-1)^2$ on the semicircle, giving $I(t) = 2\pi t e^t$ $(t < 0)$.

Solution to 5.11.4: For $t = 0$ the integral is elementary:

$$\int_{-R}^{R} \frac{dx}{(x+i)^3}dx = \frac{-1}{2(x+i)^2}\Big|_{-R}^{R} \to 0 \quad \text{as} \quad R \to \infty,$$

hence $F(0) = 0$.

For $t < 0$ the function $f_t(z) = \dfrac{e^{-itz}}{(z+i)^3}$ is bounded in the upper half-plane, and in fact is $O(|z|^{-3})$ there. We integrate f_t around the contour Γ_R consisting of the interval $[-R, R]$ on the real axis and the semicircle in the upper half-plane with center 0 and radius R ($R > 1$), oriented counterclockwise. The integral is 0 by Cauchy's Theorem [MH87, p. 152]. The length of the semicircle is πR and the integrand is $O(R^{-3})$ on it, so the contribution to the integral due to the semicircle tends to 0 as $R \to \infty$. The contribution due to the interval $[-R, R]$ tends to $F(t)$, so we conclude that $F(t) = 0$ for $t < 0$.

For $t > 0$ we integrate f_t around Γ_R^*, the reflection of Γ_R with respect to the real axis (oriented clockwise). The integrand has one singularity in the interior of Γ_R^*, a pole of order 3 at $z = -i$. The residue there is

$$\frac{1}{2!}\frac{d^2}{dz^2}(e^{-itz})\Big|_{z=-i} = \frac{-t^2 e^{-t}}{2}\,.$$

By the Residue Theorem [MH87, p. 280],

$$\int_{\Gamma_R^*} f_t(z)dz = -2\pi i\left(\frac{-t^2 e^{-t}}{2}\right) = \pi i t^2 e^{-t}\,.$$

By the same reasoning as above, the preceding integral tends to $F(t)$ as $R \to \infty$. Hence $F(t) = \pi i t^2 e^{-t}$ for $t > 0$.

Solution to 5.11.5: To see that the integral exists, notice that

$$\lim_{x\to 0}\frac{\sin^2 x}{x^2} = 1, \qquad \frac{\sin^2 x}{x^2} = O\left(\frac{1}{x^2}\right), \quad (x \to \infty)\,.$$

As the integrand is an even function, we have

$$\int_{-\infty}^{\infty}\frac{\sin^2 x}{x^2}dx = 2\int_{0}^{\infty}\frac{\sin^2 x}{x^2}dx\,.$$

Let $f(z) = \dfrac{1 - e^{2iz}}{z^2}$. Then f is analytic except for a simple pole at 0.

For $0 < \varepsilon < R$, consider the contour $C_{\varepsilon,R} = \Gamma_R \cup \gamma_\varepsilon \cup [-R, -\varepsilon] \cup [\varepsilon, R]$

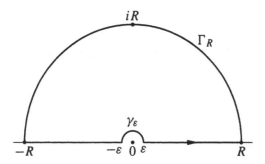

by Cauchy's Theorem [MH87, p. 152]. We have, by the Residue Theorem,

$$0 = \int_{C_{\varepsilon,R}} f(z)\,dz = \int_{\Gamma_R} + \int_{[-R,-\varepsilon]} + \int_{\gamma_\varepsilon} + \int_{[\varepsilon,R]}.$$

It is easy to see that on Γ_R we have

$$|f(z)| = \frac{|1 - e^{2iz}|}{z^2} \leqslant \frac{2}{|z|^2},$$

therefore

$$\int_{-\infty}^{\infty} \frac{1 - e^{2ix}}{x^2}\,dx = \pi i \operatorname{Res}(f, 0) = \pi i(-2i) = 2\pi.$$

Since $\sin^2(x) = \dfrac{1}{2}\Re\left(1 - e^{2ix}\right)$, we have

$$\int_0^{\infty} \frac{\sin^2 x}{x^2}\,dx = \frac{1}{4}\Re \int_{-\infty}^{\infty} \frac{1 - e^{2ix}}{x^2}\,dx,$$

so

$$\int_0^{\infty} \frac{\sin^2 x}{x^2}\,dx = \frac{\pi}{2}.$$

Solution 2. Assuming Dirichlet's Integral

$$\int_0^{\infty} \frac{\sin x}{x}\,dx = \frac{\pi}{2}$$

whose evaluation can be found in [Boa87, p. 86] and [MH87, p. 313], we can use integration by parts:

$$\int_0^{\infty} \frac{\sin^2 x}{x^2}\,dx = -\frac{1}{x}\sin^2 x \Big|_0^{\infty} - \int_0^{\infty} -\frac{1}{x}\sin 2x\,dx$$

$$= 0 + \int_0^\infty \frac{\sin y}{y}\, dy$$

$$= \frac{\pi}{2}.$$

Solution to 5.11.6: The integrand is absolutely integrable, because it is bounded in absolute value by 1 near the origin and by $1/|x|^3$ away from the origin. We have

$$\sin^3 x = -\frac{1}{8i}\left(e^{ix} - e^{-ix}\right)^3 = -\frac{1}{8i}\left(e^{3ix} - e^{-3ix} - 3e^{ix} + 3e^{-ix}\right)$$

$$= \Im\left(\frac{3}{4}e^{ix} - \frac{1}{4}e^{3ix}\right).$$

For $0 < \varepsilon < R$, consider the contour $C_{\varepsilon,R} = \Gamma_R \cup \gamma_\varepsilon \cup [-R, -\varepsilon] \cup [\varepsilon, R]$

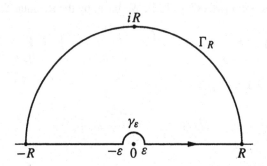

by Cauchy's Theorem [MH87, p. 152],

$$0 = \int_{C_{\varepsilon,R}} \frac{3e^{iz} - e^{3iz}}{z^3}\, dz = \int_{\Gamma_R} + \int_{[-R,-\varepsilon]} + \int_{\gamma_\varepsilon} + \int_{[\varepsilon,R]}.$$

The integral over Γ_R is bounded in absolute value by $(4R^{-3})(2\pi R)$ since $|e^{iz}|$ and $|e^{3iz}|$ are bounded by 1 in the upper half-plane, so it tends to 0 as $R \to \infty$. To estimate the integral over γ_ε, we note that

$$\frac{3e^{iz} - e^{3iz}}{z^3} = \frac{1}{z^3}\left(2 + \frac{3(iz)^2}{2} - \frac{(3iz)^2}{2} + O(z^3)\right)$$

$$= \frac{2}{z^3} + \frac{3}{z} + O(1) \qquad (z \to 0).$$

Hence,

$$\int_{\gamma_\varepsilon} \frac{3e^{iz} - e^{3iz}}{z^3}\, dz = \int_\pi^0 \left(\frac{2}{\varepsilon^3 e^{3i\theta}} - \frac{3}{\varepsilon e^{i\theta}} + O(1)\right) i\varepsilon e^{i\theta}\, d\theta$$

$$= \frac{2i}{\varepsilon^2}\int_\pi^0 e^{-2i\theta}\, d\theta + 3i \int_\pi^0 d\theta + O(\varepsilon)$$

$$= 0 - 3\pi i + O(\varepsilon) \to -3\pi \qquad \text{as} \quad \varepsilon \to 0.$$

Thus,

$$\lim_{\substack{R\to\infty \\ \varepsilon\to 0}} \left(\int_{-R}^{-\varepsilon} \frac{3e^{ix} - e^{3ix}}{x^3}\, dx + \int_{\varepsilon}^{R} \frac{3e^{ix} - e^{3ix}}{x^3}\, dx \right) = 3\pi i.$$

The integral is one-fourth the imaginary part of the preceding limit, so $\dfrac{3\pi}{4}$.

Solution to 5.11.7: Let

$$f(z) = \frac{e^{iz} z^3}{(z^2 + 1)^2} = \frac{e^{iz} z^3}{(z + i)^2 (z - i)^2}.$$

Integrate f over a closed semicircular contour $C_R = [-R, R] \cup \Gamma_R$ with radius R in the upper half plane.

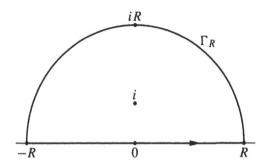

The portion $[-R, R]$ along the axis gives

$$\int_{-R}^{R} \frac{x^3}{(x^2 + 1)^2} (\cos x + i \sin x)\, dx$$

whose imaginary part is

$$\int_{-R}^{R} \frac{x^3 \sin x}{(x^2 + 1)^2}\, dx.$$

This integral converges as $R \to \infty$ to

$$\int_{-\infty}^{\infty} \frac{x^3 \sin x}{(x^2 + 1)^2}\, dx.$$

To establish the convergence, note that integration by parts gives

$$\int_{-R}^{R} \frac{x^3 \sin x}{(x^2 + 1)^2}\, dx = \left. \frac{(-\cos x) x^3}{(x^2 + 1)^2} \right|_{-R}^{R} + \int_{-R}^{R} \cos x \, \frac{d}{dx} \left(\frac{x^3}{(x^2 + 1)^2} \right) dx$$

and

$$\left| \frac{(-\cos x) x^3}{(x^2 + 1)^2} \right| \leqslant \frac{1}{|x|} \to 0 \quad \text{as} \quad x \to \infty,$$

and the second term is integrable by the Comparison Test [Rud87, p. 60], $O(1/x^2)$. The real part also converges by a similar reasoning, and since the integrand is odd, it converges to zero.

The integral of f over Γ_R is treated as follows. Let $z = Re^{i\theta}$ for $0 \leqslant \theta \leqslant \pi$. Then

$$\int_{\Gamma_R} \frac{e^{iz} z^3}{(z^2 + 1)^2} \, dz = \int_0^\pi \frac{e^{iR(\cos\theta + i\sin\theta)} z^3}{(z^2 + 1)^2} \, iRe^{i\theta} \, d\theta .$$

This is bounded above in absolute value for large R by

$$A \int_0^\pi e^{-R\sin\theta} \, d\theta$$

for a constant A and this integral tends to zero as $R \to \infty$ (Jordan's Lemma [MH87, p. 301]). Finally,

$$\int_{C_R} f(z) dz = 2\pi i \, \mathrm{Res}(f, i)$$

and since we have a second order pole, we have

$$\mathrm{Res}(f, i) = \frac{d}{dz} \left(\frac{e^{iz} z^3}{(z + i)^2} \right) \bigg|_{z=i} = \frac{1}{4e}.$$

Thus,

$$\int_{C_R} f(z) dz = \frac{\pi i}{2e}$$

and so the required integral is $\dfrac{\pi}{2e}$.

Solution to 5.11.8: Using an argument similar to the one in Problem 5.11.7 with the function

$$f(z) = \frac{e^{iz} z}{(z^2 + 1)^2},$$

we get

$$\int_{-\infty}^{\infty} \frac{x \sin x}{(x^2 + 1)^2} \, dx = \Re \left(2\pi i \, \mathrm{Res}(f, i) \right) = \Re 2\pi i \frac{d}{dz} \left(\frac{e^{iz} z}{(z + i)^2} \right) \bigg|_{z=i}$$

$$= \Re \left(2\pi i \frac{1}{4e} \right) = \frac{\pi}{2e}.$$

Solution to 5.11.9: Consider the complex integral

$$\int_{C_{\varepsilon, R}} \frac{e^{iz}}{z(z^2 + a^2)} \, dz$$

where the contour $C_{\varepsilon, R}$ is the contour described below oriented counterclockwise. The integrand has simple poles at 0 and ia.

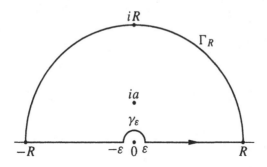

Notice that $\dfrac{\sin x}{x(x^2 + a^2)} = \Re\left(\dfrac{e^{ix}}{x(x^2 + a^2)}\right)$ for real x. By Jordan's Lemma [MH87, p. 301], we have

$$\left|\int_{\Gamma_R} \frac{e^{iz}}{z(z^2 + a^2)} dz\right| \leqslant \frac{\pi}{R} \int_{\Gamma_R} \frac{1}{(z^2 + a^2)} dz = o(1) \quad (|R| \to \infty)$$

so, at the limit, the contribution from the bigger semicircle to the integral above is zero. Using the Residue Theorem [MH87, p. 280], we have

$$\int_{C_{\varepsilon, R}} \frac{e^{iz}}{z(z^2 + a^2)} dz = 2\pi i \operatorname{Res}\left(\frac{e^{iz}}{z(z^2 + a^2)}, ia\right) + \pi i \operatorname{Res}\left(\frac{e^{iz}}{z(z^2 + a^2)}, 0\right)$$

$$= 2\pi i \frac{e^{-a}}{ia(2ia)} + \pi i \frac{1}{a^2}$$

$$= -\frac{\pi i}{e^a a^2} + \frac{\pi i}{a^2}$$

since f is even we have,

$$I = \frac{\pi}{2a^2}\left(1 - \frac{1}{e^a}\right).$$

Solution 2. Consider the function $\dfrac{e^{iz} - 1}{z(z^2 + a^2)}$ which has a removable singularity at the origin and a simple pole at ia. Again, we have $\dfrac{\sin x}{x(x^2 + a^2)} = \Re\left(\dfrac{e^{ix} - 1}{x(x^2 + a^2)}\right)$.
By the same argument as above, but with one less pole, we can use the following contour

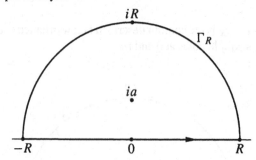

and get

$$\int_{-\infty}^{\infty} \frac{e^{ix}-1}{x(x^2+a^2)}dx = 2\pi i \text{Res} \left(\frac{e^{iz}-1}{z(z^2+a^2)}, ia \right) = \frac{\pi i}{a^2}\left(1-e^{-a}\right),$$

therefore,

$$I = \frac{\pi i}{2a^2}\left(1-e^{-a}\right).$$

Solution to 5.11.10: We have

$$\frac{\sin x}{x-3i} = \frac{(\sin x)(x+3i)}{x^2+9} = \Im\left(\frac{xe^{ix}}{x^2+9}\right) + 3i\Im\left(\frac{e^{ix}}{x^2+9}\right).$$

So

$$\int_{-R}^{R} \frac{\sin x}{x-3i}dx = \Im\int_{-R}^{R} \frac{xe^{ix}}{x^2+9}dx + 3i\Im\int_{-R}^{R} \frac{e^{ix}}{x^2+9}dx.$$

We evaluate these integrals over the contour $C_R = \Gamma_R \cup [-R, R]$:

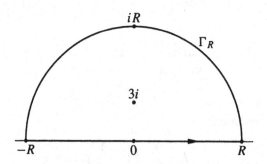

As $z^2 + 9$ has simple zeros at $\pm 3i$, by the Residue Theorem [MH87, p. 280],

$$\int_{C_R} \frac{ze^{iz}}{z^2+9}dz = 2\pi i \, \text{Res}\,(f, 3i) = e^{-3}\pi i.$$

By Jordan's Lemma [MH87, p. 301], we have

$$\left| \int_{\Gamma_R} \frac{ze^{iz}}{z^2+9}dz \right| \leqslant \int_{\Gamma_R} \frac{|z||e^{iz}|}{|z|^2-9}|dz| \leqslant \frac{\pi R}{R^2-9} = o(1) \quad (R \to \infty)$$

so, in the limit the integral along the upper half-circle contributes nothing.
We can evaluate the second integral in the same way, getting

$$\int_{C_R} \frac{e^{iz}}{z^2 + 9} \, dz = \frac{e^{-3}\pi}{3}.$$

Again, in the limit, the upper half-circle contributes zero, so

$$\lim_{R \to \infty} \int_{-R}^{R} \frac{\sin x}{x - 3i} \, dx = e^{-3}\pi.$$

Solution to 5.11.11: Consider the function f defined by

$$f(z) = \frac{e^{iz}}{z(z - \pi)}.$$

By Cauchy's Theorem [MH87, p. 152],

$$\int_{C_{\varepsilon,R}} f(z) \, dz = 0$$

where $C_{\varepsilon,R}$ is the contour $\Gamma_R \cup \gamma_0 \cup [\varepsilon, \pi - \varepsilon] \cup \gamma_\pi \cup [\pi + \varepsilon, R]$.

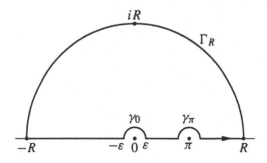

The poles of f at 0 and π are simple; therefore,

$$\lim_{\varepsilon \to 0} \int_{\gamma_0} f(z) \, dz = -\pi i \, \text{Res} \, (f(z), 0) = i$$

$$\lim_{\varepsilon \to 0} \int_{\gamma_\pi} f(z) \, dz = -\pi i \, \text{Res} \, (f(z), \pi) = i.$$

We also have, for $|z| = R$,

$$\frac{e^{iz}}{z(z - \pi)} = O\left(R^{-2}\right) \quad (R \to \infty)$$

so

$$\lim_{R \to \infty} \int_{\Gamma_R} f(z)dz = 0.$$

Taking imaginary parts, we obtain

$$\int_{-\infty}^{\infty} \frac{\sin x}{x(x - \pi)} dx = -2.$$

Solution to 5.11.12: The integral equals $\Re \int_{-\infty}^{\infty} \frac{e^{ix}}{(1 + x^2)^3} dx$. For $R > 1$ let Γ_R be the contour consisting of the interval $[-R, R]$ on the real axis and the top half of the circle $|z| = R$, with the counterclockwise orientation. The function $f(z) = e^{iz}/(1 + z^2)^3$ is analytic on and inside Γ_R except for a pole of order 3 at $z = i$. By the Residue Theorem [MH87, p. 280],

$$\int_{\Gamma_R} f(z)dz = 2\pi i \operatorname{Res}(f(z), i) .$$

Writing $f(z) = \dfrac{e^{iz}}{(z - i)^3 (z + i)^3}$, we see that

$$
\begin{aligned}
2 \operatorname{Res}(f(z), i) &= \frac{d^2}{dz^2}\left(\frac{e^{iz}}{(z + i)^3} \right)\bigg|_{z=i} \\
&= \frac{d}{dz}\left(\frac{ie^{iz}}{(z + i)^3} - \frac{3e^{iz}}{(z + i)^4} \right)\bigg|_{z=i} \\
&= \left(\frac{-e^{iz}}{(z + i)^3} - \frac{6ie^{iz}}{(z + i)^4} + \frac{12e^{iz}}{(z + i)^5} \right)\bigg|_{z=i} \\
&= e^{-1}\left(\frac{-1}{8i^3} - \frac{6i}{16i^4} + \frac{12}{32i^5} \right) = \frac{e^{-1}}{i}\left(\frac{1}{8} + \frac{3}{8} + \frac{3}{8} \right) \\
&= \frac{7}{8ie} .
\end{aligned}
$$

Hence $\int_{\Gamma_R} f(z)dz = \dfrac{7\pi}{8e}$. Now, since $\Im f$ is an odd function on $\mathbb{R}$,

$$\int_{\Gamma_R} f(z)dz = \int_{-R}^{R} \frac{\cos x}{(1 + x^2)^3} dx + \int_{S_R} f(z)dz ,$$

where S_R is the top half of Γ_R. As $R \to \infty$ the first term on the right converges to the integral we want, and the second term converges to 0 because $|f(z)| \leqslant 1/(|z|^2 - 1)^3$ in the upper half-plane, and we conclude that the value of the integral is $7\pi/8e$.

Solution to 5.11.13: Note that $1 + x + x^2 = (x + \frac{1}{2})^2 + \frac{3}{4}$, so the denominator does not vanish on the real axis. The given integral is $I = \Re \int_{-\infty}^{\infty} \frac{e^{ikx}}{1 + x + x^2}$. As $|e^{ikz}| \leq 1$ for z in the upper half-plane, and since the denominator has degree 2, we may close the contour in the upper half-plane to get $I = \Re(2\pi i R)$, where R is the sum of the residues of $f(z) = \frac{e^{ikz}}{1 + z + z^2}$ in the upper half-plane. In this domain f has just one singularity, a simple pole at $-\frac{1}{2} + i\frac{\sqrt{3}}{2}$, with residue

$$R = \frac{e^{ik(-\frac{1}{2} + i\frac{\sqrt{3}}{2})}}{2(-\frac{1}{2} + i\frac{\sqrt{3}}{2}) + 1} = \frac{e^{-\frac{k\sqrt{3}}{2}}e^{-i\frac{k}{2}}}{i\sqrt{3}}. \text{ Therefore}$$

$$I = \Re(2\pi i R) = \frac{2\pi}{\sqrt{3}}e^{-\frac{k\sqrt{3}}{2}}\cos\frac{k}{2}.$$

Solution to 5.11.14: An argument similar to the one used in Problem 5.11.11 with the contour around the simple poles $-1/2$ and $1/2$, gives

$$\int_{-\infty}^{\infty} \frac{\cos(\pi x)}{4x^2 - 1}\, dx = \Re\left(\pi i \left(\text{Res}\left(\frac{e^{\pi i z}}{4z^2 - 1}, -\frac{1}{2}\right) + \text{Res}\left(\frac{e^{\pi i z}}{4z^2 - 1}, \frac{1}{2}\right)\right)\right)$$

$$= \Re\pi i \left(\frac{i}{4} + \frac{i}{4}\right) = -\frac{\pi}{2}.$$

Solution to 5.11.15: Consider the following contour C_R and the function

$$f(z) = \frac{e^{inz}}{z^4 + 1}$$

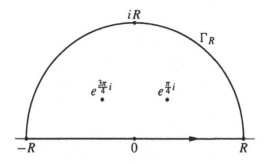

We have

$$\left| \int_{\Gamma_R} \frac{e^{inz}}{z^4 + 1}\, dz \right| \leq \int_{\Gamma_R} \frac{|e^{inz}|}{R^4 - 1}|dz| \leq \frac{\pi R}{R^4 - 1}$$

which approaches 0 when $R \to \infty$ (note that for $z \in \Gamma_R$, $|e^{inz}| = e^{-\Im nz} \leqslant 1$).
Taking the real part and using the fact that for these roots $1/z_0^3 = -z_0$, we have

$$\int_{-\infty}^{\infty} \frac{\cos nx}{x^4 + 1}\, dz = \lim_{R \to \infty} \Re \left(\int_C \frac{e^{inz}}{z^4 + 1}\, dz \right)$$

$$= \Re \left(2\pi i \left(\operatorname{Res}\left(\frac{e^{inz}}{z^4+1}, e^{\pi i/4} \right) + \operatorname{Res}\left(\frac{e^{inz}}{z^4+1}, e^{3\pi i/4} \right) \right) \right)$$

$$= -2\pi \Im \left(\frac{e^{inz}}{4z^3}\bigg|_{z=e^{\pi i/4}} + \frac{e^{inz}}{4z^3}\bigg|_{z=e^{3\pi i/4}} \right)$$

$$= \frac{\pi e^{-\frac{n}{\sqrt{2}}}}{\sqrt{2}}\left(\cos \frac{n}{\sqrt{2}} + \sin \frac{n}{\sqrt{2}} \right).$$

Solution to 5.11.16: Consider the same contour C_R as in Problem 5.11.15; which encircles two of the poles ($e^{\frac{\pi i}{4}}$ and $e^{\frac{3\pi i}{4}}$), and the function

$$f(z) = \frac{z^2 + 1}{z^4 + 1}.$$

We have

$$\int_0^{\infty} \frac{x^2 + 1}{x^4 + 1}\, dz = \frac{1}{2}\left(2\pi i \left(\operatorname{Res}\left(\frac{z^2+1}{z^4+1}, e^{\pi i/4} \right) + \operatorname{Res}\left(\frac{z^2+1}{z^4+1}, e^{3\pi i/4} \right) \right) \right)$$

$$= \pi i \left(\frac{z^2+1}{4z^3}\bigg|_{z=e^{\pi i/4}} + \frac{z^2+1}{4z^3}\bigg|_{z=e^{3\pi i/4}} \right)$$

$$= -\frac{\pi i}{4}\left((z^2+1)z\big|_{z=e^{\pi i/4}} + (z^2+1)z\big|_{z=e^{3\pi i/4}} \right)$$

$$= \frac{\pi}{\sqrt{2}}.$$

Solution to 5.11.17: We have

$$\int_{-\infty}^{\infty} \frac{\cos^3 x}{a^2 + x^2}\, dx = \int_{-\infty}^{\infty} \frac{\left(\frac{e^{ix} + e^{-ix}}{2} \right)^3}{a^2 + x^2}\, dx$$

$$= \frac{1}{8}\int_{-\infty}^{\infty} \frac{e^{3ix} + e^{-3ix} + 3e^{ix} + 3e^{-ix}}{a^2 + x^2}\, dx$$

$$= \frac{1}{4}\int_{-\infty}^{\infty} \frac{e^{3ix}}{a^2 + x^2}\, dx + \frac{3}{4}\int_{-\infty}^{\infty} \frac{e^{ix}}{a^2 + x^2}\, dx$$

$$= \frac{1}{4}I_1(a) + \frac{3}{4}I_2(a).$$

To evaluate $I_1(a)$, let Γ_R ($R > a$) be the contour consisting of the interval $[-R, R]$ on the real axis plus the semicircle in the upper half-plane with center 0 and radius R, oriented counterclockwise. By Cauchy's Integral Formula [MH87, p. 167],

$$\int_{\Gamma_R} \frac{e^{3iz}}{z^2 + a^2}\, dz = \int_{\Gamma_R} \frac{e^{3iz}}{(z+ia)(z-ia)}\, dz$$

$$= 2\pi i \left(\frac{e^{3iz}}{z+ia}\right)\Bigg|_{z=ia} = \frac{\pi e^{-3a}}{a}.$$

The contribution to the integral from the semicircle is $O\left(\dfrac{\pi R}{R^2 - a^2}\right)$ (as $|e^{3iz}| \leqslant 1$ for $\Im z \geqslant 0$), so it tends to 0 as $R \to \infty$. The contribution from the interval $[-R, R]$ tends to $I_1(a)$ as $R \to \infty$. It follows that $I_1(a) = \dfrac{\pi e^{-3a}}{a}$. Exactly the same reasoning shows that $I_2(a) = \dfrac{\pi e^{-a}}{a}$. We obtain

$$I(a) = \frac{\pi}{4a}\left(e^{-3a} + 3e^{-a}\right).$$

Solution to 5.11.18: This integral converges, by Dirichlet's Test [MH93, p. 287], since

$$\lim_{x \to \pm\infty} \frac{x}{x^2 + 4x + 20} = 0$$

monotonically and

$$\int_\alpha^\beta \sin x\, dx = O(1) \qquad (\alpha \to -\infty,\ \beta \to \infty).$$

Consider the integral

$$I = \int_{C_R} f(z)e^{iz}\, dz \quad \text{where} \quad f(z) = \frac{z}{z^2 + 4z + 20}.$$

The curve C_R is the contour

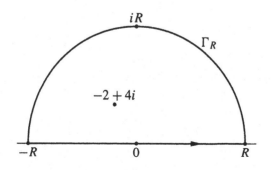

where Γ_R is the semicircle of radius $R > 0$. The integral I is equal to the sum of the residues of $g(z) = f(z)e^{iz}$ inside of C_R. f has poles at $-2 \pm 4i$, of which only $-2 + 4i$ lies inside C_R. Hence,

$$I = 2\pi i \operatorname{Res}(g, -2 + 4i) = 2\pi i \lim_{z \to -2+4i} \frac{ze^{iz}}{z - (-2 - 4i)} = \frac{\pi}{4}(-2 + 4i)e^{-4-2i}.$$

We have, by Jordan's Lemma [MH87, p. 301], when $R \to \infty$,

$$\left| \int_{\Gamma_R} g(z)\,dz \right| \leqslant \frac{R}{R^2 - 4R - 20} \int_{\Gamma_R} |e^{iz}|\,|dz| \leqslant \frac{R\pi}{R^2 - 4R - 20} = o(1).$$

So,

$$\begin{aligned}
\int_{-\infty}^{\infty} f(x) \sin x\,dx &= \Im \left(\int_{-\infty}^{\infty} f(x)e^{ix}\,dx \right) \\
&= \Im \left(\lim_{R \to \infty} \int_{C_R} f(z)e^{iz}\,dz \right) \\
&= \Im \left(\frac{\pi}{4}(-2 + 4i)e^{-4-2i} \right) \\
&= \frac{\pi}{2e^4}(2 \cos 2 + \sin 2).
\end{aligned}$$

Solution to 5.11.19: Consider the following contour avoiding the pole of the function

$$f(z) = \frac{z + ie^{iz}}{z^3}$$

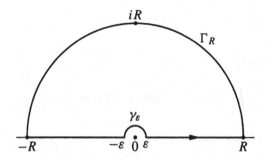

we have

$$\left| \int_{\Gamma_R} \frac{z + ie^{iz}}{z^3}\,dz \right| \leqslant \int_{\Gamma_R} \frac{|z| + |e^{iz}|}{R^3}|dz| \leqslant \frac{\pi(R + 1)}{R^2}$$

which approaches 0 when $R \to \infty$ (note that for $z \in \Gamma_R$, $|e^{iz}| = e^{-\Im z} \leqslant 1$); and around 0,

$$\frac{z + ie^{iz}}{z^3} = \frac{1}{z^3}\left(z + i\left(1 + iz + \frac{(iz)^2}{2} + \cdots \right) \right) = \frac{i}{z^3} - \frac{i}{2z} + g(z),$$

where g is analytic. Then

$$\int_{\gamma_\varepsilon} \frac{z + ie^{iz}}{z^3}\, dz = i\int_{\gamma_\varepsilon} \frac{dz}{z^3} - \frac{i}{2}\int_{\gamma_\varepsilon} \frac{dz}{z} + \int_{\gamma_\varepsilon} g(z)dz = -\frac{\pi}{2} + \int_{\gamma_\varepsilon} g(z)dz.$$

Now since

$$\left|\int_{\gamma_\varepsilon} g(z)dz\right| \leqslant \max_{|z|\leqslant\varepsilon} |g(z)|\pi\varepsilon \to 0 \text{ as } \varepsilon \to 0$$

the integral along the real axis will approach $\dfrac{\pi}{2}$. Taking the real part, one gets

$$\int_0^\infty \frac{x - \sin x}{x^3}dx = \frac{1}{2}\frac{\pi}{2} = \frac{\pi}{4}.$$

Solution to 5.11.20: Denote the given integral by I. Consider the function $f(z) = (1 + z + z^2)^{-2}$. f has two poles of order 2 at $z = (-1 \pm i\sqrt{3})/2$. We evaluate the contour integral

$$I' = \int_{C_R} f(z)\, dz.$$

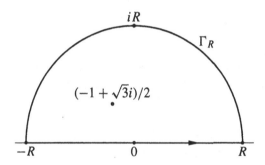

By the Residue Theorem [MH87, p. 280], for R large enough,

$$I' = 2\pi i \operatorname{Res}\left(f(z), \frac{-1 + i\sqrt{3}}{2}\right).$$

As R tends to infinity, the integral along the upper half-circle tends to 0 since the denominator of $f(z)$ has degree 4 higher than the numerator. Hence, I' tends to I as R tends to infinity. Therefore,

$$I = 2\pi i \operatorname{Res}\left(f(z), \frac{-1 + i\sqrt{3}}{2}\right)$$

$$= 2\pi i \frac{d}{dz}\left(\left(z + \frac{1 + i\sqrt{3}}{2}\right)^{-2}\right)\Bigg|_{z = \frac{-1+i\sqrt{3}}{2}}$$

$$= \frac{4\pi}{3\sqrt{3}}.$$

Solution to 5.11.21: Letting $t^2 = x$, we get

$$\int_0^\infty \frac{x^{\alpha-1}}{1+x}\,dx = 2\int_0^\infty \frac{t^{2\alpha-1}}{1+t^2}\,dt.$$

Consider the integral

$$\int_{C_{\varepsilon,R}} \frac{z^{2\alpha-1}}{1+z^2}\,dz,$$

where $C_{\varepsilon,R}$ is the contour $\Gamma_R \cup \gamma_\varepsilon \cup [-R, -\varepsilon] \cup [\varepsilon, R]$:

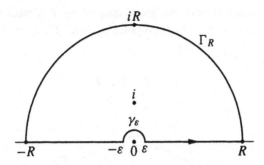

where the small circle, γ_ε, has radius ε, and the large circle, Γ_R, radius R. Since $z^{2\alpha-1}$ is defined to be $e^{(2\alpha-1)\log z}$, we choose as our branch of the logarithm that with arguments between $-\pi/2$ and $3\pi/2$ (that is, the one with the cut along the negative imaginary axis).

The integrand has a simple pole inside $C_{\varepsilon\,R}$ at i, so, by the Residue Theorem [MH87, p. 280], the integral is equal to

$$2\pi i \operatorname{Res}(f, i) = \frac{2\pi i e^{(2\alpha-1)\log i}}{2i} = -\pi i e^{\alpha\pi i}.$$

As ε and R tend to 0 and infinity, respectively, the integral on $[\varepsilon, R]$ tends to the desired integral. On the segment $[-R, -\varepsilon]$, we make the change of variables $y = -z$, getting

$$\int_{-R}^{-\varepsilon} \frac{e^{(2\alpha-1)\log z}}{1+z^2}\,dz = -e^{2\alpha\pi i}\int_\varepsilon^R \frac{e^{(2\alpha-1)\log y}}{1+y^2}\,dy,$$

so this integral tends to a constant multiple of the desired integral. On Γ_R, a calculation shows that $|z^{2\alpha-1}| = |z|^{2\Re\alpha-1}e^{-2\Im\alpha\,\arg z}$, so if we assume that $\Im\alpha \geq 0$, we have $|z^{2\alpha-1}| \leq |z|^{2\Re\alpha-1}$ and

$$\left|\int_{\Gamma_R} \frac{z^{2\alpha-1}}{1+z^2}\,dz\right| \leq \frac{\pi R^{2\Re\alpha}}{R^2-1}.$$

This tends to 0 whenever $\Re\alpha < 1$.

On γ_ε, essentially the same estimate holds:

$$\left| \int_{\gamma_\varepsilon} \frac{z^{2\alpha-1}}{1+z^2} \, dz \right| \leqslant \frac{\pi \varepsilon^{2\Re\alpha}}{1-\varepsilon^2}$$

and this tends to 0 whenever $\Re\alpha > 0$.

So we have

$$\left(1 - e^{2\alpha\pi i}\right) \int_0^\infty \frac{t^{2\alpha-1}}{1+t^2} \, dt = -\pi i e^{\alpha\pi i}.$$

Dividing through, we get

$$\frac{-\pi i e^{\alpha\pi i}}{1 - e^{2\alpha\pi i}} = \frac{\pi}{2\sin\pi\alpha}.$$

Therefore, twice this is our answer, subject to the restrictions $0 < \Re\alpha < 1$ and $\Im\alpha \geqslant 0$. However, if $\Im\alpha \leqslant 0$, we may replace α by $\bar{\alpha}$ and obtain the above equality with $\bar{\alpha}$. Then, by taking the complex conjugate of both sides, we see that we can eliminate the second restriction.

Solution 2. Considering the slightly more complex contour of integration below, we can do away with the change of variables. So consider the integral

$$\int_C \frac{z^{\alpha-1}}{z+1} dz$$

where C is

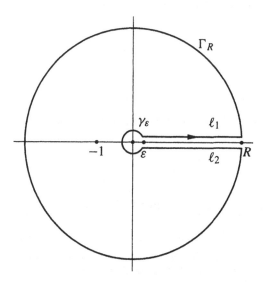

The origin is a branch point and the two straight lines are close to the x–axis. The integrand has one simple pole at $z = -1$ inside C. Thus,

$$\int_C \frac{z^{\alpha-1}}{z+1} dz = 2\pi i \text{ Res} \left(\frac{z^{\alpha-1}}{z+1}, -1 \right) = 2\pi i \lim_{z \to -1} \frac{z^{\alpha-1}}{z+1}(z+1) = 2\pi i e^{(\alpha-1)\pi i}.$$

On the other hand, the integral can be split in four integrals along the segments of the contour:

$$2\pi i \, e^{(\alpha-1)\pi i} = \int_{\Gamma_R} \frac{z^{\alpha-1}}{z+1} dz + \int_{\gamma_\varepsilon} \frac{z^{\alpha-1}}{z+1} dz + \int_{\ell_1} \frac{z^{\alpha-1}}{z+1} dz + \int_{\ell_2} \frac{z^{\alpha-1}}{z+1} dz$$

$$= \int_0^{2\pi} \frac{(Re^{i\theta})^{\alpha-1}}{Re^{i\theta}+1} i R \, e^{i\theta} \, d\theta + \int_{2\pi}^0 \frac{(\varepsilon e^{i\theta})^{\alpha-1}}{\varepsilon e^{i\theta}+1} i\varepsilon e^{i\theta} \, d\theta$$

$$+ \int_\varepsilon^R \frac{x^{\alpha-1}}{x+1} dx + \int_R^\varepsilon \frac{(xe^{2\pi i})^{\alpha-1}}{xe^{2\pi i}+1} dx.$$

Taking the limits as $\varepsilon \to 0$ and $R \to \infty$, we get

$$\int_0^\infty \frac{x^{\alpha-1}}{x+1} dx + \int_\infty^0 \frac{e^{2\pi i(\alpha-1)} x^{\alpha-1}}{x+1} dx = 2\pi i \, e^{(\alpha-1)\pi i}$$

or

$$\left(1 - e^{2\pi i(\alpha-1)} \right) \int_0^\infty \frac{x^{\alpha-1}}{x+1} dx = 2\pi i \, e^{(\alpha-1)\pi i}.$$

Hence,

$$\int_0^\infty \frac{x^{\alpha-1}}{x+1} dx = \frac{2\pi i \, e^{(\alpha-1)\pi i}}{1 - e^{(\alpha-1)2\pi i}} = \frac{\pi}{\sin \alpha\pi}.$$

Solution to 5.11.22: Making the substitution $x = \sqrt{y}$ the integral becomes

$$\frac{1}{2} \int_0^\infty \frac{y^{-1/4}}{1+y} dy$$

which is equal to $\dfrac{\pi}{\sqrt{2}}$ by Problem 5.11.21 with $\alpha = \dfrac{3}{4}$.

Solution to 5.11.23: Making the substitution $x = \sqrt[5]{y}$ the integral becomes

$$\frac{1}{5} \int_0^\infty \frac{y^{-4/5}}{1+y} dy$$

which is equal to $\dfrac{\pi}{5 \sin \frac{\pi}{5}}$ by Problem 5.11.21 with $\alpha = \dfrac{1}{5}$.

Solution 2. We integrate the function $f(z) = 1/(1+z^5)$ around the counterclockwise oriented contour Γ_R consisting of the segment $[0, R]$, the circular arc joining the points R and $R \, e^{2\pi i/5}$, and the segment $[R \, e^{2\pi i/5}, 0]$, where $R > 1$.

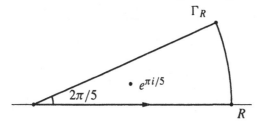

The function f has one singularity inside Γ_R, a simple pole at $e^{\pi i/5}$. The residue is given by

$$\text{Res}(f, e^{\pi i/5}) = \frac{1}{\frac{d}{dz}(1 + z^5)}\bigg|_{z=\pi i/5} = \frac{1}{5\, e^{4\pi i/5}}.$$

By the Residue Theorem [MH87, p. 280], $\displaystyle \int_{\Gamma_R} f(z)dz = \frac{2\pi i}{5\, e^{4\pi i/5}}.$

The contribution to the integral from the segment $[0, R]$ equals $\displaystyle \int_0^R \frac{1}{1+x^5}\,dx$ and the contribution from the segment $[R\, e^{2\pi i/5}, 0]$ equals

$$-\int_0^R \frac{1}{1 + (t\, e^{2\pi i/5})^5}d(t\, e^{2\pi i5}) = -e^{2\pi i/5}\int_0^R \frac{1}{1+t^5}\,dt$$

$$= -e^{2\pi i/5}\int_0^R \frac{1}{1+x^5}\,dx$$

The contribution from the circular arc is $O\!\left(\frac{R}{(R-1)^5}\right)$, so it tends to 0 as $R \to \infty$. Taking the limit as $R \to \infty$, we thus obtain

$$(1 - e^{2\pi i/5})\int_0^\infty \frac{1}{1+x^5}\,dx = \frac{2\pi i}{5\, e^{4\pi i/5}},$$

or

$$\int_0^\infty \frac{1}{1+x^5}\,dx = \frac{2\pi i}{5(e^{4\pi i/5} - e^{6\pi i/5})} = \frac{2\pi i}{5(e^{\pi i/5} - e^{-\pi i/5})}$$

$$= \frac{2\pi i}{5(2i\sin\frac{\pi}{5})} = \frac{\pi}{5\sin\frac{\pi}{5}}.$$

Solution to 5.11.24: Observing that the function is even, doubling the integral from 0 to ∞ and making the substitution $y = x^{2n}$, we get

$$\int_{-\infty}^\infty \frac{dx}{1+x^{2n}} = 2\int_0^\infty \frac{dx}{1+x^{2n}} = \frac{1}{n}\int_0^\infty \frac{y^{\frac{1}{2n}-1}}{y+1}\,dy = \frac{\pi}{n\sin\pi/2n}$$

by Problem 5.11.21.

Solution 2. The $2n^{th}$ roots of -1 in the upper half-plane are $z_k = e^{i(\frac{\pi}{2n}+\frac{k\pi}{n})}$ for $0 \leqslant k \leqslant n-1$.

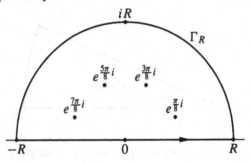

Therefore

$$\int_{-\infty}^{\infty} \frac{dx}{1+x^{2n}} = 2\pi i \sum_{k=0}^{n-1} \text{Res}\left(\frac{1}{1+z^{2n}}, z_k\right).$$

Since z_k is a simple pole of the integrand,

$$\text{Res}\left(\frac{1}{1+z^{2n}}, z_k\right) = \lim_{z \to z_k} \frac{z - z_k}{1+z^{2n}} = \frac{1}{2nz_k^{2n-1}} = -\frac{z_k}{2n},$$

giving

$$\int_{-\infty}^{\infty} \frac{dx}{1+x^{2n}} = -\frac{i\pi}{n} \sum_{k=0}^{n-1} z_k - \frac{i\pi}{n} e^{i\frac{\pi}{2n}} \sum_{k=0}^{n-1} e^{i\frac{k\pi}{n}}$$

$$= -\frac{i\pi}{n} e^{i\frac{\pi}{2n}} \frac{1 - e^{i\pi}}{1 - e^{i\frac{\pi}{n}}} = \frac{2\pi i}{n} e^{i\frac{\pi}{2n}} \frac{1}{e^{i\frac{\pi}{n}} - 1} = \frac{\pi}{n \sin\frac{\pi}{2n}}.$$

Solution 3. Let $f(z) = \dfrac{1}{1+z^{2n}}$ and consider the contour in shape of a slice of pizza containing the positive real axis from 0 to R, the arc of circle of angle π/n and radius of the circle from the end of the arc back to 0, oriented in this way.

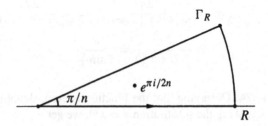

f has one singularity inside the contour, which is a simple pole at $e^{\pi i/2n}$. Therefore, the integral of f around such a contour equals, by the Residue Theorem [MH87, p. 280],

$$2\pi i \text{Res}\left(\frac{1}{1+z^{2n}}, e^{\pi i/2n}\right) = \frac{-\pi i e^{\pi i/2n}}{n}.$$

We have, then

$$\int_0^R \frac{1}{1+x^{2n}}dx + \int_0^{\pi/4} \frac{1}{1+R^{2n}e^{2ni\theta}}iRe^{i\theta}\,d\theta + \int_R^0 \frac{1}{1+r^{2n}e^{2\pi i}}e^{\pi i/n}\,dr$$

$$= \left(1 - e^{\pi i/n}\right)\int_0^R \frac{1}{1+x^{2n}}dx + iR\int_0^{\pi/n} \frac{1}{1+R^{2n}e^{2ni\theta}}e^{i\theta}\,d\theta .$$

The integral on the arc approaches zero as $R \to \infty$, therefore, we get

$$\int_{-\infty}^{\infty} \frac{1}{1+x^{2n}}dx = -2\frac{\pi i e^{\pi i/2n}}{n(1 - e^{\pi i/n})} = \frac{\pi}{n \sin \frac{\pi}{2n}} .$$

Solution 4. Consider the complex function $f(z) = \dfrac{1}{z^{2n}+1}$. Its poles are all simple and located at the points

$$c_k = e^{i\left(\frac{\pi}{2n}+2k\frac{\pi}{2n}\right)} = e^{i(2k+1)\frac{\pi}{2n}} \qquad k = 0, 1, \dots, 2n - 1$$

and since k and n are integers, the roots are never real and calling the primitive angle $\alpha = \frac{\pi}{2n}$ the ones located above the real axis are

$$c_k = e^{i(2k+1)\alpha} \qquad k = 0, 1, \dots, n - 1.$$

Computing the residue of f at these points we have:

$$\mathrm{Res}\left(\frac{1}{z^{2n+1}}, c_k\right) = \frac{1}{\frac{d}{dz}(z^{2n}+1)} = \frac{1}{2nc_k^{2n-1}} \qquad k = 0, 1, \dots, n - 1$$

so

$$\mathrm{Res}\left(\frac{1}{z^{2n+1}}, c_k\right) = \frac{1}{2n}e^{[i\frac{(2k+1)(1-2n)}{2n}\pi]}$$

$$= \frac{1}{2n}e^{[i\frac{(2k+1)}{2n}\pi + i\frac{(2k+1)(-2n)}{2n}\pi]}$$

$$= \frac{1}{2n}e^{[i\frac{(2k+1)}{2n}\pi]} \cdot e^{[-i(2k+1)\pi]}$$

$$= -\frac{e^{[i(2k+1)\alpha]}}{2n}$$

Using the upper-semi-circle as the contour of integration, we have to add the residues on all poles in the upper-half-plane. Using the fact that $\displaystyle\sum_{k=0}^{n-1} z^k = \frac{1-z^n}{1-z}$ and applying the Residue Theorem:

$$\int_{-R}^R \frac{1}{x^{2n+1}}\,dx + \int_{C_R} \frac{1}{z^{2n+1}}\,dz = 2\pi i \sum_{k=0}^{n-1} \mathrm{Res}\left(\frac{1}{z^{2n+1}}, c_k\right)$$

$$= -\frac{\pi i}{n} e^{i\alpha} \sum_{k=0}^{n-1} (e^{i2\alpha})^k$$

$$= -\frac{\pi i}{n} e^{i\alpha} \frac{1 - e^{i2\alpha n}}{1 - e^{i2\alpha}} \cdot \frac{e^{-i\alpha}}{e^{-i\alpha}}$$

$$= -\frac{\pi i}{n} \frac{e^{i2\alpha n} - 1}{e^{i\alpha} - e^{-i\alpha}} = \frac{\pi}{n} \frac{2i}{e^{i\alpha} - e^{-i\alpha}} = \frac{\pi}{n \sin \alpha} \cdot$$

Solution to 5.11.25: Let

$$f(z) = \frac{z^2}{z^n + 1} \cdot$$

The answer is $2I$, where $I = \displaystyle\int_0^\infty f(x)\, dx$.

For $R > 1$ let Γ_R be the straight line path from 0 to R followed by the arc Re^{it} for $t \in [0, 2\pi/n]$, followed by the straight line from $Re^{2\pi i/n}$ to 0.

Let $\zeta = e^{\pi i/n}$. The poles of f are ζ^{2m+1} for $m \in \mathbb{Z}$, so the only pole inside Γ_R is ζ and f has a simple pole at ζ with residue $\frac{1}{n}\zeta^{3-n}$. By the Residue Theorem, we have,

$$\int_{\Gamma_R} f(z)dz = \frac{2\pi i}{n} \zeta^{3-n} = -\frac{2\pi i}{n} \zeta^3 .$$

The integral over the first line tends to I as $R \to \infty$, over the arc of circle part it approaches 0, since the integrand is $O(R^{2-n}$ while the arc length is $O(R)$, and the integral over the last line segment tends to $-\zeta^6 I$, as the substitution $z = \zeta^2 w$ shows. Thus,

$$I - \zeta^6 I = -\frac{2\pi i}{n} \zeta^3 .$$

Now,

$$\sin \frac{3\pi}{n} = \frac{\zeta^3 - \zeta^{-3}}{2i} = \frac{\zeta^6 - 1}{2i\zeta^3}$$

so

$$2I = \frac{2\pi}{n \sin \frac{3\pi}{n}} \cdot$$

Solution to 5.11.26: We have

$$I = \int_0^\infty \frac{x}{e^x + e^{-x}} dx$$

$$= \frac{1}{2} \int_{-\infty}^\infty \frac{x}{e^x - e^{-x}} dx$$

$$= \frac{1}{2} \int_0^\infty \frac{\log u}{(u - 1/u)u} du$$

$$= \frac{1}{4} \int_{-\infty}^\infty \frac{\log |u|}{u^2 - 1} du$$

where we used $u = e^x$. Integrate

$$f(z) = \frac{\log z}{z^2 - 1}$$

over the contour $\Gamma = \Gamma_R \cup \gamma_{-1} \cup \gamma_0 \cup \gamma_1$.

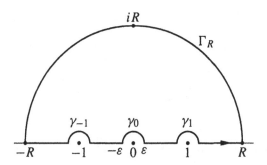

By Cauchy's Theorem [MH87, p. 152], we have

$$\int_\Gamma f(z)dz = 0$$

also,

$$\int_{\gamma_{-1}} f(z)dz = -i\pi \ \text{Res}(f, -1) = \frac{\pi^2}{2}$$

$$\int_{\gamma_1} f(z)dz = -i\pi \frac{\log 1}{2} = 0$$

$$\int_{\Gamma_R} f(z)dz \to 0 \quad \text{as} \quad R \to \infty \quad \text{since} \quad \lim_{R \to \infty} \frac{R \log R}{R^2 - 1} = 0$$

$$\int_{\gamma_0} f(z)dz \to 0 \quad \text{since} \quad \lim_{\varepsilon \to 0} \frac{\varepsilon \log \varepsilon}{\varepsilon^2 - 1} = 0.$$

So we get

$$\int_{-\infty}^{\infty} \frac{\log|u|}{u^2 - 1}du = \frac{\pi^2}{2}$$

and

$$I = \frac{\pi^2}{8}.$$

Solution to 5.11.27: The roots of the denominator are $1 \pm i\sqrt{3}$, and for large $|x|$, the absolute value of the integrand is of the same order of magnitude as x^{-2}. It follows that the integral converges. For $R > 2$, let

$$I_R = \int_{C_R} \frac{e^{-iz}}{z^2 - 2z + 4}dz,$$

where the contour C_R consists of the segment $[-R, R]$ together with the semicircle Γ_R in the lower half-plane with the same endpoints as that segment, directed clockwise.

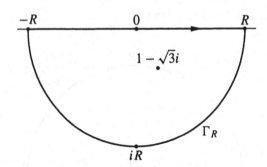

The integrand has only one singularity inside C_R, a simple pole at $1 - i\sqrt{3}$. By the Residue Theorem [MH87, p. 280],

$$I_R = -2\pi i \operatorname{Res}\left(\frac{e^{-iz}}{z^2 - 2z + 4}, 1 - i\sqrt{3}\right).$$

The residue, since we are dealing with a simple pole, equals

$$\lim_{z \to 1 - i\sqrt{3}}\left(z - 1 + i\sqrt{3}\right)\left(\frac{e^{-iz}}{z^2 - 2z + 4}\right) = \frac{e^{-iz}}{z - 1 - i\sqrt{3}}\bigg|_{z=1-i\sqrt{3}} = \frac{e^{-\sqrt{3}-i}}{-2i\sqrt{3}}.$$

Hence, $I_R = \dfrac{\pi e^{-\sqrt{3}-i}}{\sqrt{3}}$. Since $|e^{-iz}|$ is bounded by 1 in the lower half-plane, we have, for large R,

$$\left|\int_{\Gamma_R}\frac{e^{-iz}}{z^2 - 2z + 4}\,dz\right| \leq \frac{2\pi R}{R^2 - 2R - 4} \longrightarrow 0 \qquad (R \to \infty).$$

Hence,

$$\frac{\pi e^{-\sqrt{3}-i}}{\sqrt{3}} = \int_{-R}^{R}\frac{e^{-ix}}{x^2 - 2x + 4}\,dx + \int_{\Gamma_R}\frac{e^{-iz}}{z^2 - 2z + 4}\,dz \longrightarrow I \qquad (R \to \infty),$$

giving $I = \dfrac{\pi e^{-\sqrt{3}-i}}{\sqrt{3}}$.

Solution to 5.11.28: The integral converges absolutely since, for each $0 < \varepsilon < 1$, we have $\log x = o(x^\varepsilon)$ $(x \to 0+)$.

Consider the function

$$f(z) = \frac{\log z}{(z^2 + 1)(z^2 + 4)}.$$

and, for $0 < \varepsilon < R$, consider the contour $C_{\varepsilon,R} = \Gamma_R \cup \gamma_\varepsilon \cup [-R, -\varepsilon] \cup [\varepsilon, R]$.

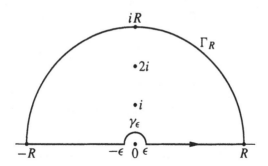

On the small circle, we have

$$\left| \int_{\gamma_\varepsilon} f(z)dz \right| \leqslant \frac{|\log \varepsilon| + \pi}{\frac{1}{2}} \pi \varepsilon = o(1) \quad (\varepsilon \to 0)$$

and on the larger one,

$$\left| \int_{\Gamma_R} f(z)dz \right| \leqslant \frac{|\log R| + \pi}{(R^2 - 1)(R^2 - 4)} \pi R = o(1) \quad (R \to \infty).$$

Combining with the Residue Theorem [MH87, p. 280] we get

$$\int_{-\infty}^{0} f(z)dz + \int_{0}^{\infty} f(z)dz = 2\pi i \left(\text{Res}(f, i) + \text{Res}(f, 2i) \right).$$

We have

$$\text{Res}(f, i) = \lim_{z \to i} f(z)(z - i) = \frac{\pi}{12}$$

and

$$\text{Res}(f, 2i) = \lim_{z \to 2i} f(z)(z - 2i) = \frac{\log 2 + \frac{\pi i}{2}}{-12i} = \frac{i \log 2}{12} - \frac{\pi}{24}.$$

We also have

$$\int_{-\infty}^{0} f(z)dz = \int_{-\infty}^{0} \frac{\log(-x) + \pi i}{(x^2 + 1)(x^2 + 4)} dx$$

$$= \int_{0}^{\infty} \frac{\log x}{(x^2 + 1)(x^2 + 4)} dx + \int_{0}^{\infty} \frac{\pi i}{(x^2 + 1)(x^2 + 4)} dx$$

and

$$\int_{0}^{\infty} f(z)dz = \int_{0}^{\infty} \frac{\log x}{(x^2 + 1)(x^2 + 4)} dx$$

using the same contour and singularities to evaluate

$$\int_0^\infty \frac{\pi i}{(x^2+1)(x^2+4)} dx$$

we get $\dfrac{\pi^2 i}{12}$, so

$$2\int_0^\infty \frac{\log x}{(x^2+1)(x^2+4)} dx + \frac{\pi^2 i}{12} = 2\pi i\left(\frac{\pi}{12} + \frac{i\log 2}{12} - \frac{\pi}{24}\right)$$

and

$$\int_0^\infty \frac{\log x}{(x^2+1)(x^2+4)} dx = -\frac{\pi \log 2}{12}.$$

Solution to 5.11.29: For $0 < \varepsilon < 1 < R$, let $C_{\varepsilon,R}$ denote the contour pictured below. Let $\log z$ denote the branch of the logarithm function in the plane slit along the negative imaginary axis.

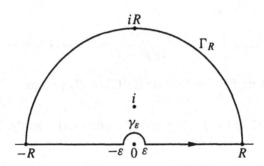

By the Residue Theorem [MH87, p. 280], since i is a simple pole of the integrand, we have

$$\int_{C_{\varepsilon,R}} \frac{(\log z)^2}{z^2+1} dz = 2\pi i \, \mathrm{Res}\left(\frac{(\log z)^2}{z^2+1}, i\right)$$

$$= 2\pi i \lim_{z\to i} \frac{(\log z)^2}{z+i}$$

$$= 2\pi i \frac{\left(\frac{\pi}{2}i\right)^2}{2i}$$

$$= -\frac{\pi^3}{4}.$$

For x negative, we have $\log x = \log|x| + \pi i$, so $(\log x)^2 = (\log|x|)^2 - \pi^2 + 2\pi i \log|x|$. Thus, the contribution to the integral from the interval $[-R, -\varepsilon]$

equals

$$\int_\varepsilon^R \frac{(\log x)^2 - \pi^2 + 2\pi i \log x}{x^2 + 1} dx = \int_\varepsilon^R \frac{(\log x)^2 - \pi^2}{x^2 + 1} dx + \int_{\Gamma_R} \frac{(\log z)^2}{z^2 + 1} dz$$
$$+ \int_{\gamma_\varepsilon} \frac{(\log z)^2}{z^2 + 1} dz + 2\pi i \int_\varepsilon^R \frac{\log x}{x^2 + 1} dx$$
$$= -\frac{\pi^3}{4}.$$

Also,

$$\left| \int_{\Gamma_R} \frac{(\log z)^2}{z^2 + 1} dz \right| \leq \frac{(\log R)^2(\pi R)}{R^2 - 1} = o(1) \qquad (R \to \infty)$$

and

$$\left| \int_{\gamma_\varepsilon} \frac{(\log z)^2}{z^2 + 1} dz \right| \leq \frac{(\log \varepsilon)^2(\pi \varepsilon)}{\varepsilon^2 - 1} = o(1) \qquad (\varepsilon \to 0).$$

In the limit we then obtain, considering the real parts,

$$2 \int_0^\infty \frac{(\log x)^2}{x^2 + 1} dx = -\frac{\pi^3}{4} + \int_0^\infty \frac{\pi^2}{x^2 + 1} dx = -\frac{\pi^3}{4} + \frac{\pi^3}{2} = \frac{\pi^3}{4}$$

and

$$\int_0^\infty \frac{(\log x)^2}{x^2 + 1} dx = \frac{\pi^3}{8}.$$

Solution to 5.11.30: For $\lambda = 0$, we have

$$\int_0^\infty (\operatorname{sech} x)^2 dx = \tanh x |_0^\infty = 1.$$

For $\lambda \neq 0$,

$$(\operatorname{sech} x)^2 \cos \lambda x = \Re \left(\frac{4e^{i\lambda x}}{(e^x + e^{-x})^2} \right)$$

so we consider the function f given by

$$f(z) = \frac{e^{i\lambda z}}{(e^z + e^{-z})^2}.$$

This function has a simple pole inside the contour

$$C = [-R, R] \cup [R, R + \pi i] \cup [R + \pi i, -R + \pi i] \cup [-R + \pi i, -R].$$

pictured below:

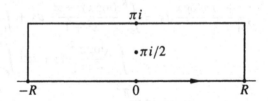

We have

$$\int_{[R+\pi i,-R+\pi i]} f(z)dz = -\int_{-R}^{R} \frac{e^{i\lambda(x+\pi i)}\,dx}{\left(-e^x - e^{-x}\right)^2}$$

$$= -e^{-\lambda\pi}\int_{-R}^{R} \frac{e^{i\lambda x}\,dx}{\left(e^x + e^{-x}\right)^2}$$

and

$$\left|\int_{[R,R+\pi i]} f(z)dz\right| = \left|\int_{0}^{1} \frac{e^{i\lambda(R+\pi ix)}\,\pi i dx}{\left(e^{R+\pi ix} + e^{-R-\pi ix}\right)^2}\right|$$

$$\leqslant \frac{\pi e^{|\lambda|\pi}}{\left(e^R - e^{-R}\right)^2} = o(1) \quad (R \to \infty)$$

similarly, we get

$$\int_{[-R+\pi i,-R]} f(z)dz = o(1) \quad (R \to \infty).$$

Therefore,

$$(1 - e^{-\lambda\pi})\int_{-\infty}^{\infty} \frac{e^{i\lambda x}\,dx}{\left(e^x + e^{-x}\right)^2} = 2\pi i\, \mathrm{Res}\left(f(z), \frac{\pi i}{2}\right) = \frac{\lambda\pi e^{-\lambda\pi/2}}{2}.$$

Thus,

$$\int_{0}^{\infty} (\mathrm{sech}\, x)^2 \cos \lambda x\, dx = \frac{\lambda\pi/2}{\sinh(\lambda\pi/2)}.$$

Solution to 5.11.31: If $b = 0$, the integral is well known:

$$\int_{0}^{\infty} e^{-x^2}\,dx = \frac{\sqrt{\pi}}{2}$$

and can be computed either by doubling it up

$$\int_{0}^{\infty} e^{-x^2}\,dx \int_{0}^{\infty} e^{-y^2}\,dy = \int_{0}^{\infty}\int_{0}^{\infty} e^{-(x^2+y^2)}\,dx\,dy$$

and converting to polar coordinates or by using the Residue Theorem applied to the function $f(z) = e^{-z^2}$ and the following contour, as shown by [Cad47, Mir49]. For the details, see [MH93, p. 321-322]. We will use this result to compute the full integral (for other values of b) below.

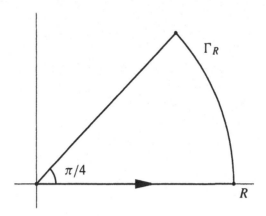

Now consider the function $f(z) = e^{-z^2}$. If $b > 0$, we will consider the integral over the rectangle contour of height b and if $b < 0$ the symmetric one below the x-axis.

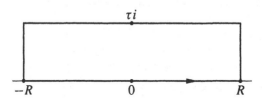

Since f is entire, the integral around the contour is zero, for any value of R. If $b < 0$ draw the contour below the x-axis. Evaluating the integral over each one of the two sides parallel to the x-axis, we have

$$\int_I f(z)\, dz = \int_{-R}^R e^{-x^2}\, dx$$

$$\int_{III} f(z)\, dz = \int_R^{-R} e^{-(x+bi)^2}\, dx$$

$$= -\int_{-R}^R e^{-(x^2-b^2)} e^{2bxi}\, dx$$

$$= -e^{b^2} \int_{-R}^R e^{-x^2} (\cos(2bx) + i\sin(2bx))\, dx$$

$$= -e^{b^2} \int_{-R}^{R} e^{-x^2} \cos(2bx)\, dx.$$

Along the vertical segments (II and IV), $|f(z)| = \left|e^{-(\pm R+iy)^2}\right| = e^{-R^2+y^2} \leqslant$ e^{-R^2} then $\left|\int_{II} f\right|$ and $\left|\int_{IV} f\right|$ are bounded by be^{-R^2}, and with b fixed, this goes to 0 as $R \to \infty$. Letting $R \to \infty$, the integral along the circuit becomes

$$\int_{-\infty}^{\infty} e^{-x^2}\, dx - e^{b^2} \int_{-\infty}^{\infty} e^{-x^2} \cos(2bx)\, dx = 0$$

Now using the pre-computed integral the result follows:

$$\int_{0}^{\infty} e^{-x^2} \cos(2bx)\, dx = \frac{1}{2} e^{-b^2} \int_{-\infty}^{\infty} e^{-x^2} \cos(2bx)\, dx$$

$$= \frac{1}{2} e^{-b^2} \int_{-\infty}^{\infty} e^{-x^2}\, dx$$

$$= \frac{\sqrt{\pi}}{2} e^{-b^2}.$$

Solution to 5.11.32: The integral is the real part of

$$I = \int_{0}^{\infty} e^{-(1+i)x^2}\, dx = \int_{0}^{\infty} e^{-\sqrt{2}e^{i\pi/4}x^2}\, dx = \int_{0}^{\infty} e^{-\sqrt{2}(e^{i\pi/8}x)^2}\, dx$$

Consider the contour Γ in shape of a slice of pizza, like the one in Solution 3 of Prob. 5.11.24, with angle $\pi/8$. By Cauchy's Theorem [MH87, p. 152],

$$\int_{\Gamma} e^{-\sqrt{2}z^2}\, dz = 0$$

On the arc the integral will approach 0 when the radius $R \to \infty$, because the integrand is bound in absolute value by $|e^{-\sqrt{2}e^{i\pi/4}R^2}| = e^{-R^2}$. Thus

$$0 = \int_{0}^{\infty} e^{-\sqrt{2}x^2}\, dz - \int_{0}^{\infty} e^{-\sqrt{2}(e^{i\pi/8}x)^2}\, d(e^{i\pi/8}x)$$

or equivalently

$$0 = 2^{-1/4} \int_{0}^{\infty} e^{-u^2}\, du - e^{i\pi/8} I$$

so $I = 2^{-1/4} e^{-i\pi/8} \sqrt{\pi}2$, and taking the real part, the integral is equals to

$$2^{-5/4}(\cos \pi/8)\sqrt{\pi} = \frac{(2+\sqrt{2})\sqrt{\pi}}{2^{13/4}}.$$

Solution to 5.11.33: Denote the line segment in $\mathbb{C}$ from z_0 to z_1 by $[z_0, z_1]$. Let $a, b > 0$, and $C_1 = [b, -a]$, $C_2 = [-a, -a - i\gamma]$, $C_3 = [-a - i\gamma, b - i\gamma]$, $C_4 = [b - i\gamma, b]$.

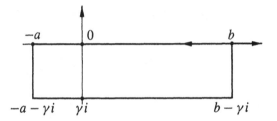

We integrate $f(z) = e^{-z^2/2}$ round the rectangular contour $= C_1 + C_2 + C_3 + C_4$. (The figure shows the case $\gamma > 0$.)

The reverse of C_1 is parametrized by $z = t$, $-a \leqslant t \leqslant b$. Thus

$$\int_{C_1} f(z)\,dz = -\int_{-a}^{b} e^{-\frac{1}{2}t^2}\,dt.$$

The curve C_2 is parametrized by $z = -a - it\gamma$, $0 \leqslant t \leqslant 1$. Thus

$$\int_{C_2} f(z)\,dz = -i\gamma \int_0^1 e^{-\frac{1}{2}(-a-it\gamma)^2}\,dt = -i\gamma e^{-\frac{1}{2}a^2} \int_0^1 e^{\frac{1}{2}t^2\gamma^2} e^{-iat\gamma}\,dt$$

so

$$\left| \int_{C_2} f(z)\,dz \right| \leqslant |\gamma| e^{-\frac{1}{2}a^2} \int_0^1 e^{\frac{1}{2}t^2\gamma^2}\,dt \to 0, \quad \text{as } a \to \infty.$$

The curve C_3 is parametrized by $z = t - i\gamma$, $-a \leqslant t \leqslant b$. Thus

$$\int_{C_3} f(z)\,dz = \int_{-a}^{b} e^{-\frac{1}{2}(t-i\gamma)^2}\,dt.$$

The opposite of C_4 is parametrized by $z = b - it\gamma$, $0 \leqslant t \leqslant 1$. Thus

$$\int_{C_4} f(z)\,dz = i\gamma \int_0^1 e^{-\frac{1}{2}(b-it\gamma)^2}\,dt = i\gamma e^{-\frac{1}{2}b^2} \int_0^1 e^{\frac{1}{2}t^2\gamma^2} e^{ibt\gamma}\,dt$$

therefore

$$\left| \int_{C_4} f(z)\,dz \right| \leqslant |\gamma| e^{-\frac{1}{2}b^2} \int_0^1 e^{\frac{1}{2}t^2\gamma^2}\,dt \to 0, \quad \text{as } b \to \infty.$$

By Cauchy's Theorem [MH87, p. 152], the integral of f round the rectangular contour is zero. Thus

$$-\int_{-a}^{b} e^{-\frac{1}{2}t^2}\,dt + \int_{C_2} f(z)\,dz + \int_{-a}^{b} e^{-\frac{1}{2}(t-i\gamma)^2}\,dt + \int_{C_4} f(z)\,dz = 0.$$

Let $a, b \to \infty$. We deduce that

$$\int_{-\infty}^{\infty} \frac{e^{-(t-i\gamma)^2/2}}{\sqrt{2\pi}}\,dt = \frac{1}{\sqrt{2\pi}} \int_{-\infty}^{\infty} e^{-\frac{1}{2}t^2}\,dt = 1$$

proving the result.

6

Algebra

6.1 Examples of Groups and General Theory

Solution to 6.1.1: 1. The set G is the set of all invertible 2×2 real matrices. If $A, B \in G$, then $|AB| = |A||B| \neq 0$ and thus $AB \in G$. Matrix multiplication is associative. The identity matrix I has determinant one, and thus belongs to G. Finally if $A \in G$ then A is invertible, with $|A^{-1}| = 1/|A| \neq 0$, and thus G is closed to inverses. This proves that G is a multiplicative group.

2. Let

$$A = \begin{pmatrix} a & b \\ c & d \end{pmatrix}, \quad B = \begin{pmatrix} 1 & 1 \\ 0 & 1 \end{pmatrix}, \quad C = \begin{pmatrix} 1 & 0 \\ 1 & 1 \end{pmatrix}.$$

Then

$$AB = \begin{pmatrix} a & a+b \\ c & c+d \end{pmatrix} \quad BA = \begin{pmatrix} a+c & b+d \\ c & d \end{pmatrix},$$

$$AC = \begin{pmatrix} a+b & b \\ c+d & d \end{pmatrix} \quad CA = \begin{pmatrix} a & b \\ a+c & b+d \end{pmatrix}.$$

The matrices B and C belong to G. If A commutes with B and with C then $c = 0$, $a = d$, $b = 0$ and $A = aI$ is a scalar matrix. On the other hand any scalar matrix commutes with every element of G. We conclude that the center of G is the set of all real 2×2 matrices aI with $a \neq 0$.

3. If A, B are orthogonal then $(AB)(AB)^t = ABB^t A^t = I$, so O is closed under multiplication. The identity matrix is orthogonal. Finally, if A is orthogonal then $A^{-1} = A^t$ and $A^{-1}(A^{-1})^t = A^t A = I$, thus O is a subgroup of G.

If $x = \begin{pmatrix} 1 & 0 \\ 1 & 1 \end{pmatrix}$ and $y = \begin{pmatrix} 0 & 1 \\ 1 & 0 \end{pmatrix}$ then x is in G, y is in O and

$$x^{-1}yx = \begin{pmatrix} 1 & 0 \\ -1 & 1 \end{pmatrix}\begin{pmatrix} 0 & 1 \\ 1 & 0 \end{pmatrix}\begin{pmatrix} 1 & 0 \\ 1 & 1 \end{pmatrix} = \begin{pmatrix} 1 & 1 \\ 0 & -1 \end{pmatrix}$$

is not orthogonal. This proves that O is not a normal subgroup of G.

4. $\mathbb{R}^* = \mathbb{R} \setminus \{0\}$ is an abelian multiplicative group. The map $\det : G \to \mathbb{R}^*$ is a group homomorphism which is clearly surjective, and nontrivial.

Solution to 6.1.2: 1. The set G is a subset of the multiplicative group $GL_3(\mathbb{R})$ of nonsingular matrices. It is easy to verify that the product of two elements of G has ones on the diagonal, and zeros below the diagonal. Moreover the identity matrix belongs to G. To verify closure to inverses, let $A \in G$. Then the characteristic polynomial of A is $(x - 1)^3$. By Cayley–Hamilton's Theorem [HK61, p. 194], $(A - I)^3 = 0$, therefore $A^{-1} = A^2 - 3A + 3I$. From the closure property, $A^2 \in G$, and it is easy to see that $A^2 - 3A + 3I$ has ones on the diagonal, and zeros below the diagonal, so $A^{-1} \in G$. Thus G is a group.

2. Let $A = \begin{pmatrix} 1 & a & b \\ 0 & 1 & c \\ 0 & 0 & 1 \end{pmatrix}$ be in the center of G, so that A commutes with every member of G. Then the products

$$\begin{pmatrix} 1 & a & b \\ 0 & 1 & c \\ 0 & 0 & 1 \end{pmatrix}\begin{pmatrix} 1 & 1 & 0 \\ 0 & 1 & 0 \\ 0 & 0 & 1 \end{pmatrix} = \begin{pmatrix} 1 & 1+a & b \\ 0 & 1 & c \\ 0 & 0 & 1 \end{pmatrix}$$

$$\begin{pmatrix} 1 & 1 & 0 \\ 0 & 1 & 0 \\ 0 & 0 & 1 \end{pmatrix}\begin{pmatrix} 1 & a & b \\ 0 & 1 & c \\ 0 & 0 & 1 \end{pmatrix} = \begin{pmatrix} 1 & 1+a & b+c \\ 0 & 1 & c \\ 0 & 0 & 1 \end{pmatrix}$$

show that $c = 0$. And the products

$$\begin{pmatrix} 1 & a & b \\ 0 & 1 & c \\ 0 & 0 & 1 \end{pmatrix}\begin{pmatrix} 1 & 0 & 0 \\ 0 & 1 & 1 \\ 0 & 0 & 1 \end{pmatrix} = \begin{pmatrix} 1 & a & a+b \\ 0 & 1 & 1+c \\ 0 & 0 & 1 \end{pmatrix}$$

$$\begin{pmatrix} 1 & 0 & 0 \\ 0 & 1 & 1 \\ 0 & 0 & 1 \end{pmatrix}\begin{pmatrix} 1 & a & b \\ 0 & 1 & c \\ 0 & 0 & 1 \end{pmatrix} = \begin{pmatrix} 1 & a & b \\ 0 & 1 & 1+c \\ 0 & 0 & 1 \end{pmatrix}$$

show that $a = 0$. Finally the products

$$\begin{pmatrix} 1 & a & b \\ 0 & 1 & c \\ 0 & 0 & 1 \end{pmatrix}\begin{pmatrix} 1 & 0 & x \\ 0 & 1 & 0 \\ 0 & 0 & 1 \end{pmatrix} = \begin{pmatrix} 1 & a & x+b \\ 0 & 1 & c \\ 0 & 0 & 1 \end{pmatrix}$$

$$\begin{pmatrix} 1 & 0 & x \\ 0 & 1 & 0 \\ 0 & 0 & 1 \end{pmatrix}\begin{pmatrix} 1 & a & b \\ 0 & 1 & c \\ 0 & 0 & 1 \end{pmatrix} = \begin{pmatrix} 1 & a & b+x \\ 0 & 1 & c \\ 0 & 0 & 1 \end{pmatrix}$$

show that the center of G is the set of all matrices

$$\begin{pmatrix} 1 & 0 & x \\ 0 & 1 & 0 \\ 0 & 0 & 1 \end{pmatrix}, \qquad x \in \mathbb{R}.$$

Solution to 6.1.3: 1. S_3 is a nonabelian group.

2. The Klein four group $V = \mathbb{Z}_2 \times \mathbb{Z}_2$ is a finite abelian group that is not cyclic.

3. $\mathbb{Z}$ is an infinite additive group; the subgroup $5\mathbb{Z}$ is a subgroup of index 5 in $\mathbb{Z}$.

4. $\mathbb{Z}_4$ and $\mathbb{Z}_2 \times \mathbb{Z}_2$ are nonisomorphic finite groups of order 4.

5. S_3 has a subgroup $H = \langle \alpha \rangle$, where α is any transposition in S_3, which is not normal in S_3.

6. The alternating group A_5 is nonabelian and has no normal subgroups other than the full group and the identity.

7. The additive group $\mathbb{Z}$ has a normal subgroup $H = 3\mathbb{Z}$; the factor group $\mathbb{Z}/3\mathbb{Z}$ has order 3, and is not isomorphic to any subgroup of $\mathbb{Z}$.

8. The left (also right) cosets of H in G form a partition of G, and for any $g \in G$, $gH = H$ iff $g \in H$. Let H have index 2 in G. Then there are two left cosets H and $G \setminus H$, and these are also the two right cosets of H in G. Fix $x \in G$. If $x \in H$ then $xH = H = Hx$. If $x \notin H$ then $xH = G \setminus H = Hx$. This proves that H is normal in G.

Solution to 6.1.4: 1. If $x, y \in G(R)$ have inverses $x', y' \in R$ respectively, then $xy(y'x') = (y'x')xy = 1$ proving that xy has an inverse. The identity 1 has itself as an inverse . If x has inverse x', then x' has inverse x. Associativity of multiplication is given. Therefore $G(R)$ is a multiplicative group.

2. Let $x = a + ib$ have inverse $y = u + iv$, with a, b, u, v integers. Then $xy = 1$, and taking moduli $|x||y| = 1$ and $|x|^2|y|^2 = 1$. Since a, b, u and v are integers, $|x|^2 = a^2 + b^2 = 1$, giving $a = \pm 1, b = 0$ or $a = 0, b = \pm 1$. The units are $1, i, -1, -i$. This group is cyclic, generated by i, and so isomorphic to $\mathbb{Z}_4$.

Solution to 6.1.5: Let G be the group of symmetries of the network and let D be the triangle with vertices P_0, P_2, and P_3.

Let α and β be $180°$ degree rotations about the midpoints of the segments $P_0 P_3$ and $P_0 P_1$, respectively, and let τ be a reflection in the line extending the segment $P_0 P_2$. Let x be a point in the plane. With this notation, it is easy to see that:

$$\alpha\beta(x) = x + (0, 1)$$
$$\alpha\tau\alpha\tau(x) = x + (2, 0).$$

Claim 1: α, β, τ generate the group G.

Proof: Note that the four triangles D, $\alpha(D)$, $\alpha\tau(D)$, and $\alpha\tau\alpha(D)$ tile a one-by-two rectangle. Applying the symmetries $\alpha\beta$ and $\alpha\tau\alpha\tau$ we may tile the plane with copies of this rectangle. Finally, any element of G which fixes D setwise must be the identity.

Note also that α, β, and τ satisfy the relations:

$$\alpha^2 = \beta^2 = \tau^2 = 1$$

$$[\alpha\beta, \tau] = 1$$

the second relation holds because the line of reflection of τ is parallel to the translation given by $\alpha\beta$.

We will show that the abstract group G' with presentation

$$G' = \; < \alpha, \beta, \tau \mid \alpha^2 = \beta^2 = \tau^2 = 1, [\alpha\beta, \tau] = 1 >$$

is isomorphic to G. We first prove that G' surjects onto G via the natural map by finding a single preimage for every element of G.

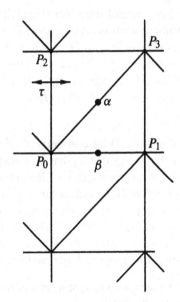

Claim 2: The set of elements $N = \{(\alpha\beta)^p(\alpha\tau)^{2q}Z\}$ in G', where p and q are integers and where $Z = 1$, α, $\alpha\tau$, or $\alpha\tau\alpha$, injects and surjects onto the elements of G.

Proof: Surjectivity follows from *Claim 1*. Injectivity is also clear as there is a bijection between N and the triangles of the network. We will refer elements of N as being in *normal form*.

Claim 3: Any word in α, β, and τ can be put in normal form using the relations of G'.

Proof: To do this we first apply the recursive algorithm below. Note that the relation $[\alpha\beta, \tau] = 1$ can be rewritten to obtain $\tau\beta = \beta\alpha\tau\alpha$ and $\tau\alpha\beta = \alpha\beta\tau$. We are given a word of the form $(\alpha\beta)^r W(\alpha, \beta, \tau)$:

Step 1: Move all β's in W to the left:

$$\beta\beta = 1$$
$$\tau\beta = \beta\alpha\tau\alpha$$
$$\alpha\beta - \text{use step 2.}$$

Step 2: Move all $\alpha\beta$'s in the new word to the left using:

$\alpha\alpha\beta$ – delete the $\alpha\alpha$ and go back to step 1.

$\tau\alpha\beta = \alpha\beta\tau$

$\beta\alpha\beta$ – go back to step one and move the β which is on the left.

This procedure can be shown to terminate by applying induction to the correct lexicographic order. Thus we have shown that any word can be put in the form $(\alpha\beta)^p W(\alpha, \tau)$, where W is a word in α and τ only. But, since $\alpha^2 = \tau^2 = 1$, we may assume that $W(\alpha, \tau) = (\alpha\tau)^r Q$, where $Q = 1, \alpha$, or τ. It is straightforward to place W in the desired normal form by considering the possible values of r and Q. It follows that G' is isomorphic to G. This group is isomorphic to the **pmg** group of [CM57, p. 45].

Solution to 6.1.6: If $a, b \in G$, then $a > 0$ and $b \neq 1$, so $a^{\log b} \in G$. Therefore, the operation $*$ is well defined.

Identity. The constant e is the identity since $a * e = a^{\log e} = a^1 = a$ and $e * a = e^{\log a} = a$.

Associativity. We have

$$(a * b) * c = c^{\log(e^{\log a \log b})} = e^{\log a \log b \log c} = a^{\log(e^{\log b \log c})} = a * (b * c).$$

Invertibility. Since $a \neq 1$, $e^{1/\log a}$ exists and is an element of G. A calculation shows that $a * e^{1/\log a} = e^{1/\log a} * a = e$.

Solution 2. The map $\log : G \to \mathbb{R} \setminus \{0\}$ is a bijection that transforms the operation $*$ into multiplication; that is, $\log(a * b) = (\log a)(\log b)$. Since $\mathbb{R} \setminus \{0\}$ forms a group with respect to multiplication, G is a group with respect to $*$.

Solution to 6.1.7: 1. Let $a \in G$. The centralizer $C_a = \{x \in G \mid xax^{-1} = a\}$ of a is a subgroup of G.

We have,

$$xax^{-1} = yay^{-1} \Leftrightarrow y^{-1}xa = ay^{-1}x \Leftrightarrow y^{-1}x \in C_a \Leftrightarrow xC_a = yC_a,$$

which shows that two conjugates xax^{-1}, yay^{-1} of a are equal if and only if the left cosets xC_a, yC_a are equal. Therefore the number of distinct conjugates of a is equal to the number of distinct cosets of C_a in G. Thus $|C(a)|$ divides $|G|$.

2. Each element $\sigma \in S_n$ can be written as a product of disjoint cycles $\sigma = \sigma_1 \ldots \sigma_k$; this representation is unique apart from the ordering of the cycles. Include the cycles of length 1, and denote the length of σ_i by r_i, which we assume

are in increasing order. The cycle pattern of σ is then the sequence $\{r_1, \ldots, r_k\}$. Thus the elements of

$$S_3 = \{(1)(2)(3), (1)(2\ 3), (2)(1\ 3), (3)(1\ 2), (1\ 2\ 3), (1\ 3\ 2)\}$$

have cycle patterns $\{1, 1, 1\}$, $\{1, 2\}$, $\{1, 2\}$, $\{1, 2\}$, $\{3\}$, $\{3\}$. In S_n two permutations are conjugate if and only if they have the same cycle pattern. The conjugacy classes in S_3 are therefore the classes $\{e\}$, $\{(1\ 2), (1\ 3), (2\ 3)\}$ and $\{(1\ 2\ 3), (1\ 3\ 2)\}$. This shows that conjugacy classes do not all have the same number of elements.

3. Denote the order of G by n. The conjugacy class of the identity has one element, namely the identity. The second conjugacy class then has $n - 1$ elements, and thus $n - 1$ divides n. We deduce that $n = 2$.

Solution to 6.1.8: Fix an element $a \in G$ different from the identity, and consider the map $\varphi : G \setminus \{e\} \to G \setminus \{e\}$ defined by $\varphi(c) = c^{-1}ac$. The map is onto, so, since G is finite, it is one-to-one. As $\varphi(a) = a$ and $a^{-2}aa^2 = a$, it must be that $a^2 = e$. Thus, all elements of G, other than the identity, have order 2. Then, if a and b are in G, we have

$$ab = a(ab)^2 b = a^2(ba)b^2 = ba$$

in other words, G is commutative, and it follows that G has order 2.

Solution 2. Since G is finite, it has an element of prime order p. Hence, every element of G, other than the identity, has order p. Since G is a p–group, it has a nontrivial central element. Therefore, all elements are central; in other words, G is abelian. Hence, G has order 2.

Solution 3. By our hypothesis, any two nonidentity elements are conjugate. Hence, there are two conjugacy classes: The class containing the identity and the class containing all the other elements of G. Letting n be the order of G, we see that the second conjugacy class must contain $n - 1$ elements. But, by the class equation, we know that the order of any conjugacy class divides the order of the group, so $(n - 1)|n$. Solving for $n > 0$, we see that the only possible solution is $n = 2$.

Solution to 6.1.9: Since G is finite, every element of G has finite order. Since any two elements of $G \setminus \{e\}$ are related by an automorphism of G, all such elements have the same order, say q. Since all powers of an element of $G \setminus \{e\}$ have order q or 1, q is prime. By Sylow's Theorems [Her75, p. 91], the order of G is a power of q. Therefore the center Z of G contains an element other than e. Since the center of G is invariant under all automorphisms, Z is the whole of G, i.e. G is abelian.

Solution to 6.1.11: Consider the group $G' = \{a^n b^m \mid 0 \leqslant n \leqslant r - 1, 0 \leqslant m \leqslant s - 1\}$. As the order of any element of G' divides the order of G', we have $|G'| = rs$. This shows that $(ab)^k$ is never the identity for $0 < k < rs$. Clearly we have $(ab)^{rs} = a^r b^s = e$, so the order of ab is rs.

Solution to 6.1.12: 1. Suppose $X \subset Y$. If $z \in C(Y)$ then $zy = yz$ for all $y \in Y$, and thus $zx = xz$ for all $x \in X$, so $z \in C(X)$.

2. Let $x \in X$. If $z \in C(X)$ then $zx = xz$; thus x commutes with each member of $C(X)$, that is, $x \in C(C(X))$.

3. Replace X by $C(X)$ in (2): we get that $C(X) \subset C(C(C(X)))$. Further, using (2) in (1), we conclude that $C(C(C(X))) \subset C(X)$.

Solution to 6.1.13: Let $g, h \in H$ and let $x \in D \setminus H$. Then

$$h^{-1}g^{-1} = xhx^{-1}xgx^{-1} = xhgx^{-1} = (hg)^{-1} = g^{-1}h^{-1}$$

and we can conclude immediately from this that H is abelian.

Let $x \in D \setminus H$. We have [D:H]=2, so $x^2 \in H$. By hypothesis, $xx^2x^{-1} = x^{-2}$ or $x^4 = 1$. Therefore, x has order 1, 2, or 4. But n is odd, so 4 does not divide the order of D, so, by Lagrange's Theorem [Her75, p. 41], x cannot have order 4. By our choice of x, $x \neq 1$, so x cannot have order 1. Hence, x has order 2.

6.2 Homomorphisms and Subgroups

Solution to 6.2.1: Let $f: \mathbb{Z}_2 \times \mathbb{Z}_2 \rightarrow S_3$ be a homomorphism. Then $(\mathbb{Z}_2 \times \mathbb{Z}_2)/\ker f \simeq \operatorname{im} f$. Denote $|\ker f|$ by k, and $|\operatorname{im} f|$ by l. Then $k = 1, 2$ or 4, and $l = 1, 2, 3$ or 6. Therefore $4/k = l$. The only solutions are $k = 4, l = 1$ and $k = 2, l = 2$. The first case is realized by the map which sends each element of $\mathbb{Z}_2 \times \mathbb{Z}_2$ to the identity of S_3. Now

$$\mathbb{Z}_2 \times \mathbb{Z}_2 = \{(0, 0), (0, 1), (1, 0), (1, 1)\},$$
$$S_3 = \{e, (1\ 2), (1\ 3), (2\ 3), (1\ 2\ 3), (1\ 3\ 2)\}.$$

Each element of $\mathbb{Z}_2 \times \mathbb{Z}_2$, not the identity, has order 2; denote these by z_1, z_2, z_3. The cycles in S_3 of length 2 have order 2; denote these by y_1, y_2, y_3. We therefore consider the map f given by

$$f(0, 0) = f(z_1) = e, \qquad f(z_2) = f(z_3) = y_1.$$

It is routine to verify that f is a homomorphism. By 'varying' the zs and the ys we find nine such homomorphisms. Thus there are ten homomorphisms from $\mathbb{Z}_2 \times \mathbb{Z}_2$ to S_3.

Solution to 6.2.2: 1. Let C_g denote the conjugacy class of g and Z_g the centralizer of g. Using the Orbit Stabilizer Theorem [Fra99, Thm. 16.16], [Lan94, Prop. I.5.1], $|Z_g| \cdot |C_g| = |G|$ for every element g. Hence $\sum_{g \in C} |Z_g| = |G|$ for every conjugacy class C, and $|X| = \sum_{g \in G} |Z_g| = c|G|$.

2. In this case $G = S_5$, so $|G| = 5! = 120$. The number of conjugacy classes is the number of partitions of 5, namely 7. So there are $7 \cdot 120 = 840$ pairs of commuting permutations.

Solution to 6.2.3: A matrix A is a solution of $x^3 = x^2$, or equivalently $x^2(x-1) = 0$, if and only if all its Jordan blocks are solution to the same equation. So each

Jordan block must have eigenvalues 0 and 1, and the possibilities are (0), (1), $\begin{pmatrix} 0 & 1 \\ 0 & 0 \end{pmatrix}$.

The conjugacy type of a matrix is determined by the multiplicity of the Jordan blocks. Let a, b, c be the multiplicities of the blocks above, respectively. Then the answer is the number of nonnegative integer solutions of $a + b + 2c = 5$. For fixed $c \in \{0, 1, 2\}$, there are $6 - 2c$ solutions to $a + b = 5 - 2c$. Thus the answer is

$$(6 - 2 \cdot 0) + (6 - 2 \cdot 1) + (6 - 2 \cdot 2) = 12.$$

Solution to 6.2.4: Let $n = [\mathbb{C}^* : H]$. By Lagrange's Theorem [Her75, p. 41], the order of any element of $\mathbb{C}^*/H$ divides n. So $x^n \in H$ for all $x \in \mathbb{C}^*$. Therefore, $(\mathbb{C}^*)^n = \mathbb{C} \subset H$.

Solution to 6.2.5: 1. The number of conjugates of H in G is $|G : N(H; G)|$ where $N(H; G) = \{g \in G \mid g^{-1}Hg = H\}$ is the normalizer of H. As $H \subset N(H; G)$, we have $|G : H| \geqslant |G : N(H; G)|$.

2. By Problem 6.4.19, there is a normal subgroup of G, N, contained in H, such that G/N is finite. By Part 1 we can find a coset $Ng \in G/N$ such that Ng is not contained in any conjugate $y^{-1}Hy/N$ of H/N in G/N. Then $g \notin y^{-1}Hy$ for any $y \in G$.

Solution to 6.2.6: We will prove a more general result. Let $G = \langle g \rangle$, $|G| = n$, and $\alpha \in \operatorname{Aut} G$. As $\alpha(g)$ also generates G, we have $\alpha(g) = g^k$ for some $1 \leqslant k < n$, $(k, n) = 1$. Conversely, $x \mapsto x^k$ is an automorphism of G for $(k, 1) = 1$. Let $\mathbb{Z}_n$ be the multiplicative group of residue classes modulo n relatively prime to n. If $\overline{k}$ denotes the residue class containing k, we can define $\Phi : \operatorname{Aut} G \to \mathbb{Z}_n$ by

$$\Phi(\alpha) = \overline{k} \qquad \text{iff} \qquad \alpha(g) = g^k.$$

It is clear that Φ is an isomorphism. As $\mathbb{Z}_n$ is an abelian group of order $\varphi(n)$ (φ is Euler's totient function [Sta89, p. 77], [Her75, p. 43]), so is $\operatorname{Aut} G$. When n is prime, these groups are also cyclic.

Solution to 6.2.7: 1. If $g^{-1}\varphi(g) = h^{-1}\varphi(h)$ for some $g, h \in G$, then $\varphi(g)\varphi(h)^{-1} = gh^{-1}$, so, by hypothesis, $gh^{-1} = 1$ and $g = h$. Thus, there are $|G|$ elements of that form, so they must constitute all of G.

2. Using Part 1, we have, for $z = g^{-1}\varphi(g) \in G$,

$$\varphi(z) = \varphi(g^{-1})\varphi^2(g) = \varphi(g^{-1})g = z^{-1}$$

so $g \mapsto g^{-1}$ is an automorphism of G, which implies that G is abelian. For any $z \in G$, $z \neq 1$, we have $z^{-1} = \varphi(z) \neq z$, so G has no element of order 2, and $|G|$ is odd.

Solution to 6.2.8: Let G be the group. If G is not abelian and a is an element not in the center, then the map $x \mapsto a^{-1}xa$ is the desired automorphism. If G is cyclic, say of order m, and n is an integer larger than 1 and relatively prime to m, then the map $x \mapsto x^n$ is the desired automorphism. If G is any finite abelian group, then, by the Structure Theorem [Her75, p. 109], it is a direct product of cyclic groups. If one of the factors has order at least 3, we get the desired automorphism by using the preceding one in that factor and the identity in the other factors. If every factor has order 2, we get the desired automorphism by permuting any two of the factors.

Solution to 6.2.9: We show that u is an isomorphism from $\ker f$ onto $\ker g$. Let $y \in \ker f$. We have

$$g(u(y)) = u(y) - u(v(u(y))) = u(y - v(u(y))) = u(f(y)) = u(0) = 0.$$

Hence, u maps $\ker f$ into $\ker g$. Let $y \in \ker f$ with $u(y) = 0$; then

$$0 = f(y) = y - v(u(y)) = y.$$

Thus, u is injective. Let $x \in \ker g$. Then $0 = g(x) = x - u(v(x))$, so $u(v(x)) = x$. Therefore, u is onto if $v(x) \in \ker f$. However, an argument identical to the first one shows that this is the case, so we are done.

Solution to 6.2.10: Given an element $\beta \in G$, by the surjectivity of h, there is an element $\gamma \in G$ such that
$$\gamma - u(v(\gamma)) = u(\beta)$$
now build $\alpha = \beta + v(\gamma)$, for this element

$$
\begin{aligned}
f(\alpha) &= \beta + v(\gamma) - v(u(\beta + v(\gamma))) \\
&= \beta + v(\gamma) - v(u(\beta) + u(v(\gamma))) \\
&= \beta + v(\gamma) - v(\gamma) \\
&= \beta
\end{aligned}
$$

showing that f is surjective.

Solution to 6.2.12: 1. Suppose this is not the case, then there exists $h_1 \in H_1 \setminus H_2$ and $h_2 \in H_2 \setminus H_1$. Since they both belong to the group $H_1 \cup H_2$, the element $h_1 h_2 \in H_1 \cup H_2$. In both cases, if $h_1 h_2 \in H_1$ we get a contradiction with $h_2 = h_1^{-1}(h_1 h_2) \in H_1$ or if $h_1 h_2 \in H_2$ we get the contradiction with $h_1 = (h_1 h_2)h_2^{-1} \in H_2$.
2. Consider the product group $G = \mathbb{Z}_2^{n-1}$, and the subgroups

$$H_i = \{(x_1, \ldots, x_{n-1}) \in G | x_i = 0\}.$$

Then $H_1 \cup \cdots \cup H_{n-1} = G \setminus \{(1, 1, \ldots, 1)\}$. Let $H_n = \{(x_1, \ldots, x_{n-1}) \in G | x_1 + x_2 = 0\}$. Then $(1, 1, \ldots, 1) \in H_n$, so $H_1 \cup H_2 \cdots H_n = G$, no H_i is contained in any other, since they are distinct groups of the same order.

Solution to 6.2.13: 1. The statement is false. Let $G = S_3 = \langle a, b \mid a^3 = b^2 = 1, ba = a^2 b \rangle$ and $H = \{1, b\}$, so that $|G : H| = 3$. Let $x = ab \in G$. Then $x^3 = ab \notin H$.

2. The statement is true. The $n + 1$ cosets $H, Ha, Ha^2, \ldots, Ha^n$ are not all distinct. There are integers $0 \leqslant i < j \leqslant n$ such that $Ha^i = Ha^j$. This means that $a^{j-i} \in H$, proving the result.

Solution to 6.2.14: For each nonzero rational s, the map $f : \mathbb{Q} \to \mathbb{Q}$ given by $x \mapsto sx$ is a bijection and satisfies $f(x + y) = f(x) + f(y)$. Thus f is an automorphism of $\mathbb{Q}$.

Conversely, let f be an automorphism of $\mathbb{Q}$. Then $f(0) = 0$, and $s = f(1)$ is a nonzero rational. The inductive step $f(n + 1) = f(n) + f(1) = ns + s$ proves that $f(n) = ns$ for $n \geqslant 1$. Further as $0 = f(n + (-n)) = f(n) + f(-n)$ we see that $f(-n) = -f(n) = -ns$ for $n \geqslant 1$. Thus $f(n) = ns$ for $n \in \mathbb{Z}$. Now let $x = m/n$ where $n > 0$ and $m \in \mathbb{Z}$. Then $ms = f(m) = f(nx) = nf(x)$. Thus $f(x) = sx$.

Solution to 6.2.15: If the origin were a limit point of Γ then Γ would be dense in $\mathbb{R}$, since Γ contains all integer multiples of its elements. But then Γ would equal $\mathbb{R}$, since it is closed, which is impossible because Γ is countable as a set $\Gamma = \{m\alpha + n\beta \mid m, n \in \mathbb{Z}\}$. Hence the origin is not a limit point of Γ.

Therefore Γ contains a smallest positive number γ. If x is in Γ and n is the largest integer such that $n\gamma \leqslant x$, then $x - n\gamma$ is in Γ, and $0 \leqslant x - n\gamma < \gamma$, implying that $x - n\gamma = 0$. Therefore $\Gamma = \gamma\mathbb{Z}$, from which the desired conclusion is immediate.

Solution to 6.2.16: Suppose H is a proper subgroup of G of finite index. Let $N = G/H$, so that N is a finite non trivial abelian group. Let p be a prime dividing the order of N, and $n \in N$ an element of order p. Suppose n is the image of the class of a/b in G. Let n_k be the image of the class of $a/p^k b$. Then n_k has order p^{k+1}, because $p^k n_k = n$, so $p^{k+1} n_k = 0$ and $d n_k$ does not equal zero for any proper divisor d of p^{k+1}. This is not possible since every element of N has order dividing $|N|$.

Solution to 6.2.17: The only homomorphism is the trivial one. Suppose φ is a nontrivial homomorphism. Then $\varphi(a) = m \neq 1$ for some $a, m \in \mathbb{Q}$. We have

$$a = \frac{a}{2} + \frac{a}{2} = \frac{a}{3} + \frac{a}{3} + \frac{a}{3} + \cdots$$

but m is not the nth power of a rational number for every positive n. For example $3/5 = 1/5 + 1/5 + 1/5$ but $\sqrt[3]{\frac{3}{5}} \notin \mathbb{Q}^+$.

Solution to 6.2.18: For $i = 1, 2$ let S_i denote the set of right cosets of H_i, G/H_i. Then G acts on S_i on the left. Hence G acts on $S_1 \times S_2$, and the stabilizer of the trivial pair (H_1, H_2) of cosets is $H_1 \cap H_2$ Hence $[G : H_1 \cap H_2]$ is the size of the G-orbit of $(H_1, H_2) \in S_1 \times S_2$, so $[G : H_1 \cap H_2] \leqslant 9$. On the other hand, H_1 and H_2 are distinct subgroups of the same index, so H_1 is not contained in H_2, and $[G : H_1 \cap H_2]$ is a proper multiple of $[G : H_1] = 3$. Thus it equals 6 or 9.

Both are possible. The index is 6 if $H_1 = \{1, (12)\}$ and $H_2 = \{1, (13)\}$ in the symmetric group $G = S_3$. The index is 9 if H_1 and H_2 are distinct subgroups of order 3 in $G = \mathbb{Z}/3 \times \mathbb{Z}/3$.

Solution to 6.2.19: By assumption, H has only finitely many right cosets, say $H, x_1 H, \ldots, x_n H$, whose union is G. Hence, K is the union of the sets $K \cap H, K \cap x_1 H, \ldots, K \cap x_n H$, some of which may be empty, say $K = (K \cap H) \cup (K \cap x_1 H) \cup \cdots \cup (K \cap x_m H)$ (the notation being so chosen that $K \cap x_j H = \emptyset$ if and only if $j > m$). If $K \cap x_j H \neq \emptyset$, then we may assume x_j is in K (since y is in xH if and only if $yH = xH$). After making this assumption, we have $K = x_j K$, so that $K \cap x_j H = x_j K \cap x_j H = x_j (K \cap H)$, whence

$$K = (K \cap H) \cup x_1 (K \cap H) \cup \cdots \cup x_m (K \cap H).$$

This shows $K \cap H$ has only finitely many right cosets in K, the desired conclusion.

Solution to 6.2.20: For each $g \in G$, let $S_g = \{a \in G \mid ag = ga\}$. Each S_g is a nontrivial subgroup of G, because $g \in S_g$ and $S_e = G$. The intersection of all S_g, $g \in G$, is the center of G. So H is a subset of the center of G.

Solution to 6.2.21: Let $N = 6^k$. Let $G_i = C_6^i \times S_3^{k-i}$ for $0 \leqslant i \leqslant k$. These groups are not isomorphic because the center of G_i is C_6^i.

Solution 2. Let $N = 2^{2k}$. Let $G_i = C_{2^i} \times C_{2^{2k-i}}$ for $0 \leqslant i \leqslant k$. These groups are not isomorphic because the exponent of C_i is 2^{2k-i}.

Solution to 6.2.22: Since G is finitely generated, $\mathrm{Hom}(G, S_k)$ is finite (bounded by $(k!)^n$), where S_k denotes the symmetric group on k numbers $1, 2, \ldots, k$. For any subgroup H of index k in G, we can identify G/H with this set of symbols, sending the coset H to 1. Then the left action of G on G/H determines an element of $\mathrm{Hom}(G, S_k)$ such that H is the stabilizer of 1. Thus, the number of such H's is, at most, $(k!)^n$.

6.3   Cyclic Groups

Solution to 6.3.1: 1. Let G be the subgroup of $\mathbb{Q}$ generated by the nonzero numbers $a_1, \ldots, a_r$, and let q be a common multiple of the denominators of $a_1, \ldots, a_r$. Then each a_i has the form p_j/q with $p_j \in \mathbb{Z}$, and, accordingly, $G = \frac{1}{q} G_0$, where G_0 is the subgroup of $\mathbb{Z}$ generated by $p_1, \ldots, p_r$. Since all subgroups of $\mathbb{Z}$ are cyclic, it follows that G is cyclic, that is, $G = \frac{p}{q}\mathbb{Z}$, where p is a generator of G_0.

2. Let $\pi : \mathbb{Q} \to \mathbb{Q}/\mathbb{Z}$ be the quotient map. Suppose G is a finitely generated subgroup of $\mathbb{Q}/\mathbb{Z}$, say with generators $b_1, \ldots, b_r$. Let G_1 be the subgroup of $\mathbb{Q}$ generated by $1, \pi^{-1}(b_1), \ldots, \pi^{-1}(b_r)$. Then $\pi(G_1) = G$, and G_1 is cyclic by Part 1. Hence, G is cyclic.

Solution to 6.3.2: The subgroup of $\mathbb{Q}/\mathbb{Z}$ generated by the coset of $1/t$ is a cyclic subgroup of order t, H_0 say. Let H be any cyclic subgroup of order t, and $r+\mathbb{Z}$ be one of its generators, where $0 < r < 1$. Then $tr = p$ is an integer, and $r = p/t$. This implies that the coset of r is in H_0, hence that H is a subset of H_0. Since H and H_0 both have order t, we have $H = H_0$, as desired.

Solution to 6.3.3: Suppose $G = \{0, g_1, g_2, \ldots\}$ is countable, and, for each $n \in \mathbb{N}$, $\{g_1, \ldots, g_n\}$ is contained in an infinite cyclic subgroup of G. Then the subgroup generated by $\{g_1, \ldots, g_n\}$, H_n, is infinite cyclic, and $H_n \subset H_{n+1}$. Let H_n be generated by h_n. We have $h_n = h_{n+1}^{q_n}$ for some nonzero $q_n \in \mathbb{Z}$. Let $r_n = q_1 q_2 \cdots q_{n-1}$, and let $\varphi_n : H_n \to \mathbb{Q}$ be the morphism verifying $\varphi_n(h_n) = 1/r_n$. Then we have

$$\varphi_{n+1}(h_n) = \varphi_{n+1}(h_{n+1}^{q_n}) = \frac{q_n}{r_{n+1}} = \frac{1}{r_n} = \varphi_n(h_n).$$

Since h_n generates H_n, $\varphi_n = \varphi_{n+1}|_{H_n}$. Thus there is a unique homomorphism $\varphi : G \to \mathbb{Q}$ such that $\varphi_n = \varphi|_{H_n}$ for all $n \in \mathbb{N}$. As each φ_n is injective and the subgroups H_n cover G, φ is an isomorphism from G onto a subgroup of $\mathbb{Q}$. The converse is clear.

Solution to 6.3.4: 1. Let $G = \langle c \rangle$. We have $a = c^r$ and $b = c^s$ for some positive odd integers r and s, so $ab = c^{r+s}$ with $r + s$ even, and ab is a square.
2. Let $G = \mathbb{Q}^*$, the multiplicative group of the rational numbers, $a = 2$, and $b = 3$. Then $ab = 6$, and none of these is a square in G.

Solution to 6.3.6: Let e be the identity in G, and let $N = \{e, a\}$ be the normal subgroup of order 2. If $x \in G$, then $x^{-1}ax \in N$ and certainly does not equal e, so it equals a. Thus, $xa = ax$ for all $x \in G$. The quotient group G/N has order p and so is cyclic. Let x be any element not in N. Then the coset of x in G/N has order p, so, in particular, the order of x itself is not 2. But the order of x divides $2p$, so it must be p or $2p$. In the latter case, G is the cyclic group generated by x. In the former case, since $xa = ax$, we have $(xa)^p = x^p a^p = a$, so $(xa)^{2p} = a^2 = e$, and xa has order $2p$, which means G is the cyclic group generated by xa.

Solution to 6.3.8: It follows that every Sylow subgroup is normal. From this it follows that if a and b are in distinct Sylow subgroups a and b commute. Indeed $aba^{-1}b^{-1}$ lies in the intersection of the two groups which is trivial. Hence, it suffices to establish the result when G has order p^n for a prime p. Let $a \in G$ be an element of largest order, say p^k. Let $H = \langle a \rangle$. If $b \in G$ and b has order $p^i \leqslant p^k$, then $b \in H$ because $|\langle b \rangle| = p^i$ and H contains the unique subgroup of G of order p^i. Thus if $c \notin H$ then c has order strictly bigger than p^k which contradicts the maximality of the order of a. Thus $H = G$ and G is cyclic.

6.4   Normality, Quotients, and Homomorphisms

Solution to 6.4.1: 1. For each $x \in G$ the map $h \mapsto xh$ is a bijection between H and the left coset xH. Thus, $|H| = |xH|$. Further, the left cosets of H form a partition of G, arising from the equivalence relation on G given by $x \sim y$ iff $x^{-1}y \in H$. Thus, the number of left cosets of H in G is given by $|G|/|H|$. The same holds for right cosets: here the partition of G is given by the relation $x \sim y$ iff $xy^{-1} \in H$. This proves the result.

2. The group of symmetries of the square is the dihedral group D_4 of order 8. Take the square with vertices $P_4 = \{1, i, -1, -i\}$ in the complex plane. Let r denote the rotation of the about the origin through the angle $\pi/2$. Then $\{1, r, r^2, r^3\}$ is a normal subgroup of D_4. Let s be the reflection in the real axis. The elements of D_4 are

$$1, r, r^2, r^3, s, rs, r^2s, r^3s.$$

Further, $s^2 = 1$, and rs is also a reflection, so $(rs)^2 = 1$ or $sr = r^{-1}s = r^3s$. We obtain the presentation

$$D_4 = \langle r, s \mid r^4 = 1 = s^2, sr = r^{-1}s \rangle.$$

Let $H = \{1, s\}$. Then $rH = \{r, rs\} \neq Hr = \{r, sr\}$.

Solution to 6.4.3: The subgroup H is normal only if $aHa^{-1} = H$ or $aH = Ha$ for all $a \in G$. Since H has only one left coset different from itself, it will suffice to show that this is true for a fixed a which is a representative element of this coset. Since H has the same number of right and left cosets, there exists a b such that H and bH form a partition of G. Since cosets are either disjoint or equal and $H \cap aH = \emptyset$, we must have that $aH = Hb$. But then $a \in Hb$, so $Hb = Ha$.

Solution to 6.4.5: The number of Sylow 2-groups is odd, and divides 112, therefore it divides 7, thus, it is 1 or 7. If there is only one Sylow 2-group, it's normal, so we may assume otherwise.

Let G act on the set of seven 2-groups by conjugation, so it maps into S_7. This is a nontrivial map, since the action is transitive. If it is not an injection, then the kernel is a nontrivial normal subgroup, so we may assume it is an injection.

Since 16 divides $|G|$ and does not divide $|A_7|$, the image of G is not entirely inside A_7. Therefore the composite of $G \rightarrow S_7 \rightarrow S_7/A_7 = \mathbb{Z}_2$ is onto, and its kernel is a normal subgroup (of index 2).

Solution to 6.4.6: The groups of order $\leqslant 6$ are $\{1\}$, $\mathbb{Z}_2$, $\mathbb{Z}_3$, $\mathbb{Z}_4$, $\mathbb{Z}_2 \times \mathbb{Z}_2$, $\mathbb{Z}_5$, $\mathbb{Z}_6$, and the symmetric group S_3. Suppose that G is a group with the mentioned property. Let S be a Sylow-2 subgroup of G. Some Sylow-2 subgroup of G must contain (a subgroup isomorphic to) $\mathbb{Z}_4$, and all Sylow 2-subgroups are conjugate, so S contains $\mathbb{Z}_4$. Similarly S contains $\mathbb{Z}_2 \times \mathbb{Z}_2$. Thus S cannot equal either, so 8 divides the order of S, which divides the order of G. On the other hand, G has subgroups of orders 3 and 5, so 3 and 5 also divide $|G|$. Since 8, 3, 5 are pairwise relatively prime, their product 120 divides $|G|$. Thus $|G| \geqslant 120$.

On the other hand, the group $G = \mathbb{Z}_4 \times \mathbb{Z}_5 \times S_3$ of order 120 has the property: $\mathbb{Z}_2$ is isomorphic to a subgroup of $\mathbb{Z}_4$, and $\mathbb{Z}_2$ and $\mathbb{Z}_3$ are isomorphic to subgroups of S_3, so $\mathbb{Z}_2 \times \mathbb{Z}_2$ and $\mathbb{Z}_6 \cong \mathbb{Z}_2 \times \mathbb{Z}_3$ are isomorphic to subgroups of G. The other desired inclusions are obvious.

Solution to 6.4.7: Denote by $h_i H$ (and $k_i K$) a coset of H (and K) and suppose

$$G = h_1 H \cup \cdots \cup h_r H \cup k_1 K \cup \cdots \cup k_s K.$$

Since all of the cosets of K are equal or disjoint and since the index of K in G is infinite, there is a $k \in K$ such that

$$kK \subset h_1 H \cup \cdots \cup h_r H.$$

Therefore, for $1 \leqslant i \leqslant s$,

$$k_i K \subset k_i k^{-1} h_1 H \cup \cdots \cup k_i k^{-1} h_r H.$$

This implies that G can be written as the union of a finite number of cosets of H, contradicting the fact that the index of H in G is infinite. Hence, G cannot be written as the finite union of cosets of H and K.

Solution to 6.4.8: 1. First, note that the multiplication rule in G reads

$$\begin{pmatrix} a & b \\ 0 & a^{-1} \end{pmatrix} \begin{pmatrix} a_1 & b_1 \\ 0 & a_1^{-1} \end{pmatrix} = \begin{pmatrix} aa_1 & ab_1 + ba_1^{-1} \\ 0 & a^{-1}a_1^{-1} \end{pmatrix}$$

which gives $\begin{pmatrix} a & b \\ 0 & a^{-1} \end{pmatrix}^{-1} = \begin{pmatrix} a^{-1} & -b \\ 0 & a \end{pmatrix}$. This makes it clear that N is a subgroup, and if $\begin{pmatrix} 1 & \beta \\ 0 & 1 \end{pmatrix}$ is in N, then

$$\begin{pmatrix} a & b \\ 0 & a^{-1} \end{pmatrix}^{-1} \begin{pmatrix} 1 & \beta \\ 0 & 1 \end{pmatrix} \begin{pmatrix} a & b \\ 0 & a^{-1} \end{pmatrix} = \begin{pmatrix} a^{-1} & -b \\ 0 & a \end{pmatrix} \begin{pmatrix} a & b + \beta a^{-1} \\ 0 & a^{-1} \end{pmatrix}$$

$$= \begin{pmatrix} 1 & \beta a^{-2} \\ 0 & 1 \end{pmatrix} \in N,$$

proving that N is normal.

By the first equation, the map from G onto $\mathbb{R}_+$ (the group of positive reals under multiplication) given by $\begin{pmatrix} a & b \\ 0 & a^{-1} \end{pmatrix} \mapsto a$ is a homomorphism whose kernel is N (which by itself proves that N is a normal subgroup). Hence, G/N is isomorphic to $\mathbb{R}_+$, which is isomorphic to the additive group $\mathbb{R}$.

2. To obtain the desired normal subgroup majorizing N, we can take the inverse image under the homomorphism above of any nontrivial proper subgroup of $\mathbb{R}_+$. If we take the inverse image of $\mathbb{Q}_+$, the group of positive rationals, we get the proper normal subgroup

$$N' = \left\{ \begin{pmatrix} a & b \\ 0 & a^{-1} \end{pmatrix} \mid a \in \mathbb{Q}_+ \right\}$$

of G, which contains N properly.

Solution to 6.4.9: $H_1 = \{m(1, 2) + n(4, 1) \mid m, n \in \mathbb{Z}\}$. By inspection, we see that a transversal for the quotient group G_1, a set containing one representative of each coset, consists of the lattice points inside the parallelogram with edges connecting $(0, 0)$, $(1, 2)$, $(5, 3)$ and the origin. A glance at the figure shows that

$$G_1 = \{(0, 0), (1, 1), (2, 1), (3, 1), (2, 2), (3, 2), (4, 2)\}$$

where we have abbreviated $(a, b) = H_1 + (a, b)$. Thus G_1 is an abelian group of order 7, isomorphic to the cyclic group $\mathbb{Z}_7$. In the same way we see that

$$G_2 = \{(0, 0), (1, 1), (1, 2), (2, 2), (2, 3), (3, 3), (3, 4)\},$$

so that G_2 is isomorphic to the cyclic group $\mathbb{Z}_7$. Thus $G_1 \simeq G_2$.

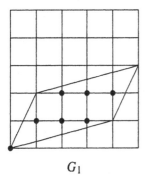

G_1

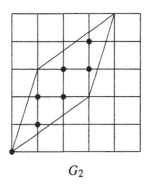

G_2

Solution to 6.4.10: Let $H = \{1, b\}$ be a normal subgroup of order 2. The group G contains an element of order 10, or G contains an element of order 5, or all elements (not 1) of G have order 2. If G has an element of order 10, or if $x^2 = 1$ for all $x \in G$, then G is abelian. When G has an element of order 5, a, let $K = \{1, a, a^2, a^3, a^4\}$. Since $|H|$, $|K|$ are coprime, we have $H \cap K = 1$. Thus $b \notin K$ and $G = K \cup Kb = \{1, a, a^2, a^3, a^4, b, ab, a^2b, a^3b, a^4b\}$. Now $\{a, ab\} = aH = Ha = \{a, ba\}$ whence $ab = ba$. It is now easy to see that G is abelian, for each element of G has the form $a^i b^j$, and $a^i b^j a^k b^l = a^{i+k} b^{j+l} = a^k b^l a^i b^j$.

Solution to 6.4.11: Let $x \in N_1$ and $y \in N_2$, since N_1 is normal, $x(yx^{-1}y^{-1}) \in N_1$ and by the same token $(xyx^{-1})y^{-1} \in N_2$. So $xyx^{-1}y^{-1} \in N_1 \cap N_2 = \{e\}$ and $xy = yx$. A similar argument works for the other pairs of subgroups, showing that elements of N_i commutes with elements of N_j for all $i \neq j$. Now suppose that $x, y \in N_1$. Since N_1 is normal in $N_2 N_3 = G$, there exist $x_2 \in N_2$ and $x_3 \in N_3$ such that $x = x_2 x_3$. By the previous remarks, y commutes with x_2 and x_3 so $xy = x_2 x_3 y = y x_2 x_3 = yx$, and we can now see that the elements of N_i commutes will all elements of N_j for all i and j, since $G = N_1 N_2$, G is abelian.

Applying the Second Isomorphism Theorem [Fra99, §3.1], we obtain,

$$N_2 \simeq \frac{N_2}{\{e\}} \simeq \frac{N_2}{N_1 \cap N_2} \simeq \frac{N_1 N_2}{N_1} = \frac{G}{N_1} = \frac{N_3 N_1}{N_1} \simeq N_3.$$

The other two isomorphisms $N_1 \simeq N_3$ and $N_1 \simeq N_2$ are obtained in the same way.

Solution to 6.4.13: Let $\{a_1 H, a_2 H, \ldots, a_{\frac{n}{m}} H\}$ be the set of distinct cosets of H. G acts on on this set by left multiplication and any $g \in G$ permutes these n/m cosets. This group action defines a map

$$\varphi : G \to S_{\frac{n}{m}}$$

from G to the permutation group on n/m objects. There are two cases to consider depending on φ being injective or not.

If φ is not injective, then $\ker \varphi$ is a normal subgroup $K \neq \{e\}$; and $K \neq G$, as well, because if $g \notin H$, $gH \neq H$; so g is not a trivial permutation.

If φ is injective, then $|\varphi(G)| = n$, and $\varphi(G)$ is a subgroup of $S_{\frac{n}{m}}$. But

$$\left[S_{\frac{n}{m}} : \varphi(G) \right] = \left| S_{\frac{n}{m}} \right| / |\varphi(G)| = \left(\frac{n}{m}\right)! / n < 2$$

So $\left[S_{\frac{n}{m}} : \varphi(G) \right] = 1$, that is, G is isomorphic to $S_{\frac{n}{m}}$, and in that case, $A_{\frac{n}{m}}$ is a nontrivial normal subgroup.

Solution to 6.4.14: Let H be a subgroup of G of index 3. G acts by left multiplication on the left cosets $\{gH\}$ of the subgroup H. This gives a homomorphism of G into the symmetric group of degree 3, the group of permutations of these cosets. The subgroup H is the stabilizer of one element of this set of cosets, namely the coset $1H$. This homomorphism cannot map onto the entire symmetric group, since this symmetric group has a subgroup of index 2, which would pull back to a subgroup of G of index 2. Thus, it must map onto the cyclic subgroup of order 3, and the group H is then the kernel of this homomorphism.

Solution to 6.4.15: For each element $g \in G$ consider the inner automorphism $\psi_g(h) = ghg^{-1}$. Defined in this way the map $g \mapsto \psi_g$ defines a homomorphism $G \to A$. It is nontrivial because G is not abelian, and injective since G is simple. Let B be its image, so $G \simeq B$. If $\alpha \in A$ and $g, h \in G$, we have

$$\alpha(\psi_g(h)) = \alpha(ghg^{-1}) = \alpha(g)\alpha(h)\alpha(g)^{-1} = \psi_{\alpha(g)}(\alpha(h)),$$

so $\alpha \circ \psi_g = \psi_{\alpha(g)} \circ \alpha$ in A. Thus, $\alpha \circ \psi_g \circ \alpha^{-1} = \psi_{\alpha(g)}$ and B is normal in A.

Solution to 6.4.16: 1. The cycles of λ_g are the right cosets $\langle g \rangle x$ of the subgroup $\langle g \rangle$ in G, so the lengths of each one is the order of g, and the number of cycles is $[G : \langle g \rangle]$.

If the order of g is odd, then each cycle has odd length, so each cycle is even, and so is λ_g. If the order of g is even, then each cycle is of even length and, therefore, odd. Also,

$$[G : \langle g \rangle] \cdot |\langle g \rangle| = |G|$$

is odd. As $[G : \langle g \rangle]$ is the number of cycles, λ_g is odd.

2. Let $\varphi : G \to \{-1, 1\}$ be defined by $g \mapsto \varphi(g) = $ sign of λ_g. φ is a morphism and, by Part 1, its kernel is N. So N is a normal subgroup of G with index 1 or 2. By Cauchy's Theorem [Her75, p. 61], G has an element of order 2, which is not in N, so N has order 2.

Solution to 6.4.17: Let $g = xyx^{-1}y^{-1}$ be a commutator. It suffices to show that conjugation by g fixes every element of N. As N is cyclic, Aut(N) is abelian, and, because N is normal, conjugation by any element of G is an automorphism of N. Let φ_x be the automorphism of conjugation by x. We have $\varphi_x \varphi_y = \varphi_y \varphi_x$. Hence, for $n \in N$, $gng^{-1} = \varphi_x \circ \varphi_y \circ \varphi_x^{-1} \circ \varphi_y^{-1}(n) = n$.

Solution to 6.4.18: We will show that if the index of N in G is not finite and equal to a prime number, then there is a subgroup H properly between N and G. Since any nontrivial proper subgroup of G/N is the image of such a subgroup, we need only look at subgroups of G/N.

Suppose first that the index of N in G is infinite, and let g be an element of G/N. If g is a generator of G/N, then G/N is isomorphic to $\mathbb{Z}$, and the element g^2 generates a proper nontrivial subgroup of G/N. Otherwise, g generates such a subgroup.

Suppose that the index of N in G is finite but not a prime number. Let p be any prime divisor of the index. By Cauchy's Theorem [MH87, p. 152], there is an element of order p in G/N. This element cannot generate the whole group, so it generates a nontrivial proper subgroup of G/N.

Solution to 6.4.19: Let the index of A in G be n. G acts by left multiplication on the cosets gA, and this gives a homomorphism into the group of permutations of the cosets, which has order $n!$. The kernel, N, of this homomorphism is contained in A, so the index of N in G is, at most, $n!$.

Solution to 6.4.20: 1. False. Consider the integers $\mathbb{Z}$ and the subgroup of the multiples of $n > 1$, then the quotient $H \simeq \mathbb{Z}_n$ is finite cyclic but $G \not\simeq H \times K$ because the product has torsion and $\mathbb{Z}$ is free abelian.

2. False. If H is a countable direct product of infinite cyclic groups and G is a free abelian group mapping onto H with kernel K, G is not isomorphic to the direct product of H and K. For if G were isomorphic to $H \times K$, H would be isomorphic to a subgroup of G. All subgroups of G are free abelian groups, but H is not.

6.5 $S_n, A_n, D_n, \ldots$

Solution to 6.5.1: The matrix $X = \begin{pmatrix} a & b \\ c & d \end{pmatrix}$ is in G if and only if $|A| = ad - bc = 1$. If $a = 1$ then $d = 1 + bc$ and X is one of

$$\begin{pmatrix} 1 & 0 \\ 0 & 1 \end{pmatrix}, \begin{pmatrix} 1 & 0 \\ 1 & 1 \end{pmatrix}, \begin{pmatrix} 1 & 1 \\ 0 & 1 \end{pmatrix}, \begin{pmatrix} 1 & 1 \\ 1 & 0 \end{pmatrix}.$$

If $a = 0$ then $bc = 1$ and X is one of

$$\begin{pmatrix} 0 & 1 \\ 1 & 0 \end{pmatrix}, \begin{pmatrix} 0 & 1 \\ 1 & 1 \end{pmatrix}.$$

Hence $|G| = 6$.

Let

$$A = \begin{pmatrix} 0 & 1 \\ 1 & 1 \end{pmatrix}, \quad B = \begin{pmatrix} 0 & 1 \\ 1 & 0 \end{pmatrix}.$$

It is easily verified that $A^3 = I = B^2$, $BA = A^2B$. The subgroup $\langle A \rangle$ has index 2 in G. Moreover, $B \notin \langle A \rangle$ since B has order 2 and elements of $\langle A \rangle$ have order 1 or 3. Therefore we have the coset decomposition

$$G = \langle A \rangle \cup \langle A \rangle B = \{I, A, A^2, B, AB, A^2B\}.$$

The relations $A^3 = I = B^2$, $BA = A^2B$ completely determine the multiplication in G.

Now S_3 has an element $\alpha = (1\ 2\ 3)$ of order 3, and an element $\beta = (1\ 2)$ of order 2. Moreover $\beta\alpha = \alpha^2\beta$. As above

$$S_3 = \{1, \alpha, \alpha^2, \beta, \alpha\beta, \alpha^2\beta\}$$
$$= \langle \alpha, \beta \mid \alpha^3 = 1 = \beta^2, \ \beta\alpha = \alpha^2\beta \rangle$$

Thus the correspondence $A^i B^j \leftrightarrow \alpha^i \beta^j$ is an isomorphism between G and S_3.

Solution to 6.5.2: Think of S_4 as permuting the set $\{1, 2, 3, 4\}$. For $1 \leqslant i \leqslant 4$, let $G_i \subset S_4$ be the set of all permutations which fix i. Clearly, G_i is a subgroup of S_4; since the elements of G_i may freely permute the three elements of the set $\{1, 2, 3, 4\} \setminus \{i\}$, it follows that G_i is isomorphic to S_3. Thus, the G_i's are the desired four subgroups isomorphic to S_3.

Similarly, for $i, j \in \{1, 2, 3, 4\}, i \neq j$, let H_{ij} be the set of permutations which fix i and j. Again H_{ij} is a subgroup of S_4, and since its elements can freely permute the other two elements, each must be isomorphic to S_2. Since for each pair i and j we must get a distinct subgroup, this gives us six such subgroups.

Finally, note that S_2 is of order 2 and so is isomorphic to $\mathbb{Z}_2$. Therefore, any subgroup of S_4 which contains the identity and an element of order 2 is isomorphic to S_2. Consider the following three subgroups: $\{1, (1\ 2)(3\ 4)\}$, $\{1, (1\ 3)(2\ 4)\}$, and $\{1, (1\ 4)(2\ 3)\}$. None of these three groups fix any of the elements of $\{1, 2, 3, 4\}$, so they are not isomorphic to any of the H_{ij}. Thus, we have found the final three desired subgroups.

Solution to 6.5.3: Let σ be a 5-cycle and τ a 2-cycle. By renaming the elements of the set, we may assume $\sigma = (1\ 2\ 3\ 4\ 5)$ and $\tau = (a\ b)$. Recall that if $(i_1\ i_2\ \cdots\ i_n)$ is a cycle and σ is any permutation, then

$$\sigma(i_1\ i_2\ \cdots\ i_n)\sigma^{-1} = (\sigma(i_1)\,\sigma(i_2)\ \cdots\ \sigma(i_n)).$$

Letting σ act repeatedly on τ as above, we get the five transpositions $(a+i\ b+i)$, $1 \leqslant i \leqslant 5$, where we interpret $a+i$ to be $a+i-5$ if $a+i > 5$. Fixing i such that $a+i-5 = 1$ and letting $c = b+i$, we see that G contains the five transpositions $(1\ c)$, $(2\ c+1)$, ..., $(5\ c+4)$. Since $a \neq b$, $c \neq 1$.

Let $d = c - 1$. Since d does not equal 0 or 5, these five transpositions can be written as $(c + nd\ c + (n + 1)d)$, $0 \leqslant n \leqslant 4$. By the Induction Principle [MH93, p. 7] the equation above shows that G contains the four transpositions

$$(1\ c + nd) = (1\ c + (n - 1)d)(c + (n - 1)d\ c + nd)(1\ c + (n - 1)d),$$

$1 \leqslant n \leqslant 4$. Since they are distinct, it follows that G contains the four transpositions $(1\ 2)$, $(1\ 3)$, $(1\ 4)$, and $(1\ 5)$. Applying the equation a third time, we see that $(i\ j) = (1\ i)(1\ j)(1\ i)$ is an element of G for all i and j, $1 \leqslant i, j \leqslant 5$. Hence, G contains all of the 2-cycles. Since every element in S_n can be written as the product of 2-cycles, we see that G is all of S_n.

Solution to 6.5.5: We use the notation $1^{\alpha_1} 2^{\alpha_2} \ldots$ to denote the cycle pattern of a permutation in disjoint cycles form; this means that the permutation is a product of α_1 cycles of length 1, α_2 cycles of length 2,.... (Thus the pattern of, say, $(8)(3\ 4)(5\ 6)(2\ 7\ 1) \in S_8$ is denoted by $1^1 2^2 3^1$.) The order of a permutation in disjoint cycles form is the least common multiple of the orders of its factors. The order of a cycle of length r is r. The possible cycle patterns of elements of S_7 are

$$1^7$$
$$1^5 2^1;\quad 1^3 2^2\quad 1^1 2^3;$$
$$1^4 3^1;\quad 1^2 2^1 3^1;\quad 2^2 3^1;\quad 1^1 3^2;$$
$$1^3 4^1;\quad 1^1 2^1 4^1;\quad 3^1 4^1$$
$$1^2 5^1;\quad 2^1 5^1;$$
$$1^1 6^1;$$
$$7^1.$$

We see that possible orders of elements of S_7 are 1, 2, 3, 4, 5, 6, 7, 10, 12.

Solution to 6.5.6: 1. $\sigma = (1234)(56789)$ has order $\operatorname{lcm}\{4, 5\} = 20$.
2. If $\sigma = \sigma_1 \cdots \sigma_k$ is written as a product of disjoint cycles σ_i of length r_i, then the order of σ is $\operatorname{lcm}\{r_1, \ldots, r_k\}$. In order that this lcm is 18, some r_i must have a factor 3^2, and some r_i a factor 2. As $1 < r_i \leqslant 9$ and $r_1 + \cdots + r_k \leqslant 9$ this is impossible.

Solution to 6.5.7: The order of a k–cycle is k, so the smallest m which simultaneously annihilates all 9-cycles, 8-cycles, 7-cycles, and 5-cycles is $2^3 \cdot 3^2 \cdot 5 \cdot 7 = 2520$. Any n–cycle, $n \leqslant 9$, raised to this power is annihilated, so $n = 2520$.

   To compute n for A_9, note that an 8-cycle is an odd permutation, so no 8-cycles are in A_9. Therefore, n need only annihilate 4-cycles (since a 4-cycle times a transposition is in A_9), 9-cycles, 7-cycles, and 5-cycles. Thus, $n = 2520/2 = 1260$.

Solution to 6.5.8: We have $1111 = 11 \times 101$, the product of two primes. So G is cyclic, say $G = \langle a \rangle$. From $1111 > 999$, it follows that a, when written as a product of disjoint cycles, has no cycles of length 1111. Therefore, all cycles of a have lengths $1, 11$, and 101. Let there be x, y, and z cycles of lengths, respectively, $1, 11$, and 101. If $x > 0$, then a has a fixed point, and this is then the desired fixed point for all of G. So assume that $x = 0$. Then $11x + 101z = 999$. It follows that $2z \equiv 9 \pmod{11}$, so $z \equiv 10 \pmod{11}$, and, therefore, $z \geqslant 10$. But then $999 = 11y + 101z \geqslant 1010$, a contradiction.

Solution to 6.5.9: For σ in S_n, let $\varepsilon_\sigma = 0$ if σ is even and $\varepsilon_\sigma = 1$ if σ is odd. Let τ denote the cycle $(n+1\ n+2)$ in S_{n+2}, and regard the elements of S_n as elements of S_{n+2} in the obvious way. The map of S_n into S_{n+2} defined by $\sigma \mapsto \sigma\tau^{\varepsilon_\sigma}$ is then an isomorphism of S_n onto its range, and the range lies in A_{n+2}.

Solution to 6.5.10: Call i and $j \in \{1, 2, \ldots, n\}$ equivalent if there exists $\sigma \in G$ with $\sigma(i) = j$. (This is clearly an equivalence relation.) For each i, the set $G_i = \{\sigma \in G \mid \sigma(i) = i\}$ is a subgroup of G, and the map $G \to \{1, 2, \ldots, n\}$, $\sigma \mapsto \sigma(i)$, induces a bijection from the coset space G/G_i to the equivalence class of i. Hence, for each i, the size of its equivalence class equals $[G : G_i]$, which is a power of p. Choosing one i from each equivalence class and summing over i, one finds that all these powers of p add up to n, since p does not divide n, one of these powers has to be $p^0 = 1$. This corresponds to an equivalence class that contains a single element i, and this i satisfies $\sigma(i) = i$ for all $\sigma \in G$.

Solution to 6.5.13: To determine the center of

$$D_n = \langle a, b \mid a^n = b^2 = 1, ba = a^{-1}b \rangle$$

(a is a rotation by $2\pi/n$ and b is a flip), it suffices to find those elements which commute with the generators a and b. Since $n \geqslant 3$, $a^{-1} \neq a$. Therefore,

$$a^{r+1}b = a(a^r b) = (a^r b)a = a^{r-1}b$$

so $a^2 = 1$, a contradiction; thus, no element of the form $a^r b$ is in the center. Similarly, if for $1 \leqslant s < n$, $a^s b = ba^s = a^{-s}b$, then $a^{2s} = 1$, which is possible only if $2s = n$. Hence, a^s commutes with b if and only if $n = 2s$. So, if $n = 2s$, the center of D_n is $\{1, a^s\}$; if n is odd the center is $\{1\}$.

Solution to 6.5.14: The number of Sylow 2-subgroups of D_n is odd and divides n. Each Sylow 2-subgroup is cyclic of order 2, since 2^1 is the largest power of 2 dividing the order of the group. By considering the elements of D_n as symmetries of a regular n–gon, we see that there are n reflections through axes dividing the n–gon in half, and each of these generates a different subgroup of order 2. Thus, the answer is that there are exactly n Sylow 2-subgroups in D_n when n is odd.

6.6   Direct Products

Solution to 6.6.3: Suppose $\mathbb{Q}$ was the direct sum of two nontrivial subgroups A and B. Fix $a \neq 0$ in A and $b \neq 0$ in B. We can write $a = a_0/a_1$ and $b = b_0/b_1$, where the a_i's and b_i's are nonzero integers. Since A and B are subgroups, $na \in A$ and $nb \in B$ for all integers n. In particular, $(a_1 b_0)a \in A$ and $(b_1 a_0)b \in B$. But

$$(a_1 b_0)a = (a_1 b_0)a_0/a_1 = a_0 b_0 = (b_1 a_0)b_0/b_1 = (b_1 a_0)b.$$

Hence, A and B have a nontrivial intersection, a contradiction.

Solution to 6.6.4: Let C_m denote the cyclic group of order m for $m \in \mathbb{N}$. By the Structure Theorem for abelian groups [Her75, p. 109], if G is a finite abelian group, there exist unique nonnegative integers $n_{p^r}(G)$ for each prime number p and each nonnegative integer r such that G is isomorphic to

$$\prod_p \prod_r C_{p^r}^{n_{p^r}(G)}.$$

If H is another abelian group, $G \times H$ is isomorphic to

$$\prod_p \prod_r C_{p^r}^{n_{p^r}(G)} \times \prod_p \prod_r C_{p^r}^{n_{p^r}(H)} \simeq \prod_p \prod_r C_{p^r}^{n_{p^r}(G)+n_{p^r}(H)}.$$

Hence, $n_{p^r}(G \times H) = n_{p^r}(G) + n_{p^r}(H)$. Now this and the fact that $A \times B$ is isomorphic to $A \times C$ yield the identities $n_{p^r}(B) = n_{p^r}(C)$ for all primes p and all nonnegative integers r. We conclude B and C are isomorphic.

Solution to 6.6.5: Let $\pi_3 : A \to G_1 \times G_2$ be the natural projection map. We have $\ker \pi_3 = \{(1, 1, g_3) \in A\}$. Let $N_3 = \{g_3 \mid (1, 1, g_3) \in \ker \pi_3\}$. Since $\ker \pi_3$ is a normal subgroup of A, N_3 is normal in G_3. Let $A' = G_1 \times G_2 \times G_3/N_3$. Since π_3 is onto, for any $(g_1, g_2) \in G_1 \times G_2$, there exists $g_3 \in G_3$ such that $(g_1, g_2, g_3) \in A$, and, thus, $(g_1, g_2, \overline{g_3}) \in A'$.

Define the map $\varphi : G_1 \times G_2 \to G_3/N_3$ by $\varphi(g_1, g_2) = \overline{g_3}$, where g_3 is such that $(g_1, g_2, g_3) \in A$. This is well defined, for if (g_1, g_2, g_3) and (g_1, g_2, h_3) are both in A, then $(1, 1, g_3 h_3^{-1}) \in A$, so $g_3 h_3^{-1} \in N_3$, which, in turn, implies that $\overline{g_3} = \overline{h_3}$. The map φ is clearly a homomorphism. Furthermore, since $\pi_1 : A \to G_2 \times G_3$ is onto, if $g_3 \in G_3$, there exist $g_1 \in G_1$ and $g_2 \in G_2$ such that $(g_1, g_2, g_3) \in A$. Thus, $\varphi(g_1, g_2) = \overline{g_3}$ so φ is onto.

Therefore, $\varphi(G_1 \times \{1\})$ and $\varphi(\{1\} \times G_2)$ are subgroups of G_3/N_3 which commute with each other. If these two subgroups were equal to one another and to G_3/N_3, then G_3/N_3 would be abelian. As G_3 is generated by its commutator subgroup, this would imply G_3/N_3 to be trivial or, equivalently, $G_3 = N_3$, which we assumed not to be the case. So we may assume that $\varphi(\{1\} \times G_2) \neq G_3/N_3$. Pick $\overline{g_3} \in G_3/N_3 \setminus \varphi(\{1\} \times G_2)$. Since $\pi_2 : A \to G_1 \times G_3$ is onto, there exists a $g_2 \in G_2$ such that $(1, g_2, g_3) \in A$. Hence, $\varphi(1, g_2) = \overline{g_3}$, contradicting our choice of $\overline{g_3}$.

Therefore, we must have that $N_3 = G_3$, so $\{1\} \times \{1\} \times G_3 \subset A$. Similar arguments show that $\{1\} \times G_2 \times \{1\}$ and $G_1 \times \{1\} \times \{1\}$ are contained in A and, thus, $A = G$.

Solution to 6.6.6: Yes, we will show that there exist a $Y \subset C$ such that $C = A+Y$. Choose $Y = X \cap C$, then $A \cap Y \subset A \cap X = 0$. Now to see the decomposition, observe that for all $c \in C \subset B$, there exist an $a \in A$ and $x \in X$ such that $c = a + x$, but $x = c - a \in C$ so $x \in C \cap X = Y$, thus showing that for all $c \in C$ there exist an $a \in A$ and $x \in Y$ such that $c = a + x$.

Solution to 6.6.7: The obvious isomorphism $F \colon \operatorname{Aut}(G) \times \operatorname{Aut}(H) \to \operatorname{Aut}(G \times H)$ is defined by

$$F(\alpha_G, \alpha_H)(g, h) = (\alpha_G(g), \alpha_H(h)).$$

Since α_G and α_H are automorphisms of G and H, it is clear that $F(\alpha_G, \alpha_H)$ is an automorphism of $G \times H$. Let us prove now that F is an isomorphism.

- F is injective: Let $F(\alpha_G, \alpha_H) = \operatorname{id}_{G \times H}$. Then $\alpha_G = \operatorname{id}_G$ and $\alpha_H = \operatorname{id}_H$ by definition of F. Hence, $\ker F$ is trivial so F is injective.

- F is surjective: Choose $\alpha \in \operatorname{Aut}(G \times H)$. Define α_G, α_H by

$$\alpha_G(g) = \pi_G(\alpha(g, \operatorname{id}_H)) \qquad \text{and} \qquad \alpha_H(h) = \pi_H(\alpha(\operatorname{id}_G, h))$$

where π_G and π_H are the quotient maps. Thus,

$$\alpha(g, h) = (\alpha_G(g), \alpha_H(h)) = F(\alpha_G, \alpha_H)(g, h).$$

Since the situation is symmetric between G and H; and G is finite, we need only show that α_G is injective. Let $\alpha_G(g) = \operatorname{id}_G$. Then $\alpha(g, \operatorname{id}_H) = (\operatorname{id}_G, h)$ for some $h \in H$. Suppose $n = |G|$. Then

$$(\operatorname{id}_G, h^n) = \alpha(g, \operatorname{id}_H)^n = \alpha(g^n, \operatorname{id}_H) = (\operatorname{id}_G, \operatorname{id}_H).$$

Hence, $h^n = \operatorname{id}_H$, so the order of h divides $|G|$. But, by Lagrange's Theorem [Her75, p. 41], the order of h also divides $|H|$, which is relatively prime to $|G|$, so the order of h is one and $h = \operatorname{id}_H$.

Solution to 6.6.8: We only have to determine where the automorphism send its generators $(1, 0)$ and $(0, 1)$, and these can be mapped into (a, b) and (c, d), respectively, if and only if,

$$a \notin p\mathbb{Z}/p^2\mathbb{Z}, \qquad c \in p\mathbb{Z}/p^2\mathbb{Z} \qquad and \qquad d \neq 0 \in \mathbb{Z}_p$$

If ψ is an automorphism that takes $(1, 0)$ to (a, b) and $(0, 1)$ to (c, d), then the (a, b) must not be killed by p, so $a \notin p\mathbb{Z}/p^2\mathbb{Z}$ and (c, d) must be killed by p, so $c \in p\mathbb{Z}/p^2\mathbb{Z}$. Moreover (c, d) should not be multiple of $p(a, b) = (pa, 0)$, so $d \neq 0$.

On the other hand, if (a, b) and (c, d) satisfy the conditions, then it is easy to find an automorphism, mapping the generators there, since (a, b) is killed by p^2 and (c, d) is killed by p. The condition on a implies that (a, b) has order p^2. If (c, d) were a a multiple of (a, b), then since $c \in p\mathbb{Z}/p^2\mathbb{Z}$, the element (c, d) would be a multiple of $p(a, b) = (pa, 0)$, which is impossible, since $d \neq 0 \in \mathbb{Z}_p$. This makes the $\#\psi(G) > p^2$ and by the Lagrange's Theorem the number has to be p^3. Thus ψ is surjective, but G is finite, so then ψ is injective and then an automorphism.

We need now to count all possibilities for a, b, c and d that satisfy the conditions. First there are $p^2 - p$ choices for a, p total for b, p for c and $p - 1$ choices for d, and these are all independent, so in total there are $p^3(p-1)^2$ automorphisms of G.

6.7 Free Groups, Generators, and Relations

Solution to 6.7.2: Suppose $\mathbb{Q}$ is finitely generated, with generators

$$\alpha_1/\beta_1, \ldots, \alpha_k/\beta_k \qquad \alpha_i, \beta_i \in \mathbb{Z}.$$

Then any element of $\mathbb{Q}$ can be written as $\sum n_i\alpha_i/\beta_i$, where the n_i's are integers. This sum can be written as a single fraction with denominator $s = \beta_1 \cdots \beta_k$. Consider a prime p which does not divide s. Then we have $1/p = r/s$ for some integer r, or $pr = s$, contradicting the fact that p does not divide s. Hence, $\mathbb{Q}$ cannot be finitely generated.

Solution 2. Suppose $\mathbb{Q}$ is finitely generated, then using the solution to Part 1 of Problem 6.3.1, $\mathbb{Q}$ is cyclic, which is a contradiction.

Solution to 6.7.3: Let A be the matrix

$$A = \begin{pmatrix} 15 & 3 & 0 \\ 3 & 7 & 4 \\ 18 & 14 & 8 \end{pmatrix}$$

that represents the relations of the group G, following [Hun96, p. 343] we perform elementary row and column operations on A, to bring it into diagonal form.

$$\begin{pmatrix} 15 & 3 & 0 \\ 3 & 7 & 4 \\ 18 & 14 & 8 \end{pmatrix} \rightarrow \begin{pmatrix} 15 & 3 & 0 \\ 3 & 7 & 4 \\ 12 & 0 & 0 \end{pmatrix} \rightarrow \begin{pmatrix} 3 & 3 & 0 \\ 3 & 7 & 4 \\ 12 & 0 & 0 \end{pmatrix} \rightarrow \begin{pmatrix} 3 & 3 & 0 \\ 0 & 4 & 4 \\ 12 & 0 & 0 \end{pmatrix}$$

$$\rightarrow \begin{pmatrix} 3 & 3 & 0 \\ 0 & 0 & 4 \\ 12 & 0 & 0 \end{pmatrix} \rightarrow \begin{pmatrix} 0 & 3 & 0 \\ 0 & 0 & 4 \\ 12 & 0 & 0 \end{pmatrix} \rightarrow \begin{pmatrix} 12 & 0 & 0 \\ 0 & 3 & 0 \\ 0 & 0 & 4 \end{pmatrix}$$

that is, $G \simeq \mathbb{Z}_{12} \times \mathbb{Z}_3 \times \mathbb{Z}_4$.

1. G is then the direct product of two cyclic groups $\mathbb{Z}_{12} \times \mathbb{Z}_{12}$.
2. G is also the direct product of the cyclic groups of prime power orders, namely $\mathbb{Z}_3 \times \mathbb{Z}_3 \times \mathbb{Z}_4 \times \mathbb{Z}_4$.
3. Using the decomposition $G \simeq \mathbb{Z}_{12} \times \mathbb{Z}_{12}$, we can see that for any element (p, q) of order 2 of G, p and q have to be either 0 or 6, so it has three elements of order 2, which are $(6, 0)$, $(0, 6)$ and $(6, 6)$.

Solution to 6.7.4: $x^5 y^3 = x^8 y^5$ implies $x^3 y^2 = 1$. Then $x^5 y^3 = x^3 y^2$ and $x^2 y = 1$. Hence, $x^3 y^2 = x^2 y$, so $xy = 1$. But then $xy = x^2 y$, so we have that $x = 1$. This implies that $y = 1$ also, and G is trivial.

Solution to 6.7.5: We start with the given relations

$$a^{-1} b^2 a = b^3 \qquad \text{and} \qquad b^{-1} a^2 b = a^3$$

that we will call first and second, respectively. From the first one, we get $a^{-1} b^4 a = b^6$ and

$$a^{-2} b^4 a^2 = a^{-1} b^6 a = (a^{-1} b^2) b^4 a = b^3 (a^{-1} b^4 a) = b^3 b^6 = b^9$$

using the relation just obtained and the second one, we get

$$b^9 = (a^{-2}) b^4 a^2 = b a^{-3} b^{-1} b^4 (a^2) = b a^{-3} b^{-1} b^4 b a^3 b^{-1} = b a^{-3} b^4 a^3 b^{-1}$$

and we conclude that

$$a^{-3} b^4 a^3 = b^9 = a^{-2} b^4 a^2$$

from which we obtain $ab^4 = b^4 a$. This, combined with the square of the first relation, $(a^{-1} b^4 a = b^6)$, gives $b^2 = 1$ and substitution back into the first shows that $b = 1$; then, substituting that into the second relation, we see that $a = 1$, so the group is trivial.

Solution 2. Using the same labels for the first and second relations, as above, we can conjugate the first one by a^2 to obtain the new relation $ab^2 a^{-1} = a^2 b^3 a^{-2}$. At the same time multiplying the second relation by b on the left and a^{-2} on the right, we get $a^2 b a^{-2} = ba$ and multiplying three copies of it back to back we get $a^2 b^3 a^{-2} = (ba)^3$. Getting these two last relations together we see that

$$ab^2 a^{-1} = (ba)^3 \qquad *$$

and similarly we can show that

$$ba^2 b^{-1} = (ab)^3 \qquad \star$$

Now consider this last relation ($\star$):

$$ba^2b^{-1} = (ab)^3$$
$$= a(ba)^2b$$
$$= a(ab^2a^{-2}b^{-1})b \qquad \text{using } (\star)$$
$$= a^2b^2a^{-2}$$
$$= (a^2b)ba^{-2}$$
$$= (ba^3)ba^{-2} \qquad \text{using second relation}$$
$$= ba(a^2ba^{-2})$$
$$= baba \qquad \text{using second relation}$$
$$= (ba)^2$$

so $ab^{-1} = ba$ and similarly we can show that $ba^{-1} = ab$. These two together combined show that $a^2 = b^{-2}$, and substituting back into the second relation, we see that $a = 1$ and the group is trivial.

Solution 3. We have

$$a^{-2}b^{-2}a^2 = a^{-1}b^{-3}a = a^{-1}b^{-1}(b^{-2}a) = a^{-1}b^{-1}ab^{-3}$$
$$= a^{-1}b^{-3}a = (a^{-1}b^{-2})b^{-1}a = b^{-3}a^{-1}b^{-1}a$$
$$b^{-2}a^{-2}b^2 = b^{-1}a^{-3}b = b^{-1}a^{-1}(a^{-2}b) = b^{-1}a^{-1}ba^{-3}$$
$$= b^{-1}a^{-3}b = (b^{-1}a^{-2})a^{-1}b = a^{-3}b^{-1}a^{-1}b$$
$$a^{-2}b^2a^2 = a^{-1}b^3a = a^{-1}b(b^2a) = a^{-1}bab^3$$
$$= a^{-1}b^3a = (a^{-1}b^2)ba = b^3a^{-1}ba$$
$$b^{-2}a^2b^2 = b^{-1}a^3b = b^{-1}a(a^2b) = b^{-1}aba^3$$
$$= b^{-1}a^3b = (b^{-1}a^2)ab = a^3b^{-1}ab$$

so we get

$$a^{-2}b^{-2}a^2b^2 = (a^{-2}b^{-2}a^2)b^2 = a^{-1}b^{-1}ab^{-3}b^2 = a^{-1}b^{-1}ab^{-1}$$
$$= a^{-2}(b^{-2}a^2b^2) = a^{-2}a^3b^{-1}ab = ab^{-1}ab$$
$$b^{-2}a^{-2}b^2a^2 = (b^{-2}a^{-2}b^2)a^2 = b^{-1}a^{-1}ba^{-3}a^2 = b^{-1}a^{-1}ba^{-1}$$
$$= b^{-2}(a^{-2}b^2a^2) = b^{-2}b^3a^{-1}ba = ba^{-1}ba$$

therefore, $ba^{-1}ba = baba^{-1}$ or $a^{-1}ba = aba^{-1}$ and $a^{-2}ba^2b^{-1} = 1$. Thus, $ba^2 = a^2b$ so the original relations give $b^2 = b^3$, $a^2 = a^3$ so $a = b = 1$ and the group is trivial.

Solution 4. By induction we show that

$$a^{-n}b^{2^nm}a^n = b^{3^nm}$$

in particular $a^{-3}b^8a^3 = b^{27}$, and then substituting the second relation in, we get $b^{-1}a^2b^8a^2b = b^{27}$, but we know from the general relation obtained by induction that $a^{-2}b^{4\cdot 2}a^{-2} = b^{18}$, so $b^{18} = b^{27}$ or $b^9 = 1$. Similarly we obtain $a^9 = 1$, now using the general relation again for $n = 9$ and $m = 1$ we get

$$b^{2^9} = a^{-9}b^{2^9}a^{-9} = b^{3^9}$$

but since $3^9 - 2^9 \equiv 1 \pmod 9$, we have $b = 1$ and the group is trivial.

Solution 5. In this one we will go further and show that the group

$$\langle a, b \, | \, a = [a^m, b^n], \ b = [b^p, a^q] \rangle$$

is trivial for all $m, n, p, q \in \mathbb{Z}$. Here $[x, y] = x^{-1}y^{-1}xy$, and for $m = p = 2$ and $n = q = 1$, we have our original group.

The two relations can be written as

$$a^{-q}b^p a^q = b^{p+1}$$

$$b^{-n}a^m b^n = a^{m+1}$$

By induction and raising the first relation to the kth-power shows that

$$a^{-q}b^{kp}a^q = b^{k(p+1)} \qquad k \in \mathbb{Z}$$

and we can write

$$b^{-k(p+1)}a^l b^{k(p+1)} = a^{-q}b^{-kp}a^l b^{kp}a^q \qquad k, l \in \mathbb{Z}$$

and in the same fashion we can obtain from the second relation

$$b^{-rn}a^{m^r}b^{rn} = a^{(m+1)^r} \qquad r \geqslant 1 \qquad \dagger$$

Now taking $k = n$, $l = m^{p+1}$ and $r = p + 1$ in the two equations above and equating the right-hand sides we get

$$b^{-np}a^{m^{(p+1)}}b^{np} = a^{(m+1)(p+1)}$$

and the relation $\dagger$ by itself is

$$b^{-np}a^{m^p}b^{np} = a^{(m+1)^p}$$

so raising this last relation to $m + 1$ and equating with the left-hand side of the previous one we end up with

$$a^{m^{(p+1)}} = a^{m^p(m+1)} = a^{m^{(p+1)}}a^{m^p}$$

which shows that

$$a^{m^p} = 1 \qquad \ddagger$$

Now $a^{m^p} = (a^m)^{m^{p-1}}$ so substituting back into the first relation we see that

$$a^{(m+1)m^{p-1}} = 1$$

that is,

$$a^{m^p} a^{m^{p-1}} = 1$$

and using ‡ again

$$a^{m^{p-1}} = 1$$

continuing in this way we eventually get to $a^{m^0} = 1$, that is $a = 1$ and the group is trivial.

Solution to 6.7.6: Since $a^2 = b^2 = 1$, every element of G can be expressed in (at least) one of the forms $ab \ldots ab$, $ba \ldots ba$, $ab \ldots aba$, $ba \ldots bab$. But $ab \ldots ab = (ab)^n$ for some $n \geqslant 0$, $ba \ldots ba = (ba)^n = (ab)^{-n}$, $ab \ldots aba = (ab)^n a$, and $ba \ldots bab = (ab)^{-n} baa = (ab)^{1-n} a$. Let H be the cyclic subgroup of G generated by ab. Then $G = H \cup Ha$, so either H is a cyclic subgroup of index 2 in G or $G = H$ is itself cyclic. In the latter case, G is either infinite cyclic or finite cyclic of even order, so G has a cyclic subgroup of index 2.

Solution to 6.7.8: We use the Induction Principle [MH93, p. 7]. For $n = 1$, the result is obvious.

Suppose $g_1, \ldots, g_n$ generate the group G and let H be a subgroup of G. If $H \subset \langle g_2, \ldots, g_n \rangle$, by the induction hypothesis, H is generated by $n - 1$ elements or fewer. Otherwise let

$$y = g_1^{m_1} \cdots g_n^{m_n} \in H$$

be such that $|m_1|$ is minimal but nonzero. We can assume, without loss of generality, that $m_1 > 0$. For any $z \in H$,

$$z = g_1^{k_1} \cdots g_n^{k_n}$$

there are integers q and r such that $k_1 = qm_1 + r$ and $0 \leqslant r < m_1$. Then the exponent of g_1 in zy^{-q} is r, and, by the choice of m_1, we obtain $r = 0$. Hence,

$$H = \langle y, K \rangle \quad \text{where} \quad K = H \cap \langle g_2, \ldots, g_n \rangle.$$

By the induction hypothesis, K is generated by, at most, $n - 1$ elements, and the result follows.

Solution to 6.7.10: By the Structure Theorem for finite abelian groups [Her75, p. 109], there are integers $m_1, \ldots, m_k$, $m_j | m_{j+1}$ for $1 \leqslant j \leqslant k - 1$, such that A is isomorphic to $\mathbb{Z}_{m_1} \oplus \cdots \oplus \mathbb{Z}_{m_k}$. We identify A with this direct sum. Clearly $m = m_k$. Therefore, S must contain the elements of the form $(0, 0, \ldots, 1)$ and $(0, \ldots, 0, 1, 0, \ldots, 1)$, where the middle 1 is in the jth position, $1 \leqslant j \leqslant k - 1$. Hence, using elements in S, we can generate all the elements of A which are zero everywhere except in the jth position. These, in turn, clearly generate A, so S generates A as well.

6.8   Finite Groups

Solution to 6.8.1: Any group with one element must be the trivial group.

By Lagrange's Theorem [Her75, p. 41], any group with prime order is cyclic and so abelian. Therefore, by the Structure Theorem for abelian groups [Her75, p. 109], every group of orders 2, 3, or 5 is isomorphic to $\mathbb{Z}_2$, $\mathbb{Z}_3$, or $\mathbb{Z}_5$, respectively.

If a group G has order 4, it is either cyclic, and so abelian, or each of its elements has order 2. In this case, we must have $1 = (ab)^2 = abab$ or $ba = ab$, so the group is abelian. Then, again by the Structure Theorem for abelian groups, a group of order 4 must be isomorphic to $\mathbb{Z}_4$ or $\mathbb{Z}_2 \times \mathbb{Z}_2$. These two groups of order 4 are not isomorphic since only one of them has an element of order 4.

Since groups of different orders can not be isomorphic, it follows that all of the groups on this list are distinct.

Solution to 6.8.2: By the Solution to the Problem 6.8.1 all groups of order up to 5 are abelian and the group of symmetries of the triangle, D_3, has 6 elements and is nonabelian.

Solution to 6.8.3: Let G be a group of order 6. By Cauchy's Theorem there exist an element a of order 3, and an element b of order 2. $H = \langle a \rangle$ is a subgroup of G and, since b has order 2 and 2 does not divide $|H|$, $b \notin H$. Therefore

$$G = H \cup Hb = \{1, a, a^2, b, ab, a^2b\}.$$

The multiplication of G is determined if we know which of the six elements of G is equal to ba. We find that

$$
\begin{array}{lll}
ba \neq 1 & \text{else} & a = b^{-1} = b, \\
ba \neq a & \text{else} & b = 1, \\
ba \neq a^2 & \text{else} & b = a, \\
ba \neq b & \text{else} & a = 1.
\end{array}
$$

Two cases remain.

Suppose $ba = ab$. Then $(ab)^n = a^n b^n$ for all $n \geq 1$. The order of ab is 2, 3 or 6. Now $(ab)^2 = a^2 b^2 = a^2 \neq 1$, and $(ab)^3 = a^3 b^3 = b \neq 1$. Hence ab has order 6. Thus G is cyclic, and $G \simeq \mathbb{Z}_6$.

Suppose $ba = a^2 b$. In this case $G \simeq S_3$. For S_3 has an element $\alpha = (1\ 2\ 3)$ of order 3, and an element $\beta = (1\ 2)$ of order 2 which satisfy $\beta\alpha = \alpha^2\beta$. Since $\langle \alpha \rangle$ has index 2 in S_3 and $\beta \notin \langle \alpha \rangle$, it follows that

$$S_3 = \{1, \alpha, \alpha^2, \beta, \alpha\beta, \alpha^2\beta\}.$$

Then the correspondence $a^i b^j \leftrightarrow \alpha^i \beta^j$ is an isomorphism between G and S_3.

Solution to 6.8.4: 1. The Klein four-group $V = \mathbb{Z}_2 \times \mathbb{Z}_2$ is a noncyclic group of order 4, by the arguments of the Solution to Problem 6.8.1 this is the only one of order 4. Alternatively to show uniqueness up to isomorphism, let G be a noncyclic group of order 4. Then the order of any non-identity element is equal to 2, whence $g = g^{-1}$ for all $g \in G$. Thus G is abelian. For if $x, y \in G$ then $xy = x^{-1}y^{-1} = (yx)^{-1} = yx$.

Let $G = \{1, a, b, c\}$, and consider ab. We see that

$$
\begin{array}{lll}
ab \neq 1 & \text{else} & b = a^{-1} = a, \\
ab \neq a & \text{else} & b = 1, \\
ab \neq b & \text{else} & a = 1.
\end{array}
$$

Hence $ab = c$. The multiplication of G is then completely determined using the fact that each element of G appears exactly once in any row or column in the multiplication table. This completes the proof of uniqueness.

2. Let $G = \{1, a, b, c\}$ be the noncyclic group of order 4. If $\varphi \in \text{Aut}(G)$, then $\varphi(1) = 1$ and $\widetilde{\varphi}$, the restriction of φ to $X = \{a, b, c\}$, is a permutation of X, i.e., $\widetilde{\varphi} \in S_3$. Moreover, for $\varphi, \psi \in \text{Aut}(G)$, $\widetilde{\varphi\psi} = \widetilde{\varphi}\widetilde{\psi}$. Thus the map $f : \varphi \mapsto \widetilde{\varphi}$ is a group homomorphism $\text{Aut}(G) \to S_3$.

If $\varphi \neq \psi \in \text{Aut}(G)$, then $\varphi(1) = \psi(1)$ and φ, ψ differ on X, so f is injective.

To show that f is surjective, let σ be a permutation of X, and let $\hat{\sigma}$ be the extension of σ to G given by $\hat{\sigma}(1) = 1$. Then $\hat{\sigma}$ is a bijection $G \to G$. It is left to show that $\hat{\sigma}$ is a homomorphism.

Let $x, y \in G$. If $x = 1$ or $y = 1$ then $\hat{\sigma}(xy) = \hat{\sigma}(x)\hat{\sigma}(y)$, since $\hat{\sigma}(1) = 1$. If $x = y$, then $\hat{\sigma}(xy) = \hat{\sigma}(x^2) = \hat{\sigma}(1) = (\hat{\sigma}(x))^2 = \hat{\sigma}(x)\hat{\sigma}(y)$, since $g^2 = 1$ for all $g \in G$. Finally for the case when x, y are distinct elements of X we note that the product of two of a, b, c is the third, i.e.,

$$ab = ba = c, \quad bc = cb = a, \quad ca = ac = b.$$

Hence $\hat{\sigma}(a)\hat{\sigma}(b) = \hat{\sigma}(c) = \hat{\sigma}(ab)$. The other cases are similar. Thus the map $f : \text{Aut}(G) \to S_3$ is an isomorphism.

Solution to 6.8.5: By the Structure Theorem for finitely generated abelian groups [Her75, p. 109], there are three: $\mathbb{Z}_8$, $\mathbb{Z}_2 \times \mathbb{Z}_4$, and $\mathbb{Z}_2 \times \mathbb{Z}_2 \times \mathbb{Z}_2$.

1. $(\mathbb{Z}_{15})^* = \{1, 2, 4, 7, 8, 11, 13, 14\}$. By inspection, we see that every element is of order 2 or 4. Hence, $(\mathbb{Z}_{15})^* \simeq \mathbb{Z}_2 \times \mathbb{Z}_4$

2. $(\mathbb{Z}_{17})^* = \{1, 2, \ldots, 16\} = \{\pm 1, \pm 2, \ldots, \pm 8\}$, passing to the quotient $(\mathbb{Z}_{17})^*/\{\pm 1\} = \{1, 2, \ldots, 8\}$ which is generated by 3, so $(\mathbb{Z}_{17})^* \simeq \mathbb{Z}_8$.

3. The roots form a cyclic group of order 8 isomorphic to $\mathbb{Z}_8$.

4. $\mathbf{F}_8$ is a field of characteristic 2, so every element added to itself is 0. Hence,

$$\mathbf{F}_8^+ \simeq \mathbb{Z}_2 \times \mathbb{Z}_2 \times \mathbb{Z}_2.$$

5. $(\mathbb{Z}_{16})^* = \{1, 3, 5, 7, 9, 11, 13, 15\} \simeq \mathbb{Z}_2 \times \mathbb{Z}_4$.

Solution to 6.8.6: Order 24: The groups S_4 and $S_3 \times \mathbb{Z}_4$ are nonabelian of order 24. They are not isomorphic since S_4 does not contain any element of order 24 but $S_3 \times \mathbb{Z}_4$ does.

Order 30: The groups $D_3 \times \mathbb{Z}_5$ and $D_5 \times \mathbb{Z}_3$ have different numbers of Sylow 2-subgroups, namely 3 and 5, respectively.

Order 40: There are two examples where the Sylow 2-subgroup is normal: The direct product of $\mathbb{Z}_5$ with a nonabelian group of order 8. Such order 8 groups are the dihedral group of symmetries of the square (which has only two elements of order 4), and the group of the quaternions $\{\pm 1, \pm i, \pm j, \pm k\}$ (which has six elements of order 4). There are also several other examples where the Sylow 2-subgroup is not normal.

Solution to 6.8.7: Consider the following eight groups of order 36:

$$\mathbb{Z}_2^2 \times \mathbb{Z}_3^2, \, \mathbb{Z}_2^2 \times \mathbb{Z}_9, \, \mathbb{Z}_4 \times \mathbb{Z}_3^2, \, \mathbb{Z}_4 \times \mathbb{Z}_9, \, \mathbb{Z}_6 \times S_3, \, S_3 \times S_3, \, \mathbb{Z}_2 \times D_{2 \cdot 9}, \, \mathbb{Z}_3 \times A_1.$$

The first four are abelian and pairwise nonisomorphic because each pair has either distinct 2-Sylow subgroups or distinct 3-Sylow subgroups. They are not isomorphic to the last four, since the latter are not abelian.

Of the last four, only $\mathbb{Z}_2 \times D_{2 \cdot 9}$ has a cyclic 3-Sylow subgroup, only $\mathbb{Z}_3 \times A_1$ has a normal 2-Sylow subgroup, and only $S_3 \times S_3$ has trivial center. Thus the last four are pairwise nonisomorphic.

Solution to 6.8.8: Let G be a group of order p^2. The conjugacy classes of G form a partition of G, as the ones associated with the equivalence relation $x \sim y$ iff there is z such that $y = zxz^{-1}$. Let Z denote the center of G. Then $x \in Z$ iff $C(x) = \{x\}$, where $C(x)$ denotes a conjugacy class. Thus $|Z|$ is the number of singleton conjugacy classes. Now $|C(g)|$ is a divisor of $|G|$, for each $g \in G$. Denote the non-singleton conjugacy classes by $C_1, \ldots, C_k$, where $k \geqslant 0$. Then $|G| = |Z| + |C_1| + \cdots + |C_k|$. If $|Z| = 1$ we obtain the contradiction that p divides 1. Thus $|Z| > 1$.

If G is cyclic then G is abelian. So we suppose that G is not cyclic. Therefore, each element of G, except 1, has order p. Select $x \neq 1$ in Z, and $y \notin \langle x \rangle$. Then $|\langle x \rangle| = p = |\langle y \rangle|$ and thus, by Lagrange's theorem, $\langle x \rangle \cap \langle y \rangle = \{1\}$. We deduce that the elements $x^i y^j$, $(1 \leqslant i, j \leqslant p)$ are distinct, and therefore $\langle x \rangle \langle y \rangle = G$. As $\langle x \rangle \subset Z$ we see that each element of $\langle x \rangle$ commutes with each element of $\langle y \rangle$.

Write $H = \langle x \rangle$, $K = \langle y \rangle$; $H \times K$ is the direct product group. We define a bijection $f : H \times K \to G$ given by $f(h, k) = hk$. This map is an isomorphism as

$$f\{(h, k)(h', k')\} = f(hh', kk')$$
$$= hh'kk'$$
$$= hkh'k' \qquad (u \in H \quad \text{commutes} \quad \text{with} \quad v \in K)$$
$$= f(h, k) f(h'k').$$

Now H and K are both isomorphic to the cyclic group $\mathbb{Z}_p$ of order p. Hence G is isomorphic to the abelian group $\mathbb{Z}_p \times \mathbb{Z}_p$, which proves that G is abelian. (In fact, this proves that a group of order p^2 is either $\mathbb{Z}_{p^2}$ or $\mathbb{Z}_p \times \mathbb{Z}_p$.)

Solution to 6.8.9: The number of Sylow 3-subgroups is congruent to 1 mod 3 and divides 5; hence, there is exactly one such subgroup, which is normal in the group. It is an abelian group of order 9. The abelian groups of this order are the cyclic group of order 9 and the direct product of two cyclic groups of order 3.

The number of Sylow 5-subgroups is congruent to 1 mod 5 and divides 9; hence, there is exactly one such subgroup, which is the (normal) cyclic group of order 5. The Sylow 3-subgroup and the Sylow 5-subgroup intersect trivially so their direct product is contained in the whole group, and a computation of the order shows that the whole group is exactly this direct product. Therefore, there are, up to isomorphism, two possibilities

$$\mathbb{Z}_9 \times \mathbb{Z}_5 , \qquad \mathbb{Z}_3 \times \mathbb{Z}_3 \times \mathbb{Z}_5 .$$

Solution to 6.8.10: Every element of G of order 7 generates a Sylow 7-subgroup. Therefore G contains more than one Sylow 7-subgroup. The number of Sylow 7-subgroups of G is congruent to 1 modulo 7, and any two of them have a trivial intersection. Hence the number must be 8 (since G, with 56 elements, is too small to accommodate 15 or more Sylow 7-subgroups). Thus G has $8 \times 6 = 48$ elements of order 7.

If P is a Sylow 2-subgroup of G, then P is contained in the complement of the set of elements of order 7, and that complement has cardinality $56 - 48 = 8$. Therefore G has only one Sylow 2-subgroup.

Since G has a unique Sylow 2-subgroup P, the subgroup P is normal in G. Let g be an element of G of order 7, and let g act on P by conjugation: $h \mapsto ghg^{-1}$ ($h \in P$). If this automorphism were the identity, then P would commute with the subgroup generated by g, and it would follow that G is the direct product of P and the subgroup generated by g. In that case G would have only 6 elements of order 7, a contradiction. Hence the automorphism of P induced by g is not the identity. That automorphism therefore has order 7 . Being a permutation of the 7-element set $P \backslash \{1\}$, it must be a cyclic permutation of that set. It follows that all elements of $P \backslash \{1\}$ have the same order. Since P contains an element of order 2, all of its nonidentity elements must have order 2.

Solution to 6.8.11: The prime factorization of the order is $80 = 2^4 \cdot 5^1$. Since there are five partitions of 4 and one of 1, by the Structure Theorem for finite abelian groups [Her75, p. 109], there are five non-isomorphic groups of order 80, which are:

$$\mathbb{Z}_2 \times \mathbb{Z}_2 \times \mathbb{Z}_2 \times \mathbb{Z}_2 \times \mathbb{Z}_5$$
$$\mathbb{Z}_2 \times \mathbb{Z}_2 \times \mathbb{Z}_4 \times \mathbb{Z}_5$$
$$\mathbb{Z}_2 \times \mathbb{Z}_8 \times \mathbb{Z}_5$$
$$\mathbb{Z}_4 \times \mathbb{Z}_4 \times \mathbb{Z}_5$$
$$\mathbb{Z}_{16} \times \mathbb{Z}_5$$

Solution to 6.8.12: If $p = 2$, then the group is either cyclic and so isomorphic to $\mathbb{Z}_4$, or every element has order 2 and so is abelian and isomorphic to $\mathbb{Z}_2 \oplus \mathbb{Z}_2$; for details see the Solution to Problem 6.8.1.

Now suppose $p > 2$ and let G have order $2p$. By Sylow's Theorems [Her75, p. 91], the p-Sylow subgroup of G must be normal, since the number of such subgroups must divide $2p$ and be congruent to 1 mod p. Since the p-Sylow subgroup has order p, it is cyclic; let it be generated by g. A similar argument shows that the number of 2-Sylow subgroups is odd and divides $2p$; hence, there is a unique, normal 2-Sylow subgroup, or there are p conjugate 2-Sylow subgroups. Let one of the 2-Sylow subgroups be generated by h.

In the first case, the element $ghg^{-1}h^{-1}$ is in the intersection of the 2-Sylow and the p-Sylow subgroups since they are both normal; these are cyclic groups of different orders, so it follows that $ghg^{-1}h^{-1} = 1$, or $hg = gh$. Since g and h must generate G, we see that G is abelian and isomorphic to $\mathbb{Z}_2 \oplus \mathbb{Z}_p$.

In the second case, a counting argument shows that all the elements of G can be written in the form $g^i h^j$, $0 \leqslant i < p$, $0 \leqslant j < 2$. Since all the elements of the form g^i have order p, it follows that all the 2-Sylow subgroups are generated by the elements $g^i h$. Hence, all of these elements are of order 2; in particular, $ghgh = 1$, or $hg = g^{-1}h$. Thus, $G = \langle g, h \mid g^p = h^2 = 1, hg = g^{-1}h \rangle$ and so G is the dihedral group D_n.

Solution to 6.8.13: By Cayley's Theorem [Her75, p. 71], every group of order n is isomorphic to a subgroup of S_n, so it is enough to show that S_n is isomorphic to a subgroup of $\mathbb{O}(n)$. For each $\sigma \in S_n$, consider the matrix $A_\sigma = (a_{ij})$, where $a_{\sigma(i)i} = 1$ and all other entries are zero. Let φ be defined by $\sigma \mapsto \varphi(\sigma) = A_\sigma$. The matrix A_σ has exactly one 1 in each row and column. Hence, both the rows and columns form an orthonormal basis of $\mathbb{R}^n$, so A_σ is orthogonal. φ maps S_n into $\mathbb{O}(n)$. Let $A_\sigma = (a_{ij})$ and $B_\tau = (b_{ij})$. Then

$$A_\sigma B_\tau = (c_{ij}) = \left(\sum_{k=1}^{n} a_{ik} b_{kj} \right).$$

An element of this matrix is 1 if and only if $i = \sigma(k)$ and $k = \tau(j)$ for some k; equivalently, if and only if $i = \sigma(\tau(j))$. Hence, $c_{\sigma(\tau(i))i} = 1$ and all the other entries are 0. Therefore, $(c_{ij}) = A_{\sigma \cdot \tau}$, so φ is a homomorphism.

If A_σ equals the identity matrix, then $\sigma(i) = i$ for $1 \leqslant i \leqslant n$, so σ is the identity permutation. Thus, φ has trivial kernel and is one-to-one hence an isomorphism.

Solution to 6.8.14: Suppose P and Q are Sylow subgroups corresponding to different primes. If $x \in P$ and $y \in Q$ then $xyx^{-1} \in Q$ (since Q is normal), therefore $[x, y] = xyx^{-1}y^{-1} \in Q$. Similarly, using the normality of P, $[x, y] \in P$. But $P \cap Q = \{1\}$ since P and Q have coprime orders, so $xy = yx$. The union of the Sylow subgroups generate a subgroup of G, H say, whose order is divisible by the order of each Sylow subgroup, therefore by the order of G, and we get $G = H$. Since G is generated by a set of commuting elements, it is abelian.

Solution to 6.8.15: Any counterexample is clearly nonabelian. In the symmetric group S_3 we have $S = A_3$, a subgroup. But if $G = S_4$, then S contains the squares of all 4-cycles, hence all transpositions. But the transpositions generate S_4, so if S were a subgroup it would equal S_4. However 4-cycles are not in S since S has no elements of order 8.

Solution 2. Let G be any simple nonabelian group of even order. Since S is obviously invariant under conjugation ($a^{-1}Sa = S$ for all a), if it were a subgroup it would be either $\{1\}$ or G. If $S = \{1\}$ then $abab = aabb$ for all a, b, implying that G is abelian, a contradiction. If $S = G$ then the squaring map is surjective, hence injective (since G is finite), contradicting the fact that a group of even order has elements of order 2. Hence S is not a subgroup.

Solution to 6.8.16: For each element $x \in X$, $G_x = \{g \in G \mid gx = x\}$ is a subgroup of G of index $|X|$, by transitivity. Since the identity is in each G_x, $\cup_{x \in X} G_x$ contains at most

$$|X|(|G|/|X| - 1) + 1 = |G| - |X| + 1$$

elements. Therefore at least $|X| - 1$ elements of G fix no elements of X.

Solution to 6.8.17: Suppose $b \in G$ is not in the center of G. Let $P = \{gbg^{-1}\}$. Then G clearly acts transitively on P. The cardinality of P equals $|G|/|C(b)|$ where $C(b)$ is the centralizer of b. Since $\{e, b\} \subset C(b)$, $b \neq e$ and $C(b) \neq G$, the result follows.

Solution 2. Since G is non-Abelian its order is not prime, so it has a nontrivial proper subgroup H (for example, a subgroup of prime order, obtained either from a Sylow Theorem [Her75, p. 91], or as a subgroup of the cyclic subgroup generated by a nonidentity element of G). Then G acts transitively by left multiplication on the left cosets of H.

Solution to 6.8.18: 1. For any set X let S_X be the group of bijections $\sigma : X \to X$. If X is a finite set with n elements then S_X is isomorphic to S_n, the group of all permutations of $\{1, 2, \ldots, n\}$.

For any group G and any element $g \in G$, define a mapping $\varphi_g : G \to G$ by $\varphi_g(x) = gx$ for all $x \in G$. For any $g, h, x \in G$, $(\varphi_g \cdot \varphi_h)(x) = \varphi_g(\varphi_h(x)) = \varphi_g(hx) = g(hx) = (gh)x = \varphi_{gh}(x)$. This shows that $\varphi_{gh} = \varphi_g \cdot \varphi_h$.

It follows that, for each $g \in G$, $\varphi_g : G \to G$ is a bijection, with inverse $\varphi_{g^{-1}}$. In other words $\varphi_g \in S_G$ for all $g \in G$, defining a map $\varphi : G \to S_G$. Since $\varphi_{gh} = \varphi_g \cdot \varphi_h$, $\varphi : G \to S_G$ is a group homomorphism. If $\varphi_g = \varphi_h$ for two elements $g, h \in G$ then $g = \varphi_g(1) = \varphi_h(1) = h$, showing that $\varphi : G \to S_G$ is injective. Therefore, G is isomorphic to the subgroup $\varphi(G)$ of S_G.

2. Since any group of order n embeds in S_n, it suffices to embed S_n into the group of even permutations of $n + 2$ objects. Let $\varepsilon : S_n \to \mathbb{Z}_2$ be the homomorphism that maps even permutations to 0 and odd permutations to 1.

Define $\theta : S_n \to S_{n+2}$ by $\theta(\sigma) = \sigma \cdot (n+1, n+2)^{\varepsilon(\sigma)}$. Since the transposition $(n+1, n+2)$ commutes with each element $\sigma \in S_n$, θ is a homomorphism, clearly injective. Since σ and $(n+1, n+2)^{\varepsilon(\sigma)}$ have the same parity, their product $\theta(\sigma)$ is even.

Solution to 6.8.19: Case 1. Suppose a and b commute. Then the elements $a^i b^j$, $0 \leqslant i < p, 0 \leqslant j < p$ are distinct, have order p (if i, j are not both zero).

Case 2. Suppose a, b do not commute. Then the $p^2 - 1$ elements $a^i b^j a^{-i}$ and a^k, $0 \leqslant i < p, 0 < j < p, 0 < k < p$ are distinct and have order p. Indeed, if $a^i b^j a^{-i} = b^k$ then we would have a homomorphism from (a) into $Aut(b)$. Since the first group has order p and the second has order $p - 1$ this morphism must be trivial, which is impossible since a and b do not commute. We have $a^i b^j a^{-i} \neq a^k$ because (a) and (b) are distinct groups.

Solution to 6.8.20: Suppose G contains an element whose order is not 2. Then the map $g \mapsto -g$ is an automorphism of G of order 2, and it follows by Lagrange's Theorem [Her75, p. 41] that $Aut(G)$ has even order. By the Structure Theorem for finite abelian groups [Her75, p. 109], it only remains to consider the groups $G = (\mathbb{Z}_2)^r$ $(r = 1, 2, \ldots)$, plus the trivial group. If G is trivial or $r = 1$, then $Aut(G)$ is trivial. If $r \geqslant 2$, then G has the automorphism

$$(x_1, x_2, x_3, x_4, \ldots, x_r) \mapsto (x_2, x_1, x_3, x_4, \ldots, x_r)$$

of order 2, so Lagrange's Theorem again shows that $Aut(G)$ has even order.

So $Aut(G)$ has odd order if and only if G is trivial or $G \simeq \mathbb{Z}_2$. In both cases $Aut(G)$ is trivial. Observe that it is *not* necessarily true that $Aut(G \times H) = Aut(G) \times Aut(H)$ even if G and H are cyclic of orders equal to distinct powers of a prime.

Solution to 6.8.21: Let p be a prime dividing n and $\varphi(n)$. The expression for $\varphi(n)$ shows that either $p^2 | n$ or there is a different prime number q such that $q | n$ and $p | (q - 1)$. If $p^2 | n$, then $\mathbb{Z}_p \times \mathbb{Z}_p \times \mathbb{Z}_{n/p^2}$ is a noncyclic group of order n. In the second case, let G be the subgroup of $GL(\mathbb{F}_q)$ of matrices of the form $\begin{pmatrix} a & b \\ 0 & 1 \end{pmatrix}$ where $a^p = 1$. Since $\mathbb{F}_q^*$ is cyclic of order $q - 1$, there are p solutions to $a^p = 1$ in $\mathbb{F}_q$. Therefore $|G| = pq$. If $a^p = 1$ and $a \neq 1$, then, $\begin{pmatrix} a & 0 \\ 0 & 1 \end{pmatrix}\begin{pmatrix} 1 & 1 \\ 0 & 1 \end{pmatrix} \neq \begin{pmatrix} 1 & 1 \\ 0 & 1 \end{pmatrix}\begin{pmatrix} a & 0 \\ 0 & 1 \end{pmatrix}$, so G is not abelian. Therefore, $G \times \mathbb{Z}_{n/pq}$ has order n and is not abelian, so it is not cyclic either.

6.9 Rings and Their Homomorphisms

Solution to 6.9.2: 1. Let $N = \begin{pmatrix} 1 & 1 \\ 0 & 1 \end{pmatrix}$. If $A, B \in R$, then $(A+B)N = AN + BN = NA + NB = N(A + B)$, so $A + B \in R$. If $A, B \in R$ then $(AB)N = A(BN) = A(NB) = (AN)B = (NA)B = N(AB)$, so $AB \in R$. And it is trivial to verify that the matrix, $-I$ also commutes with N, so it belongs to the R, implying that R is a subring.

2. A simple calculation shows that the matrix $A = \begin{pmatrix} a & b \\ c & d \end{pmatrix}$ belongs to R iff $a = a + c,\ a + b = b + d,\ c + d = d$, that is, iff A has the form $\begin{pmatrix} a & b \\ 0 & a \end{pmatrix}$. Define a $\mathbb{Q}$-algebra homomorphism $\psi : \mathbb{Q}[x] \to R$ by mapping x to $\begin{pmatrix} 0 & 1 \\ 0 & 0 \end{pmatrix}$. Clearly $\psi(x^2) = \begin{pmatrix} 0 & 1 \\ 0 & 0 \end{pmatrix}^2 = 0$, so ψ induces a homomorphism $\mathbb{Q}[x]/\langle x^2 \rangle \to R$. Since $\psi(a+bx) = \begin{pmatrix} a & b \\ 0 & a \end{pmatrix}$, this homomorphism is an isomorphism, and the result follows.

Solution to 6.9.4: Let $\varphi : \mathbb{C}^n \to \mathbb{C}$ be a ring homomorphism and $e_1 = (1, 0, 0, \ldots, 0),\ e_2 = (0, 1, 0, \ldots, 0), \ldots, e_n = (0, 0, 0, \ldots, 1)$, then $e_i e_j = 0$ for all $i \neq j$ and if $\varphi(e_1) = \cdots = \varphi(e_n) = 0$,

$$\varphi(x_1, \ldots, x_n) = \varphi(x_1, 0, \ldots, 0)\varphi(e_1) + \cdots + \varphi(0, 0, \ldots, x_n)\varphi(e_n) = 0$$

that is, φ is identically zero.

Suppose now that φ is a nontrivial homomorphism, then $\varphi(e_i) \neq 0$ for some i and in this case $\varphi(e_i) = \varphi(e_i e_i) = \varphi(e_i)\varphi(e_i)$ and $\varphi(e_i) = 1$. At the same time $0 = \varphi(e_i e_j) = \varphi(e_i)\varphi(e_j)$ we conclude that $\varphi(e_j) = 0$ for all $j \neq i$, and φ is determined by its value on the ith coordinate.

$$\begin{aligned} \varphi(x_1, \ldots, x_i, \ldots, x_n) &= \varphi(0, \ldots, x_i, \ldots, 0)\varphi(e_i) \\ &= \varphi(0, \ldots, x_i, \ldots, 0)1 \\ &= \varphi(0, \ldots, x_i, \ldots, 0) \end{aligned}$$

So for every homomorphism $\sigma : \mathbb{C} \to \mathbb{C}$ we can create n such homomorphisms from $\mathbb{C}^n$ to $\mathbb{C}$ by composing $\varphi(x_1, \ldots, x_n) = \sigma(\pi_i(x_1, \ldots, x_n))$ where π_i is the projection on the ith coordinate, and the argument above shows that all arise in this way. We observe here that is probably the best that can be done, since homomorphims from $\mathbb{C}$ to $\mathbb{C}$ cannot be easily classified.

Solution to 6.9.5: Let R contain k elements $(k < \infty)$ and consider the ring

$$S = \underbrace{R \times R \times \cdots \times R}_{k \text{ copies}}$$

Let $R = \{r_1, \ldots, r_k\}$ and $\alpha = (r_1, \ldots, r_k) \in S$. Now consider the collection of elements $\alpha, \alpha^2, \alpha^3, \ldots$ Since S is also a finite ring, by the Pigeonhole Principle [Her75, p. 127], there exist n and m sufficiently large with $\alpha^n = \alpha^m$. Coordinatewise, this means that $r_i^n = r_i^m$ for $1 \leqslant i \leqslant k$, and we are done.

Solution to 6.9.6: If $ax = 0$ (or $xa = 0$) with $a \neq 0$, then $axa = 0a$ or $a0 = 0$. If b is as in the text, then $a(b + x)a = a$, so, by uniqueness of b, $b = b + x$ and $x = 0$. Thus, there are no zero divisors.

Fix a and b such that $aba = a$. If $x \in R$, then $xaba = xa$ and, as there are no zero divisors, $xab = x$, so ab is a right identity. Similarly, $abax = ax$ implies $bax = x$ and ba is a left identity. Since any right identity is equal to any right identity, we get $ab = ba = 1$. Since $b = a^{-1}$, R is a division ring.

Solution to 6.9.7: Since $(R, +)$ is a finite abelian group with p^2 elements, by the Structure Theorem for finite abelian groups [Her75, p. 109], it is isomorphic to either $\mathbb{Z}_{p^2}$ or $\mathbb{Z}_p \times \mathbb{Z}_p$. In the first case, there is an element $x \in R$ such that every element of R can be written as nx, for $1 \leqslant n \leqslant p^2$. Since all elements of this form commute, it follows that R is abelian.

In the second case, every nonzero element must have additive order p. Let $x \in R$ be any element not in the additive subgroup generated by 1. Then it too must have additive order p. Thus, a counting argument shows that every element of R can be written in the form $n + kx$, $1 \leqslant n \leqslant p$, $1 \leqslant k \leqslant p$. Since all elements of this form commute, it follows that R is commutative.

Solution to 6.9.8: Assume R is not the zero ring, and let A be the center of R. Then $0, 1$ are distinct elements of A. Hence $\#A \geqslant 2$. The quotient of additive groups R/A has less than $8/2 = 4$ elements, so it must be cyclic. Thus there exists $x \in R$ such that $R = \{a + mx \mid a \in A, m \in \mathbb{Z}\}$. Such expressions commute (so in fact $A = R$).

The result is the best possible, the ring of upper triangular 2×2 matrices over the field of two elements is an example of a noncommutative ring of order 8, since

$$\begin{pmatrix} 1 & 0 \\ 0 & 0 \end{pmatrix}\begin{pmatrix} 0 & 1 \\ 0 & 0 \end{pmatrix} \not\equiv \begin{pmatrix} 0 & 1 \\ 0 & 0 \end{pmatrix}\begin{pmatrix} 1 & 0 \\ 0 & 0 \end{pmatrix} \quad (\text{mod } 2).$$

Solution to 6.9.9: R is trivially an additive subgroup and closed under multiplication and since $\mathbb{C}$ is a field, any subring is an integral domain, finishing the first part. Now consider two factorizations of the integer 10 in R, $10 = 2 \cdot 5$ and $10 = (1 + 3i) \cdot (1 - 3i)$. The norm $|z|^2 = a^2 = 9b^2$ of any $z \in R$ is an integer, and if $|z|^2 < 9$ then $b = 0$, so z is a real integer. This implies, in particular, that 2 has no non-trivial factorizations in R. If R were a unique factorization domain, then 2 would divide $1 + 3i$ or $1 - 3i$, but can't since $(1 \pm 3i)/2$ are not in R.

Solution to 6.9.10: One can embed R in the field of quotients of R; then the finite subgroup of R^* is a finite subgroup of the multiplicative group of the field; it is a finite abelian group, and so can be written as a direct product of $\mathbb{Z}_{p^n}$ for various primes p. If there are two such factors for the same p, then there are at least p elements of the field satisfying the equation $x^p - 1 = 0$. However, in a field, due to the uniqueness of factorization in the polynomial ring, there are, at most, n solutions to any n^{th} degree polynomial equation in one variable. Thus, in the factorization of our group, each prime p occurs at most once, therefore, any such group is cyclic.

Solution to 6.9.11: Consider the map $\alpha : R \to R$ defined by $\alpha(x) = ax$. If $\alpha(x) = \alpha(y)$ then $ax = ay$, so $a(x - y) = 0$. Since a is not a left zero divisor this implies $x = y$. Therefore $\alpha : R \to R$ is one-to-one. Since R is finite, α is also onto, so there is an element $b \in R$ such that $ab = 1$.

Similarly, using the fact that a is not a right zero divisor, there is an element $c \in R$ such that $ca = 1$. Therefore $b = (ca)b = c(ab) = c$ satisfies $ab = ba = 1$.

Solution to 6.9.12: $1 \Rightarrow 2$: There exist $v_1 \neq v_2$ such that $uv_1 = uv_2 = 1$; thus, $u(v_1 - v_2) = 0$ and u is a zero divisor.

$2 \Rightarrow 3$: Suppose that u is a unit with inverse v. If the $uv = 0$ then $w = (vu)w = v(uw) = v0 = 0$ and, therefore, u is not a left zero divisor.

$3 \Rightarrow 1$: Let v be a right inverse for u, that is, $uv = 1$. Since u is not a unit $vu \neq 1$ implying $vu - 1 \neq 0$. Now consider the element $v' = v + (vu - 1) \neq v$, and we have

$$uv' = uv + u(vu - 1)$$
$$= 1 - (uv)u - u$$
$$= 1 + u - u = 1$$

showing that u has more than one right inverse.

Solution to 6.9.13: The identity element of such a ring would belong to the additive group of the ring, which is a torsion group; thus there is some finite n such that if you add the identity element, 1, to itself n times, you get 0. In other words, the ring would have some finite characteristic n. But this implies that the additive order of every element of the ring divides n, and this is false, for example, for the element $1/(n + 1)$.

Solution to 6.9.14: Consider separately the cases: $a > 1, a = 1, a < 1$.

- If $a > 1$ then $a - 1 > 0$; multiplying this by a (which is > 0) we get $a^2 - a > 0$. Adding 1, we get $a^2 - a + 1 > 1 > 0$.

- If $a = 1$ then $a^2 - a + 1 = 1 > 0$.

- If $a < 1$, then $1 - a > 0$. If $a \neq 0$, then $a^2 > 0$; the sum of positives is positive, so $a^2 - a + 1 > 0$. If $a = 0$ then $a^2 - a + 1 = 1 > 0$.

Solution to 6.9.15: The degree 2 polynomial $x^2 + y^2 - 1$ does not factor into the product of two linear ones (since the circle $x^2 + y^2 = 1$ is not a union of two lines). This implies that the ideal $\langle x^2 + y^2 - 1 \rangle$ is prime and, thus, the ring $R = \mathbb{Q}[x, y]/\langle x^2 + y^2 - 1 \rangle$ is an integral domain.

Consider now the *stereographic projection* $(x, y) \mapsto (1, y/(x + 1))$ (at half of the speed of the standard one, in order to make the expressions simpler) of the circle from the point $(-1, 0)$ to the line $(1, t)$. It provides a homomorphism $t = y/(x + 1)$ of $\mathbb{Q}(t)$ to the field of fractions of R. The inverse homomorphism is given by the formulas $x = (1 - t^2)/(1 + t^2)$ and $y = 2t/(1 + t^2)$.

Solution to 6.9.16: For any set S of primes, let R_S be the set of rational numbers of the form a/b where a, b are integers and b is a product of powers of primes in S. It is clear that R_S contains 0 and 1, and is closed under addition, multiplication, and additive inverses. Hence R_S is a subring. If S and T are distinct sets of primes, say with $p \in S \setminus T$, then $1/p \in R_S$ but $1/p \notin R_T$, so $R_S \neq R_T$. Since there are infinitely many primes, we obtain at least $2^{\aleph_0}$ subrings in this way.

On the other hand, as $\mathbb{Q}$ is countable, its number of subsets is $2^{\aleph_0}$, therefore the set of subrings of $\mathbb{Q}$ has cardinality $2^{\aleph_0}$.

6.10   Ideals

Solution to 6.10.1: For two matrices in A we have

$$\begin{pmatrix} a & b \\ 0 & c \end{pmatrix} \begin{pmatrix} a_1 & b_1 \\ 0 & c_1 \end{pmatrix} = \begin{pmatrix} aa_1 & ab_1 + bc_1 \\ 0 & cc_1 \end{pmatrix},$$

therefore the sets of matrices of any of the forms $\begin{pmatrix} 0 & b \\ 0 & 0 \end{pmatrix}$, $\begin{pmatrix} a & b \\ 0 & 0 \end{pmatrix}$, and $\begin{pmatrix} 0 & b \\ 0 & c \end{pmatrix}$ are ideals in A. Call these ideals $\mathfrak{I}_1, \mathfrak{I}_2, \mathfrak{I}_3$, respectively. We'll show that, together with $\{0\}$ and A, these are the only ideals in A.

Consider any ideal $\mathfrak{I} \neq \{0\}$. As A contains all scalar multiples of the identity matrix, $\mathfrak{I}$ is closed under scalar multiplication. From the equalities

$$\begin{pmatrix} 1 & 0 \\ 0 & 0 \end{pmatrix} \begin{pmatrix} a & b \\ 0 & c \end{pmatrix} \begin{pmatrix} 0 & 0 \\ 0 & 1 \end{pmatrix} = \begin{pmatrix} 0 & b \\ 0 & 0 \end{pmatrix},$$

$$\begin{pmatrix} a & b \\ 0 & c \end{pmatrix} \begin{pmatrix} 0 & 1 \\ 0 & 0 \end{pmatrix} = \begin{pmatrix} 0 & a \\ 0 & 0 \end{pmatrix}, \quad \begin{pmatrix} 0 & 1 \\ 0 & 0 \end{pmatrix} \begin{pmatrix} a & b \\ 0 & c \end{pmatrix} = \begin{pmatrix} 0 & c \\ 0 & 0 \end{pmatrix},$$

it follows that $\mathfrak{I}$ contains a nonzero matrix in $\mathfrak{I}_1$, hence that $\mathfrak{I}$ contains $\mathfrak{I}_1$. Suppose $\mathfrak{I} \neq \mathfrak{I}_1$. Every matrix in $\mathfrak{I}$ is the sum of a diagonal matrix and a matrix in $\mathfrak{I}_1$. There are three cases:

- All diagonal matrices in $\mathfrak{I}$ have the form $\begin{pmatrix} a & 0 \\ 0 & 0 \end{pmatrix}$. Then $\mathfrak{I}$ contains all such matrices, since it contains a nonzero one, and we conclude that $\mathfrak{I} = \mathfrak{I}_2$.

- All diagonal matrices in $\mathfrak{I}$ have the form $\begin{pmatrix} 0 & 0 \\ 0 & c \end{pmatrix}$. Then, by a similar argument, $\mathfrak{I} = \mathfrak{I}_3$.

- $\mathfrak{I}$ contains a matrix of the form $\begin{pmatrix} a & 0 \\ 0 & c \end{pmatrix}$ with $a \neq 0 \neq c$. Since

$$\begin{pmatrix} 1 & 0 \\ 0 & 0 \end{pmatrix} \begin{pmatrix} a & 0 \\ 0 & c \end{pmatrix} = \begin{pmatrix} a & 0 \\ 0 & 0 \end{pmatrix}, \quad \begin{pmatrix} 0 & 0 \\ 0 & 1 \end{pmatrix} \begin{pmatrix} a & 0 \\ 0 & c \end{pmatrix} = \begin{pmatrix} 0 & 0 \\ 0 & c \end{pmatrix},$$

we conclude in this case that $\mathfrak{I}$ contains both $\mathfrak{I}_2$ and $\mathfrak{I}_3$, hence that $\mathfrak{I} = A$.

Solution to 6.10.2: Assume that $\mathfrak{I}$ is a nontrivial ideal. Let M_{ij} be the $n \times n$ matrix with 1 in the $(i, j)^{th}$ position and zeros elsewhere. Choose $A \in \mathfrak{I}$ such that $a = a_{ij} \neq 0$. Then, for $1 \leq k \leq n$, $M_{ki} A M_{jk}$ is a matrix which has a in the $(k, k)^{th}$ entry and 0 elsewhere. Since $\mathfrak{I}$ is an ideal, $M_{ki} A M_{jk} \in \mathfrak{I}$. The sum of these matrices is aI and so this matrix is also in $\mathfrak{I}$. However, since $\mathbf{F}$ is a field, a is invertible, so $\mathfrak{I} = M_n(\mathbf{F})$.

Solution to 6.10.3: We have see in the Solution to Problem 6.10.2 that $M_{n \times n}(\mathbf{F})$ has no nontrivial proper ideals. $M_{n \times n}(\mathbf{F})$ is an $\mathbf{F}$–vector field, and if we identify $\mathbf{F}$ with $\{a \Im \mid a \in \mathbf{F}\}$, we see that any ring homomorphism induces a vector space homomorphism. Hence, if $M_{n \times n}(\mathbf{F})$ and $M_{(n+1) \times (n+1)}(\mathbf{F})$ are isomorphic as rings, they are isomorphic as vector spaces. However, they have different dimensions n^2 and $(n + 1)^2$, respectively, so this is impossible.

Solution to 6.10.4: Since the kernel of h is a two sided ideal in R, it suffices to show that every ideal $\Im$ in R is either trivial (h is injective) or all of R (h is zero).

Assume $\Im$ is a non-trivial two sided ideal in R. Suppose $A \in \Im$ with $a_{ij} \neq 0$ for some $0 \leqslant i, j \leqslant n$. If we multiply A on the left by the elementary matrix E_{ji} we get the matrix $B = E_{ji} A$ which has only one nonzero entry ($b_{ij} = a_{ij}$). If we multiply B on the left by $(1/a_{ij})I$, we get E_{ij}, therefore $E_{ij} \in \Im$. Multiplying E_{ij} on the left and on the right by elementary matrices, we produce all elementary matrices, so these are in $\Im$. As they generate R we have $\Im = R$.

Solution to 6.10.5: Each element of $\mathbf{F}$ induces a constant function on X, and we identify the function with the element of $\mathbf{F}$. In particular, the function 1 is the unit element in $R(X, \mathbf{F})$.

Let $\Im$ be a proper ideal of $R(X, \mathbf{F})$. We will prove that there is a nonempty subset Y of X such that $\Im = \{f \in R(X, \mathbf{F}) \mid f(x) = 0 \text{ for all } x \in Y\} = \Im_Y$. Suppose not. Then either $\Im \subset \Im_Y$ for some set Y or, for every point $x \in X$, there is a function f_x in $\Im$ such that $f_x(x) = a \neq 0$. In the latter case, since $\Im$ is an ideal and $\mathbf{F}$ is a field, we can replace f_x by the function $a^{-1} f_x$, so we may assume that $f_x(x) = 1$. Multiplying f_x by the function g_x, which maps x to 1 and all other points of X to 0, we see that $\Im$ contains g_x for all points $x \in X$. But then, since X is finite, $\Im$ contains $\sum g_x \equiv 1$, which implies that $\Im$ is not a proper ideal.

Hence, there is a nonempty set Y such that $\Im \subset \Im_Y$. Let Y be the largest such set. As for every $x \notin Y$, there is an $f_x \in \Im$ such that $f_x(x) \neq 0$ (otherwise we would have $\Im \subset \Im_{Y \cup \{x\}}$) by an argument similar to the above, $\Im$ contains all the functions g_x, $x \notin Y$. But, from these, we can construct any function in $\Im_Y$, so $\Im_Y \subset \Im$.

Let $\Im$ and $\Im$ be two ideals, and the associated sets be Y and Z. Then $\Im \subset \Im$ if and only if $Z \subset Y$. Therefore, an ideal is maximal if and only if its associated set is as small as possible without being empty. Hence, the maximal ideals are precisely those ideals consisting of functions which vanish at one point of X.

Solution to 6.10.6: If $f, g \in E(U, W)$ then so is the homomorphism $f + g$ since its image on U is contained in $fU + gU$. If Y is a subspace and $h \in E(W, Y)$, then $h \circ f \in E(U, Y)$ since $fU \subseteq W + X$ for some finite dimensional subspace X and $h(f(U)) \subseteq h(W) + h(X)$. Thus $E(W, Y)E(U, W) \subseteq E(U, Y)$. From this we see that $E(U, U)$ is a ring with left ideal $E(V, U)$ and right ideal $E(U, 0)$. Also, $E(U, U) \subseteq E(V, V) = E(0, 0)$ so these latter two sets are also right and left ideals. The conclusion follows.

Solution to 6.10.7: Let $\mathfrak{I} = \langle a^n - 1, a^m - 1 \rangle$ and $\mathfrak{J} = \langle a^d - 1 \rangle$. For $n = rd$ the polynomial $x^n - 1$ factors into $(x^d - 1)(x^{r(d-1)} + x^{r(d-2)} + \cdots + x^r + 1)$. Therefore, in R, $a^n - 1 = (a^d - 1)(a^{r(d-1)} + a^{r(d-2)} + \cdots + a^r + 1)$. A similar identity holds for $a^m - 1$. Hence, the two generators of $\mathfrak{I}$ are in $\mathfrak{J}$, so $\mathfrak{I} \subset \mathfrak{J}$.

Since $d = \gcd\{n, m\}$, there exist positive integers x, y such that $xn - ym = d$. A calculation gives

$$
\begin{aligned}
a^d - 1 &= a^d - 1 - a^{d+ym} + a^{xn} \\
&= -a^d(a^{ym} - 1) + a^{xn} - 1 \\
&= -a^d(a^m - 1)(a^{y(m-1)} + \cdots + a^y + 1) \\
&\quad + (a^n - 1)(a^{x(n-1)} + \cdots + a^x + 1).
\end{aligned}
$$

Hence, $a^d - 1$ is in $\mathfrak{I}$, so $\mathfrak{J} \subset \mathfrak{I}$ and the two ideals are equal.

Solution to 6.10.8: 1. If there is an ideal $\mathfrak{I} \neq R$ of index, at most, 4, then there is also a maximal ideal of index, at most, 4, $\mathfrak{M}$, say. Then $R/\mathfrak{M}$ is a field of cardinality less than 5 containing an element α with $\alpha^3 = \alpha + 1$, namely $\alpha = a + \mathfrak{M}$. By direct inspection, we see that none of the fields $\mathbf{F}_2$, $\mathbf{F}_3$, $\mathbf{F}_4$ contains such an element. Therefore, $\mathfrak{I} = R$.
2. Let $R = \mathbb{Z}_5$, $a = 2 \pmod 5$, and $\mathfrak{I} = \{0\}$.

Solution 2. Since $R/\mathfrak{I}$ has order less than 5, two of the elements $0, 1, a, a^2, a^3$ have the same image in $R/\mathfrak{I}$. Then $\mathfrak{I}$ contains one of their differences, that is, one of

$$
1, a, a^2, a^3, a - 1, a^2 - 1, a^3 - 1, a(a - a), a(a^2 - 1), a^2(a - 1).
$$

But all these elements are units, since

$$
a(a - 1)(a + 1) = a(a^2 - 1) = 1, \qquad a^3 - 1 = a.
$$

Therefore, $\mathfrak{I}$ contains a unit, so $\mathfrak{I} = R$.

Solution to 6.10.9: Let $\mathfrak{I}$ be such an ideal. Consider $\varphi : R \rightarrow R/\mathfrak{I}$, the quotient map. Since $R/\mathfrak{I}$ is a three element ring with 1, it must be isomorphic to $\mathbb{Z}_3$. If $u \in R^*$ is a unit then so is $\varphi(u)$. Hence, $\varphi(u) = \pm 1$, and $\varphi(u^2) = 1$. As the squares of the units generate the additive group of R, this uniquely determines φ so there is, at most, one such $\mathfrak{I}$.

Solution to 6.10.10: Let $\mathfrak{M}$ be the maximal ideal, so in particular $2 \notin \mathfrak{M}$. If $a \in 1 + \mathfrak{M}$, then $a \notin \mathfrak{M}$, so the ideal (a) of R is not contained in any maximal ideal, so $(a) = (1)$, and a is a unit. By assumption, R^* is trivial, so $1 + \mathfrak{M}$ has at most one element, and $\mathfrak{M}$ has at most one element. Thus $\mathfrak{M} = \{0\}$, and R is a field. Hence $1 = \#(R^*) = \#R - 1$, and $\#R = 2$. Therefore R is the field of two elements, $\mathbb{Z}_2$, which can easily be verified to satisfy the condition.

Solution to 6.10.11: It suffices to show that if $ab - ba$ is any generator of $\mathfrak{I}$ and if c is any element of R, then $abc - bac$ is in $\mathfrak{I}$. By the definition of $\mathfrak{I}$, $a(bc) - (bc)a$

is an element of $\mathfrak{J}$. Further, since $\mathfrak{J}$ is a left ideal, $b(ca - ac) = bca - bac$ is an element of $\mathfrak{J}$. Therefore, $abc - bac = abc - bca + bca - bac$ is in $\mathfrak{J}$, and we are done.

Solution to 6.10.12: Using the direct sum decomposition, there are elements $u_i \in \mathfrak{J}_i$ such that

$$1 = u_1 + u_2 + \cdots + u_n.$$

If $a_1 \in \mathfrak{J}_1$ then

$$a_1 = a_1 \cdot 1 = (u_1 + u_2 + \cdots + u_n) = a_1 u_1 + a_2 u_2 + \cdots + a_n u_n.$$

Therefore $(a_1 u_1 - a_1) + a_2 u_2 + \cdots + a_n u_n = 0$. Since the ith summand is in $\mathfrak{J}_i$ and $R = \mathfrak{J}_1 \oplus \mathfrak{J}_2 \oplus \cdots \oplus \mathfrak{J}_n$, each summand is zero, i.e., $a_1 u_1 = a_1$ and $a_1 u_j = 0$ for $j \neq 1$. Similarly we get $a_i u_i = a_i$ and $a_i u_j = 0$ for $j \neq i$.

Solution to 6.10.13: The fact that this map is a ring homomorphism follows from the fact that the inclusion map $R \to S$ is a ring homomorphism mapping mR into mS. Let $1 = an + bm$, with integers a, b. Let $r \in R$ be in the kernel. Then $r = ms$ for some $s \in S$, so $r = (am + bn)r = m(ar + bns)$. Since R has index n in S, we have $ns \in R$ and so $r \in mR$. This shows that the map is an injection. Now suppose $s \in S$. Then $s = (am + bn)s \equiv b(ns) \bmod mS$. As $ns \in R$, the map is a surjection.

Solution to 6.10.15: We have $\mathfrak{J} = \langle i \rangle$ and $\mathfrak{J} = \langle j \rangle$, for some $i, j \in R$. Suppose first that $\mathfrak{J} + \mathfrak{J} = R$. Then $1 \in \mathfrak{J} + \mathfrak{J}$, so $1 = ri + sj$ for some $r, s \in R$. Therefore, the greatest common divisor of i and j is 1. Now $\mathfrak{J}\mathfrak{J}$ and $\mathfrak{J} \cap \mathfrak{J}$ are both ideals and, clearly, have generators ij and k, respectively, where k is the least common multiple of i and j. But the greatest common divisor of i and j is 1, so $ij = k$, and $\mathfrak{J}\mathfrak{J} = \mathfrak{J} \cap \mathfrak{J}$. Since every implication in the previous argument can be reversed, if $\mathfrak{J}\mathfrak{J} = \mathfrak{J} \cap \mathfrak{J}$, then $\mathfrak{J} + \mathfrak{J} = R$ and we are done.

6.11 Polynomials

Solution to 6.11.1: If $P(z)$ is a polynomial of degree n with α as a root, then $z^n P(1/z)$ is a polynomial of degree at most n with $1/\alpha$ as a root, multiplying by an appropriate term z^k, we have a polynomial of degree n.

Solution to 6.11.2: We have

$$x^3 + 2x^2 + 7x + 1 = (x - \alpha_1)(x - \alpha_2)(x - \alpha_3).$$

Equating coefficients we get

$$\alpha_1 + \alpha_2 + \alpha_3 = -2$$
$$\alpha_1 \alpha_2 + \alpha_2 \alpha_3 + \alpha_3 \alpha_1 = 7.$$

Therefore

$$\alpha_1^2 + \alpha_2^2 + \alpha_3^2 = (\alpha_1 + \alpha_2 + \alpha_3)^2 - 2(\alpha_1\alpha_2 + \alpha_2\alpha_3 + \alpha_3\alpha_1)$$
$$= -10.$$

For $i = 1, 2, 3$

$$\alpha_i^3 + 2\alpha_i^2 + 7\alpha_i + 1 = 0.$$

Adding these equations we get

$$(\alpha_1^3 + \alpha_2^3 + \alpha_3^3) + 2(\alpha_1^2 + \alpha_2^2 + \alpha_3^2) + 7(\alpha_1 + \alpha_2 + \alpha_3) + 3 = 0,$$

hence,

$$\alpha_1^3 + \alpha_2^3 + \alpha_3^3 = 31.$$

Solution to 6.11.3: Since ζ is a primitive seventh root of unity, we have

$$\zeta^6 + \zeta^5 + \cdots + \zeta + 1 = 0.$$

Dividing this by ζ^3, we get

$$(\zeta^3 + \zeta^{-3}) + (\zeta^2 + \zeta^{-2}) + (\zeta + \zeta^{-1}) + 1 = 0.$$

As $(\zeta + \zeta^{-1})^2 = (\zeta^2 + \zeta^{-2}) + 2$ and $(\zeta + \zeta^{-1})^3 = (\zeta^3 + \zeta^{-3}) + 3(\zeta + \zeta^{-1})$, the above equation becomes, letting $\alpha = (\zeta + \zeta^{-1})$,

$$\alpha^3 + \alpha^2 - 2\alpha - 1 = 0.$$

Solution to 6.11.4: 1. Let $x = \sqrt{5} + \sqrt{7}$. Squaring and rearranging successively, we get

$$x - \sqrt{5} = \sqrt{7}$$
$$x^2 - 2\sqrt{5}x + 5 = 7$$
$$x^2 - 2 = 2\sqrt{5}x$$
$$x^4 - 24x^2 + 4 = 0$$

This calculation shows that $\sqrt{5} + \sqrt{7}$ is a root of $f(x) = x^4 - 24x^2 + 4$.

2. If f had a linear factor, then it would have a rational root, but a calculation shows that none of $\pm 1, \pm 2$ is such a root (if p/q in lowest terms is a root of $a_n x^n + \cdots + a_0$ then, $p | a_0$ and $q | a_n$). Suppose now that for some $a, b, c, d \in \mathbb{Z}$,

$$f(x) = (x^2 + ax + b)(x^2 + cx + d).$$

Since the coefficient of x^3 in f is zero, $c = -a$, so we have

$$f(x) = (x^2 + ax + b)(x^2 - ax + d).$$

As the coefficient of x in f is zero, we get $ad - ab = 0$. If $a = 0$, then $f(x) = (x^2 + b)(x^2 + d) = x^4 + (b + d)x^2 + bd$, but the equations $bd = 4, b + d = -24$ have no integer solutions. If $b = d$, then $f(x) = (x^2 + ax + b)(x^2 - ax + b) = x^4 + (2b - a^2)x^2 + b^2$, so $b^2 = 4$ and $2b - a^2 = -24$, which also have no solutions in $\mathbb{Z}$.

Solution to 6.11.5: It is easy to see that $\sqrt{2} + \sqrt[3]{3}$ is a zero of a monic polynomial $p \in \mathbb{Z}[x]$ (use the process described on Problem 6.11.4.) If it were a rational number, it would have to be an integer, since its denominator would divide the leading coefficient of p. As $\sqrt{2} + \sqrt[3]{3}$ is strictly between 2 and 3, it must be irrational.

Solution 2. Suppose $\sqrt{2} + \sqrt[3]{3} \in \mathbb{Q}$. Then $\mathbb{Q}\left(\sqrt{2}\right) = \mathbb{Q}\left(\sqrt[3]{3}\right)$. However, this contradicts the fact that the fields $\mathbb{Q}\left(\sqrt{2}\right)$ and $\mathbb{Q}\left(\sqrt[3]{3}\right)$ have degrees 2 and 3 over $\mathbb{Q}$, respectively, since by Eisenstein Criterion [Her75, p. 160], the polynomials $x^2 - 2$ and $x^3 - 3$ are irreducible over $\mathbb{Q}$.

Solution to 6.11.6: Suppose that ω is a primitive k^{th} root of unity and that $\omega_i = \omega^i$ for $1 \leqslant i \leqslant k$. Let $P(z) = a_0 + a_1 z + \cdots + a_j z^j$ $(j < k)$; we have

$$\frac{1}{k} \sum_{i=1}^{k} P(\omega^i) = \frac{1}{k} \sum_{i=1}^{k} \sum_{r=0}^{j} a_r \omega^{ir} = \frac{1}{k} \sum_{r=0}^{j} a_r \sum_{i=1}^{k} \omega^{ir}$$

Since $\omega^k = 1$, we have $\omega^{rk} - 1 = 0$ for $1 \leqslant r \leqslant j$. Factoring and replacing 1 by ω^{rk}, we get

$$0 = (\omega^r - 1)(\omega^{rk} + \omega^{r(k-1)} + \cdots + \omega^r).$$

Since $r < k$ and ω is a primitive root of unity, $\omega^r \neq 1$. Therefore,

$$\omega^{rk} + \omega^{r(k-1)} + \cdots + \omega^r = 0.$$

Substituting this into the above equality gives

$$\frac{1}{k} \sum_{i=1}^{k} P(\omega^i) = \frac{1}{k} \sum_{i=1}^{k} a_0 \omega^{0i} = a_0 = P(0).$$

Solution to 6.11.7: By the Euclidean Algorithm, the vector space $V = \mathbb{Q}[x]/\langle f \rangle$ has dimension $d = \deg(f)$. Therefore, the infinitely many equivalence classes

$$\overline{x}^2, \overline{x}^3, \overline{x}^5, \ldots$$

are linearly dependent in V, so we can let q_i be a finite collection of rational numbers not all zero and satisfying

$$q_2 \overline{x}^2 + q_3 \overline{x}^3 + q_5 \overline{x}^5 + \cdots = 0.$$

This means that

$$q_2 x^2 + q_3 x^3 + q_5 x^5 + \cdots = f(x)g(x)$$

for some nonzero $g \in \mathbb{Q}[x]$.

Solution to 6.11.8: First, note that each polynomial $p(x)$ in $\mathbb{Z}[x]$ is congruent modulo $x - 7$ to a unique integer, namely the remainder one obtains by using the division algorithm to divide $x - 7$ into $p(x)$. (Only integer coefficients arise in the process, because the coefficient of x in $x - 7$ is 1.) If $p(x)$ lies in $\mathfrak{I}$, then so does the preceding remainder. However, the only members of $\mathfrak{I} \cap \mathbb{Z}$ are the integers that are divisible by 15. In fact, if k is in $\mathfrak{I} \cap \mathbb{Z}$, say $k = (x - 7)q(x) + 15r(x)$, then $k = 15r(7)$. Hence, we get a well defined map from $\mathbb{Z}[x]/\mathfrak{I}$ into $\mathbb{Z}_{15}$ by sending $p(x) + \mathfrak{I}$ to $k + 15\mathbb{Z}$, where k is the remainder one gets when dividing $p(x)$ by $x - 7$. The map is clearly a homomorphism. If $p(x)$ is not a unit in $\mathfrak{I}$, then the remainder is clearly not divisible by 15, from which we conclude that the map is one-to-one. The map is obviously surjective. It is, thus, the required isomorphism.

Solution 2. The map $\varphi : \mathbb{Z}[x] \to \mathbb{Z}[x]$ defined by $\varphi(p(x)) = p(x + 7)$ is a ring automorphism and it maps $\mathfrak{I}$ onto the ideal generated by x and 15. The quotient ring $\mathbb{Z}[x]/\varphi(\mathfrak{I})$ is isomorphic to $\mathbb{Z}_{15}$ under the map $p(x) \mapsto p(0) + 15\mathbb{Z}$, implying the desired conclusion.

Solution to 6.11.9: $\mathfrak{I}$ is prime. To show this we will prove that the quotient ring $\mathbb{Z}[x]/\mathfrak{I}$ is a field. Since $5 \in \mathfrak{I}$, this quotient ring is isomorphic to $\mathbb{Z}_5[x]/\langle x^3 + x + 1 \rangle$. So it suffices to show that $x^3 + x + 1$ is irreducible (mod 5). If it were reducible, it would have a linear factor, and, hence, a zero. But we can evaluate this polynomial for each $x \in \mathbb{Z}_5$ as follows:

x	$x^3 + x + 1$
0	1
1	3
2	1
3	1
4	4

Since there is no zero, the polynomial is irreducible, and the quotient ring

$$\frac{\mathbb{Z}_5[x]}{\langle x^3 + x + 1 \rangle}$$

is a field.

Solution to 6.11.12: Case 1: $p = 2$. In this case, define a map of $\mathbf{F}_2[x]$ into itself by $\varphi(1) = 1$ and $\varphi(x) = x + 1$, and extend it in the obvious way. Since constants are fixed and $\varphi(x + 1) = x$, it is clear that this is a ring isomorphism. Further, $\varphi(x^2 - 2) = (x - 1)^2 - 2 = x^2 + 1 = x^2 - 3$; we see that φ maps the ideal

$\langle x^2 - 2 \rangle$ onto the ideal $\langle x^2 - 3 \rangle$. It follows immediately from this that the two rings $\mathbf{F}_2[x]/\langle x^2 - 2 \rangle$ and $\mathbf{F}_2[x]/\langle x^2 - 3 \rangle$ are isomorphic.

Case 2: $p = 5$. By checking all the elements of $\mathbf{F}_5$, we see that $x^2 - 2$ and $x^2 - 3$ are both irreducible polynomials in $\mathbf{F}_5[x]$. Therefore, the ideals they generate are maximal and the quotient rings $\mathbf{F}_5[x]/\langle x^2 - 2 \rangle$ and $\mathbf{F}_5[x]/\langle x^2 - 3 \rangle$ are fields. The Euclidean Algorithm [Her75, p. 155] shows that each is a finite field with 25 elements. Since finite fields of the same order are isomorphic, the quotient rings in this case are isomorphic.

Case 3: $p = 11$. In this case, checking all the elements of $\mathbf{F}_{11}$ shows that $x^2 - 2$ is irreducible, but $x^2 - 3 = (x - 5)(x + 5)$ is not. Hence, the quotient ring $\mathbf{F}_{11}[x]/\langle x^2 - 2 \rangle$ is a field, whereas $\mathbf{F}_{11}[x]/\langle x^2 - 3 \rangle$ is not, so the two quotient rings are not isomorphic in this case.

Solution to 6.11.13: Since $x - 3$ is a monic, given any polynomial $r(x)$ in $\mathbb{Z}[x]$, there exist polynomials $t(x)$ and $s(x)$ such that $r(x) = t(x)(x - 3) + s(x)$ and $\deg s(x) < \deg(x - 3) = 1$. Hence, $s(x)$ is a constant, and so it is congruent modulo 7 to some α, $0 \leqslant \alpha \leqslant 6$. Hence, $r(x) - \alpha = t(x)(x - 3) + (s(x) - \alpha)$, and the right-hand term is clearly an element of $\mathfrak{I}$.

In the special case where $r(x) = x^{250} + 15x^{14} + x^2 + 5$, we have, by the Euclidean Algorithm [Her75, p. 155], $r(x) = t(x)(x-3)+\alpha$. Substituting $x = 3$, we get $r(3) = \alpha$. Since we only need to know α modulo 7, we reduce $r(3)$ mod 7 using the fact that $n^7 \equiv n \pmod 7$, getting $r(3) \equiv 3^4 + 3^2 + 3^2 + 5 \equiv 6 \pmod 7$. Hence, $\alpha = 6$ is the desired value.

Solution to 6.11.14: Let $\varphi : \mathbb{Z}[x] \to \mathbb{Z}_{13}$ be the unique ring homomorphism such that $\varphi(x) = 4$. A polynomial $\alpha(x) \in \mathbb{Z}[x]$ is in the kernel of φ if and only if $\alpha(4) \equiv 0 \pmod{13}$. This occurs if and only if $\alpha(x) \equiv (x - 4)\beta(x) \pmod{13}$ for some $\beta(x) \in \mathbb{Z}[x]$, i.e., exactly when $\alpha(x) = (x - 4)\beta(x) + 13\gamma(x)$ for some $\gamma(x) \in \mathbb{Z}[x]$, in other words if and only if $\alpha(x) \in \mathfrak{I}$.

Set $f(x) = (x^{26} + x + 1)^{73} \in \mathbb{Z}[x]$; then $f(x) - m \in \mathfrak{I}$ if and only if $\varphi(f(x) - m) = 0$, which holds if and only if $f(4) \equiv m \pmod{13}$.

By Fermat's Little Theorem [Sta89, p. 80], [Her75, p. 44], if $a \in \mathbb{Z}$ is not divisible by the prime p then $a^{p-1} \equiv 1 \pmod p$. This gives $(4^{26} + 4 + 1) \equiv (4^2 + 5) \equiv 8 \pmod{13}$, and $f(4) \equiv 8^{73} \equiv 8 \pmod{13}$. So $m = 8$ is the unique integer in the range $0 \leqslant m \leqslant 12$ such that $(x^{26} + x + 1)^{73} - m \in \mathfrak{I}$.

Solution to 6.11.15: Let $f(x) = x^3 - 2$, and denote by $\sqrt[3]{2}$ the real cube root of 2. Then $\mathbb{Q}\left(\sqrt[3]{2}\right)$ is an algebraic extension of $\mathbb{Q}$ generated by a root of f, hence isomorphic to $K = \mathbb{Q}[x]/(x^3 - 2)$. Since $\mathbb{Q}\left(\sqrt[3]{2}\right) \subset \mathbb{R}$, and $x^3 - 2$ has only one real root, $x^3 - 2$ does not factor completely over K.

Solution to 6.11.16: 1. Let K be the set of $f \in \mathbb{R}[x]$ for which $f(2) = f'(2) = f''(2) = 0$. K is an ideal in the ring $\mathbb{R}[x]$ if

(i) K is closed under addition and negation, and $0 \in K$,

(ii) gf and fg belong to K whenever $g \in \mathbb{R}[x]$ and $f \in K$.

Condition (i) is easily verified. For condition (ii), let $g \in \mathbb{R}[x]$, $f \in K$. Then $(gf)' = g'f + gf'$, $(gf)'' = g''f + 2g'f' + gf''$. We see that $(gf)(2) = (gf)'(2) = (gf)''(2) = 0$, so that gf, and also fg, belongs to K. Thus K is an ideal of $\mathbb{R}[x]$.

If K is an (nonzero) ideal of $R = \mathbb{R}[x]$, let a be a nonzero element of K of lowest degree. If $b \in K$ then there exist $q, r \in R$ such that $b = aq + r$ with $r = 0$ or $\deg(r) < \deg(a)$. Now $r = b - aq \in K$, so $\deg(r) \geq \deg(a)$. That is, $r = 0$ and $b = aq$. We conclude that $K = aR$ is generated by a. Two generators of K are associates and there is a unique monic generator. Expand in powers of $(x - 2)$:

$$a = a_0 + a_1(x - 2) + a_2(x - 2)^2 + a_3(x - 2)^3 + \cdots + a_n(x - 2)^n.$$

Then $a_0 = a(2) = 0$, $a_1 = a'(2) = 0$, $a_2 = a''(2)/2! = 0$. The monic generator is $(x - 2)^3$.

2. The function $f(x) = x^2 - 6x + 8$ satisfies $f(2) = 0 = f'(3)$. If $g(x) = x$ then $(gf)'(3) \neq 0$. Thus the stated condition does not define an ideal.

Solution to 6.11.17: Let $\varphi : \mathbb{Z}[x] \to \mathbb{Z}[x]$ be any automorphism. Since φ is determined by the value of x, every element of $\mathbb{Z}[x]$ must be a polynomial in $\varphi(x)$. In order to get x in the image of φ, we see that $\varphi(x)$ must be of the form $\pm x + \alpha$ for some constant α.

Solution to 6.11.19: The multiplicative group of units $\mathbb{Z}_p^*$ has $p - 1$ elements. Therefore, $a^{p-1} = 1$ for each $a \in \mathbb{Z}_p$, whence $a^p - a = 0$ for all $a \in \mathbb{Z}_p$. The polynomial $x^p - x$ has p distinct roots, thus factorizes as

$$x^p - x = x(x - 1)\ldots(x - (p - 1)).$$

If $f(x) = u p_i^{\alpha_1} \ldots p_k^{\alpha_k}$, where u is a unit and p_i are distinct monic irreducible polynomials, then we see that the gcd of f and g is equal to the product of the distinct linear factors of f.

Solution to 6.11.20: Let $p = u_1 a_1^{\alpha_1} \ldots a_k^{\alpha_k}$, $q = u_2 a_1^{\beta_1} \ldots a_k^{\beta_k}$ where a_i are monic irreducible in $\mathbf{F}[x]$, and $\alpha_i, \beta_i \geq 0$, u_i units of $\mathbf{F}$. If $\gamma_i = \min\{\alpha_i, \beta_i\}$ then $h = a_i^{\gamma_1} \ldots a_k^{\gamma_k}$ is the gcd of p and q over $\mathbf{F}$.

The units of $\mathbf{F}[x]$ are the nonzero constants, and remain units in $\mathbf{K}[x]$. If $a_i = \prod_j b_{ij}^{\alpha_{ij}}$ is the irreducible factorization of a_i over $\mathbf{K}$ then $p = u_1 \prod_i \prod_j b_{ij}^{\alpha_{ij}}$ is the irreducible factorization of p over $\mathbf{K}$, by uniqueness of factorization. We thus obtain the irreducible factorizations of p and q over $\mathbf{K}$ by further factorizing the a_i over $\mathbf{K}$.

If a and b are distinct irreducible polynomials over $\mathbf{F}$ then their gcd over $\mathbf{F}$ is 1. There exist polynomials s, t over $\mathbf{F}$ with $sa + tb = 1$. Thus, if $d \in \mathbf{K}[x]$ is the gcd of a and b over $\mathbf{K}$, then $d|1$, and so d is a unit of $\mathbf{K}[x]$, and $d = 1$ as d is chosen monic.

Let h, h' be the gcds of p and q over $\mathbf{F}$, $\mathbf{K}$ respectively. Then h divides h'. By the above we cannot get any extra factors over $\mathbf{K}$ in h'. Thus the gcd of p and q over $\mathbf{K}$ is the same as their gcd over $\mathbf{F}$.

Solution to 6.11.21: If the fraction p/q in lowest terms is a zero of $x^{10}+x^9+x^8+\cdots+x+1$, then $p|1$ and $q|1$, so the possible rational zeros are ± 1. A calculation shows that neither of these is a zero, so the given polynomial is irreducible over $\mathbb{Q}$. Now -1 is a zero of the second polynomial, so it is reducible over $\mathbb{Q}$.

Solution to 6.11.22: Note that $539 = 7^2 \cdot 11$, $511 = 7 \cdot 23$, and $847 = 7 \cdot 11^2$. Thus, all the coefficients except the leading one are divisible by 7, but the constant term is not divisible by 7^2. Since 7 is a prime, by Eisenstein Criterion [Her75, p. 160], the polynomial is irreducible in $\mathbb{Z}[x]$.

Solution to 6.11.23: By the Gauss Lemma [BML97, p. 85], an integral polynomial that can be factored over rationals can be factored into polynomials of the same degree over the integers. Since ± 1 are not roots (and they are the only ones possible because $a_0 = a_n = 1$), there are no linear terms and the only possible factorizations are in polynomials of degree 2.

- $(x^2 + ax + 1)(x^2 + bx + 1)$
- $(x^2 + ax - 1)(x^2 + bx - 1)$

In the first, case we get $x^4 + (a+b)x^3 + (2+ab)x^2 + (a+b)x + 1$, which implies that the coefficients of the terms of degree 1 and 3, are the same, a contradiction. The other case is analogous, showing that the polynomial is irreducible over $\mathbb{Q}$.

Solution to 6.11.24: We will use Eisenstein Criterion [Her75, p. 160]. Let $x = y + 1$. We have

$$
\begin{aligned}
x^{p-1} + x^{p-2} + \cdots + 1 &= (y+1)^{p-1} + (y+1)^{p-2} + \cdots + 1 \\
&= \frac{(y+1)^p - 1}{(y+1) - 1} \\
&= \frac{\sum_{k=0}^{p} \binom{p}{k} y^k - 1}{y} \\
&= \sum_{k=1}^{p} \binom{p}{k} y^{k-1} \\
&= y^{p-1} + p y^{p-2} + \cdots + p.
\end{aligned}
$$

Since the prime p divides all the coefficients except the first and p^2 does not divide the last, it follows that the polynomial is irreducible in $\mathbb{Q}[x]$. Therefore, the given polynomial must also be irreducible, since if it were not, the same change of variables would give a factorization of the new polynomial.

Solution to 6.11.25: Put $x = y + 1$ to get

$$f(x) = \frac{x^5 - 1}{x - 1} + 5x = \frac{(y + 1)^5 - 1}{y} + 5y + 5.$$

The coefficients of y^3, y^2, y, and 1 are integers divisible by $p = 5$ and the constant term is 10, which is not divisible by p^2. Thus, by Eisenstein Criterion [Her75, p. 160], f is irreducible over $\mathbb{Q}$.

Solution to 6.11.26: By the Gauss Lemma [BML97, p. 85], it is enough to show that $f(x)$ is irreducible over the integers. Suppose $f(x) = g(x)h(x)$, where $g(x)$ and $h(x)$ have positive degrees and integer coefficients, say

$$g(x) = b_j x^j + b_{j-1} x^{j-1} + \ldots + b_0, \qquad h(x) = c_k x^k + c_{k-1} x^{k-1} + \ldots + c_0.$$

The we can see that $b_0 c_0 = 25$, which can only happens in two ways, either $b_0 = 25$, $c_0 = 1$ or $b_0 = 5$, $c_0 = 5$.
Case 1: $b_0 = 25$, $c_0 = 1$.
Then $75 = b_1 c_0 + c_1 b_0 = b_1 + 25c_1$, so 25 divides b_1. Hence $-100 = b_2 c_0 + b_1 c_1 + b_0 c_2 = b_2 + b_1 c_1 + 25c_1$, so 25 divides b_2. Continuing in this way, we find that 25 also divides b_3 and b_4. However 16, the leading coefficient of $f(x)$, is given by

$$16 = b_4 c_1 + b_3 c_2 + b_2 c_3 + b_1 c_4 \qquad (*)$$

(since $j < 5, k < 5$), and 25 does not divide 16. Therefore Case 1 is impossible.
Case 2: $b_0 = 5$, $c_0 = 5$.
Then $75 = 5b_1 + 5c_1$, $-100 = 5b_2 + b_1 c_1 + 5c_2$. The first of these equalities tells us that if either b_1 or c_1 is divisible by 5 then both are divisible by 5. By the second equality, 5 divides $b_1 c_1$. Hence 5 divides both b_1 and c_1.

Similarly, we have $50 = 5b_3 + b_2 c_1 + b_1 c_2 + 5c_3$ and $-125 = 5b_4 + b_3 c_1 + b_2 c_2 + b_1 c_3 + 5b_4$. Exactly the same reasoning as before shows that 5 divides b_2 and c_2, again a contradiction with $(*)$, since 5 does not divide 16.

Solution to 6.11.27: Let $f(x) = x^3 + x + 2$. A calculation shows that 2 is a zero of $f(x)$ over $\mathbb{Z}_3$, but 0 and 1 are not. Hence, we get the factorization $f(x) = (x - 2)(x^2 + 2x + 2) = (x - 2)g(x)$. Clearly, 0 and 1 are not roots of $g(x)$ since they are not roots of $f(x)$; another calculation shows that 2 is not a root of $g(x)$. Hence, $g(x)$ is irreducible, and the above factorization is the desired one.

Solution to 6.11.28: Let $f = x^4 + x^3 + x + 3$. A calculation shows that f has no zeros in $\mathbb{Z}_5$, so it contains no linear factors. Consider a product of two monic quadratics. Now $3 = 3 \cdot 1 = 2 \cdot 4$. If $f = (x^2 + ax + 1)(x^2 + bx + 3)$, then, equating coefficients of powers of x, we get $a+b = 1, 4+ab = 0$ and $3a+b = 1$. This equation has no solutions in $\mathbb{Z}_5$. If $f = (x^2 + ax + 2)(x^2 + bx + 4)$, then $a + b = 1, 1 + ab = 0$ and $4a + 2b = 1$. This equation has no solutions in $\mathbb{Z}_5$. Thus $x^4 + x^3 + x + 3$ is irreducible in $\mathbb{Z}_5[x]$.

Solution to 6.11.29: 1. There are five distinct monic irreducible polynomials of degree 1: x, $x + 1$, $x + 2$, $x + 3$, $x + 4$. Multiplying these in pairs we obtain reducible monic polynomials of type p^2 or type $p_1 p_2$ with $p_1 \neq p_2$. There are $5 + \binom{5}{2} = 15$ distinct reducible monic polynomials of degree 2. The number of monic polynomials $x^2 + ax + b$ is 25. Therefore there are 10 monic irreducible polynomials of degree 2.

2. There are 5 monic irreducible polynomials of degree 1 denoted p, and 10 of degree 2 denoted q. A reducible polynomial of degree 3 has the form p^3 (5 of these), or $p_1^2 p_2$ ($5 \times 4 = 20$ of these), or $p_1 p_2 p_3$ ($\binom{5}{3} = 10$ of these), or pq ($5 \times 10 = 50$ of these). There are $5^3 = 125$ monic polynomials $x^3 + ax^2 + bx + c$ of degree 3. Thus there are $125 - 85 = 40$ monic irreducible polynomials of degree 3.

Solution to 6.11.30: $x^4 + 1 = (x^2 + 1)^2 - 2x^2 = (x^2 - \sqrt{2}x + 1)(x^2 + \sqrt{2}x + 1)$. Thus, $x^4 + 1$ is reducible over the real numbers.

The above expression is the irreducible factorization of $x^4 + 1$ over $\mathbb{R}$. The irreducible factorization over $\mathbb{Q}$ can be obtained from this by suitably grouping factors and can only be $x^4 + 1$ itself. Thus $x^4 + 1$ is irreducible over the rationals.

The field $\mathbf{F}_{16}$ has characteristic 2; that is, $1 + 1 = 0$. Thus $x^4 + 1 = (x + 1)^4$ is reducible over $\mathbf{F}_{16}$.

Solution to 6.11.31: In this solution we use the fact that a polynomial of degree 2 or 3 is reducible over a field iff it has a root there.

Decomposing $x^4 - 4$:

- over $\mathbb{R}$, we have

$$x^4 - 4 = (x - \sqrt{2})(x + \sqrt{2})(x^2 + 2),$$

where $x^2 + 2$ is irreducible.

- over $\mathbb{Z}$, we have

$$x^2 - 4 = (x^2 - 2)(x^2 + 2)$$

where both $x^2 - 2$ and $x^2 + 2$ are irreducible.

- over $\mathbb{Z}_3$, we have

$$x^4 - 4 = x^4 - 1 = (x - 1)(x^3 + x^2 + x + 1) = (x - 1)(x + 1)(x^2 + 1)$$

and $x^2 + 1$ is irreducible.

Decomposing $x^3 - 2$:

- over $\mathbb{R}$, we have

$$x^3 - 2 = (x - \sqrt[3]{2})(x^2 + \sqrt[3]{2}x + \sqrt[3]{4})$$

and $x^2 + \sqrt[3]{2}x + \sqrt[3]{4}$ is irreducible.

- over $\mathbb{Z}$, $x^3 - 2$ has no roots, and is irreducible.

- over $\mathbb{Z}_3$, we have

$$x^3 - 2 = x^3 + 1 = (x+1)(x^2 - x + 1) = (x+1)(x+1)(x+1) = (x+1)^3.$$

Solution to 6.11.32: Suppose, to the contrary, that $x^p - a$ has nontrivial factors $f(x)$ and $g(x)$ in $\mathbf{F}[x]$. Let $\mathbf{K}$ be a splitting field of $x^p - a$. Then, in $\mathbf{K}$, there are elements $a_1, \ldots, a_p$ such that $x^p - a = (x - a_1) \cdots (x - a_p)$. We may assume without loss of generality that $f(x) = (x - a_1) \cdots (x - a_k)$ and $g(x) = (x - a_{k+1}) \cdots (x - a_p)$. Therefore, $A = a_1 \cdots a_k = \pm f(0)$ and $B = a_{k+1} \cdots a_p = \pm g(0)$ are both elements of $\mathbf{F}$. Further, since the a_j's are zeros of $x^p - a$, $a_j^p = a$ for all j. Hence, $A^p = a^k$ and $B^p = a^{p-k}$. Since k and p are relatively prime, there exist integers x and y such that $kx + py = 1$. Let $r = -y$ and $s = x + y$. Then A^s / B^r is an element of $\mathbf{F}$ and

$$\left(\frac{A^s}{B^r}\right)^p = \frac{a^s k}{a^{rp - rk}} = a^{kx + py} = a.$$

Hence, a is a pth power, contradicting our assumptions. Therefore, $x^p - a$ must be irreducible over $\mathbf{F}$.

Solution to 6.11.33: Suppose $g(x)$ (or $h(x)$) is not irreducible. Then $x^4 + 1$ has a linear factor and, hence, a zero. In other words, there is an element $a \in \mathbb{Z}_p$ with $a^4 = -1$. It follows that $a^8 = 1$, and since $a^4 \neq 1$ (p is odd), a has order 8 in the multiplicative group $\mathbb{Z}_p^*$ of the field with p elements. But $\mathbb{Z}_p^*$ is a group of order $p - 1 \equiv 2 \pmod{4}$, so 8 cannot divide $p - 1$, and we have a contradiction.

Solution to 6.11.34: Let $\deg f = n$. Then the collection of all real polynomials of $\deg \leqslant n - 1$ is an m–dimensional vector space. Hence, any collection of $n + 1$ polynomials of degree $\leqslant n - 1$ is linearly dependent. By the Euclidean Algorithm [Her75, p. 155], we have

$$
\begin{aligned}
x^{2^0} &= q_0(x) f(x) + r_0(x) \quad \text{with } \deg r_0 < n \\
x^{2^1} &= q_1(x) f(x) + r_1(x) \quad \text{with } \deg r_1 < n \\
&\vdots \\
x^{2^n} &= q_n(x) f(x) + r_n(x) \quad \text{with } \deg r_n < n.
\end{aligned}
$$

The polynomials $r_0, \ldots, r_n$ are linearly dependent, so, for some $\alpha_i \in \mathbb{R}$, we have

$$\sum \alpha_i r_i(x) = 0.$$

Therefore,

$$p(x) = \sum \alpha_i x^{2^i} = f(x) \sum \alpha_i q_i(x)$$

and $f(x) | p(x)$.

Solution to 6.11.35: Fix $f \in R \setminus \mathbf{F}$ of least degree n, say. Choose polynomials $f_1, f_2, \ldots, f_{n-1}$ in R such that deg $f_j \equiv j \pmod{n}$ and such that each f_j is of least degree with this property, $1 \leqslant j \leqslant n - 1$, if such a polynomial exists, otherwise take $f_j = 0$. Let $f_n = f$. We will prove that $R = \mathbf{F}[f_1, f_2, \ldots, f_n]$. Suppose not, and fix $g \in R \setminus \mathbf{F}[f_1, f_2, \ldots, f_n]$ of least degree, and suppose deg $g \equiv j \pmod{n}$. For some $k \geqslant 0$, deg $g = \deg(f_n^k f_j)$. Hence, $g - \alpha f_n^k f_j$ is of lower degree than g for some $\alpha \in \mathbf{F}$, and, by the minimality of g, must lie in $\mathbf{F}[f_1, f_2, \ldots, f_n]$. However, this implies that $g \in \mathbf{F}[f_1, f_2, \ldots, f_n]$ as well.

6.12 Fields and Their Extensions

Solution to 6.12.1: Since the 2×2 matrices over a field form a ring, to show that R is a commutative ring with 1 it suffices to show that it is closed under addition and multiplication, commutative with respect to multiplication, and contains I. The first and the last are obvious, and the second follows almost as quickly from

$$\begin{pmatrix} a & -b \\ b & a \end{pmatrix}\begin{pmatrix} c & -d \\ d & c \end{pmatrix} = \begin{pmatrix} ac - bd & -ad - bc \\ ad + bc & ac - bd \end{pmatrix} = \begin{pmatrix} c & -d \\ d & c \end{pmatrix}\begin{pmatrix} a & -b \\ b & a \end{pmatrix}.$$

The inverse of a nonzero element in R is given by

$$\frac{1}{a^2 + b^2}\begin{pmatrix} a & b \\ -b & a \end{pmatrix}$$

which lies in R provided that $a^2 + b^2 \neq 0$. If $\mathbf{F} = \mathbb{Q}$, then $a^2 + b^2 > 0$ for a and b not both equal to 0; hence, every nonzero element of R has an inverse, so R is a field. If $\mathbf{F} = \mathbb{C}$, then the matrix

$$\begin{pmatrix} i & -1 \\ 1 & i \end{pmatrix}$$

has no inverse since $i^2 + 1^2 = 0$. Therefore, in this case, R is not a field. Similarly, if $\mathbf{F} = \mathbb{Z}_5$, we have that $2^2 + 1^2 = 0$, so there exists a noninvertible matrix in R and so R is not a field. Finally, if $\mathbf{F} = \mathbb{Z}_7$, the equation $a^2 + b^2 = 0$ has no nonzero solutions, so every nonzero element of R has an inverse and R is a field.

Solution to 6.12.2: Let R be a finite integral domain. Let $0 \neq b \in R$ and enumerate the elements of R by $c_1, c_2 \ldots, c_n$. Since R, has no zero divisors, cancellation shows that the elements bc_i are distinct. Since there are n of them, it follows that there is an element c_{i_0} such that $bc_{i_0} = 1$. Hence, c_{i_0} is the inverse of b and we are done.

Solution to 6.12.3: Since a and b are algebraic over $\mathbf{F}$, there exist integers n and m such that $[\mathbf{F}(a) : \mathbf{F}] = n$ and $[\mathbf{F}(b) : \mathbf{F}] = m$. Because b is algebraic over $\mathbf{F}$, it must also be algebraic over $\mathbf{T} = \mathbf{F}(a)$ of degree, at most, m. Hence, $[\mathbf{T}(b) : \mathbf{F}(a)] \leqslant m$, which implies

$$[\mathbf{T}(b) : \mathbf{F}] = [\mathbf{T}(b) : \mathbf{F}(a)][\mathbf{F}(a) : \mathbf{F}] \leqslant nm.$$

Therefore, $\mathbf{T}(b)$ is a finite extension of $\mathbf{F}$. $\mathbf{T}(b)$ contains $a + b$ and so contains $\mathbf{F}(a+b)$; the latter must, therefore, be a finite extension of $\mathbf{F}$, so $a + b$ is algebraic over $\mathbf{F}$.

Solution to 6.12.4: From the fact that the group is finite, we can see that all elements are in the unit circle, otherwise consecutive powers would make an infinite sequence.

All elements of the group are roots of unity (maybe of different degrees), since a high enough power will end up in 1 (the order of an element always divides the order of the group). We will prove that they are all roots of unity of the same degree. Let α be the element with smallest positive argument ($\arg \alpha \in (0, 2\pi)$). We will show that this element generates the whole group. Suppose that there is an element β that is not in the group generated by α. There is a $p \in \mathbb{N}$ such that

$$\arg \alpha^p < \arg \beta < \arg \alpha^{p+1}$$

therefore,

$$\arg(\beta \alpha^{-p}) < \arg \alpha$$

which contradicts the minimality of $\arg \alpha$. We conclude then that the group is generated by α. See also the Solution to Problem 6.12.5.

Solution to 6.12.5: Let G be a finite subgroup of $\mathbf{F}^*$ of order n. By the Structure Theorem for finite abelian groups [Her75, p. 109], there are integers $m_1 | m_2 | \cdots | m_k$ such that G is isomorphic to $\mathbb{Z}_{m_1} \times \cdots \times \mathbb{Z}_{m_k}$. To show that G is cyclic, it suffices to show that $m_k = n$. Suppose that $m_k < n$. From the structure of G we know that $g^{m_k} = 1$ for every $g \in G$. Hence, the polynomial $x^{m_k} - 1$ has n roots, contradicting the fact that a polynomial over a field has no more roots than its degree. Hence, $m_k = n$ and G is cyclic.

Solution to 6.12.6: Let $\mathbf{K}_+$ denote the additive group of $\mathbf{K}$. By the Structure Theorem on finitely generated abelian groups [Her75, p. 109], $\mathbf{K}_+ \cong \mathbb{Z}^n \oplus T$, where $n \geqslant 0$ and T is a finite abelian group. Then $2\mathbf{K}_+ \cong (2\mathbb{Z})^n \oplus (2T)$. If $n > 0$, then $2\mathbf{K}$ is an ideal of $\mathbf{K}$ not equal to $\{0\}$ or $\mathbf{K}$, but this contradicts the assumption that $\mathbf{K}$ is a field. Thus $n = 0$, and $\mathbf{K} \cong T$ is finite.

Solution to 6.12.7: Since $x^3 - 2$ is irreducible over $\mathbb{Q}[x]$, $\langle x^3 - 2 \rangle$ is a maximal ideal in $\mathbb{Q}[x]$, so $\mathbf{F} = \mathbb{Q}[x]/\langle x^3 - 2 \rangle$ is a field. Using the relationship $\overline{x}^3 = 2$, we get that every element of $\mathbf{F}$ can be written in the form $a + b\overline{x} + c\overline{x}^2$, where $a, b, c \in \mathbb{Q}$. Further, such a representation is unique, since otherwise we could find $a, b, c \in \mathbb{Q}$, not all 0, with $a + b\overline{x} + c\overline{x}^2 = 0$. On pulling back to $\mathbb{Q}[x]$, we find that $x^3 - 2$ divides $a + bx + cx^2$, a contradiction.

Consider the map $\varphi : \mathbf{F} \to \mathbf{F}$ given by $\varphi(a) = a$ if $a \in \mathbb{Q}$, and $\varphi(\sqrt[3]{2}) = \bar{x}$. Since $(\sqrt[3]{2})^3 = 2$ and $\bar{x}^3 = 2$ this extends to a ring epimorphism in the obvious way. It is also one-to-one, since if $\varphi(a+b\sqrt[3]{2}+c\sqrt[3]{4}) = 0$ then $a+b\bar{x}+c\bar{x}^2 = 0$, so $a = b = c = 0$. Hence, $\mathbf{F}$ is the isomorphic image of a field, and it is field.

Further, by the isomorphism we see that every element can be expressed uniquely in the desired form. In particular, $(1 - \sqrt[3]{2})(-1 - \sqrt[3]{2} - \sqrt[3]{4}) = 1$.

Solution to 6.12.8: Consider the ring homomorphism $\varphi : \mathbb{Z} \to \mathbf{F}$ that satisfies $\varphi(1) = 1$. Since $\mathbf{F}$ is finite, $\ker \varphi \neq 0$. Since $\mathbf{F}$ is a field, it has characteristic p, where p is a prime number. Hence, $\ker \varphi$ contains the ideal $\langle p \rangle$, which is maximal. Therefore, $\ker \varphi = \langle p \rangle$, and the image of $\mathbb{Z}$ is a subfield of $\mathbf{F}$ isomorphic to $\mathbb{Z}_p$. Identify $\mathbb{Z}_p$ with this subfield. Then $\mathbf{F}$ is a vector space over $\mathbb{Z}_p$, and it must be of finite dimension since $\mathbf{F}$ is finite. Let $\dim \mathbf{F} = r$. A counting argument shows that $\mathbf{F}$ has p^r elements.

Solution to 6.12.9: Let a be any element of $\mathbf{K}$. Then $\mathbf{L}$ contains a^2 and $(a + 1)^2$, hence it contains

$$(a + 1)^2 - a^2 - 1 = 2a .$$

Since $2 \neq 0$, $\mathbf{L}$ contains a.

If $\mathbf{K}$ is the finite field of order 2^n, then the multiplicative group of $\mathbf{K}$ is cyclic of order $2^n - 1$, an odd number. The homomorphism $a \mapsto a^2$ on this group has a trivial kernel, hence it is surjective, therefore, every element of $\mathbf{K}$ is a square. In this case, then, $\mathbf{L} = \mathbf{K}$.

To obtain a field of characteristic 2 in which the equality $\mathbf{L} = \mathbf{K}$ can fail, consider $\mathbf{K} = \mathbf{F}_2(x)$, the field of rational functions with coefficients in $\mathbf{F}_2$, the field of order 2. For $r(x)$ in $\mathbf{F}_2(x)$ we have $r(x)^2 = r(x^2)$, so that the subfield $\mathbf{L} = \{r(x^2) \mid r(x) \in \mathbf{K}\}$ is the desired counterexample.

Solution to 6.12.10: Since $\mathbf{F}$ has characteristic p, it contains a subfield isomorphic to $\mathbb{Z}_p$, which we identify with $\mathbb{Z}_p$. For $j \in \mathbb{Z}_p$, $\alpha + j$ is an element of $\mathbf{F}(\alpha)$. Using the identity $(\alpha + j)^p = \alpha^p + j^p$, we get

$$f(\alpha + j) = \alpha^p - \alpha + 3 + j^p - j = 0.$$

Therefore, f has p roots in $\mathbf{F}(\alpha)$, which are clearly distinct.

Solution to 6.12.11: The polynomial $x^{10}-2$ is irreducible over $\mathbb{Q}$, by Eisenstein's Criterion. Hence

$$\mathbb{Q}\left(\sqrt[10]{2}\right) = \mathbb{Q}[x]/\langle x^{10} - 2 \rangle$$

has degree 10 over $\mathbb{Q}$.

Let φ be an automorphism of $\mathbf{K}$. It is well known that φ acts like the identity on $\mathbb{Q}$. Let $\alpha = \sqrt[10]{2}$. Then α and $-\alpha$ are the only roots of $x^{10} - 2$ in $\mathbb{R}$ and hence in $\mathbf{K}$. Therefore, either $\varphi(\alpha) = \alpha$ or $\varphi(\alpha) = -\alpha$. In the first case φ is the identity. It is easy to see that $p(\alpha) \mapsto p(-\alpha)$ ($p \in \mathbb{Q}[x]$) defines an automorphism of $\mathbf{K}$. Thus, $\mathbf{K}$ has exactly two automorphisms.

Solution to 6.12.12: Notice that $\mathbb{Q}[x]$ is an Euclidean domain, so the ideal $\langle f, g \rangle$ is generated by a single polynomial, h say. f and g have a common root if and only if that root is also a root of h. We can use the Euclidean Algorithm to find $h = x^2 + 3x + 1$, which has roots $-3/2 \pm \sqrt{5}/2$. Therefore, f and g have exactly two common roots, both in $\mathbb{Q}(\sqrt{5})$.

Solution to 6.12.13: Let x be an element of $\mathbf{F}$ that is not in $\mathbb{Q}$. Then x satisfies an equation $ax^2 + bx + c = 0$ with a, b, and $c \in \mathbb{Q}$. Completing the square, we see that $\left(x + \frac{a}{2}\right)^2 \in \mathbb{Q}$, whereas $\left(x + \frac{a}{2}\right) \notin \mathbb{Q}$. Let $\xi = \left(x + \frac{a}{2}\right)$. As $\xi^2 \in \mathbb{Q}$, we can write it as $\frac{c^2}{d^2}m$, where $c, d, m \in \mathbb{Z}$ and m is square free (i.e., m has no multiple prime factors). As $\mathbf{F} = \mathbb{Q}(\xi) = \mathbb{Q}\left(\frac{d}{c}\xi\right)$, we get an isomorphism $\mathbf{F} \to \mathbb{Q}(\sqrt{m})$ by sending $r + s\left(\frac{d}{c}\xi\right)$ to $r + s\sqrt{m}$. The uniqueness of m follows from the fact that the elements of $\mathbf{F}$ that are not in $\mathbb{Q}$ but whose squares are in $\mathbb{Q}$ are those of the form $k\xi$ for some nonzero $k \in \mathbb{Q}$.

Solution to 6.12.14: Let $p_1 = 2$, $p_2 = 3$, $\ldots$, be the prime numbers and $\mathbf{F}_i = \mathbb{Q}(\sqrt{p_i})$. *Claim:* The fields $\mathbf{F}_i$ are pairwise nonisomorphic. Indeed, if $\mathbf{F}_i$ were isomorphic to $\mathbf{F}_j$, then there would exist $r \in \mathbf{K}_j$ such that $r^2 = p_i$. Write such an r in the form

$$r = a + b\sqrt{p_j} \qquad a, b \in \mathbb{Q}.$$

Then

$$r^2 = a^2 + b^2 p_j + 2ab\sqrt{p_j} = p_i$$

if and only if $ab = 0$. Therefore, either $p_i = c^2$ or $p_j = c^2$ for some $c \in \mathbb{Q}$, which contradicts the primality of p_i and p_j.

Solution to 6.12.15:

$$\left(\cos\frac{\theta}{3} + i\sin\frac{\theta}{3}\right)^3 = \cos\theta + i\sin\theta$$

$$\Rightarrow 3\sin\frac{\theta}{3} - 4\sin^3\frac{\theta}{3} = \sin\theta$$

$$\Rightarrow E_\theta \supset F_\theta$$

All the possibilities can occur. For example

$$\dim_{F_\theta} E_\theta = 1 \quad \text{if} \quad \theta = \frac{\pi}{2}$$

$$\dim_{F_\theta} E_\theta = 2 \quad \text{if} \quad \theta = \pi$$

$$\dim_{F_\theta} E_\theta = 3 \quad \text{if} \quad \theta = \frac{\pi}{6}$$

In the last example, $4x^3 - 3x + 1/2$ or $8x^3 - 6x + 1$ is irreducible because ± 1, $\pm 1/2$, $\pm 1/4$, and $\pm 1/8$ are not roots of the above polynomial.

Solution to 6.12.16: We first prove, using the Induction Principle [MH93, p. 7], that there are n algebraically independent ([BML97, p. 73]) real numbers. The case $n = 1$ is obvious. Suppose $\alpha_1, \ldots, \alpha_{n-1}$ are algebraically independent. Then $\mathbb{Q}(\alpha_1, \ldots, \alpha_{n-1})$ is countable, and so is the the set of numbers algebraic over it. Let α_n be transcendent over $\mathbb{Q}(\alpha_1, \ldots, \alpha_{n-1})$. Then clearly $\alpha_1, \ldots, \alpha_{n-1}, \alpha_n$ are independent too.

Let $\psi : \mathbb{Q}(t_1, \ldots, t_n) \to \mathbb{R}$ be defined by

$$\frac{p(t_1, \ldots, t_n)}{q(t_1, \ldots, t_n)} \mapsto \frac{p(\alpha_1, \ldots, \alpha_n)}{q(\alpha_1, \ldots, \alpha_n)}.$$

ψ is clearly a homomorphism. If

$$\frac{p_1(\alpha_1, \ldots, \alpha_n)}{q_1(\alpha_1, \ldots, \alpha_n)} = \frac{p_2(\alpha_1, \ldots, \alpha_n)}{q_2(\alpha_1, \ldots, \alpha_n)}$$

then $(p_1 q_2 - q_2 p_1)(\alpha_1, \ldots, \alpha_n) = 0$ and the independence of the α's gives $p_1/q_1 = p_2/q_2$, so ψ is injective, therefore it is an isomorphism over its range.

Solution to 6.12.18: We'll show that $\mathbf{K}[t^a, t^b]$ is a unique factorization domain, UFD for short, if and only if one of a and b divides the other. Let $d = \gcd(a, b)$, $a' = a/d$ and $b' = b/d$. Then $\mathbf{K}[t^a, t^b]$ is the image of $\mathbf{K}[u^{a'}, u^{b'}]$ under the injective $\mathbf{K}$–algebra homomorphism $\mathbf{K}[u] \to \mathbf{K}[t]$ mapping u to t^d, so we may reduce to the case that $\gcd(a, b) = 1$. If $a = 1$ or $b = 1$, then $\mathbf{K}[t^a, t^b] = \mathbf{K}[t]$ is a UFD. It remains to show that if $a, b > 1$ and $\gcd(a, b) = 1$, then $\mathbf{K}[t^a, t^b]$ is not a UFD.

First we show that t^a is an irreducible element of $\mathbf{K}[t^a, t^b]$. Since t is an irreducible element of the UFD $\mathbf{K}[t]$ with unit group $\mathbf{K}^*$, the only ways to factor t^a into nonunits of $\mathbf{K}[t]$ are as $(ct^m)(c^{-1}t^{a-m})$ for some $c \in \mathbf{K}^*$ and $0 < m < a$. But m and $a - m$ cannot both be nonnegative integer combinations of a and b (they would have to be positive multiples of b, and their sum then could not be a), so ct^m and $c^{-1}t^{a-m}$ cannot both lie in $\mathbf{K}[t^a, t^b]$. Thus t^a is irreducible in $\mathbf{K}[t^a, t^b]$. Similarly t^b is irreducible in $\mathbf{K}[t^a, t^b]$. Moreover t^a and t^b are not associate in $\mathbf{K}[t^a, t^b]$ because even in $\mathbf{K}[t]$ it is not true that each of them divides the other. Now t^{ab} can be factored in $\mathbf{K}[t^a, t^b]$ either as $(t^a)^b$ or as $(t^b)^a$, violating unique factorization.

As an alternative to the argument in the previous paragraph, one could note that since $\gcd(a, b) = 1$, there exist $r, s \in \mathbb{Z}$ with $ra + sb = 1$, so the fraction field of $\mathbf{K}[t^a, t^b]$ (viewed as a subfield of $\mathbf{K}(t)$) contains $t = (t^a)^r (t^b)^s$, even though $\mathbf{K}[t^a, t^b]$ obviously does not contain t if $a, b > 1$. But t is integral over $\mathbf{K}[t^a, t^b]$ because it satisfies the polynomial equation $x^a - t^a = 0$. Hence $\mathbf{K}[t^a, t^b]$ is not integrally closed (in its fraction field). UFD's are integrally closed, so $\mathbf{K}[t^a, t^b]$ is not a UFD.

Solution to 6.12.19: 1. A 2×2 matrix over $\mathbf{F}$ is invertible if and only if the first column is nonzero and the second is not a $\mathbf{F}$–multiple of the first. This gives $|\mathbf{F}|^2 - 1 = p^{2n} - 1$ possibilities for the first column of an invertible matrix, and, given the first column, $|\mathbf{F}|^2 - |\mathbf{F}| = p^{2n} - p^n$ for the second, hence the result. See also Problem 7.1.3 for a more general solution.

2. The map $\mathbf{F} \to G$ sending a to $\left(\begin{smallmatrix} 1 & a \\ 0 & 1 \end{smallmatrix}\right)$ is easily checked to be an injective group homomorphism. Its image is a subgroup S of G that is isomorphic to the additive group of $\mathbf{F}$ and that, consequently, has order p^n. By 1., this is the largest power of p dividing the order of G. Hence, S is, in fact, a p-Sylow subgroup of G. Since all p-Sylow subgroups of a finite group are conjugate (and hence isomorphic), this implies the result.

Solution to 6.12.20: Assume $g \in GL_n(\mathbf{F}_p)$ has order p^k, and let $m(x)$ be the minimal polynomial of $g - 1$. By Fermat's Little Theorem [Sta89, p. 80], [Her75, p. 44], $(x - 1)^p = x^p - 1 \pmod{p}$, and iteration gives $(x - 1)^{p^\ell} = x^{p^\ell} - 1 \pmod{p}$ for all positive integers ℓ. Since $g^{p^k} - 1 = 0$, also $(g - 1)^{p^k} = 0$, so $m(x)$ divides $(x - 1)^{p^k}$. By the same token, $m(x)$ does not divide $(x - 1)^{p^{k-1}}$. Therefore $m(x) = (x - 1)^j$ with $p^{k-1} < j \leqslant p^k$. But $j = \deg m(x) \leqslant n$, so the desired inequality follows.

For the other direction, assume first $p^{k-1} < n \leqslant p^k$. Let g be the $n \times n$ matrix with 1 in each position on the main diagonal and immediately above the main diagonal, 0 elsewhere. Then $g - 1$ is nilpotent of index n, so

$$0 = (g - 1)^{p^k} = g^{p^k} - 1,$$
$$0 \neq (g - 1)^{p^{k-1}} = g^{p^{k-1}} - 1.$$

It follows that the order of g divides p^k but does not divide p^{k-1}, so it must equal p^k.

Finally, if $n > p^k$, we can get a g of order p^k by taking the direct sum of the preceding g (corresponding to $n = p^k$) and an identity matrix of size $(n - p^k) \times (n - p^k)$.

Solution to 6.12.21: 1. We can write down the general commutator:

$$\begin{pmatrix} a & b \\ 0 & a^{-1} \end{pmatrix} \begin{pmatrix} c & d \\ 0 & c^{-1} \end{pmatrix} \begin{pmatrix} a^{-1} & -b \\ 0 & a \end{pmatrix} \begin{pmatrix} c^{-1} & -d \\ 0 & c \end{pmatrix}$$
$$= \begin{pmatrix} ac & ad + bc^{-1} \\ 0 & a^{-1}c^{-1} \end{pmatrix} \begin{pmatrix} a^{-1}c^{-1} & -a^{-1}d - bc \\ 0 & ac \end{pmatrix}$$
$$= \begin{pmatrix} 1 & -cd - abc^2 + a^2cd + ab \\ 0 & 1 \end{pmatrix}$$

The upper-right entry can be rewritten as $ab(1 - c^2) - cd(1 - a^2)$. This is 0 if $\mathbf{F}$ is the field of two or the field of three elements. Otherwise we can take $d = 0$ and c such that $1 - c^2 \neq 0$. Then, by suitably choosing a and b, we can make the upper-right entry equal any element of $\mathbf{F}$. Conclusion: the commutators are the matrices $\left(\begin{smallmatrix} 1 & x \\ 0 & 1 \end{smallmatrix}\right)$ with x in $\mathbf{F}$, except in case $\mathbf{F}$ is the field of two or the field of three elements, in which case the identity is the only commutator. In either case, the commutators form a group, which is the commutator subgroup.

2. Except in the trivial cases $p = 2$ or 3, $k = 1$, the commutator subgroup is isomorphic to the additive group of $\mathbf{F}$, which is a vector space of dimension k over the field of p elements. The fewest number of generators is thus k.

Solution to 6.12.22: The zero element of $\mathbf{F}_p$ obviously has a unique square root and a unique cube root. Let $\mathbf{F}_p^*$ denote the multiplicative group of nonzero elements of $\mathbf{F}_p$. It is a cyclic group of order $p - 1$. Since $p - 1$ is even, the homomorphism $x \to x^2$ of $\mathbf{F}_p^*$ into itself has a kernel of order 2, which means that its range has order $(p - 1)/2$. There are, thus, $1 + (p - 1)/2$ elements of $\mathbf{F}_p$ with square roots.

If $p - 1$ is not divisible by 3, the homomorphism $x \to x^3$ of $\mathbf{F}_p^*$ into itself has a trivial kernel, and so every element of $\mathbf{F}_p$ has a cube root. If 3 divides $p - 1$, then the preceding homomorphism has a kernel of order 3, so its range has order $(p - 1)/3$. In this case, there are $1 + (p - 1)/3$ elements of $\mathbf{F}_p$ with cube roots.

Solution to 6.12.23: All functions are polynomials. A polynomial with the value 1 at 0 and 0 elsewhere is $p(x) = 1 - x^{q-1}$; from this one, we can construct any function by considering sums $\sum f_i \cdot p(x - x_i)$. Thus, there are q^q such functions, and that many polynomials. Another way is to observe that all polynomials of degree, at most, $q - 1$ define nonzero functions unless the polynomial is the zero polynomial.

Solution to 6.12.24: Let $\mathbf{K}$ be that subfield. The homomorphism of multiplicative groups $\mathbf{F}^* \to \mathbf{K}^*$ sending x to x^3 has a kernel of order, at most, 3, so $|\mathbf{K}^*| \geqslant |\mathbf{F}^*|/3$, that is, $(|\mathbf{F}| - 1)/(|\mathbf{K}| - 1) \leq 3$. Also, if the extension degree $[\mathbf{F} : \mathbf{K}]$ equals n, then $n \geqslant 2$, so $|\mathbf{F}| = |\mathbf{K}|^n \geqslant |\mathbf{K}|^2$, and $(|\mathbf{F}| - 1)/(|\mathbf{K}| - 1) \geqslant |\mathbf{K}| + 1$, with equality if and only if $n = 2$. Thus, $3 \geqslant |\mathbf{K}| + 1$, which gives $|\mathbf{K}| = 2, n = 2$, and $|\mathbf{F}| = 2^2 = 4$.

Solution to 6.12.25: As A has dimension at least 2 as a vector space over $\mathbb{C}$, it contains an element a which is not in the subspace spanned by the identity element $1 \in A$. Since A has finite dimension over $\mathbb{C}$, there exists a complex number λ such that $(a - \lambda 1)x = 0$ for some nonzero $x \in A$. Let $\mathfrak{I}$ be the ideal generated by $b = a - \lambda 1$, and $\mathfrak{J} = \{x \in A \mid bx = 0\}$ the annihilator of b. We have $\mathfrak{I} \cap \mathfrak{J} = \{0\}$ since all the elements of $\mathfrak{I} \cap \mathfrak{J}$ have zero square. As the dimensions of $\mathfrak{I}$ and $\mathfrak{J}$ add up to the dimension of A we must have $A = \mathfrak{I} \oplus \mathfrak{J}$ as vector spaces over $\mathbb{C}$. Since $\mathfrak{I}$ and $\mathfrak{J}$ are ideals in A, $A = \mathfrak{I} \oplus \mathfrak{J}$ as rings. Let $1 = e + f$ with $e \in \mathfrak{I}$ and $f \in \mathfrak{J}$. Then $e^2 = e$, $f^2 = f$ and neither of them is zero or 1.

Solution to 6.12.26: Suppose that A^* has order 5. We have $-1 = 1$ in A, since otherwise A^* would have even order. Hence we have a homomorphism $\varphi : \mathbf{F}_2[x] \to A$ sending x to a generator of A^*. The kernel of φ is generated by a polynomial dividing $x^5 - 1$ in $\mathbf{F}_2[x]$, but not dividing $x - 1$, since the image of x is not 1. Now $x^5 - 1 = (x - 1)(x^4 + x^3 + x^2 + x + 1)$, and the latter is irreducible over $\mathbf{F}_2$, since otherwise it would have a factor of degree $\leqslant 2$, and $\mathbf{F}_2$ or $\mathbf{F}_4$ would contain a nontrivial fifth root of unity. Hence $\ker \varphi$ equals $(x^4 + x^3 + x^2 + x + 1)$

or $((x-1)(x^4+x^3+x^2+x+1))$ and A contains a subring $B = \mathbf{F}_2[x]/\ker\varphi$ isomorphic to $\mathbf{F}_{2^4}$ or $\mathbf{F}_2 \times \mathbf{F}_{2^4}$. In either case, $|B^*| = 15$, a contradiction.

Solution to 6.12.27: The relations imply $ab + ba = 1$. In particular $a \neq 0 \neq b$. Multiplying the equality $ab + ba = 1$ (from either side) by a gives $aba = a$, so $ab \neq 0 \neq ba$. Multiplying the same equality by b gives $bab = b$. The three equalities $ab + ba = 1$, $aba = a$, $bab = b$, imply that the four elements a, b, ab, ba generate R as a vector space over $\mathbf{K}$. It is asserted that these four elements are linearly independent. In fact, suppose u, v, w, x are in $\mathbf{K}$ and

$$ua + vb + wab + xba = 0.$$

Multiplying on the right by b, we get $uab + xb = 0$. Multiply this on the left by a to get $xab = 0$, whence $x = 0$. Thus $uab = 0$, whence $u = 0$. Similarly $v = w = 0$. Thus R has dimension 4 over $\mathbf{K}$.

Let T be the quotient of the noncommutative ring $\mathbf{K}[X, Y]$ by the two-sided ideal generated by X^2, Y^2 and $(X+Y)^2 - 1$. It follows from the above that T has dimension 4 and maps onto R, hence is isomorphic to R.

Now let $A = \begin{pmatrix} 0 & 1 \\ 0 & 0 \end{pmatrix}$ and $B = \begin{pmatrix} 0 & 0 \\ 1 & 0 \end{pmatrix}$. Then $A^2 = B^2 = 0$ and $(A+B)^2 = 1$ and the matrices A, B, AB, BA span $M_2(\mathbf{K})$. It follows from the above that $M_2(\mathbf{K}) \simeq T \simeq R$.

6.13   Elementary Number Theory

Solution to 6.13.1: Let the six people be Aline, Ana, Laura, Lucia, Manuel, and Stephanie. Fix one of them, Manuel, say. The five girls form two sets: X (Manuel's friends) and Y (the others). By the Pigeonhole Principle [Her75, p. 127], one of these sets has cardinality at least 3. Suppose it is X that contains at least three elements, say {Aline, Laura, Stephanie} $\subset X$. If Aline, Laura, and Stephanie are pairwise strangers, we are done. Otherwise, two of them are friends of each other, Stephanie and Aline, say. Then Manuel, Aline, and Stephanie are mutual friends. If the set with three or more elements is Y, a similar argument leads to the same conclusion.

Solution to 6.13.2: Suppose not. Changing signs, if necessary, we may assume that $M > 0$ where M is the maximum of the $u_{m,n}$'s. Define the set

$$A = \{(m, n) \mid u_{m,n} = M\} \subset \{2, 3, \ldots, N-1\} \times \{2, 3, \ldots, M-1\}.$$

and choose $(m, n) \in A$ with m minimal. Since $(m-1, n) \notin A$, we have

$$\frac{1}{4}(u_{m-1,n} + u_{m+1,n} + u_{m,n-1} + u_{m,n+1}) < \frac{1}{4}(M+M+M+M) = M = u_{m,n},$$

which contradicts the relation.

Solution to 6.13.3: There are $\binom{5}{3}$ possibilities for the range, so the answer is 10N, where N is the number of surjective functions from $\{1, 2, 3, 4, 5\}$ to a given 3-element set. The total number of functions $\{1, 2, \ldots, 5\} \to \{1, 2, 3\}$ is 3^5, from which we subtract $\binom{3}{2}$ (the number of 2-element subsets of $\{1, 2, 3\}$) times 2^5 (the number of functions mapping into that subset), but then (according to the Principle of Inclusion-Exclusion) we must add back $\binom{3}{1}$ (the number of functions mapping into a 1-element subset of $\{1, 2, 3\}$). Thus

$$N = 3^5 - \binom{3}{2}2^5 + \binom{3}{1}1^5 = 150$$

and the answer is $10N = 1500$.

Solution to 6.13.5: Let $x = 0.a_1a_2\ldots$ in base 3. If $a_j = 1$ for some j, choose the smallest such j, and define

$$x_- = 0.a_1a_2\ldots a_{j-1}022222\ldots, \quad x_+ = 0.a_1a_2\ldots a_{j-1}200000\ldots$$

These are the numbers from the Cantor set closest to x. Then $f(x_-) = f(x_+)$ so f is constant on $[x_-, x_+]$.

It suffices then to prove the inequality for $x = \sum a_j 3^{-j} \geqslant y = \sum b_j 3^{-j}$ with $a_j, b_j \in \{0, 2\}$. Let j_* be the smallest j with $a_j \neq b_j$. Then $|x - y| \geqslant 3^{-j_*}$. On the other hand, we have,

$$|f(x) - f(y)| = \left| \sum_{j \geqslant j_*} \frac{a_j - b_j}{2} 2^{-j} \right| \leqslant \sum_{j \geqslant j_*} 2^{-j} = 2 \cdot 2^{-j_*}.$$

Combining, we obtain,

$$|f(x) - f(y)| \leqslant 2 \cdot 2^{-j_*} \leqslant 2\left(3^{-j_*}\right)^{(\log 2)/(\log 3)} \leqslant 2|x - y|^{(\log 2)/(\log 3)}.$$

Solution to 6.13.6: If $a = 0$, the congruence has the trivial solution $x = 0$. For $1 \leqslant a \leqslant p - 1$, if $x^2 \equiv a \pmod{p}$, we have

$$(p - x)^2 = p^2 - 2xp + x^2 \equiv a \pmod{p}$$

so, for $a \neq 0$, there are two solutions of the quadratic congruence in each complete set of residues mod p. We conclude, then, that the total number is $1 + (p - 1)/2 = (p + 1)/2$.

Solution to 6.13.7: By Fermat's Little Theorem [Sta89, p. 80], [Her75, p. 44], we have, raising both sides of the congruence $-1 \equiv x^2 \pmod{p}$ to the power $(p - 1)/2$,

$$(-1)^{\frac{p-1}{2}} \equiv x^{p-1} \equiv 1 \pmod{p}$$

which implies that $(p - 1)/2$ is even, and the result follows.

Solution to 6.13.8: Let $f(n) = 2^n + n^2$. It suffices to show that $f(n)$ is composite if $n \equiv i \pmod 6$ for $i \neq 3$. If n is even, then $f(n)$ is an even number larger than 2, a composite. Let $n = 6k + 1$ for some integer k. We have

$$f(n) = 2^{6k+1} + 36k^2 + 12k + 1$$
$$\equiv (-1)^{6k}2 + 1$$
$$\equiv 0 \pmod 3,$$

so $f(n)$, being a multiple of 3 larger than 3, is a composite. For $n = 6k + 5$ we get

$$f(n) = 2^{6k+5} + 36k^2 + 60k + 25$$
$$\equiv (-1)^{6k}2^5 + 25$$
$$\equiv 57$$
$$\equiv 0 \pmod 3,$$

and again $f(n)$ is composite.

Solution to 6.13.9: 1. See the Solution to Part 1 of Problem 6.1.4.
2. The congruence $ka \equiv 1 \pmod n$ is equivalent to the equation $ka = mn + 1$ for some integer m. As all integer linear combinations of k and n are multiples of $\gcd\{k, n\}$, the first congruence has a solution iff $\gcd\{k, n\} = 1$.
3. Let φ be Euler's totient function [Sta89, p. 77], [Her75, p. 43]. As φ is multiplicative, we have

$$\varphi(n) = \varphi(p)\varphi(q) = (p - 1)(q - 1).$$

Solution to 6.13.10: Let $p(t) = 3t^3 + 10t^2 - 3t$ and $n/m \in \mathbb{Q}$; $\gcd\{n, m\} = 1$. We can assume $m \neq \pm 1$. If $p(n/m) = k \in \mathbb{Q}$, then $m|3$ and $n|k$. Therefore, we have $m = \pm 3$.
   Suppose $m = 3$. We have

$$p\left(\frac{n}{3}\right) = n\left(\frac{n^2}{9} + 10\frac{n}{9} - 1\right).$$

This expression represents an integer exactly when $n^2 + 10n = n(n + 10) \equiv 0 \pmod 9$. As $\gcd\{n, 3\} = 1$, this means $n + 10 \equiv 0 \pmod 9$, that is, $n \equiv 8 \pmod 9$.
   A similar argument for the case $m = -3$ shows that the numbers $n/(-3)$ with $n \equiv 1 \pmod 9$ produce integer values in p.

Solution to 6.13.11:

$$\binom{1/2}{n} = \frac{(\frac{1}{2})(\frac{1}{2} - 1)\cdots(\frac{1}{2} - (n - 1))}{n!} = \frac{(-1)^{n-1} \cdot 3 \cdot 5 \cdots (2n - 1)}{2^n n!}$$
$$= \frac{(-1)^{n-1} \cdot 2 \cdot 3 \cdot 4 \cdot 5 \cdots 2n}{(2^n n!)^2} = \frac{(-1)^{n-1}}{2^{2n}} \cdot \binom{2n}{n}.$$

Solution to 6.13.12: A counting argument shows that the power of 2 which divides $n!$ is given by

$$\sum_{k \geqslant 1} \left\lfloor \frac{n}{2^k} \right\rfloor,$$

where $\lfloor x \rfloor$ denotes the largest integer less than or equal to x. Since $c_n = \dfrac{(2n)!}{(n!)^2}$, to show that c_n is even it suffices to show that

$$\sum_{k \geqslant 1} \left\lfloor \frac{2n}{2^k} \right\rfloor > 2 \sum_{k \geqslant 1} \left\lfloor \frac{n}{2^k} \right\rfloor.$$

Suppose $2^r \leqslant n < 2^{r+1}$. For $k \leqslant r$, there is an r_k, $0 \leqslant r_k < 1$, such that

$$\frac{n}{2^k} = \left\lfloor \frac{n}{2^k} \right\rfloor + r_k$$

or

$$\frac{2n}{2^k} = 2 \left\lfloor \frac{n}{2^k} \right\rfloor + 2r_k$$

so

$$\left\lfloor \frac{2n}{2^k} \right\rfloor \geqslant 2 \left\lfloor \frac{n}{2^k} \right\rfloor$$

and equality holds if and only if n is a power of 2. For $k = r + 1$, we have that $\lfloor n/2^{r+1} \rfloor = 0$ while $\lfloor 2n/2^{r+1} \rfloor = 1$. Finally, for $k > r$, the terms in both sums are 0. Hence, we see that the above inequality holds. Further, we see that the left side is 2 or more greater than the right side (i.e., c_n is divisible by 4) if and only n is not a power of 2.

Solution to 6.13.13: We may assume that $n > 3$. Converting the sum into a single fraction, we get

$$\frac{n!/1 + n!/2 + \cdots + n!/n}{n!}.$$

Let r be such that $2^r | n!$ but 2^{r+1} does not divide $n!$, and s be such that 2^s is the largest power of 2 less than or equal to n. Since $n > 3$, $r > s > 0$. The only integer in $1, \ldots, n$, divisible by 2^s is 2^s. Hence, for $1 \leqslant k \leqslant n$, $n!/k$ is divisible by 2^{r-s}, and every term except 1 is divisible by 2^{r-s+1}. So

$$\frac{n!/1 + n!/2 + \cdots + n!/n}{n!} = \frac{2^{r-s}(2j + 1)}{2^r k} = \frac{2j + 1}{2^s k}$$

for some integers j and k. The numerator is odd and the denominator is even, so this fraction is never an integer.

Solution 2. Let $n \geqslant 4$. The Bertrand's Postulate asserts that there is at least one prime number in the interval $(n/2, n]$, [HW79, Chap. 22]. Let p be the largest such prime. Converting the sum into a single fraction, as above, we get

$$\frac{n!/1 + n!/2 + \cdots + n!/n}{n!}.$$

Clearly p divides the denominator and all summands in the numerator except $n!/p$, since $p > n/2$.

Solution to 6.13.14: Recall that if $p_1, p_2, \ldots$ is the sequence of prime numbers and $x = \prod p_i^{\xi_i}$ and $y = \prod p_i^{\eta_i}$, we have

$$\gcd\{x, y\} = \prod p_i^{\min\{\xi_i, \eta_i\}} \qquad \mathrm{lcm}\,\{x, y\} = \prod p_i^{\max\{\xi_i, \eta_i\}}.$$

Let

$$a = \prod p_i^{\alpha_i} \quad b = \prod p_i^{\beta_i} \quad c = \prod p_i^{\gamma_i}$$

we have

$$\gcd\{a, \mathrm{lcm}\{b, c\}\} = \prod p_i^{\min\{\alpha_i, \max\{\beta_i, \gamma_i\}\}}$$

$$= \prod p_i^{\max\{\min\{\alpha_i, \beta_i\}, \min\{\alpha_i, \gamma_i\}\}}$$

$$= \mathrm{lcm}\,\{\gcd\{a, b\}, \gcd\{a, c\}\}.$$

Solution to 6.13.15: There are nine prime numbers $\leqslant 25$:

$$p_1 = 2, \quad p_2 = 3, \quad p_3 = 5, \quad p_4 = 7, \quad p_5 = 11,$$

$$p_6 = 13, \quad p_7 = 17, \quad p_8 = 19, \quad p_9 = 23.$$

By unique factorization, for each $1 \leqslant a \leqslant 25$ there is an integer sequence $v(a) = \left(v_j(a)\right)_{j=1}^9$ with $a = \prod_{j=1}^9 p_j^{v_j(a)}$. The 10 sequences $v(a_i) \in \mathbb{Q}^9$ must be linearly dependent, so

$$\sum_{i=1}^{10} n_i v_j(a_i) = 0$$

for all j, for some rational numbers n_i which are not all 0. Multiplying by a common multiple of the denominators, we can assume that the n_i's are integers. So

$$\prod_{i=1}^{10} a_i^{n_i} = \prod_{j=1}^9 p_j^{\sum_{i=1}^{10} n_i v_j(a_i)} = 1,$$

as required.

Solution to 6.13.16: Denote the given number by n and let $n = a^{13}$. By counting digits, we see that $n < 10^{26}$, so $a < 100$. As $8^{13} = 7934527488$, we have $80^{13} < n$ and $a > 80$. Note that $n \equiv 9 \pmod{10}$. The integers $c < 10$ such that $c^k \equiv 9 \pmod{10}$ for some k are 3, 7 and 9. But $3^4 \equiv 7^4 \equiv 9^4 \equiv 1 \pmod{10}$ and $13 = 3 \cdot 4 + 1$, so $c^{13} \equiv c^{3 \cdot 4} c \equiv c \equiv 9 \pmod{10}$ so $c = 9$. Hence, $a = 89$ or $a = 99$. As 3 does not divide n, 3 does not divide a. Hence, $a = 89$.

Solution to 6.13.17: Since $17 \equiv 7 \pmod{10}$,

$$A \equiv 7^{17^{17}} \pmod{10}.$$

Since $(7, 10) = 1$, we can apply Euler's Theorem [Sta89, p. 80], [Her75, p. 43]:

$$7^{\varphi(10)} \equiv 1 \pmod{10}.$$

The numbers k such that $1 \leqslant k \leqslant 10$ and $(k, 10) = 1$ are precisely 1, 3, 7, and 9, so $\varphi(10) = 4$. Now $17 \equiv 1 \pmod 4$, so $17^{17} \equiv 1 \pmod 4$. Thus,

$$7^{17^{17}} \equiv 7^1 \equiv 7 \pmod{10}$$

and the rightmost decimal digit of A is 7.

Solution to 6.13.18: As $23 \equiv 3 \pmod{10}$, it suffices to find $3^{23^{23^{23}}} \pmod{10}$. We have $\varphi(10) = 4$, where φ is the Euler's totient function, and, by Euler's Theorem [Sta89, p. 80], [Her75, p. 43], $3^r \equiv 3^s \pmod{10}$ when $r \equiv s \pmod 4$. So we will find $23^{23^{23}} \pmod 4$. We have $23 \equiv 3 \pmod 4$, so $23^{23^{23}} \equiv 3^{23^{23}} \pmod 4$. As $-1 \equiv 3 \pmod 4$, $3^{23^{23}} \equiv (-1)^{23^{23}} \equiv -1 \pmod 4$, because 23^{23} is odd. Hence, $23^{23^{23}} \equiv 3 \pmod 4$, and $3^{23^{23^{23}}} \equiv 3^3 \equiv 7 \pmod{10}$.

Solution to 6.13.19: Let

$$N_0 = \{0, 1\} \cup \{4, 5, 6\} \cup \{11, 12, 13, 14, 15\} \cup \cdots$$
$$N_1 = \{2, 3\} \cup \{7, 8, 9, 10\} \cup \{16, 17, 18, 19, 20, 21\} \cup \cdots$$

We have

$$N_0 \cap N_1 = \emptyset, \qquad N_0 \cup N_1 = \mathbb{Z}_+$$

and, clearly, neither can contain an arithmetic progression.

Solution to 6.13.20: Consider the ring $\mathbb{Z}_{a^k-1}$. Since $a > 1$, $(a, a^k - 1) = 1$, so $a \in \mathbb{Z}_{a^k-1}^*$. Further, it is clear that k is the least integer such that $a^k \equiv 1 \pmod{a^k - 1}$, so k is the order of a in $\mathbb{Z}_{a^k-1}^*$. Hence, by Lagrange's Theorem [Her75, p. 41], k divides the order of the group $\mathbb{Z}_{a^k-1}^*$, which is $\varphi(a^k - 1)$.

Solution to 6.13.21: Let N be the desired greatest common divisor. By Fermat's Little Theorem [Sta89, p. 80], [Her75, p. 44], we have

$$n^{13} \equiv (n^6)^2 n \equiv (n^3)^2 n \equiv n^4 \equiv n^2 \equiv n \pmod 2.$$

Hence, $2|(n^{13} - n)$ for all n, so $2|N$. An identical calculation shows that $p|N$ for $p \in \{3, 5, 7, 13\}$. Since these are all prime, their product, 2730, divides N. However, $2^{13} - 2 = 8190 = 3 \cdot 2730$, so N is either 2730 or $3 \cdot 2730$. As $3^{13} - 3 = 3(3^{12} - 1)$ is not divisible by 9, $N = 2730$.

Solution to 6.13.22: Let

$$n = p_1^{k_1} p_2^{k_2} \cdots p_n^{k_n}$$

be the factorization into a product of prime powers ($p_1 < p_2 < \cdots < p_n$) for n. The positive integer divisors of n are then the numbers

$$p_1^{j_1} p_2^{j_2} \cdots p_n^{j_n} \qquad 0 \leqslant j \leqslant k_j.$$

It follows that $d(n)$ is the number of r–tuples $(j_1, j_2, \ldots, j_r)$ satisfying the preceding conditions. In other words,

$$d(n) = (k_1 + 1)(k_2 + 1) \cdots (k_n + 1),$$

which is odd iff each k_i is even; in other words, iff n is a perfect square.

Solution to 6.13.23: Only 2 and 8 do not have the required property. This can be proved in four steps as follows.

1. If the last digit of n is 0 or 5, every power of n has the same last digit, so in particular n^n does. Hence 0 and 5 have the required property.

2. Suppose n is even and not divisible by 5. Then n^n is even and, by Fermat's Little Theorem [Sta89, p. 80], [Her75, p. 44], n^n mod 5 is determined by n mod 5 and n mod 4. By the Chinese Remainder Theorem [Sta89, p. 72] (n mod 5, n mod 4) can be any pair in $\{1, 2, 3, 4\} \times \{0, 2\}$, so n^n can be 1 or 4 mod 5, and n^n has last digit 6 or 4.

3. Suppose n is odd and not divisible by 5. Then n^n is odd, and n^n mod 5 is determined by (n mod 5, n mod 4) in $\{1, 2, 3, 4\} \times \{1, 3\}$. Therefore n^n mod 5 can be anything in $\{1, 2, 3, 4\}$, and n^n has last digit 1, 7, 3, or 9.

4. Finally, each of the values 0,1,3,4,5,6,7,9 occurs infinitely often, since the arguments above show that n^n mod 10 has period 20.

Solution 2. Let $\bar{n} = n$ mod 10 denote the last digit of n. If $n = 10m + k$ with $0 \leqslant k \leqslant 9$, then

$$\overline{n^n} = \overline{k^n} = \overline{k^{10m+k}}.$$

For fixed k, we get a sequence as m increases. Each term is the image under $x \mapsto \overline{k^{10}x}$ of the previous term, so once a value in the sequence is repeated, we can predict the rest of the sequence, which will be periodic. After computing $a_k = \overline{k^k}$ and $b_k = \overline{k^{10}}$ for k up to 9, we can immediately write down, for each k, the sequence that starts with a_k (or $\overline{10^{10}}$ if $k = 0$) and then iteratively multiply by b_k and reduce modulo 10:

$$
\begin{array}{ll}
\overline{10^{10}}, \overline{20^{20}}, \ldots = 0, 0, 0, 0, 0, 0, \ldots & (b_0 = 0) \\[4pt]
\overline{1^1}, \overline{11^{11}}, \ldots = 1, 1, 1, 1, 1, 1, \ldots & (b_1 = 1) \\[4pt]
\overline{2^2}, \overline{12^{12}}, \ldots = 4, 6, 4, 6, 4, 6, \ldots & (b_2 = 4) \\[4pt]
\overline{3^3}, \overline{13^{13}}, \ldots = 7, 3, 7, 3, 7, 3, \ldots & (b_3 = 9) \\[4pt]
\overline{4^4}, \overline{14^{14}}, \ldots = 6, 6, 6, 6, 6, 6, \ldots & (b_4 = 6) \\[4pt]
\overline{5^5}, \overline{15^{15}}, \ldots = 5, 5, 5, 5, 5, 5, \ldots & (b_5 = 5) \\[4pt]
\overline{6^6}, \overline{16^{16}}, \ldots = 6, 6, 6, 6, 6, 6, \ldots & (b_6 = 6)
\end{array}
$$

$$\overline{7^7}, \overline{17^{17}}, \ldots = 3, 7, 3, 7, 3, 7, \ldots \qquad (b_7 = 9)$$

$$\overline{8^8}, \overline{18^{18}}, \ldots = 6, 4, 6, 4, 6, 4, \ldots \qquad (b_8 = 4)$$

$$\overline{9^9}, \overline{19^{19}}, \ldots = 9, 9, 9, 9, 9, 9, \ldots \qquad (b_9 = 1)$$

Thus $0, 1, 3, 4, 5, 6, 7, 9$ occur infinitely often as $\overline{n^n}$, and $2, 8$ do not occur at all.

Solution to 6.13.24: We need to count the number of solutions of $x^3 = 1$ in the ring $\mathbb{Z}_N$. By the Chinese Remainder Theorem [Sta89, p. 72], $\mathbb{Z}_N$ is isomorphic, as a ring, to $\prod_{p \in \{2,3,5,7,11,13\}} \mathbb{Z}_p$. Hence the total number of solutions is $\prod_{p \in \{2,3,5,7,11,13\}} n_p$, where n_p is the number of solutions of $x^3 = 1$ in each $\mathbb{Z}_p$. On the other hand n_p is the number of elements of order dividing 3 in the multiplicative group $(\mathbb{Z}_p)^*$. Since $(\mathbb{Z}_p)^*$ is cyclic of order $p - 1$, we have $n_p = 3$ if 3 divides $p - 1$, and 1 otherwise. Therefore, the answer is

$$n_2 n_3 n_5 n_7 n_{11} n_{13} = 1 \cdot 1 \cdot 1 \cdot 3 \cdot 1 \cdot 3 = 9.$$

7
Linear Algebra

7.1 Vector Spaces

Solution to 7.1.1: 1. Is sufficient. The evaluation map that assigns to each polynomial its value at 1 is a nonzero linear functional on the linear space of polynomials of degree at most 4. Like all such functionals, its null space, having codimension 1, has dimension 3. The four polynomials being in that 3 dimensional null space are therefore a dependent set.

2. Is not sufficient. The four polynomials $1, 1 + x, 1 + x^2, 1 + x^3$ form a counterexample.

Solution to 7.1.2: Following by induction on k, we can easily see that the function $F^{(k)}$ is a linear combination of the mn functions $f^{(i)}g^{(j)}$ for $0 \leqslant i < n, 0 \leqslant j < m$, with constant coefficients. Hence, the $mn + 1$ functions $F^{(0)}, \ldots, F^{(mn)}$ are linearly dependent over $\mathbb{C}$.

Solution to 7.1.3: 1. Every element of V can be uniquely written in the form $a_1 v_1 + \cdots + a_n v_n$, where the v_i's form a basis of V and the a_i's are elements of $\mathbf{F}$. Since $\mathbf{F}$ has q elements it follows that V has q^n elements.

2. A matrix A in $GL_n(\mathbf{F})$ is nonsingular if and only if its columns are linearly independent vectors in $\mathbf{F}^n$. Therefore, the first column A_1 can be any nonzero vector in $\mathbf{F}^n$, so there are $q^n - 1$ possibilities. Once the first column is chosen, the second column, A_2, can be any vector which is not a multiple of the first, that is, $A_2 \neq cA_1$, where $c \in \mathbf{F}$, leaving $q^n - q$ choices for A_2. In general, the ith column A_i can be any vector which cannot be written in the form

$c_1 A_1 + c_2 A_2 + \cdots + c_{i-1} A_{i-1}$ where $c_j \in \mathbf{F}$. Hence, there are $q^n - q^{i-1}$ possibilities for A_i. By multiplying these together we see that the order of $GL_n(\mathbf{F})$ is then $(q^n - 1)(q^n - q) \cdots (q^n - q^{n-1})$.

3. The determinant clearly induces a homomorphism from $GL_n(\mathbf{F})$ onto the multiplicative group $\mathbf{F}^*$, which has $q - 1$ elements. The kernel of the homomorphism is $SL_n(\mathbf{F})$, and the cosets with respect to this kernel are the elements of $GL_n(\mathbf{F})$ which have the same determinant. Since all cosets of a group must have the same order, it follows that the order of $SL_n(\mathbf{F})$ is $|GL_n(\mathbf{F})|/(q - 1)$.

Solution to 7.1.4: Suppose $v \in V$ is nonzero. Let $S_v = \{A \in \operatorname{End} V \mid Av = v\}$. Choose $w \in V$ so that $\{v, w\}$ is a basis. With respect to this basis, S_v corresponds to the matrices of the form $\left(\begin{smallmatrix} 1 & a \\ 0 & b \end{smallmatrix}\right)$ with $a, b \in F$, so $\#S_v = q^2$. We have $S_v = S_{v'}$ if v and v' are F-multiples of each other, and otherwise $S_v \cap S_{v'} = \{I\}$, since if an endomorphism fixes a basis, it is the identity. There are $(q^2 - 1)/(q - 1) = q + 1$ nonzero vectors in V modulo the action of F^*, so the total number of endomorphisms fixing some nonzero vector is $(q + 1)q^2 - q$, where the $-q$ is there to avoid counting the identity $q + 1$ times. Thus the answer is $q^3 + q^2 - q$.

Solution 2. We are asked for the set of endomorphisms having 1 as an eigenvalue. There is a bijection between this set and the set of endomorphisms of determinant zero, taking A to $A - I$. Therefore we need to count the number of solutions to $ad - bc = 0$ with $a, b, c, d \in F$.

The number of solutions with $a \neq 0$ is $(q - 1)q^2$, since for each choice of $a \in F^*$ and $b, c \in F$, solving the equation for d yields a unique solution. The number of solutions with $a = 0$ equals $q(2q - 1)$, since d is arbitrary, and there are $2q - 1$ pairs (b, c) with $b = 0$ or $c = 0$ (there are q with $b = 0$, and q with $c = 0$, but $(b, c) = (0, 0)$ has been double counted). Thus the answer is $(q - 1)q^2 + q(2q - 1) = q^3 + q^2 - q$.

Solution to 7.1.5: If p is prime then the order of $GL_2(\mathbb{Z}_p)$ is the number of ordered bases of a two-dimensional vector space over the field $\mathbb{Z}_p$, namely $(p^2 - 1)(p^2 - p)$, as in the solution to Part 2 of Problem 7.1.3 above.

A square matrix A over $\mathbb{Z}_{p^n}$ is invertible when $\det(A)$ is invertible modulo p^n, which happens exactly when $\det(A)$ is not a multiple of p. Let $\rho(A)$ denote the matrix over $\mathbb{Z}_p$ obtained from A by reducing all its entries modulo p. We have $\det(\rho(A)) \equiv \det(A) \pmod{p}$, thus

$$A \in GL_2\left(\mathbb{Z}_{p^n}\right) \quad \text{iff} \quad \rho(A) \in GL_2\left(\mathbb{Z}_p\right),$$

giving a surjective homomorphism

$$\rho : GL_2\left(\mathbb{Z}_{p^n}\right) \to GL_2\left(\mathbb{Z}_p\right).$$

The kernel of ρ is composed of the 2×2 matrices that reduce to the Identity modulo p so the diagonal entries come from the set $\{1, p + 1, 2p + 1, \ldots, p^n - p + 1\}$ and the off-diagonal are drawn from the set that reduce to 0 modulo p, that is, $\{0, p, 2p, \ldots, p^n - p\}$. Both sets have cardinality p^{n-1}, so the order of the kernel is $(p^{n-1})^4$, and order of $GL_2(\mathbb{Z}_p)$ is

$$p^{4n-4}(p^2 - 1)(p^2 - p) = p^{4n-3}(p-1)(p^2 - 1).$$

Solution to 7.1.6: The multiplicative group $\mathbf{F}_{p^n}^*$ of the field of p^n elements is cyclic; let g be a generator. Then the map $T : \mathbf{F}_{p^n} \to \mathbf{F}_{p^n}$ sending x to gx is an $\mathbf{F}_p$-linear map acting as a single cycle on the nonzero elements of $\mathbf{F}_{p^n}$. The matrix of T with respect to a basis for $\mathbf{F}_{p^n}$ over $\mathbf{F}_p$ has the desired property.

Solution to 7.1.7: Let $\mathbf{F} = \{0, 1, a, b\}$. The lines through the origin can have slopes 0, 1, a, b, or ∞, so S has cardinality 5. Let L_s be the line through the origin with slope y. Suppose $\gamma \in G$ fixes all these lines, to be specific say

$$\gamma = \begin{pmatrix} x & y \\ z & w \end{pmatrix}.$$

Then

$$\gamma L_0 = L_0$$

implies that

$$\gamma(1, 0)^t = (x, z)^t = (c, 0)^t$$

for some $c \neq 0$. Thus, $z = 0$. Similarly, the invariance of L_∞ implies $y = 0$ and of L_1 implies $x = w$. Then $\det(\gamma) = x^2 = 1$ and since $\mathbf{F}$ has characteristic 2, we must have $x = 1$ and γ is the identity.

Solution to 7.1.8: 1. $\mathbf{F}[x]$ is a ring under polynomial addition and multiplication because $\mathbf{F}$ is a ring. The other three axioms of vector addition – associativity, uniqueness of the zero, and inverse – are trivial to verify; as for scalar multiplication, there is a unit (same as in $\mathbf{F}$) and all four axioms are trivial to verify, making it a vector field.
2. To see this, observe that the set $\{1, x, x^2, \ldots, x^n\}$ form a basis for this space, because any linear combination will be zero, if and only if, all coefficients are zero, by looking at the degree on both sides.
3. An argument as above shows that

$$a_0 1 + a_1(x - a) + \cdots + a_n(x - a)^n = 0$$

only if the coefficients are all zero.

Solution to 7.1.9: For any two finite-dimensional subspaces X and Y,

$$\dim(X + Y) + \dim(X \cap Y) = \dim(X) + \dim(Y),$$

which you get by applying the Rank–Nullity Theorem [HK61, p. 71] to the quotient map $X \to (X + Y)/Y$ (in this case the dimension of the kernel of the map is $\dim(X \cap Y)$, and the dimension of the range is $\dim(X + Y) - \dim(Y))$. Thus,

$$\dim(U) + \dim(V) + \dim(W) - \dim(U + V + W)$$
$$= \dim(U) + \dim(V) - \dim(U + V) + \dim((U + V) \cap W))$$
$$\geq \dim(U) + \dim(V) - \dim(U + V)$$
$$= \dim(U \cap V)$$

Observe that it is not true for subspaces of a vector space $(U + V) \cap W = (U \cap W) + (V \cap W)$.

Solution to 7.1.10: Let Y denote the given intersection. Then Y is a subspace of V and, clearly, $W \subset Y$. Suppose that there exists a nonzero vector $v \in Y \backslash W$. Since v is not in W, a set consisting of v and a basis for W is linearly independent. Extend this to a basis of V, and let Z be the $(n - 1)$-dimensional subspace obtained by omitting v from this basis. Then $W \subset Z$, so Z is a term in the intersection used to define Y. However, v is not in Z, so v cannot be an element of Y, a contradiction. Hence, $Y \subset W$ and we are done.

Solution to 7.1.11: We use the Induction Principle [MH93, p. 7] on the dimension of V. If $\dim V = 1$, the only proper subspace is $\{0\}$, so V is clearly not the union of a finite number of proper subspaces.

Now suppose the result is true for dimension $n - 1$ and that there is a V, $\dim V = n$, with

$$V = \bigcup_{i=1}^{k} W_i,$$

where we may assume $\dim W_i = n - 1$, $1 \leq i \leq k$. Suppose that there existed a subspace W of V of dimension $n - 1$ which was not equal to any of the W_i's. We have

$$W = \bigcup_{i=1}^{n} (W \cap W_i).$$

But $\dim(W \cap W_i) \leq n - 2$, and this contradicts our induction hypothesis.

Therefore, to complete the proof, it remains to show that such a subspace W exists. Fix a basis $x_1, \ldots, x_n$ of V. For each $\alpha \in \mathbf{F}$, $\alpha \neq 0$, consider the $(n - 1)$–dimensional subspace given by

$$W_\alpha = \{a_1 x_1 + \cdots + a_n x_n \mid a_1 + \cdots + a_{n-1} + \alpha a_n = 0\}.$$

Any two of these subspaces intersect in a subspace of dimension, at most, $n - 2$, so they are distinct. Since there are infinitely many of these, because $\mathbf{F}$ is infinite, we can find W as desired.

Solution 2. Suppose that $V = \cup_{1 \leq i \leq k} V_i$. After discarding superfluous V_i's, we may assume that

$$V \neq \bigcup_{i \neq i_0} V_i \qquad \text{for all} \qquad 1 \leq i_0 \leq k.$$

Then $k \geqslant 2$, and there must be vectors v_1, v_2 in V such that

$$(1) \quad v_1 \in V_1 \setminus \bigcup_{i \neq 1} V_i \quad \text{and} \quad (2) \quad v_2 \in V_2 \setminus \bigcup_{i \neq 2} V_i.$$

Let (x_s) be sequence of distinct, nonzero elements of the field. Then, for each s, the vector $u_s = v_1 + x_s v_2$ does not lie in $V_1 \cup V_2$ (if $u_s \in V_1$, then $v_2 = (u_s - v_1)/x_s \in V_1$, contradicting (1); similarly, $u_s \notin V_2$.) It follows that, for all s, $u_s \in \bigcup_{i \neq 1,2} V_i$. Since the vectors u_s are all distinct, it follows that, for some $s \neq s'$ and $i \neq 1, 2$, u_s and $u_{s'}$ lie in V_i. But then $v_2 = (u_s - u_{s'})/(x_s - x_{s'}) \in V_i$, contradicting (2). Hence, $V \neq \bigcup_{1 \leqslant i \leqslant k} V_i$.

Solution to 7.1.12: Note first that if A and B are matrices and C is an invertible matrix, then

$$AB = BA \quad \text{iff} \quad C^{-1}ACC^{-1}BC = C^{-1}BCC^{-1}AC.$$

Also, if $D_1, \ldots, D_n$ are linearly independent matrices, so are the matrices $C^{-1}D_1 C, \ldots, C^{-1}D_n C$. We may then assume that A is in Jordan Canonical Form [HK61, p. 247].

A direct calculation shows that if $\tilde{A} = \begin{pmatrix} a & 1 & \cdots & 0 \\ & \ddots & \ddots & \\ & & & 1 \\ 0 & & & a \end{pmatrix}$ is a $k \times k$ Jordan

block, then $\tilde{A}$ commutes with $\tilde{B} = \begin{pmatrix} b_1 & b_2 & & b_k \\ & \ddots & \ddots & b_2 \\ 0 & & & b_1 \end{pmatrix}$.

Therefore, by block multiplication, A commutes with any matrix of the form

$$B = \begin{pmatrix} \tilde{B}_1 & & \\ & \ddots & \\ & & \tilde{B}_r \end{pmatrix}$$

where the $\tilde{B}_r$'s have the form of $\tilde{B}$ and the same dimension as the Jordan blocks of A. Since there are n variables in B, $\dim C(A) \geqslant n$.

Solution to 7.1.13: $\text{tr}(AB - BA) = 0$, so S is contained in the kernel of the trace. Since the trace is a linear transformation from $M_{n \times n}(\mathbb{R}) = \mathbb{R}^{n^2}$ onto $\mathbb{R}$, its kernel must have dimension, at most, $n^2 - 1$. Therefore, it suffices to show that S contains $n^2 - 1$ linearly independent matrices.

Let M_{ij} denote the matrix with a 1 in the $(i, j)^{th}$ coordinate and 0's elsewhere. A calculation shows that for $i \neq j$, $M_{ij} = M_{ik}M_{kj} - M_{kj}M_{ik}$, so M_{ij} is in S. Similarly, for $2 \leqslant j \leqslant n$, $M_{11} - M_{jj} = M_{1j}M_{j1} - M_{j1}M_{1j}$. Together, these $n^2 - 1$ matrices are clearly a linearly independent set.

Solution to 7.1.14: Let $f, g \in S$ and let r and s be scalars. Then, for any $v \in A$, $(rf + sg)(v) = f(rv) + g(sv) \in A$, since A is a vector subspace and f and g fix A. Similarly $rf + sg$ fixes B, so $rf + sg \in S$ and S is a vector space.

To determine the dimension of S, it suffices to determine the dimension of the space of matrices which fix A and B. To choose a basis for V, let A' denote a complementary subspace of $A \cap B$ in A and let B' denote a complementary subspace of $A \cap B$ in B. Then, since $A + B = V$, $r = a + b - n$ is the dimension of $A \cap B$. Further, $\dim A' = a - r$ and $\dim B' = b - r$. Take one basis in each of the spaces A', B', and $A \cap B$. The union of these bases form a basis for V. Since any endomorphism which leaves A and B invariant must also fix $A \cap B$, its matrix in this basis must have the form

$$\begin{pmatrix} * & * & * \\ 0 & * & 0 \\ 0 & 0 & * \end{pmatrix}$$

which has, at most, $a^2 + b^2 + n^2 - an - bn$ nonzero entries, so the dimension of S is $a^2 + b^2 + n^2 - an - bn$.

Solution to 7.1.15: Suppose there are scalars such that

$$a_0 x + a_1 T x + \cdots + a_k T^k x + \cdots + a_{m-1} T^{m-1} x = 0$$

applying T^{m-1} to both sides, we get, since $T0 = 0$,

$$a_0 T^{m-1} x + a_1 T^m x + \cdots + a_k T^{m-1+k} x + \cdots + a_{m-1} T^{m-1+m-1} x = 0$$

so

$$a_0 T^{m-1} x = 0$$

and $a_0 = 0$. By the Induction Principle [MH93, p. 7] (multiplying by T^{m-k-1}) we see that all $a_k = 0$ and the set is linearly independent.

Solution to 7.1.16: We may assume $\alpha_1 < \cdots < \alpha_n$. Suppose that there are constants $c_1, \ldots, c_n$ such that $c_1 e^{\alpha_1 t} + \cdots + c_n e^{\alpha_n t} \equiv 0$. If not all the coefficients vanish, we have, without loss of generality, $c_n \neq 0$. Thus,

$$c_1 e^{(\alpha_1 - \alpha_n)t} + \cdots + c_{n-1} e^{(\alpha_{n-1} - \alpha_n)t} + c_n \equiv 0.$$

Letting $t \to \infty$, and noting that $e^{(\alpha_i - \alpha_n)t} \to 0$ for $1 \leqslant i \leqslant n - 1$, we get $c_n = 0$, a contradiction.

Solution 2. This proof is by induction. Let $P(n)$ denote the statement that for all distinct real numbers $\alpha_1, \ldots, \alpha_n$, the functions $e^{\alpha_1 t}, \ldots, e^{\alpha_n t}$ are linearly independent.

Then $P(1)$ is true since $e^{\alpha_1 t} \not\equiv 0$.

Assume $P(n)$ is true for some fixed $n \geqslant 1$, and let $\alpha_1, \ldots, \alpha_{n+1}$ be $n + 1$ distinct real numbers. Suppose for scalars λ_i, $\lambda_1 e^{\alpha_1 t} + \cdots + \lambda_{n+1} e^{\alpha_{n+1} t} \equiv 0$.

Then $\lambda_1 + \lambda_2 e^{\beta_2 t} + \cdots + \lambda_{n+1} e^{\beta_{n+1} t} \equiv 0$, where $\beta_j = \alpha_j - \alpha_1, j = 2, \ldots, n+1$ are distinct nonzero real numbers. Differentiating with respect to t, we obtain $\lambda_2 \beta_2 e^{\beta_2 t} + \cdots + \lambda_{n+1} \beta_{n+1} e^{\beta_{n+1} t} \equiv 0$. By the induction hypothesis, $\lambda_j \beta_j = 0$ for $j = 2, \ldots, n+1$. Thus $\lambda_j = 0, j = 2, \ldots, n+1$. Substituting in the first equation above gives $\lambda_1 = 0$, proving $P(n+1)$.

Solution to 7.1.17: Let P be the change of basis matrix from (a_i) to (b_i). A straightforward calculation shows that $I + 2P$ is the matrix taking (a_i) to $(a_i + 2b_i)$. Now $(I + 2P)v = \lambda v$ implies that $Pv = \frac{1}{2}(\lambda - 1)v$. So if λ is an eigenvalue of $I + 2P$, then $\frac{1}{2}(\lambda - 1)$ is an eigenvalue of P, and they correspond to the same eigenvectors. The reverse also holds, so there is a one-to-one correspondence between the eigenvalues of P and those of $I + 2P$. As (a_i) and (b_i) are orthonormal bases, P is orthogonal and therefore, all the eigenvalues of P are ± 1. But this implies that the only possible eigenvalues of $I + 2P$ are 3 and -1. Hence, 0 is not an eigenvalue of $I + 2P$, so it is an invertible matrix and, thus, $(a_i + 2b_i)$ is a basis. Further, $\det P = (-1)^\alpha 1^\beta$, where α and β are the algebraic multiplicities of -1 and 1 as eigenvalues of P. Thus, $\det(I + 2P) = (-1)^\alpha 3^\beta$. Since we are given that $\det P > 0$, α is even and, thus, $\det(I + 2P)$ is positive as well. Therefore, $(a_i + 2b_i)$ has the same orientation as (a_i).

7.2   Rank and Determinants

Solution to 7.2.1: 1. Let $A = (a_{ij})$, and R_i denote the ith row of A. Let r, $1 \leqslant r \leqslant m$, be the row rank of A, and $S_i = (b_{i1}, \ldots, b_{in}), 1 \leqslant i \leqslant r$, be a basis for the row space. The rows are linear combinations of the S_i's:

$$R_i = \sum_{j=1}^{r} k_{ij} S_j, \qquad 1 \leqslant i \leqslant m.$$

For $1 \leqslant l \leqslant n$, isolating the lth coordinate of each of these equations gives

$$a_{1l} = k_{11} b_{1l} + \cdots + k_{1r} b_{rl}$$
$$a_{2l} = k_{21} b_{1l} + \cdots + k_{2r} b_{rl}$$
$$\vdots$$
$$a_{ml} = k_{m1} b_{1l} + \cdots + k_{mr} b_{rl}.$$

Hence, for $1 \leqslant l \leqslant n$ the lth column of A, C_l, is given by the equation

$$C_l = \sum_{j=1}^{r} b_{jl} K_j,$$

where K_j is the column vector $(k_{1j}, \ldots, k_{mj})^t$. Hence, the space spanned by the columns of A is also spanned by the r vectors K_j, so its dimension is less than or equal to r. Therefore, the column rank of A is less than or equal to its row rank. In exactly the same way, we can show the reverse inequality, so the two are equal.

2. Using Gaussian elimination we get the matrix

$$\begin{pmatrix} 1 & 0 & 3 & -2 \\ 0 & 1 & -4 & 4 \\ 0 & 0 & 2 & 0 \\ 0 & 0 & 0 & 2 \\ 0 & 0 & 0 & 0 \end{pmatrix}$$

so the four columns of M are linearly independent.

3. If a set of rows of M is linearly independent over $\mathbf{K}$, then clearly it is also independent over $\mathbf{F}$, so the rank of M over $\mathbf{K}$ is, at most, the rank of M over $\mathbf{F}$; but looking at the Gaussian elimination process one can see that it involves operations in the sub-field only, so the rank over both fields is the same.

Solution to 7.2.2: Let v be a nonzero right null vector of A and u a left null vector of A such that $u \cdot v = 0$. Since A is diagonalizable, we can choose a basis $\{v_1, v_2, \ldots, v_n\}$ of $\mathbb{R}^n$ consisting of eigenvectors of A. Since the null space of A has dimension 1, one of these basis vectors is a multiple of v, so we can assume $v_1 = v$. Let λ_j be the eigenvalue corresponding to v_j, $j = 2, \ldots, n$. Then $\lambda_j \neq 0$, and for $j > 1$,

$$uv_j = \frac{1}{\lambda_j} u(Av_j) = \frac{1}{\lambda_j} (uA)v_j = 0.$$

Hence u is orthogonal to each of the basis vectors $v_1, \ldots, v_n$, so $u = 0$.

Solution to 7.2.3: As $A^t AV \subset A^t V$, it is sufficient to prove that $\dim A^t AV = \dim A^t V$. We know that $\operatorname{rank} A = \dim(\operatorname{Im} A)$ and that

$$\dim(\operatorname{Im} A) + \dim(\ker A) = n$$

Similar formulas hold for $A^t A$. Therefore, it is enough to show that $\ker A$ and $\ker A^t A$ have the same dimension. In fact, they are equal. Clearly, $\ker A \subset \ker A^t A$. Conversely, take any $v \in \ker A^t A$. Then

$$0 = \langle A^t Av, v \rangle = \langle Av, Av \rangle,$$

so $\|Av\| = 0$. Hence, $v \in \ker A$ and we are done.

Solution to 7.2.4: Since $I - P - Q$ is invertible, P has the same rank as

$$P(1 - P - Q) = P - P^2 - PQ = -PQ.$$

Similarly, Q has the same rank as

$$(1 - P - Q)Q = Q - PQ - Q^2 = -PQ,$$

so P and Q have the same rank.

Solution to 7.2.5: The upper $k \times k$ square minor $A(k, k)$ of $A(m, n)$ is the Vandermonde matrix, which determinant is $\prod_{0 \leqslant i < j < k}(j - i)$. If $k \leqslant p$, this determinant is not zero (mod p), which shows that $\text{rank} A(m, n) \geqslant \min\{m, n, p\}$. Now if $A(m, n)$ has at most p distinct columns (mod p), so $\text{rank} A(m, n) \leqslant p$. Since $\text{rank} A(n, n) \leqslant \min\{n, m\}$, we have $\text{rank} A(m, n) = \min\{m, n, p\}$.

Solution to 7.2.6: Let $V = K^n$, and consider the two quotient spaces AV/A^2V and A^2V/A^3V. The first has dimension $\text{rank} A - \text{rank} A^2$, and the second has dimension $\text{rank} A^2 - \text{rank} A^3$. The matrix A induces a linear transformation of AV/A^2V onto A^2V/A^3V, so $\dim(A^2V/A^3V) \leqslant \dim(AV/A^2V)$, as desired.

Solution to 7.2.7: 1. *and* 2. Since T is symmetric it is diagonalizable, so $\mathbb{R}^n$ can be written as the direct sum of the eigenspaces of T. It suffices to show that any eigenspace has dimension, at most, 1. For if this is the case, then the kernel has dimension, at most, 1, and, by the Rank–Nullity Theorem [HK61, p. 71], T has rank at least $n - 1$, and there must be n distinct eigenspaces, so there are n distinct eigenvalues associated with them.

Let $\lambda \in \mathbb{R}$, and consider the system of equations $Tx = \lambda x$. The first equation is $a_1 x_1 + b_1 x_2 = \lambda x_1$. Since $b_1 \neq 0$, we can solve for x_2 in terms of x_1. Suppose that we can solve the first $i - 1$ equations for $x_2, \ldots, x_i$ in terms of x_1. Then, since $b_i \neq 0$, we can solve the ith equation $b_{i-1} x_{i-1} + a_i x_i + b_i x_{i+1} = \lambda x_i$ for x_{i+1} in terms of x_1. Therefore, by the Induction Principle [MH93, p. 7], we can solve the first $n - 1$ equations for $x_2, \ldots, x_n$ in terms of x_1.

The last equation, $b_{n-1} x_{n-1} + a_n x_n = \lambda x_n$, is either consistent with this or is not. If not, λ is not an eigenvalue; if it is, then λ is an eigenvalue and we have one degree of freedom in determining eigenvectors. Hence, in either case the associated eigenspace has dimension, at most, 1 and we are done.

Solution 2. 1. The submatrix one obtains by deleting the first row and the first column is upper-triangular with nonzero diagonal entries, so its determinant is nonzero. Thus, the first $n - 1$ columns of T are linearly independent.
2. By the Spectral Theorem [HK61, p. 335], [Str93, p. 235], $\mathbb{R}^n$ has a basis consisting of eigenvectors of T. If λ is an eigenvalue of T, then $T - \lambda I$ has rank $n - 1$ by Part 1, so $\ker(T - \lambda I)$ has dimension 1. Since the eigenspaces span $\mathbb{R}^n$ and each has dimension 1, there must be n of them.

Solution to 7.2.9: 1. Write the characteristic polynomial of A, $\det(A - \lambda I)$, as $(-1)^r \lambda^r + c_1 \lambda^{r-1} + \cdots + c_r$. Since the entries of A are integers, each c_k is an integer, and $c_r = \det A$. If λ is an integer eigenvalue, then $\det(A - nI) = 0$, so

$$\det A = (-1)^{r-1} n^r + c_1 n^{r-1} + \cdots + c_{r-1} n$$

showing that n divides $\det A$.
2. Under the given hypotheses, n is an eigenvalue with eigenvector $(1, 1, \ldots, 1)^t$, so Part 1 applies.

Solution to 7.2.10: Since A has rank $n-1$ it has $n-1$ linearly independent rows, and the remaining row is a linear combination of those $n-1$. Interchanging rows and corresponding columns of A, we can assume without loss of generality that rows 1 through $n-1$ of A are linearly independent. Thus, there are real numbers $c_1, \ldots, c_{n-1}$ such that, letting a_{jk} denote the entries of A, we have

$$a_{nk} = \sum_{j=1}^{n-1} c_j\, a_{jk}\,, \qquad k = 1, \ldots, n\,. \tag{$*$}$$

It is asserted that the matrix obtained from A by deletion of its last row and column has rank $n-1$. Assume not. Then there are real numbers $b_1, \ldots, b_{n-1}$, not all 0, such that

$$\sum_{j=1}^{n-1} b_j\, a_{jk} = 0\,, \qquad k = 1, \ldots, n-1\,. \tag{$**$}$$

By $(*)$ and the symmetry of A,

$$a_{jn} = \sum_{k=1}^{n-1} c_k\, a_{jk}\,, \qquad j = 1, \ldots, n\,.$$

Hence

$$\sum_{j=1}^{n-1} b_j\, a_{jn} = \sum_{j=1}^{n-1}\sum_{k=1}^{n-1} b_j c_k\, a_{jk} = \sum_{k=1}^{n-1} c_k \sum_{j=1}^{n-1} b_j\, a_{jk} = 0\,,$$

i.e., $(**)$ also holds for $k = n$, contradicting the linear independence of the first $n-1$ rows of A.

Solution to 7.2.11: We use the Induction Principle [MH93, p. 32] in the order of the matrix. If $n = 2$,

$$A = \begin{pmatrix} 1 & x_1 \\ 1 & x_2 \end{pmatrix}$$

which has determinant $(x_2 - x_1)$.

Suppose the result holds for all $k < n$, and let A be the $n \times n$ Vandermonde matrix, see the Solution to Problem 7.2.11 or [HK61, p. 125]. Treating the indeterminates $x_1, \ldots, x_{n-1}$ as *constants* and expanding the determinant of A along the last row, we see that $\det A$ is an $(n-1)^{th}$ degree polynomial in x_n, which can have, at most, $n-1$ roots. If we let $x_n = x_i$ for $1 \leqslant i \leqslant n-1$, A would have two identical rows, so $\det A$ would equal 0. Hence, the x_i's are the roots of $\det A$ as a polynomial in x_n. In other words, there exists a constant $c > 0$ such that

$$\det A = c \prod_{i=1}^{n-1} (x_n - x_i).$$

c is the coefficient of the x_n^{n-1} term, which, when we expand the determinant, is equal to the determinant of the $(n-1) \times (n-1)$ Vandermonde matrix. So, by the induction hypothesis,

$$\det A = \prod_{j < i \leqslant n-1} (x_i - x_j) \prod_{i=1}^{n-1} (x_n - x_i) = \prod_{i>j}(x_i - x_j).$$

Solution to 7.2.12: 1. As shown in the solution of Problem 7.2.11, the determinant of the matrix is

$$\prod_{i>j}(a_i - a_j)$$

which is nonzero if the a_i are all different.

2. The function f given by

$$f(x) = \sum_{i=0}^{n} \frac{(x - a_0) \cdots (x - a_{i-1}) b_i (x - a_{i+1}) \cdots (x - a_n)}{(a_i - a_0) \cdots (a_i - a_{i-1}) b_i (a_i - a_{i+1}) \cdots (a_i - a_n)}$$

has degree n and takes $f(a_i)$ into b_i. Now, if $\psi(x)$ is another such polynomial of degree n, the polynomial

$$f(x) - \psi(x)$$

has degree n with $n+1$ different roots (the a_i's), so it has to be the zero polynomial and f is unique.

Solution to 7.2.13: Define the matrix $C = A^{-1}B$, one can immediatly see that it is also unitary. And we can also verify that $A + B = A(I + C)$. Since A is unitary, its eigenvalues have absolute value 1. Multiplying them together shows that $|\det A| = 1$. If $\zeta_1, \ldots, \zeta_n$ are the eigenvalues of C, listed with their multiplicities, then $|\zeta_i| = 1$, so the eigenvalues of $I + C$ are $1 + \zeta_1, \ldots, 1 + \zeta_n$, so

$$|\det(I + C)| = |1 + \zeta_1| \cdots |1 + \zeta_n| \leqslant 2 \cdot 2 \cdots 2 = 2^n.$$

Hence

$$|\det(A + B)| = |\det A| |\det(I + C)| \leqslant 2^n.$$

Solution to 7.2.14: Consider the function $v(t) = (1, t, t^2)$. To show that $v(t_1)$, $v(t_2)$, and $v(t_3)$ form a basis for $\mathbb{R}^3$ whenever the t_i's are distinct, it will suffice to show that the matrix which has these vectors as rows has nonzero determinant. But this matrix is

$$\begin{pmatrix} 1 & t_1 & t_1^2 \\ 1 & t_2 & t_2^2 \\ 1 & t_3 & t_3^2 \end{pmatrix}$$

which is the 3×3 Vandermonde matrix, see the Solution to Problem 7.2.11 or [HK61, p. 125]. Its determinant is given by

$$(t_3 - t_2)(t_3 - t_1)(t_2 - t_1)$$

which is nonzero whenever the t_i's are distinct.

Solution to 7.2.15: Let G be the matrix with entries

$$G_{ij} = \int_a^b f_i(x)f_j(x)dx.$$

If the determinant of G vanishes, then G is singular; let a be a nonzero n–vector with $Ga = 0$. Then

$$0 = a^T Ga = \sum_{i=1}^n \sum_{i=j}^n \int_a^b a_i f_i(x)a_j f_j(x)dx = \int_a^b \left(\sum_{i=1}^n a_i f_i(x)\right)^2 dx,$$

so, since the f_i's are continuous functions, the linear combination $\sum a_i f_i$ must vanish identically. Hence, the set $\{f_i\}$ is linearly dependent on $[a, b]$. Conversely, if $\{f_i\}$ is linearly dependent, some f_i can be expressed as a linear combination of the rest, so some row of G is a linear combination of the rest and G is singular.

Solution to 7.2.16: Identify $M_{2\times2}$ with $\mathbb{R}^4$ via

$$\begin{pmatrix} a & b \\ c & d \end{pmatrix} \leftrightarrow \begin{pmatrix} a \\ b \\ c \\ d \end{pmatrix}$$

and decompose L into the multiplication of two linear transformations,

$$M_{2\times2} \simeq \mathbb{R}^4 \xrightarrow{L_A} \mathbb{R}^4 \xrightarrow{L_B} \mathbb{R}^4 \simeq M_{2\times2}$$

where $L_A(X) = AX$ and $L_B(X) = XB$.

The matrices of these two linear transformations on the canonical basis of $\mathbb{R}^4$ is

$$L_A = \begin{pmatrix} 1 & 0 & 2 & 0 \\ 0 & 1 & 0 & 2 \\ -1 & 0 & 3 & 0 \\ 0 & -1 & 0 & 3 \end{pmatrix} \quad \text{and} \quad L_B = \begin{pmatrix} 2 & 0 & 0 & 0 \\ 1 & 4 & 0 & 0 \\ 0 & 0 & 2 & 0 \\ 0 & 0 & 1 & 4 \end{pmatrix}$$

then $\det L = \det L_A \cdot \det L_B = (9 + 6 + 2(2 + 3)) \cdot (2 \cdot 32) = 2^6 \cdot 5^2$, and to compute the trace of L, we only need the diagonal elements of $L_A \cdot L_B$, that is,

$$\text{tr } L = 2 + 4 + 6 + 12 = 24.$$

Solution to 7.2.17: Let $X = (x_{ij})$ be any element of $M_3(\mathbb{R})$. A calculation gives

$$T(X) = \begin{pmatrix} x_{11} & 3x_{12}/2 & x_{13} \\ 3x_{21}/2 & 2x_{22} & 3x_{23}/2 \\ x_{31} & 3x_{32}/2 & x_{33} \end{pmatrix}.$$

It follows that the basis matrices M_{ij} are eigenvectors of T. Taking the product of their associated eigenvalues, we get $\det T = 2(3/2)^4 = 81/8$.

Solution to 7.2.18: Since the minimal polynomial of A splits into distinct linear factors, $\mathbb{R}^3$ has a basis $\{v_1, v_2, v_3\}$ of eigenvectors of A. Since $\det A = 32$, two of those, say v_1 and v_2, are associated with the eigenvalue 4, and one, v_3, is associated with the eigenvalue 2. Now consider the nine matrices E_{ij}, $1 \leqslant i, j \leqslant 3$, whose i^{th} column is the vector v_j and whose other columns are zero. Since the v_i's are linearly independent, the matrices E_{ij} are linearly independent in $M_{3\times 3}$ and form a basis of $M_{3\times 3}$. Further, a calculation shows that $AE_{ij} = \lambda_j E_{ij}$, where $\lambda_1 = \lambda_2 = 4$ and $\lambda_3 = 2$. Hence, $M_{3\times 3}$ has a basis of eigenvectors of L_A, so it follows that $\operatorname{tr} L_A = 6 \cdot 4 + 3 \cdot 2 = 30$.

Solution to 7.2.20: We have

$$\dim \operatorname{range} T = \dim M_{7\times 7} - \dim \ker T = 49 - \dim \ker T$$

so it suffices to find the dimension of $\ker T$; in other words, the dimension of the subspace of matrices that commute with A. Let E_+ be the eigenspace of A for the eigenvalue 1 and E_- be the eigenspace of A for the eigenvalue -1. Then $\mathbb{R}^7 = E_+ \oplus E_-$. A matrix that commutes with A leaves E_+ and E_- invariant, so, as linear transformations on $\mathbb{R}^7$, can be expressed as the direct sum of a linear transformation on E_+ with a linear transformation on E_-. Moreover, any matrix that can be so expressed commutes with A. Hence, the space of matrices that commute with A is isomorphic to $M_{4\times 4} \oplus M_{3\times 3}$, and so has dimension $16 + 9 = 25$. It follows that $\dim \operatorname{range} T = 49 - 25 = 24$.

Solution to 7.2.21: $m > n$: We write $T = T_1 T_2$, where $T_2 : M_{n\times m} \to M_{n\times n}$ is defined by $T_2(X) = XB$ and $T_1 : M_{n\times n} \to M_{m\times n}$ is defined by $T_1(Y) = AY$. Since $\dim M_{n\times m} = nm > n^2 = \dim M_{n\times n}$, the transformation T_2 has a nontrivial kernel, by the Rank–Nullity Theorem [HK61, p. 71]. Hence, T also has a nontrivial kernel and is not invertible.

$m < n$: We write $T = T_2 T_1$, where $T_1 : M_{n\times m} \to M_{m\times m}$ is defined by $T_1(X) = AX$ and $T_2 : M_{m\times m} \to M_{m\times n}$ is defined by $T_2(Y) = YB$. Now we have $\dim M_{n\times m} = nm > m^2 = \dim M_{m\times m}$, so T_1 has a nontrivial kernel, and we conclude as before that T is not invertible.

7.3 Systems of Equations

Solution to 7.3.1: Subtracting the first equation from the third we get $x_1 = x_7$, continuing in this fashion subtracting the kth equation from the $(k+2)^{th}$ equation we obtain

$$x_k = x_{k+6} \qquad \text{for every} \quad k$$

and we are left with a maximum of six independent parameters in the system. Solving the first and second equations, separatedly, we get

$$x_5 = -(x_1 + x_3)$$
$$x_6 = -(x_2 + x_4)$$

so four free parameters is all that is required.

Solution to 7.3.2: 1. Through linear combinations of rows, reduce the system of equations to a row-reduced echelon form, that is, a system where:

- the first nonzero entry in each nonzero row is equal to 1;

- each column which contains the leading nonzero entry of some row has all its other entries 0;

- every row which has all entries 0 occurs below every row that has a nonzero entry;

- if rows $1, \ldots, r$ are the nonzero rows and if the leading nonzero entry of row i occurs in column k_i, $i = 1, \ldots, r$, then $k_i < k_2 < \cdots < k_r$.

This new system has a number of nonzero rows $r \leqslant m < n$ and it is easy to see that it has nonzero solution.

Since the original system is equivalent to the row-reduced one, they have exactly the same solutions.

2. Let V be a vector space spanned by m vectors $\beta_1, \ldots, \beta_m$. We will show that every subset $S = \{\alpha_1, \ldots, \alpha_n\}$ of V with $n > m$ vectors is linear dependent.

Since $\beta_1, \ldots, \beta_m$ span V, there are scalars A_{ij} in the field $\mathbf{F}$ such that

$$\alpha_i = \sum_{i=1}^{m} A_{ij} \beta_i .$$

For any set of scalar $x_1, \ldots, x_n$ in $\mathbf{F}$, we have

$$x_1 \alpha_1 + \cdots + x_n \alpha_n = \sum_{j=1}^{n} x_j \alpha_j$$

$$= \sum_{j=1}^{n} x_j \sum_{i=1}^{m} A_{ij} \beta_i$$

$$= \sum_{j=1}^{n} \sum_{i=1}^{m} (A_{ij} x_j) \beta_i$$

$$= \sum_{i=1}^{m} \left(\sum_{j=1}^{n} A_{ij} x_j \right) \beta_i .$$

Since $n > m$, the linear systems of equations

$$\sum_{j=1}^{n} A_{ij} x_j = 0, \quad 1 \leqslant i \leqslant m$$

has a nontrivial solution so the set is linear dependent, proving the assertion.

Solution to 7.3.3: The answer is yes. Writing the system of linear equations in matrix form, we have $Ax = 0$, where A is an $m \times n$ matrix with rational entries. Let the column vector $x = (x_1, \ldots, x_n)^t$ be a complex solution to this system, and let V be the $\mathbb{Q}$-vector space spanned by the x_i's. Then $\dim V = p \leqslant n$. If $y_1, \ldots, y_p \in \mathbb{C}$ is a basis of V, then there is a rational $n \times p$ matrix B with $By = x$ (where y is the column vector $(y_1, \ldots, y_p)^t$). Substituting this into the original equation, we get $ABy = 0$. Since y is composed of basis vectors, this is possible only if $AB = 0$. In particular, every column of B is a rational solution of the equation $Ax = 0$.

7.4   Linear Transformations

Solution to 7.4.1: 1. We need to show that vector addition and scalar multiplication are closed in $S(E)$, but this is a trivial verification because if $v = S(x)$ and $w = S(y)$ are vectors in $S(E)$, then

$$v + w = S(x + y) \quad \text{and} \quad cv = S(cx)$$

are also in $S(E)$.

2. If S is not injective, then two different vectors x and y have the same image $S(x) = S(y) = v$, so

$$S(x - y) = S(x) - S(y) = v - v = 0$$

that is, $x - y \neq 0$ is a vector in the kernel of S. On the other hand, if S is injective, it only takes $0 \in E$ into $0 \in F$, showing the result.

3. Assuming that S is injective, the application $S^{-1} : S(E) \to E$ is well defined. Given $av + bw \in S(E)$ with $v = S(x)$ and $w = S(y)$, we have

$$\begin{aligned}
S^{-1}(av + w) &= S^{-1}(aS(x) + bS(y)) \\
&= S^{-1}(S(ax + by)) \\
&= ax + by \\
&= aS^{-1}(v) + bS^{-1}(w)
\end{aligned}$$

therefore, S^{-1} is linear.

Solution to 7.4.2: Let $\{\alpha_1, \ldots, \alpha_k\}$ be a basis for $\ker T$ and extend it to $\{\alpha_1, \ldots, \alpha_k, \ldots, \alpha_n\}$, a basis of V. We will show that $\{T\alpha_{k+1}, \ldots, T\alpha_n\}$ is a basis for the range of T. It is obvious they span the range since $T\alpha_j = 0$ for $j \leqslant k$. Assume

$$\sum_{i=k+1}^{n} c_i T\alpha_i = 0,$$

which is equivalent to

$$T\left(\sum_{i=k+1}^{n} c_i\alpha_i\right) = 0,$$

that is, $\alpha = \sum_{i=k+1}^{n} c_i\alpha_i$ is in the kernel of T. We can then write α as $\alpha = \sum_{i=1}^{k} b_i\alpha_i$ and have

$$\sum_{i=1}^{k} b_i\alpha_i - \sum_{i=k+1}^{n} c_i T\alpha_i = 0,$$

which implies all $c_i = 0$, and the vectors $T\alpha_{k+1}, \ldots, T\alpha_n$ form a basis for the range of T.

Solution to 7.4.3: Applying the Rank–Nullity Theorem [HK61, p. 71] (see Solution to Problem 7.4.2) to the map $T|_{T^{-1}(X)} : T^{-1}(X) \rightarrow X$, we get

$$\dim T^{-1}(X) = \dim \ker T|_{T^{-1}(X)} + \dim X,$$

but $\ker T|_{T^{-1}(X)} = \ker T$, so, by the same theorem, we have

$$\dim \ker T|_{T^{-1}(X)} = \dim V - \dim X \leqslant \dim V - \dim W.$$

Therefore
$$\dim T^{-1}(X) \geqslant \dim V - \dim W + \dim X.$$

Solution 2. Let Q be the projection from W onto the quotient space $Z = W/X$. The linear map $QT : V \rightarrow Z$ has null space $T^{-1}(X)$, therefore

$$\dim V - \dim T^{-1}(X) = \text{rank}\,(QT) \leqslant \dim Z = \dim W - \dim X.$$

Solution to 7.4.4: We have $A(B(\ker(AB))) = \{0\}$ implying that $B(\ker(AB)) \subset \ker A$, so $\dim B(\ker(AB)) \leqslant \dim \ker A$. Using the Rank–Nullity Theorem [HK61, p. 71], we have

$$\dim \ker(AB) = \dim B(\ker(AB)) + \dim \ker\left(B|_{\ker(AB)}\right)$$
$$\leqslant \dim \ker A + \dim \ker B.$$

Solution to 7.4.5: Let v_0 be a vector in V such that $Av_0 \neq 0$. By the second condition there is a scalar α such that $Bv_0 = \alpha Av_0$.

Suppose v is any vector in V such that Av_0 and Av are linearly independent. By the second condition there are scalars β and γ such that $Bv = \beta Av$ and $B(v_0 + v) = \gamma A(v_0 + v)$. Then

$$\alpha Av_0 + \beta Av = \gamma Av_0 + \gamma Av ,$$

which by the linear independence of Av_0 and Av implies that $\alpha = \beta = \gamma$. Hence $Bv = \alpha Av$.

Fix a vector v_1 in V such that Av_0 and Av_1 are linearly independent. Suppose v is a vector in V such that Av_0 and Av are linearly dependent, but $Av \neq 0$. Then Av_1 and Av are linearly independent, and the reasoning above shows that $Bv = \alpha Av$.

Suppose finally that v is a nonzero vector in V such that $Av = 0$. Then $A(v + v_0)$ and Av_1 are linearly independent, so the reasoning above gives $B(v + v_0) = \alpha A(v + v_0) = \alpha Av_0$, implying that $Bv = 0$ (since $Bv_0 = \alpha Av_0$). Thus $Bv = \alpha Av$ for all v in V, i.e., $B = \alpha A$.

Solution to 7.4.6: Let $v_1, \ldots, v_n$ be a basis for V such that $v_1, \ldots, v_k$ is a basis for W. Then the matrix for L in terms of this basis has the form $\left(\begin{smallmatrix} M & M \\ 0 & 0 \end{smallmatrix}\right)$, where M is a $k \times k$ matrix and N is $k \times (n - k)$. It follows that M is the matrix of L_W with respect to the basis $v_1, \ldots, v_k$. As the matrix of $1 - tL$ is $\left(\begin{smallmatrix} 1-tM & -tN \\ 0 & 1 \end{smallmatrix}\right)$, it follows that $\det(1 - tL) = \det(1 - tM) = \det(1 - tL_W)$.

Solution to 7.4.7: The trace of f equals the trace of the restriction $f|_W$ plus the trace of the induced endomorphism g of V/W. Since f maps V into W, g is zero and has trace zero.

Solution to 7.4.8: Suppose W is a k-dimensional invariant subspace for T. Any basis $\{e_1, \ldots, e_k\}$ of W extends to a basis of V, $\{e_1, \ldots, e_n\}$. The matrix of T with respect to this basis has the form $\left(\begin{smallmatrix} B & C \\ 0 & D \end{smallmatrix}\right)$, where B is the matrix of $T|_W$. It follows that the characteristic polynomial of $T|_W$ is a factor of the characteristic polynomial of T, of degree k.

Conversely, suppose $\det(a - \lambda I) = f(t)g(t)$ where $f, g \in \mathbf{F}[t]$ are both nonconstant. Then $f(A)g(A) = 0$, by the Cayley–Hamilton Theorem [HK61, p. 194], so at least one of $f(A)$, $g(A)$, is singular. Assume $f(A)$ is singular, and choose a nonzero vector $w \in \ker f(A)$. Then the subspace W spanned by

$$w, Tw, \ldots, T^{k-1}w ,$$

where $k = \deg f$, is invariant under T, nonzero, and has dimension at most $k < n$.

Solution to 7.4.9: Let $f(x) \in \mathbb{R}[x]$ denote the characteristic polynomial of T. If $f(x)$ has a nonreal root λ, let $v \in \mathbb{C}^n$ be a nonzero eigenvector with eigenvalue λ; then the 2-dimensional subspace M spanned by the real vectors $v + \bar{v}$ and $(\bar{v} - v)/i$ will be mapped into itself by T.

Otherwise all roots of the degree-n polynomial $f(x)$ are real. If $f(x)$ has two distinct real roots r_1, r_2, then there exist nonzero eigenvectors v_1 and v_2 with these eigenvalues, and we may take M to be the span of v_1 and v_2.

We are left with the case in which $f(x) = (x - r)^n$ for some $r \in \mathbb{R}$. Let v_1 be a nonzero eigenvector. Then the characteristic polynomial of the transformation T' of $\mathbb{R}^n/(\mathbb{R}v_1)$ induced by T is $(x - r)^{n-1}$, and there exists a nonzero eigenvector $\tilde{v}_2 \in \mathbb{R}^n/(\mathbb{R}v_1)$ of T', since $n > 1$. Choose $v_2 \in \mathbb{R}^n$ representing $\tilde{v}_2$. Then T maps the span M of v_1 and v_2 into itself.

Solution to 7.4.10: Let $V_i = \{v \in V \mid \chi_i(L)(v) = 0\}$, for $i = 1, 2$. Clearly, each V_i is a subspace of V with $\chi_i(L)V_i = 0$. To show that V is the direct sum of V_1 and V_2, choose polynomials a and b over $\mathbf{F}$ for which $a\chi_1 + b\chi_2 = 1$. Then $a(L)\chi_1(L) + b(L)\chi_2(L) = 1$. If $v \in V_1 \cap V_2$, then $v = 1 \cdot v = a(L)\chi_1(L) + b(L)\chi_2(L)v = a(L)0 + b(L)0 = 0$, so $V_1 \cap V_2 = \{0\}$. If $v \in V$, then by the Cayley–Hamilton Theorem [HK61, p. 194], we have $\chi(L)v = 0$. Hence, $v_1 = a(L)\chi_1(L)v$ is annihilated by $\chi_2(L)$ and, therefore, belongs to V_2. Likewise, $v_2 = b(L)\chi_2(L)v$ belongs to V_1. Since $v = v_1 + v_2$, this shows that $V = V_1 + V_2$.

Solution to 7.4.11: It is sufficient to prove that $y \notin \langle x \rangle$ and $z \notin \langle x, y \rangle$. Suppose that $y = \lambda x$ for some $\lambda \in \mathbb{Q}$. Then $z = Ty = \lambda y = \lambda^2 x$, and $x + y = Tz = \lambda^2 y = \lambda^3 x$. Hence $x + \lambda x = \lambda^3 x$, $(\lambda^3 - \lambda - 1)x = 0$ and $\lambda^3 - \lambda - 1 = 0$ since $x \neq 0$. This is a contradiction, since $\lambda^3 - \lambda - 1 = 0$ has no rational solutions. Therefore $y \notin \langle x \rangle$.

Suppose that $z = \alpha x + \beta y$ for some $\alpha, \beta \in \mathbb{Q}$. Then

$$x + y = Tz = \alpha Tx + \beta Ty = \alpha y + \beta z = \alpha y + \beta(\alpha x + \beta y) = \alpha\beta x + (\alpha + \beta^2)y \,.$$

By the independence of x, y we obtain

$$1 = \alpha\beta, \qquad 1 = \alpha + \beta^2 \,.$$

Eliminating β yields $\alpha^3 - \alpha^2 + 1 = 0$. This is a contradiction since $\alpha^3 - \alpha^2 + 1 = 0$ has no rational solutions.

Hence $x \neq 0$, $y \notin \langle x \rangle$ and $z \notin \langle x, y \rangle$, so x, y, z are linearly independent.

Solution to 7.4.14: Since the linear transformation f has rank $n - 1$, we know that $f(\mathbb{R}^m)$ is an $n - 1$–dimensional subspace of $\mathbb{R}^n$. Hence, there exist real constants $\lambda_1, \ldots, \lambda_n$, not all zero, such that

$$\sum_{i=1}^{n} \lambda_i f_i(v) = 0$$

for all $v \in \mathbb{R}^m$. The λ_i's are unique up to constant multiples. Further, this equation determines the subspace: If $w \in \mathbb{R}^n$ satisfies it, then $w \in f(\mathbb{R}^m)$.

Now suppose that the λ_i's all have the same sign, or, without loss of generality, that they are all nonnegative. Then if there existed $v \in \mathbb{R}^m$ with $f_i(v) > 0$ for all i, we would have $\sum \lambda_i f_i(v) > 0$, a contradiction. Hence, there can be no such v.

Conversely, suppose that two of the λ_i's, say λ_1 and λ_2, have different signs. Let $x_3 = x_4 = \cdots = x_n = 1$, and choose $x_1 > 0$ sufficiently large so that

$$\sum_{i \neq 2} \lambda_i x_i > 0.$$

Then there is a real number $x_2 > 0$ such that

$$\sum_{i=1}^{n} \lambda_i x_i = 0.$$

But then we know that there exists $v \in f(\mathbb{R}^m)$ such that $f(v) = (x_1, \ldots, x_n)$. Since each of the x_i's is positive, we have found the desired point v.

Solution to 7.4.15: Let $\langle \, , \, \rangle$ denote the ordinary inner product. From $d(s, t)^2 = d(s, 0)^2 + d(t, 0)^2 - 2\langle s, t \rangle$ and the hypothesis, it follows that

$$\langle \varphi(s), \varphi(t) \rangle = \langle s, t \rangle \qquad \text{for all} \quad s, t \in S.$$

Let $V \subset \mathbb{R}^n$ denote the subspace spanned by S, and choose a subset $T \subset S$ that is a basis of V. Clearly, there is a unique linear map $f : V \to V$ that agrees with φ on T. Then one has $\langle f(t), f(t') \rangle = \langle t, t' \rangle$ for all t and $t' \in T$. By bilinearity, it follows that $\langle f(v), f(v') \rangle = \langle v, v' \rangle$ for all v and $v' \in V$. Taking $v = v'$, one finds that $f(v) \neq 0$ for $v \neq 0$, so f is injective, and $f(V) = V$. Taking $v = s \in S$ and $v' = t \in T$, one finds that

$$\langle f(s), f(t) \rangle = \langle s, t \rangle = \langle \varphi(s), \varphi(t) \rangle = \langle \varphi(s), f(t) \rangle,$$

so $f(s) - \varphi(s)$ is orthogonal to $f(t)$ for all $t \in T$, and hence to all of $f(V) = V$. That is, for all $s \in S$, one has $f(s) - \varphi(s) \in V^\perp$; but also $f(s) - \varphi(s) \in V$, so $f(s) - \varphi(s) = 0$. This shows that f agrees with φ on S. It now suffices to extend f to a linear map $\mathbb{R}^n \to \mathbb{R}^n$, which one can do by supplementing T to a basis for $\mathbb{R}^n$ and defining f arbitrarily on the new basis vectors.

Solution to 7.4.16: Define $U(x) = T(x) - T(0)$, then $\|U(x) - U(y)\| = \|x - y\|$ for all x and y and, in particular, since 0 is a fixed point for U, $\|U(x)\| = \|x\|$. Squaring the first equality yields

$$\langle U(x) - U(y), U(x) - U(y) \rangle = \langle x - y, x - y \rangle,$$

or

$$\|U(x)\|^2 - 2\langle U(x), U(y) \rangle + \|U(y)\|^2 = \|x\|^2 - 2\langle x, y \rangle + \|y\|^2.$$

Canceling equal terms gives $\langle U(x), U(y) \rangle = \langle x, y \rangle$. Expanding the inner product $\langle U(x+y) - U(x) - U(y), U(x+y) - U(x) - U(y) \rangle$ and applying this equality immediately gives that $U(x+y) - U(x) - U(y) = 0$. A similar calculation shows that $U(\alpha x) - \alpha U(x) = 0$. Hence, U is linear, and $\langle U(x), U(y) \rangle = \langle x, y \rangle$, so U is in fact orthogonal. Letting $a = T(0)$ gives the desired decomposition of T.

Solution 2. An isometry T of $\mathbb{R}^2$ is a bijection $\mathbb{R}^2 \to \mathbb{R}^2$ which preserves distance. Such a map also preserves the inner product, and hence preserves angles. Euclidean geometry shows that a point in the plane is specified by its distances from any three non-collinear points. Thus an isometry is determined by its images of any three non-collinear points.

The orthogonal linear maps of $\mathbb{R}^2$ correspond to rotations about 0, and reflections in lines through 0. The rotation of $\mathbb{R}^2$ through the angle θ is the map $R: z \mapsto z e^{i\theta}$. The reflection on a line with an angle α with the x-axis is given by composing three maps: rotation through $-\alpha$, reflection in the x-axis, rotation through α. These are the maps $z \mapsto z e^{-i\alpha}$, $z \mapsto \bar{z}$, and $z \mapsto z e^{i\alpha}$, respectively. Thus reflection in that line is the map $S: z \mapsto \bar{z} e^{2i\alpha}$.

Let $T(0) = a$. The translation $W(x) = x - a$ is an isometry, and $U = WT$ is an isometry that satisfies $U(0) = 0$. Now $d(U(1), 0) = d(U(1), U(0)) = d(1, 0) = 1$ and also $d(U(i), 0) = 1$. Further if $U(i) = e^{i\theta}$, then, by preservation of angles, we see that $U(i) = \pm i e^{i\theta}$.

First the case $U(i) = iU(1)$. As $R : z \mapsto z e^{i\theta}$ has the same effect as U on the non-collinear points $0, 1, i$, we deduce that U is that rotation. Next the case $U(i) = -iU(1)$. Let $S: z \mapsto \bar{z} e^{i\theta}$ be the reflection in line of slope $\alpha = \theta/2$. We see that that U agrees with S at $0, 1, i$. Thus U is reflection in the line with slope $\theta/2$. In either case, U is an orthogonal map.

Finally, as $U = WT$ we have $T(x) = W^{-1}U(x) = a + U(x)$, where U is an orthogonal linear transformation.

Solution to 7.4.17: We use Complete Induction [MH93, p. 32] on the dimension of V. If $\dim V = 1$, then V has a single basis vector $f_1 \neq 0$, so there is $x_1 \in X$ such that $f_1(x_1) \neq 0$. Hence, the map $f \mapsto f(x_1)$ is the desired isomorphism.

Now suppose the result is true for dimensions less than n and let $\dim V = n$. Fix a basis $\{f_1, f_2, \ldots, f_n\}$ of V. Then, by the induction hypothesis, there are points $x_1, x_2, \ldots, x_{n-1}$ such that the map $f \mapsto (f(x_1), \ldots, f(x_{n-1}), 0)$ is an isomorphism of the subspace of V spanned by $\{f_1, \ldots, f_{n-1}\}$ onto $\mathbb{R}^{n-1} \subset \mathbb{R}^n$. In particular, the vector $(f_n(x_1), \ldots, f_n(x_{n-1}), 0)$ is a linear combination of the basis vectors $\{(f_i(x_1), \ldots, f_i(x_{n-1}), 0), 1 \leqslant i \leqslant n - 1\}$, so there exists a unique set of λ_i's, $1 \leqslant i \leqslant n$, such that

$$\sum_{i=1}^{n} \lambda_i f_i(x_j) = 0, \qquad 1 \leqslant j \leqslant n - 1.$$

Suppose there is no point $x \in X$ such that the given map is an isomorphism from V onto $\mathbb{R}^n$. This implies that the set $\{(f_i(x_1), \ldots, f_i(x_{n-1}), f_i(x)), 1 \leqslant i \leqslant n\}$ is linearly dependent for all x. But because of the uniqueness of the λ_i's, this, in turn, implies that for all x,

$$\sum_{i=1}^{n} \lambda_i f_i(x) = 0.$$

Hence, the f_i's are linearly dependent in V, a contradiction. Therefore, such an x exists and we are done.

Solution to 7.4.18: We argue by induction on the dimension of V.

If $\dim V = 1$ let $\{g_1\}$ be a basis of V. g is not the zero function, therefore we have $g(x_1) \neq 0$ for some $x_1 \in X$. Thus $f_1(x) = g_1(x)/g_1(x_1)$ satisfies the given condition.

Suppose now that for any given subspace W of the vector space of continuous real valued functions on X with $\dim W = n$ there exists a basis $\{f_1, \ldots, f_n\}$ for W and points $x_1, \ldots, x_n \in X$ such that $f_i(x_j) = \delta_{ij}$.

Let V be such a subspace with $\dim V = n + 1$. Let $\{g_1, \ldots, g_{n+1}\}$ be a basis for V, and let W be the subspace spanned by $\{g_1, \ldots, g_n\}$. Then there exists a basis for W, $\{h_1, \ldots, h_n\}$, and points $x_1, \ldots, x_n \in X$ with $h_i(x_j) = \delta_{ij}$.

As each h_i is a linear combination of elements of the basis $\{g_1, \ldots, g_n\}$, the set $\{h_1, \ldots, h_n, g_{n+1}\}$ is linerly independent, therefore a basis for V. Let

$$h_{n+1} = g_{n+1} - g_{n+1}(x_1)h_1 - \cdots - g_{n+1}(x_n)h_n ,$$

then $\{h_1, \ldots, h_n, h_{n+1}\}$ is linearly independent and, for each $1 \leqslant j \leqslant n$, we have

$$h_{n+1}(x_j) = g_{n+1}(x_j) - \sum_{i=1}^{n} g_{n+1}(x_i)\delta_{ij} = g_{n+1}(x_j) - g_{n+1}(x_j) = 0 .$$

As h_{n+1} is nonzero, there is $x_{n+1} \in X$ with $h_{n+1}(x_{n+1}) \neq 0$. We have $x_{n+1} \neq x_i$ for $1 \leqslant i \leqslant n$ since $h_{n+1}(x_i) = 0$. Let $f_{n+1} = h/h_{n+1}(x_{n+1})$. For $1 \leqslant i \leqslant n$ let $f_i = h_i - h_i(x_{n+1})f_{n+1}$ so that $f_i(x_{n+1}) = 0$. The set $\{f_1, \ldots, f_{n+1}\}$ is linearly independent and satisfies $f_i(x_j) = \delta_{ij}$.

Solution to 7.4.19: Since the formula holds, irrespective of the values of c_k, for the polynomials x^{2n+1}, it suffices, by linearity, to restrict to the vector space P_{2n} of polynomials of degree, at most, $2n$. This vector space has dimension $2n + 1$ and the map $P_{2n} \rightarrow \mathbb{R}^{2n+1}$ given by $p \mapsto (p(-n), p(-n + 1), \ldots, p(n))$ is an isomorphism. As the integral is a linear function on $\mathbb{R}^{2n+1}$, there exist unique real numbers $c_{-n}, c_{-n+1}, \ldots, c_n$ such that

$$\int_{-1}^{1} p(x)dx = \sum_{k=-n}^{n} c_k p(k) \qquad \text{for all } p \in P_{2n}.$$

We have

$$\int_{-1}^{1} p(x)dx = \int_{-1}^{1} p(-x)dx = \sum_{k=-n}^{n} c_k p(k) \quad \text{for all } p \in P_{2n},$$

so $c_k = c_{-k}$ by uniqueness of the c_k, and, therefore,

$$\int_{-1}^{1} p(x)dx = c_0 p(0) + \sum_{k=1}^{n} c_k \left(p(k) + p(-k)\right) \qquad \text{for all } p \in P_{2n}.$$

Setting $p = 1$, we find that

$$2 = c_0 + \sum_{k=1}^{n} 2c_k$$

so, upon eliminating c_0,

$$\int_{-1}^{1} p(x)dx = 2p(0) + \sum_{k=1}^{n} c_k \left(p(k) + p(-k) - 2p(0) \right).$$

Solution to 7.4.20: Let $\{v_1, v_2, \ldots, v_n\}$ be a basis for $\mathbb{R}^n$ consisting of eigenvectors of T, say $Tv_n = \lambda_n v_n$. Let $\{u_1, u_2, \ldots, u_n\}$ be the orthonormal basis one obtains from $\{v_1, v_2, \ldots, v_n\}$ by the Gram–Schmidt Procedure [HK61, p. 280]. Then, for each index k, the vector u_k is a linear combination of $v_1, \ldots, v_k$, say

$$u_k = c_{k1}v_1 + c_{k2}v_2 + \cdots + c_{kk}v_k.$$

Also, each v_k is a linear combination of $u_1, \ldots, u_k$. (This is guaranteed by the Gram–Schmidt Procedure; in fact, $\{u_1, \ldots, u_k\}$ is an orthonormal basis for the subspace generated by $\{v_1, \ldots, v_k\}$.) We have

$$Tu_k = c_{k1}Tv_1 + c_{k2}Tv_2 + \cdots + c_{kk}Tv_k$$
$$= c_{k1}\lambda_1 v_1 + c_{k2}\lambda_2 v_2 + \cdots + c_{kk}\lambda_k v_k.$$

In view of the preceding remark, it follows that Tu_k is a linear combination of $u_1, \ldots, u_k$, and, thus, T has upper-triangular matrix in the basis $\{u_1, u_2, \ldots, u_n\}$.

Solution to 7.4.21: Let $\dim V = k$. Relative to a basis of V, each linear operator f on V is represented by a matrix A over $\mathbf{F}$. The correspondence $f \mapsto A$ is a ring isomorphism from $L(V, V)$ onto $M_k(\mathbf{F})$. Further, f is invertible if, and only if, A is invertible. As $\mathbf{F}$ is finite, the multiplicative group of invertible matrices $GL_k(\mathbf{F})$ is a finite group, so each element in $GL_k(\mathbf{F})$ has finite order. Thus if P is invertible, there is n such that $P^n = I$.

Solution to 7.4.23: 1. The characteristic polynomial of T has degree 3 so it has at least one real root. The space generated by the eigenvector associated with this eigenvalue is invariant under T.
2. The linear transformation $T - \lambda I$ has rank 0, 1, or 2. If the rank is 0 then $T = \lambda I$ and all subspaces are invariant, if it is 1 then $\ker(T - \lambda I)$ will do and if it is 2 the image of $(T - \lambda I)$ is the desired subspace. This is equivalent to the Jordan Canonical Form [HK61, p. 247] of T being either a diagonal matrix with three 1×1 blocks or with one 1×1 and one 2×2 block, in both cases there is a 2-dimensional invariant subspace.

Solution to 7.4.24: Since A is symmetric and orthogonal, we have $A^2 = I$, therefore its minimal polynomial, μ_A, divides $x^2 - 1$, which leaves three possibilities:

- $\mu_A = x - 1$. In this case $A = I$.

- $\mu_A = x + 1$. We have $A = -I$.

- $\mu_A = x^2 - 1$. Here we consider two cases. If $\dim \ker(A - I) = 2$, there is a 2-dimensional subspace where every vector remains fixed ($\ker(A - I)$), and the complement of $\ker(A - I)$, $\ker(A + I)$, has dimension 1, is perpendicular to $\ker(A - I)$, since A is orthogonal. In this case we have a reflection about $\ker(A - I)$.

  If $\dim \ker(A - I) = 1$, there is an axis ($\ker(A - I)$) that remains fixed and $\ker(A + I)$ has dimension 2. Moreover, since A is orthogonal, $\ker(A + I) \perp \ker(A - I)$, hence for $x \in \ker(A + I)$ we have $A(x) = -x$, that is, A is a rotation by $180°$ about the axis $\ker(A - I)$.

Solution to 7.4.25: Clearly, both R and S are rotations and so have rank 3. Therefore, T, their composition, is a rank 3 operator. In particular, it must have trivial kernel. Since T is an operator on $\mathbb{R}^3$, its characteristic polynomial is of degree 3, and so it has a real root. This root is an eigenvalue, which must be nontrivial since T has trivial kernel. Hence, the associated eigenspace must contain a line which is fixed by T.

Solution to 7.4.26: Let $x = (x_1, x_2, x_3)$ in the standard basis of $\mathbb{R}^3$. The line joining the points x and Tx intersects the line containing e at the point $f = \langle e, x \rangle e$ and is perpendicular to it. We then have $Tx = 2(f - x) + x = 2f - x$, or, in the standard basis, $Tx = (2\langle e, x \rangle a - x_1, 2\langle e, x \rangle b - x_2, 2\langle e, x \rangle c - x_3)$. With respect to the standard basis for $\mathbb{R}^3$, the columns of the matrix of T are Te_1, Te_2, and Te_3. Applying our formula and noting that $\langle e, e_1 \rangle = a$, $\langle e, e_2 \rangle = b$, and $\langle e, e_3 \rangle = c$, we get that the matrix for T is

$$\begin{pmatrix} 2a^2 - 1 & 2ab & 2ac \\ 2ab & 2b^2 - 1 & 2bc \\ 2ac & 2bc & 2c^2 - 1 \end{pmatrix}.$$

Solution to 7.4.27: Since the minimal polynomial divides the characteristic polynomial and this last one has degree 3, it follows that the characteristic polynomial of T is $(t^2 + 1)(t - 10)$ and the eigenvalues $\pm i$ and 10.

Now $T(1, 1, 1) = \lambda(1, 1, 1)$ implies that $\lambda = 10$ because 10 is the unique real eigenvalue of T.

The plane perpendicular to $(1, 1, 1)$ is generated by $(1, -1, 0)$ and $(\frac{1}{2}, \frac{1}{2}, -1)$ since these are perpendicular to each other and to $(1, 1, 1)$.

Let

$$\begin{aligned} f_1 &= (1, 1, 1) \\ f_2 &= (1, -1, 0)/\sqrt{2} \\ f_3 &= \left(\tfrac{1}{2}, \tfrac{1}{2}, -1\right)/\sqrt{\tfrac{1}{4} + \tfrac{1}{4} + 1} = \left(\tfrac{1}{2}, \tfrac{1}{2}, -1\right)/\sqrt{\tfrac{3}{2}} \end{aligned}$$

we have $Tf_1 = 10f_1$ and, for $\pm i$ to be the other eigenvalues of T, $Tf_2 = f_3$, and $Tf_3 = -f_2$.

The matrix of T in the basis $\{f_1, f_2, f_3\} = \beta$ is then

$$[T]_\beta = \begin{pmatrix} 10 & 0 & 0 \\ 0 & 0 & 1 \\ 0 & -1 & 0 \end{pmatrix}$$

The matrix that transforms the coordinates relative to the basis β into the coordinates relative to the canonical basis is

$$P = \begin{pmatrix} 1 & 1/\sqrt{2} & \sqrt{6}/4 \\ 1 & -1/\sqrt{2} & \sqrt{6}/4 \\ 1 & 0 & -\sqrt{6}/2 \end{pmatrix}$$

and a calculation gives

$$P^{-1} = \begin{pmatrix} 1/3 & 1/3 & 1/3 \\ \sqrt{2}/2 & -\sqrt{2}/2 & 0 \\ 2/3\sqrt{6} & 2/3\sqrt{6} & -4/3\sqrt{6} \end{pmatrix}.$$

Therefore, the matrix of T in the canonical basis is

$$[T] = P[T]_\beta P^{-1} = \begin{pmatrix} \frac{10}{3} + \frac{1}{3\sqrt{3}} & \frac{10}{3} + \frac{13}{36}\sqrt{3} & \frac{10}{3} - \frac{2}{3\sqrt{3}} \\ \frac{10}{3} - \frac{1}{3\sqrt{3}} & \frac{10}{3} + \frac{5\sqrt{3}}{36} & \frac{10}{3} + \frac{2}{3\sqrt{3}} \\ \frac{10}{3} + \frac{\sqrt{3}}{2} & \frac{10}{3} - \frac{\sqrt{3}}{2} & \frac{10}{3} \end{pmatrix}.$$

Solution to 7.4.28: $AA^t = I$ so $|A|^2 = 1$ and indeed $|A| = 1$. Thus also $|A^t| = 1$. We have

$$|A - I| = |(A - I)|\,|A^t| = |(A - I)A^t|$$
$$= |I - A^t| = |(I - A^t)^t|$$
$$= |I - A| = (-1)^3|A - I|;$$

hence $|A - I| = 0$ and there is a nonzero v such that $Av = v$.

Solution to 7.4.29: For $n = 1, 2, \ldots$, let P_n be the space of polynomials whose degrees are, at most, n. The subspaces P_n are invariant under E, they increase with n, and their union is P. To prove E is invertible (i.e., one-to-one and onto), it will suffice to prove that each restriction $E|_{P_n}$ is invertible. The subspace P_n is of dimension $n + 1$, it has the basis $1, x, x^2, \ldots, x^n$, with respect to which the matrix of $E|_{P_n}$ is

$$\begin{pmatrix} 1 & 1 & 0 & 0 & \cdots & 0 \\ 0 & 1 & 2 & 0 & \cdots & 0 \\ 0 & 0 & 1 & 3 & \cdots & 0 \\ \vdots & \vdots & \vdots & \ddots & \ddots & \vdots \\ 0 & 0 & 0 & 0 & 1 & n \\ 0 & 0 & 0 & 0 & 0 & 1 \end{pmatrix}.$$

In particular, the matrix is upper-triangular, with 1 at every diagonal entry, so its determinant is 1. Thus, $E|_{P_n}$ is invertible, as desired. Alternatively, since deg $Ef = \deg f$, the kernel of E is trivial, so its restriction to any finite dimensional invariant subspace is invertible.

Solution 2. We can describe E to be $I + D$, where I is the identity operator and D is the derivative operator, on the vector space of all real polynomials P. For any element f of P, there exists n such that $D^n(f) = 0$; namely $n = \deg p + 1$. Thus, the inverse of E can be described as $I - D + D^2 - D^3 + \cdots$.

Specifically, writing elements of P as polynomials in x, we have $E^{-1}(1) = 1$, $E^{-1}(x) = x - 1$, $E^{-1}(x^2) = x^2 - 2x + 2$, etc.

Solution to 7.4.30: Given the polynomial $\pi(x)$, there are constants a and $r > 0$ and a polynomial $\varphi(x)$ such that $\pi(x) = x^r \varphi(x) + a$. If $\varphi(x) \equiv 0$, then $\pi(D) = aI$, it follows that the minimal polynomial of the operator $\pi(D)$ is $x - a$. If $\varphi(x)$ is not zero, then for any polynomial $f \in P_n$, by the definition of D, $(\pi(D) - aI)(f(x)) = g(x)$, where $g(x)$ is some polynomial such that $\deg g = \max(\deg f - r, 0)$. Hence, letting $E = \pi(D) - aI$, we have $e^{\lfloor n/r \rfloor + 1}(f) = 0$ for all $f \in P_n$. ($\lfloor n/r \rfloor$ denotes the greatest integer less than or equal to n/r.) The polynomial $f(x) = x^n$ shows that $\lfloor n/r \rfloor + 1$ is the minimal degree such that this is true. It follows from this that the minimal polynomial of $\pi(D)$ is $(x - a)^{\lfloor n/r \rfloor + 1}$.

Solution to 7.4.31: 1. Consider the basis $\{1, x, x^2, \ldots, x^{10}\}$ for this space of polynomials of dimension 11. On this basis the operator takes each element into a multiple of a previous one, of one degree less, so the diagonal on this basis consist of zeroes and $\operatorname{tr} D = 0$.

2. Since D decreases the degree of the polynomial by one, the equation

$$Dp(x) = \alpha p(x)$$

has no non-trivial solutions for α, and D has only 0 as eigenvalue, the eigenvectors being the constant polynomials.

Now since the 11th-derivative of any of the polynomials in this space is zero, the expression for the exponential of D becomes a finite sum

$$e^D = I + D + \frac{D^2}{2!} + \cdots + \frac{D^{10}}{10!},$$

then any eigenvector $p(x)$ of degree k of e^D would have to satisfy

$$p + p' + \frac{p''}{2!} + \cdots + \frac{p^{(10)}}{10!} = \alpha p$$

since p is the only polynomial of degree k in this sum, the eigenvalue $\alpha = 1$ and the summation can be reduced to

$$p' + \frac{p''}{2!} + \cdots + \frac{p^{(10)}}{10!} = 0$$

now p' is the only polynomial of degree $k-1$ in the sum so its leading coefficient $k.a_k$ is zero and the same argument repeated over the lower degrees shows that all of them are zero except for the constant one, making it again the only eigenvector.

Solution to 7.4.32: Let $\mathcal{P}_n$ be the vector space of polynomials with real coefficients and degrees at most n. It is a real vector space of dimension $n + 1$. Define the map $V : \mathcal{P}_n \to \mathbb{R}^{n+1}$ by

$$V p = (p(x_0), p(x_1), \ldots, p(x_n)) .$$

This map is linear, and its null space is trivial because a nonzero polynomial in $\mathcal{P}_n$ cannot vanish at $n + 1$ distinct points. Hence V is invertible. Define the homomorphism $\varphi : \mathbb{R}^{n+1} \to \mathbb{R}$ by

$$\varphi(w) = \int_0^1 (V^{-1}w)(t)dt .$$

Being a linear functional φ is induced by a unique vector $a = (a_0, a_1, \ldots, a_n)$ in $\mathbb{R}^{n+1}$:

$$\varphi(w) = \sum_{j=0}^n a_j w_j .$$

The numbers $a_0, a_1, \ldots, a_n$ have the required property.

Solution to 7.4.33: 1. If p satisfies the equations then it does not have a constant sign on I_j, $j = 1, \ldots, n$, so p has at least n zeros. If $\deg p < n$ this implies that p is the zero polynomial.
2. Let $\mathcal{P}_n$ denote the vector space of real polynomials of degree at most n. Its dimension is $n + 1$. Each I_j defines a functional φ_j in the dual space of $\mathcal{P}_n$ by

$$\varphi_j(p) = \int_{I_j} p(x)dx .$$

The dual space of $\mathcal{P}_n$ has dimension $n + 1$, so the functionals $\varphi_1, \ldots, \varphi_n$ do not span it. Hence there is a nonzero p in $\mathcal{P}_n$ such that $\varphi_j(p) = 0$ for each j, in other words, all equations hold.

7.5   Eigenvalues and Eigenvectors

Solution to 7.5.1: 1. The minimal polynomial of M divides

$$x^3 - 1 = (x - 1)(x^2 + x + 1) .$$

Since $M \neq I$, the minimal polynomial (and characteristic as well) is

$$(x - 1)(x^2 + x + 1)$$

and the only possible real eigenvalue is 1.

2.

$$\begin{pmatrix} 1 & 0 & 0 \\ 0 & \cos\frac{2\pi}{3} & \sin\frac{2\pi}{3} \\ 0 & -\sin\frac{2\pi}{3} & \cos\frac{2\pi}{3} \end{pmatrix}$$

Solution to 7.5.2: Suppose such X exists; then the characteristic polynomial for X is $\chi_X(t) = t^n$, but this is a contradiction since $X^{2n-1} \neq 0$ and $2n - 1 > n$.

Solution to 7.5.3: Since $A^m = 0$ for some m, A is a root of the polynomial $p(x) = x^m$. By the definition of the minimal polynomial, $\mu_A(t) | p(t)$, so $\mu_A(t) = t^k$ for some $k \leqslant n$, then $A^n = A^k = 0$.

Solution to 7.5.4: Let $\{\alpha, \beta\}$ be a basis for the kernel of B, and let γ be another vector so that $\{\alpha, \beta, \gamma\}$ is a basis for the three dimensional space. The representation of B with respect to this basis is

$$\begin{pmatrix} 0 & 0 & a \\ 0 & 0 & b \\ 0 & 0 & c \end{pmatrix}$$

and the characteristic polynomial is: $\chi(\lambda) = \lambda^2(c - \lambda)$, from which the two first assertions follow.

The third is false since if we take $a = 1$, $b = c = 0$ above we get a matrix all of whose eigenvalues vanish, but which is not diagonalizable since it is not similar to the zero matrix.

Solution to 7.5.5: The condition on the minimal polynomial implies that T is not the zero operator and that its square, T^2, is. Therefore, $\dim \ker T$ is at most six, and $\dim \ker T^2$ is 7. Since the nullity of the square of any linear operator is at most twice the nullity of the operator, the nullity of T is at least 4.

Each of the three remaining integers, 4, 5, 6, is possible, as is illustrated by the examples below, where $\{e_1, e_2, \ldots e_7\}$ is a basis for the space.

- $\dim \ker T = 6$. Define T by $Te_7 = e_6$, and $Te_i = 0$ for all $i < 7$.

- $\dim \ker T = 5$. Let $Te_7 = e_5$, $Te_6 = e_4$, and $Te_i = 0$ for all $i < 6$.

- $\dim \ker T = 4$. Define T by $Te_7 = e_4$, $Te_6 = e_3$, $Te_5 = e_2$, and $Te_i = 0$ for all $i < 5$.

Solution to 7.5.6: 1. Being real and symmetric, L has an orthonormal basis of eigenvectors $e_1, \ldots, e_n$. Let $\lambda_1, \ldots, \lambda_n$ be the associated eigenvalues. We can assume that $\lambda_1 = 0$ and $\lambda_i \neq 0$ for $i > 1$. Write the two vectors $v = \sum_{i=1}^n v_i$ and $x = \sum_{i=1}^n x_i$ in terms of its coordinates, then the equation $Lx + \varepsilon = v$ becomes

$\lambda_i x_i + \varepsilon x_i = v_i$ for each i, which has the unique solution $x_i = v_i/(\lambda_i + \varepsilon)$, provided that $0 < \varepsilon < \min_{i \neq 1} |\lambda_i|$.

2. Writing εx in e_i coordinates

$$\varepsilon x = \sum_{i=1}^{n} \varepsilon x_i e_i = \sum_{i=1}^{n} \frac{\varepsilon}{\lambda_i + \varepsilon} v_i e_i$$

as $\varepsilon \to 0$, all terms in the summation on the right tend to 0, except the first which approaches $v_1 e_1 = \langle v, e_1 \rangle e_1$.

Solution to 7.5.7: As $M^p = I$, the minimal polynomial of M divides $t^p - 1 = (t - 1)(t^{p-1} + t^{p-2} + \cdots + t + 1)$. Since M fixes no nontrivial vector, 1 is not an eigenvalue of M, so it cannot be a root of the minimal polynomial. Therefore, $\mu_M(t)|(t^{p-1} + t^{p-2} + \cdots + t + 1)$. Since p is prime, the polynomial on the right is irreducible, so $\mu_M(t)$ must equal it. The minimal and characteristic polynomials of M have the same irreducible factors, so $\chi_M(t) = \mu_M(t)^k$ for some $k \geq 1$. Therefore,

$$\dim V = \deg \chi_M(t) = k(p - 1)$$

and we are done.

Solution to 7.5.8: 1. Let $d = \deg \mu$. Since $\mu(T)v = 0$, the vector $T^d v$ is linearly dependent on the vectors $v, Tv, \ldots, T^{d-1}v$. Hence, $T^{d+n}v$ is linearly dependent on $T^n v, T^{n+1}v, \ldots, T^{n+d-1}v$ and so, by the Induction Principle [MH93, p. 7], on $v, Tv, \ldots, T^{d-1}v$ ($n = 1, 2, \ldots$). Thus, $v, Tv, \ldots, T^{d-1}v$ span V_1, so dim $V_1 \leq d$.

On the other hand, the minimal polynomial of $T|_{V_1}$ must divide μ (since $\mu(T|_{V_1}) = 0$), so it equals μ because μ is irreducible. Thus, $\dim V_1 \geq d$. The desired equality, $\dim V_1 = d$, now follows.

2. In the case $V_1 \neq V$, let T_1 be the linear transformation on the quotient space V/V_1 induced by T. (It is well defined because V_1 is T–invariant.) Clearly, $\mu(T_1) = 0$, so the minimal polynomial of T_1 divides μ, hence equals μ. Therefore, by Part 1, V/V_1 has a T_1–invariant subspace of dimension d, whose inverse image under the quotient map is a T–invariant subspace V_2 of V of dimension $2d$. In the case $V_2 \neq V$, we can repeat the argument to show that V has a T–invariant subspace of dimension $3d$, and so on. After finitely many repetitions, we find dim $V = kd$ for some integers k.

Solution to 7.5.9: Let v be an eigenvector in $\mathbb{C}^n$ for M with nonzero eigenvalue α. Then,

$$M A^{-1} v = A^{-1} M^2 v = \alpha^2 A^{-1} v$$

so α^2 is also an eigenvalue. Thus α^{2^k} is an eigenvalue for all nonnegative integers k. Since the set of eigenvalues is finite there exist $0 \leq k < m \in \mathbb{Z}$ such that $\alpha^{2^k} = \alpha^{2^m}$. Thus α is a root of unity.

Solution to 7.5.10: Since the matrix is real and symmetric, its eigenvalues are real. As the trace of the matrix is 0, and equal to the sum of its eigenvalues, it has at least one positive and one negative eigenvalue.

The matrix is invertible because the span of its columns has dimension 4. In fact, the space of the first and last columns contains all columns of the form

$$\begin{pmatrix} 0 \\ * \\ * \\ 0 \end{pmatrix}.$$

The span of all four columns thus contains

$$\begin{pmatrix} 5 \\ 0 \\ 0 \\ 1 \end{pmatrix} \quad \text{and} \quad \begin{pmatrix} 1 \\ 0 \\ 0 \\ 5 \end{pmatrix},$$

which together span all columns of the form

$$\begin{pmatrix} * \\ 0 \\ 0 \\ * \end{pmatrix}.$$

Since the matrix is invertible it does not have 0 as an eigenvalue. There are now only three possibilities:

- three positive and one negative eigenvalues;

- two positive and two negative eigenvalues;

- one positive and three negative eigenvalues.

A calculation shows that the determinant is positive. Since it equals the product of the eigenvalues, we can only have two positives and two negatives, completing the proof.

Solution to 7.5.11: A calculation shows that the characteristic polynomial of the given matrix is

$$-x(x^2 - 3x - 2(1.00001^2 - 1))$$

therefore one of the eigenvalues is 0 and the product of the other two is

$$-2(1.00001^2 - 1)) < 0$$

so one is negative and the other is positive.

Solution to 7.5.12: Denote the matrix by A. A calculation shows that A is a root of the polynomial $p(t) = t^3 - ct^2 - bt - a$. In fact, this is the minimal polynomial of A. To prove this, it suffices to find a vector $x \in \mathbf{F}^3$ such that $A^2 x$, Ax, and x are linearly independent. Let $x = (1, 0, 0)$. Then $Ax = (0, 1, 0)$ and $A^2x = (0, 0, 1)$; these three vectors are linearly independent, so we are done.

Solution to 7.5.13: The matrix $A + I$ is diagonalizable, being unitary, so A also is diagonalizable. It will therefore suffice to show that A has no nonzero eigenvalues.

Suppose λ is a nonzero eigenvalue of A. Then $\lambda^j + 1$ is an eigenvalue of $A^j + I$ ($j = 1, 2, 3$) and so has unit modulus. Thus $\lambda, \lambda^2, \lambda^3$ all lie on the circle $|z + 1| = 1$, so each has an argument strictly between $\frac{\pi}{2}$ and $\frac{3\pi}{2}$. But if $\frac{3\pi}{4} \leqslant \arg \lambda \leqslant \pi$ then $\frac{3\pi}{2} \leqslant \arg \lambda^2 \leqslant 2\pi$, a contradiction. And if $\frac{\pi}{2} < \arg \lambda < \frac{3\pi}{4}$ then $\frac{3\pi}{2} < \arg \lambda^3 < \frac{9\pi}{4} \equiv \frac{\pi}{4} \pmod{2\pi}$, again a contradiction. The case where λ is in the third quadrant can be handled similarly.

Solution to 7.5.14: The first direction is obvious, lets assume now that every nonzero vector is an eigenvector of A, then in particular, the vectors of the standard basis, e_i are are eigenvectors and A is diagonal. Lets assume the diagonal entries are $A_{ii} = \lambda_i$. If $\lambda_i \neq \lambda_j$, then $A(e_i + e_j) = \lambda_i e_i + \lambda_j e_j$ is not a multiple scalar of $e_i + e_j$, contradicting the hypothesis that $e_i + e_j$ is an eigenvector of A. Then all diagonal entries are equal, finishing the proof.

Solution to 7.5.15: 1. This is false. Consider $A = \begin{pmatrix} 1 & 1 \\ 1 & 1 \end{pmatrix}$ and $B = \begin{pmatrix} 1 & 1 \\ 0 & 1 \end{pmatrix}$. $AB = \begin{pmatrix} 1 & 2 \\ 1 & 2 \end{pmatrix}$ and has $(1, 1)$ as an eigenvector, which is clearly not an eigenvector of $BA = \begin{pmatrix} 2 & 2 \\ 1 & 1 \end{pmatrix}$. The condition that AB and BA have a common eigenvector is algebraic in the entries of A and B; therefore it is satisfied either by all of A and B or by a subset of codimension at least one, so in dimensions two and higher almost every pair of matrices would be a counterexample.

2. This is true. Let x be an eigenvector associated with the eigenvalue λ of AB. We have

$$BA(Bx) = B(ABx) = B(\lambda x) = \lambda Bx$$

so λ is an eigenvalue of BA.

Solution to 7.5.16: Regard A and B as linear transformations on $\mathbb{C}^n$. Then A has an eigenvalue λ. Let S_λ be the corresponding eigenspace. For v in S_λ, we have

$$A(Bv) = B(Av) = B(\lambda v) = \lambda Bv,$$

showing that S_λ is invariant under B. The linear transformation $B|_{S_\lambda}$ has an eigenvalue μ. If v is a corresponding eigenvector we have $Av = \lambda v$ and $Bv = \mu v$.

Solution to 7.5.17: We use the Induction Principle [MH93, p. 7]. As the space $M_n(\mathbb{C})$ is finite dimensional, we may assume that S is finite. If S has one element, the result is trivial. Now suppose any commuting set of n elements has a common eigenvector, and let S have $n+1$ elements $A_1, \ldots, A_{n+1}$. By induction hypothesis, the matrices $A_1, \ldots, A_n$ have a common eigenvector v. Let E be the vector space spanned by the common eigenvectors of $A_1, \ldots, A_n$. If $v \in E$, $A_i A_{n+1} v = A_{n+1} A_i v = \lambda_i A_{n+1} v$ for all i, so $A_{n+1} v \in E$. Hence, A_{n+1} fixes E. Let B be the restriction of A_{n+1} to E. The minimal polynomial of B splits into linear factors (since we are dealing with complex matrices), so B has an eigenvector in E, which must be an eigenvector of A_{n+1} by the definition of B, and an eigenvector for each of the other A_i's by the definition of E.

Solution to 7.5.20: 1. For $(a_1, a_2, a_3, \ldots)$ to be an eigenvector associated with the eigenvalue λ, we must have

$$S((a_1, a_2, a_3, \ldots)) = \lambda(a_2, a_3, a_4, \ldots)$$

which is equivalent to

$$a_2 = \lambda a_1, \quad a_3 = \lambda a_2, \quad \ldots, \quad a_n = \lambda a_{n-1}, \quad \ldots$$

so the eigenvectors are of the form $a_1(1, \lambda, \lambda^2, \ldots)$.
2. Let $x = (x_1, x_2, \ldots) \in W$. Then x is completely determined by the first two components x_1 and x_2. Therefore, the dimension of W is, at most, two. If an element of W is an eigenvector, it must be associated with an eigenvalue satisfying $\lambda^2 = \lambda + 1$, which gives the two possible eigenvalues

$$\varphi = \frac{1 + \sqrt{5}}{2} \quad \text{and} \quad -\varphi^{-1} = \frac{1 - \sqrt{5}}{2}.$$

A basis for W is then

$$\left\{ (\varphi, \varphi^2, \varphi^3, \ldots), (-\varphi^{-1}, \varphi^{-2}, -\varphi^{-3}, \ldots) \right\}$$

which is clearly invariant under S.
3. To express the Fibonacci sequence in the basis above, we have just to find the constants k_1 and k_2 that satisfy

$$\begin{cases} 1 &= k_1\varphi - k_2\varphi^{-1} \\ 1 &= k_1\varphi^2 + k_2\varphi^{-2} \end{cases}$$

which give $k_1 = \dfrac{1}{\sqrt{5}} = -k_2$. We then have, for the Fibonacci numbers,

$$f_n = \frac{1}{\sqrt{5}} \left(\left(\frac{1 + \sqrt{5}}{2} \right)^n - \left(\frac{1 - \sqrt{5}}{2} \right)^n \right).$$

Solution to 7.5.21: If $T(v) = \lambda v$ then $T^n(v) = \lambda^n v$ for all $n \geq 1$, by induction. If $f = a_0 + a_1 x + \cdots + a_n x^n$ then $f(T) = a_0 I + a_1 T + \cdots + a_n T^n$, and $f(T)(v) = a_0 v + a_1 T(v) + \cdots + a_n T^n(v) = f(\lambda)v$ so that $f(\lambda)$ is an eigenvalue of $f(T)$.

If λ is an eigenvalue of $f(T)$ then $|f(T) - \lambda I| = 0$. Factorize $f(x) - \lambda = \alpha_0(x - \alpha_1) \cdots (x - \alpha_n)$. (The case of constant $f(x)$ is trivial.) Then $f(T) - \lambda I = \alpha_0(T - \alpha_1 I) \cdots (T - \alpha_n I)$. Taking determinants we see that there is i such that $|T - \alpha_i I| = 0$. Therefore α_i is an eigenvalue of T with $f(\alpha_i) = \lambda$.

Solution to 7.5.22:

$$A = uu^t - I \quad \text{where} \quad u = \begin{pmatrix} 1 \\ \vdots \\ 1 \end{pmatrix}$$

and I is the identity matrix. If $Ax = \lambda x$, where $x \neq 0$, then

$$uu^t x - x = (u^t x)u - x = \lambda x$$

so x is either perpendicular or parallel to u. In the latter case, we can suppose without loss of generality that $x = u$, so $u^t uu - u = \lambda u$ and $\lambda = n - 1$. This gives a 1-dimensional eigenspace spanned by u with eigenvalue $n - 1$. In the former case x lies in a $(n - 1)$–dimensional eigenspace which is the null space of the rank one matrix uu^t, so

$$Ax = (uu^t - I)x = -Ix = -x$$

and the eigenvalue associated with this eigenspace is -1, with multiplicity $n - 1$. Since the determinant is the product of the eigenvalues, we have $\det(A) = (-1)^{n-1}(n - 1)$.

Solution to 7.5.23: Since A is positive definite, there is an invertible Hermitian matrix C such that $C^2 = A$. Thus, we have $C^{-1}(AB)C = C^{-1}C^2BC = CBC$. By taking adjoints, we see that CBC is Hermitian, so it has real eigenvalues. Since similar matrices have the same eigenvalues, AB has real eigenvalues.

Solution to 7.5.24: Let v be the column vector defined by $\begin{pmatrix} a & b & c & d \end{pmatrix}^t$, then the operator given, that we will call T, is literally the product of the two matrices vv^t. and the operator T applied to a vector x is:

$$Tx = (vv^t)x = v(v^t x) = v\langle v, x \rangle$$

where $\langle , \rangle$ is the standard euclidean inner product in $\mathbb{R}^4$.

Let u be the vector of norm 1 in the direction of v, that is, $u = v/|v|$, V the uni-dimensional space generated by v, and $V^\perp$ its orthogonal complement in $\mathbb{R}^4$. Then any vector in $x \in \mathbb{R}^4$ can be written as $x = x_\| + x_\perp$ where

$$x_\| = \langle u, x \rangle u$$
$$x_\perp = x - \langle u, x \rangle u$$

In this decomposition $Tu = |v|^2 u |u|^2 = |v|^2 u$ and u is an eigenvector of T with eigenvalue $|v|^2$ and multiplicity 1. For any vector in the $V^\perp$ space

$$Tx_\perp = |v|^2 u \langle u, x_\perp \rangle = 0$$

so taking any basis for this 3-dimensional space, the matrix representing T will be

$$\begin{pmatrix} a^2 + b^2 + c^2 + d^2 & 0 & 0 & 0 \\ 0 & 0 & 0 & 0 \\ 0 & 0 & 0 & 0 \\ 0 & 0 & 0 & 0 \end{pmatrix}$$

Solution to 7.5.25: $1 \Rightarrow 2$: As A is symmetric, it is similar to a diagonal matrix, $\mathrm{diag}(\lambda_1, \lambda_2, \lambda_3)$. We have $\mathrm{tr}\,A = \lambda_1 + \lambda_2 + \lambda_3$. The eigenvalues of A are the zeros of the polynomial $(\lambda - \lambda_1)(\lambda - \lambda_2)(\lambda - \lambda_3)$ and the conclusion follows.
$2 \Rightarrow 3$: Let $\{c_1, c_2, c_3\}$ be the ordered basis of $\mathbb{R}^3$ with respect to which the representation of A is $\mathrm{diag}(\lambda_1, \lambda_2, \lambda_3)$. For $W \in S$ we have $W^t = -W$, therefore, using the standard inner product in $\mathbb{R}^3$, we obtain, for $i = 1, 2, 3$,

$$\langle Wc_i, c_i \rangle = \langle c_i, W^t c_i \rangle = -\langle c_i, Wc_i \rangle = -\langle Wc_i, c_i \rangle,$$

therefore $Wc_i \perp c_i$.

L is clearly a linear self map of S, so it suffices to show that it is injective. Suppose $L(W) = 0$. Then $AW = -WA$, so we get, for $i \neq j$,

$$\begin{aligned} \langle AWc_i, c_j \rangle &= -\langle WAc_i, c_j \rangle \\ &= \langle Ac_i, Wc_j \rangle \\ &= \lambda_i \langle c_i, Wc_j \rangle \end{aligned}$$

and

$$\begin{aligned} \langle -WAc_i, c_j \rangle &= \langle AWc_i, c_j \rangle \\ &= \lambda_j \langle Wc_i, c_j \rangle \\ &= -\lambda_j \langle c_i, Wc_j \rangle \end{aligned}$$

as $\lambda_i + \lambda_j \neq 0$, we must have $Wc_j \perp c_i$. We conclude then that $Wc_i = 0$ for $i = 1, 2, 3$, therefore $W = 0$.
$3 \Rightarrow 1$: Suppose $\mathrm{tr}\,A$ is an eigenvalue. According to what we saw previously in the Part $1 \Rightarrow 2$, there is no loss in generality in assuming $\lambda_1 + \lambda_2 = 0$. Let W be defined on the vectors of the basis $\{c_1, c_2, c_3\}$ by $Wc_1 = -c_2$, $Wc_2 = -c_1$, $Wc_3 = 0$. Clearly $W \in S$, and $W \neq 0$. We have,

$$\begin{aligned} (AW + WA)c_1 &= AWc_1 + WAc_1 \\ &= -Ac_2 + \lambda_1 Wc_1 \\ &= -\lambda_2 c_2 - \lambda_1 c_2 \\ &= 0 \end{aligned}$$

so L is not an isomorphism.

Solution to 7.5.26: The characteristic polynomial of A is

$$\chi_A(t) = t^2 - (a + d)t + (ad - bc).$$

which has roots

$$t = \frac{1}{2}(a+d) \pm \frac{1}{2}\sqrt{(a-d)^2 + 4bc} = \frac{1}{2}\left(a+d \pm \sqrt{\Delta}\right).$$

Δ is positive, so A has real eigenvalues. Let $\lambda = \frac{1}{2}\left(a+d+\sqrt{\Delta}\right)$ and let $v = (x, y)$ be an eigenvector associated with this eigenvalue with $x > 0$. Expanding the first entry of Av, we get

$$ax + by = \frac{1}{2}\left(a+d+\sqrt{\Delta}\right)x$$

or

$$2by = \left(d - a + \sqrt{\Delta}\right)x.$$

Since $b > 0$, to see that $y > 0$ it suffices to show that $d - a + \sqrt{\Delta} > 0$, or $\sqrt{\Delta} > a - d$. But this is immediate from the definition of $\sqrt{\Delta}$ and we are done.

Solution to 7.5.27: It suffices to show that A is positive definite. Let $x = (x_1, \ldots, x_n)$, we have

$$\langle Ax, x \rangle = 2x_1^2 - x_1x_2 - x_1x_2 + 2x_2^2 - x_2x_3 - \cdots - x_{n-1}x_n + 2x_n^2$$
$$= x_1^2 + (x_1 - x_2)^2 + (x_2 - x_3)^2 + \cdots + (x_{n-1} - x_n)^2 + x_n^2.$$

Thus, for all nonzero x, $\langle Ax, x \rangle \geq 0$. In fact, it is strictly positive, since one of the center terms is greater than 0, or $x_1 = x_2 = \cdots = x_n$ and all the x_i's are nonzero, so $x_1^2 > 0$. Hence, A is positive definite and we are done.

Solution 2. Since A is symmetric, all eigenvalues are real. Let $x = (x_i)_1^n$ be an eigenvector with eigenvalue λ. Since $x \neq 0$, we have $\max_i |x_i| > 0$. Let k be the least i with $|x_i|$ maximum. Replacing x by $-x$, if necessary, we may assume $x_k > 0$. We have

$$\lambda x_k = -x_{k-1} + 2x_k - x_{k+1}$$

where nonexistent terms are taken to be zero. By the choice of x_k, we have $x_{k-1} < x_k$ and $x_{k+1} \leq x_k$, so we get $\lambda x_k > 0$ and $\lambda > 0$.

Solution to 7.5.28: Let λ_0 be the largest eigenvalue of A. We have

$$\lambda_0 = \max \{ \langle Ax, x \rangle \mid x \in \mathbb{R}, \|x\| = 1 \},$$

and the maximum it attained precisely when x is an eigenvector of A with eigenvalue λ_0. Suppose v is a unit vector for which the maximum is attained, and let u be the vector whose coordinates are the absolute values of the coordinates of v. Since the entries of A are nonnegative, we have

$$\langle Au, u \rangle \geq \langle Av, v \rangle = \lambda_0,$$

implying that $\langle Au, u \rangle = \lambda_0$ and so that u is an eigenvector of A for the eigenvalue λ_0.

Solution to 7.5.29: Let ξ_n be the characteristic polynomial of A_n. We have $\xi_1 = -z$, $\xi_2 = z^2 - 1$, and, for $n \geqslant 2$, $\xi_n = -z\xi_{n-1}(z) - \xi_{n-2}(z)$. Thus, by induction, ξ_n contains only even powers of z for even n, and only odd powers of z for odd n. It follows that the zeros of ξ_n are symmetric with respect to the origin. *Solution 2.* Let S_n be the diagonal $n \times n$ matrix with alternating 1's and -1's on the main diagonal. Then $S^{-1}AS = -A$, from which the desired conclusion follows.

Solution to 7.5.30: Let λ be an eigenvalue of A and $x = (x_1, \ldots, x_n)^t$ a corresponding eigenvector. Let x_i be the entry of x whose absolute value is greatest. We have

$$\lambda x_i = \sum_{j=1}^{n} a_{ij} x_j$$

so

$$|\lambda||x_i| \leqslant \sum_{j=1}^{n} a_{ij}|x_j| \leqslant |x_j| \sum_{j=1}^{n} a_{ij} = |x_i|.$$

Hence, $|\lambda| \leqslant 1$.

Solution to 7.5.31: 1. Regard S as a linear transformation on $\mathbb{C}^n$. Since S is orthogonal it preserves norms, $\|Sx\| = \|x\|$ for all x. Hence all eigenvalues have modulo 1. The characteristic polynomial of S has real coefficients, so the nonreal eigenvalues occur in conjugate pairs. There is thus an even number of nonreal eigenvalues, counting multiplicities, and the product of all of them is 1. Assuming n is odd, there must be an odd number of real eigenvalues, which can only equal 1 or -1. The product of all the eigenvalues is $\det(S) = 1$. Hence -1 must occur as an eigenvalue with even multiplicity, so that 1 occurs with an odd, hence nonzero, multiplicity.
2. If n is even then $-I_n$ is a special orthogonal matrix, yet it does not have 1 as an eigenvalue.

Solution to 7.5.32: Since A is Hermitian, by Rayleigh's Theorem [ND88, p. 418], we have

$$\lambda_{\min} \leqslant \frac{\langle x, Ax \rangle}{\langle x, x \rangle} \leqslant \lambda_{\max}$$

for $x \in \mathbb{C}^m$, $x \neq 0$, where $\lambda_{\min}$ and $\lambda_{\max}$ are its smallest and largest eigenvalues, respectively. Therefore,

$$a \leqslant \frac{\langle x, Ax \rangle}{\langle x, x \rangle} \leqslant a'$$

Similarly for B:

$$b \leqslant \frac{\langle x, Bx \rangle}{\langle x, x \rangle} \leqslant b'$$

Hence,

$$a + b \leqslant \frac{\langle x, (A+B)x \rangle}{\langle x, x \rangle} \leqslant a' + b'.$$

However, $A + B$ is Hermitian, since A and B are, so the middle term above is bounded above and below by the largest and smallest eigenvalues of $A + B$. But, again by Rayleigh's Theorem, we know these bounds are sharp, so all the eigenvalues of $A + B$ must lie in $[a + b, a' + b']$.

Solution to 7.5.33: Let $v = (1, 1, 0, \ldots, 0)$. A calculation shows that $Av = (k + 1, k + 1, 1, 0, \ldots, 0)$, so

$$\frac{\langle Av, v \rangle}{\langle v, v \rangle} = k + 1.$$

Similarly, for $u = (1, -1, 0, \ldots, 0)$, we have $Au = (k - 1, 1 - k, -1, 0, \ldots, 0)$ and so

$$\frac{\langle Au, u \rangle}{\langle u, u \rangle} = k - 1.$$

By Rayleigh's Theorem [ND88, p. 418], we know that

$$\lambda_{min} \leqslant \frac{\langle Av, v \rangle}{\langle v, v \rangle} \leqslant \lambda_{max}.$$

for all nonzero vectors v, and the desired conclusion follows.

Solution to 7.5.34: As B is positive definite, there is an invertible matrix C such that $B = C^t C$, so

$$\frac{\langle Ax, x \rangle}{\langle Bx, x \rangle} = \frac{\langle Ax, x \rangle}{\langle C^t Cx, x \rangle} = \frac{\langle Ax, x \rangle}{\langle Cx, Cx \rangle}.$$

Let $Cx = y$. The right-hand side equals

$$\frac{\langle AC^{-1} y, C^{-1} y \rangle}{\langle y, y \rangle} = \frac{\langle (C^{-1})^t AC^{-1} y, y \rangle}{\langle y, y \rangle}.$$

Since the matrix $(C^{-1})^t AC^{-1}$ is symmetric, by Rayleigh's Theorem [ND88, p. 418], the right-hand side is bounded by λ, where λ is the largest eigenvalue of $(C^{-1})^t AC^{-1}$. Further, the maximum is attained at the associated eigenvector. Let y_0 be such an eigenvector. Then $G(x)$ attains its maximum at $x = C^{-1} y_0$, which is an eigenvector of the matrix $(C^{-1})^t A$.

Solution to 7.5.35: Let $y \neq 0$ in $\mathbb{R}^n$. A is real symmetric, so there is an orthogonal matrix, P, such that $B = P^t AP$ is diagonal. Since P is invertible, there is a nonzero vector z such that $y = Pz$. Therefore,

$$\frac{\langle A^{m+1} y, y \rangle}{\langle A^m y, y \rangle} = \frac{\langle A^{m+1} Pz, Pz \rangle}{\langle A^m Pz, Pz \rangle} = \frac{\langle P^t A^{m+1} Pz, z \rangle}{\langle P^t A^m Pz, z \rangle} = \frac{\langle B^{m+1} z, z \rangle}{\langle B^m z, z \rangle}.$$

Since A is positive definite, we may assume without loss of generality that B has the form

$$
\begin{pmatrix}
\lambda_1 & 0 & \cdots & 0 \\
0 & \lambda_2 & \cdots & 0 \\
\vdots & \vdots & \ddots & \vdots \\
0 & 0 & \cdots & \lambda_n
\end{pmatrix}
$$

where $\lambda_1 \geqslant \lambda_2 \geqslant \cdots \geqslant \lambda_n > 0$. Let $z = (z_1, \ldots, z_n) \neq 0$, and $i \leqslant n$ be such that z_i is the first nonzero coordinate of z. Then

$$
\begin{aligned}
\frac{\langle B^{m+1}z, z \rangle}{\langle B^m z, z \rangle} &= \frac{\lambda_i^{m+1} z_i^2 + \cdots + \lambda_n^{m+1} z_n^2}{\lambda_i^m z_i^2 + \cdots + \lambda_n^m z_n^2} \\
&= \lambda_i \left(\frac{z_i^2 + (\lambda_{i+1}/\lambda_i)^{m+1} z_{i+1}^2 + \cdots + (\lambda_n/\lambda_i)^{m+1} z_n^2}{z_i^2 + (\lambda_{i+1}/\lambda_i)^m z_{i+1}^2 + \cdots + (\lambda_n/\lambda_1)^m z_n^2} \right) \\
&\sim \lambda_i \quad (m \to \infty).
\end{aligned}
$$

7.6   Canonical Forms

Solution to 7.6.1: The minimal polynomial of A divides $x^k - 1$ so it has no multiple roots, which implies that A is diagonalizable.

Solution to 7.6.2: Assume A^m is diagonalizable. Then its minimal polynomial, $\mu_{A^m}(x)$, has no repeated roots, that is,

$$
\mu_{A^m}(x) = (x - a_1) \cdots (x - a_k)
$$

where $a_i \neq a_j$ for $i \neq j$.
  The matrix A^m satisfies the equation

$$
(A^m - a_1 I) \cdots (A^m - a_k I) = 0
$$

so A is a root of the polynomial $(x^m - a_1) \cdots (x^m - a_k)$, therefore, $\mu_A(x)$ divides this polynomial. To show that A is diagonalizable, it is enough to show this polynomial has no repeated roots, which is clear, because the roots of the factors $x^m - a_i$ are different, and different factors have different roots.
  This proves more than what was asked; it shows that if A is an invertible linear transformation on a finite dimensional vector space over a field $\mathbf{F}$ of characteristic not dividing n, the characteristic polynomial of A factors completely over $\mathbf{F}$, and if A^n is diagonalizable, then A is diagonalizable.
  On this footing, we can rewrite the above proof as follows: We may suppose that the vector space V has positive dimension m. Let λ be an eigenvalue of A. Then $\lambda \neq 0$. We may replace V by the largest subspace of V on which $A - \lambda I$ is nilpotent, so that we may suppose the characteristic polynomial of A is

$(x - \lambda)^m$. Since A^n is diagonalizable, we must have $A^n = \lambda^n I$ since λ^n is the only eigenvalue of A^n. Thus, A satisfies the equation $x^n - \lambda^n = 0$. Since the only common factor of $x^n - \lambda^n$ and $(x - \lambda)^n$ is $x - \lambda$, and as the characteristic of $\mathbf{F}$ does not divide n, $A = \lambda I$ and, hence, is diagonal.

Solution to 7.6.3: The characteristic polynomial of A is $\chi_A(x) = x^2 - 3$, so $A^2 = 3I$, and multiplying both sides by A^{-1}, we have

$$A^{-1} = \frac{1}{3} A.$$

Solution to 7.6.4: 1. Subtracting the second line from the other three and expanding along the first line, we have

$$\det A_x = (x - 1) \begin{vmatrix} x & 1 & 1 \\ 1-x & x-1 & 0 \\ 1-x & 0 & x-1 \end{vmatrix} + (x-1) \begin{vmatrix} 1 & 1 & 1 \\ 0 & x-1 & 0 \\ 0 & 0 & x-1 \end{vmatrix}$$

$$= (x-1)^3(x+3).$$

2. Suppose now that $x \neq 1$ and -3. Then A_x is invertible and the characteristic polynomial is given by:

$$\chi_{A_x}(t) = \begin{vmatrix} t-x & -1 & -1 & -1 \\ -1 & t-x & -1 & -1 \\ -1 & -1 & t-x & -1 \\ -1 & -1 & -1 & t-x \end{vmatrix} = \begin{vmatrix} x-t & 1 & 1 & 1 \\ 1 & x-t & 1 & 1 \\ 1 & 1 & x-t & 1 \\ 1 & 1 & 1 & x-t \end{vmatrix}$$

$$= (x-t-1)^3(x-t+3)$$

Now an easy substitution shows that the minimal polynomial is

$$\mu_{A_x}(t) = (x - t - 1)(x - t - 3)$$

so substituting t by A_x, we have

$$((x-1)I_4 - A_x)((x+3)I_4 - A_x) = 0$$
$$(x-1)(x+3)I_4 - 2(x+1)A_x - A_x^2 = 0$$

multiplying both sides by A_x^{-1},

$$(x-1)(x+3)A_x^{-1} = 2(x+1)I_4 - A_x$$
$$= -A_{-x-2}$$

so

$$A_x^{-1} = -(x-1)^{-1}(x+3)^{-1}A_{-x-2}.$$

Solution to 7.6.5: The characteristic polynomial of the matrix A is $\chi_A(t) = t^3 - 8t^2 + 20t - 16 = (t-4)(t-2)^2$ and the minimal polynomial is $\mu_A(t) = (t-2)(t-4)$. By the Euclidean Algorithm [Her75, p. 155], there is a polynomial $p(t)$ and constants a and b such that

$$t^{10} = p(t)\mu_A(t) + at + b.$$

Substituting $t = 2$ and $t = 4$ and solving for a and b yields $a = 2^9(2^{10} - 1)$ and $b = -2^{11}(2^9 - 1)$. Therefore, since A is a root of its minimal polynomial,

$$A^{10} = aA + bI = \begin{pmatrix} 3a+b & a & a \\ 2a & 4a+b & 2a \\ -a & -a & a+b \end{pmatrix}.$$

Solution to 7.6.6: The characteristic polynomial of A is $\chi_A(t) = t^2 - 2t + 1 = (t-1)^2$. By the Euclidean Algorithm [Her75, p. 155], there is a polynomial $q(t)$ and constants a and b such that $t^{100} = q(t)(t-1)^2 + at + b$. Differentiating both sides of this equation, we get $100t^{99} = q'(t)(t-1)^2 + 2q(t)(t-1) + a$. Substituting $t = 1$ into each equation and solving for a and b, we get $a = 100$ and $b = -99$. Therefore, since A satisfies its characteristic equation, substituting it into the first equation yields $A^{100} = 100A - 99I$, or

$$A^{100} = \begin{pmatrix} 51 & 50 \\ -50 & -49 \end{pmatrix}.$$

An identical calculation shows that $A^7 = 7A - 6I$, so

$$A^7 = \begin{pmatrix} 9/2 & 7/2 \\ -7/2 & -5/2 \end{pmatrix}.$$

From this it follows immediately that

$$A^{-7} = \begin{pmatrix} -5/2 & -7/2 \\ 7/2 & 9/2 \end{pmatrix}.$$

Solution to 7.6.7: Counterexample: Let

$$A = \begin{pmatrix} 0 & 1 \\ 0 & 0 \end{pmatrix} = B^2 = \begin{pmatrix} a & b \\ c & d \end{pmatrix}\begin{pmatrix} a & b \\ c & d \end{pmatrix} = \begin{pmatrix} a^2+bc & ab+bd \\ ca+dc & cb+d^2 \end{pmatrix}.$$

Equating entries, we find that $c(a+d) = 0$ and $b(a+d) = 1$, so $b \neq 0$ and $a+d \neq 0$. Thus, $c = 0$. The vanishing of the diagonal entries of B^2 then implies that $a^2 = d^2 = 0$ and, thus, $a+d = 0$. This contradiction proves that no such B can exist, so A has no square root.

Solution 2. Let

$$A = \begin{pmatrix} 0 & 1 \\ 0 & 0 \end{pmatrix}.$$

Any square root B of A must have zero eigenvalues, and since it cannot be the zero matrix, it must have Jordan Canonical Form [HK61, p. 247] $JBJ^{-1} = A$. But then $B^2 = J^{-1}A^2J = 0$ since $A^2 = 0$, so no such B can exist.

Solution to 7.6.8: 1. Let
$$A = \begin{pmatrix} a & b \\ c & d \end{pmatrix}$$

then
$$A^2 = \begin{pmatrix} a^2 + bc & (a+d)b \\ (a+d)c & bc + d^2 \end{pmatrix}.$$

Therefore, $A^2 = -I$ is equivalent to the system

$$\begin{cases} a^2 + bc &=& -1 \\ (a+d)b &=& 0 \\ (a+d)c &=& 0 \\ bc + d^2 &=& -1 \end{cases}$$

if $a + d \neq 0$, the second equation above gives $b = 0$, and from the fourth, we obtain $d^2 = -1$, which is absurd. We must then have $a = -d$ and the result follows.

2. The system that we get in this case is

$$\begin{cases} a^2 + bc &=& -1 \\ (a+d)b &=& 0 \\ (a+d)c &=& 0 \\ bc + d^2 &=& -1 - \varepsilon \end{cases}$$

As above, we cannot have $a \neq -d$. But combining $a = -d$ with the first and fourth equations of the system, we get $\varepsilon = 0$, a contradiction. Therefore, no such matrix exists.

Solution to 7.6.9: Suppose such a matrix A exists. One of the eingenvalues of A would be w and the other $(1 + \varepsilon)^{1/20}w$ where w is a twentieth root of -1. From the fact that A is real we can see that both eigenvalues are real or form a complex conjugate pair, but neither can occur because none the twentieth root of -1 are real and the fact that
$$|w| = 1 \neq (1 + \varepsilon)^{1/20}$$

make it impossible for them to be a conjugate pair, so no such a matrix exist.

Solution to 7.6.10: $A^n = I$ implies that the minimal polynomial of A, $\mu(x) \in \mathbb{Z}[x]$, satisfies $\mu(x)|(x^n - 1)$. Let $\zeta_1, \ldots, \zeta_n$ be the distinct roots of $x^n - 1$ in $\mathbb{C}$. We will separate the two possible cases for the degree of μ:

- $\deg \mu = 1$. We have $\mu(x) = x - 1$ and $A = I$, or $\mu(x) = x + 1$ and $A = -I$, $A^2 = I$.

- deg $\mu = 2$. In this case, ζ_i and ζ_j are roots of μ for some $i \neq j$, in which case we have $\zeta_j = \bar{\zeta}_i = \zeta$ say, since μ has real coefficients. Thus, $\mu(x) = (x - \zeta)(x - \bar{\zeta}) = x^2 - 2\Re(\zeta)x + 1$. In particular, $2\Re(\zeta) \in \mathbb{Z}$, so the possibilities are $\Re(\zeta) = 0, \pm 1/2$, and ± 1. We cannot have $\Re(\zeta) = \pm 1$ because the corresponding polynomials, $(x-1)^2$ and $(x+1)^2$, have repeated roots, so they are not divisors of $x^n - 1$.

$\Re(\zeta) = 0$. We have $\mu(x) = x^2 + 1$ and $A^2 = -I$, $A^4 = I$.

$\Re(\zeta) = 1/2$. In this case $\mu(x) = x^2 - x + 1$. ζ is a primitive sixth root of unity, so $A^6 = I$.

$\Re(\zeta) = -1/2$. We have $\mu(x) = x^2 + x + 1$. ζ is a primitive third root of unity, so $A^3 = I$.

From the above, we see that if $A^n = I$ for some $n \in \mathbb{Z}_+$, then one of the following holds:

$$A = I, \ A^2 = I, \ A^3 = I, \ A^4 = I, \ A^6 = I.$$

Further, for each $n = 2, 3, 4$, and 6 there is a matrix A such that $A^n = I$ but $A^k \neq I$ for $0 < k < n$:

- $n = 2$.

$$\begin{pmatrix} -1 & 0 \\ 0 & -1 \end{pmatrix}$$

- $n = 3$.

$$\begin{pmatrix} 0 & 1 \\ -1 & -1 \end{pmatrix}$$

- $n = 4$.

$$\begin{pmatrix} 0 & 1 \\ -1 & 0 \end{pmatrix}$$

- $n = 6$.

$$\begin{pmatrix} 0 & 1 \\ -1 & 1 \end{pmatrix}$$

Solution to 7.6.11: Since A is upper-triangular, its eigenvalues are its diagonal entries, that is, 1, 4, and 9. It can, thus, be diagonalized, and in fact, we will have

$$S^{-1}AS = \begin{pmatrix} 1 & 0 & 0 \\ 0 & 4 & 0 \\ 0 & 0 & 9 \end{pmatrix}$$

where S is a matrix whose columns are eigenvectors of A for the respective eigenvalues 1, 4, and 9. The matrix

$$B = S \begin{pmatrix} 1 & 0 & 0 \\ 0 & 2 & 0 \\ 0 & 0 & 3 \end{pmatrix} S^{-1}$$

will then be a square root of A.

Carrying out the computations, one obtains

$$S = \begin{pmatrix} 1 & 1 & 1 \\ 0 & 1 & 1 \\ 0 & 0 & 1 \end{pmatrix} \quad \text{and} \quad S^{-1} = \begin{pmatrix} 1 & -1 & 0 \\ 0 & 1 & -1 \\ 0 & 0 & 1 \end{pmatrix}$$

giving

$$B = \begin{pmatrix} 1 & 1 & -1 \\ 0 & 2 & 1 \\ 0 & 0 & 3 \end{pmatrix}.$$

The number of square roots of A is the same as the number of square roots of its diagonalization, $D = S^{-1}AS$. Any matrix commuting with D preserves its eigenspaces and so is diagonal. In particular, any square root of D is diagonal. Hence, D has exactly eight square roots, namely

$$\sqrt{\begin{pmatrix} 1 & 0 & 0 \\ 0 & 4 & 0 \\ 0 & 0 & 9 \end{pmatrix}} = \begin{pmatrix} \pm 1 & 0 & 0 \\ 0 & \pm 2 & 0 \\ 0 & 0 & \pm 3 \end{pmatrix}.$$

Solution to 7.6.12: $n = 1$. There is the solution $X = A$.
$n = 2$. A is similar to the matrix

$$\begin{pmatrix} 0 & 0 & 1 & 0 \\ 0 & 0 & 0 & 0 \\ 0 & 0 & 0 & 0 \\ 0 & 0 & 0 & 0 \end{pmatrix}$$

under the transformation that interchanges the third and fourth basis vectors and leaves the first and second basis vectors fixed. The latter matrix is the square of

$$\begin{pmatrix} 0 & 1 & 0 & 0 \\ 0 & 0 & 1 & 0 \\ 0 & 0 & 0 & 0 \\ 0 & 0 & 0 & 0 \end{pmatrix}.$$

Hence, A is the square of

$$\begin{pmatrix} 0 & 1 & 0 & 0 \\ 0 & 0 & 0 & 1 \\ 0 & 0 & 0 & 0 \\ 0 & 0 & 0 & 0 \end{pmatrix}.$$

$n = 3$. The Jordan matrix [HK61, p. 247]

$$X = \begin{pmatrix} 0 & 1 & 0 & 0 \\ 0 & 0 & 1 & 0 \\ 0 & 0 & 0 & 1 \\ 0 & 0 & 0 & 0 \end{pmatrix}$$

is a solution.

$n \geqslant 4$. If $X^k = A$, then X is nilpotent since A is. Then the characteristic polynomial of X divides x^4, so that $X^4 = 0$, and, a fortiori, $X^n = 0$ for $n \geqslant 4$. There is, thus, no solution for $n \geqslant 4$.

Solution to 7.6.13: Suppose such a matrix A exists. Its minimal polynomial must divide $t^2 + 2t + 5$. However, this polynomial is irreducible over $\mathbb{R}$, so $\mu_A(t) = t^2 + 2t + 5$. Since the characteristic and minimal polynomials have the same irreducible factors, $\chi_A(t) = \mu_A(t)^k$. Therefore, $\deg \chi_A(t) = n$ must be even.

Conversely, a calculation shows that the 2×2 real matrix

$$A_0 = \begin{pmatrix} 0 & -5 \\ 1 & -2 \end{pmatrix}$$

is a root of this polynomial. Therefore, any $2n \times 2n$ block diagonal matrix which has n copies of A_0 on the diagonal will satisfy this equation as well.

Solution to 7.6.14: Let $p(t) = t^5 + t^3 + t - 3$. As $p(A) = 0$, we have $\mu_A(t)|p(t)$. However, since A is Hermitian, its minimal polynomial has only real roots. Taking the derivative of p, we see that $p'(t) = 5t^4 + 3t^2 + 1 > 0$ for all t, so $p(t)$ has exactly one real root. A calculation shows that $p(1) = 0$, but $p'(1) \neq 0$. Therefore, $p(t) = (t - 1)q(t)$, where $q(t)$ has only nonreal complex roots. It follows that $\mu_A(t)|(t - 1)$. Since $t - 1$ is irreducible, $\mu_A(t) = t - 1$ and $A = I$.

Solution to 7.6.16: Note that

$$A = \begin{pmatrix} 2 & 0 & 0 \\ 0 & 2 & 0 \\ 0 & -1 & 1 \end{pmatrix}$$

can be decomposed into the two blocks (2) and $\begin{pmatrix} 2 & 0 \\ -1 & 1 \end{pmatrix}$, since the space spanned by $(1\,0\,0)^t$ is invariant. We will find a 2×2 matrix C such that $C^4 = \begin{pmatrix} 2 & 0 \\ -1 & 1 \end{pmatrix} = D$, say.

The eigenvalues of the matrix D are 2 and 1, and the corresponding Lagrange Polynomials [MH93, p. 286] are $p_1(x) = (x - 2)/(1 - 2) = 2 - x$ and $p_2(x) = (x - 1)/(2 - 1) = x - 1$. Therefore, the spectral projection of D can be given by

$$P_1 = -\begin{pmatrix} 2 & 0 \\ -1 & 1 \end{pmatrix} + 2\begin{pmatrix} 1 & 0 \\ 0 & 1 \end{pmatrix} = \begin{pmatrix} 0 & 0 \\ 1 & 1 \end{pmatrix}$$

$$P_2 = -\begin{pmatrix} 1 & 0 \\ 0 & 1 \end{pmatrix} + \begin{pmatrix} 2 & 0 \\ -1 & 1 \end{pmatrix} = \begin{pmatrix} 1 & 0 \\ -1 & 0 \end{pmatrix}$$

We have

$$D = \begin{pmatrix} 0 & 0 \\ 1 & 1 \end{pmatrix} + 2\begin{pmatrix} 1 & 0 \\ -1 & 0 \end{pmatrix}.$$

As $P_1 \cdot P_2 = P_2 \cdot P_1 = 0$ and $P_1^2 = P_1$, $P_2^2 = P_2$, letting $C = \begin{pmatrix} 0 & 0 \\ 1 & 1 \end{pmatrix} + 2^{1/4} \begin{pmatrix} 1 & 0 \\ -1 & 0 \end{pmatrix} = 1 P_1 + 2^{1/4} P_2$, we get

$$C^4 = P_1^4 + \underbrace{\cdots}_{0} + (2^{1/4} P_2)^4 = P_1 + 2 P_2 = D.$$

Then

$$C = \begin{pmatrix} 2^{1/4} & 0 \\ 1 - 2^{1/4} & 1 \end{pmatrix}$$

and B is

$$\begin{pmatrix} 2^{1/4} & 0 & 0 \\ 0 & 2^{1/4} & 0 \\ 0 & 1 - 2^{1/4} & 1 \end{pmatrix}.$$

Solution to 7.6.17: It suffices to show that every element $w \in W$ is a sum of eigenvectors of T in W. Let $a_1, \ldots, a_n$ be the distinct eigenvalues of T. We may write

$$w = v_1 + \cdots + v_n$$

where each v_i is in V and is an eigenvector of T with eigenvalue a_i. Then

$$\prod_{i \neq j} (T - a_j) w = \prod_{i \neq j} (a_i - a_j) v_i.$$

This element lies in W since W is T invariant. Hence, $v_i \in W$ for all i and the result follows.

Solution 2. To see this in a matrix form, take an ordered basis of W and extend it to a basis of V; on this basis, a matrix representing T will have the block form

$$[T]_B = \begin{pmatrix} A & C \\ 0 & B \end{pmatrix}$$

because of the invariance of the subspace W with respect to T.

Using the block structure of T, we can see that the characteristic and minimal polynomials of A divide the ones for T. For the characteristic polynomial, it is immediate from the fact that

$$\det(xI - [T]_B) = \det(xI - A) \det(xI - B)$$

For the minimal polynomial, observe that

$$[T]_B^k = \begin{pmatrix} A^k & C_k \\ 0 & B^k \end{pmatrix}$$

where C_k is some $r \times (n - r)$ matrix. Therefore, any polynomial that annihilates $[T]$ also annihilates A and B; so the minimal polynomial of A divides the one for $[T]$.

Now, since T is diagonalizable, the minimal polynomial factors out in different linear terms and so does the one for A, proving the result.

Solution to 7.6.19: Let λ be an eigenvalue of A and v a vector in the associated eigenspace, A_λ. Then $A(Bv) = BAv = B(\lambda v) = \lambda(Bv)$, so $Bv \in A_\lambda$. Now fix an eigenvalue λ and let C be the linear transformation obtained by restricting B to A_λ. Take any $v \in A_\lambda$. Then, since C is the restriction of B,

$$\mu_B(C)v = \mu_B(B)v = 0,$$

so C is a root of $\mu_B(t)$. It follows from this that $\mu_C(t)|\mu_B(t)$. But B was diagonalizable, so $\mu_B(t)$ splits into distinct linear factors. Therefore, $\mu_C(t)$ must split into distinct linear factors as well and so A_λ has a basis of eigenvectors of C. As A is diagonalizable, V can be written as the direct sum of the eigenspaces of A. However, each of these eigenspaces has a basis which consists of vectors which are simultaneously eigenvectors of A and of B. Therefore, V itself must have such a basis, and this is the basis which simultaneously diagonalizes A and B.

Solution 2. (This one, in fact, shows much more; it proves that a set of $n \times n$ diagonalizable matrices over a field $\mathbf{F}$ which commute with each other are all simultaneously diagonalizable.) Let S be a set of $n \times n$ diagonalizable matrices over a field $\mathbf{F}$ which commute with each other. Let $V = \mathbf{F}^n$. Suppose T is a maximal subset of S such that there exists a decomposition of

$$V = \oplus_i V_i$$

where V_i is a nonzero eigenspace for each element of T such that for $i \neq j$, there exists an element of T with distinct eigenvalues on V_i and V_j. We claim that $T = S$. If not, there exists an $N \in S - T$. Since N commutes with all the elements of T, $NV_i \subset V_i$. Indeed, there exists a function $a_i : T \to \mathbf{F}$ such that $v \in V_i$, if and only if $Mv = a_i(M)v$ for all $M \in T$. Now if $v \in V_i$ and $M \in T$,

$$MNv = NMv = Na_i(M) = a_i(M)Nv$$

so $Nv \in V_i$. Since N is diagonalizable on V, it is diagonalizable on V_i. (See Problem 7.6.17; it satisfies a polynomial with distinct roots in $\mathbf{K}$.) This means we can decompose each V_i into eigenspaces $V_{i,j}$ for N with distinct eigenvalues. Hence, we have a decomposition of the right sort for $T \cup N$,

$$V = \oplus_i \oplus_j V_{i,j}.$$

Hence, $T = S$. We may now make a basis for V by choosing a basis for V_i and taking the union. Then A will be the change of basis matrix.

Solution to 7.6.20: 1. Take

$$A = \begin{pmatrix} 1 & 0 \\ -1 & -1 \end{pmatrix}, \qquad B = \begin{pmatrix} -1 & 1 \\ 0 & 1 \end{pmatrix}$$

they are both diagonalizable since the characteristic polynomial factors in linear terms, but $A + B = \begin{pmatrix} 0 & 1 \\ -1 & 0 \end{pmatrix}$ is not diagonalizable.

2. If $A = \begin{pmatrix} 1 & 2 \\ 0 & 2 \end{pmatrix}$ and $B = \begin{pmatrix} 1 & 0 \\ 0 & 1/2 \end{pmatrix}$ then A and B have distinct eigenvalues and are diagonalizable. However $AB = \begin{pmatrix} 1 & 1 \\ 0 & 1 \end{pmatrix}$ is not diagonalizable.

3. If $A^2 = A$ then μ_A divides $x^2 - x$, whence $\mu_A = x$ or $x - 1$ or $x(x - 1)$. It follows that A is diagonalizable, as μ_A is a product of distinct linear factors.

4. If A^2 is diagonalizable its minimal polynomial is the product of distinct linear factors: $\mu_{A^2} = (x - \lambda_1) \cdots (x - \lambda_k)$, where $\lambda_i \in \mathbb{C}$. Therefore

$$(A^2 - \lambda_1 I) \cdots (A^2 - \lambda_k I) = 0.$$

It follows that $(A - \sqrt{\lambda_1} I)(A + \sqrt{\lambda_1} I) \cdots (A - \sqrt{\lambda_k} I)(A + \sqrt{\lambda_k} I) = 0$. The minimum polynomial of A divides the polynomial

$$f = (x - \sqrt{\lambda_1})(x + \sqrt{\lambda_1}) \cdots (x - \sqrt{\lambda_k})(x + \sqrt{\lambda_k}).$$

If A is invertible then none of the λ_i is zero, and thus the factors of f are distinct. It follows that μ_A is a product of distinct linear factors and A is diagonalizable.

Solution to 7.6.21: The characteristic polynomial of A is

$$\chi_A(x) = \begin{vmatrix} x - 7 & -15 \\ 2 & x + 4 \end{vmatrix} = (x - 1)(x - 2)$$

so A is diagonalizable and a short calculation shows that eigenvectors associated with the eigenvalues 1 and 2 are $(5, -2)^t$ and $(3, -1)^t$, so the matrix B is $\begin{pmatrix} 5 & 3 \\ -2 & -1 \end{pmatrix}$. Indeed, in this case, $B^{-1}AB = \begin{pmatrix} 1 & 0 \\ 0 & 2 \end{pmatrix}$.

Solution to 7.6.23: The characteristic polynomial is $\chi_A(t) = (t - 1)(t - 4)^2$. Since the minimal polynomial and the characteristic polynomial share the same irreducible factors, another calculation shows that $\mu_A(t) = (t - 1)(t - 4)^2$. Therefore, the Jordan Canonical Form [HK61, p. 247] of A must have one Jordan block of order 2 associated with 4 and one Jordan block of order 1 associated with 1. Hence, the Jordan form of A is

$$\begin{pmatrix} 1 & 0 & 0 \\ 0 & 4 & 1 \\ 0 & 0 & 4 \end{pmatrix}.$$

Solution to 7.6.24: We have

$$|A - \lambda I| = \begin{vmatrix} 2 - \lambda & 1 & 1 \\ 1 & 2 - \lambda & 1 \\ 1 & 1 & 2 - \lambda \end{vmatrix} = \begin{vmatrix} 1 - \lambda & 1 & 0 \\ -1 + \lambda & 2 - \lambda & -1 + \lambda \\ 0 & 1 & 1 - \lambda \end{vmatrix}$$

$$= (\lambda - 1)^2 \begin{vmatrix} -1 & 1 & 0 \\ 1 & 2 - \lambda & 1 \\ 0 & 1 & -1 \end{vmatrix} = -(\lambda - 1)^3,$$

using column operations $C_1 - C_2$ and $C_3 - C_2$. The single eigenvalue of A is $\lambda = 1$ with algebraic multiplicity 3.

The eigenvectors of A are the solutions to the equation $(A - I)x = 0$. That is, $x_1 + x_2 + x_3 = 0$, or $x = (-x_2 - x_3, x_2, x_3)$. The eigenvectors corresponding to the eigenvalue $\lambda = 1$ are $x_2(-1, 1, 0) + x_3(-1, 0, 1)$, $(x_2, x_3 \in \mathbf{F}_3)$ (including the zero vector).

The characteristic polynomial of A is $(x - 1)^3$. Now

$$A - I = \begin{pmatrix} 1 & 1 & 1 \\ 1 & 1 & 1 \\ 1 & 1 & 1 \end{pmatrix}, \quad (A - I)^2 = 0.$$

The minimal polynomial of A is $(x - 1)^2$. The Jordan form has an elementary Jordan block of size 2. The Jordan form of A is therefore

$$\begin{pmatrix} 1 & 1 & 0 \\ 0 & 1 & 0 \\ 0 & 0 & 1 \end{pmatrix}.$$

Solution to 7.6.25: Combining the equations, we get $\mu(x)^2 = \mu(x)(x-i)(x^2+1)$ and, thus, $\mu(x) = (x - i)^2(x + i)$. So the Jordan blocks of the Jordan Canonical Form [HK61, p. 247] J_A, correspond to the eigenvalues $\pm i$. There is at least one block of size 2 corresponding to the eigenvalue i and no larger block corresponding to i. Similarly, there is at least one block of size 1 corresponding to $-i$. We have $\chi(x) = (x - i)^3(x + i)$, so $n = \deg \chi = 4$, and the remaining block is a block of size 1 corresponding to i, since the total dimension of the eigenspace is the degree with which the factor appears in the characteristic polynomial. Therefore,

$$J_A = \begin{pmatrix} i & 1 & 0 & 0 \\ 0 & i & 0 & 0 \\ 0 & 0 & i & 0 \\ 0 & 0 & 0 & -i \end{pmatrix}.$$

Solution to 7.6.26: 1. As all the rows of M are equal, M must have rank 1, so its nullity is $n - 1$. It is easy to see that $M^2 = nM$, or $M(M - nI) = 0$, so the minimal polynomial is $\mu_M = x(x - n)$, since the null space associated with the characteristic value 0 is $n - 1$, then the characteristic polynomial is $\chi_M = x^{n-1}(x - n)$

2. If char $\mathbf{F} = 0$ or if char $\mathbf{F} = p$ and p does not divide n, then 0 and n are the two distinct eigenvalues, and since the minimal polynomial does not have repeated roots, M is diagonalizable.

If char $\mathbf{F} = p$, $p | n$, then n is identified with 0 in $\mathbf{F}$. Therefore, the minimal polynomial of M is $\mu_M(x) = x^2$ and M is not diagonalizable.

3. In the first case, since the null space has dimension $n - 1$, the Jordan form [HK61, p. 247] is

$$\begin{pmatrix} n & 0 & \cdots & 0 \\ 0 & 0 & \cdots & 0 \\ \vdots & \vdots & \ddots & \vdots \\ 0 & 0 & \cdots & 0 \end{pmatrix}.$$

If char $F = p$, $p|n$, then all the eigenvalues of M are 0, and there is one 2-block and $n - 1$ 1-blocks in the Jordan form:

$$\begin{pmatrix} 0 & 1 & \cdots & 0 \\ 0 & 0 & \cdots & 0 \\ \vdots & \vdots & \ddots & \vdots \\ 0 & 0 & \cdots & 0 \end{pmatrix}.$$

Solution to 7.6.27: A computation gives

$$T \begin{pmatrix} x_{11} & x_{12} \\ x_{21} & x_{22} \end{pmatrix} = \begin{pmatrix} -x_{21} & x_{11} - x_{22} \\ 0 & x_{21} \end{pmatrix}.$$

In particular, for the basis elements

$$E_1 = \begin{pmatrix} 1 & 0 \\ 0 & 0 \end{pmatrix}, \quad E_2 = \begin{pmatrix} 0 & 1 \\ 0 & 0 \end{pmatrix}, \quad E_3 = \begin{pmatrix} 0 & 0 \\ 1 & 0 \end{pmatrix}, \quad E_4 = \begin{pmatrix} 0 & 0 \\ 0 & 1 \end{pmatrix}$$

we have

$$T E_1 = \begin{pmatrix} 0 & 1 \\ 0 & 0 \end{pmatrix} = E_2, \quad T E_2 = 0,$$

$$T E_3 = \begin{pmatrix} -1 & 0 \\ 0 & 1 \end{pmatrix} = -E_1 + E_4, \quad T E_4 = \begin{pmatrix} 0 & -1 \\ 0 & 0 \end{pmatrix} = -E_2.$$

The matrix for T with respect to the basis $\{E_1, E_2, E_3, E_4\}$ is then

$$S = \begin{pmatrix} 0 & 0 & -1 & 0 \\ 1 & 0 & 0 & -1 \\ 0 & 0 & 0 & 0 \\ 0 & 0 & 1 & 0 \end{pmatrix}.$$

A calculation shows that the characteristic polynomial of S is λ^4. Thus, S is nilpotent. Moreover, the index of nilpotency is 3, since we have

$$T^2 E_1 = T^2 E_2 = T^2 E_4 = 0, \quad T^2 E_3 = -2E_2.$$

The only 4×4 nilpotent Jordan matrix [HK61, p. 247] with index of nilpotency 3 is

$$\begin{pmatrix} 0 & 0 & 0 & 0 \\ 0 & 0 & 1 & 0 \\ 0 & 0 & 0 & 1 \\ 0 & 0 & 0 & 0 \end{pmatrix}$$

which is, therefore, the Jordan Canonical Form of T. A basis in which T is represented by the preceding matrix is

$$\left\{ E_1 + E_4, \ E_2, \ \frac{E_1 - E_4}{2}, \ -\frac{E_3}{2} \right\}.$$

Solution to 7.6.28: It will suffice to prove that $\ker(T_A)$, the subspace of matrices that commute with A, has dimension at least n. We can assume without loss of generality that A is in Jordan form. A $k \times k$ Jordan block commutes with the $k \times k$ identity and with its own powers up to the $(k-1)$, hence with k linearly independent matrices. Thus, the commutant of each Jordan block has a dimension at least as large as the size of the block, in fact, equality holds. By taking direct sums of matrices commuting with the separate Jordan blocks of A, we therefore obtain a subspace of matrices commuting with A of dimension at least n and, in fact, equal to n.

Solution to 7.6.29: The answer is 0 and 1. If A is diagonal with diagonal entries $1, 2, \ldots, n$ and $g(t) = (t-1)(t-2) \ldots (t-(n-1))$, then $g(A)$ has $(n-1)!$ in the lower right corner and zeros elsewhere, so $g(A)$ has rank 1. If A is the zero matrix and $g(t) = t^{n-1}$, then $g(A)$ has rank 0. We will show that these are the only possibilities for the rank of $g(A)$, even if A is allowed to have complex entries.

Since the ground field is now algebraically closed, we may conjugate A to put it in Jordan canonical form, without affecting the rank of $g(A)$. Let $A_1, \ldots, A_r$ be the Jordan blocks of A, and let $f_i(t) \in \mathbb{C}[t]$ be the characteristic polynomial of A_i for each i. Since $g(t)$ divides $\prod \xi_i(t)$ and $\deg g = n - 1$, we can factor $g(t)$ as $\prod g_i(t)$ where $g_i = \xi_i$ for all i except one. Without loss of generality the exception is $i = 1$. Then for some $k \geqslant 1$ and $\lambda \in \mathbb{C}$, A_1 is a $k \times k$ block

$$A_1 = \begin{pmatrix} \lambda & 1 & 0 & \cdots & 0 \\ 0 & \lambda & 1 & \cdots & 0 \\ \vdots & \vdots & \vdots & \ddots & \vdots \\ 0 & 0 & 0 & \cdots & 1 \\ 0 & 0 & 0 & \cdots & \lambda \end{pmatrix},$$

$\xi_1(t) = (t - \lambda)^k$, and $g_1(t) = c(t-\lambda)^{k-1}$, where $c \neq 0$. Then $g(A)$ is formed of the blocks $g(A_1), \ldots, g(A_r)$ (not necessarily full Jordan blocks), and $g(A_i) = 0$ for $i \geqslant 2$, since the characteristic polynomial of A_i divides g. On the other hand $g(A_1)$ is some matrix times

$$g_1(A_1) = c \begin{pmatrix} 0 & 1 & 0 & \cdots & 0 \\ 0 & 0 & 1 & \cdots & 0 \\ \vdots & \vdots & \vdots & \ddots & \vdots \\ 0 & 0 & 0 & \cdots & 1 \\ 0 & 0 & 0 & \cdots & 0 \end{pmatrix}^{k-1} = \begin{pmatrix} 0 & 0 & 0 & \cdots & c \\ 0 & 0 & 0 & \cdots & 0 \\ \vdots & \vdots & \vdots & \ddots & \vdots \\ 0 & 0 & 0 & \cdots & 0 \\ 0 & 0 & 0 & \cdots & 0 \end{pmatrix},$$

which is of rank 1, so the rank of $g(A_1)$ is at most 1, and the rank of $g(A)$ is at most 1.

Solution to 7.6.30: A direct calculation shows that $(A - I)^3 = 0$ and this is the least positive exponent for which this is true. Hence, the minimal polynomial of A is $\mu_A(t) = (t-1)^3$. Thus, its characteristic polynomial must be $\chi_A(t) = (t-1)^6$. Therefore, the Jordan Canonical Form [HK61, p. 247] of A must contain one 3×3 Jordan block associated with 1. The number of blocks is the dimension of the eigenspace associated with 1. Letting $x = (x_1, \ldots, x_6)^t$ and solving $Ax = x$, we get the two equations $x_1 = 0$ and $x_2 + x_3 + x_4 + x_5 = 0$. Since x_6 is not determined, these give four degrees of freedom, so the eigenspace has dimension 4. Therefore, the Jordan Canonical Form of A must contain four Jordan blocks and so it must be

$$\begin{pmatrix} 1 & 1 & 0 & 0 & 0 & 0 \\ 0 & 1 & 1 & 0 & 0 & 0 \\ 0 & 0 & 1 & 0 & 0 & 0 \\ 0 & 0 & 0 & 1 & 0 & 0 \\ 0 & 0 & 0 & 0 & 1 & 0 \\ 0 & 0 & 0 & 0 & 0 & 1 \end{pmatrix}.$$

Solution to 7.6.31: Since AB has rank 3, both A and B must also have rank 3, from which it follows that BA has rank 3. Hence the null space of BA has dimension 2; let v_1, v_2 form a basis for it.

Let e_1, e_2, e_3 be the standard basis vectors for $\mathbb{C}^3$. We have

$$ABe_1 = e_1, \qquad ABe_2 = e_1 + e_2, \qquad ABe_3 = -e_3,$$

so

$$BABe_1 = Be_1, \qquad BABe_2 = Be_1 + Be_2, \qquad BABe_3 = -Be_3.$$

Since B has rank 3, no nontrivial linear combination of Be_1, $Be_1 + Be_2$, $-Be_3$ can vanish, implying that Be_1, Be_2, Be_3 are linearly independent modulo the null space of BA. Hence v_1, v_2, Be_1, Be_2, Be_3 form a basis for $\mathbb{C}^5$. Relative to his basis, the transformation induced by BA has the matrix

$$\begin{pmatrix} 0 & 0 & 0 & 0 & 0 \\ 0 & 0 & 0 & 0 & 0 \\ 0 & 0 & 1 & 1 & 0 \\ 0 & 0 & 0 & 1 & 0 \\ 0 & 0 & 0 & 0 & -1 \end{pmatrix}$$

which is the Jordan form of BA.

Solution to 7.6.33: The ith diagonal element of AA^t is $\sum_{j=i}^n a_{ij}^2$ and the same element on $A^t A$ is $\sum_{j=1}^i a_{ji}^2$. Comparing these expressions successively for

$i = n, n - 1, \ldots 1$, we conclude that all a_{ij} with $j > i$ are zero, that is, A is diagonal.

Solution to 7.6.38: Since A is nonsingular, $A^t A$ is positive definite. Let $B = \sqrt{A^t A}$. Consider $P = BA^{-1}$. Then $PA = B$, so it suffices to show that P is orthogonal, for in that case, $Q = P^{-1} = P^*$ will be orthogonal and $A = QB$. We have

$$P^t P = (A^t)^{-1} B^t B A^{-1} = (A^t)^{-1} B^2 A^{-1} = (A^t)^{-1} A^t A A^{-1} = 1.$$

Suppose that we had a second factorization $A = Q_1 B_1$. Then

$$B^2 = A^t A = B_1^t Q_1^t Q_1 B_1 = B_1^2.$$

Since a positive matrix has a unique positive square root, it follows that $B = B_1$. As A is invertible, B is invertible, and canceling gives $Q = Q_1$.

Solution to 7.6.39: Conjugating A changes neither the convergence nor the eigenvalues, so we may assume A in Jordan canonical form, $A = \begin{pmatrix} a & 0 \\ 0 & b \end{pmatrix}$ or $A = \begin{pmatrix} a & 1 \\ 0 & a \end{pmatrix}$. In the first case $A^n = \begin{pmatrix} a^n & 0 \\ 0 & b^n \end{pmatrix}$ and $\sum A^n$ converges when the eigenvalues, a and b, have absolute value less than 1, since the entries of the sum are geometric series.

In the second case write $A = aI + N$, with N the nilpotent matrix $N = \begin{pmatrix} 0 & 1 \\ 0 & b \end{pmatrix}$. In this case $N^2 = 0$, and $A^n = a^n I + na^{n-1} N$. If $I + A + A^2 + \cdots$ converges, then the diagonal entries, a^n, of the terms A^n must converge to 0, so $|a| < 1$. Conversely, if $|a| < 1$, then $\sum a^n$ and $\sum na^{n-1}$ converge, by the Ratio Test [Rud87, p. 66], therefore, $\sum A^n$ converges.

Solution to 7.6.40: An easy calculation shows that A has eigenvalues 0, 1, and 3, so A is similar to the diagonal matrix with entries 0, 1, and 3. Since clearly the problem does not change when A is replaced by a similar matrix, we may replace A by that diagonal matrix. Then the condition on a is that each of the sequences (0^n), (a^n), and $((3a)^n)$ has a limit, and that at least one of these limits is nonzero. This occurs if and only if $a = 1/3$.

Solution to 7.6.41: Let g be an element of the group. Consider the Jordan Canonical Form [HK61, p. 247] of the matrix g in $\mathbf{F}_{p^2}^*$ a quadratic extension of $\mathbf{F}_p$. The characteristic polynomial has degree 2 and is either irreducible in $\mathbf{F}_p$ and the canonical form is diagonal with two conjugate entries in the extension or reducible with the Jordan Canonical Form having the same diagonal elements and a 1 in the upper right-hand corner. In the first case, we can see that $g^{p^2-1} = I$ and in the second $g^{p(p-1)} = I$.

7.7   Similarity

Solution to 7.7.1: A simple calculation shows that A and B have the same characteristic polynomial, namely $(x - 1)^2(x - 2)$. However,

$$A - I = \begin{pmatrix} 0 & 0 & 0 \\ -1 & 0 & 1 \\ -1 & 0 & 1 \end{pmatrix}, \qquad B - I = \begin{pmatrix} 0 & 1 & 0 \\ 0 & 0 & 0 \\ 0 & 0 & 1 \end{pmatrix}.$$

Since $A - I$ has rank 1 and $B - I$ has rank 2, these two matrices are not similar, and therefore, neither are A and B.

Solution to 7.7.3: A calculation gives

$$\det(B - zI) = z^2(z - 1)(z + 1) = \det(A - zI).$$

The matrix A is in Jordan Canonical Form, and there are only two matrices with the same characteristic polynomial as A, namely A and the diagonal matrix with diagonal entries $1, -1, 0, 0$. Since B obviously has rank 3, its Jordan form must be A, i.e., B and A are similar.

Solution to 7.7.5: The eigenvalues of A an B are either ± 1 and neither is I or $-I$, since the equation $AB + BA = 0$ would force the other matrix to be zero. Therefore, A and B have distinct eigenvalues and are both diagonalizable. Let S be such that $SAS^{-1} = \left(\begin{smallmatrix} 1 & 0 \\ 0 & -1 \end{smallmatrix}\right)$. Multiplying on the left by S and on the right by S^{-1} the relations above we see that $C = SBS^{-1}$ satisfies $C^2 = I$ and $(SAS^{-1})(SBS^{-1}) + (SBS^{-1})(SAS^{-1}) = 0$. We get

$$C = \begin{pmatrix} 0 & 1/c \\ c & 0 \end{pmatrix} \quad \text{for} \quad c \neq 0$$

and taking $D = \left(\begin{smallmatrix} ck & 0 \\ 0 & k \end{smallmatrix}\right)$ we can easily see that $T = DS$ satisfies

$$TAT^{-1} = \begin{pmatrix} 1 & 0 \\ 0 & -1 \end{pmatrix} \qquad TBT^{-1} = \begin{pmatrix} 0 & 1 \\ 1 & 0 \end{pmatrix}.$$

Solution to 7.7.7: Let $\chi(x) = x^2 - bx + d$ be the characteristic polynomial of A. Since d is the determinant of A, $d^n = 1$, but the only roots of unity in $\mathbb{Q}$ are ± 1 so $d = \pm 1$. Let α and β be the two complex roots of χ. Over $\mathbb{C}$ A is similar to a matrix M of the form

$$M = \begin{pmatrix} \alpha & 0 \\ 0 & \beta \end{pmatrix}.$$

Since the roots of the characteristic polynomial of A^n are the the nth powers of the roots of χ, it follows that $\alpha^n = \beta^n = 1$. Since ± 1 are the only real roots of unity, if α or β is real, we get that $\alpha = \pm\beta = \pm 1$ so in this case, since A is similar to M, $A^2 = I$. Now suppose α is not real. This means $b^2 - 4d < 0$ and $|\alpha| = d$. So $d = 1$ and $b = 0$ or $b = \pm 1$. If $b = 0$, α and β are fourth roots of unity so $M^4 = I$ and thus $A^4 = I$. If $b = \pm 1$, α and β are sixth roots of unity so M^6 and A^6 equal I. Thus in all cases $A^{12} = I$.

Solution 2. Alternatively, we can analyze the characteristic polynomial in the following way: $\chi(x) = x^2 - bx + d$ where $d = \det A = \pm 1$ and $b = \mathrm{trace}\, A$ is an integer. Since A has finite order, its eigenvalues are roots of unity, so have modulus 1. Since b is the sum of the eigenvalues, $|b| \leqslant 2$.

Therefore the characteristic polynomial of A is one of $x^2 \pm 1$, $x^2 \pm x + 1$, $x^2 \pm x - 1$, $x^2 \pm 2x + 1$, $x^2 \pm 2x - 1$. Note that $x^2 \pm 1$ divides $x^4 - 1$, and $x^2 \pm x + 1$ divides $x^6 - 1$, so any matrix with one of these polynomials has order dividing 12. The zeros of $x^2 \pm x - 1$ and $x^2 \pm 2x - 1$ do not have modulus 1, so these polynomials cannot occur.

A matrix A with polynomial $x^2 \pm 2x + 1$ has repeated eigenvalues ± 1. Recalling that $A^n = 1$ for some $n > 0$, A is diagonalizable (for example, since it satisfies a polynomial equation with distinct linear factors), so $A = \pm I$.

Solution 3. Since $A^n - I = 0$ the minimal polynomial μ_A of A is a divisor of $x^n - 1$. The irreducible factorization of $x^n - 1$ is given by

$$x^n - 1 = \prod_{m \mid n} \Phi_m,$$

where $\Phi_m = \prod_{\omega_i \in \Omega_m} (x - \omega_i)$, is the m-th cyclotomical polynomial, product of all the factors $(x - \omega_i)$ where w_i are the primitive m-th roots of unity. $\Omega_m = \{e^{2\pi i r/m} \mid 1 \leqslant r \leqslant m, (r, m) = 1\}$ and the numbers of elements is given by $\varphi(m)$, the Euler's totient function. Thus the irreducible factorizations of μ_A and χ_A over $\mathbb{Q}$ have the form

$$m_A = \Phi_{m_1} \cdots \Phi_{m_r}, \quad \chi_A = \Phi_{m_1}^{d_1} \cdots \Phi_{m_r}^{d_r}$$

where $r \geqslant 1$, $m_1, \ldots, m_r$ are distinct and divide n, and $d_i \geqslant 1$. It follows that $d_1 \varphi(m_1) + \cdots + d_r \varphi(m_r) = 2$.

There are two cases: solving $\varphi(m_1) + \varphi(m_2) = 2$ gives $(m_1, m_2) = (1, 1)$, $(1, 2)$, $(2, 2)$ and solving $\varphi(m) = 2$ gives $m = 3, 4, 6$.

Take the case $(1, 1)$, that is, the minimal and characteristic polynomial of A are given by

$$\mu_A = \Phi_1, \quad \chi_A = \Phi_1^2.$$

By the Cyclic Decomposition Theorem $A \sim C(\Phi_1) \oplus C(\Phi_1)$, a direct sum of companion matrices.

The other cases are similar, that is, A is similar over $\mathbb{Q}$ to one of the six matrices

$$C(\Phi_1) \oplus C(\Phi_1),\ C(\Phi_1) \oplus C(\Phi_2),\ C(\Phi_2) \oplus C(\Phi_2),\ C(\Phi_3),\ C(\Phi_4),\ C(\Phi_6)$$

which are

$$\begin{pmatrix} 1 & 0 \\ 0 & 1 \end{pmatrix}, \begin{pmatrix} 1 & 0 \\ 0 & -1 \end{pmatrix}, \begin{pmatrix} -1 & 0 \\ 0 & -1 \end{pmatrix},$$

$$\begin{pmatrix} 0 & -1 \\ 1 & -1 \end{pmatrix}, \begin{pmatrix} 0 & -1 \\ 1 & 0 \end{pmatrix}, \begin{pmatrix} 0 & -1 \\ 1 & 1 \end{pmatrix},$$

and these have orders 1, 2, 2, 3, 4, 6 respectively. (the order of $C(\Phi_m)$ is m). In all cases $A^{12} = I$.

Solution to 7.7.8: 1. Let A be any element of the group:

(i) Every element in a finite group has finite order, so there is an $n > 0$ such that $A^n = I$. Therefore, $(\det A)^n = \det(A^n) = 1$. But A is an integer matrix, so $\det A$ must be ± 1.

(ii) If λ is an eigenvalue of A, then $\lambda^n = 1$, so each eigenvalue has modulo 1, and at the same time, λ is a root of a second degree monic characteristic polynomial $\chi_A(x) = x^2 + ax + b$ for A. If $|\lambda| = 1$ then $b = \pm 1$ and $a = 0, \pm 1$, and ± 2 since all roots are in the unit circle. Writing out all 10 polynomials and eliminating the ones whose roots are not in the unit circle, we are left with $x^2 \pm 1$, $x^2 \pm x + 1$, and $x^2 \pm 2x + 1$, and the possible roots are $\lambda = \pm 1, \pm i$, and $\frac{1 \pm \sqrt{3}i}{2}$ and $\frac{-1 \pm \sqrt{3}i}{2}$, the sixth roots of unity.

(iii) The Jordan Canonical Form [HK61, p. 247] of A, J_A, must be diagonal, otherwise it would be of the form $J_A = \begin{pmatrix} x & 1 \\ 0 & x \end{pmatrix}$, and the subsequent powers $(J_A)^k = \begin{pmatrix} x^k & kx^{k-1} \\ 0 & x^k \end{pmatrix}$, which is never the identity matrix since $kx^{k-1} \neq 0$ (remember $|x| = 1$). So the Jordan Canonical Form of A is diagonal, with the root above and the complex roots occurring in conjugate pairs only.

The Rational Canonical Form [HK61, p. 238] can be read off from the possible polynomials.

(iv) A can only have order 1, 2, 3, 4, or 6, depending on λ.

Solution to 7.7.10: Let R_A and R_B be the Rational Canonical Forms [HK61, p. 238] of A and B, respectively, over $\mathbb{R}$; that is, there are real invertible matrices K and L such that

$$R_A = KAK^{-1}$$
$$R_B = LBL^{-1}.$$

Observe now that R_A and R_B are also the Rational Canonical Forms over $\mathbb{C}$ as well, and by the uniqueness of the canonical form, they must be the same matrices. If $KAK^{-1} = LBL^{-1}$ then $A = K^{-1}LB(K^{-1}L)^{-1}$, so $K^{-1}L$ is a real matrix defining the similarity over $\mathbb{R}$. Observe that the proof works for any subfield; in particular, two *rational* matrices that are similar over $\mathbb{R}$ are similar over $\mathbb{Q}$.

Solution 2. Let $U = K + iL$ where K and L are real and $L \neq 0$ (otherwise we are done). Take real and imaginary parts of

$$A(K + iL) = AU = UB = (K + iL)B$$

and add them together after multiplying the imaginary part by z to get

$$A(K + zL) = (K + zL)B$$

for any complex z. Let $p(z) = \det(K + zL)$. Since p is a polynomial of degree n, not identically zero ($p(i) \neq 0$), it has, at most, n roots. For real z_0 not one of the roots of p, $V = K + z_0 L$ is real and invertible and $A = VBV^{-1}$.

Solution to 7.7.11: The minimal polynomial of A divides $(x - 1)^n$, so $I - A$ is nilpotent, say of order r. Thus, A is invertible with

$$A^{-1} = (I - (I - A))^{-1} = \sum_{j=0}^{r-1} (I - A)^j .$$

Suppose first that A is just a single Jordan block [HK61, p. 247], say with matrix

$$\begin{pmatrix} 1 & 1 & 0 & \cdots & 0 \\ 0 & 1 & 1 & \cdots & 0 \\ \vdots & \vdots & \vdots & \ddots & \vdots \\ 0 & 0 & 0 & \cdots & 1 \end{pmatrix}$$

relative to the basis $\{v_1, v_2, \ldots, v_n\}$. Then A^{-1} has the same matrix relative to the basis $\{v_1, v_1 + v_2, \cdots, v_1 + v_2 + \cdots + v_n\}$, so A and A^{-1} are similar.

In the general case, by the theory of Jordan Canonical Form, the vector space can be written as a direct sum of A–invariant subspaces on each of which A acts as a single Jordan block. By the formula above for A^{-1}, each subspace in the decomposition is A^{-1}–invariant, so the direct sum decomposition of A is also one of A^{-1}. The general case thus reduces to the case where A is a single Jordan block.

Solution to 7.7.12: The statement is true. First of all, A is similar to a Jordan matrix [HK61, p. 247], $A = S^{-1}JS$, where S is invertible and J is a direct sum of Jordan blocks. Then $A^t = S^t J^t (S^t)^{-1}$ (since $(S^{-1})^t = (S^t)^{-1}$); that is, A^t is similar to J^t. Moreover, J^t is the direct sum of the transposes of the Jordan blocks whose direct sum is J. It will, thus, suffice to prove that each of these Jordan blocks is similar to its transpose. In other words, it will suffice to prove the statement for the case where A is a Jordan block.

Let A be an $n \times n$ Jordan block:

$$A = \begin{pmatrix} \lambda & 1 & 0 & \cdots & 0 \\ 0 & \lambda & 1 & \cdots & \vdots \\ \vdots & \vdots & \ddots & \ddots & 0 \\ 0 & 0 & \cdots & \lambda & 1 \end{pmatrix}$$

Let $e_1, \ldots, e_n$ be the standard basis vectors for $\mathbb{C}^n$, so that $Ae_j = \lambda e_j + e_{j-1}$ for $j > 1$ and $Ae_1 = \lambda e_1$. Let the matrix S be defined by $Se_j = e_{n-j+1}$. Then

$S = S^{-1}$, and

$$S^{-1}ASe_j = SAe_{n-j}$$
$$= \begin{cases} S(\lambda e_{n-j+1} + e_{n-j}), & j < n \\ S(\lambda e_{n-j+1}), & j = n \end{cases}$$
$$= \begin{cases} \lambda e_j + e_{j+1}, & j < n \\ \lambda e_j, & j = n \end{cases}$$

which shows that $S^{-1}AS = A^t$.

Solution to 7.7.14: Using the first condition the Jordan Canonical Form [HK61, p. 247] of this matrix is a 6×6 matrix with five 1's and one -1 on the diagonal. The blocks corresponding to the eigenvalue 1 are either 1×1 or 2×2, by the second condition, with at least one of them having dimension 2. Thus, there could be three 1-blocks and one 2-block (for the eigenvalue 1), or one 1-block and two 2-blocks. In this way, we get the following two possibilities for the Jordan Form of the matrix:

$$\begin{pmatrix} 1 & 0 & 0 & 0 & 0 & 0 \\ 0 & 1 & 0 & 0 & 0 & 0 \\ 0 & 0 & 1 & 0 & 0 & 0 \\ 0 & 0 & 0 & 1 & 1 & 0 \\ 0 & 0 & 0 & 0 & 1 & 0 \\ 0 & 0 & 0 & 0 & 0 & -1 \end{pmatrix}, \quad \begin{pmatrix} 1 & 0 & 0 & 0 & 0 & 0 \\ 0 & 1 & 1 & 0 & 0 & 0 \\ 0 & 0 & 1 & 0 & 0 & 0 \\ 0 & 0 & 0 & 1 & 1 & 0 \\ 0 & 0 & 0 & 0 & 1 & 0 \\ 0 & 0 & 0 & 0 & 0 & -1 \end{pmatrix}.$$

Solution to 7.7.15: Since A and B have the same characteristic polynomial, they have the same n distinct eigenvalues $l_1, \ldots, l_n$. Let $\chi(x) = (x - l_1)^{c_1} \cdots (x - l_n)^{c_n}$ be the characteristic polynomial and let $\mu(x) = (x - l_1)^{m_1} \cdots (x - l_n)^{m_n}$ be the minimal polynomial. Since a nondiagonal Jordan block [HK61, p. 247] must be at least 2×2, there can be, at most, one nondiagonal Jordan block for $N \leqslant 3$. Hence, the Jordan Canonical Form is completely determined by $\mu(x)$ and $\chi(x)$ for $N \leqslant 3$. If $\mu(x) = \chi(x)$, then each distinct eigenvalue corresponds to a single Jordan block of size equal to the multiplicity of the eigenvalue as a root of $\chi(x)$, so the Jordan Canonical Form is completely determined by $\chi(x)$, and A and B must then be similar.

Solution to 7.7.16: It will suffice to show that A is similar to a matrix of the form $\begin{pmatrix} 0 & * \\ * & B \end{pmatrix}$, where B is $(n-1) \times (n-1)$. For in that case B has the same trace as A, hence trace 0, and the argument can be iterated.

Assume, without loss of generality, that A is not the zero matrix. Since A has trace 0, it is not a scalar multiple of the identity matrix. Hence, there is a nonzero vector v in $\mathbb{C}^n$ that is not an eigenvector of A. Take a basis for $\mathbb{C}^n$ in which the first basis vector is v and the second one is Av. The matrix with respect to such a basis of the linear transformation induced by A is similar to A and has the required form.

Solution to 7.7.17: Since $M^3 = tI$, $N^3 = 0$, so it suffices to show that $N^2 \neq 0$. If $N^2 = 0$, then $M^2 = tP$ for some 3×3 matrix P with entries in $\mathbb{R}[t]$. Therefore $tPM = M^3 = tI$, hence $PM = I$, implying that $tI = M^3$ has an inverse P^3 with entries in $\mathbb{R}[t]$, a contradiction.

Solution to 7.7.19: Since the map $X \mapsto X^2$ on $M_n(\mathbb{C})$ is continuous, we have

$$B^2 = \left(\lim_{n \to \infty} A^n \right)^2 = \lim_{n \to \infty} A^{2n} = B.$$

The minimal polynomial of B divides $x^2 - x$, so the eigenvalues of B are zeros and ones. Since the minimal polynomial is squarefree, the Jordan blocks of B are of size 1, i.e., B is similar to a diagonal matrix with its eigenvalues (zeros and ones) along the main diagonal.

7.8   Bilinear, Quadratic Forms, and Inner Product Spaces

Solution to 7.8.1: For all $x \in \mathbb{R}^n$, $g(x, x) = f(2x) - 2f(x) = 4f(x) - 2f(x) = 2f(x)$. Therefore $f(x) = g(x, x)/2$ is the quadratic form associated with the bilinear form g. For each $x \in \mathbb{R}^n$, $y \mapsto g(x, y)$ is linear, so there is exactly one vector $A(x) \in \mathbb{R}^n$ such that $g(x, y) = \langle A(x), y \rangle$ for all $y \in \mathbb{R}^n$. Since g is linear in the first variable, A is a linear map.

Solution to 7.8.2: Let H be the Hermitian matrix that induces the quadratic form

$$H = \begin{pmatrix} A & B & D \\ B & C & E \\ D & E & F \end{pmatrix},$$

and let

$$G = \begin{pmatrix} A & B \\ B & C \end{pmatrix}.$$

H is positive definite if and only if all of its eigenvalues are positive. Since the product of those eigenvalues is det H, the positivity of det H is a necessary condition for H to be positive definite. Since the positive definiteness of H implies the positive definiteness of G, whose associated quadratic form is $Ax^2 + 2Bxy + Cy^2$, the positivity of det G is also a necessary condition. The positivity of A is similarly a necessary condition.

Assume that $A > 0$, det $G > 0$, and det $H > 0$, but that H is not positive definite. Then the product of the eigenvalues of H ($=$ det H) is positive, but not all of the eigenvalues are positive. Hence H must have one positive eigenvalue and two negative ones, or a single negative one of multiplicity two. It follows that H is negative definite on the two dimensional subspace spanned by the eigenvectors corresponding to the negative eigenvalues. That subspace must have a nontrivial intersection with the subspace spanned by the first two basis vectors. Hence there is a nonzero vector v in $\mathbb{R}^3$ with its last coordinate 0 such that $v^t H v < 0$. It follows that G is not positive definite. But since $A > 0$ neither is G negative definite. Thus G must have one positive and one negative eigenvalue, in contradiction to the assumption $\det G > 0$.

Solution to 7.8.3: We have

$$2x_1^2 + x_2^2 + 3x_3^2 + 2tx_1x_2 + 2x_1x_3 = (x_1, x_2, x_3) \begin{pmatrix} 2 & t & 1 \\ t & 1 & 0 \\ 1 & 0 & 3 \end{pmatrix} \begin{pmatrix} x_1 \\ x_2 \\ x_3 \end{pmatrix}$$

By the Solution to Problem 7.8.2 above, the form is positive definite if and only if the determinants

$$\begin{vmatrix} 2 & t & 1 \\ t & 1 & 0 \\ 1 & 0 & 3 \end{vmatrix} > 0 \quad \text{and} \quad \begin{vmatrix} 2 & t \\ t & 1 \end{vmatrix} > 0$$

that is, when $-1 + 3(2 - t^2) = 5 - 3t^2 > 0$ and $2 - t^2 > 0$.

Both conditions hold iff $|t| < \sqrt{\frac{5}{3}}$. For these values of t the form is positive definite.

Solution to 7.8.4: Every vector in W is orthogonal to $v = (a, b, c)$. Let Q be the orthogonal projection of $\mathbb{R}^3$ onto the space spanned by v, identified with its matrix. The columns of Q are Qe_j, $1 \leq j \leq 3$, where the e_j's are the standard basis vectors in $\mathbb{R}^3$. But

$$Qe_1 = \langle v, e_1 \rangle v = (a^2, ab, ac)$$

$$Qe_2 = \langle v, e_2 \rangle v = (ab, b^2, bc)$$

$$Qe_3 = \langle v, e_3 \rangle v = (ac, bc, c^2).$$

Therefore, the orthogonal projection onto W is given by

$$P = I - Q = \begin{pmatrix} 1 - a^2 & -ab & -ac \\ -ab & 1 - b^2 & -bc \\ -ac & -bc & 1 - c^2 \end{pmatrix}.$$

Solution to 7.8.5: 1. The monomials $1, t, t^2, \ldots, t^n$ form a basis for P_n. Applying the Gram–Schmidt Procedure [HK61, p. 280] to this basis gives us an orthonormal basis $p_0, p_1, \ldots, p_n$. The $(k + 1)^{th}$ vector in the latter basis, p_k, is a linear combination of $1, t, \ldots, t^k$, the first $k + 1$ vectors in the former basis, with t^k having a nonzero coefficient. (This is built into the Gram–Schmidt Procedure.) Hence, $\deg p_k = k$.

2. Since p'_k has degree $k - 1$, it is a linear combination of $p_0, p_1, \ldots, p_{k-1}$, for those functions form an orthonormal basis for P_{k-1}. Since p_k is orthogonal to $p_0, p_1, \ldots, p_{k-1}$, it is orthogonal to p'_k.

Solution to 7.8.6: Since p is even, it is orthogonal on $[-1, 1]$ to all odd polynomials. Hence a and b are determined by the two conditions

$$\int_{-1}^{1} p(x)\,dx = 0, \qquad \int_{-1}^{1} x^2 p(x)\,dx = 0.$$

Carrying out the integrations, one obtains the equations

$$2a + \frac{2b}{3} - \frac{2}{5} = 0, \qquad \frac{2a}{3} + \frac{2b}{5} - \frac{2}{7} = 0.$$

Solving these for a and b, one gets $a = -\dfrac{3}{35}, b = \dfrac{6}{7}$, therefore

$$p(x) = -\frac{3}{35} + \frac{6x^2}{7} - x^4.$$

Solution to 7.8.7: Let $n = \dim E$, and choose a basis $v_1, \ldots, v_n$ for E. Define the $n \times n$ matrix $A = (a_{jk})$ by $a_{jk} = B(v_k, v_j)$. The linear transformation T_A on E induced by A is determined by the relations

$$T_A v_k = \sum_j a_{jk} v_j = \sum_j B(v_k, v_j) v_j, \qquad k = 1, \ldots, n,$$

implying that $T_A v = \sum_j B(v, v_j) v_j$ ($v \in E$). It follows that $E_1 = \ker T_A$. By similar reasoning, $E_2 = \ker T_{A'}$, where A' is the transpose of A. By the Rank–Nullity Theorem [HK61, p. 71], $\dim E_1$ equals n minus the dimension of the column space of A, and $\dim E_2$ equals n minus the dimension of the row space of A. Since the row space and the column space of a matrix have the same dimension, the desired equality follows.

Solution to 7.8.8: We have, for any $x \in \mathbb{R}^n$,

$$\langle Ax, x \rangle = \langle x, Ax \rangle = \langle A'x, x \rangle \geqslant 0,$$

hence, $\langle (A + A')x, x \rangle \geqslant 0$ for all $x \in \mathbb{R}^n$. As $A + A'$ is symmetric, therefore diagonalizable, there exist an integer $k \leqslant n$, positive numbers $\lambda_1, \lambda_2, \ldots, \lambda_k$, and a basis of $\mathbb{R}^n$, $\{v_1, \ldots, v_n\}$, such that $(A + A^T)v_i = \lambda_i v_i$ for $1 \leqslant i \leqslant k$, and $(A + A^T)v_i = 0$ for $k + 1 \leqslant i \leqslant n$. The matrix representing A in this basis has the form

$$A' = \begin{pmatrix} B & C \\ -C & D \end{pmatrix}$$

where D is antisymmetric and

$$B = \frac{1}{2} \begin{pmatrix} \lambda_1 & \cdots & 0 \\ \vdots & \ddots & \vdots \\ 0 & \cdots & \lambda_k \end{pmatrix} + B',$$

where B' is antisymmetric. We have

$$x^t A' x = \lambda_1 x_1^2 + \cdots + \lambda_k x_k^2,$$

so $x \in \ker A'$ exactly when the first k components of x are zero and the last $n - k$ form a vector in $\ker C \cap \ker D$. As

$$(A')^t = \begin{pmatrix} B^t & C \\ -C & D \end{pmatrix},$$

we conclude that a vector is in the kernel of A' iff it is in the kernel of $(A')^t$.

Solution to 7.8.9: 1. Since A is positive definite, one can define a new inner product $\langle\,,\,\rangle_A$ on $\mathbb{R}^n$ by

$$\langle x, y \rangle_A = \langle Ax, y \rangle.$$

The linear operator $A^{-1}B$ is symmetric with respect to this inner product, that is,

$$\langle A^{-1}Bx, y \rangle_A = \langle Bx, y \rangle = \langle x, B^t y \rangle = \langle x, By \rangle$$
$$= \langle A^{-1}Ax, By \rangle = \langle Ax, A^{-1}By \rangle = \langle x, A^{-1}By \rangle_A.$$

So there is a basis $\{v_1, \dots v_n\}$ of $\mathbb{R}^n$, orthonormal with respect to $\langle\,,\,\rangle_A$, in which the matrix for $A^{-1}B$ is diagonal. This is the basis we are looking for; in particular, v_i is an eigenvector for $A^{-1}B$, with eigenvalue λ_i and

$$\langle v_i, v_j \rangle_A = \delta_{ij}$$

$$\langle Bv_i, v_j \rangle = \langle A^{-1}Bv_i, v_j \rangle_A = \langle \lambda_i v_i, v_j \rangle_A = \lambda_i \delta_{ij}.$$

2. Let U be the matrix which takes the standard basis to $\{v_1, \dots v_n\}$ above, that is, $Ue_i = v_i$. Since the e_i form an orthonormal basis, for any matrix M, we have

$$Mx = \sum_{j=1}^{n} \langle Mx, e_j \rangle e_j,$$

in particular,

$$U^t A U e_i = \sum_{j=1}^{n} \langle U^t A U e_i, e_j \rangle e_j$$
$$= \sum_{j=1}^{n} \langle A U e_i, U e_j \rangle e_j$$

$$= \sum_{j=1}^{n} \langle Av_i, v_j \rangle e_j$$

$$= \sum_{j=1}^{n} \delta_{ij} e_j = e_i ,$$

showing that $U' A U = I$.

Using the same decomposition for $U' B U$, we have

$$U' B U e_i = \sum_{j=1}^{n} \langle U' B U e_i, e_j \rangle e_j$$

$$= \sum_{j=1}^{n} \langle B v_i, v_j \rangle e_j$$

$$= \sum_{j=1}^{n} \lambda_i \delta_{ij} e_i = \lambda_i e_i ,$$

so $U' B U$ is diagonal.

Solution 2. Since A is positive definite $A = W' W$ for some invertible matrix W. Setting $N = W^{-1}$ we get $N' A N = W'^{-1} W' W W^{-1} = I$ and $N' B N$ is still positive definite since $\langle x, N' B N x \rangle = \langle N x, B(N x) \rangle \geqslant 0$. As $N' B N$ is positive definite there is an orthogonal matrix O such that $O'(N' B N)O = D$, where D is a diagonal matrix. The entries of D are positive, since $N' B N$ is positive definite. Let $M = N O$. Then $M' B M = O' N' B N O = D$ and $M' A M = O' N' A N O = O' I O = I$, and we can easily see that M is invertible, since N and O are.

Solution to 7.8.11: From the Criteria proved in the Solution to Problem 7.8.2 we see that the matrix is positive definite and $v' A v = N(v)$ defines a norm in $\mathbb{R}^3$. Now all norms in $\mathbb{R}^3$ are equivalent, see the Solution to Problem 2.1.6, so

$$\alpha N(v) \leqslant \|v\| \leqslant \beta N(v)$$

where α and β are the minimum and the maximum of $\|v\|$ on the set $v' A v = 1$.

We can use the Method of Lagrange Multipliers [MH93, p. 414] to find the maximum of the function

$$f(x, y, z) = x^2 + y^2 + z^2$$

over the surface defined by $\psi(x, y, z) = 6$, where

$$\psi = \frac{1}{6} \left(13x^2 + 13y^2 + 10z^2 - 10xy - 4xz - 4yz \right)$$

is the quadratic form defined by the matrix A. Setting up the equations we have:

$$\frac{\partial \psi}{\partial x} = 26x - 10y - 4z = \lambda \frac{\partial f}{\partial x} = 2\lambda x$$

$$\frac{\partial \psi}{\partial y} = 26y - 10x - 4z = \lambda \frac{\partial f}{\partial y} = 2\lambda y$$

$$\frac{\partial \psi}{\partial z} = 20z - 4x - 4y = \lambda \frac{\partial f}{\partial z} = 2\lambda z$$

$$\psi(x, y, z) = 6$$

which is equivalent to the linear system of equations

$$\begin{pmatrix} 13 - \lambda & -5 & -2 \\ -5 & 13 - \lambda & -2 \\ -2 & -2 & 10 - \lambda \end{pmatrix} \begin{pmatrix} x \\ y \\ z \end{pmatrix} = 0$$

together with the equation $\psi(x, y, z) = 6$. Now if the determinant of the system is nonzero, there is only one solution to the system, the trivial one $x = y = z = 0$ and that point is not in the surface $\psi(x, y, z) = 6$, so let's consider the case where the determinant is zero. One can easily compute it and it is:

$$-\lambda^3 + 36\lambda^2 - 396\lambda + 1296$$

and one can easily see that $\lambda = 18$ is a root of the polynomial, because it renders the first two rows of the matrix the same, factoring it completely we get:

$$(18 - \lambda)(\lambda - 6)(\lambda - 12)$$

Considering each one of the roots that will correspond to non-trivial solutions of the system of linear equations

- $\lambda = 18$. In this case the system becomes:

$$\begin{pmatrix} -5 & -5 & -2 \\ -5 & -5 & -2 \\ -2 & -2 & -8 \end{pmatrix} \begin{pmatrix} x \\ y \\ z \end{pmatrix} = 0$$

which reduces to $\begin{cases} 5x + 5y + 2z = 0 \\ 2x + 2y + 8z = 0 \end{cases}$. Therefore $z = 0$ and $y = -x$. Substituting this back in the equation of the ellipsoid, we find $x = \pm \dfrac{1}{\sqrt{6}}$, and the solution points are:

$$\pm \frac{1}{\sqrt{6}} (1, -1, 0)$$

- $\lambda = 6$. The system now is

$$\begin{pmatrix} 7 & -5 & -2 \\ -5 & 7 & -2 \\ -2 & -2 & 4 \end{pmatrix} \begin{pmatrix} x \\ y \\ z \end{pmatrix} = 0$$

and the solutions are: $x = y = z$ and the only points in this line on the ellipsoid are:

$$\pm \frac{1}{\sqrt{3}} (1, 1, 1)$$

- $\lambda = 12$. The system now is

$$\begin{pmatrix} 1 & -5 & -2 \\ -5 & 1 & -2 \\ -2 & -2 & -2 \end{pmatrix} \begin{pmatrix} x \\ y \\ z \end{pmatrix} = 0$$

and the solutions are: $y = x$ and $z = -2x$ the points in this line on the ellipsoid are:

$$\pm \frac{1}{2\sqrt{3}} (1, 1, -2) .$$

Computing the sizes of each (pair) of the vectors we obtain $1/\sqrt{3}$, 1 and $1/\sqrt{2}$, respectively; so the least upper bound is 1.

Solution 2. A, being a symmetric matrix, can be diagonalized by an orthogonal matrix S, so

$$D = S^t A S$$

where the matrix S is given by the eigenvalues of A. Following the computation of the previous solution, we get

$$S = \begin{pmatrix} 1/\sqrt{2} & 1/\sqrt{3} & 1/\sqrt{6} \\ -1/\sqrt{2} & 1/\sqrt{3} & 1/\sqrt{6} \\ 0 & 1/\sqrt{3} & -2/\sqrt{6} \end{pmatrix}$$

but in our case we only need the matrix D which is the diagonal matrix with the eigenvalues of A, that is,

$$D = \begin{pmatrix} 18 & 0 & 0 \\ 0 & 6 & 0 \\ 0 & 0 & 12 \end{pmatrix}$$

so after the change of basis given by the matrix S above, the form ψ becomes $\psi(r, s, t) = 18r^2 + 6s^2 + 12t^2$ and on these variables the ellipsoid

$$\frac{r^2}{1/3} + s^2 + \frac{t^2}{1/2} = 1$$

has semi-axis $1/\sqrt{3}$, 1 and $1/\sqrt{2}$, and the least upper bound is 1.

Solution 3. Rotating around the z-axis by 45° on the positive direction is equivalent to the change of variables

$$x = \frac{1}{\sqrt{2}} (r + s)$$

$$y = \frac{1}{\sqrt{2}} (-r + s)$$

which substituted in the equation of the ellipsoid

$$13x^2 + 13y^2 + 10z^2 - 10xy - 4xz - 4yz = 6$$

eliminates the mixed terms in xy and xz. After the substitution we end up with

$$18r^2 + 8s^2 + 10z^2 - 4\sqrt{2}sz = 6.$$

Now we rotate around the r-axis by an angle α where $\cot 2\alpha = \dfrac{8-10}{-4\sqrt{2}} = \dfrac{1}{2\sqrt{2}}$ in order to eliminate the mixed term in sz. Using trigonometric identities we find $\cos 2\alpha = \dfrac{1}{3}$, and $\cos \alpha = \sqrt{\dfrac{2}{3}}$, $\sin \alpha = \dfrac{1}{\sqrt{3}}$, so the change of coordinates can now be written as

$$s = w \cos \alpha - t \sin \alpha = \frac{1}{\sqrt{3}} \left(w\sqrt{2} - t \right)$$

$$z = w \sin \alpha + t \cos \alpha = \frac{1}{\sqrt{3}} \left(w + t\sqrt{2} \right)$$

and the final substitution renders the equation of the ellipsoid

$$18r^2 + 6w^2 + 12t^2 = 6$$

whose largest semi-axis is 1.

Solution to 7.8.12: Suppose that A has a diagonalization with p strictly positive and q strictly negative entries, and that B has one with p' strictly positive and q' strictly negative entries. Then $\mathbb{R}^n$ contains a subspace V of dimension p such that $x^t A x > 0$ for all nonzero $x \in V$, and a subspace W of dimension $n - p'$ such that $x^t B x \leqslant 0$ for all $x \in W$. Since $x^t A x \leqslant x^t B x$ for all $x \in \mathbb{R}^n$, $x^t A x \leqslant 0$ for all $x \in W$. Therefore $V \cap W = \{0\}$, so $\dim V + \dim W \leqslant n$, i.e., $p + (n - p') \leqslant n$ giving $p \leqslant p'$. Similarly $q' \leqslant q$, so $p - q \leqslant p' - q'$.

Solution to 7.8.13: Let $A = T^*T$. Then $\langle x, y \rangle = 0$ implies $\langle Ax, y \rangle = 0$. For any $x \in \mathbb{C}^n$ let $x^\perp = \{y \in \mathbb{C}^n \mid \langle x, y \rangle = 0\}$. Thus, $\langle x, x^\perp \rangle = 0$ so $\langle Ax, x^\perp \rangle = 0$, implying that $Ax \in \left(x^\perp \right)^\perp$, so $Ax = \lambda x$ for some $\lambda \in \mathbb{C}$. Since every vector is an eigenvector of A, it follows that rI for some scalar r. The constant r is a nonnegative real number since A is positive semidefinite. If $r = 0$ then $A = 0$, hence $T = 0$ and we may take $k = 0$, $U = I$. If $r > 0$ we take $k = \sqrt{r}$ and set $U = \frac{1}{\sqrt{r}}T$. Clearly k is real, and U is unitary because

$$U^*U = \frac{1}{\sqrt{k}} T^*T \frac{1}{\sqrt{k}} = \frac{1}{k}A = I.$$

As $T = kU$ we are done.

Solution to 7.8.15: Assume that such u, v exist. Then there is a 3×3 orthogonal matrix Q whose three rows are (u_1, u_2, a), (v_1, v_2, b), (w_1, w_2, c). Every column of Q is a unit vector. This implies that $a^2 + b^2 \leqslant 1$.

Conversely, assume that $a^2 + b^2 \leqslant 1$. Let c be a real number such that $a^2 + b^2 + c^2 = 1$. The vector (a, b, c) can be extended to a basis of $\mathbb{R}^3$ and by the Gram–Schmidt Procedure there exist vectors (u_1, v_1, w_1), (u_2, v_2, w_2) that together with (a, b, c) form an orthonormal basis. Thus the matrix M given by

$$\begin{pmatrix} u_1 & u_2 & a \\ v_1 & v_2 & b \\ w_1 & w_2 & c \end{pmatrix}$$

is an orthogonal, that is, $M^t M = I$. Since a square matrix is a left-inverse if and only if is a right-inverse we also have that $M M^t = I$, and the first two rows of this matrix are the desired vectors.

Solution 2. Suppose we have such u and v. By Cauchy–Schwarz Inequality [MH93, p. 69], we have

$$(u_1 v_1 + u_2 v_2)^2 \leqslant (u_1^2 + u_2^2)(v_1^2 + v_2^2).$$

Since $u \cdot v = 0$, $(u_1 v_1 + u_2 v_2)^2 = (ab)^2$; since $\|u\| = \|v\| = 1$, $1 - a^2 = u_1^2 + u_2^2$, and $1 - b^2 = v_1^2 + v_2^2$. Combining these, we get

$$(ab)^2 \leqslant (1 - a^2)(1 - b^2) = 1 - a^2 - b^2 + (ab)^2,$$

which implies $a^2 + b^2 \leqslant 1$.

Conversely, suppose that $a^2 + b^2 \leqslant 1$. Let $u = (0, \sqrt{1 - a^2}, a)$. $\|u\| = 1$, and we now find v_1 and v_2 such that $v_1^2 + v_2^2 + b^2 = 1$ and $u_2 v_2 + ab = 0$. If $a = 1$, then $b = 0$, so we can take $v = (0, 1, 0)$. If $a \neq 1$, solving the second equation for v_2, we get

$$v_2 = \frac{-ab}{\sqrt{1 - a^2}}.$$

Using this to solve for v_1, we get

$$v_1 = \frac{\sqrt{1 - a^2 - b^2}}{\sqrt{1 - a^2}}.$$

By our condition on a and b, both of these are real, so u and $v = (v_1, v_2, b)$ are the desired vectors.

7.9   General Theory of Matrices

Solution to 7.9.1: Let A be such an algebra, and assume A does not consist of just the scalar multiples of the identity. Take a matrix M in A that is not a scalar multiple of the identity. We can assume without loss of generality that M is in Jordan form, so there are two possibilities:

$$\text{(i)} \quad M = \begin{pmatrix} \lambda_1 & 0 \\ 0 & \lambda_2 \end{pmatrix} \text{ with } \lambda_1 \neq \lambda_2 \qquad \text{(ii)} \quad M = \begin{pmatrix} \lambda & 1 \\ 0 & \lambda \end{pmatrix}.$$

In the first case the matrices commuting with M are the 2×2 diagonal matrices, which form an algebra of dimension 2. Thus $\dim A \leqslant 2$.

In the second case M has a one-dimensional eigenspace spanned by the vector $\begin{pmatrix} 1 \\ 0 \end{pmatrix}$, so any matrix commuting with M has $\begin{pmatrix} 1 \\ 0 \end{pmatrix}$ as an eigenvector, i.e., it is upper triangular. Also, a matrix commuting with M cannot have two distinct eigenvalues, because then the corresponding eigenvectors would have to be eigenvectors of M (and the only eigenvectors of M are the multiples of $\begin{pmatrix} 1 \\ 0 \end{pmatrix}$). Thus, the matrices commuting with M are the matrices

$$\begin{pmatrix} \mu & \nu \\ 0 & \mu \end{pmatrix}.$$

The algebra of all such matrices has dimension 2, so $\dim A \leqslant 2$ in this case also.

Solution to 7.9.2: In what follows we will use frequently the facts that $I + A$ is invertible, which can easily be concluded from the fact that -1 is not an eigenvalue of A, as we have $\ker(A + I) = 0$, as well as that A commutes with $(I + A)^{-1}$, which can be seen by factoring $A + A^2$ in two different ways $(I+A)A = A(I+A)$ and now multiplying on the right and left by $(I + A)^{-1}$.

1.

$$\begin{aligned}
\left((I - A)(I + A)^{-1}\right)^t &= \left((I + A)^{-1}\right)^t (I - A)^t \\
&= \left((I + A)^t\right)^{-1} (I - A^t) \\
&= (I + A^t)^{-1} (I - A^t) \qquad \text{since } A^t = A^{-1} \\
&= \left(I + A^{-1}\right)^{-1} \left(I - A^{-1}\right) \\
&= \left(A^{-1}(A + I)\right)^{-1} \left(A^{-1}(A - I)\right) \\
&= (A + I)^{-1} A A^{-1} (A - I) \\
&= (A + I)^{-1} (A - I) \\
&= -(I + A)^{-1} (I - A) \\
&= -(I - A)(I + A)^{-1}
\end{aligned}$$

on the last equality we used the fact that A commutes with $(I + A)^{-1}$ and this shows that the product is skew-symmetric.

2. Now let's suppose S is skew-symmetric, then

$$\left((I - S)(I + S)^{-1}\right)^t = \left((I + S)^{-1}\right)^t (I - S)^t$$

$$= \left(I + S^t\right)^{-1} \left(I - S^t\right)$$
$$= (I - S)^{-1} (I + S)$$
$$= (I + S) (I - S)^{-1}$$
$$= \left((I - S) (I + S)^{-1}\right)^{-1}$$

that is the product is orthogonal. Now to see that it doesn't have an eigenvalue 1, observe that

$$\det\left((I - S)(I + S)^{-1} + I\right) = \det\left((I - S + I + S)(I + S)^{-1}\right)$$
$$= \det\left(2I\,(I + S)^{-1}\right)$$
$$= 2^n \det\left((I + S)^{-1}\right)$$
$$\neq 0$$

3. Suppose the correspondence is not one-to-one, then

$$(I - A)(I + A)^{-1} = (I - B)(I + B)^{-1}$$

$$(I + A)^{-1}(I - A) = (I - B)(I + B)^{-1}$$

$$(I - A) = (I + A)(I - B)(I + B)^{-1}$$

$$(I - A)(I + B) = (I + A)(I - B)$$

simplifying we see that $A = B$.

Solution to 7.9.4: If $x, y \in \mathbb{R}^n$ then $x^t P y$ is a scalar, so $(x^t P y)^t = x^t P y$, that is, $y^t P^t x = x^t P y$. We conclude, from the hypothesis, that $y^t P^t x = -y^t P x$ for all x, y. Therefore $y^t (P + P^t) x = 0$. Write $A = P + P^t = (a_{ij})$. Then, $a_{ij} = e_i^t A e_j = 0$, where the e_i's are the standard basis vectors, proving that $A = 0$, that is, P is skew-symmetric.

Solution to 7.9.5: We will use a powerful result on the structure of real normal operators, not commonly found in the literature. We provide also a second solution, not using the normal form, but which is inspired on it.

Lemma (Structure of Real Normal Operators): Given a normal operator A on $\mathbb{R}^n$, there exists an orthonormal basis in which the matrix of A has the form

$$
\begin{pmatrix}
\begin{pmatrix} \sigma_1 & \tau_1 \\ -\tau_1 & \sigma_1 \end{pmatrix} & & & & & \\
& \begin{pmatrix} \sigma_2 & \tau_2 \\ -\tau_2 & \sigma_2 \end{pmatrix} & & & & \\
& & \ddots & & & \\
& & & \begin{pmatrix} \sigma_k & \tau_k \\ -\tau_k & \sigma_k \end{pmatrix} & & \\
& & & & \lambda_{2k+1} & \\
& & & & & \ddots & \\
& & & & & & \lambda_n
\end{pmatrix}
$$

where the numbers $\lambda_j = \sigma_j + i\tau_j$, $j = 1, \ldots, k$ and $\lambda_{2k+1}, \ldots, \lambda_n$ are the eigenvalues of A.

The proof is obtained by embedding each component of $\mathbb{R}^n$ as the real slice of each component of $\mathbb{C}^n$, extending A to a normal operator on $\mathbb{C}^n$, and noticing that the new operator has the same real matrix (on the same basis) and over $\mathbb{C}^n$ has basis of eigenvectors. A change of basis, picking the new vectors as the real and imaginary parts of the eigenvectors associated with the imaginary eigenvalues, reduces it to the desired form. For details on the proof we refer the reader to [Shi77, p. 265-271] or [HS74, p. 117].

The matrix of an anti-symmetric operator A has the property

$$
a_{ij} = \langle Ae_i, e_j \rangle = \langle e_i, A^*e_j \rangle = \langle e_i, -Ae_j \rangle = -\overline{\langle Ae_j, e_i \rangle} = -\bar{a}_{ji}.
$$

Since anti-symmetric operators are normal, they have a basis of eigenvalues, these satisfy the above equality and are all pure imaginary.

Thus, in the standard decomposition described above all eigenvalues are pure imaginary, i.e., $\sigma_1 = \cdots = \sigma_k = \lambda_{2k+1} = \cdots = \lambda_n = 0$ and the decomposition in this case is

$$
\begin{pmatrix}
\begin{pmatrix} 0 & \tau_1 \\ -\tau_1 & 0 \end{pmatrix} & & & & & \\
& \begin{pmatrix} 0 & \tau_2 \\ -\tau_2 & 0 \end{pmatrix} & & & & \\
& & \ddots & & & \\
& & & \begin{pmatrix} 0 & \tau_k \\ -\tau_k & 0 \end{pmatrix} & & \\
& & & & 0 & \\
& & & & & \ddots & \\
& & & & & & 0
\end{pmatrix}
$$

which obviously has even rank.

Solution 2. Consider $\hat{A}$ the *complexification* of A, that is, the linear operator from $\mathbb{C}^n$ to $\mathbb{C}^n$ with the same matrix as A with respect to the standard basis. Since A is skew-symmetric, all its eigenvalues are pure imaginary and from the fact that the characteristic polynomial has real coefficients, the non-real eigenvalues show up in conjugate pairs, therefore, the polynomial has the form

$$\chi_A(t) = t^k p_1(t)^{n_1} \cdots p_r(t)^{n_r},$$

where the p_i's are real, irreducible quadratics.

From the diagonal form of $\hat{A}$ over $\mathbb{C}$ we can see that the minimal polynomial has the factor in t with power 1, that is, of the form

$$\mu_A(t) = \mu_{\hat{A}}(t) = t p_1(t)^{m_1} \cdots p_r(t)^{m_r}.$$

Now consider the Rational Canonical Form [HK61, p. 238] of A. It is a block diagonal matrix composed of blocks of even size and full rank, together with a block of a zero matrix corresponding to the zero eigenvalues, showing that A has even rank.

Solution to 7.9.6: Since A is real and symmetric, there is an orthogonal matrix U such that $D = U^{-1}AU$ is diagonal, say with entries $\lambda_1, \ldots, \lambda_n$. Let $E = U^{-1}BU$, then $\operatorname{tr} AB = \operatorname{tr} DE = \sum_{i=1}^n \lambda_i e_{ii}$. If A and B are both positive semi-definite, then so are D and E, therefore $\lambda_i \geqslant 0$ and $e_{ii} \geqslant 0$, giving $\operatorname{tr} AB \geqslant 0$.

If A is not positive definite then $\lambda_i < 0$ for some i. Let E be a diagonal matrix with $e_{ii} = 1$ and all other entries zero. Then $B = U^{-1}EU$ is positive semi-definite and $\operatorname{tr} AB < 0$.

Solution to 7.9.7: Since A is symmetric it can be diagonalized: Let

$$A = QDQ^{-1}$$

where $D = \operatorname{diag}(d_1, \ldots, d_n)$ and each d_i is nonnegative. Then

$$0 = Q^{-1}(AB + BA)Q = DC + CD$$

where $C = Q^{-1}BQ$. Individual entries of this equation read

$$0 = (d_i + d_j)c_{ij}$$

so for each i and j we must have either $c_{ij} = 0$ or $d_i = d_j = 0$. In either case,

$$d_i c_{ij} = d_j c_{ij} = 0$$

which is the same as

$$DC = CD = 0.$$

Hence, $AB = BA = 0$.

Example:

$$A = \begin{pmatrix} 1 & 0 \\ 0 & 0 \end{pmatrix} \qquad B = \begin{pmatrix} 0 & 0 \\ 0 & 1 \end{pmatrix}.$$

Solution 2. Since A is symmetric, it is diagonalizable. Let v be an eigenvector of A with $Av = \lambda v$, then

$$A(Bv) = -BAv = -\lambda Bv$$

that is, Bv is an eigenvector of A with eigenvalue $-\lambda$.

Using one of the conditions we get $\langle A\,Bv,\,Bv \rangle \geq 0$ but on the other hand $\langle A\,Bv,\,Bv \rangle = -\lambda \langle Bv,\,Bv \rangle \leq 0$, so either $\lambda = 0$ or $Bv = 0$. Writing A and B on this basis, that diagonalizes A, ordered with the zero eigenvalues in a first block we have

which implies that $AB = 0$ and similarly that $BA = 0$.

Solution to 7.9.8: Assuming first that A are B are invertible, we have

$$\begin{aligned} A(A + B)^{-1}B &= (A^{-1})^{-1}(A + B)^{-1}(B^{-1})^{-1} \\ &= (B^{-1}(A + B)A^{-1})^{-1} \\ &= (B^{-1} + A^{-1})^{-1}. \end{aligned}$$

and the same reasoning shows that $B(A + B)^{-1}A = (B^{-1} + A^{-1})$.

In the general case, for some $\delta > 0$ the matrices $A + \lambda I$ and $B - \lambda I$ will be invertible for $0 < |\lambda| < \delta$. Then by what has already been established,

$$(A + \lambda I)(A + B)^{-1}(B - \lambda I) = (B - \lambda I)(A + B)^{-1}(A + \lambda I).$$

Taking the limit as $\lambda \to 0$, we get the desired equality.

Solution to 7.9.9: 1. No, for example take $A = \begin{pmatrix} 1 & 0 \\ 1 & 1 \end{pmatrix}$.

$$AA^t = \begin{pmatrix} 1 & 1 \\ 1 & 2 \end{pmatrix} \neq \begin{pmatrix} 2 & 1 \\ 1 & 1 \end{pmatrix} = A^t A.$$

2. True. If the columns of A form an orthonormal set then $A^t A = I$, transposing the product $AA^t = (A^t A)^t = I$ and now the rows form an orthonormal set.

Solution to 7.9.10: The characteristic polynomial of M_1 is $z^2 - 7z + 10$, with roots 5, 2. Hence M_1 is similar to $\begin{pmatrix} 5 & 0 \\ 0 & 2 \end{pmatrix}$, so $(M_1^n)_{n=1}^\infty$ is bounded away from 0 but is not bounded.

The characteristic polynomial of M_2 is $z^2 - z + 1$, with roots $\left(1 \pm i\sqrt{3}\right)/2$, both of unit modulus. Hence M_2 is similar to a unitary matrix, and $(M_2^n)_{n=1}^{\infty}$ is both bounded and bounded away from 0.

The characteristic polynomial of M_3 is $z^2 - z + 0.7$, with roots $(1 \pm i\sqrt{1.8})/2$, each of modulus less than 1. Hence M_2 is similar to a diagonal matrix with each diagonal entry of modulus less than 1, implying that $(M_3^n)_{n=1}^{\infty}$ is bounded but not bounded away from 0.

Solution to 7.9.13: 1. The matrix $A - I$ has the property that the sum of the entries in each column is equal to 0. The row operation $R_1 \to R_1 + \cdots + R_n$ applied to $A - I$ results in a matrix whose first row is zero. Hence $|A - I| = 0$. Thus 1 is an eigenvalue of A and there is an eigenvector $x \neq 0$ such that $Ax = x$.

2. Let $A = \begin{pmatrix} a & b \\ c & d \end{pmatrix}$ with a, b, c, $d > 0$. The characteristic polynomial of A is $\lambda^2 - (a + d)\lambda + (ad - bc)$. The discriminant is $(a + d)^2 - 4(ad - bc) = (a - d)^2 + 4bc > 0$. Therefore, A has two distinct real eigenvalues λ_1, λ_2. Now $\lambda_1 + \lambda_2 = a + d > 0$. Thus A has at least one positive eigenvalue λ, and there is an eigenvector $y \neq 0$ such that $Ay = \lambda y$.

Solution to 7.9.14: Let $Y = AD - BC$. We have

$$\begin{pmatrix} A & B \\ C & D \end{pmatrix} \begin{pmatrix} D & -B \\ -C & A \end{pmatrix} = \begin{pmatrix} AD - BC & -AB + BA \\ CD - DC & -CB + DA \end{pmatrix} = \begin{pmatrix} Y & 0 \\ 0 & Y \end{pmatrix}.$$

If Y is invertible, then so are $\begin{pmatrix} Y & 0 \\ 0 & Y \end{pmatrix}$ and X.

Assume now that X is invertible, and let v be vector in the kernel of $Y : (AD - BC)v = 0$. Then

$$\begin{pmatrix} A & B \\ C & D \end{pmatrix} \begin{pmatrix} Dv \\ -Cv \end{pmatrix} = 0 = \begin{pmatrix} A & B \\ C & D \end{pmatrix} \begin{pmatrix} -Bv \\ Av \end{pmatrix}$$

implying that $Dv = Cv = Bv = Av = 0$. But then $X\begin{pmatrix} v \\ v \end{pmatrix} = 0$, so, by the invertibility of X, $v = 0$, proving that Y is invertible.

Solution to 7.9.15: Let $A \in GL_2(\mathbb{C})$ be a matrix representing a. Then $A^n = \lambda I$ for some $\lambda \in \mathbb{C}^*$. We may assume, without loss of generality, that $A^n = I$. Since the polynomial $x^n - 1$ has distinct roots, A is diagonalizable, and its eigenvalues must be the n-roots of 1. Now conjugating and dividing by the first root of 1, we may assume that $A = \begin{pmatrix} 1 & 0 \\ 0 & \zeta \end{pmatrix}$. If for some $m \geqslant 1$, $A^m = sI$ with $s \in \mathbb{C}^*$, then, comparing upper left hand corners we see that $s = 1$. Since the order of a is exactly n, the previous sentence implies that A has order exactly n, so ζ is a primitive n-root of 1.

In the same way we may represent b by a matrix that is conjugate to $B = \begin{pmatrix} 1 & 0 \\ 0 & \zeta' \end{pmatrix}$ for some primitive n-root of 1, ζ'. Then ζ' is a power of ζ, so B is a power A, and consequently b is conjugate to a power of a.

Solution to 7.9.16: If $\det B = 0$, then $\det B \equiv 0 \pmod 2$. Hence, if we can show that $\det B \neq 0$ over the field $\mathbb{Z}_2$, we are done. In the field $\mathbb{Z}_2$, $1 = -1$, so

B is equal to the matrix with zeros along the diagonal and 1's everywhere else. Since adding one row of a matrix to another row does not change the determinant, we can replace B by the matrix obtained by adding the first nineteen rows of B to the last row. Since each column of B contains exactly nineteen 1's, B becomes

$$\begin{pmatrix} 0 & 1 & 1 & \cdots & 1 & 1 \\ 1 & 0 & 1 & \cdots & 1 & 1 \\ \vdots & \vdots & \vdots & \ddots & \vdots & \vdots \\ 1 & 1 & 1 & \cdots & 0 & 1 \\ 1 & 1 & 1 & \cdots & 1 & 1 \end{pmatrix}.$$

By adding the last row of B to each of the other rows, B becomes

$$\begin{pmatrix} 1 & 0 & 0 & \cdots & 0 & 0 \\ 0 & 1 & 0 & \cdots & 0 & 0 \\ \vdots & \vdots & \vdots & \ddots & \vdots & \vdots \\ 0 & 0 & 0 & \cdots & 1 & 0 \\ 1 & 1 & 1 & \cdots & 1 & 1 \end{pmatrix}.$$

This is a lower-triangular matrix, so its determinant is the product of its diagonal elements. Hence, the determinant of B is equal to 1 over $\mathbb{Z}_2$, and we are done.

Solution 2. In the matrix modulo 2, the sum of all columns except column i is the ith standard basis vector, so the span of the columns has dimension 20, and the matrix is nonsingular.

Solution to 7.9.17: Let C be the set of real matrices that commute with A. It is clearly a vector space of dimension, at most, 4. The set $\{sI + tA \mid s, t \in \mathbb{R}\}$ is a two-dimensional subspace of C, so it suffices to show that there are two linearly independent matrices which do not commute with A. A calculation show that the matrices $\begin{pmatrix} 0 & 1 \\ 1 & 0 \end{pmatrix}$ and $\begin{pmatrix} 0 & 1 \\ -1 & 0 \end{pmatrix}$ are such matrices.

Solution to 7.9.18: Let

$$A = \begin{pmatrix} a & b \\ c & d \end{pmatrix} \quad \text{and} \quad X = \begin{pmatrix} x & y \\ z & w \end{pmatrix}.$$

If $AX = XA$, we have the three equations: $bz = yc$, $ay + bw = xb + yd$, and $cx + dz = za + wc$.

- $b = c = 0$. Since A is not a multiple of the identity, $a \neq d$. The above equations reduce to $ay = dy$ and $dz = az$, which, in turn, imply that $y = z = 0$. Hence,

$$X = \begin{pmatrix} x & 0 \\ 0 & w \end{pmatrix} = \left(x - \frac{a}{a-d}(x-w) \right) I + \left(\frac{x-w}{a-d} \right) A.$$

- $b \neq 0$ or $c \neq 0$. We can assume, without loss of generality, that $b \neq 0$, as the other case is identical. Then $z = cy/b$ and $w = x - y(a-d)/b$. Hence,

$$X = \frac{1}{b}\begin{pmatrix} bx - ay + ay & by \\ cy & bx - ay + dy \end{pmatrix} = \left(\frac{bx - ay}{b}\right)I + \frac{y}{b}A.$$

Solution to 7.9.19: 1. Assume without loss of generality that $\lambda = 0$ (otherwise replace J by $J - \lambda I$). Let $e_1, \ldots, e_n$ be the standard basis vectors for $\mathbb{C}^n$, so that $Je_k = e_{k-1}$ for $k = 2, \ldots, n$, $Je_1 = 0$. Suppose $AJ = JA$. Then, for $k = 1, \ldots, n-1$,

$$Ae_k = AJ^{n-k}e_n = J^{n-k}Ae_n, \tag{$*$}$$

so A is completely determined by Ae_n. If $Ae_n = \begin{pmatrix} c_n \\ \vdots \\ c_1 \end{pmatrix}$, then

$$A = \begin{pmatrix} c_1 & c_2 & c_3 & \cdots & c_{n-1} & c_n \\ 0 & c_1 & c_2 & \cdots & c_{n-2} & c_{n-1} \\ 0 & 0 & c_1 & \cdots & c_{n-3} & c_{n-2} \\ \vdots & \vdots & \vdots & \ddots & \vdots & \vdots \\ 0 & 0 & 0 & \cdots & c_1 & c_2 \\ 0 & 0 & 0 & \cdots & 0 & c_1 \end{pmatrix}.$$

In other words, A has zero entries below the main diagonal, the entry c_1 at each position on the main diagonal, the entry c_2 at each position immediately above the main diagonal, the entry c_3 at each position two slots above the main diagonal, etc. From $(*)$ it follows that every matrix of this form commutes with J. The commutant of J thus has dimension n.

2. Consider a $2n \times 2n$ matrix in block form $\begin{pmatrix} A & B \\ C & D \end{pmatrix}$ where A, B, C, D are $n \times n$. A simple computation shows that it commutes with $J \oplus J$ if and only if A, B, C, D all commute with J. The commutant of J, as a vector space, is thus the direct sum of 4 vector spaces of dimension n, so it has dimension $4n$.

Solution to 7.9.20: The characteristic polynomial

$$\chi_A(x) = (2 - x)((2 - x)^2 - 2)$$

has distinct roots $2 - \sqrt{2}, 2, 2 + \sqrt{2}$, therefore A is diagonalizable, and, as B commutes with A, B is diagonalizable over the same basis. Let T be the transformation that realizes these diagonalizations,

$$TAT^{-1} = \text{diag}\,(2, 2 + \sqrt{2}, 2 - \sqrt{2}), \qquad TBT^{-1} = \text{diag}\,(\alpha, \beta, \gamma).$$

The linear system

$$aI + b\,\text{diag}\,(2, 2 + \sqrt{2}, 2 - \sqrt{2}) + c\,\text{diag}^2\,(2, 2 + \sqrt{2}, 2 - \sqrt{2}) = \text{diag}\,(\alpha, \beta, \gamma)$$

always has a solution, as it is equivalent to the Vandermonde system

$$\begin{pmatrix} 1 & 2 & 2^2 \\ 1 & 2+\sqrt{2} & (2+\sqrt{2})^2 \\ 1 & 2-\sqrt{2} & (2-\sqrt{2})^2 \end{pmatrix} \begin{pmatrix} a \\ b \\ c \end{pmatrix} = \begin{pmatrix} \alpha \\ \beta \\ \gamma \end{pmatrix}$$

with nonzero determinant, see the Solution to Problem 7.2.11 or [HK61, p. 125]. Applying T^{-1} on the left and T on the right to the above equation we get

$$B = aI + bA + cA^2.$$

Solution 2. The characteristic polynomial

$$\chi_A(x) = (2-x)((2-x)^2 - 2)$$

has distinct roots $2 - \sqrt{2}$, 2, $2 + \sqrt{2}$, that we will call λ_1, λ_2 and λ_3, and let $\Lambda = \text{diag}(\lambda_1, \lambda_2, \lambda_3)$, then if $AB = BA$ and $A = T\Lambda T^{-1}$ then $\Lambda(T^{-1}BT) = (T^{-1}BT)\Lambda$ and since the λ_i are distinct, $T^{-1}BT$ must be diagonal. Say $T^{-1}BT = \text{diag}(b_1, b_2, b_3)$, choose a quadratic polynomial p such that $p(\lambda_i) = b_i$, for $i = 1, 2, 3$, for example

$$p(\lambda) = b_1 \frac{(\lambda - \lambda_2)(\lambda - \lambda_3)}{(\lambda_1 - \lambda_2)(\lambda_1 - \lambda_3)} + b_2 \frac{(\lambda - \lambda_1)(\lambda - \lambda_3)}{(\lambda_2 - \lambda_1)(\lambda_2 - \lambda_3)} + b_3 \frac{(\lambda - \lambda_1)(\lambda - \lambda_2)}{(\lambda_3 - \lambda_1)(\lambda_3 - \lambda_2)}$$

Then $p(A) = Tp(\Lambda)T^{-1} = T\text{diag}(b_1, b_2, b_3)T^{-1} = B$.

Solution 3. Let

$$B = \begin{pmatrix} \alpha & \beta & \gamma \\ \delta & \varepsilon & \zeta \\ \eta & \theta & \iota \end{pmatrix}.$$

Solving for $AB = BA$ leads to a system of 9 equations in 9 unknowns with a solution of the form

$$\begin{pmatrix} \alpha & \beta & \gamma \\ \beta & \alpha+\gamma & \beta \\ \gamma & \beta & \alpha \end{pmatrix}.$$

Solving the system

$$\begin{pmatrix} \alpha & \beta & \gamma \\ \beta & \alpha+\gamma & \beta \\ \gamma & \beta & \alpha \end{pmatrix} =$$

$$a\begin{pmatrix} 1 & 0 & 0 \\ 0 & 1 & 0 \\ 0 & 0 & 1 \end{pmatrix} + b\begin{pmatrix} 2 & -1 & 0 \\ -1 & 2 & -1 \\ 0 & -1 & 2 \end{pmatrix} + c\begin{pmatrix} 5 & -4 & 1 \\ -4 & 6 & -4 \\ 1 & -4 & 5 \end{pmatrix}$$

we obtain

$$a = \alpha + 2\beta + 3\gamma$$
$$b = -\beta - 4\gamma$$
$$c = \gamma$$

thus,

$$B = (\alpha + 2\beta + 3\gamma)I + (-\beta - 4\gamma)A + \gamma A^2 .$$

Solution to 7.9.21: 1. Take $A = \begin{pmatrix} 0 & 1 \\ 0 & 0 \end{pmatrix}$ and $B = \begin{pmatrix} 0 & 0 \\ 1 & 0 \end{pmatrix}$, a multiplication shows that $A^2 = B^2 = 0$ but $(A + B)^2 = I$, so $A + B$ is not nilpotent.

2. If A is nilpotent then neither ± 1 are eigenvalues of A, otherwise 1 would be eigenvalue of A^{2k} for all k, which is a contradiction, so $\ker(I \pm A) = 0$, that is, $I \pm A$ are invertible.

3. Let k_A and k_B be exponents that take A and B into zero, respectively. Since A commutes with B

$$(A + B)^{k_A + k_B} = \sum_{i=0}^{k_A + k_B} \binom{k_A + k_B}{i} A^i B^{k_A + k_B - i}$$

and one of the powers on the right is always zero, and so is the whole expression.

Solution to 7.9.23: Define the norm of a matrix $X = (x_{ij})$ by $\|X\| = \sum_{i,j} |x_{ij}|$. Notice that if B_k, $k = 0, 1, \ldots$, are matrices such that $\sum \|B_k\| < \infty$, then $\sum B_k$ converges, because the convergence of the norms clearly implies the absolute entrywise convergence.

In our case, we have $B_k = A^k$. The desired result follows from the fact that $\sum \|A\|^k / k!$ converges for any matrix A.

Solution to 7.9.24: Note that, if A is an invertible matrix, we have

$$Ae^M A^{-1} = e^{AMA^{-1}}$$

so we may assume that M is upper-triangular. Under this assumption e^M is also upper-triangular, and if $a_1, \ldots, a_n$ are M's diagonal entries, then the diagonal entries of e^M are

$$e^{a_1}, \ldots, e^{a_n}$$

and we get

$$\det\left(e^M\right) = \prod_{i=1}^{n} e^{a_i} = e^{\sum_{i=1}^{n} a_i} = e^{tr(M)}.$$

Solution to 7.9.26: It is easy to see that

$$\langle X, Y \rangle = tr(XY^*)$$

defines an inner product on the space of $n \times n$ complex matrices, in fact, it is the standard inner product under the identification with $\mathbb{C}^{n^2}$. The inequality is then the Cauchy–Schwarz Inequality [MH93, p. 69] for the norm.

Solution to 7.9.27: 1. Let A and B be $n \times n$ real matrices and k a positive integer. We have

$$(A + tB)^k = A^k + t \sum_{i=0}^{k-1} A^i B A^{k-1-i} + O(t^2) \quad (t \to 0)$$

where $O(t^2)$ is composed of terms with a power of t higher than 1. Hence

$$\lim_{t \to 0} \frac{1}{t} \left((A + tB)^k - A^k \right) = \sum_{i=0}^{k-1} A^i B A^{k-1-i} .$$

2. We have

$$\frac{d}{dt} \operatorname{tr} (A + tB)^k \Big|_{t=0} = \operatorname{tr} \frac{d}{dt} (A + tB)^k \Big|_{t=0} .$$

By definition,

$$\frac{d}{dt} (A + tB)^k \Big|_{t=0} = \lim_{t \to 0} \frac{1}{t} \left((A + tB)^k - A^k \right) .$$

Using the previous part of the problem we get

$$\frac{d}{dt} \operatorname{tr} (A + tB)^k \Big|_{t=0} = \operatorname{tr} \sum_{j=0}^{k-1} A^j B A^{k-j-1} = \sum_{j=0}^{k-1} \operatorname{tr} (A^j B A^{k-j-1}) .$$

Solution to 7.9.28: 1. Let $\alpha = \frac{1}{n} \operatorname{tr}(M)$, so that $\operatorname{tr}(M - \alpha I) = 0$. Let

$$A = \frac{1}{2}(M - M^t), \qquad S = \frac{1}{2}(M + M^t) - \alpha I.$$

The desired conditions are then satisfied.
2. We have

$$M^2 = A^2 + S^2 + \alpha^2 I + 2\alpha A + 2\alpha S + AS + SA.$$

The trace of M is the sum of the traces of the seven terms on the right. We have $\operatorname{tr}(A) = \operatorname{tr}(B) = 0$. Also,

$$\operatorname{tr}(AS) = \operatorname{tr} \left((AS)^t \right) = \operatorname{tr} \left(S^t A^t \right) = \operatorname{tr}(-SA) = -\operatorname{tr}(SA),$$

so $\operatorname{tr}(AS + SA) = 0$ (in fact, $\operatorname{tr}(AS) = 0$ since $\operatorname{tr}(AS) = \operatorname{tr}(SA)$). The desired equality now follows.

Solution to 7.9.29: We must show that 0 is the only eigenvalue of A. Suppose the contrary. We may assume that A is in Jordan form, since the trace is preserved by similarity, in particular that A is upper-triangular. Let $\lambda_1, ..., \lambda_p$ be the distinct nonzero eigenvalues of A, each λ_j repeated m_j times in the diagonal of A. Then the nonzero entries in the diagonal of A^k are $\lambda_1^k, \ldots, \lambda_p^k$, repeated $m_1, \ldots, m_p$ times, respectively. We have

$$0 = \operatorname{tr} A^k = \sum_{j=1}^{p} m_j \lambda_j^k, \quad \text{for } k = 1, \ldots, n.$$

Therefore $M(m_1 \; m_2 \; \cdots \; m_p)^t = 0$ where the matrix M is

$$
\begin{pmatrix}
\lambda_1 & \lambda_2 & \cdots & \lambda_p \\
\lambda_1^2 & \lambda_2^2 & \cdots & \lambda_p^2 \\
\lambda_1^3 & \lambda_2^3 & \cdots & \lambda_p^3 \\
\vdots & \vdots & \ddots & \vdots \\
\lambda_1^k & \lambda_2^k & \cdots & \lambda_p^k
\end{pmatrix}.
$$

But

$$
\det M = \lambda_1 \lambda_2 \cdots \lambda_p \det
\begin{pmatrix}
1 & 1 & \cdots & 1 \\
\lambda_1^2 & \lambda_2^2 & \cdots & \lambda_p^2 \\
\lambda_1^3 & \lambda_2^3 & \cdots & \lambda_p^3 \\
\vdots & \vdots & \ddots & \vdots \\
\lambda_1^k & \lambda_2^k & \cdots & \lambda_p^k
\end{pmatrix}
$$

which is nonzero since the λ's are nonzero and the other factor is a Vandermonde determinant. We got a contradiction, therefore A has no nonzero eigenvalues.

Solution to 7.9.30: Consider the function f defined on the unit disc by $f(z) = (1 + z)^{1/r}$, and its Maclaurin expansion [MH87, p. 234] $f(z) = \sum_{k=0}^{\infty} c_k z^k$. We have $f(z)^r = 1 + z$. Let the matrix A be defined by

$$A = \sum_{k=0}^{\infty} c_k N^k,$$

a finite sum, because A is nilpotent. The computation of A^r involves the same formal manipulations with power series as does the computation of $f(z)^r$. If we replace z by N everywhere in the latter computation, we arrive in the end at the equality $A^r = I + N$.

Solution to 7.9.31: We will prove the equivalent result that the kernel of A is trivial. Let $x = (x_1, \ldots, x_n)^t$ be a nonzero vector in $\mathbb{R}^n$. We have

$$Ax \cdot x = \sum_{i,j=1}^{n} a_{ij} x_i x_j$$

$$= \sum_{i=1}^{n} a_{ii} x_i^2 + \sum_{i \neq j} a_{ij} x_i x_j$$

$$\geqslant \sum_{i=1}^{n} x_i^2 - \sum_{i\neq j} |a_{ij}x_ix_j|$$

$$\geqslant \sum_{i=1}^{n} x_i^2 - \sqrt{\sum_{i\neq j} a_{ij}^2 \sum_{i\neq j} x_i^2 x_j^2}$$

$$> \sum_{i=1}^{n} x_i^2 - \sqrt{\sum_{i\neq j} x_i^2 x_j^2}$$

$$\geqslant \sum_{i=1}^{n} x_i^2 - \sqrt{\sum_{i=1}^{n} x_i^2 \sum_{j=1}^{n} x_j^2}$$

$$\geqslant 0.$$

So $\langle Ax, x \rangle \neq 0$ and $Ax \neq 0$, therefore, the kernel of A is trivial.

Solution to 7.9.32: Suppose x is in the kernel. Then

$$a_{ij}x_i = -\sum_{j\neq i} a_{ij}x_j$$

for each i. Let i be such that $|x_i| = \max_k |x_k| = M$, say. Then

$$|a_{ii}|M \leqslant \sum_{j\neq i} |a_{ij}||x_j| \leqslant \sum_{j\neq i} |a_{ij}|M$$

so

$$\left(|a_{ii}| \sum_{j\neq i} |a_{ij}|\right) M \leqslant 0.$$

Since the therm inside the parenthesis is strictly positive by assumption, we must have $M = 0$, so $x = 0$ and A is invertible.

Solution to 7.9.33: It will suffice to prove that $\ker(I - A)$ is trivial. Let $x = (x_1, x_2, \cdots, x_n)^t$ be a nonzero vector in $\mathbb{R}^n$, and let $y = (I - A)x$. Pick k such that $|x_k| = \max\{|x_1|, \ldots, |x_n|\}$. Then

$$|y_k| = |x_k - \sum_{j=1}^{n} a_{kj}x_j|$$

$$\geqslant |x_k| - \sum_{j=1}^{n} |a_{kj}||x_j|$$

$$\geqslant |x_k| - \sum_{j=1}^{n} |a_{kj}||x_k|$$

$$= |x_k| \left(1 - \sum_{j=1}^{n} |a_{kj}| \right) > 0 .$$

Hence, $y \neq 0$, as desired.

Solution 2. Let $\alpha < 1$ be a positive number such that, for all values of i, we have $\sum_{j=1}^{n} |a_{ij}| \leqslant \alpha$. Then,

$$\sum_{j,k} |a_{ij} a_{jk}| = \sum_{j} \left(|a_{ij}| \sum_{k} |a_{jk}| \right) \leqslant \alpha \sum_{j} |a_{ij}| \leqslant \alpha^2 .$$

And so, inductively, the sum of the absolute values of the terms in one row of A^n is bounded by α^n. Thus, the entries in the infinite sum

$$I + A + A^2 + A^3 + \cdots$$

are all bounded by the geometric series $1 + \alpha + \alpha^2 + \cdots$, and so are absolutely convergent; thus, this sum exists and the product

$$(I - A)(I + A + A^2 + \cdots) = I$$

is valid, so that the inverse of $I - A$ is this infinite sum.

Solution to 7.9.34: 1. If A is symmetric then, by the Spectral Theorem [HK61, p. 335], [Str93, p. 235], there is an orthonormal basis $\{e_1, e_2, \ldots, e_n\}$ for $\mathbb{R}^n$ with respect to which A is diagonal: $Ae_j = \lambda_j e_j$, $j = 1, \ldots, n$. Let x be any vector in $\mathbb{R}^n$. We can write $x = c_1 e_1 + \cdots + c_n e_n$ for some scalars $c_1, \ldots, c_n$, and have $Ax = \lambda_1 c_1 e_1 + \cdots + \lambda_n c_n e_n$. Moreover,

$$\|x\|^2 = c_1^2 + \cdots + c_n^2$$
$$\|Ax\|^2 = \lambda_1^2 c_1^2 + \cdots + \lambda_n^2 c_n^2$$
$$\leqslant \max\{\lambda_1^2, \ldots, \lambda_n^2\}(c_1^2 + \cdots + c_n^2) = M^2 \|x\|^2 ,$$

which is the desired inequality.

2. The matrix $\left(\begin{smallmatrix} 0 & 1 \\ 0 & 0 \end{smallmatrix} \right)$ gives a counterexample with $n = 2$. Its only eigenvalue is 0, yet it is not the zero matrix.

Solution to 7.9.35: 1. Suppose λs not an eigenvalue of A, so $\mu(A) \neq 0$. Write

$$\mu(z) = \mu(\lambda) + (z - \lambda)g(z)$$

with g a polynomial of degree $k - 1$. Then

$$0 = \mu(A) = \mu(\lambda)I + (A - \lambda I)g(A) ,$$

so the polynomial $\mu_\lambda(z) = -g(z)/\mu(\lambda)$ has the required property.

2. It will suffice to show that the polynomials $\mu_{\lambda_1}, \ldots, \mu_{\lambda_k}$ are linearly independent, for then they will form a basis for the vector space of polynomials of degree $\leqslant k - 1$, so some linear combination of them will equal the constant polynomial 1. Suppose $a_1, \ldots, a_k$ are complex numbers such that $\sum a_j \mu_{\lambda_j} = 0$. Then $\sum a_j (A - \lambda_j I)^{-1} = 0$. Multiplying the last equality by $\prod (A - \lambda_i I)$, we find that $q(A) = 0$, where

$$q(z) = \sum_{j=1}^{k} a_j \prod_{i \neq j} (z - \lambda_i),$$

a polynomial of degree less than k. Therefore $q = 0$. Hence, for each j,

$$q(\lambda_j) = a_j \prod_{i \neq j} (\lambda_j - \lambda_i) = 0,$$

so $a_j = 0$. This proves the desired linear independence.

Part III

Appendices

Appendix A
How to Get the Exams

A.1 On-line

Open a browser of your choice on the URL

$$\texttt{http://www.booksinbytes.com/bpm}$$

You can then proceed to explore the set of exams or download them. You will also find new sets of problems and updates to the book. When this page changes, you can do a search on the words `Berkeley Problems in Mathematics` and you should be guided to a possible new location by one of the net search engines.

To suggest improvements, comments, or submit a solution, you can send e-mail to the authors. If you are submitting a new solution, make sure you include your full name, so we can cite you, if you so wish.

A.2 Off-line, the Last Resort

Even if you do not have access to the net, you can reconstruct each of the exams using the following tables. This method should be used only as a last resort, since some of the problems have been slightly altered to uniformize notation and/or correct some errors in the original text.

Spring 77	Summer 77	Fall 77	Spring 78	Summer 78	Fall 78
1.4.8	7.1.8	7.6.21	1.6.10	6.1.3	1.2.9
1.5.4	5.7.6	7.3.2	2.1.2	6.12.1	1.6.24
5.1.5	6.11.4	7.9.25	5.6.9	7.9.13	2.2.44
5.7.5	7.9.23	7.4.21	5.11.5	7.3.2	5.10.27
6.12.3	5.1.2	6.13.9	6.11.11	5.7.2	5.7.8
6.1.1	1.1.33	5.9.1	6.12.3	5.5.10	3.1.6
7.2.12	7.6.22	5.11.24	2.3.1	3.1.2, 1.4.11	6.4.1
7.7.14	3.1.9	3.2.3	3.1.5	4.1.19	7.6.26
3.2.2	7.4.28	1.1.2	7.6.23	4.1.11	7.8.13
2.2.22	1.1.23	1.6.2	7.9.9	2.2.30	6.2.1
5.2.6	2.2.24	3.1.3	5.10.3	5.2.7	2.2.21
6.13.6	7.4.31	4.1.7	5.2.16	5.6.8	2.2.29
7.9.22	5.2.5	5.3.1	3.1.4	6.11.18	2.2.31
7.9.21	6.3.1	5.1.1	6.4.10	7.7.8	5.3.2
5.6.31	4.1.17	2.2.25	6.11.30	7.5.19	5.4.11
5.11.15	5.10.26	7.4.1	7.5.16	7.6.34	7.7.2
6.7.1	6.11.10	6.1.2	2.3.2	3.3.1	6.2.14
6.11.16	5.8.30	6.11.1	7.2.1	2.1.6	6.12.4
2.2.23	6.13.7	7.5.1	1.5.5	1.1.26	7.6.3
1.2.6	7.3.1	5.8.1	2.2.35	2.2.40	4.1.24

Spring 79	Summer 79	Fall 79	Spring 80	Summer 80	Fall 80
2.2.27	7.5.10	5.8.2	1.7.1	7.4.27	1.1.35
7.8.14	6.11.20	5.4.13	1.6.2	7.6.15	7.7.1
6.5.5	2.2.20	6.2.8	2.2.19	7.4.12	5.6.3
4.2.5	6.2.6	6.11.13	5.10.8	6.2.5	6.4.2
3.1.11	1.4.4	7.9.5	5.4.6	3.4.3	5.11.21
5.8.16	5.7.8	7.5.2	6.4.13	5.10.10	6.9.1
5.11.16	5.6.35	7.5.20	6.13.8	5.3.5	1.7.6
7.7.13	2.1.3	1.3.13	7.6.20	4.1.8	6.5.1
7.7.4	7.4.23	1.1.34	7.9.17	1.3.21	2.2.17
6.11.19	3.4.2	3.4.2	3.1.8	2.2.11	2.2.42
6.8.11	7.7.6	4.3.1	2.3.6	6.8.18	5.9.2
6.4.11	6.13.10	4.1.20	7.2.18	2.2.18	6.8.3
3.4.1	7.1.3	1.6.12	6.5.11	6.11.13	1.3.11
2.2.45	6.6.4	3.2.6	7.6.16	7.7.5	7.7.9
7.7.2	5.2.10	5.5.5	7.1.11	7.8.7	3.4.8
5.2.9	5.2.17	5.11.21	2.2.8	1.6.20	7.9.7
5.10.5	2.2.32	7.6.35	6.5.6	4.3.3	1.6.12
1.6.32	7.4.11	7.9.3	3.4.19	5.8.6	5.8.3
7.9.4	1.6.25	6.10.2	5.4.4	5.8.32	4.2.6
1.4.25	3.2.4	6.7.3	5.3.4	3.1.9	6.1.4

Spring 81	Summer 81	Fall 81	Spring 82	Summer 82	Fall 82
2.3.7	1.1.9	5.11.15	5.7.5	7.6.23	3.1.10
7.1.15	6.2.7	3.4.11	4.1.12	5.11.24	6.11.13
6.9.14	4.1.6	6.5.10	7.9.26	4.3.4	5.10.2
3.4.4	1.5.6	7.9.2	2.2.28	6.7.5	7.2.1
6.11.31	7.2.19	1.3.12	1.6.22	5.8.21	1.6.27
5.8.30	4.1.25	1.7.8	1.1.15	1.4.8	7.5.21
5.11.8	5.10.13	6.11.4	6.7.4	7.2.17	5.11.14
1.6.11	6.11.21	5.8.4	5.11.15	6.13.12	6.4.8
7.5.25	1.7.2	7.2.16	7.6.23	5.2.11	1.1.31
1.6.6	7.5.20	2.2.47	6.3.5	5.1.11	5.1.9
5.10.24	1.1.14	7.9.12	5.8.33	5.6.12	1.1.10
7.6.4	6.9.13	5.9.1	4.1.22	6.11.21	7.7.6
1.3.19	6.2.10	2.2.10	7.4.2	1.7.3	1.6.9
4.1.23	1.4.28	3.4.12	1.8.2	7.8.8	7.4.22
6.10.14	7.5.7	5.4.17	7.1.3	5.4.15	6.6.6
1.1.3	1.6.26	6.4.11	6.4.14	7.9.21	3.4.2
7.6.36	5.5.5	1.2.13	1.6.21	2.2.14	5.11.6
2.2.36	7.7.10	7.7.10	5.6.35	1.1.29	4.3.5
6.1.7	5.8.25	1.1.7	3.1.9	7.5.7	5.4.2
1.8.1	3.2.6	6.4.10	6.11.23	2.2.41	6.7.3

Spring 83	Summer 83	Fall 83	Spring 84	Summer 84	Fall 84
1.5.13	6.13.16	5.11.30	1.5.23	6.4.3	6.2.13
7.2.8	5.7.11	6.10.3	6.1.10	6.11.22	7.9.27
5.3.3	7.6.25	2.2.26	1.4.12	7.4.14	1.6.1
6.1.5	1.4.1	1.3.17	1.1.20	7.5.26	5.11.19
3.1.12	1.3.18	7.5.22	7.5.18	5.10.14	3.1.15
1.2.5	7.1.17	5.8.7	1.3.7	5.8.26	6.13.14
6.2.11	1.5.23	6.6.1	5.8.8	2.2.39	7.2.11
6.13.13	6.6.5	7.4.24	3.4.5	1.5.8	7.5.24
2.1.8	5.7.12	3.2.8	7.3.4	3.4.7	5.6.36
5.11.5	1.7.7	1.5.11	6.10.15	4.2.8	1.4.14
7.6.37	7.6.24	1.4.14	7.7.10	6.5.2	6.8.2
5.6.13	1.2.11	5.10.1	3.1.14	7.1.3	7.4.25
1.6.28	7.9.11	3.4.10	1.4.23	7.9.18	1.6.7
6.7.8	6.5.12	7.6.19	6.10.5	7.1.10	5.11.20, 5.10.26
5.10.5	5.4.3	5.9.3	5.6.37	6.11.27	3.4.9
7.8.10	5.6.15	2.2.46	7.6.2	1.1.16	6.11.32
5.11.11	7.6.14	6.12.2	1.4.24	5.2.12	7.6.26
7.1.3	3.1.13	2.2.34	1.2.8	4.3.7	7.4.30
3.1.9	5.4.5	6.11.24	1.4.2	3.2.5	5.6.17
4.1.10	1.4.27	4.1.26	6.7.9	5.11.18	1.1.17

Spring 85	Summer 85	Fall 85	Spring 86	Fall 86	Spring 87
1.4.3	7.6.8	5.11.2	7.4.26	1.6.16	4.1.16
6.1.11	1.1.25	5.8.17	1.7.4	5.1.4	2.2.15
5.1.3	7.6.38	6.11.29	5.10.15	2.3.4	5.6.16
5.2.14	6.4.4	6.12.4	1.4.18	5.8.22	6.11.36
6.11.28	6.11.2	5.8.12	6.9.3	7.6.27	1.6.31
7.5.32	5.11.21	7.5.33	7.4.13	3.1.18	5.4.22
5.11.31	1.5.7	3.1.16	3.4.13	6.12.13	3.2.9
3.2.13	1.6.13	1.5.22	5.8.11	1.4.22	5.8.13
1.6.29	1.4.6	7.8.11	6.4.9	5.11.28	7.3.4
7.5.12	5.8.16	1.6.4	6.9.4	1.1.5	6.8.13
6.12.7	6.3.4	6.5.4	5.11.1	7.2.1	1.4.10
5.11.21	7.2.3	5.7.1	1.8.3	4.1.15	1.5.15
5.8.18	6.11.6	5.4.14	7.5.17	2.2.4	2.2.26
1.5.24	6.10.3	7.5.15	4.1.5	7.8.15	6.8.4
6.11.3	5.4.8	1.5.9	7.4.16	6.11.24	5.6.29
1.6.30	1.2.10	6.12.12	6.11.12	3.1.19	4.2.10
1.4.13	5.8.31	4.3.8	3.4.14	6.5.3	3.3.3
7.5.23	3.4.2	6.1.12	5.2.18	5.10.17	7.6.19
6.12.8	1.3.1	7.8.9	6.7.2	6.13.1	6.12.5
5.4.21	6.12.10	1.5.10	5.10.16	1.5.12	5.10.28

Fall 87	Spring 88	Fall 88	Spring 89	Fall 89	Spring 90
1.1.1	1.4.19				
1.6.3	6.13.18	6.9.5	1.3.25	6.7.10	1.4.5
1.5.16	7.3.3	5.6.30	7.5.3	1.4.11	7.6.1
2.2.37	5.6.2	3.4.16	1.2.15	7.6.33	5.8.23
7.6.6	6.1.13	7.6.30	5.8.16	5.4.22	6.10.9
6.6.7	7.6.13	1.5.18	6.4.16	6.4.17	1.3.15
7.5.34	6.10.7	1.2.14	7.1.13	4.1.13	7.2.14
6.12.1	5.10.33	6.8.5	3.4.20	7.6.18	5.11.9
5.10.29	1.5.21	5.8.27	5.4.17	1.5.23	6.2.4
5.4.18	5.6.18	7.5.32	6.4.7	6.11.35	1.2.3
7.6.17	6.5.7	6.8.12	7.9.14	1.3.5	6.8.6
7.7.10	1.4.20	5.7.13	6.9.6	6.13.20	5.7.16
6.2.17	6.2.8	6.11.34	5.10.18	5.7.14	7.5.27
6.10.2	5.4.9	7.1.12	6.5.13	7.2.3	1.4.16
5.6.23	7.6.5	6.7.6	2.2.9	4.1.18	7.1.14
5.8.8	7.4.17	2.2.12	7.9.16	6.7.7	5.3.6
5.9.5	5.4.19	5.10.4	5.8.34	5.8.2	6.13.21
3.1.1	7.6.10	5.11.26	6.10.8	7.1.11	1.5.10
4.3.2	6.13.19	4.3.6	2.1.4	1.1.27	5.1.7
1.1.30	1.7.5				

Fall 90	Spring 91	Fall 91	Spring 92	Fall 92	Spring 93
1.1.18	6.8.1	6.2.8	6.3.1	7.7.1	1.3.16
5.10.19	5.6.35	1.2.12	7.6.11	6.11.9	7.9.33
6.10.11	6.13.22	5.10.20	5.6.25	5.8.5	5.11.7
1.5.3	6.9.7	7.9.28	4.1.14	3.2.6	6.7.4
7.5.35	7.2.9	1.4.16	2.2.33	6.2.19	3.2.11
5.5.2	2.3.8	5.4.10	6.11.33	4.2.12	7.7.14
6.4.18	5.4.20	3.4.17	6.13.15	5.10.21	1.6.22
1.5.14	7.2.7	1.3.20	5.11.6	7.5.8	6.8.9
7.8.4	2.1.5	6.3.6	5.8.15	5.6.20	5.8.28
7.6.23	6.6.3	7.1.3	7.6.12	6.5.14	1.1.28
1.2.2	1.3.22	4.2.11	5.2.19	7.8.7	7.4.29
6.2.9	1.3.24	5.6.24	7.5.28	1.6.18	5.10.6
5.4.13	5.11.10	7.5.30	2.3.12	5.11.9	6.9.13
6.11.5	3.4.6	2.3.9	6.11.4	3.1.20	3.1.7
7.4.19	6.1.8	6.11.8	6.5.8	6.4.8	6.12.1
1.3.26	7.7.11	4.1.27	1.3.3	2.2.8	2.2.1
6.1.6	5.6.32	2.2.16	1.1.19	5.9.4	6.4.14
5.5.11	1.4.21	5.5.13	5.6.26	1.3.6	5.4.16

Fall 93	Spring 94	Fall 94	Spring 95	Fall 95	Spring 96
4.2.4	4.1.5	1.3.23	1.6.14	6.2.22	1.3.2
6.2.17	7.9.34	7.5.11	7.6.40	5.1.10	4.1.5
5.2.4	5.10.26	5.10.32	5.10.25	5.2.15	5.10.26
7.2.21	6.4.19	6.2.20	6.12.19	7.7.10	5.8.10
3.2.10	3.3.5	3.3.4	3.1.21	6.9.15	7.6.7
6.6.4	7.7.12	7.9.31	6.10.13	3.2.7	7.3.3
2.2.45	5.1.8	2.2.38	1.1.4	5.6.36	6.11.25
7.6.1	6.11.17	6.12.15	7.4.6	7.9.32	6.13.17
5.11.27	5.6.27	5.11.29	5.8.24	1.3.10	6.12.14
1.1.13	1.5.19	2.1.1	1.6.8	6.12.5	1.1.22
6.9.10	7.4.20	7.7.14	7.4.15	5.7.10	1.1.32
5.10.23	5.6.28	5.2.2	5.10.31	1.5.17	2.2.43
6.8.6	6.12.23	6.12.22	6.5.10	5.2.1	5.6.1
3.1.3	3.4.18	1.1.12	3.3.6	6.11.7	5.6.35
7.5.10	1.3.5	7.2.20	6.12.24	1.4.29	7.9.7
1.6.23	1.1.21	2.3.10	4.3.4	7.7.15	7.5.22
7.8.5	6.8.6	6.9.12	7.4.10	7.2.15	7.1.7
5.6.34	5.9.6	5.6.21	5.3.8	1.1.36	6.6.7

Fall 96	Spring 97	Fall 97	Spring 98	Fall 98	Spring 99
1.6.19	1.3.21	1.3.4	5.8.19	4.1.5	4.1.11
1.2.10	4.2.7	1.4.4	5.8.20	3.3.2	1.6.5
5.11.22	1.5.14	5.5.4	4.3.8	5.11.21	1.4.9
5.5.12	5.5.8	5.11.24	1.5.2	5.4.12	5.11.21
7.4.23	5.11.33	5.4.17	6.10.1	7.8.1	5.6.19
7.7.5	7.4.18	7.1.16	6.3.2	7.4.4	7.1.1
7.6.41	7.5.15	7.8.12	7.5.16	7.9.6	7.5.5
6.13.11	6.8.11	6.4.12	6.13.4	6.9.11	7.7.17
6.13.14	6.10.4	6.8.8	7.9.10	6.8.14	6.1.9
1.1.24	1.2.14	1.1.8	1.1.12	1.5.1	1.1.11
1.1.31	3.2.1	1.1.17	7.8.2	2.2.2	1.4.26
5.8.8	5.11.5	2.1.7	2.3.3	2.3.5	5.8.29
5.8.1	5.6.30	5.3.7	5.7.3	5.5.7	5.2.8
5.4.3	7.9.24	5.11.13	1.4.7	5.8.35	7.6.32
7.6.9	7.2.4	7.1.13	7.8.6	7.5.15	7.7.18
7.9.20	6.12.25	7.7.7	5.6.33	7.5.4	7.9.36
6.11.14	7.1.5	6.7.6	7.7.16	7.1.3	6.12.17
6.8.18	6.4.20	6.3.7	7.9.30	6.12.16	6.6.2

Fall 99	Spring 00	Fall 00	Spring 01	Fall 01	Spring 02
7.4.3	7.7.3	7.4.7	7.1.4	7.6.29	7.1.9
4.2.2	4.2.9	4.2.3	1.2.5	1.2.16	1.2.1
6.10.12	6.2.16	6.12.27	6.10.10	6.8.20	6.5.9
5.5.3	5.8.34	5.10.11	5.11.23	5.10.22	5.10.9
2.2.40	1.3.9	1.4.15	6.2.15	5.8.9	6.13.3
7.4.8	7.9.35	6.10.6	7.5.31	7.2.10	7.4.32
6.8.16	1.4.17	1.2.4	1.6.17	2.2.13	4.2.1
5.10.7	6.9.16	7.1.6	6.8.15	6.13.23	6.12.26
3.4.15	5.10.30	5.6.14	5.6.11	5.5.14	5.5.6
1.2.7	4.1.28	7.8.3	7.6.28	7.9.1	7.6.31
7.6.19	7.5.29	4.1.9	4.1.21	1.1.37	5.5.9
5.6.6	6.2.18	6.2.21	6.5.7	6.12.21	6.8.10
6.12.11	5.8.14	5.7.7	5.2.13	5.6.22	5.7.15
4.1.14	7.4.33	7.4.9	3.1.17	5.6.5	1.3.14
7.9.29	2.2.5	1.1.6	7.2.6	7.5.13	7.9.19
5.7.4	7.7.19	6.3.8	4.1.4	3.2.12	1.5.25
6.3.3	5.11.3	5.6.7	6.11.26	6.12.18	6.12.20
1.5.20	6.8.19	6.9.8	5.7.9	5.11.4	5.11.17

Appendix B
Passing Scores

The passing scores and data presented here go back to the first Preliminary Examination, which was held on January 1977.

Date		Minimum passing score	# of students taking	# of students passing	% passing the exam
Fall	03	63/120	23	13	56.5%
Spring	03	63/120	20	15	75.0%
Fall	02	64/120	41	24	58.5%
Spring	02	67/120	28	17	60.1%
Fall	01	65/120	41	22	53.6%
Spring	01	65/120	15	5	33.3%
Fall	00	65/120	38	27	71.0%
Spring	00	64/120	11	5	45.4%
Fall	99	64/120	35	25	71.4%
Spring	99	66/120	12	8	66.7%
Fall	98	66/120	29	16	55.1%
Spring	98	71/120	15	9	60.0%
Fall	97	64/120	41	28	68.3%
Spring	97	70/120	10	5	50.0%
Fall	96	80/120	24	17	70.8%
Spring	96	84/120	17	13	76.5%
Fall	95	64/120	41	19	46.3%
Spring	95	65/120	10	4	40.0%
Fall	94	71/120	28	16	57.1%
Spring	94	70/120	11	5	45.5%
Fall	93	79/120	40	28	70.0%
Spring	93	69/120	22	17	77.3%
Fall	92	71/120	53	34	64.2%

Date		Minimum passing score	# of students taking	# of students passing	% passing the exam
Spring	92	58/120	27	13	48.1%
Fall	91	66/120	66	42	63.6%
Spring	91	60/120	43	21	48.8%
Fall	90	71/120	89	50	56.2%
Spring	90	68/120	47	24	51.0%
Fall	89	71/120	56	18	32.1%
Spring	89	66/120	31	16	51.6%
Fall	88	70/120	44	22	50.0%
Spring	88	74/140	25	16	64.0%
Fall	87	81/140	29	21	72.4%
Spring	87	73/140	46	31	67.4%
Fall	86	93/140	24	13	54.2%
Spring	86	75/140	37	25	67.6%
Fall	85	89/140	23	16	69.6%
Summer	85	90/140	16	11	68.8%
Spring	85	97/140	19	12	63.2%
Fall	84	99/140	26	18	69.2%
Summer	84	82/140	10	6	60.0%
Spring	84	82/140	35	24	69.0%
Fall	83	92/140	16	11	68.8%
Summer	83	86/140	14	7	50.0%
Spring	83	81/140	26	17	65.4%
Fall	82	84/140	15	8	53.0%
Summer	82	88/140	23	17	73.9%
Spring	82	90/140	15	12	80.0%
Fall	81	82/140	15	10	66.7%
Summer	81	65/140	21	14	67.0%
Spring	81	106/140	32	24	75.0%
Fall	80	99/140	23	17	74.0%
Summer	80	82/140	24	12	50.0%
Spring	80	88/140	30	17	56.7%
Fall	79	84/140	13	7	53.8%
Summer	79	101/140	15	7	46.7%
Spring	79	97/140	23	18	78.3%
Fall	78	90/140	17	12	71.0%
Summer	78	89/140	21	12	57.1%
Spring	78	95/140	32	19	59.4%
Fall	77	809140	13	12	92.0%
Summer	77	90/140	20	18	90.0%
Spring	77	90/140	42	29	69.0%

Appendix C
The Syllabus

The syllabus is designed around a working knowledge and understanding of an **honors** undergraduate mathematics major. Any student taking the examination should be familiar with the material outlined below.

Calculus

Basic first- and second-year calculus. Derivatives of maps from $\mathbb{R}^m$ to $\mathbb{R}^n$, gradient, chain rule; maxima and minima, Lagrange multipliers; line and surface integrals of scalar and vector functions; Gauss', Green's and Stokes' theorems. Ordinary differential equations; explicit solutions of simple equations.

Classical Analysis

Point-set topology of $\mathbb{R}^n$ and metric spaces; properties of continuous functions, compactness, connectedness, limit points; least upper bound property of $\mathbb{R}$. Sequences and series, Cauchy sequences, uniform convergence and its relation to derivatives and integrals; power series, radius of convergence, Weierstrass M-test; convergence of improper integrals. Compactness in function spaces. Inverse and Implicit Function Theorems and applications; the derivative as a linear map; existence and uniqueness theorems for solutions of ordinary differential equations; elementary Fourier series. Texts: [Ros86], [MH93], [Bar76], [Rud87].

Abstract Algebra

Elementary set theory, e.g., uncountability of $\mathbb{R}$. Groups, subgroups, normal subgroups, homomorphisms, quotient groups, automorphisms, groups acting on sets, Sylow theorems and applications, finitely generated abelian groups. Examples: permutation groups, cyclic groups, dihedral groups, matrix groups. Basic properties of rings, units, ideals, homomorphisms, quotient rings, prime and maximal ideals, fields of fractions, Euclidean domains, principal ideal domains and unique factorization domains, polynomial rings. Elementary properties of finite field extensions and roots of polynomials, finite fields. Texts: [Lan94], [Hun96], [Her75].

Linear Algebra

Matrices, linear transformations, change of basis; nullity-rank theorem. Eigenvalues and eigenvectors; determinants, characteristic and minimal polynomials, Cayley-Hamilton Theorem; diagonalization and triangularization of operators; Jordan normal form, Rational Canonical Form; invariant subspaces and canonical forms; inner product spaces, hermitian and unitary operators, adjoints. Quadratic forms. Texts: [ND88], [HK61], [Str93].

Complex Analysis

Basic properties of the complex number system. Analytic functions, conformality, Cauchy-Riemann equations, elementary functions and their basic properties (rational functions, exponential function, logarithm function, trigonometric functions, roots, e.g., $\sqrt{z}$). Cauchy's Theorem and Cauchy's integral formula, power series and Laurent series, isolation of zeros, classification of isolated singularities (including singularity at ∞), analyticity of limit functions. Maximum Principle, Schwarz's Lemma, Liouville's Theorem, Morera's Theorem, Argument Principle, Rouché's Theorem. Basic properties of harmonic functions in the plane, connection with analytic functions, harmonic conjugates, Mean Value Property, Maximum Principle. Residue Theorem, evaluation of definite integrals. Mapping properties of linear fractional transformations, conformal equivalences of the unit disc with itself and with the upper half-plane. Texts: [MH87], [Ahl79], [Con78].

References

[Ahl79]   L. V. Ahlfors. *Complex Analysis, an Introduction to the Theory of Analytic Functions of One Complex Variable.* International Series in Pure and Applied Mathematics. McGraw-Hill, New York, 1979.

[AM95]   G. L. Alexanderson and D. H. Mugler, editors. *Lion Hunting & Other Mathematical Pursuits, A Collection of Mathematics, Verse, and Stories by Ralph P. Boas, Jr.*, volume 15 of *The Dolciani Mathematical Expositions.* Mathematical Association of America, Washington, DC, 1995.

[Bar76]   R. G. Bartle. *The Elements of Real Analysis.* John Wiley & Sons Inc., New York, 1976. Also available in Spanish [Bar82].

[Bar82]   R. G. Bartle. *Introducción al análisis matemático.* Limusa, México, 1982. English original in [Bar76].

[BD65]   W. E. Boyce and R. C. DiPrima. *Elementary Differential Equations and Boundary Value Problems.* John Wiley & Sons Inc., New York, 1965. Also available in Spanish [BD79].

[BD79]   W. E. Boyce and R. C. DiPrima. *Ecuaciones diferenciales y problemas con valores en la frontera.* Limusa, México, 1979. English original in [BD65].

[BML61]   G. Birkhoff and S. Mac Lane. 現代代数学概論. 白水社, 東京, 1961. English original in [BML97].

[BML63] G. Birkhoff and S. Mac Lane. *Algebra Moderna*. Manuales Vicens-Vives. Vicens-Vives, Barcelona, 1963. English original in [BML97].

[BML97] G. Birkhoff and S. Mac Lane. *A Survey of Modern Algebra*. AKP classics. A. K. Peters, Wellesley, Mass., 1997. Previously published by The Macmillan Co. in 1941, also available in Japanese [BML61] and Spanish [BML63].

[BN82] J. Bak and D. J. Newman. *Complex Analysis*. Undergraduate Texts in Mathematics. Springer-Verlag, New York, 1982.

[Boa64] R. P. Boas, Jr. Yet Another Proof of the Fundamental Theorem of Algebra. *Amer. Math. Monthly*, 71:180, 1964. Also published in [AM95].

[Boa87] R. P. Boas, Jr. *Invitation to Complex Analysis*. The Random House/Birkhäuser Mathematics Series. Random House, New York, 1987.

[Cad47] J. H. Cadwell. Three integrals. *Math. Gazette*, 31:239–240, 1947.

[Caj69] F. Cajori. *An Introduction to the Theory of Equations*. Dover, New York, 1969. Previously published by The Macmillan Co. in 1904.

[Car60] C. Carathéodory. *Funktionentheorie*, volume 8 & 9 of *Lehrbücher und Monographien aus dem Gebiete der exakten Wissenschaften*. Birkhäuser, Basel, 1960. Also available in English [Car83].

[Car61] H. P. Cartan. *Théorie élémentaire des fonctions analytiques d'une ou plusieurs variables complexes*. Hermann, Paris, 1961. Also available in English [Car63b], German [Car66], Japanese [Car65], and Russian [Car63a].

[Car63a] H. P. Cartan. Элементарная теория аналитических функций одного и нескольких комплексных переменных. Изд-во. иностран. лит-ры., Москва, 1963. French original in [Car61].

[Car63b] H. P. Cartan. *Elementary Theory of Analytic Functions of One or Several Complex Variables*. Addison-Wesley, Reading, Mass., 1963. French original in [Car61].

[Car65] H. P. Cartan. 複素函数論. 岩波書店, 東京, 1965. French original in [Car61].

[Car66] H. P. Cartan. *Elementare Theorie der analytischen Funktionen einer oder mehrerer komplexen Veränderlichen*. Hochschultaschenbucher. Bibliographisches Institut, Mannheim, 1966. French original in [Car61].

[Car83]   C. Carathéodory. *Theory of functions of a complex variable*. Chelsea, New York, 1983. German original in [Car60].

[CB95]   R. V. Churchill and J. Brown. *Variable compleja y aplicaciones*. Aravaca and McGraw-Hill, Madrid, 1995. English original in [Chu48].

[Chu48]   R. V. Churchill. *Complex Variables and Applications*. McGraw-Hill, New York, 1948. Also available in Chinese [Chu90], Portuguese [Chu75], and Spanish [Chu66, CB95].

[Chu66]   R. V. Churchill. *Teoría de funciones de variable compleja*. McGraw-Hill, New York, 1966. English original in [Chu48].

[Chu75]   R. V. Churchill. *Variáveis complexas e suas aplicações*. McGraw-Hill do Brasil, São Paulo, 1975. English original in [Chu48].

[Chu90]   R. V. Churchill. 舒姆龍諾姆, volume 178 of 曉園. 曉園出版社有限公司, 台北市, 1990. English original in [Chu48].

[CM57]   H. S. M. Coxeter and W. O. J. Moser. *Generators and relations for discrete groups*. Springer-Verlag, Berlin, 1957. Also available in Russian [CM80].

[CM80]   H. S. M. Coxeter and W. O. J. Moser. *Порождающие элементы и определяющие соотношения дискретных групп*. Наука, Москва, 1980. Translated from the third English edition by V. A. Čurkin, With a supplement by Yu. I. Merzlyakov. English original in [CM57].

[Coh93]   P. M. Cohn. *Algebra*. John Wiley & Sons Ltd., Chichester, 1993. English original in [Coh95].

[Coh95]   P. M. Cohn. *Algebra*. John Wiley & Sons Ltd., Chichester, 1995. Also available in French [Coh93].

[Con78]   J. B. Conway. *Functions of One Complex Variable*, volume 11 of *Graduate Texts in Mathematics*. Springer-Verlag, New York, 1978.

[Fra87]   J. B. Fraleigh. *Algebra Abstracta*. Addison-Wesley Iberoamericana, Wilmington, Delaware., 1987. English original in [Fra99].

[Fra99]   J. B. Fraleigh. *A First Course in Abstract Algebra*. Addison-Wesley, Reading, Mass., 1999. Also available in Spanish [Fra87].

[Hal63]   P. R. Halmos. *Конечномерные векторные пространства*. Гос. изд-во физико-матем. лит-ры, Москва, 1963. English original in [Hal74].

[Hal65]   P. R. Halmos. *Espacios vectoriales finito-dimensionales*. Compañia Editorial Continental, S.A, México, 1965. English original in [Hal74].

584   References

[Hal74]   P. R. Halmos. *Finite Dimensional Vector Spaces*. Undergraduate Texts
           in Mathematics. Springer-Verlag, New York, 1974. Previously pub-
           lished by Princeton Univ. Press in 1942 and Van Nostrand in 1958,
           also available in Russian [Hal63] and Spanish [Hal65].

[Her70]   I. N. Herstein. *Tópicos de Álgebra*. Editora da Universidade de
           São Paulo / Editora Polígono, São Paulo, 1970. English original in
           [Her75].

[Her75]   I. N. Herstein. *Topics in Algebra*. Wiley, New York, 1975. Previ-
           ously published by Blaisdell, Vikas Publ. and Xerox College Publ. in
           1964, also available in Chinese [Her87], German [Her78], Portuguese
           [Her70], and Spanish [Her90].

[Her78]   I. N. Herstein. *Algebra*. Physik Verlag, Weinheim, 1978. English
           original in [Her75].

[Her87]   I. N. Herstein. 代數特論, volume 127 of 曉園. 曉園出版社有限
           公司, 台北市, 1987. English original in [Her75].

[Her90]   I. N. Herstein. *Algebra moderna: Grupos, anillos, campos, teoría de
           Galois*. Editorial F. Trillas, S.A., México, 1990. English original in
           [Her75].

[HK61]   K. Hoffman and R. A. Kunze. *Linear Algebra*. Prentice-Hall Mathe-
           matics Series. Prentice-Hall, Englewood Cliffs, NJ, 1961. Also avail-
           able in Portuguese [HK79] and Spanish [HK73a, HK73b].

[HK73a]   K. Hoffman and R. A. Kunze. *Algebra Lineal*. Ediciones del Castillo,
           S.A., Madrid, 1973. English original in [HK61].

[HK73b]   K. Hoffman and R. A. Kunze. *Algebra Lineal*. Prentice-Hall His-
           panoamericana, México, 1973. English original in [HK61].

[HK79]   K. Hoffman and R. A. Kunze. *Algebra Linear*. Livros Técnicos e
           Científicos Editora S.A., Rio de Janeiro, 1979. English original in
           [HK61].

[HS74]   M. W. Hirsch and S. Smale. *Differential equations, dynamical sys-
           tems, and linear algebra*, volume 60 of *Pure and Applied Mathematics*.
           Academic Press, New York, 1974. Also available in Japanese [HS76],
           Spanish [HS83], and Vietnamese [HS79].

[HS76]   M. W. Hirsch and S. Smale. 力学系入門. 岩波書店, 東京, 1976. En-
           glish original in [HS74].

[HS79]   M. W. Hirsch and S. Smale. *Phu'o'ng Trình Vi Phân Hệ Dộng Lự'c
           Và Dại Sô Tuyên Tính*. Nhà Xuât Bản Dại Học Và Trung Học Chuyên
           Nghiệp, Hà Nội, 1979. English original in [HS74].

[HS83]   M. W. Hirsch and S. Smale. *Ecuaciones diferenciales, sistemas dinámicos y álgebra lineal*, volume 61 of *Alianza Universidad Textos*. Alianza Editorial, Madrid, 1983. English original in [HS74].

[Hun96]   T. W. Hungerford. *Algebra*, volume 73 of *Graduate Texts in Mathematics*. Springer-Verlag, New York, 1996. Previously published by Holt, Rinehart & Winston in 1974.

[HW58]   G. H. Hardy and E. M. Wright. *Einführung in die Zahlentheorie*. R. Oldenbourg, Münich, 1958. English original in [HW79].

[HW79]   G. H. Hardy and E. M. Wright. *An introduction to the theory of numbers*. The Clarendon Press Oxford University Press, New York, fifth edition, 1979. Also available in German [HW58]. This book does not contain an Index, you can find one at: http://www.utm.edu/research/primes/notes/hw_index.html.

[Kra90]   S. G. Krantz. *Complex Analysis: The Geometric Viewpoint*, volume 23 of *Carus Mathematical Monographs*. Mathematical Association of America, Washington, DC, 1990.

[Lan77]   S. Lang. *Algebra*. Colección ciencia y técnica : sección matemáticas y estadística. Aguilar, Madrid, 1977. English original in [Lan94].

[Lan78]   S. Lang. *Algebra*. Addison-Wesley, Reading, Mass., 1978. English original in [Lan94].

[Lan84]   S. Lang. *Algebra*. Państwowe Wydawnictwo Naukowe (PWN), Warsaw, 1984. English original in [Lan94].

[Lan94]   S. Lang. *Algebra*. Addison-Wesley, Menlo Park, CA, 1994. Also available in French [Lan78], Polish [Lan84], and Spanish [Lan77].

[Lim82]   E. L. Lima. *Curso de Análise*, volume 2 of *Projeto Euclides*. Instituto de Matemática Pura e Aplicada, Rio de Janeiro, 1982.

[LR70]   N. Levinson and R. M. Redheffer. *Complex Variables*. Holden-Day Series in Mathematics. Holden-Day, San Francisco, 1970. Also available in Spanish [LR75].

[LR75]   N. Levinson and R. M. Redheffer. *Curso de variable compleja*. Reverté, Barcelona, 1975. English original in [LR70].

[MH87]   J. E. Marsden and M. J. Hoffman. *Basic Complex Analysis*. W. H. Freeman, New York, 1987. Also available in Spanish [MH96].

[MH93]   J. E. Marsden and M. J. Hoffman. *Elementary Classical Analysis*. W. H. Freeman, New York, 1993.

[MH96]   J. E. Marsden and M. J. Hoffman. *Análisis básico de variable compleja*. Editorial F. Trillas, S.A., México, 1996. English original in [MH87].

[Mir49]   L. Mirsky. The probability integral. *Math. Gazette*, 33:279, 1949.

[MVL78]   C. Moler and C. Van Loan. Nineteen dubious ways to compute the exponential of a matrix. *SIAM Rev.*, 20(4):801–836, 1978. Also available in Chinese [MVL82].

[MVL82]   C. Moler and C. Van Loan. Nineteen dubious ways to compute the exponential of a matrix. *Yingyong Shuxue yu Jisuan Shuxue*, 5:1–22, 1982. English original in [MVL78].

[ND86]   B. Noble and J. W. Daniel. *Álgebra Linear Aplicada*. Prentice-Hall do Brasil, Rio de Janeiro, 1986. English original in [ND88].

[ND88]   B. Noble and J. W. Daniel. *Applied Linear Algebra*. Prentice-Hall, Englewood Cliffs, NJ, 1988. Also available in Portuguese [ND86] and Spanish [ND89].

[ND89]   B. Noble and J. W. Daniel. *Algebra lineal aplicada*. Prentice-Hall Hispanoamericana, México, 1989. English original in [ND88].

[NZ69]   I. M. Niven and H. S. Zuckerman. *Introducción a la teoría de los números*. Limusa, México, 1969. English original in [NZM91].

[NZM91]   I. M. Niven, H. S. Zuckerman, and H. L. Montgomery. *An Introduction to the Theory of Numbers*. John Wiley & Sons Inc., New York, 1991. Also available in Spanish [NZ69].

[PMJ85]   M. H. Protter and C. B. Morrey Jr. *Intermediate Calculus*. Undergraduate Texts in Mathematics. Springer-Verlag, New York, 1985.

[Ros86]   M. Rosenlicht. *Introduction to Analysis*. Dover, New York, 1986. Previously published by Scott, Foresman & Co. in 1968.

[Rud66]   W. Rudin. *Princípios de análisis matemático*. McGraw-Hill, New York, 1966. English original in [Rud87].

[Rud71]   W. Rudin. *Princípios de Análise Matemática*. Ao Livro Técnico, Rio de Janeiro, 1971. English original in [Rud87].

[Rud76]   W. Rudin. *Основы математического анализа*. Изд-во. Мир, Москва, 1976. English original in [Rud87].

[Rud79]   W. Rudin. 分析學探源. 聯經出版事業公司, 台北市, 1979. English original in [Rud87].

[Rud80]   W. Rudin. *Analysis*. Physik Verlag, Weinheim, 1980. English original in [Rud87].

[Rud87]   W. Rudin. *Principles of Mathematical Analysis*. International Series in Pure and Applied Mathematics. McGraw-Hill, 1987. Also available in Chinese [Rud79], French [Rud95], German [Rud80], Portuguese [Rud71], Russian [Rud76], and Spanish [Rud66].

[Rud95]   W. Rudin. *Principes d'analyse mathématique*. Ediscience international, Paris, 1995. English original in [Rud87].

[San79]   D. A. Sanchez. *Ordinary Differential Equations and Stability Theory; An Introduction*. Dover, New York, 1979. Previously published by W. H. Freeman & Co. in 1968.

[Sán88]   J. Sándor. *Geometriai egyenlőtlenségek*. Dacia Könyvkiadó, Cluj-Napoca, 1988.

[Shi69]   G. E. Shilov. *Математический анализ: Конечномерные линейные пространства*. Наука, Москва, 1969. Also available in English [Shi77].

[Shi77]   G. E. Shilov. *Linear Algebra*. Dover, New York, 1977. Previously published by Prentice-Hall in 1971. Russian original in [Shi69].

[Sta89]   H. M. Stark. *An Introduction to Number Theory*. MIT Press, Cambridge, Mass., 1989. Previously published by Markham Pub. Co. in 1970.

[Str80]   G. Strang. *Линейная алгебра и ее применения*. Изд-во. Мир, Москва, 1980. English original in [Str93].

[Str81]   K. R. Stromberg. *Introduction to Classical Real Analysis*. Wadsworth, Belmont, California, 1981.

[Str82]   G. Strang. *Algebra lineal y sus aplicaciones*. Fondo Educativo Interamericano, México, 1982. English original in [Str93].

[Str90]   G. Strang. 線性代數及其應用. 南開大學出版社, 天津, 1990. English original in [Str93].

[Str93]   G. Strang. *Linear Algebra and its Applications*. Wellesley-Cambridge Press, San Diego, 1993. Previously published by Academic Press in 1976 and Harcourt, Brace, Jovanovich in 1980, also available in Chinese [Str90], Russian [Str80], and Spanish [Str82].

[Yue96]   Feng Yuefeng. Proof Without Words: Jordan's Inequality. *Math. Mag.*, 169:126, 1996.

Index

Problem Books in Mathematics *(continued)*

Algebraic Logic
by *S.G. Gindikin*

Unsolved Problems in Number Theory (2nd ed.)
by *Richard K. Guy*

An Outline of Set Theory
by *James M. Henle*

Demography Through Problems
by *Nathan Keyfitz and John A. Beekman*

Theorems and Problems in Functional Analysis
by *A.A. Kirillov and A.D. Gvishiani*

Exercises in Classical Ring Theory (2nd ed.)
by *T.Y. Lam*

Problem-Solving Through Problems
by *Loren C. Larson*

Winning Solutions
by *Edward Lozansky and Cecil Rosseau*

A Problem Seminar
by *Donald J. Newman*

Exercises in Number Theory
by *D.P. Parent*

Contests in Higher Mathematics:
Miklós Schweitzer Competitions 1962–1991
by *Gábor J. Székely (editor)*